MAN'S PHYSICAL WORLD

i

SECOND EDITION

MAN'S PHYSICAL WORLD

JOSEPH E. VAN RIPER

Professor of Geography
State University of New York
at Binghamton

Cartographic design by
MEI-LING HSU

Department of Geography
University of Minnesota

McGRAW-HILL BOOK COMPANY

New York   St. Louis   San Francisco
Düsseldorf   Johannesburg   Kuala Lumpur
London   Mexico   Montreal
New Delhi   Panama   Rio de Janeiro
Signapore   Sydney   Toronto

Front Cover Collage Photo Credits

Smoke stacks. *Photo by permission of Minneapolis Times. Appears in Horizon: "The Challenge of Chemistry," McGraw-Hill Book Company.*

Ordway/Earth Science/Van Nostrand.

Lightning striking the Washington Monument. *Courtesy of the American Museum of Natural History.*

World's largest telescope in Arecibo, Puerto Rico, was dedicated in 1964. *Courtesy, Commonwealth of Puerto Rico.*

Aerial view of Moriches Inlet, Long Island, N.Y. *Lounsbury and Ogden/Earth Science/Harper & Row.*

Back Cover Photo Credit

Apollo 11 view of the moon. *Photo courtesy of NASA, Manned Spacecraft Center, Houston, Texas.*

This book was set in Medallion by York Graphic Services, Inc., printed on permanent paper by Federated Lithographers-Printers, Inc., and bound by The Book Press, Inc.
The designer was J. E. O'Connor.
The editors were Janis Yates and Helen Greenberg.
John F. Harte supervised production.
Graphic materials were prepared in the Cartographic Laboratory of the Department of Geography at the University of Minnesota under the supervision of Dr. Mei-Ling Hsu and Mrs. Patricia Burwell.

MAN'S PHYSICAL WORLD

Library of Congress Catalog Card Number 73-139566

07-067156-7

234567890 FLBP 7987654321

CONTENTS

atmosphere system / 4.6 Mechanisms to adjust for imbalances in the global energy system

TEMPERATURE 114

4.7 Astronomical factors in surface temperature patterns / 4.8 Atmospheric factors in surface temperature patterns / 4.9 Thermodynamic factors in the distribution of temperatures / 4.10 Adiabatic heating and cooling / 4.11 Lapse rates and temperature inversions / 4.12 Vertical zonation of temperature changes in the upper atmosphere / 4.13 Land and water contrasts in surface temperatures / 4.14 Other land surface conditions influencing temperatures / 4.15 Relativity in the human significances of temperatures

Study Questions 139

ATMOSPHERIC MOISTURE 141

5.1 Terminology in atmospheric moisture / 5.2 Measurement of atmospheric moisture / 5.3 Variations in vapor pressure (water vapor content) / 5.4 Variations in relative humidity / 5.5 Phase changes of water and their role in heat exchange

CLOUDS AND FOG 149

5.6 The development of clouds / 5.7 The development of fog / 5.8 Cloud types

PRECIPITATION 155

5.9 Forms of precipitation / 5.10 The processes of precipitation / 5.11 Stability and instability as factors in precipitation / 5.12 Initial lifting mechanisms for precipitation / 5.13 The global precipitation pattern / 5.14 Seasonal precipitation patterns / 5.15 The distribution and meaning of snow cover / 5.16 Precipitation reliability

THE HYDROLOGIC CYCLE 173

5.17 The hydrologic cycle / 5.18 The global precipitation-evaporation balance / 5.19 The disposition of precipitation excess over the continents / 5.20 Runoff from the land surface / 5.21 Evapotranspiration

Study Questions 186

ATMOSPHERIC PRESSURE 188

6.1 Terminology and measurement / 6.2 Causes for changes in atmos-

PREFACE

This is a new volume. About 80 percent of it has been rewritten, and had it not been for the impatience of both the author and publisher, especially the former, the remainder would have been altered, some of the new material reworked, and the book appeared months later. This preface was prepared immediately after the last term was placed in the Glossary (Appendix D), and its purpose is to explain how this edition differs from the first and why.

The first and most fundamental change is one of approach. In the first edition, a deliberate attempt was made to minimize process and to attempt to present the reader with an attitude by which to view the distribution of "physical" phenomena and patterns on the earth surface. This attitude was reflected in the *principles of natural distributions,* through which it was hoped that the student of nature could comprehend the infinite complexity of pattern and change without bewilderment. These so-called principles are retained in this volume for the same reason but are not given as prominent a part. Instead, after teaching the material for a couple of years, relinquishing the course to one of his colleagues as any textbook writer-teacher should do, and after considerable soul-searching, the author became convinced that the understanding of processes, or the application of energy in altering forms and patterns, should be the goal of physical geography. Thus it behooved him as a teacher to introduce, in the portrayal of man's physical world, some of the basic processes by which the surface of this world is modified.

In order to present some of these processes, two things had to be done. First, space had to be provided. The author agrees with his editors that there has been an unfortunate trend upward in the size of college textbooks, to the growing despair of students who have to buy and read them. A good textbook should be a springboard for learning, not an encyclopedia from which to attempt to find all the answers to questions posed. It was thus decided to maintain the book at approximately the size of the first edition. Three chapters with their respective topics were removed, as follows: (1) a chapter on mineral resources, mainly because the concept of mineral resources is much less related to physical conditions than to economic and technological developments; (2) a chapter on water resources for similar reasons, although considerations of the hydrologic cycle were retained and expanded in the chapter on atmospheric moisture; and (3) the chapter on oceanography. The omission of this last chapter disturbed the author more than that of the others because he felt that the physical geography of the sea is woefully neglected in most introductory courses dealing with the face of the earth. However, in contacting many of his teaching friends in the profession who used the first edition, he found no one who had ever reached that chapter in their classes.

A new chapter is devoted to the earth as a planet and as a member of the solar system. This is not only because of our current interest in the exploration of nearby space but also because the earth itself has shrunk so much, due to our growing travel speed, that a unified look at our planet as an entity seems appropriate. The treatment of the problems of location, scale, and map representation has been expanded into a separate chapter.

Two chapters have been added to those dealing with climate, one mainly to introduce the global energy system as the driving force in nearly all earth processes, and also to introduce the systems concept. The other chapter discusses the major climatic regions of the world. The Koeppen-Geiger classification system used again may seem antiquated to some teachers, but there are serious problems of application in any classification, and the author sincerely believes, after experimentation, that the system used is a good compromise for introductory classes. The Thornthwaite climatic classification, which is presented in Appendix C, may be more functional in many ways, but disregarding the large amount of time taken to work out its water budget for weather stations (unless a computer is handy), any classification that indicates most of northern continental Europe and eastern Britain as a dry, subhumid region leaves much to be desired!

The coverage of landforms has been grossly altered with the treatment of process and a special chapter added on earth materials and the relativity of the time factor. An attempt has been made to weave a bit of geotectonics into the global landforms chapter to indicate that the flow of energy even involves the irregular crustal movements of the earth.

The vegetation chapters have been changed the least, but have been touched here and there by the new concepts. The boldest innovation in the book has been the author's attempt to introduce the relatively new principles of soil classification under the Seventh Approximation. At first, the author attempted to present the suborders of world soils, but, following the wise advice of the chief of the world soil geography unit of the U.S. Department of Agriculture, decided to retreat to the main orders of global distributions. It is hoped that the last-minute changes in the soils chapters will not prove confusing. The new classification is logical and useful, but as in so many areas of interest, there has been so little time for thorough learning, and so much to do!

There are many hazards in presenting capsuled or miniature introductions to a variety of earth sciences within a single volume. One is the risk of duplication, as the result of which the physical geographer may make few friends by skimming the "cream" from four or five introductory earth science courses. Another is the danger of superficiality, in which no real understanding can be achieved. The author has intentionally tried to avoid both of these hazards. He has found what he believes to be a powerful integrating tool in the relatively new idea of systems analysis. It is still new to him, and the trained analyst should be patient with his early attempts to use it. The idea of the interaction between phenomena in earth space is an ancient one in geography, and the last principle of natural distributions presented in the first edition—that of the tendency among elements of nature (plus man) to develop an equilibrium of sorts—lies behind the principle of the trend toward a steady state within systems.

The implications of energy flow through a system, with its goal of reducing obstacles to that flow and reducing entropy, are exciting when one begins to apply them to the complexity that one sees on the surface of the earth. The author, having attempted to apply this concept to the physical world of man, is like a young man with his first automobile—a bit overwhelmed, not quite certain yet how to use it, puzzled by some of its intricacies, aware of some of its dangers if misused, but convinced it is a vehicle that will take him where he wants to go if he drives carefully.

Processes of any kind, whether they include the erosion of mountains, the mechanics of a thunderstorm, the slow formation of soil horizons, or the progressive clothing of a bare surface by plants, are alike in a single fundamental way: they involve the transformation of forms and things by energy flows, virtually all of which are derived from the sun. The interrelationship of almost all processes to the operation of the earth-atmosphere energy system is real; it is not simply a concept. The use of the systems concept is relatively new in geography, but it is finding a warm reception throughout all branches of the subject. Both the student and the teacher are invited to "try it on for size."

The reader will find that the systems approach gradually becomes weaker in the latter portion of the book. This is not because the author believes it to have less relevance to the study of the geography of vegetation or soils, but rather reflects some of his own inadequacies and the pressure of time. He sincerely hopes that some younger physical geographer will integrate a systems approach to the global patterns of the biosphere. It would be difficult to consider the energy role played by plants apart from that of animals, including man. The disposition of the energy budget, or what happens to the surplus energy of a locality where radiation input exceeds that of output, is a fine integrating theme for understanding the organization of earth space.

The author has become humbler with age. The facts of areal differentiation and similarity that used to be piled in orderly rows in his mind now seem far less important than they used to be. The beauty of the overall scheme of things, on the other hand, grows in magnificence, and he has become properly awed and respectful. The present volume still contains too many terms, too many facts, and the author hopes that teachers and students alike will not pay undue attention to them. The Glossary is included primarily to make the book more readable without an effort at memory-stuffing. At the same time, it is satisfying for anyone to be able to recognize a kame, a barrier beach, a cirque, a granite, or a halophyte.

These may not be the most important things in life, but they can add to one's appreciation of the physical landscape. We live on an amazing planet, however, and we should not lose sight of the forest for the trees.

The reader's attention is called to the contrast between the front cover of this book and the one at the back. The one in front shows energy in action—driving things, moving things, changing things. It represents the dynamic world in which we live. The interaction of energy, materials, forms, and information within the earth-atmosphere energy system has enabled man to make this earth a potentially comfortable place on which to live, but it also threatens our very existence. The neglect of understanding man's role in the global energy system can be perilous, as we are beginning to discover. The rear cover shows the dead, inert surface of the moon, bathed by the same energy as the earth, composed of materials not much different from those that make up the earth crust, but a far different orb from the beautiful, hospitable earth that promises, and threatens, so much.

The author is again most indebted to his wife, who helped him immeasurably with her patience and forbearance when the priorities for her husband's time became confused and intolerable. He also is grateful to his professional colleagues, both in his own department and around the country, who graciously offered him ideas and constructive criticisms. Special mention should be given to Bill Kemp, a junior colleague now at McGill University, who reduced the generation gap and shared ideas within the framework of a course in cultural geography, which led to some of the new concepts that appear herein. Blessed be the young who share their fine fresh minds with the old, who find it difficult to think in new terms. Grateful acknowledgment also is extended to Nicolay Timofeeff, who advised at several critical points and who made possible the new climatic data in Appendix C. The staff of the Cartographic Laboratory at the University of Minnesota, especially Dr. Mei-ling Hsu and Patricia Burwell, has been most expert and patient in handling the author's idiosyncracies. Their skill will add much to the effectiveness of his presentation. Finally—to the many unknown critics who read part or all of the manuscript—your help and encouragement has meant much.

Joseph E. Van Riper

CHAPTER 1
THE FIELD OF PHYSICAL GEOGRAPHY

The Unity of Geography

The general field of geography has been defined in many ways by many people. There appears, however, to be a growing consensus among workers now in the field that its focus should be the *science of spatial interaction*, that is, the study of the processes by which forms and phenomena interact through their spatial arrangements on or near the earth surface, especially as to how and why these arrangements change in time, and with an emphasis on their human implications.

1-1 THE HISTORICAL DUALITY OF PHYSICAL AND HUMAN GEOGRAPHY Geography for many centuries included all knowledge pertaining to the description of the earth; it has sometimes been referred to as the "mother of the natural sciences." The explosion of knowledge that followed the discoveries and explorations of the fifteenth and sixteenth centuries led to a search for order within the mass of accumulated data concerning the earth. Some of this resulted in a growing specialization of interest and the development of several new sciences. The field of geography joined in this trend toward increased compartmentization of knowledge. The entire trend culminated in the work of Varenius (Bernard Varens), who lived in the first half of the seventeenth century. He proposed a two-part division of geography. The first was *general* or *universal geography*, which dealt with the overall characteristics of the physical world and all those properties that were interrelated on a global scale. The second division was termed *special geography*, which was to include a regional description of the earth surface. Man as a species of the animal kingdom was included in general geography, but his imprint on the surface was to be included in special geography. Varenius died at twenty-eight, having completed only the first volume, *Geographia Generalis*, which became the standard geographic reference work for over a century. During this period geography became a field with a double focus: one largely within the physical sciences that dealt systematically with the arrangement of the physical world, and the other a comparative regional focus in which the arrangement of human activities was emphasized.

During the nineteenth and early twentieth centuries, attempts were made to unite the two approaches through the search for direct causal relationships between human behavior and the physical world. *Environmental determinism* was stimulated by the work and popularity of Charles Darwin, who clearly demonstrated the impact of natural selection on the physiological characteristics of living creatures.

The expanding capabilities of man to harness the inanimate energy in nature and to modify the physical world to suit his needs during the past 100 years or more showed the fallacy of viewing mankind as a passive tool in the hands of a dynamic and dominant natural environment. Today there is a growing awareness that the face of the earth is one world, not two, and that the focus of geography should represent that unity. With all of man's capabilities, he still remains a small part of the total environment at the earth surface, part of a gigantic, integrated system through which surges a powerful flow of energy derived from the sun. Unless we understand this flow and the implications of the detours and side channels into which we are dipping, we may be faced with expulsion. In our search for ever greater comforts, we may be sealing our ultimate fate.

Human ecology, or the interrelationship between man and his total environment, is a critical and demanding approach that is much too broad to be the focus of any one branch of knowledge. Specialization in spatial arrangements is perhaps the geographer's main contribution to ecological understanding. In this connection, geography cannot specialize to the point of considering separately human and nonhuman elements when dealing with spatial interaction. This book is designed for a course in physical geography. The title implies that the so-called physical world is the home of man, but man is a dynamic part of that world, a new force that within little more than 200 years has changed from a dependent to an independent variable in the gigantic earth-atmosphere energy system. Recently man has begun to tap the mainstream of solar energy directly and to attempt the controlled use of atomic fission and fusion. This is an awesome responsibility and a dangerous one, because the evidence of his actions during these past 200 years in handling the fossil fuels and the chemical elements at his disposal indicates that he has many things to learn regarding their wise use. His technologies have surpassed his wisdom in understanding the long-term consequences of his actions.

1-2 THE INTERMEDIATE POSITION OF GEOGRAPHY BETWEEN THE PHYSICAL AND SOCIAL SCIENCES The geographer, by tradition, has been forced to consider both human and nonhuman factors in his search to comprehend the processes of spatial interaction. The duality that blurred his focus in earlier decades and centuries and led to many easy and careless ideas concerning man's relationship to his physical environment later became an asset. Specialization in all of the sciences has had a divergent effect upon their relationship to each other. Recently there has been a growing awareness of the importance of considering man in his total environmental context and understanding the ties that bind man's aspirations to the limits and potentialities of nature. The connection is extremely subtle. Kenneth Boulding, a distinguished social scientist, pleads for an understanding of the whole and decries the ever-increasing specialization of scientific research. The complexity of factors that make up the total environment of man becomes greater every decade, and the whole has become ever more elusive.

The digital, high-speed computer has become an important tool in the geographer's search for spatial interrelationships, a tool that perhaps will be as significant in the future as the map. The map is a model, a simplified picture of spatial reality, without which the complex patterns in the arrangement of things on earth would be incomprehensible. The computer also can be used to simulate reality by translating spatial patterns into digital, or numerical, sequences. Once spatial patterns are simulated on a controlled scale, and with certain built-in assumptions, the computer may test various hypotheses regarding the operation of variables in processes that could alter these patterns. Just as a computer may test the capacity of an aircraft wing strut to withstand certain stresses, so may it test the assumptions made regarding the relationships between slope and stream velocity, between the rate of rock weathering and moisture/temperature ratios, or the impact of urban growth on air pollutants. The computer frees the geographer from many of the intuitive judgments which he formerly had to make in assessing the meaning of interrelationships. It also forces upon him the use of a quantitative language that cannot be misunderstood by others. It will not replace the art of geographical description, which still is as essential in education as the analytical procedures of elementary science classes. Precision of measurement and clarity of language in geography cannot help but bring the physical and social sciences closer together.

The growing awareness that man is dangerously interfering with the operation of the total environment is a further reason for the importance of the intermediate position of geography between the physical and social sciences. Water and air pollution are only beginning to concern people, and we are just beginning to realize some of the effects of the synthetic environments caused by the use of chemical fertilizers, pesticides, and detergents. The biotic balances on earth are highly vulnerable and should be treated with great care. At present, man is eliminating animal species from the face of the earth at a rate of about one per year, some with casual disdain, some with a few fleeting regrets, and some with a growing sense of guilt and concern. Measurements of spatial patterns through accurate mapping, the construction of various types of analogue models of reality, and accurate, statistical correlations of spatial data through computer use are among the most important tools of the modern geographer, and some day these might be vital in indicating the routes and rate in the fall of man. Perhaps they might even enable man to reject once and for all, as Lynn White, Jr.,[1] has so ably suggested, the ancient Christian belief that the sole objective of nature is to serve mankind. Such a belief should appropriately be filed under "Famous last words."

1-3 THE UNITY OF GEOGRAPHY Geography is primarily interested in the organization of earth space and why certain arrangements of forms and phenomena evolved through time. We know that man is not arranged on earth like coins in a box and that the physical environment is not an inert stage on which the human drama takes place. The constant interaction between man and his total environment requires that geography not be simply a study of the physical environment on the one hand and human behav-

[1]Lynn White, Jr., "The historical roots of our ecologic crisis," *Science*, vol. 155, 1967, pp. 1203–1207.

Figure 1-1 Areal interrelationships. The sharp boundary
between the pines on the left and the barren, rocky hillside on the
right is partially explained by a sharp change from sandstone on the
left to limestone on the right, and along a *fault,* or rock-slippage
plane. Man has accentuated the contrast through centuries of serious
overgrazing by animals, in which only the pines are protected for
their shade and nuts. Pines cannot grow on the rocky limestone because their roots require easy lateral spread. View near Barouk,
Lebanon. [Van Riper]

ior on the other. The specific things with
which geography deals are the same as those
studied by many other sciences. The difference lies mainly in its focus, which is spatial
interaction (see Figure 1-1). This does not
prevent specialization in the field, however.
If a geographer wishes to specialize in the
interaction of climate, soils, and vegetation,
he may call himself a physical geographer,
but if he is a competent scientist, he will
never neglect man as one of his variables.

There are few places on earth where man has
not left an imprint of his activities.

Physical Geography

1-4 DEFINITION OF PHYSICAL GEOG-
RAPHY AND ITS EDUCATIONAL ROLE
Physical geography is a branch of the main
discipline which emphasizes the spatial interaction of the elements in the natural envi-

ronment. Man is a part of physical geography because his actions influence the patterns and interplay of such factors as climate, landforms, and soils, but he is not the center of the stage. This specialization within the discipline is largely an educational device, and one that has flourished for historical reasons. On the research frontiers of geography today, the schism between physical and human geography rarely exists. Nevertheless, there is value in such a focus on the introductory level for reasons brought out in the following subsections.

The complexity of human ecology. Human ecological problems are becoming more evident and alarming each year, and demands for action programs are beginning to be heard. The relationship between man and his environment, however, is extremely complex, and on the whole, scientists are working within much too narrow an area to be effective in this problem zone. One question is illustrative: Should we turn to electricity for all surface transportation in the United States to reduce air pollution?

A satisfactory answer to this question requires some extremely difficult analyses involving a host of contributory factors and value judgments. For example, generating the required enormous increases in electrical power undoubtedly would result in increased pollutants, either from hydrocarbon combustion plants or from those employing atomic fission. Would we be merely trading small, mobile pollutant sources for giant stationary ones? What would be the cost of redesign and retooling within the automobile industry? What implications would there be throughout the global petroleum industry? What international repercussions would there be if other countries refused to pay the price of conversion and increased their own rates of pollution? Who would pay for the

cost of conversion? What other alternatives might there be? Should there be a strict rationing of personal travel of all kinds? The questions soon leave the fields of the engineer, the atmospheric scientist, and the economist and begin to involve the accountant, the politician, the lawyer, and even the philosopher and moralist. Pollution of our air, our water, and even our soils is not simply an engineering problem. Its complexity requires a many-sided approach and involves every citizen. A major part of the approach, as with many other ecological problems, is understanding some of the interrelationships between elements within the nonhuman environment as revealed by changes in patterns through time and in their spatial context. This is one of the major objectives of physical geography.

The manageable parameters of physical geography. One of the principal reasons why the physical and natural sciences have been able to make such remarkable progress in the development of theory and the use of prediction is that the variables, or parameters, are much more measurable and controllable than those that influence human behavior. The behavior of large numbers of water molecules under varying conditions of temperature and pressure can be predicted with high probability. The same cannot be said yet regarding individual human reactions to the same variables. For this reason, it is appropriate that an important introduction to human ecology summarize those aspects of the spatial interaction where research has been more productive. Physical geography, then, is a useful base for human ecology because it deals with more manageable parameters.

The contributions of the physical and natural sciences. The reservoir of knowledge accumulated in the physical sciences is vast and available partly for reasons stated in the previous paragraph and partly for reasons of research investment priorities. Screening and tapping that reservoir for material pertinent to general geography thus is an important educational tool.

The task of descriptive generalization. Besides attempting to increase the understanding of interrelationships between physical forms and phenomena within their spatial arrangements, physical geography has the further task of making meaningful to man the infinite variety of physical differences throughout the earth surface. Careful examination of climate, vegetation, or soils will often show noticeable differences every few feet. For some purposes these minute differences from place to place are significant, but for global understanding, they become chaotic. Any description of spatial arrangements of the elements of global physical geography requires generalizations of the details that are present in the real world. The development of techniques to make such generalizations valid, consistent, and appropriate is a kind of model building, a simplified version of reality, but one that can be visualized, measured, and used for the development of theory. Depicting areal relationships through map models is one of the oldest tools of geography, and, while map making is a technique and not a goal in itself, in physical geography it plays an essential part in making sense out of the infinite variety of things on the face of the earth.

1-5 COVERAGE OF THE ELEMENTS OF PHYSICAL GEOGRAPHY IN THIS TEXT The specific elements of physical geography covered in this text are: (1) location (2) scale (3) solar energy (4) climate and weather (5) landforms and surface configuration (6) vegetation and (7) soils. The first two, location and scale, differ from the others in being concepts rather than measurable, tangible features. They are included as essential in physical geography, however, because they are involved in all considerations of earth space, or area, which in turn are part of any geography study. The list is not inclusive, partly because of space limitation and partly because of the lesser competence of the author in other aspects of physical geography—such as zoogeography and oceanography—although aspects of both are discussed in appropriate sections.

Of the major variables that can be considered in the physical[2] environment of man, two of them constitute independent variables. That is, their major features are the result of processes whose origins lie outside the general environment of the earth surface. The first of these is *solar energy,* the generating source for all processes of change on earth. The second is the solid crust of the *lithosphere,* or rock sphere, whose composition and arrangement, in some places at least, are related to internal processes inherited from the formative period of the earth, or to rearrangements of material deep within the interior.

Systems Theory in Geography

1-6 THE NATURE OF SYSTEMS One of the most useful integrating concepts that has arisen during the last few decades, and one that will appear many times in this volume, is that of *systems.* It is a concept that was first developed in engineering in connection with input-output energy models for ma-

[2]Again, it should be emphasized that the physical environment can never be regarded as completely independent of man, since he is an active variable and has modified the face of the earth almost everywhere.

chinery design and now has many applications in the theoretical development of science. It is beginning to have special significance in geography. For this reason, a brief treatment of what systems are and how they may be used is included in this introductory chapter.

A *system* is a set or aggregate of interrelated forms or objects, together with their related attributes or qualities. All systems have three important elements: (1) an internal energy flow; (2) a set of related objects or materials; and (3) signal receiving and transmitting mechanisms of varying complexity inherent in the materials, which react to changes in the flow of energy the materials encounter. This last feature of systems is sometimes loosely referred to as *information,* because some kind of sensory apparatus must be present that can detect energy fluctuations and initiate changes in the materials so as to adjust to the altered flow. This use of the word *information* should not be considered as synonymous with communicable knowledge. All materials on earth, for example (solids, liquids, and gases), react to temperature changes or to variations in the exposure to radiant energy. The differing reactions of each material are expressed in terms such as temperature or melting point. These do not involve *knowledge* as the term is generally used, yet such materials can act as sensors to gain "information" about the flow of energy and, in turn, to translate this information into changes in form, position, or state.

All of the elements in the system are interrelated and tend toward establishing as smooth a flow of energy as possible, a condition which has been termed a *steady state.* A change in any element within a steady-state system will result in a corresponding alteration of the entire system to accommodate the change, and will tend toward arriving at a new steady state.

Two types of systems are recognized: *closed systems,* in which the flow of energy is confined within the system and never crosses the borders of the system, and *open systems,* in which there is an input of energy into the system from one or more points and a corresponding outflow. An example of a closed system would be the celestial mechanics of the solar system. The planets and their satellites move around the sun without being powered from the outside. There is a slight running down of the system, but it was started billions of years ago and has been operating ever since. We are not particularly concerned with closed systems, because the important ones in physical geography are open systems, or rather a series of subsystems that belong to a single, huge earth-atmosphere system that receives solar energy, uses it to perform work, and finally loses it through reradiation to outer space.

An oil-burning heating system for a home constitutes a simple model of an open system. The energy input is the heat that is released by the combustion of oil in the furnace. The materials in the system include the oil, furnace, flues or ducts, radiators, and thermostat. The information sector of the system centers on the thermostat and its electrical connections to the oil-input valves and ignition apparatus. Like all open systems, the total output of energy equals the total input within narrow limits. If heat were to accumulate at any place within the system for an indefinite period, the entire system would break down. Either the furnace would melt or the house would be in flames. The steady state of the system is the temperature tolerance range set by the thermostat. As soon as the temperature in the room drops below a given level, the thermostat sensor notes it and sends an impulse to the furnace calling for an open valve and ignition. Once the room temperature exceeds the maximum tolerance level, the information network shuts down the input until loss of heat from the room into the air outside the house cools the air inside to the lower signal level.

The global energy system that concerns us is much more complex than that of the oil heating system, and the analogy breaks down at many points. For example, in the earth-atmosphere system, the input remains remarkably constant, and fluctuations in energy flow are caused primarily by the many detours and work patterns within the flow stream. The basic principle, however, remains the same. Input equals output, and the informational pattern and reaction apparatus of the related materials tend toward a steady state. The materials that make up the earth-atmosphere system include everything on earth and in the atmosphere. All work done on earth, which involves changing energy from one form to another, all motion, and all life processes are the changes brought about by solar energy as it passes through the system. The informational sector is as complex as the materials present and the various manifestations of energy. The subsystems, each of which might be considered as a smaller regulatory mechanism, have their own set of related objects and their informational and adaptive apparatus. The alternate evaporation and condensation of water with intervening transportation between warm seas and clouds on a global scale, often referred to as the *hydrologic cycle,* is one of the largest and most significant subsystems of the global energy system. It will be treated in detail in Chapter 5, but the reader might stop at this point and compare for himself the operation of the hydrologic cycle with that of a kitchen refrigerator, another example of an open system. The only difference is that the water contains its own sensory and reactive apparatus, whereas the refrigerator requires an electric current and a pump to regulate the temperature. Also, a special kind of heat exchanger is generally used within the refrigerator's circulatory system.

One last feature of open systems should be mentioned. This relates to the second law of thermodynamics, which states that, while energy can never be destroyed by use, its capacity for performing work decreases each time it performs work. In open energy systems, the more indirect the energy flow becomes, or the more work it performs, the greater becomes the deterioration in the "quality" of energy. A general term for this decrease in the capacity of energy to perform work is *entropy.* The smoother and more direct the flow of energy between input and output, the more order there is in the system, and therefore less entropy. The energy flow in open systems tends to remove obstacles to its smooth, direct flow or to reduce the amount of entropy within the system. This condition of minimum entropy, which is set by the characteristics of the system itself, is the steady state toward which all systems aim. A comparison of the moon and the earth reveals at once that the two have entirely different systems, despite the fact that they both are bathed in solar energy and are at about the same distance from the sun. The lack of water and air on the moon makes it possible for solar energy to enter the lunar system and perform a bit of work heating the surface, which a short time later reradiates energy back into space. The entropy in the lunar system is extremely small, because the solar radiation does little more than heat the surface before being reradiated. The swirl of motion that goes on around us here on earth, from the sweep of the planetary winds to the flow of rivers or the passage of planes overhead, constantly reminds us that the passage of energy through our great system is highly indirect. Our global energy system has great disorder and produces a large amount of entropy, but if it were not so, it would not be a fit place in which to live.

The complex patterns found on the earth surface and of special concern to the geographer are the result of the energy flow indirectly through the system and its subsystems, and the various reactions of materials to this flow. The concept of systems thus gives us an excellent base from which to study the interrelationships of environmental factors within their areal context. The study of systems and subsystems may take different forms. It may be done through an examination of the quantitative changes that take place in the objects or materials involved, to show how these are interrelated on a global scale. One example is measuring the net radiation balance, or the difference between energy input and output locally, in order to determine the amount of surplus that could be diverted into other energy forms, or the deficiencies that must be overcome by energy importation. Our global system also may be studied through the creation of models, or symbolic representations of reality, but on a simple and manageable basis, and where variables may be manipulated in various ways to find their effects. Figure 4-4 is a highly simplified model of the global system; it is a useful analytical device.

Probably the most important aspect of the growing knowledge of our complex, global energy system and subsystems is the understanding of the processes of change in the spatial arrangements of its constituent materials. Geometric arrangements of these materials and the direction of change in these geometries are of great concern to geographers. Understanding the reaction of materials to the informational signals provided by the fluctuations in energy flow is a useful way of discovering the effects of human intervention in the system.

The history of mankind has been one of greater and greater human control over the flow of energy across the face of the earth. Nearly all of this diversion has resulted in increasing the disorder and entropy within the system. The concentrated focus of energy diversions into cities is a good example. For this reason, the use of general systems theory has more than educational value. Geography has a real contribution to make here, because spatial geometric relationships can be measured and described much more accurately than the devious wanderings of the energy flow itself. A systems approach gives order to the multitude of variables in human ecology.

The concept of systems will be referred to throughout this text, either explicitly or indirectly. In some cases, as in the general presentation of the major global energy flow, the relationships with systems will be clearly defined and measured. In others, the implications of systems will be found in various statements of processes or trends toward or away from "equilibrium."

The Principles of Natural Distribution

Systems analysis provides a powerful tool for the development of theory and integration in geography, but it does not entirely remove a problem that faces each geographer as he deals with spatial relationships, especially the areal patterns involving the elements of physical geography. This is the problem of pattern complexity.

The task of characterizing the physical differences and similarities from place to place on earth is formidable. Fully as difficult is the task of avoiding misrepresentations in generalizing the content of areas on a global scale. Exceptions can be found for nearly any areal generalization on this scale. The equatorial areas are warm, but snowfields and glaciers crown the summits of Andean mountains in Ecuador. The arctic areas generally are cold, but temperatures of 90°F and over occasionally occur within the Arctic Circle. Soils in the rainy tropics generally have a low carrying capacity for agricultural use, yet the rich volcanic soil of Java supports some of the densest agricultural populations on earth. The scale limitations of any map mask the details that cannot be presented, yet for some purposes, these omitted details may be exceedingly important.

In order to indicate the ever-present limitations of areal generalizations, four basic concepts or principles of natural distributions are presented in the following paragraphs. They are referred to frequently throughout this book and therefore should always be kept in mind. They might be considered as a summary of the qualifications that must be made in any characterization of areas and as a frame of reference that, if clearly understood and utilized, can aid greatly in bringing some order out of the chaotic maze of physical differences throughout the earth.

1-7 THE PRINCIPLE OF GRADED LIKENESSES AND INFINITE DIFFERENCES IN NATURAL FORMS AND AREAS *No two forms in nature, and no two areas on the earth surface, are exactly alike, although similarities appear that permit classification of form or area. The degree of dissimilarity observed varies directly with the closeness of scrutiny. Conversely, similarities become more obvious at broader scales of observation.*

The term *cattle* is used to denote a classification of animals that have certain recognizable similarities of form. Different breeds of cattle are also recognized, such as Jersey, Holstein, and Hereford. Jersey cows have definite characteristics that are common to all of their breed, and to most people, all Jersey cows look alike. A farmer with a herd of these animals, however, or a trained judge at a county fair evaluating the qualities of Jerseys, can easily detect significant differences. This general principle of graded likeness amid an infinite variety of differences holds true for clouds as well as cows, trees as well as birds, and soils as well as mountains.

From the street, the front lawns of an urban residential area appear to be similar, yet close examination on one's knees reveals that there are appreciable differences not only between separate lawns but also within any one lawn. Minor differences in available moisture, in soil texture, in exposure to sunlight or shade, or in acidity produce distinguishable differences not only in the types and associations of grasses and weeds but also within accompanying elements of the total environment, including the microclimate and microfauna.

The monkeys, snakes, trees, and landforms of the Amazon Basin in South America show distinct differences from those of the Congo Basin in Africa, and there are easily distinguishable associations within different parts of the Amazon Basin. In contrast to the Mississippi River drainage basin, however, the environmental associations of the Congo and Amazon Basins exhibit many comparable characteristics. They both have monkeys, they both have trees which remain green the year around, their soils have comparable features and properties, and their climates are generally warm and humid throughout the year. The existence of broad similarities within the infinite variety of forms from place to place makes possible the regional approach in geography.

1-8 THE PRINCIPLE OF AREAL TRANSITIONS *The change in characteristics from one area to another is always gradational,* *although the rate of change may vary. Usually the transition becomes more gradual with closeness of scrutiny.* There are few lines or linear boundaries in nature. Even such apparent ones as mean sea level or the edge of a river represent zones of change that fluctuate with variations in water level. Maps are the geographer's main tool as he seeks to represent the orderliness and regularities in nature, yet maps almost always involve the use of boundary lines of one kind or another. Such lines are generalizations of areal transitions and sometimes depend as much on the arbitrary definition and scale of the user as on the natural conditions being shown.

To illustrate this principle, suppose we wished to determine the area covered by pine forests in the United States. Whether an area had a pine forest or not would depend in part on how we defined a pine forest. Is it a forest composed entirely of pine trees, or is it one where a certain percentage of the trees are pines? If it is one where 50 percent or more of the trees are pines, how large an area are the measurements to cover—will they include areas whose size is less than 1 acre, 1 square mile, or 100 square miles? Our "line" separating pine forests from non-pine forests will obviously represent a zone of transition, which will vary in width depending first on the preciseness of definition and second on the detail of observation. Finally, when we present a picture of the distribution of pine forests in the United States, unless we use a map ratio of 1:1,[3] an obvious impossibility, our line on a map will not represent a line at all, but an area, which will become wider as

the map's scale of presentation becomes smaller. We cannot assume, furthermore, that the zone of transition is going to be equally wide along all boundaries at the same scale of generalization in definition, observation, and presentation. The zone of transition at the base of a hill separating the hill from the nonhill area is not going to be the same width around the hill if the slopes are uneven. Where the slope increases in steepness, the zone of transition becomes narrower, regardless of how we define the bottom of a hill. An interesting problem in the relation of transitions to definitions involves where to draw a line separating a valley from an adjacent mountain. Does the side of the valley begin at the mountaintop? Does the mountain begin at the bottom of the adjacent valley?

1-9 THE PRINCIPLE OF CONTINUOUS ALTERATION OF AREAL CHARACTERISTICS WITH TIME *Because of the continuous irregular flow of energy within the earth-atmosphere system, both long- and short-term changes are taking place among the forms and phenomena that make up the environmental complex of any area. Short-term changes usually are best observed in the microfeatures, whereas long-term changes are reflected in the larger forms and phenomena. The tempo, or rate of change, varies.*

A perfectly static environment does not exist in nature. Each hour and each second bring changes of some sort to every spot on earth—changes that range from the whisper of air movement around a grass stem and the busy work of soil microorganisms to the thunder of avalanches or the continuous reduction of slopes by the gradational processes of erosion, weathering, and deposition. In presenting and interpreting the patterns of areal differences, distributional patterns should always be placed within an appropriate time continuum or scale. Just as many

[3]A map ratio represents the ratio of the size of the map projection to the earth area which it represents. A ratio of 1:10,000 signifies, for example, that 1 inch on the projection represents 10,000 inches within the area mapped. It should be remembered, however, that because of the limitations of all map projections in distorting a curved globe surface onto a flat plane, linear scale ratios are never completely correct, although they may be considered thus on large-scale maps of small areas.

differences from place to place can be dismissed as unimportant in a portrayal of graded likenesses in geographic description, so also many microchanges that occur in the environmental complex from time to time can be omitted.

In general, the most recognizable short-term changes are those associated with the weather. The changes that have the greatest amplitude and the slowest rate of change generally are those associated with crustal alterations, the great earth movements that raise and lower the continental blocks. The total pattern of change in any environment may be likened to the pattern of waves on the sea. There are small waves and large waves, some of which are fairly regular in their recurrences and others whose periodicity is vague and indefinite. Sometimes the sea of change is calm, with only tiny ripples marking miniature alterations, but at other times it is a confused mass of storm waves, representing periods of turbulent environmental change, with all the dynamic environmental forces passing rapidly from one extreme to another. Some seas are quieter than others. Similarly, some areas on earth tend to be more stable than others in regard to nearly all features of change. The task of presenting geographic patterns in their continuity of change is an essential part of any geographical description and interpretation. Change is a feature of all types of natural distributions.

1-10 THE PRINCIPLE OF TRENDS TO-WARD A STEADY STATE IN RELATED ENVIRONMENTAL ELEMENTS *Although change continually takes place within the many elements that make up areas or regions, periods of relative stability are marked by mutual adjustments within these elements and a tendency toward a steady-state condition. Long- and short-term balances may be noted, since the time required for significant changes varies widely with different elements in the environment.*

Most elements within a regional environmental complex have some effect on each of the others, although some are more influential than others. Few of them are inert or passive. Climate helps determine whether the vegetation of an area will be a forest or tundra, lays its stamp on landforms and soils, and even affects some of the characteristics of certain mineral resources. Similarly, climate itself is changed to a certain degree by many of the other environmental elements. Mountains divert winds and intercept moisture. Forests and paved city streets differ considerably in their effect as heating surfaces, and farmers long have recognized "cold" soils and "warm" soils in which soil texture and water content influence the soil climate. Even rocks and minerals are not without their influence, and the chemical composition and fertility of soils may be closely related to the composition of the underlying rocks and their included minerals. Prospecting for large low-grade copper and cobalt deposits has been done by flying over an area and noting subtle differences in the color of vegetation.

This interaction between the elements in an environmental complex tends to produce a mutual adjustment, or a tendency to adjust to each other. Absolute equilibrium, or a completely balanced environment, is impossible, because a certain amount of change in one or more of the elements is inevitable. Nevertheless, relatively stable periods can be recognized in which, for all practical purposes, an equilibrium may exist. Some changes are measured in thousands or even millions of years; hence, in the span of human affairs such changes may be relatively unimportant.

To illustrate, imagine an environment which consists of a hillside covered with a fully mature, virgin forest. The kinds of trees and shrubs present, the underlying soil, the animal life, the forest climate, the drainage, and the microlife within the soil have all reached a mutual adjustment which is relatively stable. Geologic processes, however, eventually will alter the slope of the hillside. Long-term climatic changes slowly affect the composition of life forms, including even major species of trees. As the slope gradually recedes, drainage of surface water is not so rapid, and the soil hydrology changes. With increasing age also, the soil becomes deeper. All these changes are noticeable, however, only on a time scale of hundreds or thousands of years. To a human observer, the environment is stable, and on every hand lie the evidences of mutual adjustment between the coexisting elements in the region. To be sure, short-term changes also occur. A tree is blown over during a sudden windstorm; a landslide may produce a sudden gash in the slope; an exceptionally wet year may influence the microlife forms, and so on. Although each of these changes sets off a train of minor readjustments, they do not play a significant role in the overall regional equilibrium.

Now let us alter one of the major elements in the environmental association on our hillside—for example, by having man enter the environment, remove the forest, and cultivate the slope. The resultant changes will be immediate and rapid, and the entire equilibrium will become altered. As long as man continues his cultivation, the environmental factors will tend to operate against the establishment of the old equilibrium, mainly because a new dynamic element has been introduced. Remove man, and a long chain of readjustments will take place, aimed at eventually restoring the former equilibrium.

Just as climates experience short- and long-term periods of stability, environments also develop short- and long-term relative equilibria. The short-term equilibria, however, are developed mainly in the micro-features of the environment, whereas the larger and more dominant features illustrate the long-term balances. A single unusually cold winter will generally have little effect in altering the species of trees or in decreasing the number and type of larger animals present in a forest, but such a winter may produce a noticeable effect on the general association of small ground plants and animals for several years.

Recognition of the tendency of environments either to move toward relative equilibria under the influence of mutual interaction and adjustment or to move away from such equilibria because of sudden changes in one or more environmental factors is a useful concept in human affairs. Land management programs require such knowledge in order to recognize, develop, or maintain that stability which is the most rewarding to man.

The Relativity of Similarities and Differences in Nature

A careful study of the principles of natural distributions given in the preceding sections may well leave the reader with the impression that nature is in a state of total confusion and chaos and that the task of accurately picturing and interpreting areal differences and similarities is a fruitless one. The role of the physical geographer in unraveling the complexities of nature is an important one, and it has a dual aspect.

First, as a scientist, the physical geographer must attempt to be as accurate as possible in his recognition of reality in areal complexes, with all their infinite variety, subtle changes, and complex interrelationships. In seeking an understanding of nature, he cannot dismiss arbitrarily any variable as unimportant, whether the process involved takes one year or several thousand. As his tools of observation improve and the records of environmental changes accumulate, his factual knowledge grows and, with it, his understanding. New explanations are given and old ones discarded.

Second, he has a responsibility to construct appropriate generalizations of the areal differences and similarities throughout the earth surface. In all walks of life there is need for knowledge concerning areas—what they contain, how and why one area is different from another, and where similarities in selected environmental elements occur. Maps are used by nearly everyone, and maps are one of the principal results of applied geographic generalizations.

These two functions of geography offer a continuous challenge, but the results are more rewarding than the complexities of environments would appear to indicate. Order may be perceived in the infinite variety of forms and phenomena contained within earth space, and this order appears at different levels or scales of observation.

Order often becomes apparent only after a great number of different features discovered by small-scale observation have been reexamined in broader terms. If we could shrink ourselves so as to be able to penetrate within the nucleus of a large atom such as uranium, we should find ourselves in a seemingly chaotic world of electrons, protons, neutrons, mesons, and many other minute "forms," some of which appear and disappear constantly. Each of these tiny forms is susceptible to so many interrelationships and outside forces that orderliness appears incomprehensible. If we could observe a molecule of water in a glass among billions of other molecules, however, we should see a certain degree of orderly behavior and arrangement in the constituents of the atom,

a consistency that was not apparent at the lower level. Confusion at the molecular level appears to lie mainly in the behavior of the molecules of water themselves. In the glass of water, the molecules are dashing here and there at varying speeds, bumping into one another and showing no more consistency in their movements than do individual bees in a huge swarm. Some molecules in the water break through the surface of the liquid with a burst of speed and fly out into the air; others cling to the sides of the glass, seemingly resting a moment, only to be knocked from their perch under the bombardment of their speeding companions, after which their vacant resting places are seized by still other molecules. Yet, despite all this random behavior of individual water molecules, we have formulated definite laws concerning the behavior of large numbers of water molecules in a liquid state, and these laws have an extraordinarily high degree of statistical reliability.

Each increase in the breadth or scope of observation results in a greater relative orderliness within the multitude of forms and behavior patterns at lower levels. At the level of human observation, water in a drinking glass has properties that appear to be absolutely predictable, and man puts these properties to his own use. The fact that some of the molecules of water are escaping constantly from the surface of the liquid into the air or that others are clinging to the sides of the glass does not concern a person as he tilts the glass to his mouth and drinks. At his scale, the variant behavior of molecules has no significance.

One of the responsibilities of the physical geographer is to perceive the different degrees of order among the composite natural features of areas at different levels of observation and to establish usable generalizations at appropriate levels. Global maps of iron ore deposits may be highly important and reliable in assessing the relative potential industrial strength of nations, but they would be almost useless to a geologist who wishes to prospect for iron ore in Afghanistan. Similarly, the small details of a large-scale hydrographic chart would be of little use to a pilot of an airplane traveling along the coast at 10 miles per minute. Out of the analysis and description of minute areal differences and similarities come the valid generalizations that are useful at broader levels. The geographer's descriptive task is like that of an ant crawling over a large Persian rug. Only by intelligently recording the changes in colors according to distance and position and scaling the changes down to perceptible size could the ant ever comprehend the complete design of the rug. Many of the minute differences in nature about us from place to place are tiny segments of a complex design. These segments begin to have meaning only when seen together and translated into the proper scale or frame of reference.

As we examine the physical makeup of the earth in this volume, the complexity of the natural designs should become readily apparent. The principles of natural distributions discussed in the preceding sections testify to the complexity of areal differences and change, yet at the same time they indicate that concepts of relative reality can be formed and can serve extremely useful ends. While it is true that no two spots on the earth are exactly alike in every detail, it is likewise true that within certain limits, prescribed by the purpose of the study, some spots are alike and may be grouped together in order to form a *region*.

1-11 GEOGRAPHIC GENERALIZATIONS

When a geographer draws a line on a map representing a boundary between two areas

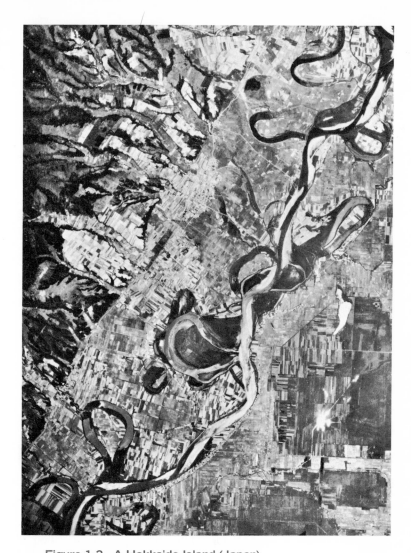

Figure 1-2 A Hokkaido Island (Japan) landscape—a tiny fraction of the world mosaic to be described and interpreted. What processes are altering this landscape? Is the present, relatively straight stream course the result of human intervention? What are some of the results of this on stream dynamics? As the forests are removed from the hill slopes (upper left), what influence does this have on erosion rates? Such questions are pertinent and typical of those asked by geographers as they view the earth surface. [U.S. Air Force photo (MATS)]

that are unlike *in some selected respects,* he is making three kinds of generalizations: (1) he chooses a classification that recognizes a certain degree of similarity of form or behavior within the area on each side of the line; (2) he generalizes the place of change between one region and another; and (3) he generalizes the changes that continually take place with time, presenting a pattern that is relatively typical only for the period of time appropriate to the purposes of his study.

Generalizations of form, area, and time are standard procedures for geographers. Each generalization is relative to the purpose and scope of the study, and its accuracy must be measured in terms of the scale on which it was made. Of the three kinds of generalizations, geographers specialize in the one dealing with differences from place to place. The physical geographers borrow heavily from the other natural sciences for classifications of form and for knowledge of processes that produce changes in form with time. Other geographers depend greatly on the social sciences for similar kinds of generalizations of form and time. Geography is the correlator of all the sciences with respect to areal content and relationships. It depends on the other sciences for basic data and techniques that must be utilized in any geographic study, but it also has its own tools and techniques, aimed specifically at understanding areal differences and similarities at levels or scales suitable for different human purposes. The other sciences, in turn, borrow the knowledge, tools, and techniques of geography in pursuing their own studies.

Study Questions

1. What is the central focus of geography?
2. Explain the relationship between the focus of geography and the importance of maps as geographic tools.
3. Enumerate some ways in which man may alter the following "natural" elements in the physical environment:
   - a. Vegetation
   - c. Climate
   - b. Soils
   - d. Stream flow
4. What is a system?
5. Using a kitchen refrigerator as a simplified model of a system, show how each of the following are integrated in its operation:
   - a. Energy input and output (note at least three sources)
   - b. Information control
   - c. Material alteration
6. What is the relationship between systems analysis and the study of human ecology? Define *ecology.*
7. Why has geography been an important subject in the study of human ecology?
8. Present your own examples to illustrate each of the principles of natural distributions.
9. What is meant by the "scaled orderliness" of nature?

Chapter 1: References

Ackerman, Edward A., *Geography as a Fundamental Research Discipline,* University of Chicago, Department of Geography, Research Paper No. 53, Chicago, 1958.

Ad Hoc Committee on Geography, Earth Sciences Division, National Research Council, *The Science of Geography,* National Academy of Sciences, Washington, D.C., 1965.

Broek, Jan O. M., *Geography: Its Scope and Spirit,* Charles E. Merrill Books, Inc., Columbus, Ohio, 1965.

Cook, Ronald, and James Johnson (eds.), *Trends in Geography,* Pergamon Press, New York, 1969.

Hartshorne, Richard, *Perspective on the Nature of Geography: A Survey of Current Thought in the Light of the Past,* Association of American Geographers, Washington, D.C., 1946.

James, Preston E., and C. F. Jones (eds.), *American Geography: Inventory and Prospect,* 2d ed., Syracuse University Press, 1964.

Taylor, T. Griffith (ed.), *Geography in the Twentieth Century; A Study of Growth, Fields, Techniques, Aims, and Trends,* 3d ed., Philosophical Library, Inc., New York, 1957.

Warntz, William, *Geographers and What They Do,* Franklin Watts, Inc., New York, 1964.

———, *Geography Now and Then; Some Notes on the History of Geography in the United States,* American Geographical Society Research Series, No. 25, American Geographical Society, New York, 1964.

PART 1

THE PLANET EARTH:

ITS PLACE IN THE SOLAR SYSTEM; ITS DIMENSIONS AND PORTRAYAL ON MAPS

The relationship between the inhabited earth and the rest of the universe has occupied a central place in human curiosity for centuries. A few centuries before the Christian era, such considerations were an important part of the discipline of geography. Most of the early models of the universe, however, remained speculative until the last few centuries, when instrumentation freed man from his dependence wholly on speculative logic in viewing his universe. The astronomer now has placed the earth in a humbling perspective: a relatively small agglomeration of matter, arranged by gravity and rotation into roughly a spherical shape; one orb among billions ranging in size from giant spheres a million times larger than the earth to tiny particles of cosmic dust; an orb that is placed in space so vast as to be almost meaningless in terms of earthbound measurements; and an orb that, like its eight companion planets, is bound in an elliptical orbit around a focal sun from which earth derives its input of energy.

Until recently, humans were confined to the surface of our planet, and it still seems huge, despite our increased speed of travel. Geography as an organized body of knowledge has had from its beginnings the task of investigating the earth surface and making meaningful the patterns that develop thereon. Maps and globes are the major tools of the geographer. As reducing instruments they bring the immensity of earth dimensions down to the level of human perception. This part of the text introduces the reader to the earth as a planetary entity within the solar system, to the complex methods of determining position, direction, and time, and to some of the technical problems connected with reproducing the global surface on flat maps.

CHAPTER 2
THE PLANET EARTH:
ITS FORM AND RELATIONSHIPS WITHIN THE SOLAR SYSTEM

The enormity of space, with its solar systems, star clusters, galaxies, and supergalaxies, deflates the ego of man as he contemplates his position and that of the relatively tiny globule of solids on which he lives. Some idea of this vastness may be comprehended by the fact that the nearest star beyond our own solar system, *Alpha Centauri*, lies about 3.68 light-years from us. A *light-year* is approximately 6 trillion miles, or the distance one could reach in a year, traveling at the speed of light—186,000 miles per second. Our most powerful telescopes have observed enormous galaxies of stars more than 300 million light-years away, and these far distant star masses are receding at incredible speeds. The stars that we see on a clear night—and we can view only a tiny fraction of those in the heavens—seem crowded in space, yet they are far apart. If we were to shrink our sun to the size of a tennis ball, and the distance to Alpha Centauri in the same proportion, the two stars would lie more than 4,600 miles apart.

Our sun (*Sol*) is an ordinary star among the millions of others that make up our own galaxy and the few hundred or so that we can distinguish by eye on a clear night. It is neither unusually large nor small, and it is about average among the stars that we have measured as to density, temperature, and composition. Even our galaxy is but one of thousands that occupy only a small fraction of total space. Lest we humble ourselves unduly, however, in contemplating our human position in this macrocosm of time and space, we may take pride in our ability to perceive and explain this "big picture." This perhaps makes us a bit godlike in the realization that—astronomical observational tools at our disposal and rumored UFO (Unidentified Flying Object) sightings notwithstanding—we have yet to receive any clear evidence of any other sentient and rational beings in space.

We need spend no more time in describing the arrangement of heavenly bodies beyond our solar system. The geography of outer space belongs to the domain of the astronomer. We must, however, treat our nearby neighbors in space, because they involve conditions that are, or may be, important to man, and some of them have significant influences on earth. After a brief look at our solar system, we will examine the form and shape of the earth in detail and investigate the methods by which we determine position and time on earth. The quality of position underlies all aspects of geography, and it is wholly appropriate that we understand clearly the reference bases by which it is determined.

The Solar System

2-1 THE NINE PLANETS The nine planets that orbit around the sun are unlike in several respects. Table 2-1 presents the major features of each. Some of the more significant facts concerning the solar system may be summarized as follows:

1. The size of the planets generally increases with distance from the sun as far as Jupiter, after which it tends to decrease. Mars is a noticeable exception.

2. Venus and the earth are nearly alike as to mass, size, and average density. The largest planets have the lowest density.

3. The orbits of most of the planets, while elliptical, are nearly circular. The highest eccentricity (variation from a true circle) occurs on Mercury and Pluto, the planets nearest to and farthest from the sun, respectively.

4. All of the revolutionary motions (movements around the sun) are counterclockwise, if viewed from a point in space above or to the north of the various ecliptics.

5. The rotational motions (turning on an axis) are from west to east, except for Uranus, which rotates from east to west, and Venus, which may have a slight retrograde rotation.

6. The planetary atmospheres vary in both

Figure 2-1 A spiral galaxy. Messier 31, one of thousands of galaxies that make up the universe. Many of them have this spiral, disk-like form. Our own galaxy—sometimes referred to as the Milky Way system—is roughly comparable to this distant galaxy, measures some 100,000 light-years in diameter, and contains about 30 thousand million stars, of which our own sun is about average in size and brightness. Such is the vastness of outer space. [Lowell Observatory]

amount and composition, but only the earth seems to have appreciable free oxygen. Ammonia (NH_3) and methane (CH_4) are major constituents of the largest planets.

Since man is beginning to leave his planetary home to explore the solar system, it is appropriate to compare the conditions that prevail on the surfaces of the other members of that system.

Mercury, the smallest of the planets, with a mass only about 5 percent that of the earth, is too small to hold much of a gaseous atmosphere, especially since it lies relatively

Table 2-1 Planets of the Solar System

	MD	D	M	SG	R_v	R_o	E	A
Mercury	.39	3,100	.054	5	.24	88 (?)d.	.21	(?)
Venus	.72	7,700	.81	5	.62	250 (?)d.	.007	CO_2, H_2O
Earth	1.00	7,900	1.00	5.52	1.00	24 h.	.017	N, O, H, CO_2, H_2O, etc.
Mars	1.52	4,216	.11	4	1.9	24.6 h.	.09	CO_2, H_2O (tr.)
Jupiter	5.2	89,000	318.	1.3	11.9	9.8 h.	.05	NH_3, CH_4, H, HCL, etc.
Saturn	9.5	74,000	95.	.7	30.	10.2 h.	.06	CH_4, NH_3, H
Uranus	19.2	30,878	15.	1.3	84.	10.8 h.	.05	CH_4, H
Neptune	30.1	28,000	17.	2.2	164.	16 h.	.009	CH_4
Pluto	39.5	3,600(?)	.07(?)	(?)	248.	6.4 d.	.25	(?)

The symbols used in the table are as follows: MD = mean distance to the sun in astronomic units (93 million miles); D = diameter in miles; M = mass (earth mass as 1); SG = density (1 is density of water); R_v = period of revolution around the sun (sidereal period) in years; R_o = period of rotation on axis (hours or days); E = eccentricity (variation of elliptical orbit from a circle, or major axis minus minor axis divided by major axis); A = atmospheric composition.

near the sun. Solar heating is intense, with temperatures rising well over 1000°F on the sunlit side. We have never had a close look at the surface of Mercury because of the competition of direct sunlight and the small size of the planet.

Venus is similar to the earth in mass and density. Despite the fact that it approaches nearer the earth than any of the other planets, astronomers have never seen its surface because of its thick cloud cover. Photospectrograms reveal large amounts of carbon dioxide, even in the outer atmosphere, and much water vapor, but no trace of free oxygen. Probes initiated by the space rocket Mariner II from only 25,000 miles above the surface indicated: (1) surface temperatures of up to 600–800°F; (2) little difference in temperature between day and night in the outer atmosphere, which remains at about −100°F; (3) no magnetic field as on earth; (4) no break in the thick cloud cover. Water probably occurs only in the vapor state, owing to the high temperatures. Data transmitted from Soviet space probes in 1969 verified the earlier conclusions.

Mars, of all the planets, has surface condi-

tions most similar to those on earth, yet falls far short of being hospitable to man. The atmosphere is extremely rarified, consisting largely of carbon dioxide in a thin film near the surface. Temperature conditions are more like those of the moon than of earth, fluctuating from below 0° at night to 130–150° in sunlight.

Photos taken by Mariner IV, a 1964 space probe, and Mariners VI and VII in 1969, revealed that the surface of Mars resembles closely that of the moon, with nearly all surface irregularities consisting of meteoric impact craters. The crater rims are not so high or so rugged as on the moon, and there are no fissure cracks or winding rills as are found on the lunar surface. Local relief generally is less than 1,000 feet. There is evidence of slightly greater weathering and material transport on the Martian surface than on the moon, but landforms typical of terrestrial deserts are totally absent despite the aridity of the surface environment. The waxing and waning of the white polar caps indicate a distinct hemispheric seasonal temperature variation, as on earth. There is little sign of daily changes in the margins of these polar caps, however, which indicates that they are fairly deep. It is now believed that they are composed of frozen carbon dioxide.

Despite some seasonal changes in coloration on the surface that were believed to have been the result of some low life forms, no chlorophyll has been detected, and there is only the slightest trace of water vapor. The virtual absence of free oxygen and nitrogen probably precludes the existence of life forms similar to those on earth. The lack of a distinct magnetic field and the sparse atmosphere tend to expose the Martian surface to dangerous cosmic and solar radiation and proton bombardments. Mars, like the moon, is no haven for population surpluses from earth.

Jupiter and *Saturn* are giant planets that are alike in having densities that average much less than that of the earth. The density of Saturn, for example, is less than that of water. Much of the total mass of these planets consists of gases, especially methane, ammonia, and molecular hydrogen. Droplets of condensed methane and ammonia form the dense cloud layers that mask any possible solid surface. The well-known "rings" of Saturn are now believed to consist largely of small crystals of ammonia. The rapid rotational speed of these planets, with periods less than half that on earth, has flattened the polar regions and bulged the equatorial zones to a much greater degree than on earth. Prevailing temperatures within the atmospheres are far below zero.

Uranus, Neptune, and *Pluto,* the outer planets, are too far from the sun to receive much solar radiation. Methane is an important constituent in the atmosphere of the first two because of their large size, which is able to hold this lightweight but apparently common primal compound of carbon and hydrogen. Little is known of the tiny, distant planet Pluto, which was discovered in 1930,

except for its highly eccentric orbit. Indirect evidence indicates a low mass and density.

Sol. The sun is much larger than any of the planets, including giant Jupiter. To illustrate, the diameter of the sun could encompass 109 earths, laid side by side. Like other stars, the sun emits both energy and matter, produced in the solar interior by processes of nuclear fusion and fission. Enormous amounts of energy are released that escape into space within a wide range of electromagnetic waves, which collectively are termed *solar radiation.* This spectrum of radiant energy, which is the driving force behind nearly all processes on earth, will be discussed in Section 4-1. Also streaming out into space from the sun are the building blocks of matter from which the atomic elements are constructed: the subatomic particles that include negatively charged electrons and positively charged protons, all moving outward at high velocities. Giant storms within the solar atmosphere are largely responsible for the discharge of these particles, which have great penetrating power. Fortunately, the upper atmosphere and the earth's magnetic field help trap or screen out most of these potentially lethal particles, but enough pass through at times to impair or even interrupt surface electrical communications on earth. Trapped protons oscillating at high velocity in the earth's magnetic field form dangerous radiation belts far above the surface.

2-2 THE MOON AND TIDES The moon is the first nonterrestrial body to be visited by man. The enormous expenditure of energy, material, and man-hours in order to reach the moon has become one of the sagas of our times. The distance, about 240,000 miles, seems short, but the difficulties of escaping from the earth's gravitational pull and the precision required for navigation are prodigious. Long before the first moon visit in 1969, astronomers and other scientists knew much concerning the moon. Detailed photographs permitted it to be mapped with great precision.

The moon presents an exceedingly hostile environment to man. There is no atmosphere, no water, no life, no movement. Temperatures range from well over 150°F in full sunlight to "space cold"—near absolute zero—in the deepness of the lunar night. Level areas are rare, and the entire surface is pockmarked with circular craters ranging from tiny pits a few inches across to gigantic craters 50 to 70 miles in diameter and with jagged rims 6,000 feet high (see Figure 2-2). Most of these circular craters are believed to have been caused by the impact of meteors that must have had a wide range of sizes. Some of the larger craters, however, have unmistakable features of ancient volcanoes, including distinct lava flows. Some of the high ridges and swells do not seem to be related to volcanic or meteoric causes, but the general surface contains no pronounced separation into high continental blocks and low basins, as on earth.

The surface of the moon is covered with loose, unconsolidated and unassorted debris composed of rock material that has solidified from a molten state. Despite the great age of the lunar surface rock material (about 3 billion years), there is no trace of chemical alteration or weathering, indicating that the generally sterile environment of the lunar surface has not changed noticeably since the rocks were formed. Among some of the puzzling features on the lunar surface are deep, sinuous chasms or rills and long, straight fissures that radiate from some of the larger plains or *marae*. Analyses of rock samples brought back to earth by the Apollo astronauts indicate: (1) a consistent great age of the surface material; (2) a common igneous origin (solidification from a molten state); (3) no new chemical elements; (4) common chemical elements similar to those in rocks of the earth crust; (5) some metallic elements, such as titanium, that are somewhat more abundant than in the surface rocks of earth.

Despite its inhospitality to man, the moon has some important potentialities. Unmanned moon stations could act as communications relay stations to replace present relay satellites, such as Telstar, which are small and

Figure 2-2   Oblique view of the far side of the lunar surface, photographed in 1969 from Apollo II.   The large central crater (International Astronomical Union #308) has a diameter of about 50 statute miles. The meteor-impact craters are about the only irregularities seen. This is an extremely ancient surface and represents the record of meteoric hits accumulated over billions of years. The earth must have been subject to the same if not greater bombardment. The main difference is that the lunar scars are unhealed because of the absence of water and air. [NASA]

liable to fluctuations in their orbit or power supply. The moon also could provide an admirable observational platform for space observation free from the atmospheric refraction that plagues earth-bound astronomical observers. Automatic scanners of the earth *albedo,* or amount of sunlight reflected from earth, could possibly aid greatly in forecasting long-range weather changes. Precision measurements of position on earth could be aided by moon-based radar or laser beams, and eventually the moon may be used as a marshaling station for further space travel and exploration. We have only fragmentary evidence concerning the mineralogical composition of the moon, and its surface material may mask important metallic resources.

Apart from its potential significance to man, the moon has an immediate and continuous influence on life on earth through its gravitational influence on tides. These rhythmic changes in sea level on earth form one of the most interesting natural phenomena. The flooding and ebbing of ocean waters along coasts have long attracted man's attention and have been important in the lives of coastal dwellers in many ways. The periodicity of tides, though relatively simple and regular in some coastal areas, is exceedingly complex in others, and accurate predictions of tides generally have had to await the accumulation of records obtained from tidal gauging stations throughout the world over a period of many years.

Causes of tides. The most important cause of tides is the gravitational attraction of the moon. The point on the earth surface nearest the moon thus experiences the greatest gravitational pull. Theoretically, a person standing at this point weighs slightly less than he would if the moon were not present. The lunar gravitational pull, however, is ex-

tremely small, amounting to only about one nine-millionth that of the earth, owing to the much smaller mass of the moon and to its distance from the earth. As Kuenen has noted, a man's weight would vary only by the equivalent of one drop of perspiration. Nevertheless, despite its relative weakness, the lunar attraction has an appreciable effect on the large mass of the earth and its ocean waters, and creates a distinct bulge in both. The ocean waters are more susceptible to warping than the solid crust of the earth. However, careful measurements indicate that the lunar pull affects the solid crust as well, and tidal bulges ranging from 9 to 12 inches have been observed in the interior of continents.

The tidal bulge on the earth at the point nearest the moon is matched by another at the antipode, the point on the earth farthest from the moon (see Figure 2-3). This opposite bulge is due to the comparatively low lunar gravitational force acting at this greater distance. Every particle of material in or on the earth is warped in position slightly in the direction of the moon. The amount of warping decreases with increased distance from the moon. The crust of the antipodal position, therefore, lags behind the rest of the earth in tidal warping, and the result is an apparent bulge. Actually, it is not a matter of the antipodal bulge being pulled out from the earth, but of the earth being pulled away from the bulge.

The zone midway between the two bulges, the intermediate position between the gravitational deformation on one side of the earth and the apparent bulging resulting from inertia on the other, is marked by a trough of low sea level, or low tide. This trough would encircle the globe like the two bulges if the oceans were continuous.

Diurnal tides. As the earth rotates on its

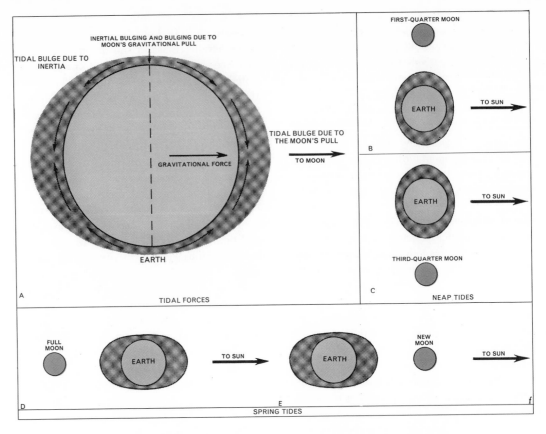

Figure labels (as printed in the diagram):

INERTIAL BULGING AND BULGING DUE TO
MOON'S GRAVITATIONAL PULL

TIDAL BULGE DUE TO
INERTIA

TIDAL BULGE DUE TO
THE MOON'S PULL

GRAVITATIONAL FORCE

TO MOON

EARTH

A   TIDAL FORCES

FIRST-QUARTER MOON

EARTH   TO SUN

B

EARTH   TO SUN

THIRD-QUARTER MOON

C   NEAP TIDES

FULL
MOON   EARTH   TO SUN   EARTH   NEW
MOON   TO SUN

D   E   SPRING TIDES   f

axis, every point on the earth surface is nearest a bulge and trough twice each day. This produces two high tides and two low tides every 24 hours, or a change in tide approximately every 6 hours. If the moon's orbit around the earth were exactly within the equatorial plane of the earth, there would be no difference in the range of daily tides at any one place (disregarding the effects of shoreline configuration, etc.). The moon's orbit, however, generally is oblique to the earth's equatorial plane, and this obliquity changes slightly. Hence, points on the earth surface vary in their nearness to the respective tidal bulges; the heights of the two daily high tides (see Figure 2-4) also vary. Sometimes, and in some places, one daily high tide

may be so much stronger than the other that there appears to be only a single daily high and low tide.

The time of occurrence of daily tides changes each day, owing to the moon's motion in its orbit around the earth. Because the moon rises approximately 45 minutes later each day, each of the two daily high tides occurs about 20 minutes later than the corresponding high tide of the previous day.

Fortnightly tides. A noticeable variation in the range (difference) between the daily high and low tides takes place according to the phases of the moon. The highest range occurs during the full- and new-moon phases. The tides during these two phases are known as

The Planet Earth: Its Place in the Solar System; Its Dimensions and Portrayal on Maps

Figure 2-3 Diagram of solar and lunar tidal influences. The influence of the moon alone is indicated in A. In B and C, the sun and the moon are pulling at right angles to each other, hence the low neap tides. The higher spring tides occur at new or full moon, when the pull of the sun parallels that of the moon.

spring tides, an unfortunate choice of term, because such tides have no relationship to the spring season. The lowest fortnightly tidal range occurs during the first- and third-quarter phases. The tides during these low-range periods are known as *neap tides* (neap = scanty).

The cause for these fortnightly changes in the range between high and low tide is the sun's gravitational attraction. Although the sun is thousands of times larger than the moon and the earth combined, it also lies some 400 times farther away than the moon. Consequently, its gravitational pull is only about one-fifth that of the moon. The effect of the sun is most noticeable in comparing the tidal range of periods when the sun and moon are "pulling" together (as in the full- and new-moon phases) with the tidal range when the two are pulling at right angles to each other (as in the quarter phases). The solar influence is diagramed in Figure 2-3. A slight variation in spring and neap tides also

Figure 2-4 Influence of the angle of the moon's ecliptic on the periodicity of tides. In (A) the moon's ecliptic (plane of its orbit) and the earth equatorial plane are in the same plane, and the daily tidal period everywhere consists of two high tides and two low tides, each 6 hours apart. More frequently, however, the moon's ecliptic, as in (B), is inclined somewhat to the earth equatorial plane, thus producing irregular tidal periods that vary with latitude. Note the similarity between this and the effect of the declination of the earth axis on the length of day and night (see Figure 2-17).

can be noted, depending on the relationship between the plane of the moon's orbit and that of the earth. The highest spring tides occur during solar and lunar eclipses, when the earth, sun, and moon are exactly in line.

Effects of shoreline configuration on tides. The tidal bulges undergo many changes during their daily journey around the earth, changes that are brought about by the shape and configuration of coastlines. An incoming tidal bulge may be channeled and concentrated by wide-mouthed and tapering estu-

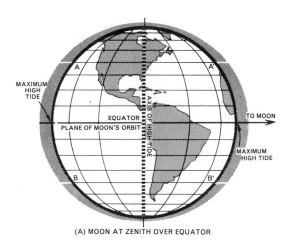

(A) MOON AT ZENITH OVER EQUATOR

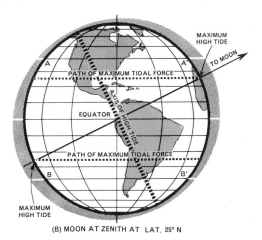

(B) MOON AT ZENITH AT LAT. 25° N

Figure 2-5 Removing fish from nets at low tide in the Bay of Fundy region. The high tidal ranges in the area create many problems, but perhaps none as unusual as this. [National Film Board of Canada]

aries that open directly toward it. The Bay of Fundy, for example, has spring tides that range up to 50 feet (see Figure 2-5). In such areas, where there is rapid shoaling near the mouth of an estuary, the incoming tide may move up the estuary and develop a steep wavefront known as a *tidal bore.* The most famous are those in the Bay of Fundy, Hangchow Bay in China, and the lower Amazon River. In Hangchow Bay, the bore may be as much as 20 feet high and may proceed up the estuary at a speed of as much as 12 miles per hour. Marco Polo called attention to this unique phenomenon. *Tidal currents,* resulting from the channeling of tides, can be hazardous to navigation in the vicinity of islands or narrow straits. The junction of tidal currents from opposite directions sometimes produces areas of especially turbulent water known as *rip tides* or *tidal falls.* The navigational charts of waters in the vicinity of many Pacific islands note such areas as hazards to small vessels.

Tides sometimes reflect from shorelines,

The Planet Earth: Its Place in the Solar System; Its Dimensions and Portrayal on Maps

producing a complex pattern of "echo" tides. The exposure, or angle, of the coastline is reflected in variations in tidal periods. Coastlines that are highly irregular, such as those in the Indonesian archipelago, may have tidal patterns that differ widely in both time of occurrence and range within short distances. The complexity of tidal patterns caused by shoreline configuration is illustrated by the tidal chart for the North Sea in Figure 2-6.

Several other factors influence tidal patterns to a greater or lesser degree. These include the deflective force of the earth rotation, variations in atmospheric pressure, and the changeable strength and direction of prevailing winds. Of all the factors involved in tidal behavior, the easiest to predict are those associated with lunar or solar positions. These alone, however, though of great importance, may be overshadowed locally by conditions of site and weather. There is no substitute for accurate tidal observations accumulated over many years.

2-3 THE PLANET EARTH Seen from space, the planet earth appears as an unusually smooth, spherical body, somewhat hazy in outline owing to its surrounding atmosphere. Clouds which are arranged in huge spiral swirls mask a relatively large part of the surface, and the brown to green continents and bright blue oceans appear here and there between the gleaming white cloud surfaces. The color television transmissions from Apollo spacecraft brought this glorious view to millions of earth-bound viewers. Compared with the colorless, dead mass of the moon, the earth truly appears to be hospitable even from far out in space.

Among the planets, the earth is unique in ways other than its external appearance. The average density of the earth, 5.6 times that of water, is the greatest of all the planets (see Table 2-1), even exceeding that of Mercury

and Venus, which are nearer the sun. A second unique feature of the earth is its high content of the lighter gases, especially hydrogen, nitrogen, oxygen, and water vapor. No other planet has more than a trace of molecular oxygen (O_2), although Venus apparently has a large amount of combined oxygen contained within the carbon dioxide and water vapor that saturate its dense, hot, lower atmosphere.

Figure 2-6 Detail of tides in the North Sea. [After Chapin]

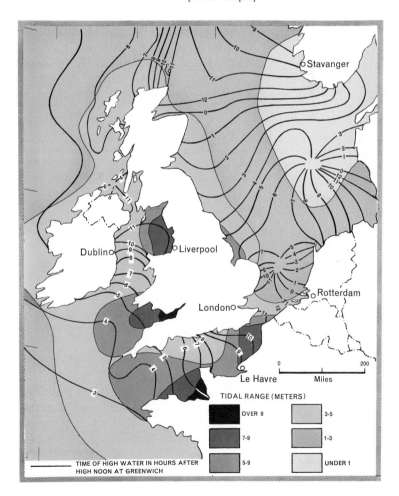

TIDAL RANGE (METERS)

OVER 9	3-5
7-9	1-3
5-9	UNDER 1

TIME OF HIGH WATER IN HOURS AFTER HIGH NOON AT GREENWICH

The earth's relatively high density, combined with exactly the right distance from the sun, enable it to hold its lighter gases and to have a temperature range within which water may occur in all of its three phases: solid, liquid, and gas. The importance of water in making our planet livable should not be underestimated. As will be seen later, it is a natural heat exchanger and air conditioner that maintains a safe temperature range on earth and prevents accumulations of energy, not only for the entire planet but for the inner workings of all living things that inhabit the earth. The beautiful blue of the oceans as seen from space is rightfully a symbol of the earth's hospitality.

The average specific gravity of the earth is approximately 5.6 times that of water. Considering the fact that the average density of rocks in the crust is about 2.7, the interior must be composed of material that is much denser than that at the surface. Seismologists have determined through the behavior of earthquake waves passing through the earth that the interior of the earth is arranged in layers much like an onion, with a small solid core contained within an extremely hot, dense, liquid zone occupying the innermost third (see Figure 2-7). This is surrounded by successive layers of progressively less dense material toward the outer surface. The outer crust of relatively lightweight rocks is extremely thin, varying from 3 miles beneath the oceans to 30–40 miles beneath parts of the continental blocks.

The atmosphere, like the solid part of the earth, is also arranged into concentric zones, with the lightest atmospheric constituents predominating at the uppermost levels. The atmosphere near the surface consists of a mixture composed almost entirely of nitrogen (78.084 percent) and oxygen (20.946 percent), with small quantities of carbon dioxide (.033 percent) and traces of rare gases such as argon, neon, krypton, and xenon. Other airborne materials, such as water vapor and dust particles, are variable in their proportions, but they still constitute only a tiny portion of the total mass and are rarely found more than 7 miles above the surface. The proportions of the gases remain remarkably constant up to an elevation of about 44 miles, although the total quantity or mass rapidly decreases with elevation, 99 percent lying below an elevation of 25 miles and 50 percent below $3\frac{1}{2}$ miles. Above 44 miles, the heavier constituents begin to disappear and the lighter elements, especially hydrogen and helium, become more common. The effect of powerful shortwave solar and cosmic radiation is much greater at these high elevations, and entirely new gases are formed out of the hydrogen, nitrogen, and oxygen atoms. Among such new compounds are nitrous oxides, ammonia, formaldehyde, and deu-

Figure 2-7 The major interior divisions of the earth.

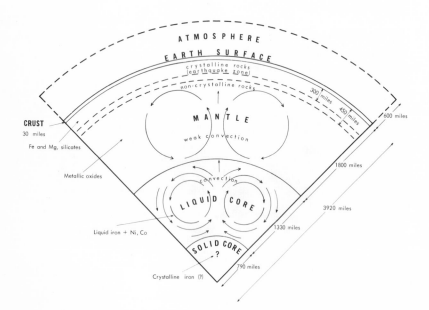

terium oxides. Still farther into space, at some 600 miles above the surface, the air is so rarified that it would be considered a good vacuum on earth.

There has been much speculation concerning the origin of O_2 in the atmosphere. A prevalent theory[1] suggests that this vital element could not have existed in our atmosphere until the earth had cooled to near its present temperature, and that it probably originated with the evolution of plant life. Prior to that time, oxygen probably occurred in the atmosphere mainly within carbon dioxide or water vapor. The atmosphere of Venus, laden with both carbon dioxide (CO_2) and water vapor, may well be similar to that of the earth prior to the development of its plant cover, and when the temperature of the surface was still too high to permit the condensation of water vapor into the liquid phase.

2-4 AGE OF THE EARTH Debate concerning the age of the earth has narrowed during the past 50 years, and there now is a general consensus that it is at least $4\frac{1}{2}$ billion years old. The amount beyond this still is uncertain. The main evidence for such great age consists of careful quantitative measurements of identifiable by-products of radioactive decomposition of certain elements in rocks. By knowing the rate of atomic breakdown and measuring the amount of the by-products produced as compared with the amount of unaltered material, a rough estimate of rock age can be made. The procedure is similar to the better-known method of dating organic material by measuring the quantities of *carbon 14,* an isotope of carbon that is produced only by radioactivity, except that the latter is not useful in dating carbon-

[1]Helmut E. Landsberg, "The Origin of the Atmosphere," *Scientific American*, vol. 189, no. 2, pp. 82–86, August, 1953.

aceous material much older than about 50,000 years.

The highlights of earth history, as recorded in the rocks of the earth crust, are presented in Table 2-2. Geologists and others continue to make corrections and additions to the record, but as in human history, there always will be many gaps or pages missing in the record. The length of geologic time is difficult to imagine. If a watch were to run at the rate of a year for each half-second, it would continue for over 70 years in order to cover the minimum time estimated since the formation of the earth crust. The length of time since earliest human recorded history would be represented by slightly more than the last half-hour, and man would have appeared on earth in roughly his present form only about 10 days previous. The Grand Canyon would have been excavated mostly within the last 2 to 3 weeks. The vastness of geologic time is almost as incomprehensible to man as the enormity of outer space.

Earth Motions and Earth-Sun Relationships

2-5 REVOLUTION, THE ZODIAC, AND OUR CALENDAR SYSTEM We have already seen that, like the other planets, the earth moves in an elliptical orbit with the sun at one of its elliptical foci. This motion is termed *revolution* and should not be confused with *rotation*, which refers to the turning of the earth on its axis. The revolution of the earth around the sun has several important implications for man. First, the elliptical orbit implies that the earth varies in its distance to the sun. In early January it reaches its nearest point (*perihelion*), a distance of about 91.5 million miles from the sun. In early July (*aphelion*) it is farthest from the sun, a distance of about 94.5 million miles

Table 2-2 The Geological Time Scale

Millions of years	Eras	Periods		Development of life forms
1 —		Quarternary	Recent Pleistocene	Ice age; development of civilization
	CENOZOIC (Recent life)	Tertiary	Pliocene	Proto-humans and humans
			Miocene	Rise of grazing animals
			Oligocene	First elephants
50 —			Eocene	First horses and camels
			Paleocene	Rise of early mammals
100 —	MESOZOIC (Middle life)	Cretaceous		Extinction of giant reptiles; rapid spread of flowering plants
150 —		Jurassic		First flowering plants; dominance of the giant reptiles
200 —		Triassic		First giant reptiles
	PALEOZOIC (Early life)	Permian		First conifers; first mammals
250 —		Pennsylvanian		First reptiles; extensive coal deposits from ferns, club mosses, etc.
		Mississippian		Abundant insect life
300 —		Devonian		The first land vertebrates (amphibians); abundant fish life
		Silurian		First land plants
350 —		Ordovician		First fish and corals
400 — 450 —		Cambrian		Abundant marine invertebrates. Multicellular organisms well advanced
	PRECAMBRIAN			
500 —	PROTEROZOIC (Former life)	Killarnean		Development of simple-celled marine life: bacteria, sponges, worms, algae
		Keweenawan		
1,000 —		Huronian		
	ARCHEOZOIC (Ancient life)	Algoman		Earliest-known life: marine fungi and algae (2,600–3,200 million years)
		Temiskaming		
		Laurentian		
		Keewatin		
3,200 —		Coutchiching		
	AZOIC (Without life)			

(see Figure 2-8). The difference of 3 million miles makes our northern hemisphere winters slightly warmer and our summers cooler than if the earth orbit were a perfect circle, but the contrast with the southern hemisphere is hidden by the greater heating and cooling caused by contrasting proportions of land and sea. Another result of the elliptical orbit is that the earth travels faster in its orbit when it is nearer the sun. Kepler's law of areas, which states that *the line joining a planet to the sun sweeps out equal areas in equal times* (see Figure 2-8), is a recognition of the fact that the orbital speed represents a vector that is the result of two velocities pulling at right angles to each other: one produced by the gravitational pull of the sun, and the other a transverse velocity imparted to the earth by some external force. Many astronomers believe that the latter was caused by the near passage of another sun at or about the time of the birth of the solar system. This varying speed of revolution has a side effect of influencing the rate by which the length of day changes at different times of the year.

Proof of the earth's rotation and revolution is not obvious, and it is not surprising that most people early in history, including careful observers of the heavens, believed that the heavenly bodies all turned in orbits around the earth. Acceptance of *heliocentricity* of the solar system proved stubborn, even after it was clearly demonstrated by Copernicus in the sixteenth century. One such proof, which was recognized at an early date but was misunderstood for many centuries, was the apparent motion of the sun along a 360° circuit through various stellar groups, or constellations, during a repetitive cycle having a period of about $365\frac{1}{4}$ days. Actually, the sun does not move through this path, but merely seems to, as the earth moves along its orbit (as shown in Figure 2-9). The year,

or an apparent circuit of 360° by the sun, is divided into twelve sky areas or stellar groups termed *constellations,* each of which covers a segment about 30° long along the solar progression path and 12° wide. The *zodiac* refers to this band that circumscribes the heavens and includes the twelve zodiacal constellations through which the sun appears to move during the course of a year.

Figure 2-8 **The orbit of the earth.** The shape of the earth orbit is an ellipse, resulting from the interaction between transverse velocity and radial velocity. Kepler's second law states that arc segments of equal area within the orbital plane represent equal periods of time. Thus, if the area of the arc segment *AB*-sun equals that of *ED*-sun, the time elapsed between *AB* along the earth orbit equals that of *ED*. For this reason, the earth travels faster when it is near the sun than when it is far from the sun.

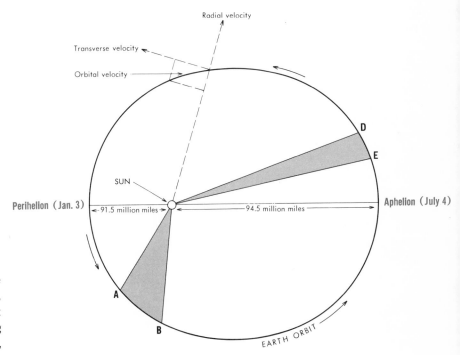

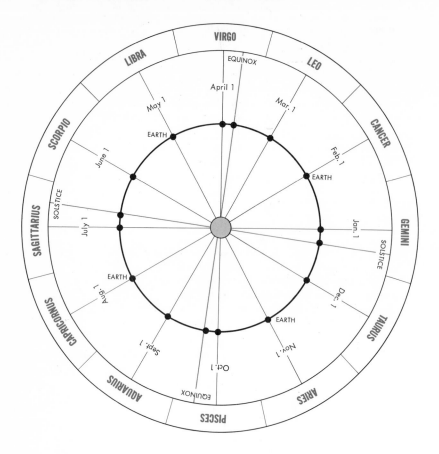

Figure 2-9 The zodiac. The signs of the zodiac represent constellations or star groups in the heavens through which the sun appears to move. The time during the year represented by each zodiacal position or "house" is the period when the sun rises within that particular constellation. In using the diagram, obtain the appropriate constellation by rotating the diagram until the sun lies to the right of the earth for a particular date. A continuation of the sight line to the sun toward the outer rim of the zodiac will indicate the zodiacal time for that date. To illustrate, on March 1 the sun rises in the constellation Aquarius. On July 1, it rises in Gemini. Complicating the zodiacal pattern is the precession of the equinoxes, in which the entire zodiacal wheel slowly rotates, completing a circuit about every 28,500 years.

The recognition of the varying position of the zodiacal divisions at certain places in the heavens at certain times of the year, and the progression of the solar path through these constellations during the year, was clearly developed by Babylonian stargazers 2,000 years before the time of Christ. The *year,* representing the annual progression of the sun's path, and the twelve zodiacal divisions of it, or *months* (which also roughly correspond to the lunar revolutionary periods), are ancient divisions of time that appeared in many unrelated cultures and do not belong solely to our own historical traditions. The zodiacal positions long have served as a natural calendar used to guide the time of seeding or

harvesting. The different but regular paths of nearby and easily visible planets (Gr. *plano-ros* = wandering) against the stellar backgrounds served as indicators of longer time periods. Great historical events were associated with the conjunction of certain planets and zodiacal positions, and an early belief in cyclical history led to the pseudoscience of *astrology,* which features an attempt to predict human events from a careful examination of planetary patterns.

Extreme accuracy in determining the length of the year is important for many reasons, including split-second timing of navigational instruments such as the chronometer; the accurate positioning of geodetic survey reference points and lines; and in astronomical observations of many kinds. The principal reference period is the *tropical year* or *year of the seasons,* which is the length of time the sun takes to complete its apparent journey through the zodiacal constellations and as observed from the earth. This period is 365.2422 mean solar days, or 365 days, 5 hours, 48 minutes, and 45.68 seconds. A different measurement unit, the *sidereal year,* from which star time is calculated for use by astronomers, is slightly longer than the tropical year, but it need not concern the geographer, whose eyes are fixed on earth.

Calendars that accommodate the solar day, tropical year, and lunar changes long have challenged man's ingenuity because of the lack of any simple even-numbered relationship between the sun, moon, and earth, and also because of differences in the accuracy of observation. Our own calendar is based primarily on a $365\frac{1}{4}$-day year, established in 45 B.C. by Julius Caesar. In this calendar, a day is added every fourth year (leap year). As we have seen, however, the tropical year is slightly less than a quarter of a day, and by the sixteenth century, the spring equinox, or the spring date at which the sun stands directly overhead at noon at the equator, had moved ahead in the year and appeared early in April rather than at its traditional date of around March 22. On the advice of his astronomer, Pope Gregory VIII arbitrarily moved the date back to the proper date in March and proclaimed that henceforth on the even century years, or those divisible by 100, no extra leap year day was to be added except on those divisible by 400. For example, 1700, 1800, and 1900 were not leap years, but 2000 will be. The new Gregorian calendar was not readily accepted, some people refusing it because of the days out of which they felt they had been cheated, and others who were against any papal proclamation. Russia, for example, did not adopt the Gregorian calendar until 1918, after the Bolshevik Revolution. In time, however, it became generally accepted, and though not perfect, should accommodate human needs for another 5,000 years or so.

Another calendar, still in somewhat limited use in the Islamic world, is a strict lunar calendar of 12 lunar revolutions per year, each consisting of about 29 days. The Islamic year, which is dated from the Hegira, or flight of Mohammed from Mecca to Medina, thus is entirely unrelated to the tropical year, being some 17 days shorter. The fasting month of Ramadan, for example, slowly progresses through the seasons from year to year at the rate of about 17 days per tropical year.

Lest we congratulate ourselves unduly for modernizing our calendar, it should be noted that both the ancient Greeks and the Mayans of Central America recognized the tropical year to be slightly less than $365\frac{1}{4}$ days, and the latter had devised an interlocking lunar and solar calendar that provided for a tropical year of 365.24 days as well as a 19-tropical-year cycle involving both the sun and the moon. In this cycle, 235 lunar revolutions, or 235 full moon phases, take place before the specific moon phase reoccurs at the same date of the tropical year. The Mayans used the more easily determined lunar calendar for their ordinary daily affairs but kept track of historical events and anniversaries by means of a notational calendar based on the tropical year.

2-6 ROTATION AND OUR TIME SYSTEM
The earth rotates on its *axis,* which is a straight line connecting the north and south poles and passing through the center of the earth. The period of rotation by definition equals one day, which in turn is divided into 24 hours, each consisting of 60 minutes, with 60 seconds per minute. Careful measurements reveal that slight changes in the length of the period of rotation take place regularly from time to time, caused by variations in lunar and planetary relationships. An extremely high tide, caused by the junction of solar and lunar gravitational forces, for example, can produce a slight slowing of the rotational speed. Our official day, as set forth

from observatories, is an average of these variations and is termed a *mean solar day.* The direction of earth rotation is from west to east, which explains why the sun and other heavenly bodies appear to rise in the east and set in the west. Likewise, as the earth spins on its axis, the circumference line that separates night from day, termed the *terminator,* or *circle of illumination,* will seem to an observer on earth to be passing westward as time elapses (see Figure 2-10).

The time of day at any place on earth is roughly correlated with the position of the sun in its apparent path across the sky from east to west. Unfortunately, at any given moment, an observer to the east and one to the west would not observe the sun at the same place in the sky. *Sun time* is the time of day that is determined by the position of the sun in its apparent sky course. Noon sun time thus becomes the time at which the sun stands at the meridian, the highest point in its sky path (see Figure 2-10). The only places

Figure 2-10 The relationship between sun time and longitude.

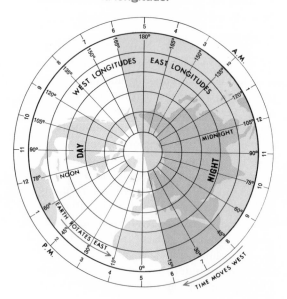

Figure 2-11 International standard time zones. [From H.O. Hydrographic Chart No. 5192]

that have the same sun time are those located along the same *meridian,* an earth semi-circumference that connects the two poles. As indicated earlier, there is a slight variation in the period of rotation during the year; consequently, our time system is based on an average of *mean solar time.*

The mean solar time at locations to the east of any reference point always is ahead of or faster than at the reference point because of the direction of the earth rotation from west to east. Locations to the west will have mean solar times that are behind or slower than the reference point at any given moment. An accurate comparison of sun time between any local position and that of an arbitrary reference meridian is the principle behind the calculation of *longitude* (see Section 2-12). The various systems of time zones used around the world represent attempts to provide rough north-south zones with a uniform designation of time. The global time system will be treated at greater length in Section 2-7.

2-7 STANDARD TIME ZONES, THE INTERNATIONAL DATE LINE, AND THE EQUATION OF TIME The basis for the global standard time zones is that all places throughout a zone of longitude 15° side and centering on a meridian that is a multiple of 15 (30, 45, 60, etc.) have a local or *standard time* equivalent to the mean solar time at the appropriate central meridian. Thus, our Central Standard time zone in the United States centers on the 90° W meridian, and if Greenwich time (often referred to as UT— Universal time or Zulu time) were 12 o'clock noon, St. Louis time would be 6 A.M. Unfortunately, the arrangement of continents, countries, and major surface communication lines rarely fits into neat longitudinal zones.

The Planet Earth: Its Place in the Solar System; Its Dimensions and Portrayal on Maps

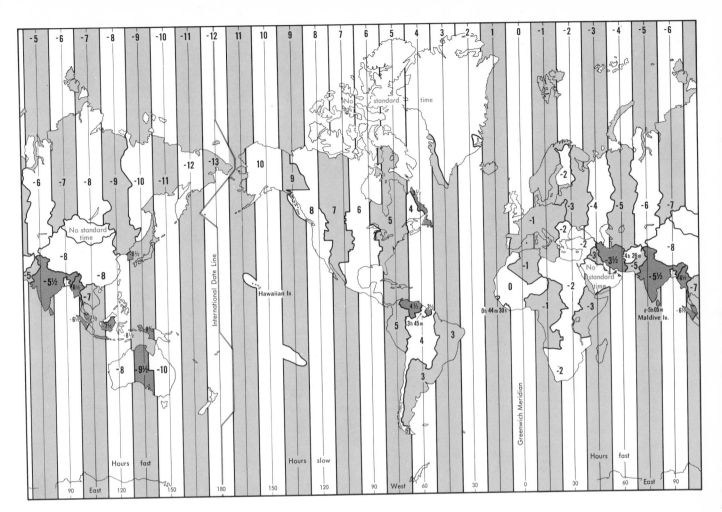

For local convenience, then, the time zone boundaries have been adjusted to the east and to the west in most zones. Figure 2-11 shows the time zones for the world. Numerous variations from the global theoretical system can be observed. Note how, with the exception of Britain and Portugal, all of northwest Europe eastward to the Soviet Union lies in the first time zone east of Greenwich. Several small countries that lie near the time zone margins base their local standard times on the half hour so that they may be nearer to mean solar time. This is especially common in southern Asia. A few countries use a local time that is based on the sun time at their capital city; they include Liberia, Afghanistan, and some of the Pacific island groups. Saudi Arabia officially is not tied to world standard time since its time is based on the Muslim lunar calendar, although clocks based on standard time are in use in many parts of the country. The Soviet Union has advanced all of its time zones one hour faster than standard time, hence remains on what we term *daylight saving time* throughout the year. The faster daylight sav-

Figure 2-12 The analemma. [With permission from John Wiley & Sons]

ing time is popular in many parts of the United States during the summer season, especially in urban areas. Farmers are less enthusiastic about it, since animals do not easily alter their daily rhythms.

People traveling around the world from east to west must set their watches back one hour each time they cross a time zone, or every 15° of longitude. If they continued this throughout their entire trip, they would return one day behind the local calendar. In fact, Magellan's ship log illustrated this puzzling point, in that although it recorded each day of sunrise and sunset, the log was one day short upon the return to the port of departure. Similarly, anyone going around the world from west to east would find his calendar one day ahead of the departure point. In order to avoid confusion in business transactions globally, an arbitrary line has been designed as the *international date line* (see Figure 2-11), where a change in days takes place. It is located at or near the 180° meridian, diverging from it in a few places to include groups of islands. Any traveler crossing the international date line from east to west adds a day because he has been "losing" time; a traveler from west to east loses a day because he has been "gaining" time. A flight passenger, for example, leaving Fiji at 7 A.M. on Friday and flying to Honolulu, would reach the latter about 2:30 P.M. on Thursday after a flight of $5\frac{1}{2}$ hours. Between the two there are 2 hours of standard time, plus the international date line change of a day.

Converting sun time to standard time, or vice versa, involves the use of an *equation of time* that gives the variation between *mean solar time,* from which standard time is derived, and *apparent solar time,* or the sun time actually observed at any local point. The *analemma* shown in Figure 2-12 contains a graphic representation of the equation of time. Note that at two opposite times of the year (September to December and April to June), apparent solar time is ahead of mean solar time, and at the other times of the year it is behind mean solar time. This variation is the net result of two separate factors: (1) the variation in orbital speed and (2) the apparent movement of the sun in an ecliptic plane rather than the equatorial plane from which our time is calculated.

2-8 PARALLELISM, THE SEASONS, AND PRECESSION The earth axis does not lie vertical to the *ecliptic,* or plane of the earth orbit around the sun. Instead, it is at an angle of $23\frac{1}{2}$° from the vertical, as shown in Figure 2-13. Further, the axis remains parallel to itself at all places along the earth orbit, or at all times of the year (see Figure 2-14). The implications of this parallelism are important. One effect is to fix the *north celestial pole,* or the zenith point above the North Pole, at the same spot among the background of stars throughout the year, or about a degree from Polaris, sometimes referred to as the North Star. Away from the North Pole, Polaris always indicates an approximately northerly direction of sight.

The most important effect of parallelism, combined with earth revolution and rotation, is the change of seasons it produces on earth. A close examination of Figure 2-14 will reveal the following:

1. The *terminator,* or *circle of illumination,* the line separating night from day, while it always bisects the equator, moves through a zone 23.5° wide surrounding both poles.
2. The result of (1) (eliminating the effect of atmospheric refraction) is that the poles

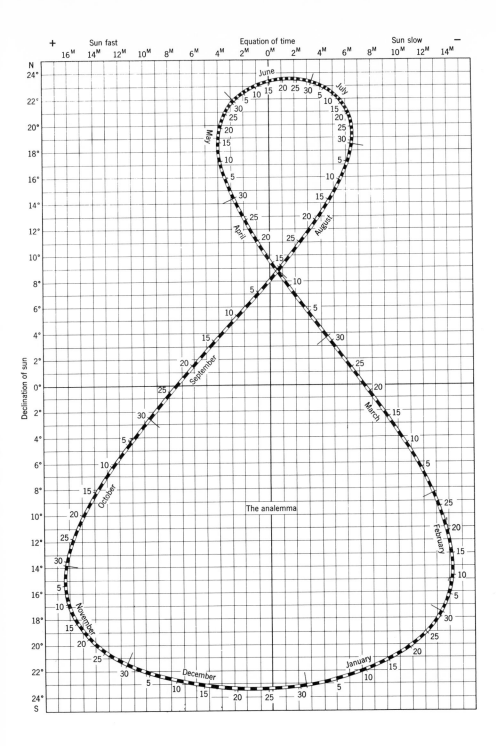

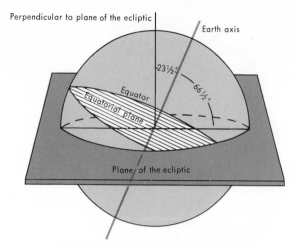

Perpendicular to plane of the ecliptic

Earth axis

23½°

66½°

Equator

Equatorial plane

Plane of the ecliptic

Figure 2-13 Angular relationship between the earth axis and the plane of the ecliptic.

have 6 months of darkness and 6 months of daylight. The length of day and night at the equator is 12 hours throughout the year.

3. The length of day exceeds 12 hours during half the year at all locations away from the equator and is less than 12 hours during the remainder of the year. The difference between the duration of night and day increases with latitude.

4. The *Arctic* and *Antarctic Circles,* located 23.5° from the North and South Poles, respectively, delimit the polar zones of possible 24-hour daylight and darkness.

5. The vertical radiation from the sun passes through a zone on earth that is 47° wide, from the *Tropic of Cancer,* $23\frac{1}{2}$° north of the equator, to the *Tropic of Capricorn,* $23\frac{1}{2}$° south of the equator.

If the earth axis had been vertical to the ecliptic at all times of the year, or had always presented the same angle toward the sun throughout the year, there would have been no change in the length of day and night or any alteration in the directness of solar radiation during the year—the two major direct causes of summer and winter.

Four critical times of the year have long been recognized as marking the seasonal changes on earth. Two of them, termed the *vernal* (spring) *equinox* (Latin: equal night) and the *autumnal equinox,* occur on or about March 22 and September 22, respectively. Because of slight variations in the orbital path of the earth from year to year, caused by external gravitational forces, the exact time of the equinoxes does not occur on the same day each year but varies back and forth over a 1- to 3-day period. The two equinoxes represent the neutral points in the seasons, when the days and nights are of equal length throughout the world, when the sun lies directly overhead at noon at the equator, and when it splits the horizon at the two poles (see Figure 2-14B). The angle which the sun makes to the zenith at noon anywhere on these dates is equal to the latitude of that place.

The other two critical dates, termed *solstices* (Latin: sun's stance), occur on or about June 21 and December 21. These are termed the summer and winter solstices, respectively, in the northern hemisphere. The seasons are reversed in the southern hemisphere. On these dates, shown diagrammatically in Figure 2-14, the days and nights are longest and the vertical rays of the sun are aligned 23.5° north and south of the equator, or at the two tropics. Although these solstices are termed the middle of summer and winter, they do not mark the periods of maximum summer or minimum winter heating. As will be seen in Chapter 7, there is a noticeable lag between the receipt of solar radiation and

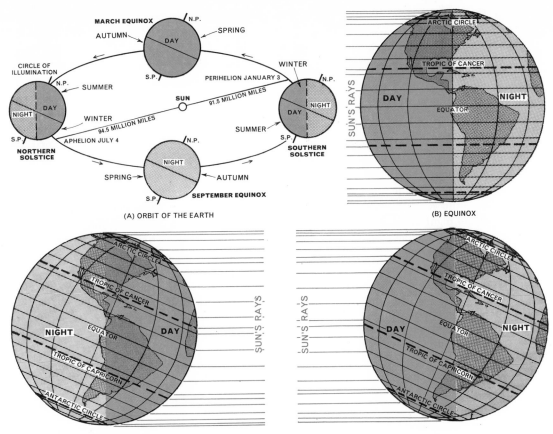

(A) ORBIT OF THE EARTH

(B) EQUINOX

(C) NORTHERN HEMISPHERE SUMMER SOLSTICE

(D) NORTHERN HEMISPHERE WINTER SOLSTICE

its ultimate role in heating the atmosphere.

Several ancient peoples, including the Egyptians, Babylonians, and Mayans, carefully noted the time of the equinoxes and solstices by observing the changing position of solar rays in passing through a small aperture into a darkened room, or by noting changes in the length of shadows at noon. Of special significance to them, as with many other cultures, was the determination of the vernal equinox, which marked the beginning of the summer season. This also marked the beginning of a new growth period, the time for planting, a time for welcoming the "return of the sun," and a natural time for festive celebrations and various forms of fertility rites. In their careful dating and solar observations, the Greeks and Mayans noted an interesting phenomenon—that over a long period of years, the rising sun at the time of the vernal equinox shifted position slightly against the background of stars. This has been termed the *precession of the equinoxes.* Hipparchus, the famous Greek astronomer and geographer at the library in Alexandria, noted this some two centuries before Christ. At that time, the vernal equinoctial sunrise took place in the constellation of *Aries,* the goat, but since then has moved westward into the constellation *Pisces,* the fish (see Figure

2-9). The complete circuit of the zodiac by the equinoxes requires 285 centuries. This precession also influences the position of the north celestial pole. At present, as noted earlier, it is only about a degree from Polaris, the North Star, and is slowly reducing that distance. A century from now, it will have reached its nearest point, about half a degree away, and then will slowly recede from Polaris. Some 18,000 years from now, our polar star will be one of the brightest stars in the heavens: *Vega,* in the constellation *Lyra.* Finally, in the year A.D. 30,500, Polaris again will be the night guide for people living in the northern hemisphere.

An interesting side light to precession is the complication that it has made in astrology. This pseudoscience is extremely old. Early in its history, certain times of the year were associated with zodiacal positions, and these, in turn, were given certain human attributes. Most horoscopes that attempt to predict the dominance of human qualities based on the time of birth still utilize zodiacal positions that are more than a thousand years old. A person born in early November, for example, is noted as having been born "under the influence of" or "in the house of" *Scorpio,* the crab, although at present the rising sun in early November has shifted slightly and is now in *Libra,* the scales. Because of precession, the zodiacal wheel has turned almost a full constellation since the time the zodiacal positions originally were defined.

The cause of precession involves a somewhat complicated interaction of forces, including the rotation of an oblate (flattened) spheroid, solar gravitation, and the tilting of the earth axis away from the perpendicular to the ecliptic. The effect is somewhat similar to that of the slow turning of the axis of a rapidly spinning top (see Figure 2-15). It has no effect on the angular relationship between the earth and the sun.

Earth Measurements: Form and Location

2-9 SHAPE AND SIZE OF THE EARTH A view of the earth as seen from the moon

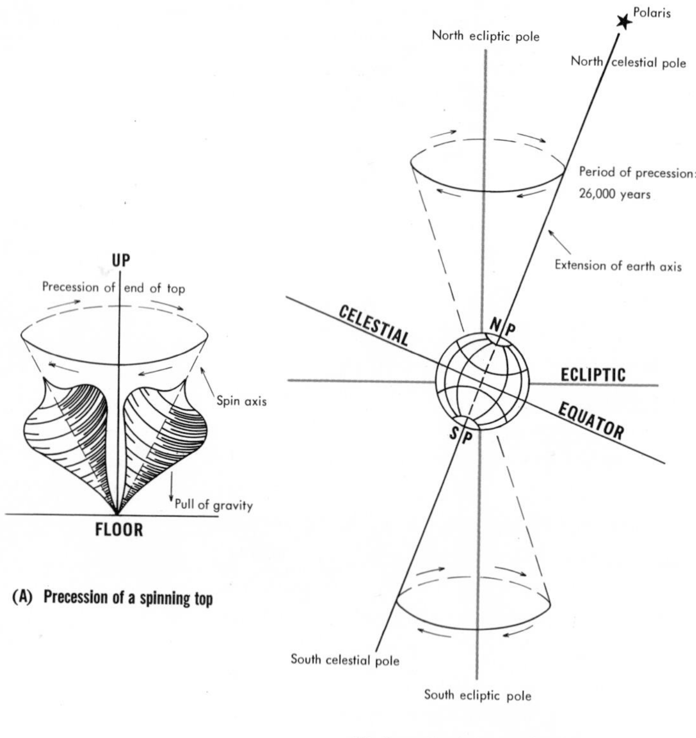

(A) Precession of a spinning top

(B) Precession of the celestial poles

Figure 2-15 Precession of the celestial poles. The spinning of the earth on its axis (rotation), operating in opposition to the gravitational pull of the sun and moon, produces a slow migration of the celestial poles along a circular path against the background of stars. The period for this circuit is about 26,000 years. The precession is analogous to the circular movement of the upper end of a spinning top, as in *A*. This precession is caused by the interaction between the spin of the top and the gravitational pull of the earth.

reveals the terrestrial planet to be a remarkably smooth, spherical body. Careful measurements on earth, however, indicate that there are many distortions from a perfect sphere. One such variation from a true sphere is caused by the rotation of the earth, which has produced a slight flattening at the poles and a bulging near the equator. The difference between the polar and equatorial diameters is about 27 miles, or a little more than .3 of 1 percent. If the earth were reduced to a circle having the diameter of a normal classroom blackboard (about 3 feet), this variation from a true sphere would be contained within the width of a chalk line and would not be noticeable from the back of the room.

Despite the relatively small distortion from a true sphere, measurements continue to be made in order to obtain greater accuracy in delineating the exact shape and size of the earth. Extreme precision in this is needed for many reasons, including the guidance of intercontinental missiles, land survey reference systems, and the instrument control in gravimetric measurements. The later are widely used in geophysical prospecting for mineral deposits. Two methods are currently being used in the attempt to sketch the accurate shape and size of the earth. One of them utilizes extremely precise measurements of the variation of a plumb bob line or line of gravity from the *astronomical latitude*.[2] Figure 2-16, which greatly exaggerates the oblateness of the earth, illustrates the method. On a perfect sphere of uniform material, the line of gravitational force, represented by the plumb bob line, will

point to the center of the earth and be the same as the earth radius at that point. In any *ellipsoid*, or flattened sphere, however, the center of mass is not at the center; hence, a gravity line will deviate from the radius except at the poles and the equator, and a vertically extended plumb bob will not point to the true astronomical latitude. If the exact latitude of a point is known, star tables will give the precise location in the heavens of the appropriate astronomical latitude at any moment of time. The difference between the line of gravity and the astronomical latitude

Figure 2-16 Measurement of earth oblateness. Two spheroids are shown, one a perfect sphere and the other highly flattened. In the perfect sphere, the angle between the plumb bob zenith (Up) and the ecliptic (astronomical latitude) always equals the true latitude, and the plumb bob points to the earth center, regardless of latitude. With flattening, the plumb bob does not point to the center of the spheroid (except at the poles and equator). The degree of flattening (oblateness) is measured by noting the difference between the astronomical latitude and the true latitude.

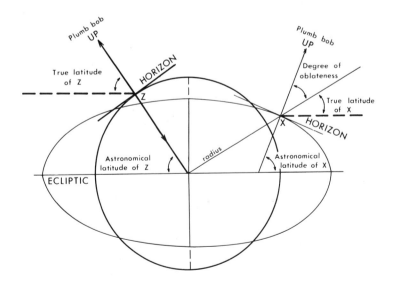

[2]The *astronomical latitude* of a locality is the angle that its zenith line (extension of the earth radius) makes to the celestial equator, or outward extension of the earth equatorial plane. It is represented by an imaginary line drawn around the heavens, paralleling the celestial equator. The zenith point completes this circuit once every 24 hours.

gives a measure of the variation of the earth surface from a true sphere at that point.

A second method involves careful triangulation, utilizing a point in space, such as an earth satellite, and extremely precise control points on earth. By such precise triangulation control, slight variations in the length of the outer circumference can be located. All of our new measurements indicate that the exact shape of the earth, or *geoid,* although never far from a true sphere, is a highly complex surface with slight ridges and depressions in many places, quite apart from the better-known flattening at the poles and bulging at the equator. Further, the irregularities that appear to be great to the earth inhabitants, such as the continental mountain ranges and the deep clefts of the oceanic trenches, are minute in comparison with the entire spheroid. Were only these irregularities involved, the earth surface would be relatively as smooth as a billiard ball.

The vital feature in the search for accuracy in the shape of the geoid is the identification of the true horizontal, or plane at right angles to the earth radius. This must always parallel the surface of the geoid. The exact determination of the geoidal surface at any point is a difficult and time-consuming task that involves precise astronomical observations. For practical reasons, mapping procedures cannot wait for the accumulation of new geoidal measurements. A substitute procedure has been devised. It involves creating a hypothetical ellipsoidal model of the earth which utilizes all of the best geoidal measurements available but averages or smooths them out into a true ellipsoid or flattened sphere, and one that has the same total surface area as the true geoid with all of its bumps and depressions. Tables can easily be prepared from the model, indicating the degree of variation from a true sphere, and hence a true horizon, at any position on earth. Such deter-

minations are reasonably accurate and are the best possible in many locations.

Several model ellipsoids have been constructed during the past century. One of the newest, that of Irene Fischer, utilizes recent measurements obtained from artificial satellite triangulation. Future refinements undoubtedly will continue to be made. Meanwhile, it is obvious that it would be impractical to change all basic land survey maps each time minor corrections were made in the ellipsoidal models, and most parts of the world continue to use maps based on ellipsoids that were developed more than 50 years ago. Table 2-3 presents the major ellipsoids that are in use today throughout the world as bases for the construction of a universal military grid mapping system. North American military grids still are based on the ellipsoid developed by A. R. Clarke of the English Ordnance Survey in 1866, while British and European grids are based on the international spheroid developed by J. F. Hayford of the U.S. Coast and Geodetic Survey in 1909.

2-10 LOCATION OF POINTS ON EARTH: LATITUDE AND LONGITUDE There must be fixed, recognizable base points, base lines, or base areas in any system of locating points, from which directions and distances can be plotted. The relationship of position on earth to position with respect to the background of stars and other luminous bodies in the heavens and to the facts of earth rotation and revolution provides us with a reference system for location and direction that is almost as old as man. Navigation using star sightings is at least 3,000 years old. Lately, however, we have utilized the principle of stellar and solar positioning as a means of establishing a rectangular network structured to the ellipsoidal shape of the earth from which the position of any point on earth can be rela-

Table 2-3 Global Ellipsoids Used in Military Mapping			
Ellipsoid	Equatorial radius (meters)	Polar radius (meters)	Flattening ratio $\dfrac{ER - PR}{ER}$
Everest (1830)	6,377,276	6,356,075	.003324
Bessel (1841)	6,377,397	6,356,079	.003343
Clarke (1866)	6,378,206	6,356,584	.003390
Clarke (1880)	6,378,249	6,356,515	.003408
Hayford (1909)	6,378,388	6,356,912	.003367

SOURCE: "The Universal Grid Systems," U.S. Army TM 5-241, U.S. Air Force TO 16-1-233. U.S. Government Printing Office, Washington, D.C., 1951.

tively easily determined. This is commonly referred to as the *geographic grid* and involves the angular measurements termed *latitude* and *longitude.*

The two most convenient reference points on the earth are the two poles that represent the two ends of the axis about which the earth rotates from west to east. The position of the poles, determined by their relationship to the celestial poles, helps to determine a reference line, the *equator,* which is a *great circle* of the earth and midway between the two poles. A great circle may be said to represent the outer arch of the earth surface. Any plane that bisects the earth sphere would form a great circle where it intersects the outer surface of the earth. The shortest angular distance between any point on earth and the equator is termed its *latitude.* The terms *north* and *south latitude* are used to designate measurements north or south of the equator, respectively. The measurement is given in units of a quadrant of a circle, that is, in degrees, minutes, and seconds, totaling no more than 90°. The latitude of the equator thus is 0° and that of the North Pole 90° N.

Lines are used on maps and globes to connect different positions having the same latitude; they are known as *parallels.* If we examine the imaginary lines on earth that represent equal latitude (see Figure 2-18), we discover the following facts:

1. Only one parallel, the equator, is a great circle.
2. The parallels gradually diminish in

Figure 2-17 An infinite number of great circles may be located on the earth surface.

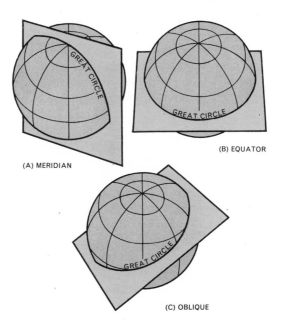

(A) MERIDIAN

(B) EQUATOR

(C) OBLIQUE

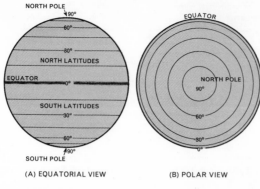

Figure 2-18 Equatorial and polar view of parallels and latitudes.

length around the earth as latitude increases, finally becoming a point at the poles.

3. The length of a degree of latitude varies only slightly in passing from the equator to the poles. At the equator, owing to the slight bulging of the geoid, it is 68.704 miles. This figure increases with latitude to 69.407 at the poles, owing to the flattening at high latitudes.

Longitude is the angular distance on the earth surface measured 180° east and west of some arbitrary great half-circle that connects the two poles. The imaginary lines that connect points of equal longitude on earth are known as *meridians*. The arbitrary meridian chosen as the reference line from which to measure longitude is known as the *prime meridian*. Different prime meridians were in use throughout the world for many years, and major nations usually selected the meridian that passed through their capital as their prime meridian. At present, nearly all of these have been abandoned, and through common agreement, the meridian that formerly was used by the British Empire became the adopted prime meridian. For convenience, this meridian passed through the site

of the Royal Observatory at Greenwich, England, a suburb of London. Although the observatory no longer is in Greenwich, having been forced to move by the expansion of the city with its vibratory interferences in astronomical observations, the meridian continues to retain its original location. A brief examination of the pattern of meridians (see Figure 2-19) reveals the following:

1. Meridians are halves of great circles.
2. They intersect the equator and all parallels at right angles.
3. They merge at the poles and are farthest apart at the equator.
4. The value of a degree of longitude, measured along the parallels of latitude, varies from about 69.172 miles at the equator to 0 at the poles.

The combination of the angular measurements of latitude and longitude gives us a system whereby points on the earth can be located with considerable accuracy. For example, the designation lat 35°30′25″ N and long 165°15′35″ E will locate a point within 101 feet of its exact position on earth. For greater accuracy, decimals of seconds can be used. Our finest instruments today can pinpoint geographic location down to within 10 feet. Obviously, such accuracy is not possible to an observer who is taking astronomical readings with a hand sextant on the deck of a rolling and pitching ship.

2-11 DETERMINATION OF LATITUDE
The geographic grid system of locating points on earth would be impractical if it required direct measurement of distances from the equator or the prime meridian. Such a system must be adaptable to on-the-spot observation. The direct measurement of latitude from observations at a given point is based on the principle of locating oneself with reference

to observable points that may be considered as fixed in space beyond the earth: stars that have known positions with reference to the earth equatorial plane.

The basic principle of latitude determination is illustrated in Figure 2-20. Assume that Polaris (the North Star) is directly over the North Pole. An observer at point X observes Polaris to be 40° above the horizon,[3] or at angle Z. Plane geometric relationships indicate that this angle is equal to angle XOE, assuming that ON and XP are parallel. This assumption may be made because of the great distance of Polaris from the earth. Since angle XOE is the angular distance of point X from the equator, it also represents the latitude of X. *The angle of elevation of Polaris, then, equals the latitude at the point of observation.* As seen earlier in this chapter, however, Polaris actually is not directly above the North Pole but lies about one degree from it, hence appears to move in a small circle of its own as the earth turns on its axis (see Figure 2-21). Compensation for this small apparent orbit of Polaris in making more accurate positioning involves noting the exact position of Polaris in its apparent circuit and consulting star tables that indicate the relationship between latitude and star locations for given times of the day and year. These star tables[4] are exceedingly important in celestial navigation and are carefully checked periodically at astronomical observatories. Alterations are made when appropriate.

[3]A crude artificial horizon may be simply established with a bubble level in a sextant or some other instrument used to determine elevation angles. A slight error is involved here owing to the oblateness of the earth, but it is not compensated unless extreme accuracy is desired.

[4]The major publication that contains essential information concerning star and solar declinations (elevation angle above the celestial equator) and other data is the *Air Almanac*, issued three times each year and produced jointly by the American and British governments. It is distributed in the United States by the Superintendent of Documents.

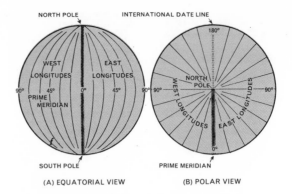

Figure 2-19   Equatorial and polar view of meridians and longitudes.

Figure 2-20   Determination of latitude, using Polaris.

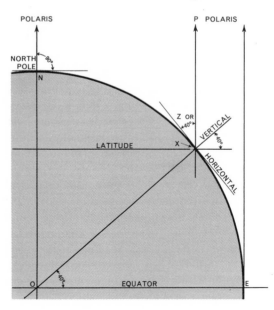

Figure 2-21 Time exposure photograph of the night sky near the north celestial pole. This photo shows the apparent path taken by the stars in the northern sky as they seem to turn about the north celestial pole. The North Star (Polaris) is not the innermost spot of light at the center, but the first, fairly bright arc to the lower left of center, indicating that Polaris does not lie directly above the North Pole on earth. [Lowell Observatory]

Figure 2-22 Determination of latitude using solar observation. At the equinoxes, the latitude equals the reciprocal angle of the sun's elevation, or the angle which it makes to the zenith. This is $90° - 51° = 39°$ at Washington, $51°$ being the angle which the sun makes above the horizon at noon on March 22 or September 22. At the solstices, the same reciprocal angle equals the latitudinal distance of the observer from the tropic where the sun is vertical. On June 21, for example, the reciprocal angle at Washington equals $90° - 74\frac{1}{2}° = 15\frac{1}{2}°$ north of the Tropic of Cancer, thus yielding a latitude of $15\frac{1}{2}° + 23\frac{1}{2}° = 39°$.

The determination of position using star sightings of Polaris has a number of weaknesses. At low latitudes it lies near the horizon. Hence, sightings there would be influenced markedly by refraction, or the bending of light when passing through our atmosphere. Accurate sightings should not be used on stars that lie below the upper half of the sky dome. Since Polaris is not visible south of the equator, or in the daytime, its use is limited to clear night readings at relatively high latitudes. The same principle, however, can be used in connection with observations of other easily identifiable stars. All that is needed is a measurement of a particular star's position in its apparent sky path and a comparison with latitudinal positions for any given moment of the year as presented in the star tables. For greater precision, the positioning of three different stars generally is used, each in a different segment of the upper sky arc, and the average of the latitude calculations taken. This tends to minimize slight human errors in handling the instruments.

Latitude determinations also can be made during the daytime. They involve a method similar to star sightings but using solar elevation angles instead (see Figure 2-22). The discussion of earth-sun relationships in Section 2-8 indicated that the sun slowly changes its apparent daily circuit across the sky through a zone 47° wide during the course of a year, shifting northward and southward with the seasons. For example, the sun stands directly at the zenith at noon sun time on March 22 (approximately), as seen by an observer located at the equator. As the year progresses, to the same observer the sun's circuit would appear to decline into the northern half of the sky, reaching a low point exactly 23.5° north of the zenith on June 21, then slowly climb again, reaching the zenith at the autumnal equinox in September, and

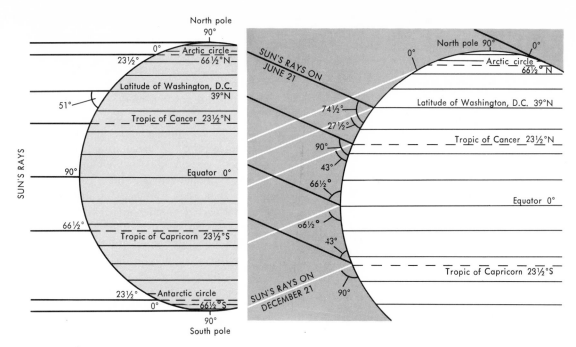

A. At the Equinoxes

B. At the Solstices

decline southward until the winter solstice on or about December 21, after which it would move northward again.

Similarly, an observer at 10° N at noon on the vernal equinox would observe the sun to be 10° south of the zenith, and an observer at the same moment who observed the sun to be 50° north of the zenith would be located at 50° S lat. In short, the angle which the sun makes to the zenith when it reaches its meridian (the highest point in its sky path) indicates the angular distance of observer from the sun's vertical location at that moment. This latter position is termed the *solar declination* and varies from $23\frac{1}{2}°$ N to $23\frac{1}{2}°$ S. Solar tables presented in published documents like the *Air Almanac* provide the data for such declinations throughout the year. A crude determination of the solar declination is frequently printed on the surface of pub-

lished globes in the form of an asymmetric figure 8, termed an *analemma* (see Figure 2-12). To illustrate the procedure involved in solar determination of latitude, the following is presented.

An observer on February 21, 1960, "shooting" the sun by means of a sextant, notes that as the sun reached its highest point (noon apparent sun time), the angle of elevation of the solar disk reads 67°36′14″. He also notes, by comparing his sun time with that of Greenwich (obtained from an accurate clock or chronometer), that he is at longitude 18°45′10″ W (see Section 2-12). His first calculation is to find the angle that the sun makes to the zenith, as follows:

$$\begin{array}{r} 89°59′60″ = 90° \\ -67°36′14″ \\ \hline 22°23′46″ \end{array}$$

He therefore is 22°23′46″ distant from the place on earth where the solar rays are vertical at that moment (the solar declination). He also notes that the sun is in the northern half of the sky, thus placing him south of the solar declination. Consulting his *Air Almanac,* he notes that the solar declination for noon at the meridian of 18°45′10″ W and on February 21, 1960, is 10°58′14″ S. His own latitude thus becomes:

$$
\begin{array}{r}
10°58′14″ \\
+\,22°23′46″ \\
\hline
32°81′60″ \text{ S or: } 33°22′00″ \text{ S lat}
\end{array}
$$

Observations during daylight hours other than at solar noon can be taken by calculating the exact position of the sun in its arc across the sky (determining true sun time) and determining what its elevation would be at noon. As with star observations, carefully prepared and periodically revised tables of the solar declinations year after year are essential to accurate navigation. The factors that influence irregularities in the earth orbit around the sun (which, in turn, affect solar declinations) include variations in orbital speed, precession of the equinoxes, gravitational effects of other planets singly or in conjunction, and lunar-solar relationships.

2-12   DETERMINATION OF LONGITUDE
The determination of longitude is based on the relationship between sun time and the period of the earth rotation. If the earth rotates completely on its axis once every 24 hours, it must turn 15° of longitude every hour (see Figure 2-10). If an observer finds that the sun is at its highest point, or "at the meridian," at a time when his chronometer, which has been set to read prime meridian time, reads 1 P.M., then he is 1 hour, or 15°, west of the prime meridian, since the earth rotates from west to east. His longitude thus is 15° W and is based on a comparison of local sun time with sun time at the prime meridian. The chronometers that are used to record Greenwich time are corrected—frequently down to the last second—by

means of radio signals sent out from astronomical observatories. Observations for the determination of longitude can be made at times other than solar noon. Solar time can be measured at any time during the day by noting the position of the sun with respect to its course across the sky. Also, solar time can be determined at night, utilizing star (sidereal) time, by calculating from the position of certain stars in their apparent path across the heavens and then translating this into solar time by means of star tables. As noted earlier, a sidereal day differs from a solar day by about 4 minutes.

The basic principle underlying the determination of longitude has been known for more than 2,000 years, yet its accurate measurement proved to be a remarkably stubborn problem until the eighteenth century. Fairly accurate latitude measurements were possible long before this. The difficulty lay not in the solar or star observations, but in the lack of an accurate timepiece, essential in comparing meridional sun times. The invention of an accurate timekeeping device had to await the development of the compensatory balance wheel combined with an escapement lever, which was introduced by Harrison in England in 1735 after a long period of sponsored research in western Europe.

Another problem lay in the difficulty of obtaining precision in astronomical readings when the various observational objects (stars or the sun) are in constant movement in their apparent orbits across the sky, owing to earth rotation. A difference of one second in a star sighting would mean a difference of about a third of a mile in longitude at the equator. Rotational speed, of course, decreases toward the poles. The coordination of star or sun meridional transects, then, with chronometer readings has been a troublesome problem in precise longitudinal measurements.

Distances and Directions

2-13 THE EARTH AND OUR MEASUREMENTS OF LENGTH Setting appropriate standards for measuring lengths or distances on earth has been in constant evolution throughout history and continues today. Some cultures, such as the Eskimo, measured distances by means of units of travel time. Most peoples, however, selected some arbitrary unit of length, one that could easily be divided or multiplied to yield flexibility. Many units, such as the *cubit* (Biblical) or *stadium* (Greek), were in common usage for centuries and were later discarded. Our English system of length, used throughout all English-speaking countries today, is one of the most arbitrary and impractical systems ever devised. Its divisions are in the following illogical mathematical sequence: 12 (inches), 3 (feet), $5\frac{1}{2}$ (yards), 4 (rods), 10 (chains), 8 (furlongs), 1 (mile). For our purposes in dealing with the earth surface, the English or *statute mile* warrants explanation. Originally it was derived from a basic Roman distance unit, the *milia passuum,* or thousand paces (double steps), which was the distance covered by a Roman legion in a thousand paces during a normal march, or a distance of 4,859 feet. This was correlated with the size of the earth, and 62 Roman *milia* was given as the length of 1 degree of latitude by Claudius Ptolemy in the second century A.D.[5] Scholars in the fourteenth century altered this figure to 60 *milia* per degree in order to conform better to the generally accepted size suggested earlier by Pausodonius. Later, following the voyages of exploration, when the earth was seen to be much larger than had been believed, two alternatives were faced:

to increase the size of the old English mile or to establish a new unit based entirely on a fraction of the earth circumference. Both were followed.

The first alternative was faced by adjusting the English mile to fit both the English field system and the earth arc. The standard length of the English field was the *furlong,*[6] or 660 feet. Eight furlongs, or 5,280 feet, became our statute mile, and this agreed closely with 69 miles per degree of earth arc.

English sea navigators needed a more accurate unit for the measurement of sea distances and one that could be more readily converted into degrees of a great circle. So, they developed what was termed the *nautical mile,* or the distance subtended by one minute of earth arc. Ocean speeds were given in *knots,* which are nautical miles per hour. Since there are 60 nautical miles per degree of earth arc instead of 69 statute miles, the nautical mile exceeds the length of a statute mile by about 15 percent. As careful measurements of the earth yielded greater and greater accuracy, it became clear that the value of a nautical mile could vary, depending upon the particular great circle that was used for reference. For many years the official nautical mile in the United States was based on an average length for a degree of latitude, derived from the Clarke hypothetical spheroid of 1866, a distance of about 6,080.2 feet. In the interest of combining accuracy and world standardization, the United States in 1954 officially recognized an arbitrary international figure of an even 1,852 meters, or 6,076.10+ feet, for the length of a nautical mile. The British continue to use the slightly

[5]A most useful summary of the history of earth measurement systems is contained in William M. McKinney, "Measuring the Earth," *The Journal of Geography,* vol. LXIV, no. 8, November, 1965.

[6]The *furlong,* or *long furrow,* equaled 10 chains of 66 feet each. It is interesting to note here that the *acre,* our basic unit of area, was derived from a measurement of length, in that it was defined as the amount of English field that could be ploughed by a yoke of oxen in one day, or an area 66 feet wide and a furlong in length.

more accurate figure of 6,080 feet, but this is not easily translated into the metric system.

A third type of mile is the *geographical mile,* which is based on the length of a minute of earth arc measured along the equator, assuming the Clarke spheroid of 1866. It is rarely used and measures 6,087.08 feet. The three official mile units in use in the United States thus are as follows:

1 statute mile = 5,280 feet
1 nautical mile = 6,076.1 feet
1 geographical mile = 6,087.08

The metric system originated in France during the eighteenth century. In 1791, following a long period of investigation sponsored by the French government, a new unit of measurement was proposed that would consist of one ten-millionth of a quadrant, or quarter of the earth circumference (polar). An extremely careful measurement of a section of a meridional arc in western Europe (between Barcelona and Dunkirk) was made, and a platinum bar of the required length, the *metre* or *meter,* was deposited in the French government vaults in 1799. Corrections in the value of the length of the meter continued to be made, and as precision grew, a meter became more difficult to define in terms of a reference bar of metal. Alloys of platinum and iridium were tried to reduce variations in the reference meter due to temperature changes, and air conditioning was used to reduce this possible error further. Finally, in 1960, an arbitrary figure close to the corrected meter was selected that could be duplicated in laboratories anywhere. This is a length that is 1,650,763.73 times the wavelength of the orange-red portion of the spectrum of krypton 86, one of the inert gases in the atmosphere.

2-14   DETERMINATION OF DIRECTION
Direction relates to the orientation of a line.

Figure 2-23   Changing direction along great circles.   The heavy black lines show the location of a great circle route on two types of map projections. In the left-hand diagram (*A*), although the great circle is in a straight line, the angle at which it crosses the meridians changes. In (*B*) the great circle is plotted on Mercator's projection. Constant compass directions are shown here by straight lines, while the great circle is an arc. Instead of changing direction constantly to follow the great circle route exactly between New Orleans and Istanbul, the pilot of a plane will fly a series of loxodromes (or rhumb lines). When he reaches the end of one flight leg (rhumb line), he knows how much to alter his compass course to start the new leg.

It may be related to some easily recognized local surface feature, such as "downtown," "toward the river," and "up the mountain." More commonly, however, it is oriented to the four cardinal points of the compass—north, east, south, and west—and as such is useful on both local and global scales. These four cardinal directions originated in antiquity and probably was related to the direction of prevailing winds. Geographic direction on the earth surface, therefore, is expressed in the angular measurement of lines with reference to those leading toward the cardinal points. These points, in turn, are related to the global system of parallels and meridians.

Two different methods are used to express directional measurements quantitatively. The first is termed *bearing,* in which the measurement is confined to a single quadrant and is made from any one of the four cardinal points. A bearing of N 35° W, for example, signifies a direction toward the north and 35° west of north. The same bearing could also be written W 55° N, but the smaller of the two complementary angles generally is used. A second method involves the entire circuit of 360°, always proceeding clockwise from

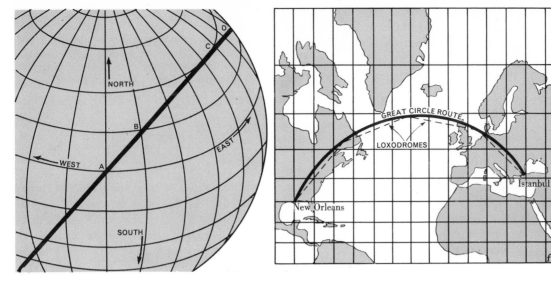

(A) CHANGING DIRECTIONS ALONG A GREAT CIRCLE   (B) PLOTTING A GREAT CIRCLE ROUTE

the meridian or north line. Such measurements are termed *azimuth angles.* A line of direction toward the east would have an azimuth angle of 90°, and one toward the west an azimuth of 270°. Azimuth angles are becoming more widely used than bearings because of less possibility of confusion. A bearing of N 35° W, for example, is not the same as W 35° N.

Any directional line on earth involves a bearing or azimuthal measurement of a great circle. The reason for this is that directions on a sphere are expressions of the shortest distance across the spherical surface between two points. This is also the nature of a great circle. Along such a line, unless along a meridian or the equator, angular directions change constantly. In Figure 2-23, for example, the direction from A to D is not the same as from C to D, even though C is on the same great circle. Conversely, a line on the earth surface that follows a constant compass direction is not a straight line. Rather, it is a constantly curving line which spirals in to-

ward the poles but theoretically never reaches them. Such a line, termed a *loxodrome* or *rhumb line,* connecting two points on a map, is a chord of a great circle. It has great value as a navigational tool. A ship captain or plane navigator, seeking to travel along a great circle route between two points, will find it exceedingly difficult to adjust his direction constantly unless he had assistance from a computer. He may approximate the great circle route, however, by sailing or flying a series of loxodromes that connect points along the plotted great circle and altering his direction at each leg or chord. As long as he maintains a constant bearing or azimuth along a particular loxodrome, he is certain to reach the end point of the course leg.

There are many methods for determining directions at any point on earth. Some of them are crude but useful, involving the direction of prevailing winds, the rising and setting of the sun, the paths of bird migrations, the direction of ocean currents, or even the direction of prevalent growth of mosses or lichens on trees and stones. Star observations were used effectively by early desert travelers at night in Asia and Africa and by Polynesian sea navigators. Precision in determining stellar azimuth angles has involved sophisticated tools and techniques that have proven to be useful in establishing accurate land survey reference meridians and parallels. The most common instrument for determining direction, however, is the magnetic compass, a simple device consisting of a slender piece of magnetized steel mounted on a pivot so that it remains aligned parallel to the earth magnetic field at all times, assuming no local interference.

The earth is a huge magnet with all of the properties of magnets, including two poles or centers from which the respective forces of magnetic attraction and repulsion diverge. These magnetic poles, unlike the poles at the ends of the earth axis, are neither directly opposite each other nor fixed in position. The north magnetic pole presently is located on Bathurst Island in the archipelago of islands north of Canada, and the south magnetic pole is located near the margin of Antarctica, south of New Zealand. The pattern of the magnetic field is neither fixed nor symmetrical. The reasons for its variability are not entirely clear, but it has been suggested that it is related to convectional movements of material within the liquid portion of the earth interior.

Despite its wide use, the magnetic compass has a number of disadvantages as a tool for precise measurements of direction. First, the magnetic needle parallels the earth magnetic field, the lines of which may differ widely from true north and south or from any other cardinal point. Second, since the magnetic field shifts and is not symmetrical, the rate of change in *magnetic declination,* or the deviation of compass direction from true

north or south, is not consistent in any given direction; furthermore, it changes appreciably over the course of years. Near the magnetic poles, the variation in magnetic declination is especially great, both from place to place and with the passage of time. Third, the magnetized needle of a compass is attracted by any nearby magnetic material, such as bodies of magnetic ore or objects of magnetized steel. Many a hunter has wondered at the erratic behavior of his compass, diverted from its true reading by a possibly magnetized zipper of a hunting jacket or a knife sheathed to his belt.

Careful measurements of the earth magnetic field are continually being made in order to plot the magnetic declination for different parts of the earth surface. Maps showing the distribution of magnetic declination (see Figure 2-24) are kept up to date in order to serve the numerous users of the magnetic compass. Many maps, especially topographic or large-scale maps of small areas, indicate the magnetic declination by means of a double arrow indicating true north and magnetic north, and with the angle designated and the year or record.

Other types of compasses are replacing the magnetic compass, particularly for airplane and ship navigation. Two of these are the *gyrocompass* and *sun compass.* The former utilizes the principle of a gyroscope to maintain a constant compass direction. When its

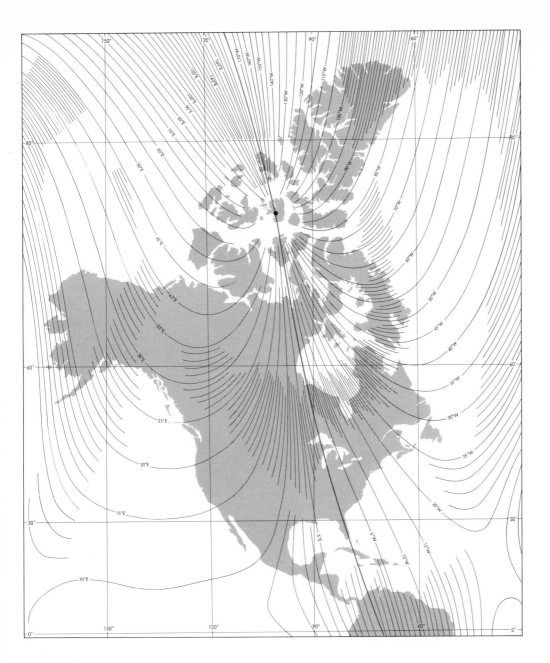

wheel is rotated with its axis oriented in a set direction, and if the spin is maintained, the axis will maintain its direction within narrow limits, regardless of the motion of the gyroscope pedestal. It is frequently used on large ships. The sun compass probably is the most accurate compass of all, since it calibrates direction by utilizing the relationship between the sun's elevation, geographic location, and the time of day or year. It is used primarily for navigation in polar areas, where the errors associated with the use of other compasses are magnified to the extent that these instruments become undependable. The sun compass, however, is not so easy to operate as other compasses, and it cannot be used whenever solar observation becomes impossible.

2-15   DETERMINATION OF POSITION USING DIRECTIONAL AIDS   Direction can be used not only to describe the orientation of any line connecting two points on the earth surface but also to locate the position of points. The principle involves solving the properties of a simple triangle, as shown in Figure 2-25. Once the sides and angles of a single triangle are known, sight lines can be extended to locate basic control points along

a network for hundreds of miles cross-country (see Figure 2-25B). This principle is used in the basic geodetic survey control system of locating points accurately by means of triangulation. The U.S. Coast and Geodetic Survey and the U.S. Geological Survey have established several triangulation networks extending from coast to coast and crisscrossing the country. First-order accuracy required for these intercontinental triangulation networks must close all triangles within one second, and lengths must agree with earth measurement within one foot in 25,000, or about $4\frac{1}{2}$ miles. Such accuracy necessitates the use of precision instruments, night observations to avoid refraction errors caused by convectional air currents, and special blinker lights for positioning.

The same principle can be used to establish the position and trajectory of an object in motion. New aids to navigation utilize short-wave radio waves. The *radio direction finder* is a radio shortwave receiver equipped with a special rotational antenna. It is not really a compass, since it does not give bearing or azimuth readings. Rotating the antenna produces variations in the strength of the signal

A.

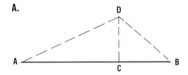

B.

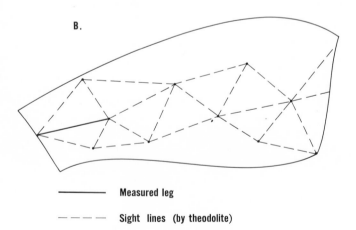

———— Measured leg

– – – – Sight lines (by theodolite)

Figure 2-25   Principles of survey triangulation.   Only the careful ground measurement of one leg of a triangle is necessary to set up an entire cross-country control survey. In (*A*), for example, angular sight lines from *A* and *B* locate point *D*. Lengths *AD*, *DC*, and *DB* can be obtained by solving trigonometrically the sides and hypotenuses of right angles. To illustrate, tan angle *CAD* = *DC*/*AC*; hence, *DC* = *AC* (tan angle *CAD*). The tangent function for any angle between 0 and 90° can be obtained from a set of trigonometric tables. Using this principle in (*B*), once the initial leg *AB* is measured, only angular measurements are necessary to carry the survey control for long distances.

received from a known transmitter. The direction from which the clearest signal is obtained thus indicates the direction in which the transmitter lies. The intersection of directions from two or more known radio transmitters can thus establish position. A special application of this technique is the establishment of radio signal beams that constitute paths between airports for use by planes. A special signal identifies the center of the beam, and a different signal is used along the beam edges. The narrowness of the beam indicates the proximity of the transmitting station. Air navigation charts, available for nearly the entire world, give the location and spread of these radio beacon beams.

Radar consists of transmitting shortwave radio waves by means of a special antenna which concentrates the microwaves into a very narrow beam. Carefully timed, synchronized radar impulses, when received and interpreted, give not only the direction but also the distance of an object from the radar transmitter. Radar beams are reflected from dense objects, and especially strong signals are reflected from metallic surfaces. By rotating both the transmission beam and a directional receiver antenna, the positions of recognizable echo surfaces are plotted on a radar receiver as small blips of light and calibrated for distance and elevation. The course and position of planes and ships can thus be accurately plotted. Accuracy and effectiveness decrease with distance, however. Determination of position by the intersection of special long-range radar beams (*loran*) is used in long-distance plane and ship navigation. Short-range (*shoran*) transmitters and receivers are used in plotting distance and position within a limited area. They can locate harbor outlines after dark

or in dense fog and spot other ships nearby even when invisible to the eye. Shoran was widely used during World War II for locating approaching planes, ships, and submarines and for spotting bombing targets from above clouds or after dark during air raids. More recent adaptations involve observing the approach of severe storms, such as tornadoes and hurricanes. A small, relatively inexpensive model is used even to locate fish in lakes and rivers for technologically minded fishermen.

A special application of loran has recently been developed for establishing position in connection with orbiting earth satellites. Radar signals are sent from two or more broadcasting sites, whose exact position has been carefully determined, to an orbiting satellite that relays them back immediately. The exact position and trajectory of the satellite in its orbit are continually being calculated by computers at the ground stations. A ship or plane whose position is desired sends a directional signal to the satellite during its passage overhead, and immediately it is locked into an extremely accurate triangulation system. As long as the ship or plane can communicate with any one of the widely spaced control stations and can contact the satellite by means of a radar beam, its position can be quickly determined at any given moment. In 1966 a high-altitude satellite, one of a series of SCOR (Sequential Collation of Ranges) satellites, was placed into orbit 2,500 miles above the earth surface; it has since been providing triangulation for distances on earth of up to 2,000 miles. This will enable intercontinental triangulation surveying networks to be tied together into a global survey system, and thus add to the growing accuracy of measuring the face of the earth.

Study Questions

1. In what ways is the earth unique among the planets in the solar system?

2. Venus is a planet roughly similar to the earth and whose distance to the sun is closer to that of the earth than that of the other planets. What other factor besides distance helps to explain why its surface is so much warmer than that of the earth? What might be the impact of introducing a variety of plant life that could tolerate a temperature of 600°F?

3. Both the moon and Mars are pockmarked with meteorite-impact craters. Why are these relatively rare on earth? What evidence exists to support the hypothesis that the earth and moon were formed at about the same time?

4. Why are ocean tides unusually high during full-moon or new-moon phases? Eliminating the wind factor, are there times when the tide at full moon is higher than usual? Explain.

5. What factors influence the number and height of tides at various coastal positions around the earth?

6. What is the zodiac? Using the chart in Figure 2-10, determine the zodiacal division in which September 1 is located. Why does the zodiacal "wheel" turn, and what is its period of rotation?

7. Using the analemma in Figure 2-12, answer the following:
   a. On December 4, at what latitude is the sun vertical?
   b. Does the earth travel faster or slower in the northern-hemisphere summer as compared with the winter months? Explain.
   c. On October 12, actual sun time would be how many minutes faster or slower than mean sun time?
   d. At what time (Central Standard time) would the sun be highest in the heavens on March 8 at Chicago (long $87\frac{1}{2}°$ W)?

8. Explain why the linear value of a degree of latitude is exactly twice that of a degree of longitude at lat 60° N.

9. On the summer solstice (June 21) at noon, sun time, an observer measures the sun to be 67°15′45″ above the northern horizon. He also notes from his chronometer that UT (Greenwich) time is 9:15 A.M. What is his latitude and longitude?

10. Why were most of the early maps distorted in scale much more from east to west than from north to south?

11. Using the map of isogonic lines shown in Figure 2-24, determine how far the magnetic compass varies from true north in New York City. Would the compass there point east or west of true north?

12. Describe the technique of finding a ship's position using an orbiting satellite.

The Planet Earth: Its Place in the Solar System; Its Dimensions and Portrayal on Maps

CHAPTER 3
MAPS AND PROJECTIONS

Maps are the principal tools of the geographer. Many other specialists make maps and use them in their work, but in no other profession do they play such an important role. Summarily, *maps are two-dimensional representations of selected features of the earth surface and at reduced scales.* They are important in geography primarily because they bring to the scale of human comprehension a picture of generalized reality regarding the arrangement of forms and phenomena on the surface of the earth. In Chapter 1 we learned that no two spots on earth are exactly alike, that distributions of forms and phenomena are transitional in their areal expression and are in a constant state of change at highly variable rates. Making sense out of this chaotic jumble is one of the geographer's principal tasks and has been since long before the Christian era. Maps are almost as old as recorded history, have a wide range of uses, and consequently exhibit a wide variety of forms, scales, and sizes. In this chapter we shall investigate the construction and properties of maps and map projections, indicate some of the ways by which the curved surface of the earth may be warped onto a flat surface to preserve selected properties, and discuss briefly the problems that are associated with surveying and cartography.

The History of Maps and Map Distortion

The earliest-known map is a Babylonian clay tablet about 4,500 years old that shows the location of an ancient land estate somewhere in what is now northern Iraq. The Tigris and Euphrates Rivers are clearly indicated, and rude symbols distinguish the mountain borders of this early settled plain. This ancient map also is oriented by symbols designating the four cardinal directions—north, east, south, and west—indicating how ancient these reference points are. Knowledge of location, direction, and areal patterns is necessary to meet even the most rudimentary requirements of mobility, selective land use, or plain curiosity as to what lies beyond their immediate horizons.

Simple pictorial maps of coastal outlines have been found on walrus tusks carved by early Eskimos. Polynesian navigators, long before having any contact with whites, used ingenious string or stick maps on which knots or attached shells represented islands, and the strings themselves portrayed the directional patterns of prevailing currents and waves (see Figure 3-1). Early explorers in the New World often reported the ability of the North American Indian to draw remarkably accurate maps of trails, rivers, lakes and mountain ranges to scale by sketching in the dust, using a stick stylus—an art that was part of the male adult training.

The evolution of the art of map making

Figure 3-1   A stick chart used by Marshall Islanders for ocean navigation.   The sticks represent the prevailing direction of waves and currents. Island positions were marked by placing shells at appropriate places on the frame. [Bernice P. Bishop Museum]

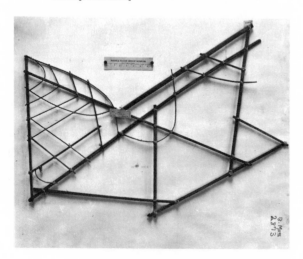

(cartography) in our own culture has been continuous, and many ages have left their imprint on our present techniques. While there has always been an active demand for maps, they were usually the exclusive property of navigators or the wealthy until the invention of the printing press. Today they are cheaper, better, and more widely used than ever before.

The Greek map makers devised a crude system of latitude and longitude to locate places two centuries before the Christian era. *Hipparchus,* a Greek astronomer and geographer of the second century B.C., is credited with being the first to use lines connecting points having the same length of day at the solstices as reference lines for north-south measurements. He termed these *klimata,* from which our word "climate" is derived. Accurate grids of longitude and latitude, however, were long in coming, and as noted in Chapter 2, the problem of measuring longitude accurately was particularly difficult until the invention of an accurate portable timekeeping device in the eighteenth century. *Claudius Ptolemy,* another Alexandrian geographer, who lived during the second century A.D., became the first to use a mathematically derived grid to represent a true map projection, and although his data were not always correct and his instruments were crude, he knew what he was doing. Centuries later, his works formed a firm base for map making in the Renaissance.

As man's mobility increased, so did his need for more accurate maps covering larger areas. When the curvature of the earth became apparent, a new problem in map making appeared, one that has remained with cartographers ever since. This unsolvable problem is that a spherical surface cannot be converted into a plane surface without some degree of distortion. In order for all or even part of a spherical surface to lie flat contiguously, it must either be compressed or stretched, or both, at various places and in various amounts. Because of this, the space characteristics of distance and direction—and consequently, of shape and area on the earth—cannot all be transferred onto one map without alteration. When such transfers are attempted, distortions of one or more of these characteristics occur. It is possible to represent selected characteristics truly on a map, but in so doing, distortions for some other properties may be increased. Anyone who uses maps to any appreciable extent should be trained to recognize the kind, amount, and distribution of the distortion inherent in any map representation of the earth surface.

The departure of the earth spherical surface from a plane is small when only limited areas are represented (see Figure 3-2). Thus, large-scale maps, or those at a scale of one or more inches per mile, can be considered free of distortion except when extremely precise measurements are required. When maps of large areas are involved, however, such as those of continents or the world, the distortion errors inevitably are great.

To illustrate the property of distortion, suppose we wished to portray the surface of an orange on a flat plane. If we photographed it or drew a perspective sketch of it, it might be recognized as an orange, but only one side could be shown. Furthermore, the shape would be distorted, and only our visual habit of perceiving things in perspective would enable us to recognize it as a sphere. We also could spread the entire surface of the orange on a flat plane by peeling it and cutting it into segments termed *gores.* In fact, the surfaces of commercial globes are printed in such segments while they still are on a flat plane, and these segments faithfully represent the total area of the globe. The total surface spread out thus may show true area,

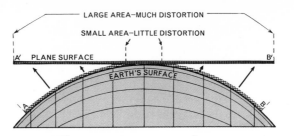

Figure 3-2 Departure of a spherical surface from a plane surface.

but it in no way resembles the shape of the globe (see Figure 3-3).

Map Projections

There are many ways by which representations of the earth surface may be laid out on a plane surface. None of them are perfect in all respects, but when they are consistent and definable in their approach, they are termed *map projections.* A map projection thus can be defined *as an orderly or systematic arrangement of the earth grid on a plane surface,* the earth grid being the framework of parallels and meridians that enclose and define the various sections of the earth surface. The term *projection* is derived from an early stage in cartography in which

Figure 3-3 Globe gores.

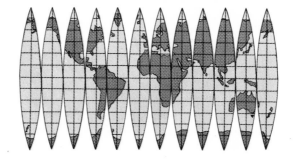

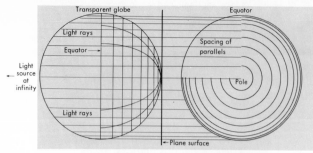

A. Geometric perspective of an orthographic projection

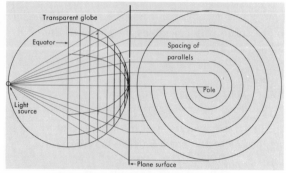

B. Geometric perspective of a stereographic projection

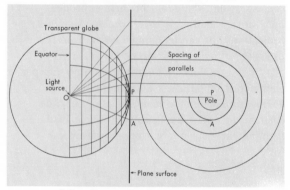

C. Geometric perspective of a gnomonic projection

Figure 3-4 Geometric perspective of three projections.
The grid lines of the basic globe in these three projections are projected upon a tangential plane surface from three different light sources: at infinity, at the antipode of the point of tangency, and at the center of the globe. Each projection may be plotted graphically, as shown in this figure, or it may be done mathematically using trigonometric functions. In the gnomonic projection (C), for example, tan (angle AOP) = AP/OP. If angle $AOP = 30°$ and OP = the radius of the basic globe, AP = tan $30°$ (radius OP), or .5775 r.

the *gnomonic,* where the point of projection lies at the center of the sphere. Geometric projections may be used on different types of surfaces besides a plane. For example, the projection may be onto the surface of a cone or a cylinder, both of which may be "developed" or spread out into a plane. While the pure geometric projections still are used occasionally for certain purposes, most of them have had their parallels and meridians mathematically "adjusted" so as to provide some unique map property that may be desirable. There is almost no limit to the variety of map projections that can be devised, and hundreds of them have been developed since the first geometric projection was developed 2,000 years ago. Only a few dozen are in common use today; they appear to suffice for the needs of most present-day maps.

The flexibility of projections is surprisingly great, and they can be made to do many things through selective distortion. Centrifugal areal distortion (exaggeration increasing outward from a central point, as in the gnomonic projection) was used effectively by the German cartographers of the Hitler regime during the 1930s in order to show the German need for *lebensraum* (living space) in the world. Their maps portrayed a tiny central Germany surrounded by a gigantic ring of

the earth grid was geometrically "projected" onto a plane surface. Figure 3-4 shows three types of geometric projections which are differentiated according to the point of projection: (1) the *orthographic,* where it lies at infinity and the projection rays are parallel; (2) the *stereographic,* where it lies at the antipode from the point of tangency; and (3)

potential enemies, especially an enormous peripheral British Empire. While map projections, like statistics, never lie if they are made correctly, they may be misleading unless their properties are correctly understood. The use of conic projections, with their curved parallels for maps of the United States, has misled many a hopeful on quiz programs when faced with the question: "Locate the most northerly and southerly points in the United States." Few people would acknowledge offhand that Bemidji, Minnesota, lies on the same parallel as St. Johns, Newfoundland, or Zurich, Switzerland. The misuse of the Mercator projection to show global political units in many older school atlases has resulted in generations of schoolchildren believing in the grossly exaggerated spread of the arctic territories within Canada, Greenland, and the Soviet Union. On this projection, Greenland appears as about twice the size of South America, when actually the latter is nine times the size of Greenland.

3-1 PROPERTIES OF MAP PROJECTIONS

A map property is a characteristic possessed by a projection which gives it value for certain purposes. Most such properties are attempts to duplicate on a plane surface the spatial characteristics of the earth surface, such as *distance* (or *scale*), *direction, shape,* and *area*. Of these, area is the only one which can truly be represented without qualification *on an entire map*. True distances, true directions, and true shapes, as properties of map projections, have more limited uses. For example, true distances can be consistent throughout a map only when radiating from one central point, and there is only one projection which can represent even this limited definition of true distance—the *azimuthal equidistant* projection. A scale of miles has little meaning on any world map other than the one mentioned above, unless it is designated as suitable only for limited portions of the map. Many world maps, for example, use a different scale of miles for each parallel. Directions, likewise, can be true consistently only from a central point, because true directions imply a great-circle path (the shortest distance) between two points on the earth surface. Even on those projections having the property of true direction from a central point, there will be errors in direction from all other points. Similarly, shapes can be shown as true only for small areas. Over larger areas, the shapes will be distorted because of distortions in both distance and direction.

The four major properties of map projections are as follows:

1. *Equivalence* (equal area) is a property in which the ratio between areas on the map projection is the same as the ratio between areas on the earth surface. This property is maintained by compensating for the extension of scale in one direction by an equivalent reduction of scale in the opposite direction. Figure 3-5 indicates a wide variety of shapes, each of which is equal in area. With each change in shape, extension in one direction has been matched by reduction in another. Equivalence on a projection can usually be recognized by comparing the areas between different pairs of meridians between the same two parallels. All such areas should be the same. Further, the areas between meridians and parallels should decrease progressively toward the poles. Equivalence as a property is important whenever areal distributions are to be accurately presented, as on maps showing the distribution of forests or areas under cultivation. Table 3-2 lists the major equivalent projections in common use.

Areas sometimes are intentionally distorted on maps in order to present a visual picture of relative areal importance. For ex-

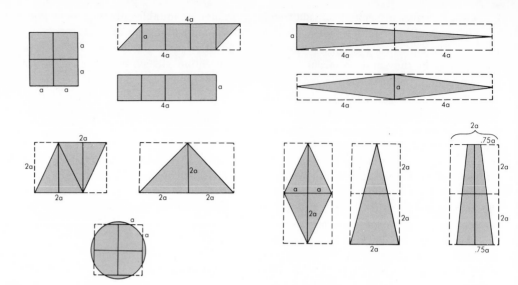

ample, if the various countries of Europe were enlarged in exact proportion to their populations or their amount of foreign trade, the result would look peculiar but might be highly effective as a graphic device (see Figure 3-19A).

2. *Conformality* (true angular relations) is the property whereby the *compass rose*[1] is the same throughout the entire projection, just as it is on the earth. All parallels and meridians intersect each other at 90° angles, and for each point on the map, every intervening direction appears to be at the appropriate angle with the cardinal directions. Bearings, therefore, may be said to be true. It should be kept in mind that while all conformal projections have 90° angles at all meridian and parallel intersections, not all projections that preserve these angles are conformal. Bearings must also be true. Another feature of conformality is that the shapes of re-

[1]The term "compass rose" was derived from the appearance of the maritime magnetic compass dial, in which the points or arrows of the major and minor compass bearings gave the entire dial somewhat the appearance of a rose with its radiating petals of unequal length.

stricted areas are faithfully preserved. A small circle, square, or triangle anywhere on the globe will be shown in its true shape (but not area) on the conformal projection. The areal distortion of conformal projections, however, has the effect of distorting the shapes of large areas, such as continental masses. Conformality as a property is important mainly in navigation, where ease in plotting location and true compass bearings from any point on the map is essential. Most air navigation charts and ocean hydrographic charts utilize conformal projections.

3. *Azimuthality* as a map property preserves true directions from a central point to any other point on the globe. By construction it may also be adjusted so as to preserve true scale or distance from the center to any point on the map (*azimuthal equidistant*). It may be made equivalent, and even conformal, although the extreme distortion of shape near the periphery usually limits the azimuthal equivalent maps to hemispheric use. Azimuthal maps are usually used in maps of polar areas and also where global rela-

The Planet Earth: Its Place in the Solar System; Its Dimensions and Portrayal on Maps

Figure 3-5 Equivalence preserved through compensation in shape adjustment. Each of the shaded figures is equivalent in area. This illustrates how equivalence may be preserved by graphic means. Each one also may be expressed mathematically. To illustrate, the diameter of a circle having an area of $4a^2$ can be solved as follows:

$$\pi R^2 = 4a^2 \qquad R^2 = 1.27a^2$$
$$R^2 = 4a^2/3.1416 \qquad R = 1.127a$$

tionships to a specific point are important. The azimuthal equidistant projection can be centered on any point. Therefore, postal centers and airfields use it to determine the location of postal and freight rate zones or the effective covering territory of planes having a certain cruising radius.

Some interesting azimuthal projections have been devised, utilizing intentional progressive distortions of scale outward from the center. The Swedish cultural geographer Hagerstrand, for example, has devised an azimuthal, logarithmic-distance scale on which to plot the outward migration of people and the diffusion of ideas from a central point. His reasoning was that the meaning of distance becomes less with increasing distance. Hence, in plotting distance as a variable from a central point, scale should be compressed outward from the center. Hagerstrand used a logarithmic scale, in which each successive unit represented ten times the distance of the previous unit, measured outward from a central point. Compressed azimuthal scales might also be used to adjust the increasing distance factor to decreasing costs of transportation (long-haul versus short-haul freight rates).

4. *Compromise* is a map property by which total, not selective, distortion is kept to a minimum. As a result, none of the map properties are without some distortion. A compromise projection does not have equivalence and is not conformal. It does have distortion of scale (as on all maps), and directions are incorrect everywhere. However, none of the distortions are excessive, and many of the compromise projections are much easier to construct than the equivalent ones. Compromise projections are becoming more widely used for general illustrative material where general global impressions and not accuracy are desired. Such maps are excellent to show the general course of isotherms (lines of equal temperature) or isohyets (lines of equal rainfall), the general relationships between ocean basins and continents, and for general atlas use. Several of the better-known compromise projections are listed in Table 3-2.

In addition to these four properties, there are many others of lesser general importance that may be especially desirable for certain purposes. Examples include true latitudinal scale, true longitudinal scale, true scale along any selected line, lines of constant compass direction (loxodromes) shown as straight lines, great circles as straight lines, and deformation reduced to a minimum along one or two lines.

More than one true property may be preserved on the same projection. As indicated above, azimuthal projections may be equidistant, equivalent, and even conformal. The Mercator projection is not only conformal but also shows true bearings as straight lines. The gnomonic projection combines azimuthality with the property of showing all great circles as straight lines. It is important to note, however, that no projection can be both conformal and equivalent. These are mutually exclusive properties. Further, no map projection can preserve true scale or maintain true direction from more than one, or at the most, two points.

3-2 THE EVALUATION AND SELECTION OF PROJECTIONS As has been noted earlier, map projections can be selected to represent some of the properties of a spherical surface either singly or in limited combinations, but never all of them. The choice of a projection would always be on the basis of its comparative strengths and weaknesses with respect to the spherical properties that are most needed for the map's use. The general properties of a spherical surface grid are outlined below. Each can easily be checked against a particular projection by using a ruler, a pair of dividers, and a protractor. With practice, most of them can be judged by eye.

1. Parallels are parallel to each other.
2. Parallels decrease in length from the equator to the poles.
3. Parallels are spaced evenly along any meridian.
4. Meridians converge toward the poles.
5. Meridians are spaced evenly along any one parallel.
6. The spacing between meridians decreases from the equator toward the poles.
7. The spacing of the meridians along the equator is approximately the same as the spacing of the parallels.
8. The spacing of the meridians along the 60th parallel is one-half that along the equator.
9. The area of any grid space between two parallels and two meridians should be the same at any such spacing between the same parallels.
10. The shape of any grid space between two parallels and two meridians should be the same at any such spacing between the same parallels.
11. The meridians and parallels always intersect each other at 90° angles.
12. The compass rose is applicable at any point.

Figure 3-6 The four major properties of projections. All areas are in true proportion to those on the globe in *equivalent* projections, as in *A*. The property of *azimuthal* projections (*B*) is that all lines radiating from the center of the projection have true global direction or bearing. In *conformal* projections (*C*) all shapes for areas of small size are consistent with those on the globe. The circle at Y and the triangle at X represent areas having the same shape as on the globe. *Compromised* projections (*D*) may be distorted in several ways, but the distortion generally is not extreme for any one property, and in many cases the ease of construction may be a determining factor.

13. The scale of distance from any point is the same in any direction.
14. Except at the equator and along any meridian, the shortest line between any two points changes constantly in direction.

Four different projections are shown in Figure 3-6. The analysis of each of them with respect to the spherical grid properties shown above would yield the following:

The *Mollweide homolographic projection* agrees with the spherical grid with respect to properties 1, 2, 4, 5, 6, 7, and 9. Property 3 appears to be satisfied but is not. The discrepancy would show up in a larger drawing. While properties 2 and 6 are satisfied, the gradient of change from the equator to the poles is not exactly proportional to that of the sphere. Since longitudinal distances are not quite true, the spacing between the parallels has been slightly adjusted mathematically to preserve the property of equivalence.

The *Azimuthal Orthographic projection* agrees with the spherical grid with respect to properties 1, 2, 4, and 6. It has the special property of true directions from the central point. This is the perspective view of a globe and presents the illusion of a globular surface.

The *Mercator projection* agrees with the spherical grid with respect to properties 1, 5, 7, 9, 10, 11, and 12. It should be noted that, although

The Planet Earth: Its Place in the Solar System; Its Dimensions and Portrayal on Maps

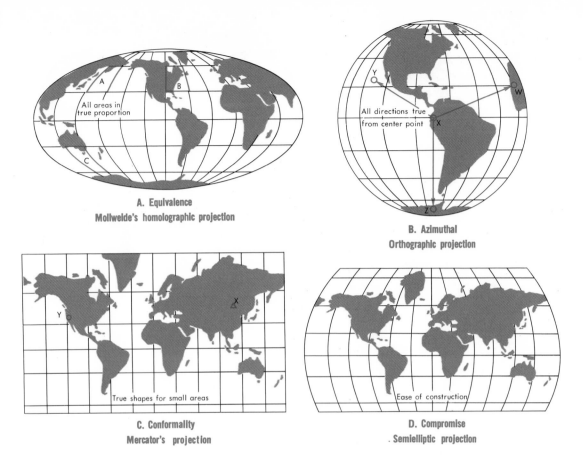

A. Equivalence
Mollweide's homolographic projection

All areas in true proportion

B. Azimuthal
Orthographic projection

All directions true from center point

C. Conformality
Mercator's projection

True shapes for small areas

D. Compromise
. Semielliptic projection

Ease of construction

property 9 is satisfied, the projection is far from equivalent because of the lack of properties 2, 3, and 4.

The *Compromise Semielliptical (Eckert) projection* conforms with respect to properties 1, 2, 4, and 6. In the case of (2) and (6), the decrease is not proportional. In (4), while the meridians tend to converge toward the poles, the convergence is not complete, and the North and South Poles are shown as a line.

Projections are highly versatile and can be altered in many ways to suit a particular purpose. For example, if an equivalent projection is desired, it may be modified in order to favor some areas over others with respect to angular or shape deformation. A conformal projection such as the Mercator need not always use the equator as a standard parallel; it can be rotated 90° to follow a particular meridian. Such a transverse form of the Mercator is well adapted as a base on which to hang the right-angle coordinate system of a military grid. A world map that is to show the location of coal deposits, where area is not especially important and where most of the data will be confined to middle latitudes, should utilize a projection that will exaggerate the mid-latitude areas somewhat, permitting the inclusion of more data.

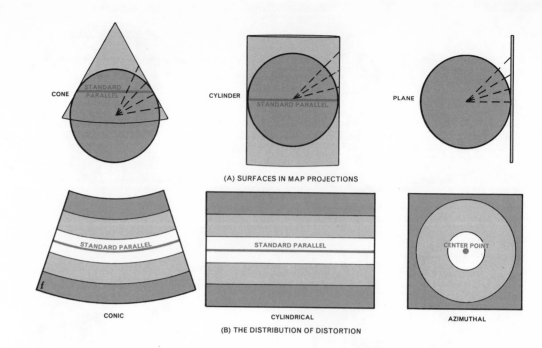

CONE · STANDARD PARALLEL

CYLINDER · STANDARD PARALLEL

PLANE

(A) SURFACES IN MAP PROJECTIONS

STANDARD PARALLEL

CONIC

STANDARD PARALLEL

CYLINDRICAL

CENTER POINT

AZIMUTHAL

(B) THE DISTRIBUTION OF DISTORTION

3-3 THE CONSTRUCTION OF PROJEC-TIONS For the purpose of accuracy, most projections employed by professional map makers are constructed by using mathematical formulas to obtain the necessary precise spacing of meridians and parallels. In most cases, these formulas express the geometric transfer of the spherical earth grid onto a plane surface or the surface of a cone or cylinder. The latter two surfaces can then be flattened into a plane without any subsequent alteration of the arrangement of the meridians or parallels (see Figures 3-4 and 3-7). The term *map projection* is derived from this process of "projecting" the earth grid onto these surfaces. The methodology can be graphically demonstrated by taking a transparent globe with the grid lines etched on it and, by means of a light, projecting the shadows of the grid onto a plane surface, as shown in Figure 3-4. The principle of using mathematical formulas to construct projec-

tions based on geometric relationships is illustrated in Figure 3-4C. Solving the trigonometric equation for the adjacent side of a right triangle will provide the exact measurement of the parallel radii to be used in the projections. These projections may also be constructed graphically, as shown in the figure. Note that, in the gnomonic projection, the parallels become farther apart as one leaves the point of tangency (the North Pole) and that it would be impossible to indicate the equator, since by construction, it would be at infinity. In the stereographic projection, (Figure 3-4B), the point of projection is based at the antipode, whereas in Figure 3-4A it lies at infinity. The three projections shown in Figure 3-4 are azimuthal projections in which the plane of tangency touches at the North Pole. A comparison of five different polar azimuthal projections is shown in Figure 3-8. Note that the only difference between them lies in the spacing of the parallels.

The polar azimuthal equidistant and equivalent projections shown in Figure 3-8 are not geometrically projected, and like many others, they are developed in a more arbitrary, although no less precise, manner. In fact, many of the map projections derived from theoretical geometric projections *and named after the surfaces involved* are mathematically adjusted or modified to achieve some desirable characteristic. In Figure 3-8, for example, the equidistant projection is produced simply by spacing the parallels in proportion to those on the globe and making them concentric. In the equivalent polar azimuthal projection, the spacing of the parallels becomes reduced somewhat toward the outside in order to compensate for the fact that the meridians diverge at a greater rate than on the globe.

One of the best-known mathematically adjusted projections is the Mercator Modified Cylindrical, developed by Gerhardt Kramer (also known as Mercator) in 1569 as a tool for mariners. For centuries, ship captains had known that by following certain compass bearings, they could sail between two ports. Only in the best-known areas, such as the Mediterranean Sea, did maps show sailing directions reasonably well, and these were the result of centuries of trial and error. What was needed was a map grid on which any straight line would have the property of true directional bearing; that is, if one followed the angular course indicated by the line, all points located along the line would be touched. Mercator recognized that in order to do this, all angles on the map had to be right angles. In searching for a projection with right angles, he went to the gnomonic cylindrical projection as a useful beginning. In this projection (see Figure 3-9), the parallels and meridians are straight lines and cross at right angles. The spacing of the parallels, however, is troublesome, since they widen rapidly away from the parallel of tangency. Mercator next realized that if he could space the parallels apart in exactly the same ratio as the meridians were spread apart, the projection would be conformal, and all bearings from any point or points on the map would be true. Figure 3-10 illustrates this point. Since the amount of distortion of meridional spacing at the poles is infinite, there can be no 90° parallel. The excessive distortion at high latitudes generally restricts the use of the equatorial Mercator to latitudes under 80°. The mathematical formula that describes the exact ratio between parallel meridians and converging meridians, thus prescribing

Figure 3-8 Composite of five polar azimuthal projections.

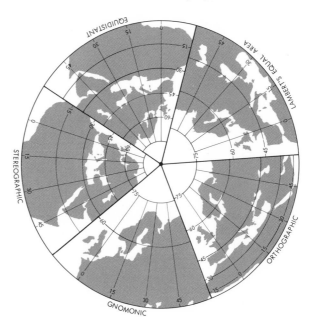

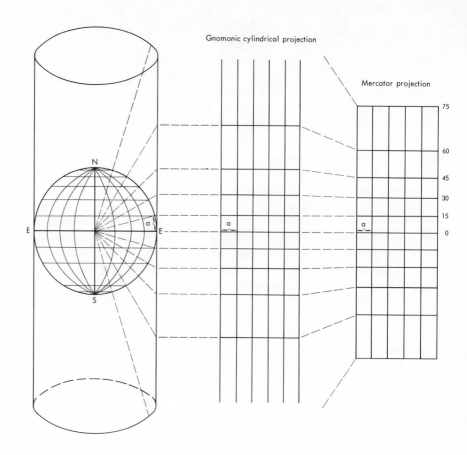

Gnomonic cylindrical projection

Mercator projection

the spacing of parallels, is translated into the data in Table 3-1.

In grouping projections on the basis of their construction, the following four major types occur. The first three obviously are named after the surfaces involved in the geometric projection.

1. Plane projections, which are also termed

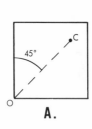

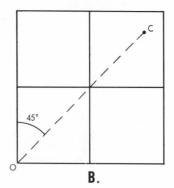

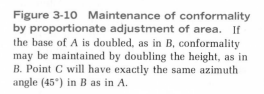

Figure 3-10 **Maintenance of conformality by proportionate adjustment of area.** If the base of *A* is doubled, as in *B*, conformality may be maintained by doubling the height, as in *B*. Point *C* will have exactly the same azimuth angle (45°) in *B* as in *A*.

Figure 3-9 Derivation of the Mercator projection from the gnomonic cylindrical. The Mercator projection was derived as a modification of the gnomonic cylindrical projection. The spacing of the meridians is preserved, but the spacing of the parallels is adjusted mathematically so as to provide conformality (see Figure 3-10). On the basic globe, the meridians should converge toward the poles. To compensate for this progressive convergence, the Mercator projection spaces the parallels in exactly the same ratio. Note that while area is distorted only slightly near the equator, it becomes progressively greater toward the poles.

azimuthal, because they all have the property of true direction from a central point.

2. Conic projections.

3. Cylindrical projections, also termed rectangular, because of the shape given to world maps on this type of projection.

4. Miscellaneous projections, including all those projections—such as oval projections—in which no geometric projection is involved.

The four major groups of projections are, in turn, subdivided according to the prop-

erties of individual projections. For example, there are equivalent, conformal, and compromise conic projections. Similar variations in properties are, to a greater or lesser degree, also true of the other groups.

A further difference results from variations in centering, or locating, individual projections. Plane projections may be centered on points, but conic and cylindrical projections are centered on lines, usually parallels. Miscellaneous projections are centered on points or lines, depending on the specific projection. Variations in centering will alter the appearance of the grid but will not alter the basic properties of the map. In this way a plane, equal-area projection, for example, may be centered on the North Pole, some point on the equator, or any position in between, and it will still retain the property of equivalence (see Figure 3-11A).

In theory, the points and lines on which projections are centered represent points or lines where the surfaces of the projections contact, are tangent to, or sometimes even secant to the earth surface (see Figure 3-11B). Such points are termed *center points,* and the lines are termed *standard lines.* If, as is most frequently the case, the line is a parallel, it is termed a *standard parallel.* Of major significance, from the point of view of distortion, is the fact that at these points or lines the projection is true in all respects, and distortion is at a minimum in the immediate vicinity. All places equally distant from the center have equal distortion on point-centered projections, and on most of the projections with standard parallels, places equally distant from that parallel have equal distortion. There are some significant exceptions to the latter generalization, however, particularly among the miscellaneous projections, which may have standard meridians in addition to standard parallels.

Table 3-2 lists some of the more common

Table 3-1 Mercator Projection		
Distances of Parallels from the Equator		
Latitude	*Minutes of longitude at equator**	*Multiples of 15° at equator*
0	000.000	.00
15	904.422	1.00
30	1,876.706	2.07
45	3,013.427	3.33
60	4,507.133	4.98
75	6,947.761	7.68
*From Deetz and Adams.		

Table 3-2 Summary of Map Projections

Projection	Maximum feasible area	Properties	Distortion	Uses
Cylindrical				
Mercator's	World	Conformal; lines of constant compass directions shown as straight lines	Increases poleward	Navigation and in transverse form for military mapping
Cylindrical equal-area	World	Equivalent	Increases north and south from equator or standard parallels	Mapping distributions where areal size relationships are important
Miller's	World	Compromise	Increases north and south from two standard parallels	Mapping distributions where areal size relationships are not paramount
Gall's stereographic	World	Compromise	Increases north and south from two standard parallels	Mapping distributions where areal size relationships are not paramount
Plane chart	World	Compromise; easy to construct	Increases north and south from equator or standard parallels	Mapping distributions where areal size relationships are not paramount
Plane (azimuthal)				
Stereographic	Hemisphere	Conformal; true directions from center point; all circles shown as circles	Distortion increases from center outward	Navigation; mapping distributions where positions are important
Azimuthal equal-area	Hemisphere usually; can show world	Equivalent; true directions from the center point	Distortion increases from center outward	Mapping distributions where areal size relationships are important and for areas less than world size
Azimuthal equidistant	Hemisphere usually; can show world	Distances and directions true from center point	Distortion increases from center outward	Measuring distances from a center point to all other points
Gnomonic	Less than hemisphere	All great circles as straight lines	Distortion increases very rapidly from the center outward	Navigation; to determine great-circle routes
Orthographic	Hemisphere	Compromise; gives visual appearance of the earth	Distortion increases from center outward	Pictorial representations of the earth
Conic				
Simple	Less than hemisphere	Compromise; easy to construct	Increases north and south from a standard parallel	Mapping distributions where relationships of areal size or position are not paramount

Table 3-2 Summary of Map Projections (continued)

Projection	Maximum feasible area	Properties	Distortion	Uses
Conic (continued)				
Alber's equal-area	Less than hemisphere	Equivalent; very little distortion for area, size, and shape of U.S.	Increases north and south from two standard parallels	Mapping distributions where areal size relationships are important at continental or lesser size
Lambert's conformal	Less than hemisphere	Conformal; very little distortion for area size and shape of U.S.	Increases north and south from standard parallels	Mapping distributions where positional relationships are important; weather maps; navigation
Bonne's	Less than hemisphere	Equivalent	Increases away from standard parallel and central meridian	Mapping distributions where areal size relationships are important for "square-shaped" areas the size of Europe or less
Polyconic	Less than hemisphere	Compromise; easy to construct	Increases away from central meridian	Large-scale military mapping
Miscellaneous				
Mollweide's homolographic	World	Equivalent	Increases away from intersection of lat 40° N & S and central meridian	Mapping distributions where areal size relationships are important
Sinusoidal	World	Equivalent	Increases away from equator and central meridian	Mapping distributions where areal size relationships are important. Good for South America and Africa
Homolosine	World	Equivalent; interrupted; minimizes other distortions; uses best parts of homolographic and sinusoidal	Generally increases away from equator and central meridian	Mapping distributions where areal size relationships are important and where interruptions will not complicate presentation
Eckert series: nos. 2, 4, 6	World	Equivalent	Increases away from equator and central meridian	Mapping distributions where areal size relationships are important
Van der Grinten	World	Compromise; easy to construct	Increases in all directions away from the intersection of central meridian and equator	Mapping distributions where areal size relationships are not paramount

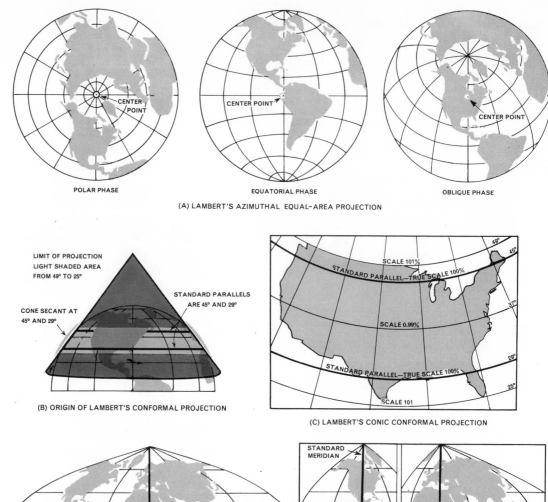

POLAR PHASE

EQUATORIAL PHASE

OBLIQUE PHASE

(A) LAMBERT'S AZIMUTHAL EQUAL-AREA PROJECTION

LIMIT OF PROJECTION LIGHT SHADED AREA FROM 49° TO 25°

STANDARD PARALLELS ARE 45° AND 29°

CONE SECANT AT 45° AND 29°

(B) ORIGIN OF LAMBERT'S CONFORMAL PROJECTION

SCALE 101%

STANDARD PARALLEL—TRUE SCALE 100%

SCALE 0.99%

STANDARD PARALLEL—TRUE SCALE 100%

SCALE 101

(C) LAMBERT'S CONIC CONFORMAL PROJECTION

STANDARD PARALLEL

STANDARD MERIDIAN

(D) SINUSOIDAL PROJECTION

STANDARD MERIDIAN

STANDARD PARALLEL

STANDARD MERIDIAN

(E) INTERRUPTED AND CONDENSED SINUSOIDAL PROJECTION

Figure 3-11   Examples of some of the more commonly used map projections.

The Planet Earth: Its Place in the Solar System; Its Dimensions and Portrayal on Maps

map projections, grouping them on the basis of construction but also indicating such features as their properties, common usage, and the maximum area possible to map on each. To reiterate, all map projections distort the features they represent in a number of ways. However, the smaller the area represented on a map, the smaller the amounts of distortion will be; very large-scale maps can be considered distortion-free for most purposes.

3-4 THE ANGULAR DEFORMATION OF GLOBAL EQUIVALENT PROJECTIONS AND ADJUSTMENTS FOR IT Equivalent (equal-area) projections are among the most useful projections for geographers when they are dealing with global patterns such as those treated in this volume. All global equivalent projections, however, suffer from great distortion of shape, angles, and scale, especially away from the equator and the mid-meridians. Different projections are selective in the distribution of such distortions. For example, in comparing the Sinusoidal projection (Figure 3-11D) with the Mollweide Homolographic (Figure 3-6A), we should note that the angular distortion of the former is greater at high latitudes but less at low latitudes. Professor J. Paul Goode, of the University of Chicago, developed a combination of the two early in the present century which he termed the *homosoline.* The *Goode's Homolosine* projection is shown in Figure 3-12. It may easily be recognized by the slight break in the curved parallels at 40°, where the two projections were joined. Other systems for reducing the angular distortion at high latitudes consist of stretching out the poles as lines and compensating for the longitudinal linear distortion by reducing the spacing of the parallels. The Eckert series was constructed by using a polar parallel that is half the distance of the equator. The fourth and the equivalent one in this series (the Eckert IV) is shown in Figure 3-12A. One of the newest equivalent global projections is the *Equivalent Polar Quartic,* developed at the Cartographic Laboratory at the University of Wisconsin. This projection (see Figure 3-12C) uses a polar parallel one-third the length of the equator. Most of the global equivalent projections are collectively termed *oval projections* because of their curved meridians. In general the distortion of angles and scale increases away from the equator and the mid-meridian.

Another and highly effective way to reduce the angular distortion on global equivalent projections is to interrupt their continuity by splitting them into sections, each of which has its own mid-meridian. These are known as *interrupted* projections. Their use might be likened to splitting the skin of an orange in order to spread it out on a flat surface. Global maps that deal with continental patterns are not harmed by splitting the projection in the ocean areas. Similarly, oceanic maps, such as those of ocean currents, can utilize mid-meridians where the split occurs on the continents. Figure 3-12E shows an *interrupted homolosine.* Note the reduction in the angular distortion that has taken place as the result of interruption. There is no limit to the number of interruptions that can be made, although, of course, the continuity of the map suffers with each interruption.

It is also possible to remove sections of the ocean areas on a global projection if their inclusion would serve no useful purpose. Space can thus be saved in order to show more continental details. Such projections are known as *condensed projections.* Figure 3-11E is an example of a sinusoidal projection that is both condensed and interrupted. Figure 3-12D does the same to a polar quartic. Such techniques are useful in reducing the inevitable distortion in shape when portraying the globe on a map.

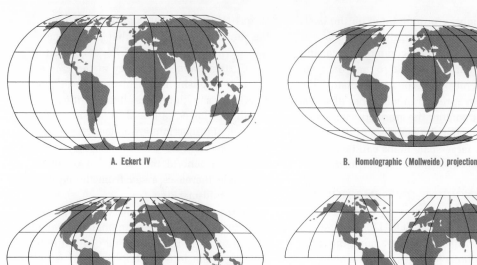

A. Eckert IV

B. Homolographic (Mollweide) projection

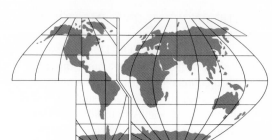

C. Flat polar quartic projection

D. Interrupted and condensed flat polar quartic equal area

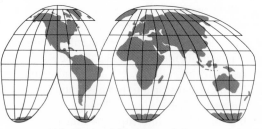

E. Goode's interrupted homolosine projection

Map Scale

Maps are reducing instruments, and the process of mapping always involves a reduction in scale, whereby patterns that are too broad to be seen at a glance are reproduced in miniature for easy comprehension. Scale, therefore, is of prime consideration in our use of maps and mapping procedures and is set apart in this chapter for special examination.

3-5 MAP SCALE AND PURPOSE Map scales must be highly flexible in order to serve a wide variety of purposes and to portray areas ranging from the entire earth to a small census enumeration district within a city. They also must be adaptable to a great range of detail. The scale of a map is much more than the numerical designation of proportionate length. It also sets limits on the amount of detail that can be shown and the level at which the mapping generalization takes place. To illustrate, a double-page world map such as Figure 14-1 in this textbook has a scale of about 1:107,000,000, or 1

The Planet Earth: Its Place in the Solar System; Its Dimensions and Portrayal on Maps

inch to about 1,700 miles. Such a scale, however, is true only along the 35th parallel. Elsewhere, scale is distorted considerably. At normal reading distance (18 inches) recognition of shape becomes impossible at less than about .01 inches. This means, therefore, that areas less than 290 square miles in size would be too small to have their shapes recognized. A general rule of thumb is that for illustrative book use, no areas should be distinguished on the final printed page that are less than about $\frac{1}{16}$ of an inch in at least one dimension and have a total area less than $\frac{1}{250}$ of a square inch. This would eliminate areas on a 1:107,000,000 global map that were smaller than about 10,000 square miles, or areas the size of the state of Maryland. Obviously maps at a scale such as this must either eliminate a large amount of detail or exaggerate the size of some things in order to present generalized reality. A line on a map at this scale, drawn to represent a road, railroad, or river, actually represents a strip of area to scale that is 40 to 50 miles wide. Distributions shown on scales such as this often are comparable to caricatures in which certain features are exaggerated for the sake of recognition. Conversely, large maps of small areas can show much greater detail. On a scale of 1 inch per mile, using the same recognition ratio as above, $\frac{1}{250}$ square inch would indicate areas approximately 2.6 acres in size, or an area of 200 × 560 feet.

The scale of a map must be related not only to its size and the area it represents but also to its purpose. The scale of a soil map sufficiently large to distinguish properties needed for advising farmers on the use of their land,

usually at scales of 1:20,000 or more, could not possibly be used to show the variability of soils within a state. If we assume a county size of 20 miles by 20 miles, the soil map for such a county at a scale of 1:20,000 would measure about 5 × 5 feet. A soil map of the United States, therefore, must be an entirely different type of map because of scale limitations. It must be limited to soil characteristics that are highly generalized and in which many local soil properties are eliminated. Global maps of soils or vegetation must show similarities consistent over areas several thousand square miles in size. Such maps usually show striking correlations with climate, not because the latter is the only determinant of soils but because the other soil-forming factors have much more local impact and could not be shown at global scales. Poor drainage, for example, is significant in soil formation and vegetation associations, and poor drainage is found throughout the world, but inadequately drained areas the size of the state of Maryland are rare anywhere in the world. At the same time, it is unlikely that a county soil map would indicate variations resulting from climatic conditions.

The form of presentation also is related to the scale of the map. A map that is to be read from the back of a classroom or auditorium must be at a much different scale than one on the page of a book. The size of the lines and symbols also is a function of scale for such maps. Robinson has prepared a useful visibility scale that is presented in Table 3-3. Lettering or symbols on a wall map to be used in a 60-foot lecture hall, for example, would not be readable unless they were at least .42 inch high or wide.

Again, it should be emphasized that scale on maps is never completely accurate, because it is impossible to lay out a spherical surface on a plane without stretching or

Table 3-3 Approximate Minimum Symbol Size for Viewing from Various Distances

Viewing distance	Size (width)
18 inches	.01 inch
5 feet	.03 inch
10 feet	.07 inch
20 feet	.14 inch
40 feet	.28 inch
60 feet	.42 inch
80 feet	.56 inch
100 feet	.70 inch

SOURCE: Arthur H. Robinson, "Elements of Cartography," 2d ed., John Wiley & Sons, Inc., New York, 1960, p. 226.

tearing. Scale may be made true in certain directions, along certain lines, and outward in all directions from one or two points, but a map can never be said to be completely true to scale. Further, the distortion in scale becomes greater as the represented area becomes greater. Each parallel of latitude on most equivalent global maps used as textbook illustrations has a different scale of miles, and such maps should not be used to find the accurate airline distance between points such as Pittsburgh and Miami. For that, one should use a map based on the oblique equidistant projection centered on Pittsburgh or Miami, or some map that has the property of true scale along or near the 80th meridian, such as a transverse Mercator. For large maps of small areas, such as the

U.S. Geological Survey topographic maps, at a scale of approximately 1 inch per mile, the scale error is negligible, and airline distances can be scaled off confidently within the permitted map error tolerance of $\frac{1}{50}$ inch, or a distance of 100 feet anywhere on the area represented.

3-6 SCALE REPRESENTATIONS Map scale may be expressed in different ways. *Numerical scale* expresses map scale by means of a simple ratio, such as 1:1,000,000. This indicates that one unit of length on the map represents one million such units in the area to be represented. In order to translate this ratio into miles per inch, divide 1,000,000 by the number of inches in a mile:

$$\frac{1,000,000}{5280 \times 12''} = \frac{1,000,000}{63,360}$$

$$= 15.78 \text{ miles per inch}$$

Scale may also be represented as a *representative fraction,* such as $\frac{1}{1,000,000}$. *Graphic scale* is the graphic representation of linear scale. Graphic scales frequently include metric equivalents, as shown in Figure 3-13.

It is often useful to classify different scales. The simplest classification distinguishes between large-scale, medium-scale, and small-scale maps. Geographers have given specific names to these three and have assigned value limits to them, as is shown in Table 3-4. *Large-scale* maps, therefore, have relatively large representative fractions (small denomi-

Table 3-4 Classification of Map Scale

Scale classification	Map ratio	Lower limit of areas to be distinguished
Topographic (large-scale)	1:500,000 and below	under $\frac{1}{4}$ sq mi
Chorographic (medium-scale)	1:500,000 to 1:5,000,000	$\frac{1}{4}$ to 25 sq mi
Geographic or global (small-scale)	1:5,000,000 and above	above 25 sq mi

The Planet Earth: Its Place in the Solar System; Its Dimensions and Portrayal on Maps

nators) and are about 8 miles or less per inch. They are used to show relatively small areas. *Small-scale* maps show large areas, have small representative fractions (large denominators), and are about 80 miles or more per inch. If we assume that measurements below $\frac{1}{16}$ inch become difficult to distinguish, delimited areas should never be below $\frac{1}{16}$ square inch or $\frac{1}{250}$ inch in size. Large-scale maps, therefore, should be used whenever areas of less than $\frac{1}{4}$ square mile are to be indicated. The relationship between map size, the area to be represented, and map scale can be illustrated by the following problem:

Problem: To determine the map ratio for a map of the United States to be shown on one page of a book and at a size of 6 × 4 inches.

Solution:

1. Approximate the size of the United States = 3,000 miles by 2,200 miles.
2. Determine map ratio of long dimension:

    3,000 (mi) × 5,280 (ft) × 2 ($\frac{1}{2}$ ft)

    $\qquad\qquad$ = 31,600,000 half-feet

    A map ratio of 1:31,600,000 would suffice to show the east-west dimension of the United States within 6 inches.

3. At this scale, determine whether 4 inches would approximate the north-south dimension:

$$4 \times 31,600,000 = \frac{126,400,000}{12} \text{ (ft)}$$

$$= \frac{10,533,333}{5,280} \text{ (mi), or 1,995 miles}$$

Conclusion: A map scale of 1:31,600,000, therefore, would be able to satisfy the space limitations, except that the north-south dimensions would have to be slightly greater than 4 inches but well within the dimensions of the page limitations.

Figure 3-13   A graphic scale.

Map *linear scale* should never be confused with map *area scale;* the latter is the square of the former. A map at a linear scale of 1 inch per mile covers an area that is four times as great as a map at the scale of 2 inches per mile.

Mapping Procedures

The preparation of maps, from the collection of raw field data to the finished map, involves many steps. Our discussion of these will be grouped into three parts: (1) the preparation of the basic locational framework of the map; (2) the collection of field data; and (3) the translation of the data into map symbolism. The first two will be treated in this section of the chapter; the last will be reserved for the last section.

3-7   PREPARATION OF THE PLANIMETRIC BASE   It is important for any map to relate the area shown to the global grid in some way and to establish a system of relative location so that the lines, points, or areas may be shown on the completed map. An air photograph may contain some of the features of a map, but it is not strictly a map, since it has not yet been related to an earth location grid. The first stage in mapping is to prepare a flat locational base, sometimes referred to as a *planimetric map* (Gr.: *planus* = flat; *metros* = to measure), that locates the major recognizable features in the area in their proper horizontal relationship (direction and scale) and ties them into some

(A) GLOBAL (1:10,000,000)

(B) CHOROGRAPHIC (1:1,000,000)

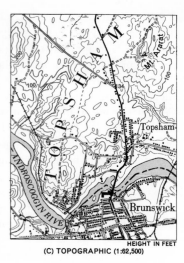
(C) TOPOGRAPHIC (1:62,500)

form of earth locational grid, possibly to the geographic grid by means of parallels or meridians, or to the Public Land Survey System by means of section and township lines, or to a national or international military grid system. Planimetric base maps at various scales already have been prepared for many parts of the United States and for selected areas of the world, and are available from governmental agencies. They contain the layout of roads, streams, lakes, rural buildings, city streets, and forested and poorly drained areas. Elevations usually are given for selected reference points such as road intersections, but no attempt is made to present the features of surface irregularities. Distances on the map are scaled according to horizontal measurement and do not consider distance up and down slopes.

In the past, the preparation of planimetric maps involved a large amount of field survey time and considerable expense. Today, airborne sensing devices, including special automatic cameras, radar, and infrared sensors, collect data relating to the earth surface rapidly and relatively cheaply. Such *imagery*

(the visible record), however, must be supplemented by ground control whose accuracy varies depending on the uses to which the maps will be put.

Ground control is established by laying out an accurate base line somewhere within the area to be mapped. The base line must be straight, both ends should be visible from each other, and the bearing or azimuth of the line must be carefully determined. Usually a straight stretch of road or railroad is selected as the base line, because measurements should not be hindered by vegetation or uneven ground. The length of the line may vary, but generally it is between 4 and 12 miles, depending on visibility and the spread of the triangulation to follow. The azimuth of the line is determined by star sightings at night, generally using Polaris if possible. The length of the line is determined with care. First-order geodetic surveys use special alloy-coated steel tape or wire that is kept off the ground and maintained at a given tension. Temperature records are kept to compensate for tape expansion or contraction, and observations often are taken at night

Figure 3-14 Global, chorographic, and topographic scales. [(A) Bartholomew World Map Series; (B) International Map of the World AMS Sheet NK 19; (C) USGS Bath 15-minute Quadrangle.]

using special signal lights to avoid refraction caused by convectional air currents. Although the official minimum standard of acceptable error in U.S. government first-order surveys is 1 in 25,000, accuracy usually is far greater than that, and tolerances may fall to as low as 1 in 500,000, or 1 inch in about 8 miles. Such accuracy is extremely important for some kinds of surveys, such as transcontinental triangulation networks, because over long distances such minute errors projected by triangulation can be cumulative. Ground control for general topographic mapping need not be as accurate, and a satisfactory base line can be established using only a steel tape and a *theodolite.* This instrument is a small telescope mounted on a tripod that can be carefully leveled and used to measure both horizontal and vertical angles.

Once the base line is established, special observation stations are selected along the line and at points where visibility is good. Triangulation techniques (see Section 2-16) are used to locate easily recognized reference points in the area such as hilltops, church steeples, and road intersections. Extension of the ground control by triangulation, therefore, can locate reference points throughout the area for ground control.

Vertical control involves establishing correct elevations of various reference points through the area with relation to a given datum plane, usually mean sea level. The same reference points determined by horizontal control may be used for vertical control, although frequently they may differ. Vertical control points usually are found along highways or at the summits of prominent hills and mountains. Vertical control points already are available throughout most parts of the United States through the extensions of the intercontinental triangulation network. First-, second-, and third-order horizontal control points in the United States, determined by government surveyors, are marked by permanent concrete or metal points surmounted by a brass disk that gives its latitude and longitude. These are known as *monuments.* First- and second-order vertical control stations are similarly marked and are known as *bench marks* (see Figure 3-15).

Once the reference points are established and marked on an appropriate grid in their proper horizontal relationship, the rest of the

Figure 3-15 Mounting a bench mark as a permanent reference point for future areal surveys. The brass disk will be set into cement. The exact latitude and longitude and elevation above mean sea level are given on the disk. [U.S. Geological Survey]

planimetric detail is filled in, either by field survey or, more recently, by aerial photography. Air photographs are invaluable for picking up the detail of irregularities in the shape of rivers or lakes, roads, and forest patterns. They must be related to ground control, however, because of inherent errors of scale distortion that increase radically from the center of the photographs and because of errors caused by slight variations in the survey plane's tilt or elevation. Improvements in remote sensing techniques have been enormous during the past two decades. Special fine-grained films adapted to preservation of minute detail make it possible to photograph from extremely high elevations, thus decreasing the scale of the photographs and increasing the area of acceptable distortion tolerance. Using radar and infrared imagery to select certain terrain details for high resolution has been an important breakthrough in the field of inventory mapping using remote sensing apparatus. Infrared sensors that are extremely sensitive to selected wavelengths of heat radiation are especially valuable in the delineation of roads, streets, bare ground, and certain vegetation patterns. Earth satellites also are being used to collect terrain information. They will be increasingly valuable in the preparation of planimetric maps for the entire world without years of expensive field surveys. Our facilities for collecting and recording field data on photos and transferring them to maps have now outreached our capacities for using them.

3-8 COLLECTION OF MAP DATA The task of any map is to portray *selected* features on the earth surface for presentation at reduced scales on a flat surface. As indicated in the discussion of the principles of natural distributions (see Chapter 1), the variability of forms and phenomena from place to place on the earth surface is infinite and subject to constant alteration. All mapping projects thus are somewhat selective and subjective. The selectivity depends on the scale, the purpose of the project, and a host of related factors, including cost, tools, and time available. One of the most difficult aspects of field mapping is to know what *not* to record, whether it is a feature of areal difference or similarity too small to be shown on the final map of presentation or a feature that may be irrelevant in the overall objective. Such judgments are involved even in the relatively simple mapping procedure of producing planimetric bases. In tracing the course of a stream, for example, the limitations of map scale must require some generalization of detail. Not every minute irregularity in the stream channel can be shown. Similarly, if one is tracing the boundaries of forest land, the transitional nature of such borders always involves generalizations of the *place of change*.

Change in time also plays an important role in the subjective nature of field mapping. This appears at all scales, ranging from the influences of tides and crustal deformation on precise determinations of bench mark elevations during first-order geodetic triangulation surveys to the periodic seasonal alteration of field cover in a land use survey. Maps always must be dated and interpreted within the parameters of change.

Map data collection takes many forms, usually depending on the particular forms and phenomena to be plotted. A partial classification of types of distribution follows:

Continuous distributions. Such distributions consist of the presence or absence of something that has continuity of presence areally. These may be single-unit distributional features, such as a field of oats or an elevation above 600 feet, or multiple-unit distributions

Figure 3-16 Map of land use illustrating the fractional system of land inventory.

The fractional system enables the map to contain a large amount of qualitative areal data through the use of fractions, each digit of which represents a qualitative rating. In the illustration the fractions of the field survey have been separated, with the numerator shown in color, referring to physical characteristics and the denominator in black indicating the type of land use and its quality rating.

NUMERATOR: PHYSICAL CHARACTERISTICS

1st digit: slope class
2d digit: drainage
3d digit: degree of erosion
4th digit: stoniness of soil
letter (*a*, *b*, *c*, etc.): soil type

Example: 3434*f* = steeply rolling (9–15° slope); droughty and excessively drained; presence of sheet erosion and incipient gullies; stony soils; soil type: Bellfontaine gravelly loam.

DENOMINATOR: RURAL LAND USE

1st digit: major land use (cropped, pasture, forest, idle, etc.)
2d letter: specific land use (wheat, rotation pasture, birch-maple forest)
3d digit: quality rating of land use

Examples: 1*f*2 = cornfield of moderate quality.
3*a*-*b*1 = forest land consisting of both oak-hickory and elm-ash-maple association of merchantable timber.
2*b*2 = permanent cleared pasture land of moderate quality.

that are arbitrarily defined as homogeneous units, such as a climatic or soil type. Air photographs are exceedingly useful in this type of mapping, when the patterns are visible and recognizable from the air. One useful device in mapping multiple land use patterns that exhibit continuous distribution is the *fractional* system, illustrated in Figure 3-16. A newer system of map data collection involves plotting the presence or absence of a form or phenomenon having continuous distribution within small quadrats (sample survey squares) and recorded on tabulating

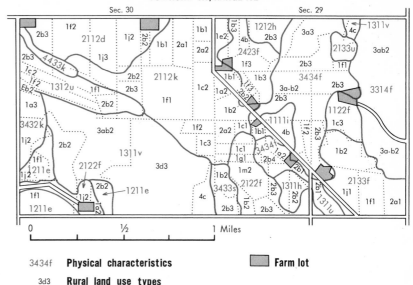

TOWNSHIP 9N; RANGE 18E

3434f **Physical characteristics**

3d3 **Rural land use types**

Farm lot

cards by means of a portable card punch, using grid coordinates as locational identification. The completeness of detail areally depends on the closeness of the rectangular grid. The advantage of this system is that it can be programmed for computer analysis in the preparation of statistical factor correlations, statistical summaries, or even in the machine preparation of map patterns. The method is diagramed in Figure 3-17. A special adaptation of such a method to receive, store, and retrieve coded weather stations throughout the United States is being used to make weather maps that are entirely machine-produced, complete with fronts, isobars and isotherms.

The newest system for the collection of continuous distribution data is still in its infancy. It consists of the rapid machine-scanning of remote-sensing imagery. The scanning device is directed to select certain colors or tones from such sources as air photographs and to translate these into computer programming for various purposes, or

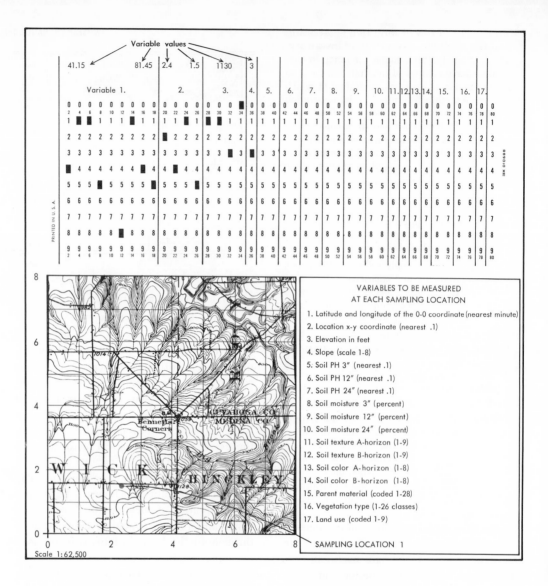

to be transferred directly onto maps of selected data. Some day, possibly, a rapid census of land use for any area on earth, complete with maps, can be prepared from high-resolution satellite imagery and performed almost entirely by machines. Inductive generalizations then can be programmed precisely and the intuitive judgments removed from map compilation. Such devices will not replace maps—they merely will produce better ones at vastly greater speeds and with less expense.

Discontinuous but stationary distributions. Such distributions include forms and phenomena that occupy points or lines, and whose areal expression on the map is so small as to be relatively insignificant. Examples on large-scale maps include buildings, roads, political boundaries, fences, streams, and mines. Small-scale maps include towns and cities in this category. Such forms normally appear on planimetric base maps and always are important map features: they serve as reference locational features and do not detract much from areal patterns, and they are centers of streams of action, indicators of the dynamics of areal integration. Examining the pattern of streams and rivers reveals much concerning the processes and direction of land sculpture, just as the pattern of roads and dwellings indicates the flow of energy into and out of an area in the economic process of human occupation. Such features are important parts of the anatomy of areas and are of special interest to geographers. Mapping procedures for such forms are relatively simple, and most of them are recognizable from air photographs, the geographer's major tool for the storage and retrieval of map data.

Discontinuous distributions in motion. Many features on the earth surface are in constant motion, and their general presence may be useful to map. Most animal life, including man, is of this type. Maps showing the distribution of population usually indicate the place of residence or the sleeping place, yet except during his daily rest period, man is in constant motion, like mountain goats or codfish. Mapping such distributions and their common routes necessitates special facilities and techniques for gathering information. When large numbers are involved, sample counts are useful. Indirect methods also are valuable tools. Persistent, repeated movement over the same route generally leaves distinctive marks on the route itself. Just as the characteristics of a stream bed and channel may reveal information concerning the flow of water through it, so too the paths of men and rabbits may reveal their numbers and habits of movement.

Perhaps the best illustration of this type of distribution in physical geography is the global pattern of air circulation. Weather stations throughout the world collect information concerning air movement throughout the lower part of the atmosphere, using special recording instruments. As a result, maps showing the streamlines of airflow at various levels above the earth surface are prepared daily and are used in weather forecasting. Similar maps for the oceanic circulation also are prepared periodically, although the sampling stations are much fewer.

Distributions of relative presence. It may be essential to map the relative presence of a particular form or phenomenon areally. To illustrate, suppose a slope map is desired in an area of hilly terrain. The entire area has slopes of all degrees. By arbitrarily dividing average area slope into classes (0–5 percent, 5–10 percent, etc.) and plotting the distribution of each grouping, some idea of relative slope presence and variation may be presented. Other examples include density of population and ratios of various kinds, such as cultivated area per square mile, relative humidity, or per capita income.

3-9 SAMPLING   The surface of the earth is so vast, so complex, and so changeable that the collection of map data often must involve sampling of some kind. This is especially true today, when our tools and techniques for area correlations are well adapted to handle enormous masses of quantitative data. Further, our demands for data have far exceeded our capacities in time, money, and personnel to produce them by direct measurement. The underdeveloped world cannot wait for detailed land inventory mapping, and shortcuts must be designed to provide useful tools for project planning. Areal sampling is not new, and it may take many forms. A topographic surveyor mapping surface relief does not measure the elevation of every point in a given area. Instead, he selects several easily identifiable points, such as a hilltop, a river or road junction, or a building corner, measures their elevations directly by barometric reading or theodolite control, and interpolates visually between points. A timber-estimating specialist often can calculate the quantity of commercial lumber contained in a square mile of forest within a few thousand board feet and within a few hours, not by counting and measuring each tree but by sampling and inspecting along selected traverses across the area. Climatic maps use a network of weather stations that sample the areal patterns of weather and climate. Just as opinion polls can be used with remarkable results to estimate political trends, so too can soil surveyors accurately interpolate the pattern of soil differences and similarities of observation at various scales, using sample borings placed at the right places.

One of the principal difficulties with sampling is human bias when the sampling sites are chosen by personal inspection. A soil surveyor may deliberately avoid sampling in the center of a swamp and assume a uniformity that is not there in order to keep his feet from getting wet. A wildlife enumerator may avoid making counts in an area he suspects contains some deterrent factor, thus saving time. Preconceived results often lead the investigator into subconscious errors of judgment. To avoid such subjective errors in the selection of sample measurements, numerous devices have been developed. Sampling techniques will differ depending on the kind of measurement to be made. Traffic counts, designed to construct a map of traffic flow in an area, will differ greatly in method and locational principles when compared with sampling devices employed in land use studies.

Figure 3-18 presents four types of sampling that can be adapted either for various types of point sampling, such as the location of soil test borings and the placing of thermometers for microclimatic studies, or for area sampling, such as the location of quadrats in plant counts or land use measurements. Linear sampling, another technique, which involves the enumeration of changes along a series of controlled transects, is a highly useful device for obtaining statistical estimates of areal variation. The accuracy of any sampling tool increases with the number of samples, and an important part of sampling is to obtain an acceptable margin of error within the limitation of available time and effort. Statistical tools have been developed to indicate the relationship between sample size, frequency, and probable percentage of error.

Cartographic Representation

Maps are recording instruments designed to present selected locational data at reduced scales for human use. All maps have some aspect of utility, and their construction must reflect the purpose for which they were de-

Figure 3-18 Methods of determining points for area sampling. In *A*, 36 points are located, each in the center of an arbitrary grid square. In *B*, 36 points are selected at random, and each point is located by a double selection: first, a random choice between 1 to 90, to determine the horizontal position; second, a random choice between 1 to 40, to determine the vertical position. In *C*, the same method is used as in *B*, except that the number of choices is altered so that more samples are taken within one part of the map than another. In *D*, a selection of numbers between 0 and 9 is chosen at random to determine the numbering of the horizontal and vertical squares. The position of the points is determined by their selected coordinates. The first digit represents the first horizontal column; the second indicates the vertical column.

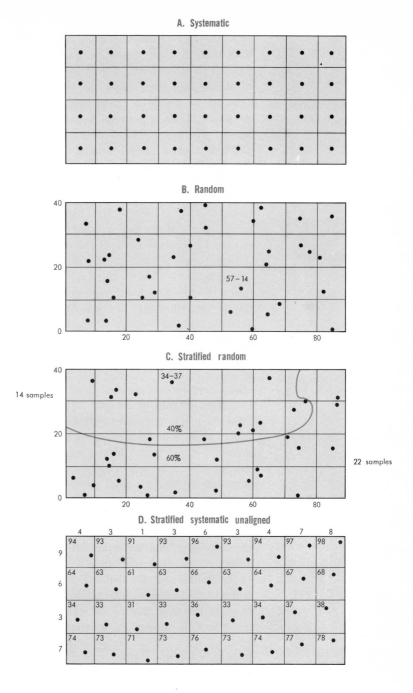

signed. Complete reality obviously cannot be shown on a map; hence, maps, as well as their constituent parts, are symbols of reality. Just as mathematicians use a wide variety of symbols to represent abstract ideas such as nothingness, equality, summation, numerical series, and infinity in the necessity for accurate treatment of logical thought, cartographers also use a wide variety of symbols to represent abstractions of location, quantification, relative presence, shape, and motion in their task of describing the content of earth space.

3-10 MAP SYMBOLS There must be a wide choice of map symbols because of the enormous variety of forms and phenomena present on the earth surface. The ideal symbol should be readily recognizable for what it represents, and any possible misunderstanding should be clarified in the map legend. Common usage and man's frequent need for maps have led to the development of certain basic map symbols that require no explanation. Blue lines on a map are readily recognized as streams or rivers; the double parallel lines representing roads are rarely confused with railroads, customarily shown with their symbolic crossties; and the fine parallel lines bordering a shoreline indicate to almost any map reader the side of the coastline on which the water lies. Similarly, the dashed lines of political boundaries, the regular grid lines of the basic map projection, and the complex pattern of concentric contour lines are almost never misunderstood, even by the rare or casual user of maps. However, the *legend,* usually shown as a boxed segment of the map and containing the identification of the map symbols, is one of the most important parts of the map when the symbolism is not perfectly clear.

Map symbols used to show quantities or ratios in their locational aspects are much less familiar than conventional topographic symbols. One of the more frequently used symbols in this connection is the *isoline* (Gr.: *isos* = equal), or line of equal value. One group of isolines, the *isarithms* (Gr.: *arithmos* = number), is especially widely used. Examples of isariths include *isotherms,* or lines of equal temperature expressed numerically; *isobars,* or lines of equal barometric pressure; *isobaths,* or lines of equal ocean depth; *isohyets,* or lines of equal rainfall, and *contours,* or lines of equal elevation above mean sea level. Isarithms are valuable to cartographers because they allow such maps to show variations in the gradients of areal change. When the isarithms are close together, the gradient of change is steep. One of the limitations of isarithmic maps is the width of the isarithmic interval. No variation in value less than that of the isarithmic interval can be shown. For example, within an area shown on a temperature map as lying between the isotherms of 50° and 60°, there is a margin of error that varies from 0 to slightly less than 10°. Further, since such maps show continuous distributions, their

accuracy must vary with the spacing of data-recording positions. Again, using isothermal maps to illustrate, a map of U.S. temperatures based on 2,000 weather stations requires interpolation over a large part of the total area. It is well known that under certain topographic conditions, variations of 10° to 15° may be experienced within a few miles, and unless the network of recording stations is unusually close-knit, such variations would not appear on the maps. Determining the proper density of measurement sites to give a desired reliability in isarithmic representation is an important statistical procedure. Such techniques are particularly significant when the interpolation cannot be done visually, as when sketching contours between bench marks in the field. Proper interpolation involves a set of analytical procedures that are important but beyond the scope of this chapter. The reader is referred to an excellent article on this subject contained in the publication *Economic Geography.*[2]

Another group of isolines include *isopleths* (Gr.: *plethron* = fullness), which represent lines of equal ratio or proportion. Examples include population maps with lines of equal density of population per square mile, maps of valuation per square mile, or maps of production income per square mile of cultivated cropland. Simpler quantitative maps may use graduations in quantity per reporting unit, such as a map of state per capita income by counties or the average production of wheat per acre for national units on a world map. *Cartograms* are graphic devices for showing quantitative data on maps. One interesting type of cartogram uses an intentional distortion of area to show degrees of importance. Figure 3-19 shows three separate

[2]J. Ross Mackay, "Some Problems and Techniques in Isopleth Mapping," *Economic Geography,* vol. 27, pp. 1–9, 1951.

Figure 3-19 Some examples of cartograms. Cartograms are graphic maps in which certain properties are distorted proportionally to show quantitative data. In *A*, for example, the area of each country in southeast Asia is proportional to its population. In *B*, the main approaches to a city are scaled in width to indicate the volume of traffic, the dots showing the location of the traffic count stations. *C* represents the environs of Seattle, with the land areas shaded. On the left, the land area and scale are roughly proportional, with the black lines representing commuting times from the city center. On the right, the commuting times are concentric, and the land area and scale have been distorted proportionately. [*A* after Broek and Webb; *C* after Bunge]

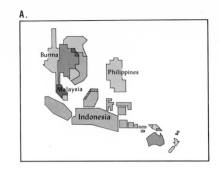

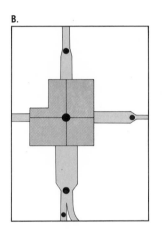

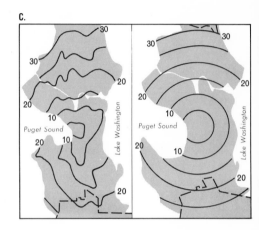

cartograms for different types of distributional data.

One type of map used to show quantity and distribution is the dot map, in which each dot represents a certain numerical value, such as a given number of people, acres of corn, or number of cows. A well-designed dot map of distributions must be carefully prepared, so that there will be sufficient dots to indicate areas of greater or lesser abundance but not so many that it would be extremely difficult to judge quantities within a given area.

3-11 REPRESENTATION OF LAND SURFACE RELIEF Probably the most challenging problem faced by cartographers has been the representation of surface relief, or the "upness" and "downness" of the land surface, upon the flat plane of the map. It is of sufficient importance to warrant separate treatment here. The earliest-known map, an ancient Babylonian clay tablet more than 4,000 years old, attempted to show the mountains encircling the Tigris-Euphrates river basin by means of a series of inverted lunes, or half-circles. Over the years, techniques have improved, but essentially they include three

basic forms: contours, hachures, and shading. The best examples of the art today include combinations of two or more of these. Each will be discussed briefly in subsequent paragraphs.

Contours. Of the three devices for representing surface configuration, the most widely used and perhaps the most adaptable for many different purposes is the contour map. Essentially, this type of map indicates comparative elevations by means of contour lines, which are lines that connect points of equal elevation above a given datum plane, usually mean sea level, or the level about which the ocean tides oscillate. The spacing between the contours varies on different

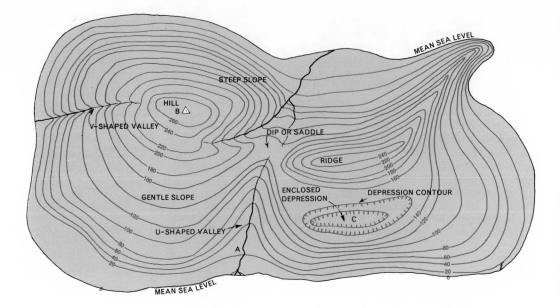

maps, depending on the degree of elevation detail desired, the amount of variability in elevation to be shown, and the scale of the map.

Figure 3-20 represents a contour map of a small island, on which the following features are labeled:

1. A central hill, steeper on one side than the other.
2. A U-shaped valley, with a gently sloping lower portion.
3. A steep-sided, narrow, V-shaped stream valley with a steep gradient through its length.
4. A ridge which contains a dip or saddle.
5. A surface depression surrounded on all sides by higher ground.

The coastline of the island at mean sea level constitutes the zero-foot contour line. If the island were submerged through a vertical distance of 20 feet, the new shoreline would correspond to the position of the 20-foot contour line prior to submergence. The amount of vertical difference in elevation represented between any two contour lines is known as the contour interval. In Figure 3-20 this interval is given as 20 feet.

An examination of the contour map in Figure 3-20 reveals a number of characteristics common to all contour maps. These may be listed as follows:

1. Contours are closed, concentric circles.
2. Contours never cross one another.
3. Every contour has an upside and a downside. The latter is on the outside of the closed circles.
4. The highest elevation lies within the innermost contour.
5. Contours become closer together as the slope of land increases.
6. Contours bend upstream when crossing a flowing stream, and the amount of bend decreases as the gradient of the stream increases.
7. The width of valleys is indicated by the

The Planet Earth: Its Place in the Solar System; Its Dimensions and Portrayal on Maps

Figure 3-20   A hypothetical contour map of an island, illustrating different topographic forms.

breadth of contour curves in their course across a valley.

8. A dip, or saddle, is indicated by a gap between two adjacent closed contours.

Closed depressions on a land surface, that is, depressions which are surrounded by higher ground on all sides, may be shown by special depression contours, which are marked by small lines or hachures on their downsides.

Most contour maps show features other than contours—especially elements of hydrology, such as lakes, rivers, or marshes, and selected cultural features, such as towns, roads, and railroads. Contour maps that contain reasonably accurate portrayals of the major surface features in an area are known as *topographic maps*. Most countries have programs for the governmental preparation of topographic maps, at least for the most important parts of their land areas. The standard topographic maps for the United States are those prepared by the U.S. Geological Survey. They are prepared at various scales, although most of them have a map ratio of 1:62,500, or slightly more than 1 inch to 1 mile in linear scale. The student should familiarize himself with the interpretation and use of such topographic maps.

Hachures.   Hachures are short segments of lines that always are drawn downslope. Increasing the length and width of the lines and decreasing the space between them give the impression of steeper slopes. The gradation is carefully correlated with slope. For example, a slope of 45° would appear as solid black, with the hachures running together, while flat terrain would contain no hachure

lines. Figure 3-21 contains an example. The technique was invented by an Austrian army officer and has been widely used by European cartographers until recently. It is rarely used today, however, except in the preparation of some wall maps, because of the skill and labor required. The difficulty of using hachuring is illustrated by a classic but unverified story familiar to many geographers and cartographers, which relates to an artisan in one of the great German map-making concerns who did nothing but draw hachures, yet was still an apprentice when he was fifty-five years old.

Shading.   Shading is the technique of showing surface irregularities by means of assuming an oblique lighting. Its simplest form would be to photograph a carefully constructed model of the area to be represented, with illumination from the side. An artist-cartographer, using careful positioning of his shading, can depict surface irregularities in much the same way as he would the recognizable contours of a human figure in a painting. Shading may be done by fine line work, or by gradations of gray to black tones. Figure 3-21 illustrates some of the excellent combinations of shading and contouring that appear on the newer topographic maps prepared by the U.S. Geological Survey.

A widely used terrain map that might be likened to a line caricature is the *physiographic diagram*. This is a line drawing of the land surface which accentuates certain features of the land surface, especially over large areas, so as to feature particular aspects of shape, genesis, and composition. As indicated earlier, many of the best topographic maps prepared today use combinations of contouring, hachuring, and shading in order to present the illusion of three dimensions without sacrificing the accuracy of scale mapping.

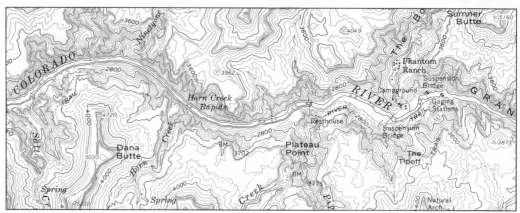

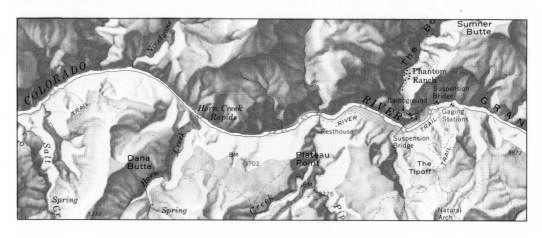

Figure 3-21 Three examples of techniques for showing relative relief. Hachuring, contouring, and side shading are illustrated in these three views of sections of the Grand Canyon. [Courtesy of Federated Lithographers-Printers, Inc. Contour data from U.S. Geological Survey]

Study Questions

1. Explain why maps can never reproduce the surface of the earth without some distortion. What kinds of distortions may take place?

2. Distinguish between an orthographic, a stereographic, and a gnomonic projection. Which one has the greatest distortion of scale, and why?

3. What are the four major properties of map projections? Give examples of map uses where each of these properties is of particular value.

4. Why was the Mercator Modified Cylindrical projection so enormously popular after its invention? What major disadvantage did it have? Can the Mercator projection show the North Pole? Explain. What is a Transverse Mercator projection?

5. Using Table 3-2, select an appropriate projection for each of the following uses:

    a. A map that will indicate the towns and cities covered by airline distance zones from a central point.

    b. A weather map of the United States.

    c. A map showing the world distribution of forest land.

    d. A navigational map from which compass bearings may be plotted from one point to another.

    e. An atlas map of Africa.

    f. A map of a small city.

    g. A world map of temperatures (area not important).

    h. A pictorial representation of the earth for use with a news story.

    i. A base for large-scale topographic mapping.

6. Explain why azimuthal equidistant projections have an unusual application in the postal service.

7. A scale of 1:10,000,000 represents about how many miles to the inch?

8. What map scale would you use to show a state on a map approximately 2×3 feet in size, assuming the state to measure about 300 miles from north to south and 500 miles from east to west?

9. Explain the principle of triangulation in the construction of a planimetric control base.

10. Prepare a coded system for collecting land use data in an urban area directly on tabulating cards using a Porta-Punch (see Figure 3-18).

11. Explain why stratified sampling might be a useful technique for positioning thermometers in determining micropatterns of temperature during a cold winter week in a section of the northern Appalachians. Where might it be more useful to have a denser sampling pattern?

Part 1: References

Birch, Thomas W., *Maps, Topographical and Statistical,* 2d ed., The Clarendon Press, New York, 1964.

Brown, Lloyd A., *The Story of Maps,* Little, Brown and Company, Boston, 1949.

Chamberlin, Wellman, *The Round Earth on Flat Paper,* The National Geographic Society, Washington, D.C., 1947.

Compton, Robert R., *Manual of Field Geology,* John Wiley & Sons, Inc., New York, 1962.

Deetz, Charles H., and O. S. Adams, *Elements of Map Projections with Applications to Map and Chart Construction,* 5th ed., U.S. Government Printing Office, Washington, D.C., 1944.

Department of the Army, *Topographic Surveying,* TM 5-234, U.S. Government Printing Office, Washington, D.C., 1953.

———, *Map Reading,* FM 21-26, U.S. Government Printing Office, Washington, D.C., 1956.

Greenhood, David, *Mapping,* rev. ed., University of Chicago Press, Chicago, 1964.

———, *Down to Earth: Mapping for Everybody,* Holiday House, Inc., New York, 1951.

Harrison, L. C., *Sun, Earth, Time, and Man,* Rand McNally & Company, Chicago, 1960.

Johnson, W. E., *Mathematical Geography,* American Book Company, New York, 1907.

Lobeck, Armin K., *Block Diagrams and Other Graphic Methods Used in Geology and Geography,* 2d ed., Emerson-Trussell Book Company, Amherst, Mass., 1958.

Monkhouse, Francis J., and Henry R. Wilkinson, *Maps and Diagrams, Their Compilation and Construction,* E. P. Dutton & Co., New York, 1952.

Paget, Thornton, and Lou Williams Page (eds.), *The Origin of the Solar System,* The Macmillan Company, New York, 1966.

Raisz, Erwin J., *Principles of Cartography,* 3d ed., McGraw-Hill Book Company, Inc., New York, 1962.

Robinson, Arthur H., and Randall S. Sale, *Elements of Cartography,* 3d ed., John Wiley & Sons, Inc., New York, 1969.

Robinson, Arthur H., *The Look of Maps,* The University of Wisconsin Press, Madison, Wis., 1952.

Steers, J. A., *An Introduction to the Study of Map Projections,* University of London Press, London, 1957.

U.S. Naval Observatory, *The Air Almanac,* published annually, U.S. Government Printing Office, Washington, D.C.

Wylie, C. C., *Astronomy, Maps, and Weather,* Harper and Bros., New York, 1942.

PART 2
CLIMATE

The ocean of air which flows everywhere about us constitutes the most changeable feature in our physical environment. Day by day and hour by hour, fluctuations occur in its movement, its pressure, and its content of heat and moisture. We speak of the momentary conditions of the atmosphere as *weather,* and in parts of the world that experience great day-by-day fluctuations, weather becomes an important topic of conversation and a significant factor in daily decisions. The broad generalizations of weather conditions are known as *climate.* The pattern of climates in the world plays a major role in the kinds of crops we grow, the kinds of trees that make up our forests, the amount of our yearly budgets spent on heating or cooling our homes, and the type of clothing we wear.

The geography of climates exemplifies the basic principles or concepts of natural distributions described in Chapter 1. Examples of each of these principles as applied to climatic characteristics come readily to mind.

The dynamics of weather and climate bring into sharp focus the operation of the global energy system. Whereas the energy that streams toward us from the sun affects *all* processes that occur on earth, it is most direct and most apparent in its influence on the lower atmosphere. Further, only a relatively small part of this energy is diverted into such processes as the erosion of mountains, the alteration of rocks, the grinding of glaciers, and life itself. Most of it is swirled around in curved trajectories within the lower atmosphere before slipping out again into outer space. To illustrate, it has been estimated that all of the energy locked within all of the fossil fuels remaining beneath the earth surface amounts to only about 2 percent of the annual supply that is received by the earth-atmosphere system. Only rarely does this prodigious energy system make its powers known, as in the violence of a hurricane or tornado.

It is appropriate that the examination of world climates be linked closely with the operation of the global energy system, and that this precede the consideration of the other major physical environmental elements in this text. The processes involving landforms and the development of vegetation and soil patterns are the result of relatively minor detours in the continuous but irregular flow of energy across the face of the earth.

Five chapters are devoted to climate in this text. The first three deal largely with the various processes that link climate to the global energy system. Included are the nature of the entire energy system; the local differences in the net surpluses and deficiencies of radiation; the various ways of evening out the flow of energy, including the hydrologic cycle with its energy-exchange mechanisms including evaporation, precipitation, and stream runoff; the planetary wind and ocean-current circulatory system that contains the three-dimensional channeling of energy flows; atmospheric pressure differentials that direct the horizontal and vertical movements of air and water, and the more violent reactions that are manifested in storms and frontal disturbances.

The last two chapters deal with the major regional divisions of climate at the earth surface, presenting and interpreting its differences, similarities, and implications for humans.

CHAPTER 4
THE GLOBAL ENERGY FLOW AND TEMPERATURE PATTERNS

The planet earth is bathed in a continuous flow of energy emanating from the sun. This energy flow is the source of all life processes and almost all forms of motion and energy on earth. It is vital to man since it keeps him warm, provides his food, permits him the blessings of sight and movement, drives his machines, waters his fields, and gives him the very energy of existence. Without this energy flow, our planet would be just another inert mass of solid material, lost in the cold darkness of outer space. It would have but a tiny energy supply leaking out from its interior—generated by the slow radioactive decay of a few elements stored within it, or perhaps left from the residual heat of the violent formative period·billions of years ago. Certainly it would not be the home of man or of any other life form as we know it. The supply of energy we receive from the sun is so important to us that we should know more about it: what it is, what it does, and what happens to it.

Probably the most obvious aspect of the energy flow is that its intake varies in different places on earth and at different periods of time. The most noticeable results of this irregularity in inflow appear in the distributional patterns of temperature on earth. We all know that the equatorial regions are warmer than the polar regions, that the seasonal variations of temperature become more pronounced with increasing latitude, and that some years are warmer and some cooler than normal. The explanations for these and many other world temperature patterns will concern us in this chapter. We shall learn that there is an extremely complicated yet efficient system of energy exchange that acts as an automatic regulator to prevent too great an accumulation of energy in some places and too rapid an outflow and loss of energy in others. As humans, we are most fortunate in having a built-in thermostat to regulate our global heating system. The fluctuations of weather everywhere are but the moving mechanisms for rearranging energy patterns, especially the areas that exhibit net radiation surpluses and deficiencies.

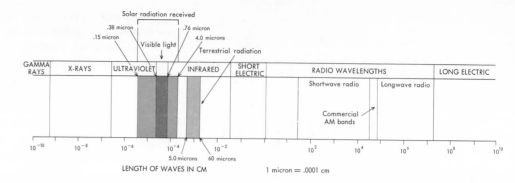

Solar radiation received

.38 micron .76 micron
.15 micron 4.0 microns
Visible light
Terrestrial radiation

GAMMA RAYS	X-RAYS	ULTRAVIOLET		INFRARED	SHORT ELECTRIC	RADIO WAVELENGTHS		LONG ELECTRIC

Shortwave radio   Longwave radio

Commercial AM bands

10^{-10} 10^{-8} 10^{-6} 10^{-4} 10^{-2} 10^{2} 10^{4} 10^{6} 10^{8} 10^{10}

5.0 microns   60 microns

LENGTH OF WAVES IN CM   1 micron = .0001 cm

The Global Energy Supply

4-1 SOLAR RADIATION The energy that
is emitted from the sun and from other stars
is a special type of energy that has the unique
property of being transmitted through space
by means of *electromagnetic*[1] waves. This
energy is termed *radiant energy,* or simply
radiation.

Electromagnetic waves vary greatly in
wavelength. Figure 4-1 presents the enormous range of known wavelengths of radia-
tion, extending from 10^{-10} to 10^{10} centimeters.
The wavelength of radiation is a function of
the temperature of the transmitting body,
higher temperatures producing shorter
wavelengths. The extremely short wavelengths of gamma radiation, for example, can
be produced only under the enormously high
temperatures generated by nuclear reactions,
such as those found in the solar interior or
in the intense flash of a thermonuclear explosion. The surface of the earth, on the other
hand, produces much longer wavelengths
because its temperatures are much lower
than those of the sun.

Solar radiation is made up of wavelengths

[1]Electromagnetic waves are so named because they are
produced by electrons as they pass from a higher to a
lower energy level within the outer atomic structures, and
also because they are bent or distorted under the influence
of magnetic fields. Their wavelengths are measured from
crest to crest.

ranging from about .15 to 5 microns (1 micron = 1/10,000 centimeter). This variation is
the result of varying temperatures on the
sun's outer, incandescent surface. The average temperature of the sun is about 10,000°F.
Hence, the maximum radiation output occurs
at wavelengths within the blue-green sector
of visible light, centering at .475 micron.
Sunlight appears to be white because it is
composed of an association of colors, ranging
from red to violet. The visible light sector
includes wavelengths of from .38 to .76 micron, with red at the long end of the spectrum. Terrestrial and atmospheric thermal
radiation, on the other hand, cannot be seen
by the human retina, because it lies entirely
within the infrared sector, ranging from 5 to
60+ microns, with a maximum between 15
and 20 microns. This position of the terrestrial radiation spectrum is related to the
average temperature of the entire earth plus
its atmosphere, which averages about 65°.

Not all of the solar radiation coming toward us reaches the earth surface. Some of
the ultraviolet (wavelengths below visible
light) radiation is intercepted by oxygen (O_2)
and ozone (O_3) molecules high above the
earth surface (see Section 4-2). This high-
elevation absorption makes it difficult to obtain accurate measurements of variations in
the total amount of solar energy received.
Scientists are interested in artificial earth

Figure 4-1   The electromagnetic wave spectrum.   Of the total electromagnetic wave spectrum, only a relatively small portion (between .15 and 4.0 microns) forms the bulk of the solar radiation that reaches the earth. The visible range, which we term *light,* has an even narrower range, between .38 and .76 micron. Note that most of the terrestrial (earth) radiation is in a somewhat longer wavelength band than that of the incoming solar radiation.

satellites and high-altitude rockets partly because these instruments are possible means of obtaining more accurate information on this variable. Solar radiation is measured in terms of an arbitrary unit termed the *solar constant,* which is the number of *calories*[2] a standard flat surface receives per square centimeter per minute when it is held perpendicular to the sun and when the earth is at its mean distance from the sun (approximately 93 million miles). The solar constant assumes no loss by atmospheric absorption and is related solely to solar radiation output. Present indications are that the amount of the solar constant is approximately 1.94 and that it varies only from 1 to 3 percent of this figure. A few authorities feel that long-term fluctuations in this constant may be responsible for the broader swings of global temperatures and that observations of the solar constant are not yet sufficiently refined or abundant to reveal significant trends. Present evidence, admittedly inadequate, appears to doubt fluctuations in solar energy output sufficiently great to produce significant changes in the average global temperature patterns.

The laws of thermodynamics indicate that

[2]A *calorie* is the amount of heat energy needed to raise the temperature of one gram of water one degree centigrade at sea level when the air temperature is 15°. The abbreviation *k cal,* for *kilocalorie,* represents a thousand calories. The calories generally used for measuring food intakes are kilocalories.

energy can never be destroyed or consumed. Absorbed radiation may be changed in form, but following such changes, when reverting to radiation, the resultant electromagnetic waves are lengthened. The process of absorption involves changing radiant energy into some other kind of energy. Usually this involves an exchange to *sensible heat,* or the energy of molecular motion that can be measured by a thermometer and recorded as *temperature.* Different substances vary in their capacity to absorb wavelengths of radiation and to convert this into sensible heat. Gases, for example, may be highly selective and absorb only a few scattered wavelengths. In general, more dense materials tend to be more efficient absorbers as well as transmitters, that is, in converting sensible heat back to radiation. The earth can be considered to be a *black body,* or perfect receiver for solar radiation, and can retransmit all of it following absorption, but, of course, at longer wavelengths in the infrared range.

Solar radiation may be transformed into other types of energy than sensible heat. Plants, for example, have the ability to change solar radiation directly into chemical energy, as in the formation of sugars and starches by photosynthesis or the formation of carbonates by microplant forms in the sea. A large part of the solar radiation also is turned into *latent heat,* as in the evaporation of water or the melting of ice. The energy is in the form of molecular motion (heat), but no change of temperature is involved. The importance and mechanics of this energy-exchange process will be treated more fully in the following chapter (see Section 5-5). The mechanical energy of winds, running water, and ocean currents also represents diversions of solar radiation, but usually these have been transformed from heat sources rather than directly from radiation absorption.

Eventually, all forms of energy within the earth-atmosphere system are converted into sensible heat and still later into final infrared radiation before leaving the system. No appreciable quantity of energy has been added along the way (our combustion of fossil fuels notwithstanding) or subtracted. The only difference between the energy that leaves and the energy that enters is that the outgoing radiation has become longer in wavelength and is somewhat less capable of performing work. The amount of entropy it has gained (see Chapter 1) is a measure of the amount of work it has performed during its stay in the system. This can be seen graphically by comparing the wavelengths of solar and terrestrial radiation shown in the top (A) graph in Figure 4-2.

4-2 COMPOSITION OF THE ATMOSPHERE AND ITS ROLE IN ENERGY EXCHANGE The earth atmosphere is a complex mixture of gases. The composition is given in Table 4-1.

Only three of the atmospheric gases shown in Table 4-1 are significantly involved in energy exchange. These are water vapor, carbon dioxide, and ozone, generally in that order. The first two are among the largest and

Table 4-1 Composition of the Earth Atmosphere (percentage of mass)	
nitrogen (N)	78.09%
oxygen (O)	20.95
argon (Ar)	.93
carbon dioxide (CO_2)	.03 (variable)
water vapor (H_2O)	trace (variable)
ozone (O_3)	trace (variable)
hydrogen (H)	trace
various inert gases, including neon, xenon, krypton, helium, etc.	traces

heaviest of the gaseous molecules and tend to be concentrated in the lower and densest part of the atmosphere. Ozone, the lightest of the three, has a zone of concentration at intermediate levels where it is formed. All three are subject to considerable variation in quantity. Nitrogen, though the most abundant gaseous element in the air, like the other inert gases, has no role in energy exchange, except in minute traces as ammonia. Oxygen has only a minor role high in the upper atmosphere. The three most active gases in the atmosphere are treated separately in order to clarify the atmosphere's role in the global energy balance.

Ozone. Ozone is a triatomic molecule (O_3) of oxygen. Most atmospheric oxygen is in the diatomic (O_2) form. O_2 is a molecule large enough to intercept and absorb the shortest wavelengths of solar energy arriving in the upper atmosphere. If sufficiently energized, the diatomic oxygen molecule breaks into single oxygen atoms that are highly unstable. These single atoms may unite with O_2 molecules to form the triatomic or ozone form. Ozone is much more efficient than oxygen in absorbing ultraviolet radiation. Its absorption bands range from one end of the solar spectrum to the other, but most of the individual bands are narrow. The narrow absorption bands in the infrared sector are not especially effective, because they coincide with the wider and more efficient bands of water vapor and CO_2. High in the atmosphere, however, in a zone roughly between 15 and 35 miles above the surface, where it is produced by shortwave radiation, ozone absorbs a large part of the ultraviolet radiation below about .3 micron. This absorption increases the temperature of this zone, as illustrated in Figure 4-18. The ozone screening effect is important to life on earth, because a strong inflow of such shortwave ra-

Figure 4-2 Absorption of solar and terrestrial radiation by the atmosphere (after Kondrat'yev). (*A*) illustrates the calculated range of electromagnetic radiation of two black bodies (perfect radiators) at temperatures equivalent to the average measured for the sun and for the earth atmosphere at the surface.

The solar radiation has a wider range and shorter wavelengths. Compare the three graphs [(*A*), (*B*), and (*C*)] and note that oxygen (O_2) and ozone (O_3) do most of the absorption at high elevations; water vapor absorbs most of the incoming radiation near the surface; and carbon dioxide near the surface absorbs most of the longer terrestrial radiation. The unshaded areas in (*B*) and (*C*) represent the "windows" through which radiation passes without being absorbed.

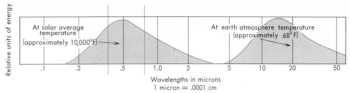

A. Distribution of energy released by black body radiation at solar and earth temperatures

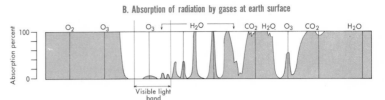

B. Absorption of radiation by gases at earth surface

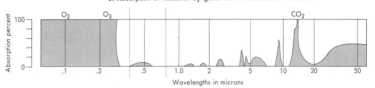

C. Absorption of radiation by gases at an elevation of 7 miles

diation directly to the surface could have harmful results, including severe skin burns, destruction of many useful microlife forms, destruction of plant leaves, and many others. Once formed, the ozone molecules may be separated into monoatomic or diatomic oxygen by further absorption of ultraviolet radiation. Being heavier than O_2, some of the ozone drifts slowly downward, reaching the earth in areas where air subsidence is taking place. Various oxidizing processes there remove the loose "companion" oxygen atom from the ozone, altering the latter to the more normal diatomic form. A study of Figure 4-2A and 4-2B reveals that a small part of the ultraviolet radiation (between about .3 and .4 micron) is not absorbed and passes through to reach the surface. This is the beneficial penetrating radiation that produces sunburn when we are exposed too long to full sunlight without previous conditioning, but which is also important in many biological processes, as in the production of certain vitamins, the maintenance of proper balance in the growth of viruses and bacteria, photosynthesis, and the production of the skin tan that is sought by beach lovers.

Water vapor. Most of the water vapor content in the atmosphere is found in the lowest reaches, although small traces have been observed 8 to 10 miles above the surface. Water vapor and its associated liquid form (cloud droplets) undoubtedly constitute the most important variable within the atmosphere in the disposition of radiation, despite the relatively small quantity involved. The gaseous molecule has the ability to absorb radiation at several bands of electromagnetic wavelengths and to transform such radiation into sensible heat. Figure 4-2B indicates that water vapor absorbs almost all the incoming radiation, except for the short ultraviolet range. It should be noted, however, that the absorption bands in the solar radiation zone are relatively narrow, with wide gaps or "windows" between. Water vapor also has absorption bands within the infrared sector, expecially between 20 and 40 microns. The influence of water vapor decreases rapidly with elevation and is relatively unimportant above 7 miles (see Figure 4-2C).

Liquid water, contained in the droplets of clouds, probably is even more important in the global energy balance than the vapor phase. Denser than the atmospheric gases, it is more efficient as an absorber and transmitter of radiation. The droplet surfaces also are good reflectors of solar radiation, and it has been estimated that somewhat more than two-thirds of all the reflected radiation is bounced off the surfaces of clouds. The phase changes between the gaseous, liquid, and solid states of water also are highly significant as heat-exchange mechanisms within the atmosphere. These mechanisms, which involve latent heat, will be treated in more detail in the next chapter (see Section 5-5).

Carbon dioxide. Carbon dioxide, like water vapor, is confined largely to the lowest levels of the atmosphere. Its principal role lies in its absorption of terrestrial thermal radiation. The main absorption bands are centered at 2.7, 4.3, 10.0, and 14.7 microns. The first three are relatively ineffective, because they lie in a zone of weak solar or terrestrial radiation (see Figure 4-2A and 4-2B). The broad band between 12.9 and 17.1 microns, however, with its center at 14.7, is extremely important, since it coincides closely with the zone of maximum earth radiation in the infrared sector.

CO_2, along with oxygen, is an essential part of the animal-plant carbon balance on earth, with plants taking in CO_2, synthesizing sugars and starches by photosynthesis, and giving off surplus oxygen. Animals, in turn, absorb oxygen, use it to burn or oxidize starches and sugars to obtain energy, and emit surplus CO_2. Imbalances between plant and animal life on earth can lead to changes in the atmospheric CO_2 content. Other such variations may result from volcanic eruptions (increase), a decrease in ocean volume by continental glaciation (increase), the formation of coal, hydrocarbon, or limestone deposits (decrease), and the combustion of carbon compounds (increase). A general lowering of global temperatures would almost automatically produce an increase in atmospheric CO_2 and a resultant corrective increase in absorbed radiation.

The role of CO_2 as a regulatory material in the global energy system has been under considerable scrutiny by scientists. A complex system of CO_2 transformations from one form to another appears to constitute a significant subsystem affecting long-run atmospheric radiation balances and temperature fluctuations. The discussion of "Limestones" (p. 365) reveals some of this complexity, as does Figure 10-6. It now is known that the interchange of CO_2 between the air and the oceans directly amounts to about 200 million tons each year. Additional amounts are ex-

Figure 4-3 Increase in carbon dioxide content in the atmosphere; 1860–1960 (after Callendar). The rise in carbon dioxide content during this period may well be the result of human activity, but the amount (about 35 parts per million of mass) is not alarming.

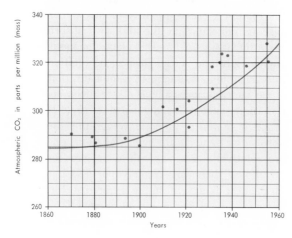

changed between the atmosphere and rock surfaces, and between the atmosphere and the total biota, or world of living things. Figure 4-3 shows that during the last century the average CO_2 concentration has increased from about 285 parts per million to 335. This may well be related to the increase in the combustion of fossil fuels and the clearing of forests, but the significance of this increase is not yet clear. Certainly the amount is not startling, and there are several mechanisms for reducing much greater concentrations.

The warming trend that has been noticeable throughout most of the northern hemisphere during the past 150 years has been attributed by some to the increase in CO_2 caused by fossil-fuel combustion. A slight drop in average temperatures since 1940, however, although irregular in area and amount, indicates that no direct correlation yet exists between temperatures and CO_2 concentrations in the air. Further, the increased absorption of radiation by CO_2 could well be offset by an increase in the earth albedo caused by a rise in the quantity of atmospheric solids (smoke, etc.) caused from the same combustion. Air pollution is a factor in the energy balance.

Atmospheric solids. Atmospheric solids include ice crystals, salt crystals blown into the air during ocean storms, pollen, and dust particles from many sources, as volcanic eruptions, meteoric falls, desert windstorms, and smoke and combustion by-products. The solids logically are concentrated in the lower portion of the atmosphere. They are not as significant as water vapor, liquid water, carbon dioxide, or ozone in the absorption of solar and terrestrial radiation, but they still exert an important influence and may play a dominant role when exceptionally concentrated. Their major function is to reflect radiation, both incoming solar radiation and

outgoing thermal radiation. Dusts and some of the larger gaseous molecules are responsible for the scattering of solar radiation or diffuse radiation that gives the sky its blue color at midday and the deep reds and yellows at sunrise and sunset.[3] Their scattering action may be compared with that of the piles of an ocean pier, which permit large waves to pass but which intercept and reflect the small ones lapping against their surfaces. Not all of the scattered radiation is lost to space. Some of it, comprising the diffused "skylight," is reflected toward the earth.

Unusually high concentrations of atmospheric solids have varied implications, including a blanketing effect in checking radiation inflow and outgo, the irritation of human respiratory tracts in city smog zones, and serious interference with visibility.

4-3 THE GLOBAL ENERGY SYSTEM A remarkable feature of the planet earth is that, despite the many changes, irregular receipts and diversions in the flow of solar radiation reaching the earth and its atmosphere, the temperature averages for the earth as a whole have varied only within relatively narrow limits over the billions of years since the first life forms appeared on earth. There have been several periods with warmer and colder temperatures than at present during this time, but it is unlikely that the average temperatures of the planet have ever varied more than 20° to 30°F. It has been estimated, for

[3]The blue of a clear sky is caused by the selective scattering of the radiation near the short end of the visible wavelengths. Astronauts and others who have risen far into the rarefied upper atmosphere have noted the gradual change in color of the sky from blue through violet to black. Selective scattering is a function of the diameter of the reflecting particles, blue from the smaller particles and red from the larger ones. The latter are nearer the earth: Hence, solar radiation near sunset and sunrise passes through the densest portion of the sky with more chance of reflection by larger particles. Unusually brilliant sunsets also are observed when the concentration of dust is greater, as in deserts or following explosive volcanic eruptions.

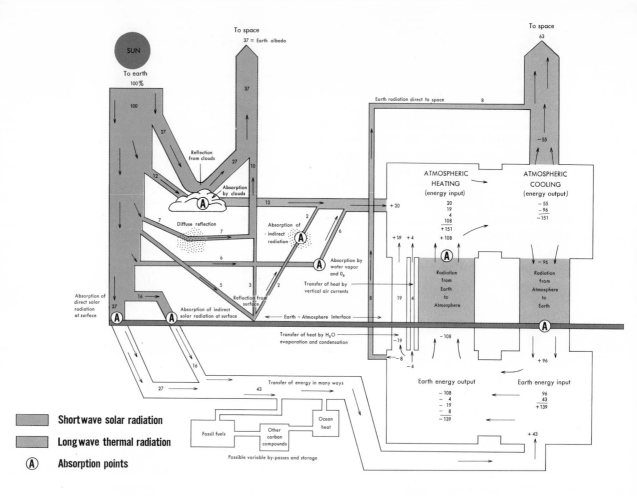

example, that a decrease of only 10°F in the average annual temperature of the earth and its atmosphere would suffice to induce continental glaciation. Considering the great extremes of temperature found within our solar system, such variations are minute. The reason for this small range of global temperature is that the earth and its atmosphere constitute an open energy system, in which the output of energy must balance the input within narrow limits and in which mechanisms exist to correct imbalances in the energy flux or flow. Some of these corrective devices already have been indicated. Our

most careful measurements reveal that the same amount of radiation leaves the earth as enters in the form of solar radiation, with only slight and temporary exceptions.

The total energy flux system of the earth is highly complex, and there are many alternative paths, detours, and even blind alleys within its structure. A generalized and simplified flow diagram for the system is presented in Figure 4-4. Values for different sections also are shown in statistical form in Table 4-2. The highly variable nature of certain sections of the flux system, and difficulties in accurately sampling the massive-

Figure 4-4 The earth energy flux system. Shortwave solar radiation input is shown on the left. Points of absorption and reflection of this radiation are indicated. Balanced atmospheric and terrestrial heat flow following absorption are shown on the right. The total outflow of longwave radiation to space, combined with the total reflection (albedo), equals the total input, expressed in percentage of total energy quantity. The values of longwave radiation percentages from earth to atmosphere and from atmosphere to earth (108 and 96, respectively) are high because there is considerable back radiation, in which the same energy is exchanged more than once. The net inflow and outflow, however, must balance within narrow limits.

ness of this enormous quantity of energy, have led to different values being assigned to different sectors of the system by various researchers. Note, however, that the discrepancies are not great. The flow diagram shown in Figure 4-4 has incorporated the values presented by the first of the four columns in Table 4-2. It is highly likely that within 5 to 10 years, the new instruments recording data from artificial satellites may yield more precise figures for some sections of the flow chart.

The climatic pattern of the earth in all of its complexity is tied closely to the mechanisms of the energy flux. Energy must flow freely in many different forms to accommodate needed adjustments in the system. Reference will be made many times to the energy system in succeeding chapters of this text, so it is important that its operation be clarified. A summary and an explanation of the system are included in the remainder of this section.

Reflection of incoming radiation (earth albedo). Just as moonlight consists of reflected solar radiation bouncing off the bar-

ren lunar surface, so does the earth shine with reflected radiation. This energy does not enter the earth-atmosphere energy system in any way, and it is reflected back into space unchanged. The best reflecting surfaces by far are the upper surfaces of clouds, and it is estimated that for the earth as a whole, some 27 percent of the incoming radiation is reflected in this way. The glaring brightness above an overcast sky is well known to air travelers. Another 7 percent is reflected back into space after having been reflected back and forth (*diffuse radiation*) among the surfaces of tiny particles in the air. Only about 5 percent of the shortwave radiation is reflected from the earth surface, and of this, 2 percent is absorbed by the atmosphere before it can return to space. Good reflecting surfaces on earth include bare ground, snow and ice, the wind-rippled surfaces of water bodies, and city streets and rooftops.

The constant scanning of the earth by satellite cameras (see Figure 5-5) is designed mainly to measure cloud cover, which in turn may be useful in making long-range weather forecasts. The role of the hydrologic cycle—the alternate evaporation and condensation of water—is highly complex within the global energy system. Its direct impact on the albedo by influencing the amount of cloud cover is important and might be likened to an input regulator valve.

Absorption and disposition of shortwave solar radiation. The processes of absorption of radiation have already been discussed (see Section 4-1). Of the 63 percent of the solar radiation that is not reflected, only about 20 percent is absorbed directly by the atmosphere, mainly by water droplets in clouds and by water vapor. The "windows" between the absorption bands of water vapor (see Figure 4-2B) are responsible for so much shortwave radiation passing directly through

Table 4-2 Average Annual Energy Flux of Earth (in percent)

	Source of Estimates			
	Budyko, Yidun, and Berlyand	Houghton	Moller	Bauer and Phillips
SHORTWAVE RADIATION				
Received at uppermost part of the atmosphere	100	100	100	100
Reflected Solar Radiation into Space:				
1. from cloud surfaces	27	25	27	30
2. from particles (diffuse radiation)	7	9	6	8
3. from earth surface	3			3
Total:	37	34	33	41
Absorbed Solar Radiation:				
Atmospheric:				
1. by clouds	12	10	11	15
2. by atmospheric gases incoming solar radiation	6	9	3	
reflected from earth	2			1
Total atmospheric absorption:	20	19	14	16
Absorption by earth surface:				
1. direct solar radiation	27	24	11	27
2. diffuse solar radiation	16	23	34	16
Total:	43	47	45	43
LONGWAVE RADIATION				
Atmospheric Thermal (Infrared) Radiation:				
1. into space	55	66	48	50
2. to earth	96	105		96
Total:	151	171		146
Terrestrial Thermal Radiation:				
1. directly into space	8		17	8
2. to atmosphere	108			112
Total	116	119		120
Net radiation of earth surface (emission minus back radiation)	20	14	23	24
OTHER ENERGY TRANSFERS				
Earth to atmosphere by evaporation and condensation of water:	19	23		23
Earth to atmosphere by turbulence, etc.:	4	10		4

SOURCE: K. Y. Kondrat'yev, *Radiative Heat Exchange in the Atmosphere*, Pergamon Press, New York, 1965, p. 346.

the atmosphere to reach the earth. As also indicated earlier, some of the solar radiation that is absorbed is used to evaporate cloud droplets, but most of it is altered into sensible or thermal heat, which in turn is reradiated as infrared radiation. In any case, the 20 percent enters the atmospheric sector of the global energy system, as indicated in the flow chart of Figure 4-4.

The 43 percent of radiation that is absorbed by the earth plays a much more diverse role than that in the atmosphere. Yet, as was indicated earlier, a large amount of it evaporates water, and most of the rest produces sensible heat that is ultimately re-radiated like that in the atmosphere. It is interesting to note that the fluctuations in the flow of energy at the earth-atmosphere interface that most concern man directly are the tiny bypasses and side eddies. Man is rarely conscious of the enormous flow that passes around him. To him, the important energy sources are the fossil fuels, the rush of mountain streams, the forests, and the fields of grain. Powerful as he is, man has only scratched the surface of the solar energy supply that ebbs and flows around him.

Absorption and disposition of longwave thermal radiation. The surface of the earth and the atmosphere both transmit thermal radiation in the infrared range following the production of sensible heat. While the earth surface absorbs shortwave and thermal radiation alike, the atmosphere is far more efficient in absorbing infrared radiation than the incoming solar radiation, largely because of the important absorption bands of CO_2. In fact, the atmosphere absorbs all of the terrestrial radiation except for a narrow band of wavelengths between 8 and 12 microns. This "window" accounts for the 8 percent of thermal radiation that passes directly from the earth to outer space without entering the atmospheric sector (see Figure 4-4). The much greater efficiency of the atmosphere in absorbing thermal radiation has often inappropriately[4] been termed the *greenhouse effect.*

The perhaps puzzling figure of 151 percent

shown in Figure 4-4 within the atmospheric sector of the system can be explained by the fact that the same energy can be exchanged several times between the atmosphere and the earth surface, with an energy exchange taking place each time. This multiple exchange is termed *back radiation.* The disposition of this 151 percent is indicated, with 55 percent passing into outer space and 96 percent being transmitted back to the earth surface. Much of this 96 percent keeps us warm at night when there is no solar input.

The input of the atmospheric sector comes first from the absorption of solar radiation (20 percent). Another 108 percent is received from terrestrial radiation, much of which, again, is back radiation. Some sensible heat is shifted into the atmosphere by air currents (4 percent), while the condensation of water vapor in the cloud zone adds another 19 percent.

The terrestrial energy exchange "boxes" are shown near the bottom of Figure 4-4. Intake totals 139 percent, 96 percent of which is atmospheric thermal radiation. Another 43 percent is derived from the absorption of solar radiation. This 139 percent of total intake is matched by 139 percent of outgo, 108 percent as terrestrial thermal radiation into the atmosphere, 8 percent passing directly through the atmosphere into space, 19 percent as latent heat transfer, and 4 percent as heat removed to the atmosphere by vertical air currents.

One energy intake and outflow is not shown on the diagram, because the total amount is so small. This is the heat energy that is generated within the earth interior as the result of the breakdown, or "decay," of radioactive substances. The total amount is much less than 1 percent of the intake of solar radiation and thus is not considered here in the total earth-atmosphere energy exchange system.

[4]Careful experimentation has revealed that the major purpose of glass in a greenhouse is to prevent the sensible heat carried by the air within the greenhouse from passing by conduction or convection into the outside air. Differential absorption of radiation by the glass apparently plays only a minor role.

In examining the operation of the entire system, it should be apparent that the alteration of any part of this highly complex yet integrated heat system should result eventually in the establishment of corrective measures. A pronounced decrease in global cloud cover for an extended period of time, for example, should lead to an increase in the total amount of radiation reaching the surface. This could, among many other possible reactions, raise surface temperatures that could increase evaporation rates, convection, precipitation, and the corrective increase in cloud cover.

The energy system, as shown in Figure 4-4, does not represent any particular day or place on earth. Instead, it represents the steady equilibrium state which the entire planet tends to develop. When the details of differences in climate from place to place on earth are examined later in this part of the text, it will be noted that imbalances in the system are the rule, rather than the exception, and that numerous fluctuations in weather conditions from place to place and from time to time are essential to the checks and balances necessary to ensure the maintenance of the system.

4-4 GEOGRAPHIC VARIATIONS IN THE RADIATION BALANCE AT THE EARTH SURFACE   The total energy balance of the earth and the interchange of energy between the atmosphere and the earth surface as a whole have been described. The radiation balance at the earth surface, or within the realm of human life, however, is rarely in a state of equilibrium at any one place or time. Determining the input and output of energy accurately for any one point on earth is extremely difficult, because energy is highly elusive, changing rapidly in form and moving freely with winds and air currents. Data for the distribution of the radiation balance on a global basis are based largely on theoretical calculations and supported by occasional careful observations at widely separated points. Like the various estimates given by authorities for the global energy flux, as illustrated in Table 4-2, there are differences in the estimates for the surface pattern of radiation balances, but the differences are not highly significant. Most researchers in the field agree on the general characteristics of the geographic pattern.

An example of a careful measurement of the radiation balance is given in Figure 4-5, which shows the variation of the different elements involved in the radiation balance at a single site, over several days of observations. The measurements were taken in August at Tashkent, in the interior of Asia, a typical desert location, when skies were clear, and over a bare surface of sand. Note on the chart that:

1. The net radiation balance for the 24-hour period clearly is positive (more input of radiation than output).

2. The negative balance is greatest shortly after sunset and decreases slowly until it becomes positive shortly after sunrise.

3. The positive radiation balance reaches a maximum at noon.

4. Terrestrial radiation is greatest at sunrise and lowest between 1 and 2 P.M., when ground temperatures are highest.

The net surplus of radiation (input exceeds output) must be removed in some way. The bare, desert surface at Tashkent indicates that conversion to chemical energy or to latent heat by evaporation would be minimal. Some of it, of course, is turned into sensible heat, and the temperature of the air at midday is much higher than at sunrise. The surplus is prevented from accumulating, however, by sensible heat flux, in which the

Figure 4-5 Diurnal variation of radiation-balance components. These curves were obtained as averages of several days of careful measurements under cloudless skies and over a bare sand surface during August at Tashkent, in the Soviet Union. Note that the radiation balance is negative during the night and positive during the day, and that the total radiation balance for the day is positive, indicating that more radiation enters than leaves. Most of the surplus energy leaves as heat in convectional air motion. In more humid regions, more of it would leave via evaporation of water. [After Aizenshtat and Zuyev]

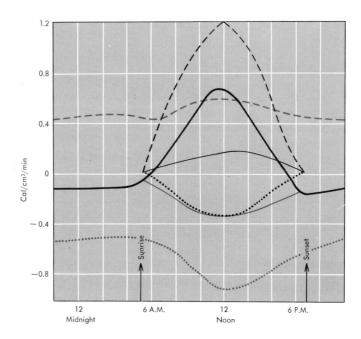

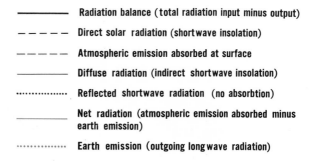

———— Radiation balance (total radiation input minus output)

– – – – – Direct solar radiation (shortwave insolation)

– – – – – Atmospheric emission absorbed at surface

———— Diffuse radiation (indirect shortwave insolation)

·············· Reflected shortwave radiation (no absorbtion)

———— Net radiation (atmospheric emission absorbed minus earth emission)

·············· Earth emission (outgoing longwave radiation)

heated air is swirled upward by convection (see Section 4-9). This helps to explain why deserts are windy places. In other kinds of climates, other effective means of energy transfer are more effective.

Figure 4-6 shows a map of the annual distribution of the radiation balance at the earth surface for the entire earth. The most apparent generalization is that *the entire earth surface exhibits an annual positive value for the radiation balance, the balance increasing from the poles to the equator and being greater over oceans than over the continents for equivalent latitudes.* There may be a small annual negative balance over large areas of permanent ice snow, as over Antarctica or Greenland, but this has not been verified. The extreme difference in radiation balance between land and water surfaces is shown in Figure 4-6 by the sudden break in the isarithms along coastlines, as along the coast of northeastern Brazil. Water absorbs radiation much more efficiently than most land surfaces (the albedo being lower) and reradiates it more slowly, owing to greater penetration with depth. Low-latitude deserts tend to have somewhat lower radiation surpluses than might be expected for their latitudes, their inputs of solar radiation, and

high temperatures. This is because of the higher reflection of incoming solar energy over bare ground. The depressed values for the Sahara Desert in North Africa in Figure 4-6 are illustrative. Convection and turbulence also help to remove sensible heat energy from ground level.

Seasonal measurements indicate that continental areas poleward of 40° have negative radiation balances at the surface in the winter. This is related to the high reflection of solar radiation from snow-covered surfaces during the short days and to the great loss of terrestrial radiation during the long, clear, winter nights.

4-5 GEOGRAPHIC VARIATIONS IN THE RADIATION BALANCE ABOVE THE EARTH AND FOR THE EARTH-ATMOS-PHERE SYSTEM If the earth surface has a positive radiation balance, at least on an annual basis, there must be a compensating negative balance in the atmosphere in order to maintain an equilibrium between total input and output for the system as a whole. High-altitude measurements indicate that this is true. Energy is transferred from the surface to the atmosphere in several ways other than radiation (as indicated in Figure 4-4), and the ultimate conversion into outbound radiation yields the net negative radiation balances of the atmosphere.

Table 4-3 shows the components of the radiation balance for the *combined* earth-atmosphere system. Note that the annual negative balances at the higher latitudes considerably exceed the total of the positive earth-atmosphere system in the lower latitudes. The two are not balanced because of the difference in areas involved. The area of the zone between 0° and 10° is much greater than that between 80° and 90°.

The transfer of both latent heat and sensible heat from the surface to the atmosphere, and the subsequent reversion to outgoing thermal radiation, constitute the principal means of reducing the net radiation balances at the surface. The atmosphere varies in its capacity to drain away energy by radiation:

1. The loss of energy by radiation without a cloud cover is at or just below the *tropopause* (see Section 4-12).
2. Radiation loss when there is a cloud cover tends to be greatest immediately above the clouds.
3. Radiation loss in the upper air tends to

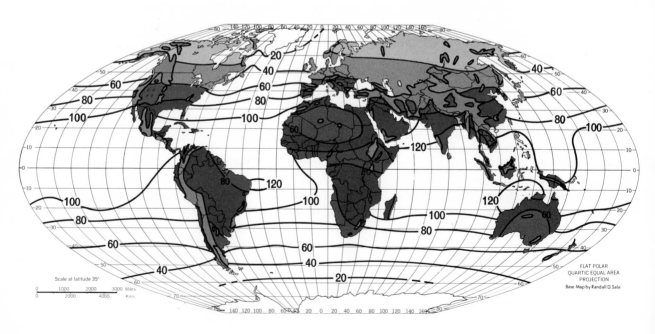

Table 4-3 Variation with Latitude of Daily Totals of the Radiation Balance of the Earth-Atmosphere System (calories/cm² /24 hr)

Latitude	January *ISR − A	AR + ER	RB	July ISR − A	AR + ER	RB	Year ISR − A	AR + ER	RB
90–80 N	0.0	292	−292	421	420	1	156	356	−200
80–70	0.0	305	−305	448	429	19	184	367	−183
70–60	13	325	−312	502	452	50	240	388	−148
60–50	72	349	−277	524	457	67	294	403	−109
50–40	148	366	−218	563	467	96	358	416	−58
40–30	234	392	−158	618	473	141	435	432	3
30–20	365	420	−55	595	461	134	491	440	51
20–10	458	441	17	535	433	102	518	438	80
10–0	513	429	84	498	422	76	518	426	92
0–10 S	545	424	121	469	426	43	518	425	93
10–20	611	436	175	424	428	−4	528	432	96
20–30	648	447	201	338	412	−74	491	430	61
30–40	648	446	202	225	392	−167	428	419	9
40–50	592	442	150	145	379	−234	361	410	−49
50–60	511	434	77	74	357	−283	278	396	−118
60–70	469	430	39	18	331	−313	214	380	−166
70–80	442	422	20	0	309	−309	171	366	−195
80–90	445	417	28	0	298	−298	156	358	−202

SOURCE: After N. A. Bagrov and quoted in K. Y. Kondrat'yev, *Radiative Heat Exchange in the Atmosphere*, Pergamon Press, New York, 1965, p. 342.

Explanation of symbols:
  $(ISR − A)$ = Incoming solar radiation minus albedo.
  $(AR + ER)$ = Outgoing radiation to space from both atmosphere (A) and earth (E).
  RB = Radiation balance of the combined earth-atmosphere system.

Figure 4-6 Distribution of annual totals of the radiation balance at the earth surface, in k cal/cm²/yr. Note that the values are positive everywhere except perhaps over the polar ice fields (not shown), and that the values over the oceans are consistently higher than over the adjacent land areas. The low-latitude oceans have the highest surpluses. [After Budyko, Beryland, and Zubenck]

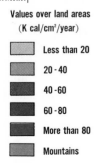

Values over land areas
(K cal/cm²/year)

Less than 20
20 - 40
40 - 60
60 - 80
More than 80
Mountains

be greatest in lower latitudes.

4. Radiation loss from the lower atmosphere is exceedingly irregular, owing to variable energy transfer mechanisms.

Any consideration of the general operation of the global energy balance must involve the transport of energy both horizontally and vertically.

4-6 MECHANISMS TO ADJUST FOR IMBALANCES IN THE GLOBAL ENERGY SYSTEM Table 4-3 and Figure 4-6 indicate the great inequalities that are found in the global energy system. Surpluses and deficiencies in the net radiation balance not only occur in different places horizontally and vertically but also change daily and seasonally in the same place. Yet—the system functions; the pattern of temperatures does not change much; the output of radiation generally matches the input within extremely narrow limits; and the inequalities in the net radiation balance are compensated. In order to maintain the global energy balance, there must be several types of energy transfer and transport to carry surplus energy from the major areas of accumulation, the *heat sources,* to those portions of the earth-atmosphere system that are more efficient in draining off surplus energy in the form of outward radiation, the *heat sinks.* The major heat sources are the low-latitude oceans, while the major heat sinks are the snow- and ice-covered surfaces of winter hemisphere continents and the upper part of the lower atmosphere. Some of the energy transport mechanisms have been previously mentioned. We shall summarize them here.

Wind. Wind is *horizontal* airflow and is capable of transporting both sensible and latent heat from the heat sources of the low latitudes to the cooler high latitudes. It is not restricted to the earth surface, and it appears at many levels within the lower and intermediate levels of the atmosphere. Wind carries energy not only directly as heat, but also in the form of mechanical energy. Violent storms result from sudden energy transfers, usually involving latent heat.

Convection, eddy diffusion, and turbulence (see Section 4-9). Vertical air currents bear surplus heat (both latent and sensible) upward from the surface, sometimes only a few feet, but other times for thousands of feet.

Ocean currents. A circulation system, roughly similar in principle to the hot water heating system in a home, exists in the world's oceans. It is not nearly as efficient or effective as air circulation, but it carries surplus heat away from the low-latitude oceans toward the heat sinks at high latitudes.

Temperature

The overall operation of the global energy system (the general mechanisms whereby inequalities in the radiation balance are compensated) and the various forms by which the energy is manifested or through which work has been done were treated in the first division of this chapter. We now will concern ourselves with the geography of *temperature.* Temperature is *a measurement of the quantity of sensible heat contained within a given mass.* Air temperature, as an indicator of the sensible heat that flows around our bodies, has a much more continuous and personal relationship to our everyday lives than measurements of the other types of energy. We sense relatively small variations in it and alter our lives accordingly almost every day by making adjustments in our food, clothing, and shelter. We should keep in mind, however, that temperature does *not* measure all of the heat present. For example, it does not measure the latent heat contained in the water vapor present, which usually is far greater than the sensible heat present. At the same time, temperature is one of the manifestations of radiation balances and is a clear indicator of the general condition of the energy balance locally.

The Fahrenheit and Celsius (centigrade) scales are the systems generally used for expressing temperature differences. The Fahrenheit scale is not so convenient in many ways as the Celsius, which is based on degrees that are 1/100 the difference in temperature between freezing and boiling with respect to water at sea level. Despite its awkwardness, the Fahrenheit scale customarily is used in English-speaking countries; it will be used throughout this text except where the Celsius scale is specifically designated. The conversion from Fahrenheit to Celsius or vice versa can be done easily through the use of the basic formula

$$\frac{C}{F - 32°} = \frac{5}{9}$$

This may be transposed as

$$F = \frac{9}{5} C + 32°, \text{ or } C = \frac{5}{9}(F - 32°)$$

For convenience, a graphic conversion scale is presented in Figure 4-7. Many scientists now are using the Kelvin scale, which is similar to the Celsius except that it begins at absolute zero ($-253°C$), the point at which all molecular motion ceases.

The range of temperatures at the surface of the earth is extremely small compared

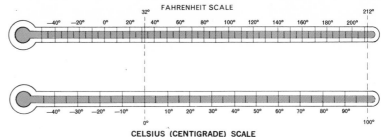

Figure 4-7 Graphic conversion scale; Celsius (centigrade) and Fahrenheit.

with the range encountered in outer space. The absolute range between the highest and lowest temperatures ever recorded at standardized weather stations on earth is about 250°F, and it rarely exceeds 100°F over most of the earth. Small as our earthly temperature range may be, it is of prime importance to life on earth.

Figures 4-8 and 4-9 indicate the generalized

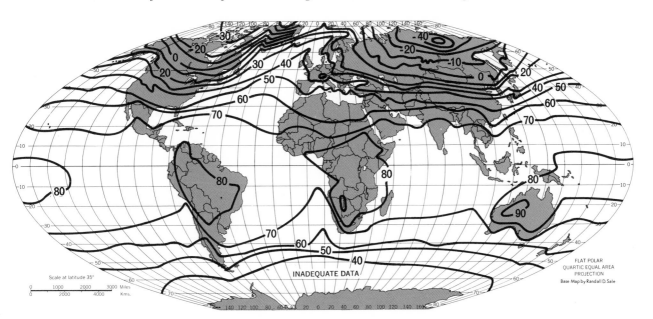

Figure 4-8 Sea-level temperatures for January in °F. [After Haurwitz and Austin]

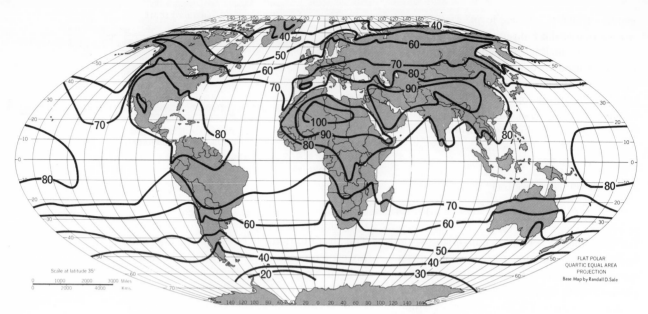

Figure 4-9 Sea-level temperatures for July in °F. [After Haurwitz and Austin]

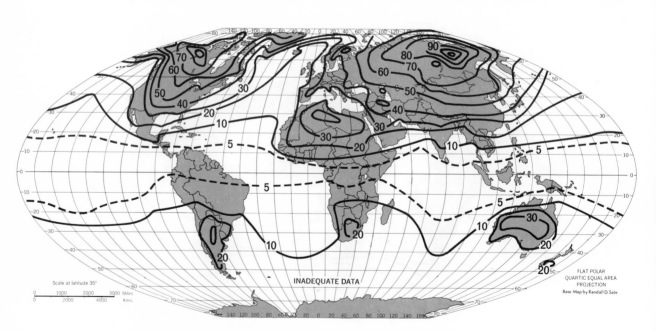

Figure 4-10 World seasonal range of temperatures in °F. [After Haurwitz and Austin]

patterns of global *isotherms* (lines connecting points of equal temperature) for the earth surface, and with temperature readings equated to sea level. Sea-level temperatures are calculated by increasing observed temperatures 3.6°F for every 1,000 feet of station elevation. This figure corresponds to the normal average *lapse rate,* or the decrease in temperature that takes place as one passes from lower to higher elevations. This adjustment is made to compensate for the complexity introduced by differences in elevation. Many of the major generalized patterns of world temperatures can be inferred from the maps showing mean temperatures for January and July, and the one giving the mean annual range of temperature (see Figures 4-8, 4-9, and 4-10).[5]

Many factors are responsible for differences in temperature patterns from place to place and from time to time. In order to interpret these patterns, it is necessary to examine the major contributory factors separately. After completing the chapter, the reader should review and analyze the patterns on the surface temperature maps.

4-7 ASTRONOMICAL FACTORS IN SURFACE TEMPERATURE PATTERNS *Output of solar energy.* As was indicated in

Section 4-1, variations in the output of solar energy may have some effect on general global temperatures, but recent measurements of the solar constant taken far above the surface do not indicate any appreciable changes. Admittedly, such measurements have been taken for only a few years, but theories about the processes of energy generation in the solar interior suggest possible fluctuations only within narrow limits.

Distance of the earth from the sun. The intensity of solar energy received at the outer portions of the atmosphere varies slightly at different positions of the earth in its orbit around the sun, owing to differences in distance to the sun. The maximum difference in distance amounts to about 3 million miles. Although measurable in terms of the solar constant, this factor does not appear to be important in present surface atmospheric temperature patterns. The earth is nearest the sun in January and farthest from it in July. Hence, if this factor were significant, the northern hemisphere winters should be warmer and the summers cooler than those in the southern hemisphere. Actually the converse is true, owing to the greater proportion of land area than ocean area in the northern hemisphere and the greater heating and cooling of land as compared with water. It is likely, nevertheless, that our winters are warmer north of the equator and our summers cooler than they would be if the earth were equidistant from the sun at all seasons.

Directness of solar radiation. The directness of solar radiation forms one of the most important variables in explaining the increase in seasonal temperature variation and the decrease in temperature with increasing latitude. The effect of directness of solar radiation is illustrated in Figure 4-11. Direct radiation is concentrated over a smaller surface area than oblique radiation and is therefore

---

[5]The terms *average* and *mean* signify the same thing, namely, the sum of all the observations divided by the number of observations. The terms *average annual temperature* and *mean annual temperature* are synonymous. One apparent contradiction should be explained. In determining generalizations of daily temperatures, instead of taking a large number of readings during the day and averaging them, an average of the maximum and minimum temperatures is used, the data being obtained from a special type of thermometer capable of recording the maximum and minimum reached during a given period. The *mean daily temperature,* therefore, is the combined highest and lowest temperatures for a daily period divided by 2. A *mean monthly temperature,* however, is the average of all the mean daily temperatures for a given month, over the period of years for which records are available. It is not the highest and lowest for the month divided by 2. Sometimes the term *average mean monthly temperature* is used to emphasize the fact that it is the average over many years, and not the mean monthly temperature for a particular year.

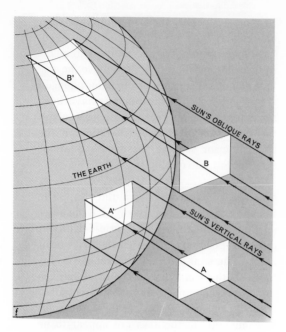

Figure 4-11 Effect of angle of insolation on receipt of solar energy. A given amount of solar energy, represented by the area of *A* and *B*, yields different amounts at the surface of the earth. The amount per unit of area at *B'* is noticeably less than at *A'*.

more intense per unit of area. Quantitatively, the intensity of solar radiation varies with the sine of the angle at which it strikes the surface. Two further considerations influence the effect of directness of solar radiation; they involve atmospheric influences. The more oblique the rays of the sun, the greater is the distance through the atmosphere which the radiation must pass—hence, the greater the opportunity for loss through atmospheric reflection and absorption. As was indicated in Section 2-8 and Figure 2-14, the directness of solar radiation at the surface is related to the tilting of the earth axis and to its parallelism within the earth orbit. During the course of a year, beginning at noon on the summer solstice (northern hemisphere), the

sun's perpendicular "rays" migrate from the Tropic of Cancer to the Tropic of Capricorn and back again.

Directness, then, is a major factor in the amount of solar energy received at the surface (*insolation*). At any one point, it is greatest at noon sun time on the summer solstice and decreases toward sunrise and toward the winter solstice. Cloud cover, of course, is a variable that may interfere with the normal influence of the radiation angle. Surface temperature patterns related to directness of radiation include the general decrease toward the poles, the higher temperature near midday, and the seasonal variation.

Length of day. The amount of solar radiation received varies also with the length of day. If the axis of the earth were not inclined from the perpendicular to the ecliptic, all parts of the earth would have a 12-hour day and a 12-hour night throughout the year. Owing to this inclination, however, the length of day varies greatly, both latitudinally and seasonally. The equator is the only place on earth where the length of day and night remains equal throughout the year. The poles, on the other hand, experience days and nights that are each about 6 months long.[6]

The latitudinal variation in the length of day, combined with seasonal variations in the directness of radiation, explains why the seasons become more pronounced with increasing latitude. One might expect that the long period of continuous daylight near the poles would make the intake of insolation there higher than it is at the equator during the summer and would thus cause higher temperatures. Factors preventing this include the lower angle of the sun, the greater thick-

[6]The period of sunlight at the poles is longer by about 15 days than the period when the sun is below the horizon because of the bending (*refraction*) of light as it passes through the atmosphere.

Table 4-4 Average Amount of Direct Solar Radiation Received at the Upper Limit of the Atmosphere in the Northern Hemisphere, in cal/cm^2/min (after Baur and Philipps)

Date	0–10°	10–20°	20–30°	30–40°	40–50°	50–60°	60–90°
Dec. 21	0.549	0.465	0.373	0.274	0.173	0.079	0.006
Mar. 21	0.619	0.601	0.563	0.509	0.441	0.358	0.211
June 21	0.579	0.629	0.664	0.684	0.689	0.683	0.703
Sept. 23	0.610	0.592	0.556	0.503	0.435	0.353	0.208

SOURCE: By permission from B. Haurwitz and J. M. Austin, *Climatology*, McGraw-Hill Book Company, New York, 1944.

ness of the atmosphere through which the radiation must pass, the high albedo over ice and snow, and the latent heat required to melt and evaporate snow or ice or to evaporate water. All of these combine to reduce sensible heat and thus keep down air temperatures in the polar regions.

Calculations have been made of the average amount of solar radiation that is received at the top of the atmosphere at different latitudes and at different times of the year in the northern hemisphere. These data, which include the albedo, are given in Table 4-4. Note that the greatest amount of solar radiation at any one time for any latitude occurs on June 20 between 60° and the North Pole. This is the result of the continuous daylight for 24 hours at this solstice. At all other times of the year, the radiation received

increases toward the equator. The great contrasts between Tables 4-4 and 4-5 show the importance of the atmosphere in influencing the amount of radiation received at the surface. The length of day and directness of radiation are important factors in net radiation balances and also in surface temperature patterns, but there are other important considerations, as the contrasts between Tables 4-4 and 4-5 suggest.

4-8 ATMOSPHERIC FACTORS IN SURFACE TEMPERATURE PATTERNS The influence of the atmosphere on incoming solar radiation has already been discussed in Section 4-2. Reflection and absorption by oxygen, ozone, and water vapor are responsible for only about 43 percent of the incoming shortwave radiation reaching the

Table 4-5 Average Amount of Direct Solar Radiation Received at the Earth Surface if Cloudiness and Turbidity (Transparency) Are Considered, in cal/cm^2/min (after Baur and Philipps)

Date	0–10°	10–20°	20–30°	30–40°	40–50°	50–60°	60–90°
Dec. 21	0.164	0.161	0.134	0.082	0.036	0.013	0.001
Mar. 21	0.191	0.224	0.206	0.161	0.116	0.096	0.055
June 21	0.144	0.170	0.216	0.233	0.183	0.159	0.133
Sept. 23	0.170	0.162	0.201	0.183	0.131	0.079	0.028

SOURCE: By permission from B. Haurwitz and J. M. Austin, *Climatology*, McGraw-Hill Book Company, New York, 1944.

earth surface. This is a total quantity, however, and there are considerable local variations in the atmospheric influence. Careful study of Tables 4-4 and 4-5 reveals several of these regional differences. For example, note that the pattern of a general increase in direct solar radiation toward the equator in Table 4-4 changes considerably at the earth surface. Except for the winter solstice, the maximum solar radiation at the surface (Table 4-5) is highest between 10° and 30° lat, even during the equinoxes, when the sun is at the zenith at the equator and the days and nights are of equal length throughout the world. The high percentage of clear skies in the subtropical deserts is largely responsible for this. The difference between the receipts of radiation at the outer limits of the atmosphere and those at the surface also is graphically shown in Figure 4-12.

A further role of the atmosphere on surface temperature patterns is in helping to cause a lag in maximum temperatures recorded at the standard measuring height above the ground ($4\frac{1}{2}$ feet) behind the period of maxi-

Figure 4-12 Annual march of insolation. [After Trewartha]

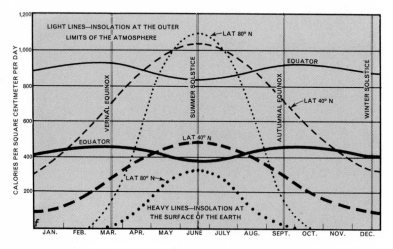

mum insolation, both daily and seasonally. Despite the fact that the maximum input of radiation from the sun as well as the maximum net radiation balance takes place around noon, the warmest time of day is usually between 1:30 and 2:30 P.M. Similarly, the warmest time of the year usually is not at the summer solstice, but a month or two beyond. The reasons for these lags are as follows:

1. On a daily basis, while solar radiation input is greatest at noon, terrestrial radiation is lowest then, and sensible heat in the lower air is derived from both. Terrestrial radiation is low at midday (see Figure 4-5), because much of the absorbed radiation is performing work in the form of sensible or latent heat. Slowly the sensible heat is converted into terrestrial radiation, which reaches a maximum just before sunrise. The maximum absorption of both solar and terrestrial radiation thus must take place sometime between noon and sunset. The same reaction takes place seasonally.

2. The processes of transferring heat en-

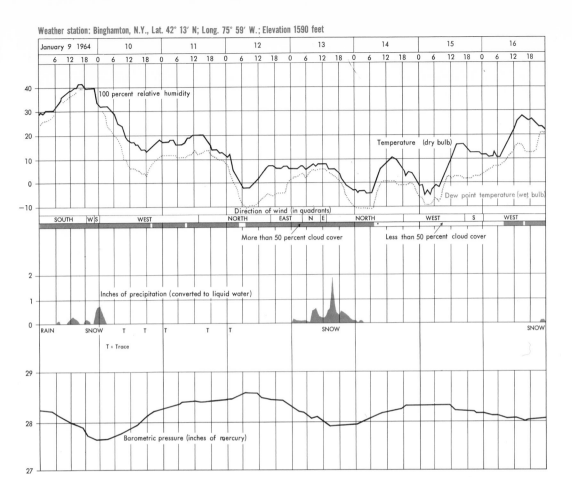

Weather station: Binghamton, N.Y., Lat. 42° 13′ N; Long. 75° 59′ W.; Elevation 1590 feet

ergy vertically from the surface to standard recording levels, though only a few feet, take time. Air is not a good conductor of heat, and the complex processes of conduction and eddy diffusion (see Section 4-9) require time for air mixing to take place. Evidence for this is that ground surface temperatures under clear skies have their daily maxima close to noon.

3. Diffuse radiation, or insolation reflected from atmospheric particles reaching the earth, has its maximum in early afternoon.

The influences of major heating agents in the air on air temperatures near the surface

are treated briefly in the following subsections.

Cloud cover. The variation in amount of cloud cover is by far the most important atmospheric variable in surface air temperatures, just as it is in the total radiation balance. As was discussed previously, clouds are important reflectors and absorbers of both solar and terrestrial radiation. The main effect of a cloud cover on surface temperatures is to reduce temperature extremes, both daily and seasonally. A direct relationship is illustrated by Figure 4-13. Regional influences of cloud cover also are indirectly indicated by several of the temperature charts shown in Chapters 7 and 8. In Figure 7-4, for example, the maximum temperatures occurring prior to the summer solstice at Poona is mainly the result of clear skies. At the onset of the summer monsoon (in late May), clouds and rain reduce the temperature extremes during the day. In areas where frost is a problem, people watch the skies for signs of frost danger. Calm, clear nights are most likely to result in frosts because of the rapid loss of sensible heat to outer space through its change into radiant energy.

Water vapor. Water vapor is an atmospheric variable that operates in much the same way as clouds, except that its effect is not nearly so great. Its principal role lies in its capacity to absorb radiation from the sun. Frequently, high humidity is associated with cloudiness, and the two factors reinforce each other in reducing surface temperature extremes. It is possible, however, to have a high water vapor content under cloudless skies.

Carbon dioxide. Being a highly mobile gas, CO_2 spreads evenly throughout the lower atmosphere and thus is not a significant factor in areal or temporal patterns of surface temperatures. Locally, in the immediate vicinity of cities, the atmospheric content of CO_2 might be sufficiently high to raise air temperatures slightly. Generally, however, other factors, such as fuel combustion and solid air pollutants, have more to do with urban-rural temperature differences than CO_2.

Dust. Small, solid particles in the atmosphere tend to increase the earth albedo, that is, to increase the reflection of radiation and thus reduce temperature extremes near the surface. It is a relatively minor factor in global surface temperature patterns, but, like

carbon dioxide, may be significant in long-term global temperature fluctuations. The blanket of smoke that accompanies certain atmospheric conditions over cities during the winter may have an appreciable influence in retarding the radiation of heat generated by space heating in cities.

4-9 THERMODYNAMIC FACTORS IN THE DISTRIBUTION OF TEMPERATURES
Temperatures within the atmosphere may be influenced by air movements vertically and horizontally and by types of heat exchange other than the direct change of radiation into sensible heat. All of these may be grouped under the general heading of *thermodynamic* factors, since they involve mechanical air movements of one kind or another. The principal ones include *conduction, convection, advection, eddy diffusion, turbulence,* and the *phase changes of water.* Most of these will be treated briefly in this section, although the phase changes of water involving conversion of latent heat (see Section 5-5) and wind transport (advection) as adjusters of energy disequilibria and temperature control will be discussed at greater length in later chapters.

Conduction. Conduction is the process of transmitting sensible heat within a substance or between materials by contact. Since sensible heat is the kinetic energy of molecular motion, the impact of molecules may transmit such energy directly from one molecule to another, much like the impact of a cue ball on the other balls in a game of billiards. The flow of sensible heat may occur between one part of a substance and another within the same material (*isotropic heat flow*)—such as the passage of sensible heat downward into a rock whose surface is absorbing radiation—or it may take place across the contact between two unlike bodies having different temperatures. The objective of conduction is to equate the amount of sensible heat in contiguous materials. The efficiency and speed of conduction vary directly with density. Air, for example, is a much poorer conductor of sensible heat than the solid surface of the earth. The use of air as insulation in homes and garments is a good example. When the surface of the earth is heated by the absorption of radiation, the transfer of heat to the air above by conduction affects a layer of air only a few inches thick. Similarly, at night, when surface temperature drops below that of the air above owing to radiation loss, a reverse transfer takes place, with the air losing sensible heat by conduction to the earth from a thin zone immediately above.

Convection. Convection is a circulation of air involving a vertical transfer of sensible heat away from a local heating source (see Figure 4-14). Its principle is similar to that of the circulation of air in a hot-air heating system. It requires unequal heat sources, or areas on the surface that contain more sensible heat than others. The air above is heated by conduction, expands and becomes less dense, whereupon the cooler, denser air adjacent moves in, displacing the warmer air upward. A cell of rising air is matched by descending air to replace the air moving toward the heating source along the ground. The downward movement generally is much broader than that of the rising air column. The height of convectional air currents depends largely on the intensity of heating and the contrast in temperature with surrounding areas. An extremely hot desert surface may generate convectional updrafts reaching heights of 10,000 feet. Convection over oceanic islands rarely is more than 200 to 4,000 feet, and mid-latitude city roofs and pavements will generate currents rising 7,000 to

8,000 feet on a warm summer day. Convection currents occur in many sizes and heights, ranging from cells hundreds of feet across and thousands of feet high to tiny circulation movements measured in inches. The process is important in mixing sensible heat through the lower levels of the atmosphere and lessening local areas of heat accumulation. It is greater over continents than over oceans, greater in the summer than winter, greater in sunny than cloudy areas, and greater over bare surfaces.

Turbulence and eddy diffusion. When wind blows over an irregular surface, turbulent flow is caused, with an eddying effect involving upward-spiraling vortexes. Some-

Figure 4-14 Heating the atmosphere by the processes of conduction, convection, and advection.

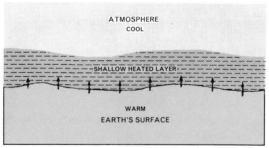

A. Conduction

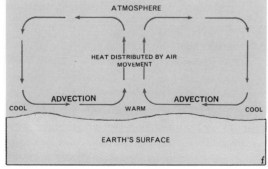

B. Convection and advection

times the turbulence is caused simply by friction against the surface and does not require a layer of heated air near the surface. In other cases, it is strongly aided by such a layer. Unlike convection, this type of airflow does not require unequal heating surfaces, nor is it a straight upward or downward movement. The more or less circular eddies range in size from tiny whirls an inch or so in diameter to huge spirals a hundred miles or more across. Hurricanes and tornadoes are huge diffusion eddies that have been intensified in speed because of special conditions. Turbulence and eddy diffusion are especially important over rough mountainous terrain and also over oceans. The trade wind zone of low-latitude oceans, where the constant strong winds maintain the sea surface in a tangled confusion of waves 8 to 10 feet high, has an extremely high incidence of eddy diffusion. It is a much more effective means of removing the sensible heat from the sea surface than either convection or conduction. An additional source of vertical turbulent flow is air masses with different temperatures and densities that move over or against each other.

Advection Advection is the horizontal transport of heat energy by wind (see Figure 4-14). The work of winds and their causes will be treated more fully in Chapter 6. As a dynamic factor in global temperature conditions, however, advection warrants special treatment here. Changes in temperature brought about by changes in horizontal wind flow are well known to everyone, and early in history, winds were characterized by temperature and strength, as well as by their prevailing direction. The association of wind direction with changes in temperature is especially noticeable in the mid-latitudes, which experience strong contrasts between the cold air masses from high latitudes and

warm winds from the tropics. The contrasts in winter temperatures between Labrador and the British Isles, both at about the same latitude, are largely due to the difference in the temperatures of their respective prevailing winds. Both locations receive air from westerly directions, but the winter air in Labrador comes from the frozen interior of the North American continent, while most of the winter winds in Britain have blown over thousands of miles of warm ocean.

4-10 ADIABATIC HEATING AND COOLING It is important at this point to discuss a few of the temperature effects caused by the dynamics of air movements and their resulting heat exchanges. A particularly important observation concerns the relationship between air density and air temperature. Physicists have known for centuries that when air expands, it cools, and when it is compressed, its temperature rises. In neither case is there any increase or decrease in heat energy. Expansion spreads the same heat energy over a greater volume, so that the measurement of heat per unit of area (temperature) becomes less. When upward or downward air currents occur, whether by convection, eddy diffusion, or turbulence, changes of temperature occur in the rising air masses. The decrease in temperature occurring in dry air that results from an upward movement and consequent expansion (there being less air pressing on it from above) amounts to 5.6°F per 1,000 feet. This is known as the *dry adiabatic rate,* and the process is known as *adiabatic cooling.* As will be discussed more fully in Section 5-11, when condensation of water vapor occurs during the ascent of an air mass, heat is released, and thus the adiabatic rate of temperature drop is retarded. This lower rate of adiabatic cooling is termed the *wet adiabatic rate.* While the dry adiabatic rate always remains the same within the lower troposphere, the wet adiabatic rate varies, depending on the amount of condensation taking place. It occasionally becomes as low as 2° to 2.5° and usually is between 4° and 5°. Descending air, since it is being compressed, has its temperature increase adiabatically at the dry rate.

4-11 LAPSE RATES AND TEMPERATURE INVERSIONS Adiabatic heating and cooling should not be confused with the normal decrease in temperature that a person encounters when moving upward in the lower part of the atmosphere. The tops of high mountains always are colder than their bases. Termed the *lapse rate,* or *vertical temperature gradient,* this decrease in temperature in free air per 1,000 feet of elevation averages about 3.6°F. This average figure, however, may vary considerably under special atmospheric conditions. In some parts of the world, it is much more stable than in others. The relationship between the dry adiabatic rate and the normal lapse rate is shown graphically in Figure 4-15.

The main reason for the decrease of temperature in free air with increasing elevation is the decrease in air density. The density of air at any elevation equals the total weight of the column of air lying above it. The higher one is above sea level, the less atmospheric mass there is above. Dense air can receive heat by conduction or by absorption of radiation more efficiently than less dense air. Further, the greater concentration of the heating agents (such as water vapor, CO_2, and cloud particles) in the air at lower levels tends to increase the absorption of radiation and the production of sensible heat at lower levels.

Variations in the lapse rate from the normal 3.6° per 1,000 feet can be caused by several factors. A lapse rate lower than normal will result from conditions that tend to warm

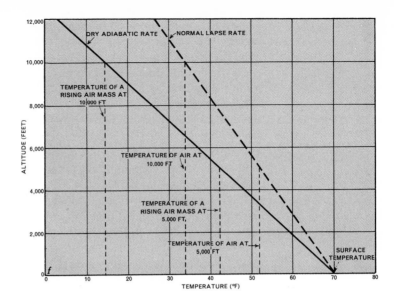

Figure 4-15 Comparison of normal lapse rate and dry adiabatic rate.

the upper air or that cool the air below. The advection or horizontal importation of a cold air mass can produce this, as can the subsidence of upper air that is being warmed adiabatically. An increase in the lapse rate is produced more frequently by surface heating. Air temperatures immediately above the hot rock and sand surfaces of low-latitude deserts may show extremely steep lapse rates for a short distance, although convectional currents usually tend to mix this heated surface air with the cooler air above. Griffiths[7] reports the following two remarkable lapse rates for a site in southern Saudi Arabia:

160°F at the surface, 100°F at 4 feet, thus producing a lapse rate of 2,800°F;

170°F at the surface, 120°F at 2 inches, thus producing a lapse rate of 55,000°F.

[7]John F. Griffiths, *Applied Climatology,* Oxford University Press, New York, 1966, p. 13.

A seasonal change in lapse rate can be observed under certain conditions, especially in areas that experience distinct seasonal differences in precipitation. Father Rey, of the Jesuit observatory at Ksara in Lebanon, has observed temperature differences for years at selected stations in that country ranging from sea level to 9,000 feet. His monthly averages are shown in Table 4-6. Upper-air *radiosonde* (balloon-borne recorder and transmitter) observations indicate that the seasonal decline of the lapse rates from winter to summer in the Mediterranean region probably is due to upper-air subsidence, induced by the migration northward of the subsident air associated with subtropical high-pressure cells.

A reversal of lapse rate sometimes occurs, in which the air temperatures increase instead of decrease with increasing elevation. Such conditions are termed *temperature inversions* when they occur in the lower portion of the atmosphere. Such inversions frequently appear as a sharp temperature boundary in the air, similar to the surface boundary between cold and warm water in an inland lake in mid-latitudes during summer months. Temperature inversions may occur at various elevations and usually are found within a relatively thin air stratum. More frequently, they represent a sudden interruption in the lapse rate, above which the lapse rate is resumed (see Figure 4-16B).

The most common example of a temperature inversion is caused by *air drainage* (see

Figure 4-16 Diagram showing two types of temperature inversions. The first (*A*) is caused by radiational cooling on calm, clear nights. The second (*B*) is an upper-air inversion caused by subsidence.

Figure 4-17). On a calm, clear, cold night within an area of rolling to rough terrain, the layer of air that is cooled by conduction with the cold ground surface becomes more dense because of the loss of heat and slips down the slope under the pull of gravity. In a large valley, the chilled air may accumulate to a depth of a hundred feet or more on a winter night. The upper boundary of this cold, heavy air is a typical inversion boundary, since temperatures above are higher than those beneath. The inversion boundary often can be clearly seen. Smoke and other combustion gases discharged from chimneys within the valley may rise up to the inversion boundary, but because the air above may be warmer, such gases cannot pass beyond; they will accumulate immediately beneath the inversion layer. The concentration of noxious gases beneath an inversion boundary is one of the principal reasons for some of the smog hazards in such localities as the Los Angeles Basin and the industrial valleys of western Pennsylvania. Condensation of water vapor in the cool lower air in the form of typical lowland fogs also may reveal the sharp inversion layer produced by air drainage. Air drainage may be an important site consid-

Table 4-6 Lebanon: Monthly Lapse Rate per 1,000 Feet in °F											
Jan.	Feb.	Mar.	Apr.	May	June	July	Aug.	Sept.	Oct.	Nov.	Dec.
4.87	4.45	3.63	3.41	1.91		no data		2.87	3.50	4.54	5.49

SOURCE: Jean Rey, *Carte Pluviometrique du Liban,* Observatoire de Ksara, Ksara, Lebanon, 1955, p. 16. It is interesting to note that the rainy season in Lebanon corresponds closely to the months when the local lapse rate exceeds the global normal rate.

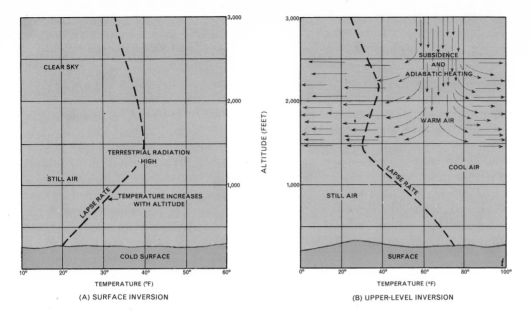

(A) SURFACE INVERSION

(B) UPPER-LEVEL INVERSION

eration in locating fruit orchards or other horticultural plants in areas of rough terrain where frost may be a hazard. Preferred sites are the hill slopes that are slightly above the customary local inversion layer. The air moving downslope, because it *is* moving, may prevent the accumulation of frost on the radiating surfaces of the plantings.

Upper-air inversions may occur from several hundred to several thousand feet above the surface. Upper-air subsidence already has been mentioned as a possible cause. Upper-air inversions also may be produced as the result of an importation of a wedge of extremely cold air that slips beneath a much warmer air mass. The discontinuity between the two generally is not as noticeable as in the lower-air inversions. A further cause for upper-air inversions may be the advection of horizontal sheets of warm air aloft, imported from lower latitudes. Upper-air temperature inversions and changes in air lapse rates aloft are extremely important in making predictions of precipitation. Inversions at any level tend to check vertical passages of air across them, hence tend to check the rising air streams that are an inevitable part of the rain-making process. This aspect will be treated more fully in Section 5-10. A useful general rule is that *low lapse rates and temperature inversions tend toward air stability, while high lapse rates promote instability and vertical air currents.*

Figure 4-17 Air drainage. Air that is chilled by contact with a cold land surface becomes more dense, flows downhill, and collects in surface depressions, thus forming local temperature inversions.

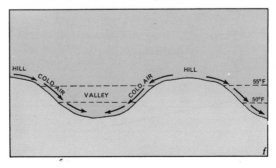

4-12 VERTICAL ZONATION OF TEMPERATURE CHANGES IN THE UPPER ATMOSPHERE The tendency to encounter lower temperatures with higher altitudes, as expressed by the normal lapse rate, does not hold true for the entire atmosphere. Instead, there appear to be four distinct zones, within each of which the lapse rates are markedly different. Figure 4-18 indicates these zones as observed in the northern hemisphere midlatitudes. The exact levels at which they occur may differ slightly with latitude. The four zones include the *troposphere, stratosphere, mesosphere,* and *thermosphere* in order, proceeding upward from the earth surface.

Troposphere. The troposphere is the lowest zone and is characterized by the presence of

a normal lapse rate of 3.6°F per 1,000 feet of elevation, subject to local differences as described in the previous section. The top of the troposphere is known as the *tropopause.* For many years the tropopause was believed to be a fairly uniform surface that sloped downward from a height of about 11 miles above the equator to about 5 miles over the poles. Upper-air observations indicate now that distinct breaks occur in it (see Figure 4-19) and that the various segments of it undulate upward and downward at different times. The tropopause surfaces tend to rise with an increase in both temperature and atmospheric pressure at the earth surface. The steepest changes in elevation of the tropopauses take place in the mid-latitudes between 30° and 40°, and the breaks in the tropopause tend to shift seasonally, being farther poleward in summer than in winter. Note also in Figure 4-19 that the tropical tropopause tends to overlap that of the midlatitudes. The edges of the tropopause segments, especially that around the Arctic tropopause, tend to be undulating, roughly following the pattern of storm tracks, or the boundaries between warm and cool air at the surface. It seems clear that the height of the tropopauses is related to the upthrust of air currents that carry surplus heat energy away from the surface. The height above the equator produces the interesting fact that whereas the higher temperatures at the surface are at low latitudes, the highest temperatures at the

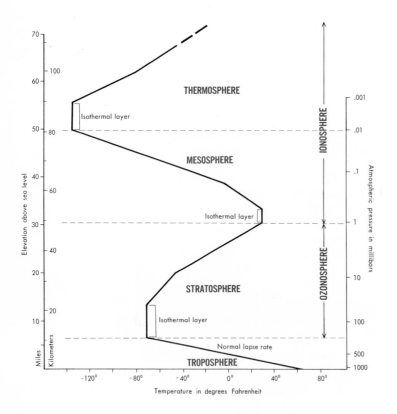

Figure 4-18 The temperature zones of the atmosphere and the standard temperature profile for the northern hemisphere mid-latitudes. The zigzag line descending through the center of the diagram represents the changes of temperature with elevation. Note that zones of little change (isothermal layers) separate the upper atmosphere. The temperature profile has been set as a standard by international agreement.

tropopause are near the poles. The role of the breaks in the tropopause is not clearly known, but they may be related to possible corridors of air interchange between the troposphere below and the stratosphere above.

Stratosphere. The stratosphere represents a thermal zone extending from the tropopause to an elevation of about 30 miles above the earth. As indicated in Figure 4-18, the lower part of the stratosphere exhibits little if any change in temperature for several miles, after which the temperature rises gradually up to 20 miles, then more steeply to the *stratopause,* or the top of the stratosphere. The rise in temperature within the stratosphere is believed to be caused primarily by an increase in the absorption of short wavelength radiation by ozone and oxygen molecules. The zone that contains most of the atmospheric ozone is closely related to the thickness of the stratosphere. It was formerly believed that the stratosphere was without much air motion, but rocket and satellite observations indicate winds of different strengths. There is little evidence, so far, however, of the kind of vertical air currents in the stratosphere that characterize the troposphere. While temperatures within the stratosphere reach those comparable to temperatures on some parts of the earth surface, it should not be assumed that a human astronaut in space at these elevations would find the same degree of body comfort as on earth. The air is so rarefied at these elevations that his body would absorb only a tiny amount of heat by conduction.

Mesosphere and thermosphere. The temperature again drops with increasing elevation immediately above the narrow isothermal zone of "warm" air that marks the stratopause. This new temperature gradient remains fairly constant up to about 50

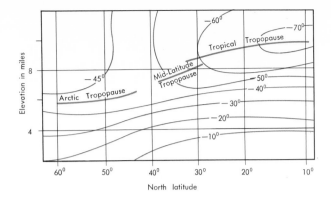

Figure 4-19 General arrangement of tropopauses and vertical temperature gradient (°F) in the northern hemisphere.
[After Rumney]

miles, where the bottom of another isothermal layer marks the *mesopause,* separating the *mesosphere* from the *thermosphere.* Within the thermosphere, as far as is now known, the velocity of molecules continues to increase indefinitely into outer space. The two outer zones sometimes are combined into a general zone known as the *ionosphere,* so named because of the ionization of gaseous molecules at these high elevations.

This ionization produces, among other effects, the *aurora borealis,* or "northern lights," located at elevations of 300 to 600 miles. Several distinct subzones of ionization intensity and concentration of free electrons have been recognized in this upper atmospheric zone, but they are not directly related to temperature distributions. Finally, space rockets have revealed a double layer of high-energy radiation, the Van Allen Belt, surrounding the earth beyond the ionosphere and shaped like a doughnut, being open above both polar regions.

4-13 LAND AND WATER CONTRASTS IN SURFACE TEMPERATURES The temperature of air near the earth surface is influenced not only by different characteristics of the atmosphere or by solar relationships, but also by different surface conditions of the earth below. Major global differences in surface materials and in their capacity for heating and cooling are well illustrated by the contrast between air temperatures over large areas of land and water. Land surfaces experience a greater and a more rapid rise in temperature following the receipt of solar energy than do water surfaces. Conversely, they tend to lose this heat faster and to a greater degree than water. Four major factors are responsible for this.

1. *Evaporation.* A large amount of heat energy is used in the evaporation of water from ocean surfaces, and it serves as the principal way of removing the surplus radiant energy that accumulates over the low-latitude oceans. It is, however, strictly a cooling process that tends to reduce the amount of sensible heat, hence to lower temperatures. A subsidiary effect of evaporation also is the downward movement of surface water that is caused by increased salinity, thus mixing the water near the surface with the cooler water below.

2. *Transparency.* Water, being transparent, permits solar energy to penetrate well beneath the surface, although the longer wavelengths of back radiation from the atmosphere are absorbed only within the top few inches. On land, however, radiation affects only the top inch or so of rock or soil. A given amount of incoming radiation thus will raise the temperature of a given amount of land surface much more rapidly and to a greater degree than it will the same amount of water surface. For the same reason, the loss of heat through outgoing radiation is faster and greater over land than over water.

3. *Movement.* Vertical mixing and the horizontal transfer of water by waves and currents can distribute heat energy over wide areas and great depths. This means that the incoming solar energy can influence a much larger mass per unit of area over oceans than over continents, and thus temperature extremes are lessened.

4. *Specific heat.* It takes only about one-fifth to one-fourth as much energy to raise the temperature of a given volume of land solids (rock, soil, vegetation, etc.) as it does to heat an equivalent amount of water. Conversely, an equivalent amount of heat loss lowers the temperature of a given volume of land more than it does the same amount of water. However, this factor is much less important than the first three, in explaining land and water temperature contrasts.

The slower rate of heating and cooling of water as compared with land results in smaller temperature ranges both daily and seasonally. Sometimes there is also a greater lag in the maximum and minimum temperatures beyond the solstices. Because of the larger proportion of land in the northern hemisphere, temperature ranges north of the equator are greater than those of equivalent latitudes in the southern hemisphere. The seasonal land-ocean contrasts are illustrated by the map of mean annual ranges of temperature (Figure 4-10) and the data contained in Table 4-7.

The regional contrasts in water temperatures in the oceans, although ranging only from 32 to about 90°F, are sufficiently great to be significant factors in world temperature patterns. Ocean currents, which carry water over long distances, and vertical movements, such as the upwelling of cold water from great depths, exhibit pronounced water *temperature anomalies,*[8] which in turn influence

---

[8]A *temperature anomaly* is a temperature that differs from the norm for the particular latitude. Positive and negative anomalies correspond to higher and lower differences, respectively.

Jan.	Feb.	Mar.	Apr.	May	June	July	Aug.	Sept.	Oct.	Nov.	Dec.	Range
					CONTINENTAL STATIONS							
					Pine Bluff, Arkansas (lat 34° N)							
42*	45	54	64	72	79	82	81	76	63	53	45	40
					Srinagar, Kashmir (lat 34° N), elev. 5,204 ft.							
31	33	45	56	64	70	73	71	64	53	44	36	41
					Rosario, Argentina (lat 33° S)							
77	75	71	65	58	52	52	54	59	63	69	79	25
					Semipalatinsk, U.S.S.R. (lat 50° N)							
3	4	14	37	56	67	71	67	56	38	20	9	68
					OCEANIC STATIONS							
					Mogador, Morocco (lat 32° N)							
57	59	60	63	65	67	68	68	68	67	63	59	11
					Evangelist Islands, Chile (lat 51° S)							
47	47	46	45	41	40	37	39	40	42	43	45	10
					Scilly Island, Great Britain (lat 50° N)							
46	45	46	49	52	57	60	61	59	54	50	47	15

*Minimum and maximum temperatures are underlined.

air temperatures and result in irregularities in the latitudinal course of isotherms as they cross oceans and continents.

Figure 4-20 shows the main negative and positive temperature anomalies for both the summer and winter seasons. Differences within the continents are largely caused by a variation in cloud cover, while differences over the oceans are caused mainly by ocean currents. Comparing Figures 4-20A and 4-20B with the map of warm and cool ocean currents (see Figure 6-27) indicates why the most pronounced negative anomalies occur off the west coasts of continents between 15° and 30° during the summer season, and poleward of 40° off the east coasts at both seasons. The greatest positive anomalies over oceans occur off the west coasts of continents poleward

of 30° (especially in winter), in the major sea inlets within the tropics and subtropics (such as the Red Sea), and in the sunny portions of the subtropics. The greatest positive water anomaly in the world is located off the Norwegian coast during the winter season, when the warm water originating in the Gulf Stream bathes this coast. This helps to explain why barley and potatoes can be cultivated along the edges of the Norwegian fjords well north of the Arctic Circle.

4-14 OTHER LAND SURFACE CONDITIONS INFLUENCING TEMPERATURES
The contrasts in temperature that can be encountered on a hot summer day between the paved parking lot of a shopping center, a grassy park in the next block, the shaded

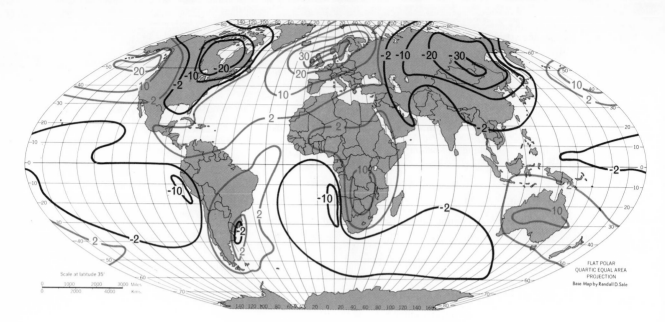

A. Global temperature anomalies for January (°F)

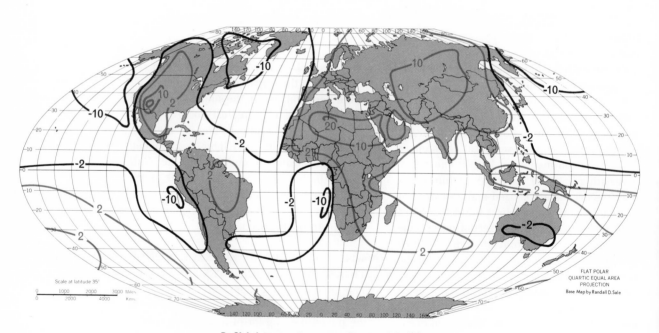

B. Global temperature anomalies for July (°F)

132

Figure 4-20 Global temperature anomalies. These two maps indicate seasonally those areas that are warmer or colder than the average for their respective latitudes. The larger continents tend to show the greatest extremes at both seasons. Note especially the influence of the Gulf Stream and Japanese Current, combined with westerly winds, in producing the extremely high positive anomalies in northwestern Europe and the northwest coast of North America during January. The cold, offshore currents along the west coasts of continents in the subtropics produce strong negative anomalies. The January anomalies are more pronounced than those of July. Careful study of these maps will suggest many interesting problems.

aisles beneath the trees in a nearby forest arboretum, or the broad expanse of a newly cut hay field in the adjacent suburbs indicate that many local conditions of vegetation and other reflecting surfaces appreciably influence temperature conditions close to the surface. The study of microclimates can be exceedingly complex. We cannot begin to treat all of the features of microthermal conditions, but there are some surface conditions that have important regional and even global temperature implications. They will be treated briefly in the following subsections.

Ice and snow. As was indicated earlier, a covering of ice and snow exerts a powerful drain on the heat content of the air that lies over it. The main reason is the extremely high albedo of such surfaces. Fresh-fallen snow may reflect as much as 90 to 95 percent of the incoming solar radiation. The slight amount absorbed is soon lost. The continued radiation of heat energy from the surface during the long winter nights, and the heat energy that is used in melting and evaporating the snow and ice, removes not only the small quantity of energy that is absorbed during the daylight hours but also the heat energy that is brought in by winds from

lower latitudes. Earlier it was noted that at the surface, the net radiation balance was negative in the winter hemisphere poleward of 40° (see Section 4-4). This is largely the result of snow cover. Geological evidence indicates that cold, snow-covered polar regions are not inevitable on earth. In fact, it indicates that much warmer high latitudes were more usual throughout most of earth history. If the global temperatures were to change sufficiently to remove the great ice caps of Greenland and Antarctica and to reduce greatly the amount of area subject to winter snowfall, the high latitudes could maintain surface temperatures at much higher levels than at present through increased radiation absorption. Whereas now the surface polar areas are important heat sinks, during the winter, in the above situation, the vertical transfer of heat into the upper atmosphere would be forced to increase in order to relieve the added positive radiation balance at the surface. Regardless of long-range implications, snow cover is an important factor in regional temperatures. "Open winters" (winters with little snow) generally are mild winters in the mid-latitudes, and a snow cover frequently will make a difference of from 8° to 10° within similar air masses.

Vegetation. The various degrees of lightness and darkness on a black and white print of an air photograph indicate that there are extremely wide differences in the reflection of solar radiation on the earth surface, and these differences could have some effect on local temperature patterns. Most water, unless highly rippled and when the light is at a low angle, appears black. This indicates a low albedo that may be as low as 2 to 4 percent. Forests also tend to be dark, but they differ between the black northern spruce forests and the lighter shades of tropical forests. Bare ground has a high albedo, and freshly plowed soil appears almost white, reflecting as much as 30 percent of the incoming radiation. While the albedo of vegetation has some effect on temperatures, there are several others that can be even more important, including transpiration, photosynthesis, and turbulence. *Transpiration* is the exhalation of water vapor by plants. It serves several purposes, but principally as a temperature regulatory apparatus, much like human perspiration, which helps to maintain a tolerable skin temperature under heat stress. Plants vary greatly in the amount of transpiration, but large individual trees in warm climates with a plentiful supply of water have been known to transpire as much as 500 gallons of water per day. Transpiration over the broad expanses of tropical forests is enormous. Other diversions of absorbed energy besides latent heat include the chemical energy bound up in plant material through photosynthesis. The irregular surface of forests tends to aid turbulence and eddy diffusion, thus helping to mix air and prevent the accumulation, or too great loss of, sensible heat at the surface.

The form taken by trees in a forest frequently is the result of mutual adjustments between the net radiation balance and the vegetation. To illustrate, the dark, conical spires of spruce and fir in the northern forests of North America and Eurasia have several functions: (1) their dark coloration decreases the amount of energy reflected; (2) the spire form increases the total surface exposed to solar radiation; (3) the angle of exposure is made more direct with respect to the incoming low-angle radiation; (4) the narrow tops produce a highly irregular forest surface to increase turbulence, thus bringing sensible heat to the surface where heat loss through radiation is greatest; (5) snow slips off the sloping branches so as to ensure that the needles have access to the small amount of incoming solar radiation. Tropical forests have a design which is aimed at almost opposite objectives.

Man. Human activities are not a significant short-term variable in global patterns of temperature distribution, although there is some debate as to whether or not man's increased use of fossil fuels has led to the general global warming trend that has characterized the past century or so, at least until very recently. The influence of humans on net radiation balances and air temperatures is much more apparent locally. Cities, for example, generally are between 5 and 10° warmer than adjacent rural areas in both winter and summer, although the difference is greatest when advection influences are minimal.

An important reason for the urban "heat islands" is related to the role of *evapotranspiration* (evaporation plus plant transpiration) as a mechanism of heat transfer. There is little evapotranspiration from a surface covered with pavement or rooftops, and that portion of the net radiation balance that normally is used to produce latent heat accumulates as sensible heat and thus raises temperatures. This is the major reason why one can fry an egg on the sidewalks of New York during a hot, sunny day in July, a feat impossible on the grassy sward of Central Park. There are other reasons, however, including the smaller unit surface for absorbing radiation on a roof as compared with the irregular surface of a field. Space heating is still another factor. The production of sensible heat within buildings during the winter and its escape into the outer air, combined with the blanketing effect of city smog to retard outward radiation, also helps to keep temperatures in the cities higher than those in the suburbs on cold winter days. It is interesting to note in this connection that most cities in the United States showed a noticeable and sudden decline in their temperature records sometime between 1920 and 1940, when the Weather Bureau recording stations generally were shifted from rooftop positions downtown to the municipal airports located in outlying areas.

4-15   RELATIVITY IN THE HUMAN SIGNIFICANCES OF TEMPERATURES   The human impact on any one environmental factor is never an isolated variable whose significance can be measured or interpreted apart from the context of other physical or cultural variables, and temperature is no exception. Our bodily sensations of warmth or coldness, for example, cannot be related directly to a given scale of temperature alone. Speed and duration of wind, relative humidity, sunshine, amount and type of clothing, metabolic rates, and human conditioning and attitudes all play a role in how a particular temperature feels to humans. A splendid example of this is recorded by Charles Darwin in his *The Diary of the Voyage of H.M.S. Beagle*, where he describes with astonishment the sight of natives of Tierra del Fuego, at the southern tip of South America, going about their daily routines naked in a sleet storm, relatively unconcerned about the weather conditions surrounding them. *Sensible temperature* is a general term used to describe the temperature that the body feels. It should not be confused with *sensible heat*, which is measured directly by thermometers and recorded as temperature.

The human body is a complicated mechanism equipped with an elaborate regulatory apparatus for maintaining operative-surface and internal temperature balances. Under normal conditions, the skin surface of the body has a temperature that is maintained at about 91.4°F, while the internal temperature is maintained at about 98.6°. Heat may be added to both the surface and internal organs of the body in various ways, or it may be removed. Many of the heat-exchange mechanisms of the atmosphere are also utilized by the body to gain or lose heat, including absorption, reflection, and reradiation of radiant energy; evaporation and condensation of water; conduction, convection, and advection; and conversion into and out of chemical and mechanical energy (metabolism).

To illustrate some of the adjustments used by the body to accommodate a continued loss of heat from the surface of the body, beginning with a neutral or comfortable state, the following physiological changes are brought into play. The list is far from inclusive, and there are many side effects not mentioned.

1. *Contraction of skin capillaries and associated skin muscles* ("goose pimples") in order to erect an insulating layer between the atmosphere and the circulatory system.

2. *Increased involuntary muscular activity* (shivering) to generate internal heat. This is a temporary increase in the metabolic rate designed to convert chemical energy into both mechanical energy and heat.

3. *Withdrawal of blood from extremities to maintain operative temperatures of vital organs.* If continued, this may involve withdrawal from surface portions of the brain, inducing coma. (Note that the hands and feet are always the first to freeze.)

4. *Decrease in general body internal temperature.* One of the more serious effects of this is the inability of lung tissue to warm the intake of cold, subfreezing air taken into the lungs. Below a certain point, serious and rapid destruction of lung tissue can take place, producing death in much the same way as smothering (cutting off the oxygen intake).

Whenever the body heat loss exceeds that of the metabolic rate (generation of body energy by oxidation) and continues at temperatures below 32°, freezing of tissues is inevitable. Fortunately, the addition of more layers of clothing always reduces the rate of heat loss.

When heat is being added to the surface of the body at temperatures above those of the body, the following mechanisms are brought into operation:

1. *Dilation of the skin capillaries* brings more internal heat faster to the surface to be lost by convection, radiation, or advection. Some people react more readily in this direction than others, and minor skin hemorrhages with accompanying tissue irritation ("prickly heat") may result.

2. *Evaporation of body fluids at the skin surface* (perspiration) uses body heat which is added to the latent heat of the surrounding air.

3. *Cell dehydration and destruction* may take place if perspiration continues too far without adequate replacement.

4. *Fever and a series of pathological conditions,* including extreme fatigue, prostration, severe hemorrhaging, coma (heat stroke), and possible death may occur with a continued increase in body heat.

Man is a highly adaptive animal and has demonstrated that with proper precautions he can work and thrive in almost any environment on earth. Under extremely cold conditions he insulates himself with added layers of clothing. The high air temperatures of low-latitude deserts are more difficult. The water intake requirements under extremely high evaporation rates and high temperatures, combined with physical exertion, are enormous and rapid. Experiments in the Sahara Desert have revealed that the loss of body fluids may be so rapid as to precede the development of strong thirst sensations. For continued maintenance of efficiency, physical labor in the hot, low-latitude deserts should be accompanied by prescribed and carefully regulated regimes of water and salt intake. The latter is important because it always is a part of the body solutions and is necessary in the internal chemical balance. Man has been able to exist outdoors in Antarctica at temperatures below −120°F, and he can perform manual labor in the Sahara Desert with temperatures rising to 130°F. Despite knowledge, however, people continue to die each year from both heat and cold. Some of the classifications of sensible temperature are shown in Figures 4-22 and 4-23. The factors that are involved are discussed in the next two sections.

Temperature and wind. The influence of temperature, combined with wind (illustrated by Figure 4-21), has been measured by U.S. military investigators who were interested in designing clothing to protect against wind chill or wind warmth. *Wind chill* may be defined as the loss of heat energy to the atmosphere from exposed skin surfaces, and is measured, as in Figure 4-21, by the number of 1000 calories that are removed from a square meter per hour. *Wind warmth* is the net gain in heat from the atmosphere added to the body. An average human has a total surface area of about 1.7 square meters. The classifications are self-explanatory. A further modification should be made in the use of Figure 4-21. In bright sunlight, the chilling effect of wind is reduced by the absorption of solar radiation by the skin. For each hour of bright sun, the wind chill factor should be reduced by 200 k cal/m²/hr. A feature of the chart indicates that the physiological impact of wind increases with a reduction

Figure 4-21 Wind chill factor. The values of the curved lines represent the loss in heat measured in k cal/m²/hr by a neutral skin having a temperature of 91.4°F. In bright sunshine, the losses are reduced by insolation absorption at a rate of about 200 k cal/m²/hr. Note that air temperatures that might feel pleasant at 55° would feel very cool under wind velocities of 25 miles per hour. [U.S. Air Force Research and Development Command, *Handbook of Geophysics,* The Macmillan Company, New York, 1961]

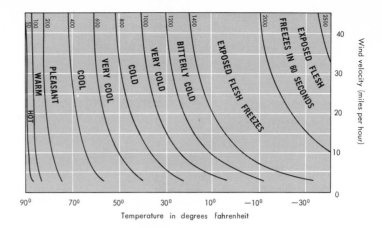

in temperature. Wind velocities accompanied by high temperatures, on the other hand, seem to have little effect on sensible temperatures. Under the latter conditions, relative humidity of the air plays a much more important role than wind velocities, because body heat loss is affected mainly by evaporation rates.

Temperature and relative humidity. Engineers interested in heating and ventilating have carried on experiments for years in the attempt to find the best combination of factors in the atmosphere that will produce comfort for most people. Such experiments have been concerned mostly with indoor conditions, where the control of the environment is much easier, and have mainly tested American adults. Wind is not a significant

Figure 4-22 Graph of comfort index. This nomograph draws boundaries to subjective degrees of comfort as felt by American office workers. The variables are temperature and humidity. The latter can be measured in different ways. See Section 5-1 for definitions of vapor pressure, wet bulb, and relative humidity. To illustrate the use of this chart, a temperature of 80°F (bottom horizontal line) would feel comfortable (Index = 0) with a relative humidity of 20 percent, warm (Index = +1) with a relative humidity of 60 percent, and hot (Index = +2) with a relative humidity of 80 percent. [After Terjung]

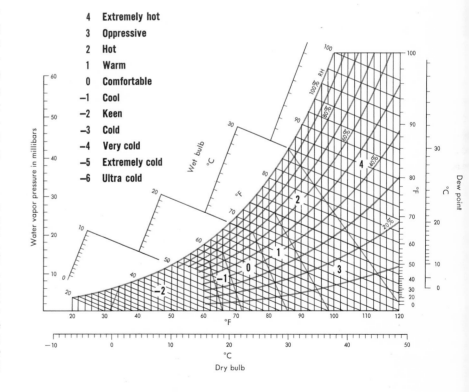

4	**Extremely hot**
3	**Oppressive**
2	**Hot**
1	**Warm**
0	**Comfortable**
−1	**Cool**
−2	**Keen**
−3	**Cold**
−4	**Very cold**
−5	**Extremely cold**
−6	**Ultra cold**

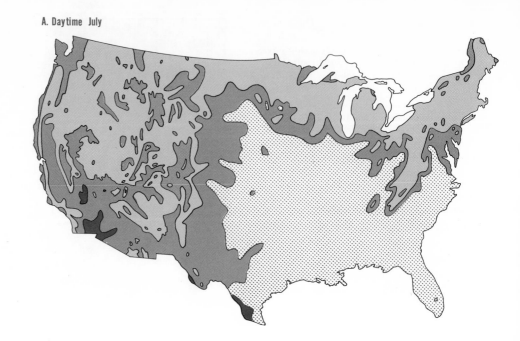

A. Daytime July

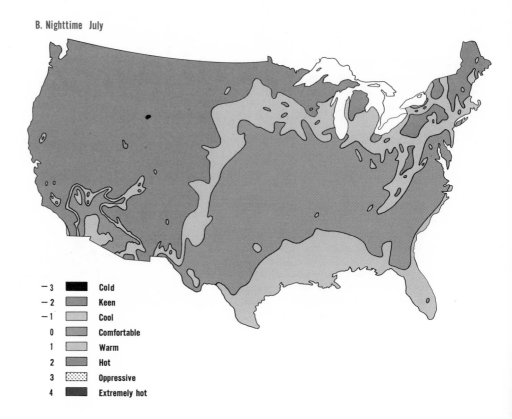

B. Nighttime July

− 3		Cold
− 2		Keen
− 1		Cool
0		Comfortable
1		Warm
2		Hot
3		Oppressive
4		Extremely hot

138

factor indoors, but, as has been known for
hundreds of years, the moisture content of
air plays an important role in sensible temperatures. Here the important variable
accompanying temperature is *relative humidity,* which represents the percentage of moisture contained in the air as compared with
the total amount that the air could hold at
a given temperature.

Figure 4-22 is a classification of various
combinations of temperature and relative
humidity developed by W. H. Terjung.[9] An
increase in relative humidity tends to reduce
the efficiency of evaporation, the principal
cooling mechanism for the skin, and also to
decrease the insulating effect of clothing.
Thus, with increasing relative humidity, the
comfort classification boundaries for the

warmer climates are shifted toward lower
temperatures (lines sloping upward and to
the left in Figure 4-22). An increasing relative
humidity has a reverse influence with respect
to the cold climates, where an increase in
humidity makes the air feel colder and shifts
the comfort index boundaries toward the
warmer temperatures (lines sloping upward
and toward the right in Figure 4-22).

An application of the comfort index in
classifying areas of the United States is given
in Figures 4-23A and B, which show the comfort index of the United States for both daytime and nighttime conditions in July, using
the same index values as shown in Figure
4-22. Terjung also has attempted more complex climatic classifications, combining such
variables as wind chill, day and night combinations, sunshine, and metabolic rates.
There are some limitations to all of the many
attempts to plot climates based on sensible
temperatures. Clothing and human conditioning can be meaningful factors, but extremely difficult to quantify. Basal metabolism rates can be changed significantly to
make the internal heat-adjusting mechanisms
more efficient with respect to special environmental stresses. People accustomed to
warmer climates feel comfortable at higher
temperatures than those who have spent
most of their lives in cool climates.

[9]Werner H. Terjung, "Physiological Climates of the Cotermi-
nous United States: A Bioclimatic Classification Based on
Man," *Annals of the Association of American Geography,*
vol. 56, p. 148, March, 1966.

Study Questions

1. a. What wavelengths of solar radiation are the most dangerous and
      why?
   b. Of the following, which one forms the bulk of solar radiation: ultra-
      violet, visible light, infrared?
   c. Are reddish stars hotter or colder than bluish stars? Why?
2. a. What is *sensible heat* and how does it differ from *latent heat*?
   b. What process changes latent heat into sensible heat?
   c. What is the solar constant, and why is it constant?
3. Describe and explain the role of the following gases in the earth-

atmosphere energy system: oxygen, ozone, water vapor, nitrogen, and carbon dioxide.

4. Examine the top curve of Figure 4-2. How would the curve of radiation from the moon differ from that of the earth, shown to the right in the diagram? Why?

5. Why is the sky blue? Why are desert sunsets famous for their coloration?

6. Arrange the following in order of their annual net radiation balance, with the highest as number 1:
   a. 60,000-foot elevation at 5° N lat
   b. Sahara Desert
   c. Amazon Basin, Brazil
   d. Tahiti Island, central Pacific Ocean
   e. 10,000-foot elevation at the North Pole
   f. Central Antarctica

7. Explain how the energy from each of the following was derived from the sun:
   a. An electric bulb   d. Waves along a rocky coast
   b. Calories in a steak   e. Niagara Falls
   c. A hot-air furnace   f. A human heartbeat

8. If the entire earth surface has a positive net radiation balance for the entire year (see Figure 4-6), why does not the temperature continue to rise?

9. Examine Table 4-3 closely and answer the following:
   a. Using the combined earth-atmosphere radiation balance, where and at what season is the greatest latitudinal rate of change in the balance?
   b. At your latitude, does the greatest latitudinal change take place in summer or winter? Is this a mild or stormy season? Explain.
   c. If the entire earth-atmosphere has a balanced energy input and output, why do the positive values not equal the negative values in the extreme right-hand column?
   d. Using the annual averages in the net radiation balances (right-hand column), explain the statement that "The so-called temperate zone is the most intemperate zone on earth."

10. What are some of the mechanisms for transporting energy from one place to another within the earth-atmosphere energy system?

11. City streets reflect much more solar radiation than a large municipal park. Why, then, are the parks considerably cooler than the streets on a hot summer day?

12. What is the comfort index of the following combinations? (See Figure 4-22.)
    a. Temperature: 40°F; relative humidity: 10 percent
    b. Temperature: 40°F; relative humidity: 80 percent
    c. Temperature: 80°F; relative humidity: 20 percent
    d. Temperature: 80°F; relative humidity: 90 percent

CHAPTER 5
ATMOSPHERIC MOISTURE, PRECIPITATION, AND THE HYDROLOGIC CYCLE

The role of water in the climatic environment of man is both vital and complex. Its function as an energy exchanger within the global energy flow system was treated briefly in the preceding chapter. Now we shall discuss the global water supply system in more detail. The water molecule (H_2O) is a remarkable substance. It is the only material that exists as a crystalline solid, a liquid, and a gas within the normal temperature ranges on earth. It is present in all protoplasm, the universal substance in living cells, and it serves many purposes in the vital life processes of humans. A glance at any global map of population distribution indicates that the most restrictive environmental factor for human life on earth is either the lack of liquid water or the overabundance of it, that is, within the deserts and oceans of the world. It is important, therefore, that we understand the role played and the patterns traced by water as it moves into and out of the atmosphere, doing its work in the global energy system.

Atmospheric Moisture

The water content within the atmosphere, as on the surface of the earth, can exist as a solid (the tiny ice or snow particles that make up the mass of clouds at high elevations), as a liquid (the droplets of water in most clouds and fog at lower levels), and as a gas. The largest quantity by far is in the invisible vapor state. Water constitutes only a tiny proportion of the total mass of the atmosphere (.0026%) and nearly all of it is confined within the first 18,000 feet of the troposphere. The total quantity, however, still is great and totals about 1.5×10^{14} tons. The water content in the atmosphere near sea level varies widely and is largely dependent on air temperature. Colder regions consistently have less water content per volume than warmer ones. If volume of water vapor is related to volume of air, the range at the earth surface is roughly between 4 and 5 percent in the warmest areas and .5 percent over the cold, frozen polar regions. It is a strange paradox that the air over the Sahara Desert contains a much greater amount of water vapor than the air over the northern Atlantic Ocean.

The total quantity of water contained within the earth atmosphere remains fairly constant in spite of the huge amounts that evaporate from the sea and land surfaces or that fall to the earth in the form of rain or snow.

5-1 TERMINOLOGY IN ATMOSPHERIC MOISTURE

It is appropriate at this point to introduce some of the terminology associated with moisture in the air.

Vapor pressure. The content of water vapor in the atmosphere can be expressed in several ways. The measure most widely used by meteorologists is *vapor pressure*, or that part of the general atmospheric pressure that is due to the contained water vapor. Every molecule of water vapor, as well as those of the other gases in the atmosphere, has a certain amount of kinetic energy which it may direct upon a surface to exert pressure. The total vapor pressure, therefore, is proportional to the number of molecules per unit volume. It is expressed in the same way as atmospheric pressure, that is, in inches of mercury, or in *millibars*.[1] Over the surface of a body of water, molecules of water vapor are passing continually from the water to the air, and some are returning. The number returning to the water increases as the vapor content (or vapor pressure) in the air increases. When the number of returning water molecules equals the number passing into the air, the air space is said to be *saturated*, and except for special conditions, no further evaporation can take place. This condition is termed *saturation vapor pressure*. If the temperature of the saturated air is raised, the capacity of the air space to hold vapor increases, evaporation again occurs, and the vapor pressure rises until a new saturation point is reached. The fluctuation of saturation vapor pressure with changes of temperature is shown in Table 5-1.

Absolute humidity. This is a measurement of water vapor content that is rarely used by meteorologists because of its rapid variability. It refers to the weight, or mass, of water vapor contained within a given volume of air, as expressed, say, in grains per cubic foot. Its variability results from the fact that the volume of air contracts and expands freely with changes in temperature unless it is confined within a given space.

Specific humidity. A more satisfactory measurement of water vapor content than absolute humidity is *specific humidity*, or the

Table 5-1 Saturation Vapor Pressure, in Inches of Mercury at Various Air Temperatures (Assuming Constant Air Pressure)

Air temperature, °F	Saturation vapor pressure (inches of mercury)
0	0.038
10	0.063
20	0.103
30	0.164
40	0.247
50	0.360
60	0.517
70	0.732
80	1.022

[1] A *millibar* is a unit of force that exerts a pressure of one-thousandth of a *bar*, which, in turn, is equivalent to a pressure of one million *dynes* per square centimeter. A *dyne* is the amount of force needed to accelerate one gram of mass one centimeter within one second. The pressure exerted by one inch of mercury in a barometer is equivalent to 34 millibars. Mean sea-level atmospheric pressure is 1,013.2 millibars, or 29.92 inches of mercury.

This same graph could be expressed in terms of vapor pressure saturation; the measuring units would be inches or millimeters of mercury (see Table 5-1) or millibars. The 1,000 millibars shown along the vertical scale refer to normal atmospheric pressure near sea level.

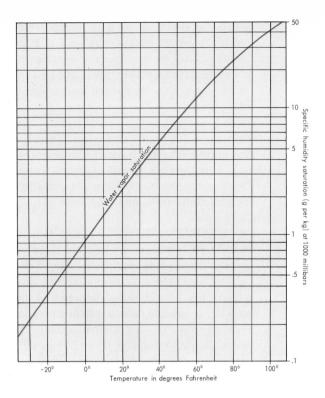

weight of water vapor contained within a given weight of air, expressed, for example, in grams per kilogram. Given appropriate equivalency tables, specific humidity can be easily calculated from data secured from standard weather instruments, even those that are carried aloft in weather balloons. Water vapor saturation also can be expressed in terms of specific humidity, as shown in Figure 5-1. Note that the trend is logarithmic, and hence the *saturation specific humidity* increases much more than in a simple arithmetic proportion with an increase in temperature. A similar relationship is revealed in Table 5-1. The capacity of air to hold water vapor at a given temperature is somewhat less over a surface of ice than over a water surface when the air temperature is below freezing (see Figure 5-12). Evaporation can take place from ice directly to the vapor state, but the process is slower than the evaporation of liquid water and more energy is required. Translating the ratio between specific humidity saturation and specific humidity into simple terms, warm air holds much more water vapor than cold air.

Relative humidity. This is the humidity measurement that is familiar to most people. It is *not* a direct measurement of moisture content, but rather, a ratio between the amount of water vapor in the air and the maximum amount that can be held at saturation at a given temperature and pressure.

More simply, relative humidity is either vapor pressure divided by saturation vapor pressure or specific humidity divided by saturation specific humidity. Hence, it is always expressed as a percentage. At saturation, the relative humidity is 100 percent. Relative humidity is much more related to *sensible humidity,* or the humidity which is felt by humans, than the other measurements. A cold wind with a relative humidity of 95 percent may feel much more wet and damp than a hot wind that contains three times as much vapor, but whose relative humidity is only 40 percent. The two important variables in relative humidity are moisture content and temperature. Pressure has a much smaller effect.

Dew point temperature. The previous paragraph indicated that, with a given amount of vapor in the air, there is an inverse relationship between temperature and relative humidity. Cooling air raises the relative humidity until the saturation point (100 percent relative humidity) is reached and condensation takes place. This temperature at saturation vapor pressure is termed the *dew point temperature,* or simply *dew point.* Given the dew point temperature and the air temperature, it is possible to calculate the vapor pressure, vapor pressure saturation, and relative humidity from empirical formulas.

5-2 MEASUREMENT OF ATMOSPHERIC MOISTURE

Various devices are used to measure atmospheric moisture. One of the simplest devices to measure relative humidity is a *hair hygrometer,* which is based on the tendency for human hair to lengthen proportionately with increased relative humidity. A newer *electric hygrometer* is much more efficient, operates at much lower temperatures, can be carried aloft by airplanes and balloons, but needs considerable adjustment. It is based on the principle of measuring the change in conductivity of a metal conductor that is coated with a chemical substance which gains or loses water obtained from the air as the relative humidity rises and falls. When a hygrometer is connected to a continuous recording device, it is termed a *hygrograph.*

The most widely used device is the *sling psychrometer,* which consists of two standard thermometers mounted on a base that is attached to a specially mounted handle, so that the assembly can be whirled through the air. One of the thermometer bulbs is covered with a thin layer of muslin which is soaked with water. When whirled, the evaporation of the water from the muslin uses heat which cools the thermometer bulb. When the temperature ceases to drop, the difference between the wet-bulb and dry-bulb readings is termed the *wet-bulb depression.* From a set of Smithsonian Meteorological Tables, the relative humidity, vapor pressure, and dew point temperature may be determined following the psychrometer readings. Variations of the sling psychrometer principle have been adapted for mechanical operation, whereby the two thermometers are rotated by an electric motor, and a thermocouple device attached to each enables both dry-bulb and wet-bulb readings to be recorded on a dial.

5-3 VARIATIONS IN VAPOR PRESSURE (WATER VAPOR CONTENT)

Vapor pressure varies noticeably during the day and night, and the characteristic pattern of diurnal changes over land is different than over oceans. Characteristically, over landmasses, a double maximum and minimum occur. As mentioned earlier, the water content of air tends to increase with temperature. The lowest daily vapor pressure thus tends to occur shortly before sunrise, when temperatures normally are lowest. In the forenoon, under the influence of a rising temperature and increased evaporation and transpiration, the vapor pressure rises to a first but slightly lower maximum. Once the surface is heated and air mixing by convection and turbulence increases, some of the evaporated water is removed to higher levels and is replaced by drier air. A secondary minimum thus occurs during the early afternoon, during the time of maximum turbulence and convection. Thereafter, toward evening, the vapor pressure rises slowly to its highest point near sunset. After sunset, cooling again tends to reduce the intake of water vapor, and the vapor pressure slowly declines to its presunrise low. Over large water bodies, the air is less likely to be subject to unequal heating, and daily temperature ranges are much less. Eddy diffusion is highly operative as an air-mixing process over the tropical oceans subject to steady and strong trade winds, but there is little diurnal variation in it, despite the greater opportunity for evaporation over the water surface. Oceans therefore tend to have a more constant diurnal vapor pressure and a single maximum and minimum related directly to temperatures. The importation or advection of new air masses having different vapor pressures owing to their source areas and courses, and several other factors, such as cloudiness, may alter the daily pattern of vapor pressure, on land as well as over the oceans.

Figure 5-2 shows that vapor pressure has distinct geographic patterns on the earth surface that vary latitudinally and seasonally. A general decrease in vapor content with increasing latitude is indicated, corresponding to the direct influence of temperature. As indicated in Figure 5-2, the wider seasonal range in vapor pressure north of the equator is related to the greater temperature range, owing to a greater proportion of land to sea. It is noteworthy to see in Figure 5-2 that the great subtropical deserts of the world do not depress the vapor pressure at their latitudes. These deserts are not dry because of lack of water in the air. On the contrary, they often show much higher vapor pressures than the rainy climates of northwestern Europe. The high daytime temperatures in the desert result in the evaporation of water even from far below the soil surface. These climates are arid because the mechanical requirements for precipitation are lacking and because the relative humidity is low (see Figure 5-3), not because there is little water in the air. The vapor pressure may be high and still be far below saturation vapor pressure in areas of high temperatures.

Vapor pressure also influences the quantity of liquid water that is released when condensation takes place. The higher vapor pressure of tropical air tends to produce

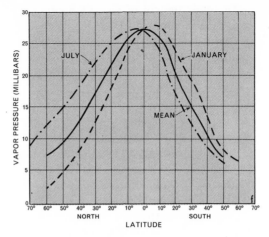

Figure 5-2 Diagram of seasonal changes in vapor pressure with latitude. Vapor pressure is greater in lower latitudes and in summer because of the greater capacity of warmer air to hold moisture. Note the greater seasonal variation north of the equator, the result of greater continentality. Note also that the arid subtropics do not show up on the vapor pressure curves. [After Haurwitz and Austin]

rapid condensation and heavy rains, whereas the precipitation of cool areas rarely is heavy, because cool air cannot hold much water vapor (see Figure 5-25).

5-4 VARIATIONS IN RELATIVE HUMIDITY As stated earlier, relative humidity is not a measure of the water vapor content of the air but a ratio between vapor pressure and saturation vapor pressure at a given air temperature and pressure. Relative humidity thus increases as temperature drops (vapor pressure remaining constant) or increases with the addition of water vapor (temperature remaining constant). It is an important aspect of climates, because it indicates how near the air space is to saturation, hence the general susceptibility of the air space to condensation with a drop in temperature.

The daily pattern of relative humidity in most places corresponds closely with temperature, usually being highest just before sunrise and lowest during the warmest part of the day, or in early afternoon. In the mid-latitudes, however, an abrupt shift in wind direction and the importation of a different air mass may result in dramatic changes in relative humidity (see Figure 4-13).

A highly generalized global pattern of relative humidity is shown in Figure 5-3. Mean figures for latitudinal zones during July and January and also for the year are shown on separate curves. Data for the high latitudes beyond 70° N and 60° S are too incomplete to be reliable. Examination of the curves reveals the following generalizations and explanations:

1. The highest mean maximum relative humidity lies at the equator. This is a reflection of the high evaporation rates in an area plentifully supplied with water and where daytime temperatures consistently remain in the 70s and 80s throughout the year. The low-latitude maximum shifts slightly north and south of the equator, corresponding to seasonal shifts in the pressure and thermal belts.

2. The minima are located roughly between lat 20° and 40° in each hemisphere. The minimum within the northern hemisphere is lower than that in the southern hemisphere. These minima correspond to areas of subtropical high atmospheric pressure that are characterized by subsiding dry air from above and that are sunny and extremely hot during the summer months. Since summer heating over the northern hemisphere continents is greater than that south of the equator, the minimum relative humidity is lower in the northern hemisphere and is located somewhat farther poleward.

3. The rise in relative humidity at latitudes greater than 30° and 40° is mainly the result of decreased temperatures.

4. High latitudes experience a maximum relative humidity in winter and a minimum in summer, owing to seasonal temperature conditions. The converse is true at low latitudes, where there are only slight seasonal temperature variations but where humidity varies under the influence of subsidence of dry air from above. The influence of such subsidence in lowering relative humidity is felt in low latitudes mainly during the "winter" months.

5. The slight drop in the relative-humidity curves beyond lat 60° N may be caused by the lack of water-evaporating surfaces due to a frozen ocean and the lesser amount of importation of water vapor from low latitudes by winds. Note that the drop is greatest in winter. Winter temperatures also are somewhat higher toward the Arctic Ocean.

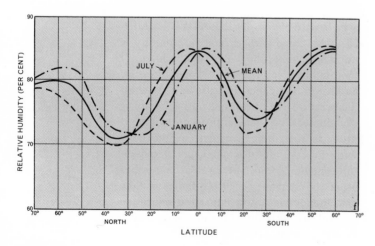

Figure 5-3 Distribution of relative humidity by latitude and by seasons. Note that, unlike Figure 5-2, the dry subtropics produce major declines in the latitudinal curves. The equatorial peak is related to high evaporation. The polar peaks are related to low temperatures and, of course, to a low net radiation balance. [After Haurwitz and Austin]

The mean data shown in Figure 5-3 do not indicate the differences within each latitudinal zone produced by continental and oceanic surfaces. In general, seasonal ranges in relative humidity are appreciably lower over oceanic surfaces than over continental areas.

5-5 PHASE CHANGES OF WATER AND THEIR ROLE IN HEAT EXCHANGE
Water, like most other substances, may occur as a solid, a liquid, or a gas. These three physical states, or *phases*, represent different stages of equilibrium with respect to the flow of heat energy about them. Each chemical substance has its own range of *phase equilibrium*, or the tendency to maintain the same phase within a given range of temperature (pressures remaining constant). However, as was stated earlier, water is the only substance that occurs in all of its three phases within the normal temperature and pressure ranges of the atmosphere at or near the earth surface. Let us now examine the three phases of water and see how each is related to heat energy.

Latent heat of fusion. Below 32°F (0°C), water is in equilibrium as ice, a crystalline solid.[2] Ice contains sensible heat, but in the crystalline phase, this occurs as molecular vibration, a to-and-fro motion from a central point which is fixed in relation to adjacent molecules in the crystal lattice. The temperature of ice is an expression of the velocity of this vibratory motion. As with other solids, there is a maximum limit to which the temperature of ice can be raised and still remain a solid. Beyond that point, a certain amount of energy must be added to break the molecular bonds of the solid state. This is termed

the *latent heat of fusion,* and different substances require different amounts of heat energy to accomplish the phase change. The term *latent heat* is used because the heat is hidden in all liquids, since no temperature change is involved in its acquisition during the melting or fusion process. Water is unique in having an extremely high latent heat of fusion: about 80 calories per gram at 32°F and at sea level pressure. This is the highest of any naturally occurring substance on earth.

Latent heat of vaporization and condensation. Once in the liquid phase, water utilizes sensible heat energy to raise its temperature at the rate of 1 gram calorie per 1°C at one atmosphere of pressure (sea level). This is termed the *specific heat* of water. The specific heat of ice and water vapor is half this amount. The rise in temperature continues until the upper limit of the liquid phase equilibrium is reached at 212°F (100°C). In order to change liquid water into vapor, the bond that holds water molecules together in a fluid phase must be broken. This bond is much tighter than that of crystallization, and the energy required to vaporize water is enormous. The *latent heat of vaporization* of water at sea level pressure amounts to 540 calories per gram at the boiling point and about 600 calories per gram at the freezing point.[3] A rough value at normal air temperatures is about 590 calories. This figure, like that of the latent heat of fusion of water, is greater than that of any other substance in natural form on earth. Thus, water is the most efficient substance in nature to act as a heat energy exchanger. Every gallon of water evaporated from the surface of a tropical ocean removes the heat equiva-

---

[2]Water may continue to remain liquid in tiny droplets even when temperatures drop below freezing, as in clouds at high elevations. This is related to the unusual conditions of surface tension in the tiny droplets and is known as *supercooling.*

[3]Vaporization need not, of course, require an air temperature of 212°F, but the total energy required is the same. For example, the evaporation of a gram of water at 10°C would require 90 calories *plus* the latent heat of vaporization (an amount somewhere between 540 and 600 calories).

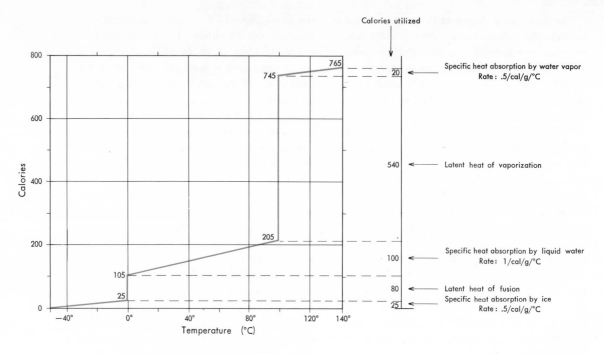

Calories utilized

Specific heat absorption by water vapor
Rate : .5/cal/g/°C

Latent heat of vaporization

Specific heat absorption by liquid water
Rate : 1/cal/g/°C

Latent heat of fusion
Specific heat absorption by ice
Rate : .5/cal/g/°C

——————— Increase in calories with changes in temperature

lent of more than 122 pounds of coal. Condensing the same water vapor within the cloud zone thousands of feet above the earth surface releases this same latent heat. The heat of vaporization thus is a cooling process for the surface from which the heat energy was derived. A good example of this cooling process is the chilling of one's body when stepping out of a hot shower bath. The cooling is caused by the rapid evaporation of the film of warm water left on the skin. This is also why the human body depends on the evaporation of perspiration as its principal means of lowering body surface temperatures when necessary. The heat released upon condensation is termed the *latent heat of condensation.*

Latent heat of sublimation. It is possible to vaporize frozen water (ice) directly without passing through the liquid phase. This process is known as *sublimation,* and the heat energy required is known as *the latent heat of sublimation.* Its quantity is 676 calories per gram at 32°F. In order for sublimation to take place, however, the dew point temperature must be below 32°F. Therefore, it tends to take place over a snow- or ice-covered surface on cold days when the air is unusually dry. The process is extremely slow. Another environment where sublimation takes place is at high elevations within the atmosphere, where cold, dry conditions prevail. The reverse process, the phase change directly from water vapor to frozen ice or snow crystals, also is possible and also is referred to as sublimation. A common example is the white frost that covers vegetation surfaces on a freezing autumn night (see Figure 5-11) or the frost that accumulates on the inside of a window pane of a heated room when the outdoor temperature is well below freezing.

To summarize the global effects of water phase changes, evaporation from the planetary surface removes heat energy and functions as a cooling process, whereas condensation adds sensible heat energy to the air and thus raises temperatures. Much of the net radiation surplus is thus removed from the surface as latent heat to higher altitudes or higher latitudes via water vapor, where it is released by condensation in the form of sensible heat or mechanical energy. This atmospheric heat exchange system, and the significance of latent heat in the system may be likened to the operation of a mechanical kitchen refrigerator.

Clouds and Fog

Clouds are air spaces that contain sufficient amounts of condensed water in the form of droplets to make these spaces somewhat opaque to light. *Fog* is a cloud at ground level. The great variety of clouds and their association with different weather types make them interesting as well as important items in our study of world climatic patterns. Their most important role, however, as we learned earlier, lies in the operation of the global energy budget system, where they absorb and reflect solar and thermal radiation and play a role in the heat exchange mechanism of condensation. The importance of keeping track of cloud cover is testified by the high expense incurred by the U.S. government since 1960 in maintaining a series of earth satellites to provide an almost con-

tinuous photographic record of cloud patterns over a wide swath around the world (see Figure 5-5). The televised cloud patterns and associated data supplied by the Tyros series of earth satellites have been made available to weather forecasters throughout the country for their use. Such information is especially valuable in planning airplane flight routes over long distances.

5-6 THE DEVELOPMENT OF CLOUDS
The development of clouds requires condensation of water vapor by cooling. Lowering the temperature of air decreases its capacity to hold water in vapor form. If cooling is sufficient to lower the temperature to the dew point, further cooling will cause some of the water to lose its latent heat of vaporization and to condense into droplets of liquid water. If the droplets are sufficiently large and numerous to limit visibility, the air space becomes a cloud. Two further requirements must be met, however. First, condensation into a droplet requires the presence of tiny particles to act as nuclei upon which the condensation takes place. Such nuclei are abundant in the air, being composed of dust and smoke particles, plant pollen grains, sea salt crystals resulting from the evaporation of ocean spray blown from the tops of ocean waves, and even meteoric dust from outer space. When temperatures are above freezing, there is never a shortage of nuclei to act as the cores of condensation droplets. Some of the larger water-soluble particles in the air, such as certain sea salts, are *hygroscopic*, in that they attract water molecules to themselves even when the relative humidity is below 100 percent. Sufficient quantities of such particles may produce a *wet haze*, which can occur with relative humidities of 75 to 80 percent.

When temperatures are below freezing, however, the condensation process is quite

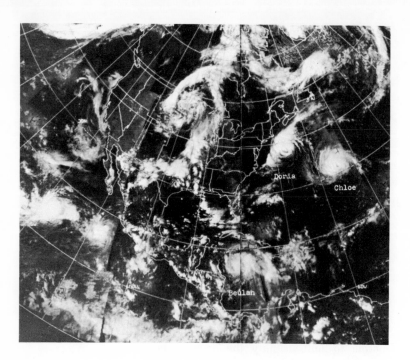

Figure 5-5 Cloud-cover photo mosaic assembled from a series of photos taken at an elevation of about 700 miles by ESSA V, an experimental weather satellite. This remarkable photo mosaic, taken and assembled on September 14, 1967, illustrates the usefulness of the weather satellite program. Circling the earth, the relatively low-level satellites plot cloud patterns within strips that cover more than a third of the earth surface. Once the photo mosaic "maps" have been compiled, they are sent to an automatic transmitting satellite positioned over the Pacific and at an altitude of 22,300 miles. The photo maps then are retransmitted and may be picked up by weather station recorders throughout a large part of the world for use in weather forecasts. Cloud patterns show as white on the photos. The three cloud swirls with names are tropical hurricanes, among the first of the 1967 season. These destructive storms can thus be watched closely. The large swirl centering over Montana and the Dakotas is a typical extratropical cyclone, or low-pressure center. Frontal disturbances can be seen over the Aleutians and near Iceland as linear bands of clouds with likely precipitation. Small scattered white spots in the tropics are individual rainstorms typical of warm, humid air. Note the line of them along the west coast of Central America. Note also the counterclockwise swirl of the storm centers, a feature characteristic of the northern hemisphere. [NASA]

different. Now condensation requires particular kinds and sizes of nuclei, usually relatively large, water-insoluble particles such as clay or dust particles blown into the air during dust storms. Sea salt and small water droplets, by far the most common nuclei present in the air, cannot be used, because the hexagonal ice crystals must have a solid water-insoluble base on which to grow. Small ice crystals appear to be the most effective nuclei. Smoke and other combustion particles generally are too small to be effective. If there are not sufficient numbers of suitable nuclei, it is possible for the relative humidity of the air to exceed 100 percent, a condition which is termed *supersaturation*. Because freezing temperatures are found above about 17,000 feet even in the tropics and may occur at ground level in mid-latitude during winters, supersaturation is not a rare phenomenon in the atmosphere, and at times it may be a factor in preventing the development of clouds. This is the special condition that is utilized by man in *cloud seeding*[4] to produce rapid condensation in supersaturated air.

The clouds that are composed of ice particles resulting from condensation at temperatures below freezing have a different appear-

[4]*Cloud seeding* is the introduction of tiny particles that have a special efficiency as condensation nuclei when temperatures are below freezing. Among the most commonly used substances are silver iodide and dry ice (frozen CO_2). Crystals of these substances are pulverized and either blown into the air by ground generators or carried aloft and spread by airplanes. The purpose of cloud seeding is to produce precipitation where otherwise it would not occur. Such artificial rain-making has been of dubious value except in some mountain areas and when conditions are unusually favorable.

ance than those formed of water droplets. The former are likely to be less opaque and to produce much more diffuse or scattered light reflection because of the many crystal facets that act as reflection surfaces. Clouds composed entirely of ice crystals do not cast shadows. The high, wispy *cirrus* clouds through which sunlight freely passes are of this type. Clouds also may consist of mixtures of liquid droplets and ice or snow particles. The fuzzy, wispy tops of many thunderstorm clouds indicate that sublimation is taking place below the freezing point at these towering heights.

Once clouds are formed, the sizes of water droplets and the total mass of liquid water present are exceedingly small. Table 5-2 presents some striking comparisons between the size of air molecules, condensation nuclei, cloud droplets, and raindrops. It will be noted, for example, that the nuclei of condensation vary greatly in size and that while there are many small nuclei, large ones are much fewer. Some giant nuclei, especially sea salt fragments, are as large as most cloud particles. Note also the wide range between the size and mass of raindrops and those of clouds. This indicates that the processes involved in rain-making are different than those of ordinary cloud formation. The data in Table 5-2 are rough approximations of average conditions. There is a range on either side of the figures given, yet the table does not represent a continuum with arbitrarily chosen divisions. Not all clouds produce rain, and in most clouds the condensation droplets usually show a remarkably narrow range of size and spacing. When water vapor condenses at temperatures above freezing, the droplets grow rather quickly up to about 10^{-3} centimeters, after which further growth is slow and requires special conditions. Another fact revealed in Table 5-2 is that, despite the variation in size of condensed water particles, the total mass per unit of air space (assuming constant air pressure) remains relatively the same. This does not refer to total water content, including water vapor, but only to the condensation forms. Thus, as the water particles grow in size, there are fewer of them.

A further feature of cloud droplets following their formation at temperatures above freezing is that if the cooling continues to

Table 5-2 The Range of Size, Mass, and Frequency in Atmospheric Water Forms

Particle	Diameter (cm)	Mass (grams)	Quantity in 1 cubic meter of air at sea level pressures — No. of particles	Mass (grams)
Air molecules	10^{-8}*	10^{-22}	10^{25}	1,000
Nuclei of condensation	10^{-5} to 10^{-3}	10^{-15} to 10^{-9}	10^{9} to 10^{3}	10^{-6}
Cloud droplets	10^{-3}	10^{-9}	10^{8} to 10^{9}	.1 to 1
Drizzly rain	10^{-2} to 50^{-2}	10^{6}	10^{6}	1
Raindrops	10^{-1} to 40^{-1}	10^{-3}	10^{3}	1

SOURCE: Reprinted from *Weather and Climate,* by R. C. Sutcliffe, F.R.S., by permission of W. W. Norton & Company, Inc., © 1966 by R. C. Sutcliffe.

*For readers who are not familiar with the use of positive and negative powers of 10, a simple rule to remember is that the number of the power refers to the number of zeros that follow the integer 1. Thus $10^2 = 100$, $10^4 = 10,000$, and $10^{-3} = \dfrac{1}{1,000}$.

below the freezing point, the water droplets may continue to retain their liquid droplet form without turning to ice if there are insufficient insoluble nuclei to act as a base for ice crystal growth. Theoretically, the lower limit of liquid droplets in nature is about $-40°F$, and there are many reports of clouds and fog retaining their droplets at temperatures well below freezing. Most airplane pilots are familiar with the hazards of *supercooling,* which is the occurrence of liquid water droplets having temperatures below freezing. When the plane flies through supercooled clouds, the droplets quickly flash into ice upon being smashed into its leading edges. One of the most important meteorological

services of the weather forecasters is the notification to pilots of the freezing levels within clouds.

Supercooled fog is a similar phenomenon close to the ground. Motorists who drive through a thick fog blanket at temperatures below freezing know how soon ice can accumulate on windshields. A more familiar expression is shown by the beautiful patterns of rime that collect on the trees that border the edge of an unfrozen river during an extremely cold, calm night. Relatively warm air above the water rises and soon condenses as it leaves the warmer water surface. The "smoke" of river mist soon becomes supercooled, and when it finally drifts against the trees that border the river, the particles of supercooled water flash into ice (see Figure 5-6).

The *contrails,* or condensed water vapor trails produced by planes flying at high elevations are examples of man-made clouds. They are produced in more than one way. In some cases, the passage of the plane through a supercooled, supersaturated air space may be sufficient to disturb the delicate equilibrium state and begin the condensation process. Often in such cases, the particles will slowly evaporate and the equilibrium become reestablished. In others, if temperatures are below $-40°F$, the ejected water vapor resulting from combustion immediately produces ice particles, which form clouds that persist for hours.

There is only one main cause for the cooling that forms clouds above ground level: the

Figure 5-6 Rime forming on trees from supercooled river mist, Chenango River, Broome County, N.Y. The temperature was below zero (F) at the time this photo was taken. Tiny droplets of condensed vapor rising from the river immediately flash into ice as they touch the cold solid surfaces of the trees. [Van Riper]

adiabatic cooling associated with rising air. The loss of heat by radiation in the upper air may be rapid enough at times to cool the air sufficiently, but wind usually mixes air enough to prevent the condensation.

5-7 THE DEVELOPMENT OF FOG There are two main causes for fog: (1) chilling of warm, moist air moving over a cold surface, and (2) air drainage following radiational cooling of a ground surface. The first type is known as *advection fog* and is by far the most widespread. When warm, moist air moves over a cold surface, it loses heat by conduction. Since air is a poor conductor, conduction cooling normally would affect only a few inches, but air turbulence produced by wind in contact with the ground surface mixes this chilled surface air with air above, producing a zone of cooled air within which condensation takes place. Such fog is particularly common over cold ocean currents during the summer in high and mid-latitudes. In the subtropics, the upwelling of cold water off west coasts (California and Peru) may cause advectional cooling during all seasons. Advection fog thus is most frequent where temperature gradients are unusually steep. While most advection fog is marine in origin, it may occur on land, in which case it occurs most often in winter. A warm winter wind blowing from the Gulf of Mexico northward over a recently frozen and snow-covered Midwest area can produce a large advective fog blanket that may cover much of the United States and be hundreds of feet thick. The famous London winter fog is also of this type. Figure 5-7 shows an advection fog drifting landward from the cold water off the California coast.

Lowland or radiation fog is associated with air drainage. While air rarely loses heat by radiation rapidly enough to cause condensation, the land surface is a good radi-

Figure 5-7 Advection fog drifting inland from the Pacific Ocean in northern California. The cold water off this coast removes heat from the air lying above it, lowering its temperature and causing water vapor condensation. Moving onshore during the morning hours, such fog is almost a daily occurrence along this coast. [Van Riper]

ating body and loses heat by radiation quickly on clear, cold nights. As with advective cooling, the air close to the ground is chilled and some of the contained water vapor condenses. Flowing down hill slopes, the thin wisps of fog may accumulate in lowlands to form thick blankets. Mid-latitude hilly areas are especially susceptible to such lowland fog during clear, calm, cool nights in the autumn and spring. It is not as common in the winter, because the air then contains little water vapor. It is relatively rare in the summer, because nighttime cooling rarely is great enough or occurs long enough to cause lowland fogs. At such latitudes, nights are much shorter than at other times of the year.

Ice fog is composed of small ice crystals formed by sublimation of water vapor di-

Figure 5-8 Typical cumulus clouds near Aukland, New Zealand. Note the even base of these clouds, indicating the level at which condensation is taking place in the rising air currents. [Van Riper]

rectly into the solid state. It is most common when temperatures are below −40°F. It can be man-made and may be a troublesome feature at airfields, where planes eject large quantities of water vapor into extremely cold air. Fairbanks, Alaska, has experienced this difficulty during very cold winter periods. Moisture from human breath may also immediately sublimate at such low temperatures.

5-8 CLOUD TYPES Everyone has observed the many types of cloud forms, and long before meteorology became one of the recognized natural sciences, people learned to associate certain cloud types with different weather patterns and to use them for short-term weather forecasting. Also, anyone who has gazed at cloud patterns recognizes further that although no two clouds are exactly alike, there are certain repetitive features that enable clouds to be classified into distinct types. As early as the eighteenth century, three main genera of clouds were recognized and given Greek names that have been retained today: *cumulus* (a heap or pile), *cirrus* (a lock of hair), and *stratus* (a layer). Meteorologists today use a standard list of ten genera, mostly made up of combinations of these three types, plus many species or subtypes. The cloud species represent special forms that the major genera assume. One example shown in Figure 5-9 is *Cumulus fractus*, which consists of scattered cumulus clouds that have irregular, shredded edges and tops owing to strong winds.

Figure 5-9 Cumulus fractus clouds with wind shear aloft in the trade wind zone east of Hawaii. Strong westerly winds from the left at elevations above about 8,000 feet are interrupting the upward movement of the cumulus clouds below. Normally in this sector the shallow trade-wind flow near the surface is southwesterly. These upper winds sometimes are termed *antitrades*. [Van Riper]

Precipitation

Precipitation includes the condensation forms that have enough mass to overcome the buoyancy of upward-moving air streams and fall to the earth in the form of rain, snow, hail, and sleet. While clouds may be the major global adjusters of energy imbalances, precipitation can be considered as a short-term, intense form of energy exchange. Precipitation, moreover, is one of the global climatic elements important to man, since it yields his major source of fresh water, a basic requirement for nearly all terrestrial plant and animal life. The total precipitation that falls on the surface is not accurately known, but estimates range from 25 to 40 inches per year. One of the most accepted figures is 34 inches. Amounts differ widely from place to place, however, as is shown on the world precipitation map (see Figure 5-18). Some areas may not experience a drop of rain in from 10 to 20 years, whereas others have extremely heavy rainfalls. The highest average annual total is found on Kauai Island in the Hawaiian group. Over a period of 40 years, this station has averaged a total of 471.8 inches per year. Even more impressive are some of the heavy seasonal rains in the tropics. Cherrapunji, in northeastern India, for example, averages more than 98 inches for the month of July and in one year (1861) totaled 366 inches for that single month!

The total energy involved in global precipitation is enormous. One inch of rain falling on a single acre involves a heat transfer equivalent to the amount of energy released in burning 15 tons of high-quality coal.

5-9 FORMS OF PRECIPITATION Precipitation may take several forms, and each has its own process of development and distribution. So, it is appropriate to distinguish them early in our discussion.

Drizzle or drizzly rain. This is a fine settling of small raindrops, usually ranging from .1 to .5 millimeters (.0025 to .02 inches). Although small, the drops are numerous. The direction of movement is highly variable and tends to follow air currents easily. The drops descend to earth vertically only in unusually calm air.

Rain. Rain is the precipitation of liquid drops that are usually 5 to 10 times the size of drizzle drops, generally from 1 to 4 millimeters. Larger drops tend to be disrupted as the result of pressure against their leading edges during descent. The typical form taken by most raindrops is kidney-shaped. Small raindrops are possible, but if so, they are much more widely spaced than in drizzle.

Snow. Snow is the precipitation of crystalline water, mainly in the form of hexagonal, starlike forms. At times, however, the flakes may be in hexagonal prisms or platelets. Larger flakes are formed when crystallization takes place just below freezing. Snow formed in extremely cold air is likely to be of the much smaller, prismatic forms. The tendency of flakes to mat together in clusters depends on the number of tiny particles of liquid water, some of which are usually attached to the crystal faces but not yet frozen. The amount of such water decreases as the temperature drops. Snow varies considerably in its liquid water equivalency per volume, but a rough figure (U.S. average) used is 10 inches of snow per inch of meltwater.

Glaze, or freezing rain. This is composed of liquid rain drops that, having passed through a cold air strata on their way to earth, are supercooled and flash into ice as soon as they strike some solid object that destroys their surface tension. It is responsible for heavy damage to trees and overhead lines when it

Figure 5-10 Ice accumulated on a high-tension electric power line following a winter glaze storm in New York State. The weight of the ice averaged 7 pounds per foot of wire. Broken lines such as this are only part of the cost of such storms. [Courtesy, U.S. Weather Bureau and N.Y. Power and Light Co.]

accumulates in large amounts (see Figure 5-10).

Sleet. Sleet is composed of pellets of clear ice resulting from the freezing and solidification of liquid raindrops that have passed through a cold air stratum on their way to earth. In Europe, the term also is used to designate a mixture of rain and snow.

Grapple, or granular snow. This is composed of small, opaque, soft, snowlike pellets that are the equivalent of frozen drizzle.

Hail. Hail consists of frozen raindrops, but it differs from sleet in being larger and infrequently having a layered structure. The larger forms may reach the size of hen's eggs or baseballs and often consist of alternate layers of clear ice and soft, opaque, snowy material.

Dew, frost, and rime are not true forms of precipitation but are condensations formed on ground surfaces. *Dew* is condensed water vapor that has collected on ground surfaces. It usually is the result of radiational cooling. A drifting advection fog may cover surfaces with heavy dew. *Frost* is similar to dew, but the condensation takes place at temperatures below freezing, and it is *not* frozen dew. *Hoar frost* is an unusually heavy coating of white frost crystals (see Figure 5-11). *Rime* is the beautiful, sparkling cover of ice crystals that is caused by the contact of a solid surface with supercooled fog droplets. Supercooled advection fog may produce spectacular displays of ice crystals that coat all exposed surfaces.

5-10 THE PROCESSES OF PRECIPITATION The entire story of the processes involved in precipitation is not yet complete, but much has been learned during the past 10 to 20 years. The formation of rain is far more complicated than the formation of clouds. It now is clear that special conditions must be present to produce the abnormal growth of some water particles to a mass sufficiently great to overcome air buoyancy. Two principal methods are involved; the first is responsible for most precipitation in middle and high latitudes, and the second is operative in some rainfall at low latitudes.

The supercooling method. It was noted in the section on cloud development that when temperatures drop below freezing, existing droplets of cloud water can continue to exist as liquid water well below the freezing point (supercooling); further, that, lacking adequate solid, insoluble nuclei for crystal development, the air space could produce supersaturation, or a relative humidity of more than 100 percent. Under such conditions, the presence of a relatively few large insoluble nuclei might result in exceedingly rapid condensation. Once ice crystals begin their growth, they strongly attract water molecules to themselves. The reason for this is that the saturation vapor pressure over ice at temperatures below freezing is less than over liquid water (see Figure 5-12). The crystal surfaces, therefore, are able to extract water vapor from the air that the liquid drops were not able to draw to themselves. Further, even the molecules of water bound to the water droplets may evaporate in the vicinity of the solid surfaces and attach themselves to the crystal faces. The snow crystals thus formed not only grow much more rapidly than the water droplets, but there is no upper limit to their growth in mass. Large snow masses begin their descent, and upon reaching elevations where air temperatures are above freezing, they melt and form raindrops that are much larger than those elsewhere in the adjacent cloud mass. The differential speed in descent between large and small droplets, owing to the difference in mass, inevitably results in collisions in which the larger drops gain at the expense of the smaller ones. The larger drops also may collect small droplets in their wake, owing to the partial vacuum created behind. With strong updrafts, the larger drops may gain much mass with only a slight downward movement. There is strong evidence that most rain requires supercooling in clouds that lie within the freezing zone. The supercooling method of producing precipitation is illustrated in Figure 5-13.

The coalescence-during-descent method. Some rain has been observed to fall from

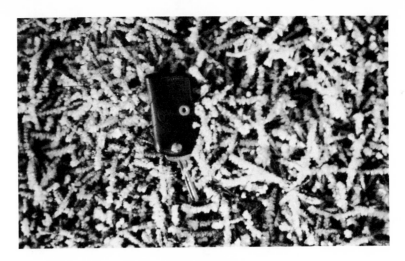

Figure 5-11 Hoar frost on blades of grass—an example of sublimation. The crystals of ice grew by addition directly from the vapor phase. [Van Riper]

Figure 5-12 Differences in vapor pressure saturation over ice and over water. As the temperature drops below freezing, the vapor pressure saturation over water becomes greater than that over ice, and as the diagram indicates, the difference is greatest near 10°F. This means that in a supercooled cloud of water droplets at or below the dew point, snow crystals will collect, condensing water vapor molecules that the water droplets cannot attract.

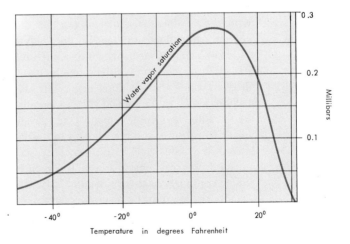

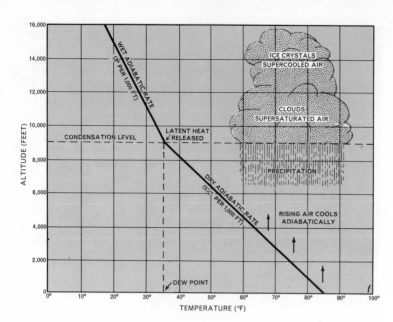

Figure 5-13 The process of precipitation by supercooling.

as a heavy downpour of rain. The coalescent process undoubtedly is also effective during the latter stages of rain produced by supercooling. The only difference is in getting the mass of cloud drops past the cloud equilibrium size. Without supercooling, the larger condensation particles must have much more time to grow, hence the requirement of much stronger and longer updrafts of air. In the supercooling process, drizzle may take place from clouds at low elevations. A steep lapse rate of 5°F per 1,000 feet, a ground temperature of 50°, and a dew point of 45° could produce supercooled clouds with bases under 4,000 feet.

The formation of snow does not require the amount of condensation necessary to produce rain. When condensation occurs below freezing and the proper nuclei are present for ice crystal growth, snow crystals will grow rapidly and drift to earth, not having melted on the way down. With low temperatures at low elevations, the relative humidity is likely to be high. Hence, a smaller lifting and consequent cooling is required for condensation below the freezing point. Although requirements for snow are much less stringent than for rain, the quantity of water involved also is much less. Continental interior locations in the upper mid-latitudes often will have many more days of precipitation (snow) during winter than summer, but the total quantity of precipitation will be much less. Snowfall is likely also to have a much more irregular distribution on a local level than rainfall, because snowfall is triggered much more easily.

Hail develops when raindrops are caught in a thick zone of rapidly ascending supercooled and supersaturated air. The violent updrafts of air maintain sufficient buoyancy to support the freezing drops in the growth medium. Once in descent, they continue to grow even more rapidly. Recycling of the

clouds in which temperatures measure at above freezing. Such rain is associated with towering cumulonimbus clouds in tropical and subtropical areas that have rapid ascending and descending vertical air columns. Theory suggests that certain giant nuclei that are near the size of drizzle drops to begin with in the upper reaches of a rising air column can collect sufficient liquid condensation to begin falling. During their fall, these unusually large droplets collect smaller cloud droplets along their leading edge, and also behind, owing to the partial vacuum created by their downward movement. Once reaching a critical size, which is about 4 to 5 millimeters, the large drops are split by air resistance into several separate particles, each of which continues to grow independently. Unusually strong updrafts also may carry such drops upward, later to fall and collect cloud droplets, until finally they reach the ground

process may take place when especially violent updrafts may sweep the frozen particles back into the upper cloud zone again.

5-11 STABILITY AND INSTABILITY AS FACTORS IN PRECIPITATION Rising air is a basic cause for the development of both clouds and precipitation. We will now focus on the mechanics for air uplift. There are three main initial ways in which air is forced to rise from lower to higher elevations: (1) *orographic lifting*, caused by the movement of air up the side of a physical obstacle, such as a mountainside; (2) *convectional lifting*, caused mainly by unequal heating of earth surfaces; included also is uplift by eddy diffusion, which is a closely allied phenomenon (see "Turbulence and eddy diffusion," p. 123); and (3) *convergent lifting*, caused by the convergence of air masses. In addition to the initial impetus supplied by the above mechanisms, another critical factor influencing lifting is the relative *instability* of the air being lifted. Since air instability is a factor in virtually all precipitation, regardless of the initial cause of lifting, it will be discussed first.

Air masses vary widely in their tendency to rise. If an air mass resists ascent, it is a *stable* air mass. If it tends to move upward easily, it is said to be *unstable*. The principal factor in instability is the relationship between the adiabatic rate of cooling and the lapse rate (see Figure 5-14). Since the adiabatic rate of cooling in dry air remains at 5.6°F per 1,000 feet, the dominant variable in the continued ascent of dry air is the lapse rate once the air is given an upward impetus. Although the normal lapse rate is 3.6°F per 1,000 feet, variations from this may occur (see Section 4-11). If the lapse rate is less than the adiabatic rate, the rising air sooner or later will be cooled to the same temperature as the surrounding air, and the rising will be

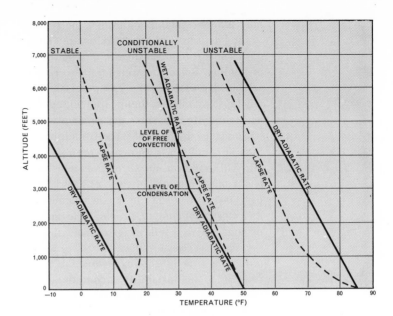

Figure 5-14 The relationship between adiabatic and lapse rates in stability and instability.

checked. This condition is termed *stability*. If the lapse rate is greater than the adiabatic rate, however, the rising air will continue to rise, and *absolute instability* thus is established. In summary, low lapse rates promote stability and are best exemplified by cool air underlying warm air; high lapse rates promote instability and are exemplified by cool air overlying warm air. If lapse rates exceed about 19°F per 1,000 feet, vertical movement can begin without any initial lifting mechanism. When this occurs, the air mass is said to be *mechanically unstable*. Such steep lapse rates are fairly common for short distances above the ground when the surface is being heated excessively by the absorption of solar radiation, but they are exceedingly rare at heights of more than 1,000 feet. Most lapse rates above 400 to 500 feet remain around 3° to 4° per 1,000 feet.

The water vapor present in an air mass

may influence stability through the release of the latent heat of condensation; the effect of this heat is to lower the *wet adiabatic rate* of cooling. Since the wet adiabatic rate is seldom depressed below 2.5° per 1,000 feet, any lapse rate below that figure will indicate an inability to rise, since the air soon would be cooled to the level of the surrounding air. Such a condition is known as *absolute stability*. If, however, an air mass is moist and has a lapse rate of between 2.5° and 5.6°F per 1,000 feet, it may be made unstable by its water vapor content, *provided the amount of heat released following condensation as the air mass rises is sufficient to depress the adiabatic rate of cooling below the lapse rate*. Since this instability is conditioned by the amount of water vapor in the air, it is termed *conditional instability*. It is represented by the center graph in Figure 5-14. Up to a level of 3,000 feet in the illustration given, the dry adiabatic rate exceeds the lapse rate. If this situation were to continue, sooner or later the adiabatic cooling would depress the temperature of the rising air to the level of that of the surrounding air, and rising would stop. At 3,000 feet, however, condensation is taking place, and now the lapse rate becomes greater than the adiabatic rate. The rising air continues to be warmer than the surrounding air, and it will rise indefinitely until the supply of water vapor begins to decrease and the adiabatic rate begins to rise. Extremely warm, humid air, once it begins to condense, may keep the column of air rising to thousands of feet in height.

The streams of moving air that pass over the earth surface may encounter different conditions that can influence both stability and instability. The following examples of induced stability and instability in North America will illustrate this statement.

1. The air stream that enters the southern United States from the Gulf of Mexico has been passing over a warm body of water since its descent somewhere in the subtropical North Atlantic. The moisture content of its lower levels thus has been greatly increased, even though the air mass has not gained much temperature since its descent. The high moisture content of the lower air stratum makes it conditionally unstable. Thus, it is extremely sensitive in summer to (1) lifting by the unequal heating surfaces of the hot, southern states; (2) orographic lifting along the flanks of the Appalachian highlands; and (3) lifting by possible convergence with cooler air masses moving southward from Canada.

2. When this type of air mass continues northeastward and encounters the cold water of the Labrador Current off the New England and southeastern Canadian coasts, its lower layer becomes chilled, hence more stable (owing to a depressed lapse rate). Only a strong, cold air mass is able to force it upward sufficiently to reach a level where the conditional instability could take over the process of ascent to high levels and consequent precipitation. The cooling below may well produce condensation into the cloud droplets of a typical advection fog.

5-12　INITIAL LIFTING MECHANISMS FOR PRECIPITATION　Section 5-11 noted the three main types of impetus that start air on an upward movement: *orographic*, *convectional*, and *convergent*. Each of these will now be treated briefly as it influences the patterns of precipitation.

Orographic lifting. This concerns air that is moving horizontally but which is forced upward by some terrain obstacle. It is diagramed in Figure 5-15. The requirements for orographic precipitation are the same as they are for all precipitation, namely: (1) sufficient uplift to lower the temperature of the rising air mass to the dew point; (2) enough vapor pressure in the rising air mass to lower the adiabatic rate following condensation to an amount below the lapse rate, so as to maintain an updraft; and (3) requisite upper air conditions, such as suitable nuclei, etc. The only distinctive feature of orographic precipitation is the influence of terrain in the initial uplift of air. Most of the heaviest rainfall in the world involves orographic factors, such as the high records accumulated at Kauai and Cherrapunji that were mentioned earlier. The pronounced differences in precipitation between the windward and leeward slopes of the mountains in the northwestern United States and California are due, partially at least, to orographic influences. How much of the rain on the windward slopes of mountains around the world is the direct result of slope alone, however, is not so clear. Conditional instability of the air may be much more important in many mountain rains than the rate or amount of forced ascent, and the initial updraft may have had only a triggering effect.

The best examples of orographic rain supplemented by conditional instability are found in the tropics, where warm, moist air, carried by steady winds, is intercepted by mountains. Many high mountainous islands in the tropics bear an almost continuous cloud cap (see Figure 7-3), and such caps are features of mountains throughout most of the humid tropics. Orographic lifting, furthermore, is generally more effective in the distribution of snowfall than of rain. As stated earlier, snowfall does not require nearly so great an uplift as rain. Hence, local differences in snowfall usually are much larger than those for rain and are likely to be more closely linked with surface irregularities. Mention already has been made of the heavy snows on the windward slopes of hills east and southeast of the Great Lakes (see Figure 5-16), yet these same areas do not exhibit comparable rainfall differences during the

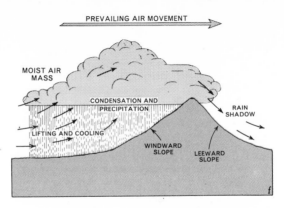

Figure 5-15 Orographic precipitation.

summer period. The heaviest snowfalls everywhere, like the heaviest rainfalls, are on windward slopes. The record in the United States is held by a station far up the western slope of Mount Rainier in Washington, with 1,000 inches during the 1955–1956 winter season.

Convectional lifting.   Convection, eddy diffusion, and turbulence all are possible and somewhat related causes for surface air to rise. Their causes were treated in Section 4-9. Without the influence of conditional instability, however, it is doubtful if any of these initial causes of air uplift could produce rain. First, they rarely lift air by themselves very far above the surface. Vertical air currents caused by irregular surface heating rarely exceed 7,000 feet, and if so, they usually are found in extremely hot desert areas where the relative humidity is extremely low. Nevertheless, when an air mass has a high vapor pressure and a temperature near the dew point, only a small initial uplift may be needed to reach the condensation point. Once condensation begins, unless the local lapse rate is abnormally low, the release of latent heat will produce the generating force for continued and even accelerated uplift. Frequently in the humid tropics, the bases of

Figure 5-16 Snow on Mount Mansfield, Vermont. The heavy orographic snowfall of the New England uplands provides ideal conditions for winter skiing. The grotesque forms are spruce trees laden with snow. [Standard Oil Company of N.J.]

cumulus and cumulonimbus clouds lie between 2,000 and 3,000 feet, well within the height range of convectional currents. Convectional rainfall then, is: (1) almost always associated with warm, tropical, or subtropical air, because only warm air can contain sufficient moisture to "run the engine"; (2) usually in the form of heavy showers of short duration and often associated with thunderstorms, and (3) more frequent over land areas in the afternoon and evening, when convection and lapse rates are likely to be greatest.

The daily period for maximum "convectional" rainfall is likely to occur more frequently at night than during the day over warm, tropical seas. This is related more closely to *eddy diffusion* than to true convection. Eddy diffusion is an important cause of spiraling updrafts over windblown water areas and has little relationship to surface temperature differences. The role of eddy diffusion as a rain producer over water, however, depends mainly on the diurnal variation in lapse rate. After sunset, the loss of radiation from the air 2,000 to 5,000 feet above the sea surface is much greater than near the surface, where the air continues to receive heat by conduction and low-level turbulence from the water surface. The reverse is true over land, where radiational cooling is greatest at ground level at night. Thus, the steepest lapse rates and most favorable conditions for rain occur at night over water and in the afternoon and evening over land.

Convergent lifting. This occurs whenever surface air streams (winds) converge toward the same air space. When this happens, part of the converging air is displaced upward. It would be exceedingly uncommon for such air streams, coming from different directions, to have exactly the same temperature. Invariably one is warmer than the other, and being warmer, is lighter and hence is displaced upward. The amount of horizontal mixing and the sharpness and steepness of the discontinuity (separation surface) between the converging air streams depend on the amount of temperature contrast. Sharp contrasts in temperature produce little mixing and sharp discontinuity surfaces. The line of contact along the separation plane between air masses (actually a zone of variable width) and the surface is termed a *front* (see Figures 6-24 and 6-25). Such lifting of the warmer air is termed *frontal lifting* or *frontal displacement.* The amount and steepness of the frontal lifting can vary greatly; hence, precipitation patterns are likely to vary accordingly. Frontal characteristics are treated in greater detail in Section 6-21. Convergence may take place at times, even within the same air stream. If some external factor, such as

a slight pressure change, forces a huge air stream hundreds of miles wide and thousands of miles long to kink or meander slightly in its course, more air converges in one portion than in another, and a broad but shallow upward displacement may take place. Given a slight angular deflection and sufficient water vapor pressure, the large convergence, if lifted into the cloud zone, may be developed into a huge diffusion eddy hundreds of miles in diameter and capable of releasing large amounts of precipitation for protracted periods. If such giant eddies shrink in size, they become more concentrated and violent. Many of our most destructive hurricanes begin as broad, shallow convergences in the tropical oceans. Their development will be treated in greater detail in Section 6-26. Some convergent precipitation also may be produced as the result of orographic effects. The heavy rainfall in the Assam region of northeastern India is partially caused by the convergence of air in the pocket formed by the sharp northeast curve of the Himalaya Mountain front. The heavy rainfall, although heaviest on the foothills, continues well out in front of the mountains and undoubtedly is the result of convergence.

5-13 THE GLOBAL PRECIPITATION PATTERN A study of the world pattern of precipitation reveals that it is somewhat similar to the water supply system of municipalities. The major sources of world precipitation, corresponding to the reservoirs, wells, or other municipal sources, are the oceans, particularly the warm ocean surfaces of low latitudes, although evaporation supplies water vapor to the atmosphere from nearly all global surfaces. Municipal systems usually require pumps to lift the water to heights so that gravity can distribute it to the consumer, and so too does the global system. The global pump, however, is a far more complicated apparatus than the ones in our local water pumping stations. The global pump is the process of evaporation, operated by solar energy, and as we have noted previously, the amount of energy used to lift water into the atmosphere is enormous. Carrying our analogy further, the water once released under pressure is led through various conduits to the consumer. During this process, the energy of pumping is finally released when the family faucet is turned and the water stands in a pan. In our global system, the conduits are the general air streams of the atmospheric circulation. These "rivers of air" are not perfect closed systems, as are the mains of a city water system, but they frequently are remarkably constant and deliver definite amounts fairly regularly to different parts of the world. Condensation and precipitation might be likened to turning the taps in a municipal home. One major difference must be indicated in our analogy, however. Unlike the water distribution system of a municipality, which is for the most part relatively horizontal, the global conduits are three-dimensional, delivering fresh water latitudinally, longitudinally, and vertically.

A generalized diagram of the total annual precipitation by latitude is shown in Figure 5-17. The principal features may be summarized and interpreted as follows:

Decrease of precipitation with increasing latitude. This is related mainly to the general decrease in temperatures and the resultant decrease in water vapor pressure (see Figure 5-2). It also is related to the high positive radiation surplus at the surface at low latitudes and to the low surplus at high latitudes (see Figure 4-6). The high rainfall in the equatorial zone is a major way of transferring heat from the surface to the upper air, where it is reradiated into space.

Offset of maximum precipitation north of the equator. This is caused mainly by the increased opportunity for convergence of surface winds (see Section 6-14, Intertropical Convergence, and Figures 6-11, 6-12, and 6-18). This convergence zone remains north of the equator during the entire year in some of the ocean areas.

Decrease in precipitation between 15° and 30°. This is a reflection of the great subtropical desert areas that center in this zone, which in turn are largely related to a zone of air subsidence, surface air divergence, low lapse rates, and low relative humidity. Were it not for appreciable precipitation in the eastern parts of continents and the western parts of oceans in this zone, the averages would be much lower.

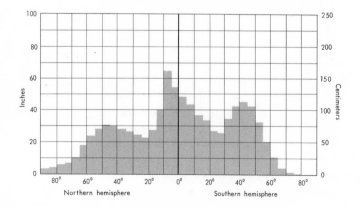

Figure 5-17 Average global precipitation by 5° latitudinal zones. The process of precipitation, with its accompanying vertical movement of air and heat-exchange mechanism, is an effective means of removing surplus energy from the earth surface. Net radiation surpluses are greatest over the low-latitude oceans; hence, precipitation is greatest in these latitudes. The demand to remove surplus surface energy is least at high latitudes; hence, precipitation is low there. The dip in the profile in the subtropics is caused by the subsidence of air in the subtropical, anticyclonic, high-pressure cells. [After Möeller]

Increase in precipitation between 30° and 50°. This is the major zone of interchange between warm surface winds from low latitudes and cold air from high latitudes. Frontal convergence is at a maximum in this zone, but summer convectional precipitation also contributes to the increase. This zone south of the equator shows somewhat higher values than that north of the equator, largely because of the greater extent of oceanic surfaces, and steep average temperature gradients and strong frontal convergences between the Antarctic continent and the southern oceans. The storm zone in southern oceans at latitudes 40 to 60° S has been noted since the days of Magellan.

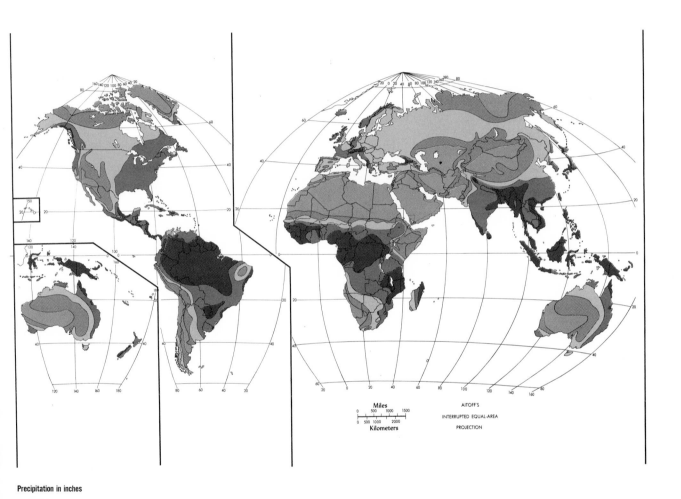

Precipitation in inches

Under 8
8-15
15-30
30-60
Over 60

Figure 5-18 World distribution of average annual precipitation.

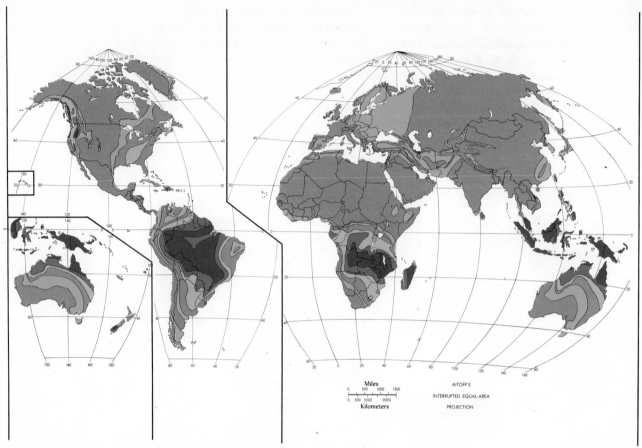

Figure 5-19 Average world precipitation for January.

Precipitation in inches

- Under 1
- 1-2
- 2-4
- 4-6
- 6-10
- Over 10

The greater rainfall at high latitudes north of the equator as compared with high south latitudes. The arctic zone is marine, while the antarctic zone consists of a thick continental ice cap. Antarctica has lower temperatures, lower vapor pressure, and lower pre-

cipitation than the areas within or bordering the Arctic Ocean.

A map of average annual rainfall (see Figure 5-18) shows that there are significant longitudinal variations in total rainfall that are not revealed in Figure 5-17. The two most significant exceptions are (1) the humid eastern sides of the continents in the subtropics as compared with the west coasts (see Decrease in Precipitation between 15° and 30°);

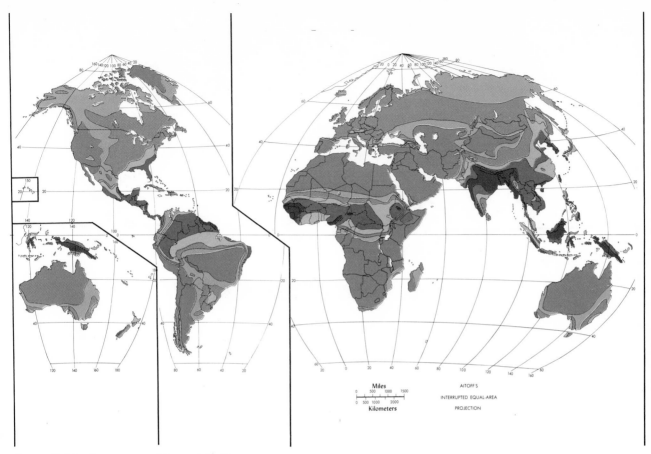

Figure 5-20   Average world precipitation for July.

Precipitation in inches

	Under 1
	1-2
	2-4
	4-6
	6-10
	Over 10

and (2) the greater rainfall on the west coasts of continents between 40° and 60° as compared with the east coasts.

5-14   SEASONAL PRECIPITATION PATTERNS   Figures 5-19 and 5-20 illustrate the essential details of precipitation distribution during winter and summer. It will be noted that in some localities there are wide seasonal differences in precipitation, whereas in others there are consistently humid or dry conditions during both seasons. The areas having alternate wet and dry seasons are transitional, lying latitudinally between a consistently dry and a consistently wet climate and coming under the influence of each

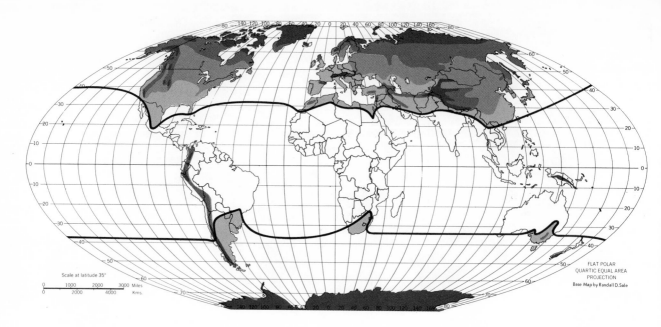

with seasonal shifts in winds, pressures, temperatures, and fronts. All of the climatic belts, except perhaps those associated with polar ice caps, shift position according to changes in the directness of solar radiation. Other strong seasonal contrasts in precipitation are caused by seasonal reversals of winds such as monsoons (see Section 6-15).

5-15 THE DISTRIBUTION AND MEANING OF SNOW COVER The global distribution of snow cover is important in worldwide climatic patterns principally because of its influence on the earth albedo, or the amount of solar energy that is radiated back into space without being absorbed into heat at the surface. There is a close, but not perfect, correlation between the distribution of snow cover and the negative radiation balance at the earth surface. As was noted earlier, temperatures are reduced appreciably with a snow cover. The boundary between severe winters and mild winters in the Koeppen climatic classification, to be used later in this text (see C and D boundary on the climatic map endpaper), closely follows the areas that are snow-covered for at least 60 days.

Figure 5-21 shows the duration of snow cover for various lengths of time as well as the absolute limit beyond which snow does not occur. It will be noted that whenever snow occurs equatorward of 20–25°, it is generally associated with increased elevation. Note, for example, the equatorward bending of the snowless line onto the highlands in southern Brazil, central Mexico, the high veldt country of South Africa, and the highlands of southern China and northern Burma and Indochina. Nearly all of the high mountain areas shown darkest have snow at elevations that lower with increasing latitude, both as to the occurrence of snow and the elevation of the permanent snow line. Table 5-3 gives a rough approximation of the elevation of the snow line at different latitudes. A *snow line* is a line beyond which a snow cover persists. In mountain areas as well as lowland areas its position varies from

Figure 5-21   Average duration of snow cover.

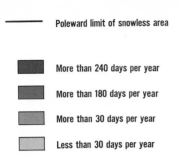

——————   Poleward limit of snowless area

More than 240 days per year

More than 180 days per year

More than 30 days per year

Less than 30 days per year

season to season and also with the quantity of snowfall, exposure to solar radiation, and advectional influences. The snow line in the Cascade Mountains of the northwestern United States sometimes lies 1,000 feet lower in elevation on the western (windward) slopes than on the eastern (lee) slopes. The upper limit of the snow line given in Table 5-3 indicates the approximate elevation of permanent snow, ranging from near sea level at high latitudes to 20,000 feet in the subtropics. The lower elevation of the upper snow line (permanent snow) near the equator, as compared with the subtropics, is due to the increase in precipitation in the equatorial zone. The continental influence of greater seasonal temperature ranges and lesser snowfall is responsible for the much wider range of the snow line north of the equator than south of it. Areas bordering the Arctic Ocean have an upper snow line at high latitudes in the northern hemisphere, in contrast to the ice-bound Antarctic continent.

Measurements of snow depths are important for many reasons. First, they are needed in the precipitation totals for global study, and also as indicators of snow melt and runoff estimates during the subsequent spring

Latitude	Highest elevation (ft and m)	Lowest elevation (ft and m)	Range (ft)
80–90° N	165 ft (50 m)	0 ft	165
70–80° N	3,280 ft (1,000 m)	984 ft (300 m)	2,300
60–70° N	6,560 ft (2,000 m)	1,800 ft (550 m)	4,760
50–60° N	10,500 ft (3,200 m)	2,625 ft (800 m)	7,885
40–50° N	13,450 ft (4,100 m)	4,600 ft (1,400 m)	9,850
30–40° N	20,000 ft (6,100 m)	11,500 ft (3,500 m)	8,500
20–30° N	19,700 ft (6,000 m)	16,075 ft (4,900 m)	3,625
10–20° N	15,430 ft (4,700 m)	15,100 ft (4,600 m)	330
0–10° N	15,100 ft (4,600 m)	14,770 ft (4,500 m)	330
0–10° S	18,370 ft (5,600 m)	14,770 ft (4,500 m)	3,600
10–20° S	20,000 ft (6,100 m)	16,400 ft (5,000 m)	3,600
20–30° S	20,000 ft (6,100 m)	15,100 ft (4,600 m)	4,900
30–40° S	14,770 ft (4,500 m)	5,250 ft (1,600 m)	9,520
40–50° S	7,220 ft (2,200 m)	2,300 ft (700 m)	4,920
50–60° S	3,940 ft (1,200 m)	1,640 ft (500 m)	2,300
60–70° S	0	0	0

Table 5-3   Typical Elevations of the Snow Line Above Sea Level

SOURCE: Gentilli, J., *A Geography of Climate*, The University of Western Australia Press, Nedlands, W.A., 2d ed., 1953, p. 107.

and summer seasons. Most parts of the world that are characterized by humid winters and arid summers, such as California and the Mediterranean Sea area, depend heavily on winter snows for their summer irrigation water. Various techniques have been devised for surveying the depth and water equivalency of snow cover in the critical watersheds of the western United States. One of the most interesting new devices utilizes a source of gamma ray radiation, such as radioactive cobalt, placed beneath the surface. The measurement of the dampening effect of an increase in snow cover on the radiation is recorded by a self-recording Geiger counter mounted above the surface. The dampening effect is also proportionate to the water equivalency of the snow, and thus is more revealing than snow depth alone. The more traditional method is by special snow survey crews, who measure both depth and weight of cores driven through the snow to the surface below (see Figure 5-22). While snow

averages about 10 inches per 1 inch of rainfall in the United States, the water equivalency of snow may range from .01 to .15 inch. Snow that has been unusually compacted or on which rain has fallen may attain higher figures.

A map of snow depth on a global scale is not included because of the extreme variability that occurs within short distances. As was indicated in discussing the processes of precipitation (see Section 5-10), snow requires far less demanding conditions than rainfall, and frequently, relatively slight terrain differences will cause major differences in snowfall within a few miles. As with rainfall, the heaviest snowfalls occur as orographic precipitation. Average snowfall for the United States is shown in Figure 8-19. In Europe, the heaviest falls and depths are found in the Scandinavian and Scottish uplands and in the Alps. The Alps are sufficiently high to have depths of snow occasionally exceeding 100 inches. The only area in all of Eurasia that experiences heavy snows at low to moderate elevations is located in northwestern Japan, where cold winter winds from Asia pick up much water vapor and become unstable in passing over the Sea of Japan. Elsewhere, the heavy snows are confined to high mountains. The Soviet Union, with its broad expanse of winter-girt land, has relatively moderate snow depths, except perhaps in the high mountains of eastern Asia. Rarely are there depths of more than 30 to 50 inches in this country noted

Figure 5-22 Snow surveyors south of Bozeman, Montana. SCS technicians trained as snow surveyors weighing the snow tube and core of snow taken vertically from the snow pack to determine the amount of water contained in the snow. Such data are important in forecasting stream runoff during the following spring and summer months. [U.S.D.A. Soil Conservation Service]

for its severe winters. Heavy snowfalls and persistent depths of more than 100 inches require a water vapor content that cannot be obtained in cold air.

5-16 PRECIPITATION RELIABILITY Reliability is an important consideration in the geography of precipitation. Long-term averages have long-term meanings, but there are many people whose lives may be influenced by short-term fluctuations. There are, of course, many parts of the world where an abundance of rainfall can be relied upon and others that are consistently dry. As a general rule, however, precipitation, like other aspects of weather, is variable and averages rarely occur. The indirect evidence[5] of many research efforts indicate that long-term periods of dry and unusually wet periods have duration periods measured in hundreds and thousands of years. Like temperature, precipitation varies in so many ways that different measures of rainfall reliability must be used for different purposes. A rice farmer in southern India probably would wish to know the general probability that the monsoon rains would begin within a critical span of two to three weeks and would last for a certain length of time. The total amount may not be important either for a single day or for the entire season. An engineer planning the construction of a flood control reservoir would wish to know the probability of individual rains of different intensities for different periods in order to plan his reservoir so as to take care of a given safety factor and within a stipulated cost.

Again we return to the principles of natural distribution. No two spots on earth are exactly alike in their pattern of precipitation occurrences. The closer one examines the

[5]Among the indirect clues for long-term precipitation variations are: pollen identification in bogs, tree-ring analyses, deposition rates in lake sediments, and glacial advances and retreats.

records, and the denser the network of recording stations, the greater become the differences that can be perceived, not only from place to place but also from time to time. There are both long-term and short-term changes, and no clearly discernable pattern that can be used for reliable long-term predictions has yet been discovered. The correlations between fluctuations in precipitation with changes in temperature, radiation balances, wind direction, stability and instability, and humidity indicate that precipitation clearly is part of the exceedingly complex global mechanism of energy flow and that the global energy budget is maintained in diverse ways. It is entirely likely that the most reliable precipitation figure, if it could ever be accurately determined, would be the total global precipitation. Figure 5-23 shows the percentage fluctuations of precipitation away from the long-run averages. In comparing this map with the distribution of average annual precipitation (Figure 5-18), note that precipitation reliability measured in this way increases with increasing dryness. There is little meaning to long-term averages in the desert, where rain may fall once or twice a year or at irregular intervals 2 to 10 years apart. One heavy thunderstorm over a desert may influence the average for many years.

Figure 5-24 shows comparative precipitation data for two widely spaced stations: Beirut, Lebanon, and Tacubaya, Mexico. The individual yearly fluctuations are given for Beirut but not for Tacubaya, although the amplitudes between the highest and lowest rainfalls for the period of record are similar. Despite the irregularity of precipitation from year to year, the reliability is fairly good between 30 and 40 inches in the case of Beirut and between 25 and 35 inches at Tacubaya. The 10-year moving averages indicate that both stations have experienced wet and dry

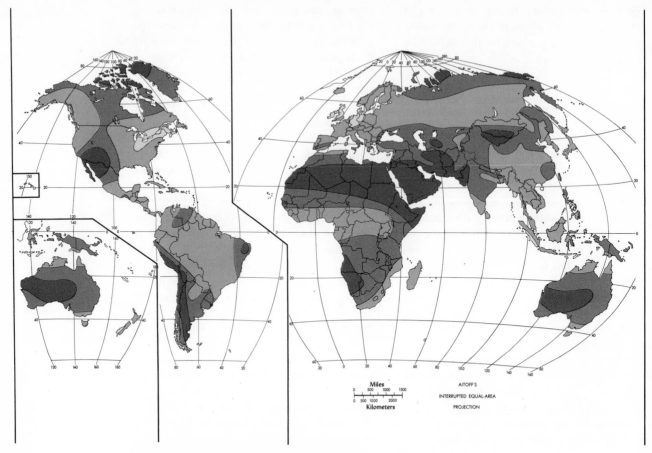

Figure 5-23 Distribution of rainfall reliability. The percentages represent the expected amount of deviation each year from the long-range average annual precipitation. [After Biel, Van Royen, Trewartha et al.]

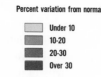

Percent variation from normal

- Under 10
- 10-20
- 20-30
- Over 30

cycles lasting about 50 years. It should be noted however, that these long-run cycles are almost exactly reversed at the two stations, indicating that such long-range cycles are not worldwide in scope. It also is interesting to observe that at Beirut, the average annual rainfall (35.5 inches) was recorded during only one year (1918) in the 85 years of record. Averages rarely occur as individual records, and this is true nearly everywhere.

Figure 5-25 presents a different type of measure of precipitation irregularity. The charts indicate average intensity-duration frequencies for different lengths of time at three extremely different climatic stations in the United States. To illustrate the use of the nomographs, at Miami, the records indicate that there is likely to be a rainfall with an intensity of 2 inches per hour that will last for 2 hours sometime every 2 years. Once

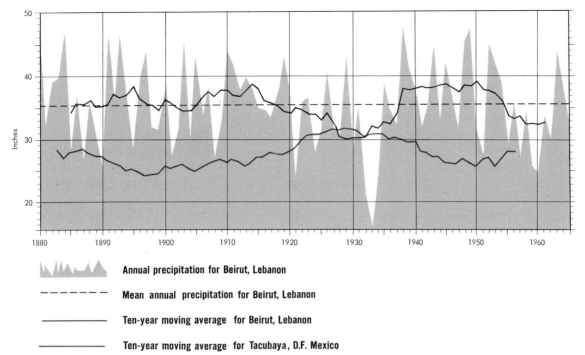

 Annual precipitation for Beirut, Lebanon

------- Mean annual precipitation for Beirut, Lebanon

———— Ten-year moving average for Beirut, Lebanon

———— Ten-year moving average for Tacubaya, D.F. Mexico

Figure 5-24 Long-term fluctuations in rainfall. Note that despite the rapid changes in precipitation amounts from year to year in Beirut, the amounts are fairly reliable between 30 to 40 inches. The use of a 10-year moving average smooths out the year-by-year variations so as to identify long-range trends. The contrast in the moving-average curves for Beirut and Tacubaya is striking and suggests that long-term trends in precipitation are regional in scope rather than global.

has an intermediate position, because its summer temperatures are much warmer than those in Seattle but not so great as those in Miami. The 2-year curve can be considered as the usual pattern for rain intensities in each case. Note that the periods of most intensive rainfall are of short duration at all three stations.

The Hydrologic Cycle

Water plays such an important role in the global energy flow system and is so essential to life on earth that an inventory should be presented, containing estimates as to how much is present on earth, in storage, or in transit, moving from place to place within the atmosphere or along the surface of the earth as rivers and streams. Since water is the principal regulator of the global energy

every 100 years, a two-hour rain will yield 3.5 inches per hour. Similarly, a rainfall at a rate of 10 inches of rain per hour could fall within a period of 5 minutes sometime within 100 years. The charts indicate that Seattle, Washington, is characterized by precipitation far less intense than that at either Miami or Boston. The cooler air of summer at Seattle cannot hold the moisture sufficient to match the heavier rains at Miami. Boston

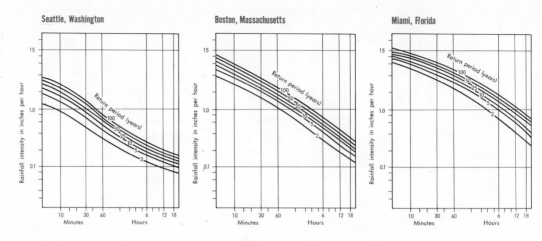

flow system, is the earth supplied with enough to take care of the imbalances that might possibly take place? Before presenting this inventory, a warning should be given that all of the figures used are estimates. No one has accurately measured all of the water in the oceans or in the great ice caps, and many rivers and streams are not regularly measured. Also in the form of crude estimates is the appraisal of the water contained in the atmosphere and the amount that falls as rain or is evaporated from the surface, despite a worldwide network of recording stations. Almost three-quarters of the earth surface is composed of oceans, where the only record-

ings are obtained from scattered island stations or from ships' logs during their passages. We are only beginning to be able to measure the amount of water that is transpired by plants, an important part of the water that is returned to the atmosphere from the land surfaces. Authorities differ with respect to certain parts of the balance sheet, but the rough proportions given in tabular form in Table 5-4 probably are not far from the truth. Several significant features stand out in this table and can be summarized as follows:

1. The total water supply of the earth is enormous, although all but a small proportion is in the form of sea water.

2. The polar ice caps and glaciers constitute a huge storage of fresh water, a quantity that, if melted, would raise the level of the oceans by about 150 feet, enough to flood most of the largest cities in the world.

3. The total amount of water in storage beneath the surface soil is almost a quarter of the total supply of fresh water.[6]

[6]An unknown quantity of the deep groundwater probably contains sufficient dissolved salts so as not to qualify as fresh water. Some of it is ancient sea water that was trapped in sediments as they were deposited on the sea floor millions of years ago.

Table 5-4 Distribution of the Total World Water (after Wolman)		
Oceans	286,230,000 cu mi	(96.7%)
Fresh water	9,770,000 cu mi	(3.3%)
Distribution of Fresh Water		
Ice caps and glaciers	7,327,500 cu mi	(75.00%)
Groundwater below 2500 ft	1,327,000 cu mi	(13.60%)
Groundwater above 2500 ft	1,074,700 cu mi	(11.00%)
Rivers and lakes	32,000 cu mi	(.32%)
Atmospheric moisture	3,400 cu mi	(.03%)
Soil moisture	5,000 cu mi	(.05%)
	9,769,600	(100.00%)

Figure 5-25 Point rate–duration-frequency curves of precipitation for Seattle, Boston, and Miami. These curves indicate the probability of occurrence of rainfalls with varying intensity and duration. In Seattle (A), for example, a 2-inch rain lasting for about 18 minutes is likely to occur sometime during a 100-year period. The heaviest rain to be expected during any 2-year period, however, would be only slightly more than an inch. Even a 1-inch rain during such an interval would last only for about 8 minutes. A comparison of the curves for Seattle, Boston, and Miami indicates the much greater intensities of rain in the lower latitudes. [U.S. Weather Bureau]

4. The average amount of water contained in the atmosphere at any one time is a tiny fraction of the total quantity of water present on earth.

The reader is reminded that Table 5-4 represents an attempt to estimate the average arrangement of the earth water supply at a given time. It does not show the shift of water from one place to another, which is the major concern of this portion of the chapter. The table shows, however, that the total water supply of the earth should be sufficient for almost any conceivable energy irregularities that might occur in the global flux of solar energy and should provide ample water for human use. The human problems of water supply are concerned not with the giant totals, but with micropatterns of the right kind of water in the right amounts in the right places and at the right times. The table also should be translated into terms of energy equivalents. Hidden in it are huge potential energy drains or surpluses. To illustrate, let us calculate the amount of energy needed to melt the estimated water that is stored in the polar ice and glaciers.

1 cu mi = 4.14 cu km

1 cu km = 10^{15} cu cm or 10^{15} grams of water

Melting 1 cu km of ice requires
$$10^{15} \times 80 \text{ cal} = 8 \times 10^{13} \text{ k cal}$$

Melting 1 cu mi of ice
$$= 4.14 (8 \times 10^{13}) \text{k cal}$$
$$= 33.12 \times 10^{13} \text{ k cal}$$

Total ice (see Table 5.4)
$$= 733 \times 10^4 \text{ cu mi}$$

Total energy needed: $733 \times 33.12 (10^{17})$
$$= 24,276 \times 10^{17} \text{ k cal}$$

The total energy received annually by the earth-atmosphere system from the sun averages about $14,140 \times 10^{17}$ k cal. Thus, melting the ice caps and glaciers of the world would require an amount of energy equivalent to the total received from the sun for over a year and a half. We know further that, less than 20,000 years ago, the extent of the ice probably was 10 times that at present. These polar ice masses have not always been present. What were the causes for these large detours in the hydrologic cycle? Do they represent a steady state or equilibrium in the global energy balance that is different from that over most of geologic time? The earth is most fortunate to have its enormous water supply that can accommodate such fluctuations in the energy flux. It is able to regulate both global and local long-term and short-term energy irregularities.

The amount of water contained in the atmosphere also represents energy values that on a short-run basis appear to be extremely large. The 3,400 cubic miles of water in the atmosphere at any one time, if all of it were in vapor form, would require energy amounting to 76×10^{17} k cal, or an amount equal to about a 2-day supply of energy from the sun. Or, extending this to account for the estimated total precipitation (approximately 34 inches of water covering the entire earth surface), the energy equivalent would be about 2,500 × 10^{17} k cal, or about one-sixth of the annual solar energy receipts. This agrees closely with the figure of 19 percent given in the diagram of global energy balance (Figure 4-4).

5-17   THE HYDROLOGIC CYCLE   Figure 5-26 is a diagram that represents what happens to water on earth. There is no attempt here to show the quantities at each position in the cycle. The *hydrologic cycle* represents that portion of the global water supply that is dynamic and moving from place to place because of changes in the energy that is being exerted upon it. The term *hydrologic system*

might have been used as well, because the quantities involved are closely correlated with the total energy that is being expended. The cycle is not a perfect closed system, however, because input and output do not remain constant, and from time to time, quantities of water are being bypassed out of the system into storage, sometimes for a long period of time. Figure 5-27 shows the hydrologic cycle in pictoral form. It indicates that water lifted from the surface by evaporation eventually finds its way back to the surface, and that the main surface supplier and ultimate return goal is the ocean. The routing is not direct, and water may take many devious routes until the cycle is completed. The general operation of the cycle appears to be relatively simple in its overall pattern, but the precise mechanics of its operation have given rise to a separate science, *hydrology*, the study of the behavior of water in its various aspects on earth. We cannot begin to deal with the intricacies in the operation of the hydrologic cycle, but we can treat briefly some of its more significant global aspects.

5-18   THE GLOBAL PRECIPITATION-EVAPORATION BALANCE   The atmosphere has a distinct limit to the amount of water that it can contain at any one time. The average amount contained has been estimated at about 3,400 cubic miles of water, most of which is in vapor form. If this were to be entirely condensed and to fall as rain, it would produce only about one inch covering the entire earth surface. There must, therefore, be a constant influx via evapora-

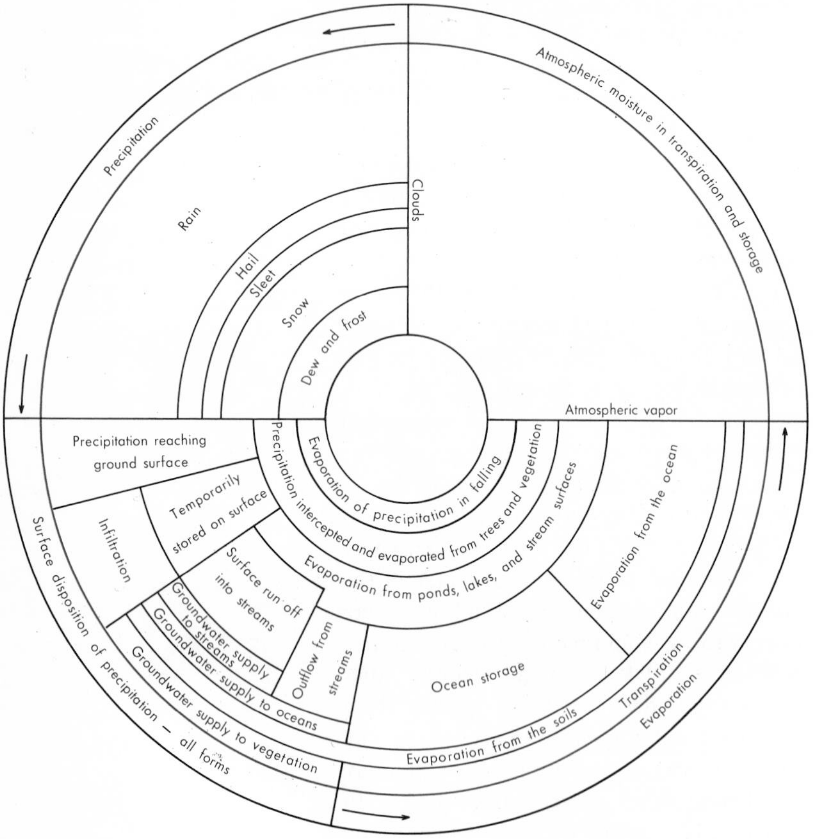

Figure 5-26   The hydrologic cycle; a qualitative representation.   [R. E. Horton, "The Field, Scope and Status of the Science of Hydrology," *Transactions, American Geophysical Union*, vol. 12, 1931]

tion and outgo via precipitation in order to produce the total world average annual precipitation of about 34 inches. In general, the total amount of evaporation equals the total amount of precipitation for the world as a whole. This does not mean, however, that this balance is divided evenly in all places on earth. We know that evaporation greatly exceeds precipitation over the warm tropical seas, and also that precipitation greatly exceeds evaporation in the storm belts of the mid-latitudes, where the warm air of the tropics meets the cold, dry winds from the polar reaches. The arrangement of the precipitation-evaporation imbalances by latitudinal zones is shown by the diagram in Figure 5-28. The diagram shows that strong imbalances take place at times within a short latitudinal distance. Note, for example, that between 0 and 10° N and between 10° and 20° N, there is an abrupt change from an excess of precipitation on both land and sea to a large excess of evaporation over precipitation, all of which is accounted for by the high oceanic evaporation. At this point, the reader should be able to deduce the direction of the air circulation system that can accommodate the transfer of energy both latitudinally and vertically. In the following chapter, we shall see how the global circulation system reacts to the surface inequalities in energy surpluses and deficiencies. We now are perceiving but one aspect of the energy exchange mechanism.

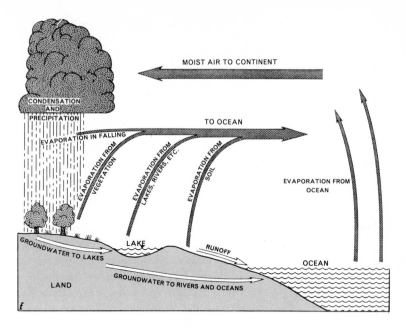

Figure 5-27   The hydrologic cycle.

Figure 5-28   The water balance of the earth by latitudinal zones. This diagram shows surpluses or deficiencies of precipitation and evaporation; it does not indicate total amounts of precipitation and evaporation. Note that the surpluses of precipitation by zones equal the surpluses of evaporation, indicating that for the earth as a whole, evaporation equals precipitation, despite wide variations latitudinally. [After Wuest]

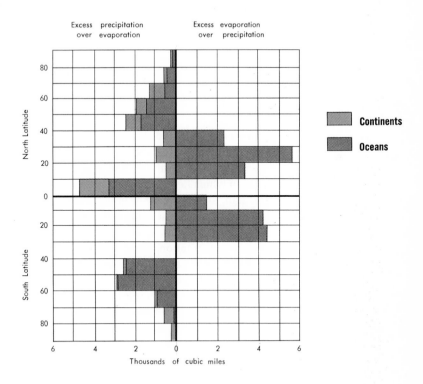

In further examining Figure 5-28, we may observe a net surplus of precipitation over evaporation poleward of 40° N and S which corresponds with the negative radiation balance of the earth-atmosphere system poleward of 40° (see Table 4-3). The earth-atmosphere radiation balance, however (see Table 4-3), shows its highest surplus between 0° and 10° N, in a zone where precipitation at the surface greatly exceeds evaporation. This apparent paradox is partially explained by the fact that the release of latent heat by precipitation is far above the surface, where it leaves by reradiation. Near the surface, the regions of the oceans around 10° N and S are breeding places for tropical storms that might be considered as safety valves for the removal of heat surpluses that accumulate too swiftly for the ordinary processes of evaporation, vertical circulation, and precipitation to remove. Another storm belt is found in the mid-latitudes which corresponds to a sharp change from a net radiation surplus to a net radiation deficiency (see Table 4-3). This will be examined in more detail in Chapter 8.

Note also in Figure 5-28 that all continental areas show surpluses of precipitation. None of them shows surpluses of evaporation. Again we must realize that we are dealing with bands of latitude that extend completely around the world, and that there are significant longitudinal differences. Even the bands within which lie the Sahara, the Arabian, the Iranian, and many other great deserts show continental surpluses of precipitation. Figure 5-17 indicates that these zones still contain much precipitation, located mainly in the eastern parts of the continents. The continental surpluses of precipitation supply our rivers, lakes, springs, groundwater, and total freshwater supply. The oceans equatorward of 40° are the principal sources of our continental supplies, and solar energy lifts this water vertically and moves it poleward to the

Figure 5-29 Average disposition of precipitation in the United States. [Abel Wolman, "Water Resources, a Report to the Committee on Natural Resources of the National Academy of Science," National Research Council, *Publication 1000-B*, Washington, D.C., 1962]

population centers in the humid continental areas of the mid-latitudes. Fortunately for us, land areas have features more favorable for lifting moist air masses than the flat surfaces of the oceans.

5-19 THE DISPOSITION OF PRECIPITATION EXCESS OVER THE CONTINENTS The disposition of water over the continents has been one of the major concerns of hydrologists. Among their many duties in this respect have been the determination of the following: (1) rates, amounts, and duration of precipitation and estimates of regional totals; (2) inflow to rivers, including surface runoff and groundwater seepage; (3) diversion of river water in transit and ultimate discharge rates; (4) evaporation from different types of soils and under different degrees of water saturation; (5) plant transpiration, an especially difficult measurement, since each plant has its own capacity for water intake and transpiration; (6) seepage into the subsoil *aquifers*[7] and the variations in groundwater levels at various depths; and (7) the role of man in disposing of water for agricultural, industrial, and municipal uses.

The balance sheet of the disposition of water over the continents is extremely difficult to prepare, and we are not yet able to describe it except in extremely general terms.

---

[7]An *aquifer* is a water-bearing stratum. The quality of an aquifer depends on its *porosity*, or the total amount of air space in which to hold water, and its *permeability*, or the ease by which water moves through the aquifer, a condition that is related to the size and connectivity of the openings.

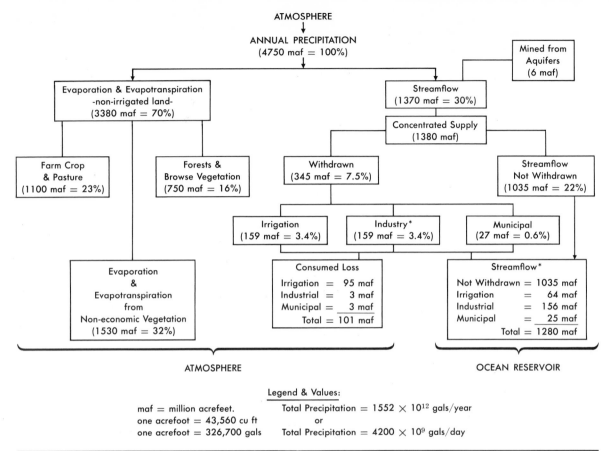

ATMOSPHERE
↓
ANNUAL PRECIPITATION
(4750 maf = 100%)

Mined from
Aquifers
(6 maf)

Evaporation & Evapotranspiration
-non-irrigated land-
(3380 maf = 70%)

Streamflow
(1370 maf = 30%)

Concentrated Supply
(1380 maf)

Farm Crop
& Pasture
(1100 maf = 23%)

Forests &
Browse Vegetation
(750 maf = 16%)

Withdrawn
(345 maf = 7.5%)

Streamflow
Not Withdrawn
(1035 maf = 22%)

Irrigation
(159 maf = 3.4%)

Industry*
(159 maf = 3.4%)

Municipal
(27 maf = 0.6%)

Evaporation
&
Evapotranspiration
from
Non-economic Vegetation
(1530 maf = 32%)

Consumed Loss

Irrigation	=	95 maf
Industrial	=	3 maf
Municipal	=	3 maf
Total	=	101 maf

Streamflow*

Not Withdrawn	=	1035 maf
Irrigation	=	64 maf
Industrial	=	156 maf
Municipal	=	25 maf
Total	=	1280 maf

ATMOSPHERE

OCEAN RESERVOIR

Legend & Values:

maf = million acrefeet.
one acrefoot = 43,560 cu ft
one acrefoot = 326,700 gals

Total Precipitation = 1552×10^{12} gals/year
or
Total Precipitation = 4200×10^{9} gals/day

* The same water may be reused at points spaced along a single stream.

One of the most carefully devised calculations of water disposal of precipitation for a large area is the one prepared by Wolman for the United States. It is shown in Figure 5-29. Since the United States is fairly representative of the mid-latitudes, with both arid and humid climates and a large, active population using water heavily for many purposes, Figure 5-29 should be examined carefully. It illustrates that 70 percent of the precipitation is disposed of by evaporation and transpiration by plants, and that of all the activities of man, by far the greatest water diversion is transpiration from his cultivated

fields and pastures. Much attention has been drawn to the problems of water needs by industry, municipalities, and irrigation, yet the consumed loss from the available supply of all these combined is only 101 million acre-feet, or slightly more than 7 percent of the total stream flow in the United States. The diagram indicates that there is sufficient stream flow to provide for a great increase in human use of runoff. The water supply problem is one of distribution and quality, not quantity.

In order to comprehend some of the intricacies in the distributional patterns of differ-

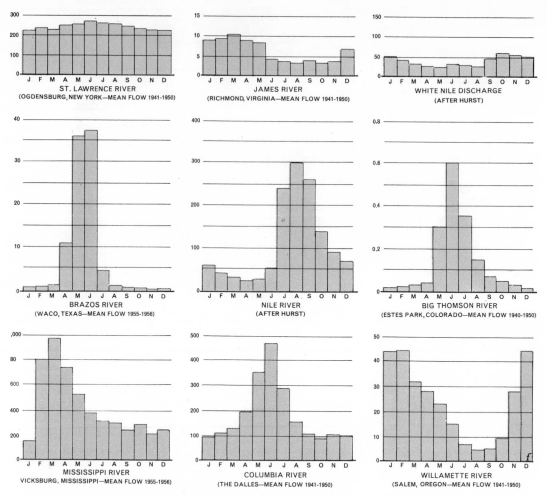

ST. LAWRENCE RIVER
(OGDENSBURG, NEW YORK—MEAN FLOW 1941-1950)

JAMES RIVER
(RICHMOND, VIRGINIA—MEAN FLOW 1941-1950)

WHITE NILE DISCHARGE
(AFTER HURST)

BRAZOS RIVER
(WACO, TEXAS—MEAN FLOW 1955-1956)

NILE RIVER
(AFTER HURST)

BIG THOMSON RIVER
(ESTES PARK, COLORADO—MEAN FLOW 1940-1950)

MISSISSIPPI RIVER
VICKSBURG, MISSISSIPPI—MEAN FLOW 1955-1956)

COLUMBIA RIVER
(THE DALLES—MEAN FLOW 1941-1950)

WILLAMETTE RIVER
(SALEM, OREGON—MEAN FLOW 1941-1950)

*NOTE: ALL DATA ARE IN 1,000 CUBIC FEET PER SECOND

ent aspects of the hydrologic cycle on land, we shall treat only two of the principal components in the last two sections of this chapter: runoff and evapotranspiration.

5-20   RUNOFF FROM THE LAND SUR-FACE   *Runoff* may be defined as that part of the precipitation discharged from a drainage basin by means of either permanent or intermittent stream flow. Some of it is supplied directly by surface flow that has not yet penetrated the soil, while some of it may represent water that is added to the streams by groundwater seepage. The first is likely to take place largely during rainfall or snow-melt periods and will fluctuate widely in amount. The second, or subsurface flow added to streams, is much more regular in both amount and duration. There are many factors that influence runoff besides the obvious ones of precipitation and evaporation amounts. A few of these include: intensity,

Figure 5-30 Graphs of stream-flow regimes for selected rivers.

type and seasonal distribution of precipitation, vegetative cover, slope, size of drainage basins, soil permeability and porosity,
form, length, and trajectory of stream channels, and possible diversions such as reservoir storage, irrigation, and other human
uses, and deep aquifer infiltration. The regimes of rivers reveal that some rivers have
remarkably even flows while others vary
from mere trickles to raging torrents. Figure
5-30 indicates some of this variety in stream
flow.

The geographic pattern of runoff on a
global scale has been estimated by L'vovich, a
Russian hydrologist. His data, translated into
English units, are shown in Table 5-5. The
average runoff of 10.5 inches represents
slightly less than a third of the total average
precipitation for the world. This agrees
closely with the data in Figure 5-29. Runoff
in the United States is about 30 percent of
the total average precipitation. Table 5-5 indicates, however, that some areas of the
world are much better supplied with runoff
than others. The favorable position of North
and South America among the continents in
this respect is related to their smaller percentage of arid and semiarid regions. Average
annual rainfall is the major factor in the gross
features of runoff distribution. Indicative of
this are the large areas of internal drainage
in Asia, Africa, and Australia, which coincide

Table 5-5 World Distribution of Runoff								
	Atlantic slope		Pacific slope		Regions of interior drainage		Total land area	
Continent (or other area)	Area, thousands of sq mi	Runoff, in.	Area, thousands of sq mi	Runoff, in.	Area, thousands of sq mi	Runoff, in.	Area, thousands of sq mi	Runoff, in.
Europe (including Iceland)	3,073	11.7	—	—	661	4.3	3,734	10.3
Asia, (including Japanese and Philippine Islands)	4,626	6.4	6,422	11.8	5,273	0.66	16,321	6.7
Africa (including Madagascar)	5,110	14.0	2,109	8.6	4,291	0.54	11,510	8.0
Australia (including Tasmania and New Zealand)	—	—	1,634	5.5	1,441	0.24	3,075	3.0
South America	6,041	18.7	519	17.5	381	2.6	6,941	17.7
North America (including West Indies and Central America)	5,657	10.8	1,914	19.1	322	0.43	7,893	12.4
Greenland and Canadian Archipelago	1,499	7.1	—	—	—	—	1,499	7.1
Malayan Archipelago	—	—	1,012	63.0	—	—	1,012	63.0
Total or average	26,006	12.4	13,610	15.5	12,369	0.82	51,985	10.5

SOURCE: After L'vovich and presented in W. B. Langbein and J. V. B. Wells, "Annual Runoff in the United States,"
U.S. Geological Survey Circular # 52, June, 1949.

closely with areas of dry land. Another reason South America has such a high average rate of runoff is that the maximum breadth of the continent occurs at precisely the latitudes of highest rainfall: 10° on either side of the equator. The discharge of the Amazon River represents an enormous quantity of almost untouched runoff. Its size can be judged by Table 5-6, which lists the ten largest rivers of the world in terms of runoff. Note that there is little relationship between discharge and the length or size of the drainage basin. For example, the Nile is longer than the Amazon, but its discharge is only slightly less than 6 percent of the latter. The Mississippi-Missouri, much of which flows through humid areas, is roughly equivalent in length to the Amazon but has only 9 percent of the latter's discharge.

The continental distribution of runoff, as shown in Table 5-5 and the totals for the major rivers (see Table 5-6), still masks great differences that occur from place to place and from time to time. Figure 5-31 shows how widely runoff can vary within the United States. The total runoff for the United States averages about 8.5 inches, yet Figure 5-31 indicates that there are broad areas that range far less and far more than the national average. This is to be expected almost everywhere on earth because of the many variations in the total environment that influence runoff. The more we examine the patterns, the more complex becomes the picture of reality, regarding both patterns in space and patterns in time. It is likewise useful sometimes to examine the effect of micropatterns (such as the effects of a growing suburban community, with its increasing area of streets, sidewalks, and rooftops) on the runoff pattern of small watersheds less than 10 square miles in size, following rains of various intensities from year to year, if serious problems are to be avoided.

5-21 EVAPOTRANSPIRATION The return of water from the land to the atmosphere by direct evaporation and through transpiration by plants is one of the major bypasses in the hydrologic cycle. *Evapotranspiration*, as this return of water to the atmosphere is termed, is of great significance in the study of hy-

---

Table 5-6 The World's Largest Rivers (subject to revision)

River	Average annual discharge (ft/cu sec, thousands)	Drainage area (per sq mi, thousands)	Length (mi)
Amazon (S. Amer.)	7,200	2,772	3,900
La Plata-Parana (S. Amer.)	2,800	1,198	2,450
Congo (Africa)	2,000	1,425	2,900
Yangtze (Asia)	770	750	3,100
Ganges-Brahmaputra (Asia)	707	793	1,800
Mississippi-Missouri (N. Amer.)	620	1,243	3,892
Yenisei (Asia)	610	1,000	3,550
Mekong (Asia)	600	350	2,600
Orinoco (S. Amer.)	600	570	1,600
Mackenzie (N. Amer.)	450	682	2,525
Nile (Africa)	420	1,293	4,053

SOURCE: "Principal Rivers of the World," *The Military Engineer*, vol. 50, no. 337, September–October, 1948.

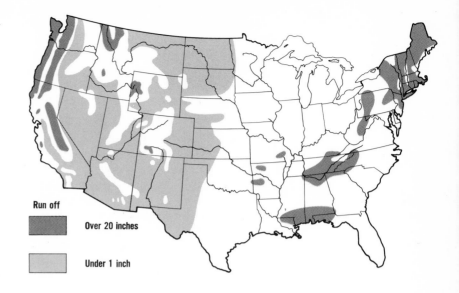

Figure 5-31 Areas of extremes in runoff in the United States. Although the average runoff in the United States is approximately 8 inches, there are extensive areas that have considerably more or less than the average.

Run off

Over 20 inches

Under 1 inch

drology, since it reduces the amount of fresh water added to the continents by a global average of about 70 percent. Whenever evapotranspiration exceeds precipitation, soil moisture reserves will be drawn upon, and the freshwater supply may become insufficient for normal plant growth. The determination of *potential evapotranspiration,* or the overall potential for evapotranspiration, is highly useful in planning water management programs ranging from individual farms to large regional continental divisions.

There have been many attempts to derive basic equations which yield an adequate measurement of potential evapotranspiration. Some of these are theoretical, using variables that logically are involved. Others are empirical, that is, derived from direct measurements, utilizing relatively simple measurements of a few major variables and combining others into empirical constants. The usefulness of the various equations depends on the degree of accuracy desired, the scale of applicability, and the quantity and type of data that can be obtained.

There is a long list of factors that influence the amount of water involved in evapotranspiration, yet the most important ones are relatively few. The conditions that permit the air to absorb more molecules of water at the surface (evaporation) than pass from the air to the surface generally are no different when plant transpiration is present than when it is not, with a few possible minor exceptions. Many of the terms used for this portion of the hydrologic cycle, including *evaporation,* *evapotranspiration, potential evaporation,* and *potential evapotranspiration,* all have

nearly the same value when considered as averages over a long period of time.

Two major factors affect evapotranspiration: (1) the relationship between the vapor pressure of the air and that of the evaporating surface, and (2) the net radiation balance. An evaporating surface, whether a plant leaf or bare ground, will yield more water under conditions of lower relative humidity than higher, and where there is more surplus energy available to be used to produce latent heat. Related to these two factors are: the amount of sensible heat present (measured by temperature); the amount of precipitation (the supply factor); the latitude, which influences the directness of radiation and length of day; and the degree of cloudiness.

Among the relatively minor factors are wind velocity and air turbulence at ground level, which remove accumulations of both water vapor and sensible heat at the evaporating surface. Soil moisture retention is another minor factor that becomes more significant under arid conditions. As soils become drier, evaporation requires more energy

to overcome the molecular attraction between the water molecules and the surfaces to which they are attached. Soils having small openings, such as clays, retain their soil moisture more than coarse-textured sands under dry conditions.

While the type of vegetation cover does not have a major influence on evapotranspiration, it has some minor influences. The height and irregularity of the vegetation affect wind velocities and turbulence. Plants that have a deep root system, such as alfalfa, can tap deep-seated water supplies and yield more transpiration than shallow-rooted field crops such as peas, although the difference is more significant in dry regions.

Three different ways of estimating potential evapotranspiration for planning purposes follow.

Inflow-outflow-storage method. This involves measuring directly the segments in the water balance equation: *precipitation minus evapotranspiration and storage equals runoff,* or $P - (E + S) = R$. Precipitation is measured by rain gauges, storage by keeping close watch of fluctuations in the *groundwater table*[8] by checking the water level in wells that are adequately spaced through the area. Runoff is measured by stream flow recording instruments. There are many sources of error in this method and results are crude.

The Thornthwaite method.[9] This is the most widely used method and involves the

[8]The *groundwater table* is the upper limit of the zone of saturation in the subsoil, below which all of the soil openings are filled with water.

[9]Those interested in the determination and application of the Thornthwaite equation are referred to C. W.

Figure 5-32 Evapotranspiration in inches in the United States. [With permission of the American Geographical Society. Source: C. W. Thornthwaite, "An Approach Toward a Rational Classification of Climate," *The Geographical Review*, vol. 38, 64, 1948]

use of a series of empirical formulas derived from measurements in evaporating test plots termed *lysimeters*. These were placed in several locations in the United States and Mexico. The formulas take into consideration temperature, precipitation, latitude, vegetation height and spacing, and soil retention. If the minor factors of vegetation and soils are removed, a rough measure of potential evapotranspiration may be derived from monthly averages of temperature and precipitation that are available at any weather recording station in the world. The calculations of the empirical formulas are time-consuming but can be aided by nomographs or computer programming. Thornthwaite utilized his methods in a classification of world climates on the basis of both temperature and moisture regimes.

The Penman method.[10] This method of determining potential evapotranspiration differs from the Thornthwaite method in being based entirely on solving an equation that utilizes all of the important variables in

the water balance. Its major handicap is that it requires the direct measurement of the net radiation balance, which in turn depends on either the use of expensive recording instruments or an exceedingly long, complex formula. The basic formula is expressed as $e = H + Ez$, in which e = daily evapotranspiration in mm; H = the net radiation balance or heat budget; and Ez = the drying capacity of air. The last is the most complex part of the equation and has a long derivation formula of its own.

Potential evapotranspiration and the water balance or budget. Regardless of how potential evapotranspiration is determined, its usability can be illustrated graphically, as in Figure 5-33. Earlier it was noted that precipitation minus runoff and storage equals evapotranspiration. This expression of the water balance represents the stable operation of the landward portion of the hydrologic cycle. The hydrologic cycle assures that the balance is maintained except for occasional diversions that were mentioned earlier in the

Thornthwaite, "An Approach Toward a Rational Classification of Climate," *Geographic Review*, January, 1948, pp. 55–94; and C. W. Thornthwaite and J. R. Mather, "Instructions and Tables for Computing Potential Evapotranspiration and the Water Balance," *Publications in Climatology*, vol. X, no. 3, Drexel Institute of Technology Laboratory of Climatology, Centerton, N.J., 1957.

[10]The factors leading to the development of the Penman equation, an evaluation of other methods of determining evapotranspiration, details of the Penman equation and many other relationships between vegetation and water supply are contained in a useful but somewhat technical little book: H. L. Penman, *Vegetation and Hydrology*, Technical Communication No. 53, Commonwealth Bureau of Soils, Harpendun; Commonwealth Agricultural Bureaux, Farnham Royal, Bucks, England, 1963.

Figure 5-33 Precipitation and potential evapotranspiration at three selected stations in the United States. [After Thornthwaite]

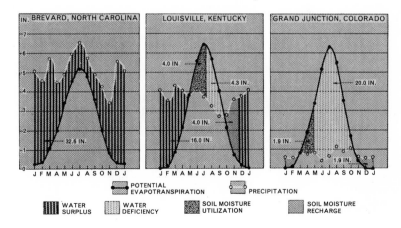

chapter. The graphs in Figure 5-33 were derived from the Thornthwaite equation, but they could just as easily have substituted values obtained from the Penman equation or others.

The water budget chart for Brevard, North Carolina, in Figure 5-33 shows that precipitation exceeds potential evapotranspiration in every month. Thus, there is a surplus of water in every month for runoff and for the replenishment of reservoirs both above and below the land surface. The chart for Louisville, Kentucky, indicates that precipitation exceeds potential evapotranspiration between October and May, and that during this period a total surplus of 16 inches of rain is available for runoff or groundwater addition. During the summer months at Louisville, however, a deficiency of 8.3 inches of water exists, 4 inches of which[11] is offset by water stored in the soil during the wet season. This leaves a seasonal total deficiency of 4.3 inches, occurring in July, August, and September. In October and November, the water

supplies in the soil at Louisville are recharged with water totaling 4 inches, after which surpluses once more appear. As indicated in the chart for Grand Junction, Colorado, the water reserves in the soil never become fully recharged, and no surpluses occur in any season for surface runoff. Agriculture must be supplemented by irrigation at nearly all times of the year.

Evapotranspiration, for the most part, detracts from the freshwater resources of the world, but at the same time, our agriculture would be impossible without it. Further, under certain unusual conditions, there may be a reverse process, in which moisture passes directly from vapor to water at the surface. This is dew, and while it is normally a short-term phenomenon that adds little to plant supply, certain areas support life because of it. Regular advection fog blown as stratus clouds from the sea onto the arid slopes of the Andes Mountains in Peru provides a meager water supply within a narrow cloud zone that nourishes a sparse vegetative cover. This cover, in turn, is the basis of a limited pastoral economy which is impossible on the arid slopes both above and below the cloud zone. Dew drip from trees may be a noticeable factor in the water balance but is rare in occurrence.

[11]The figure of 4 inches is the average amount of water that can be stored within soils and within the reach of plants. Unusually coarse-textured soils would tend to reduce this figure, and heavy clays would raise it. In later studies, it was found that 4 inches of water retention was more appropriate for fine sandy soils and that a retention amount of 8 inches was more typical of loams.

Study Questions

1. a. Assuming an air temperature of 50°F and a specific humidity of 5 grams per kilogram, how many more grams of water vapor per cubic meter could be added to the air before saturation?
   b. Assuming the same conditions, how many pints of water would there be in a room having a capacity of 17 cubic millimeters, assuming 2.2 pints per kilogram? How many more could be added?
   c. How many pints of water could be added if the windows and doors were shut and the temperature of the room was raised to 72°F?

2. Explain why the air over the Sahara contains more water vapor than that over Scotland.

3. Explain why vapor pressure normally is lowest during the early afternoon and highest near sunset.

4. How many calories are required to evaporate (sublimate) 20 grams of snow that has a temperature of $-20°C$?

5. Assuming 13,000 Btu per ton of coal, and 4 k cal per Btu, how many tons of coal would be required to produce the equivalent heat energy of 1 inch of rain over 1 square mile? *Note*: assume 590 cal/cm^3 for latent heat of vaporization, 1 gram = 1 cu cm of water, 16.4 cu cm = 1 cu in.; 144 sq in. = 1 sq ft; 27,878,400 sq ft = 1 sq mi.

6. Why does the formation of rain require much more restricted conditions than those for cloud development? What are these conditions?

7. Assuming a normal lapse rate, at what elevation will freezing temperatures be found, assuming a sea level temperature of 80°F?

8. Explain conditional instability and its role in precipitation.

9. Give some reasons for variations in lapse rates. Why does mechanical instability often occur during the daytime in the low-latitude deserts?

10. Relate the generalized pattern of global precipitation to the three-dimensional distribution of net radiational balances.

11. Why does terrain have a greater effect on the distribution of snowfall than that of rainfall?

12. Explain the role of the hydrologic cycle as a subsystem governing the flow of energy in the global energy system.

13. Explain the connection between the 34 inches of average annual precipitation around the globe and the 19 percent of the total annual radiation transferred into the atmosphere by water.

14. What is the water balance equation? What are some methods of measuring different parts of the equation? Is it synonymous with the hydrologic cycle?

15. Discuss the role of each of the following in evapotranspiration:

   a. net radiation balance   d. latitude
   b. relative humidity   e. vegetation
   c. sensible heat   f. soil texture

CHAPTER 6
ATMOSPHERIC PRESSURE, THE GLOBAL AIR AND WATER CIRCULATION, AIR MASSES AND STORMS

This chapter deals with the movement of air and water at the earth surface and within the troposphere. The transportation of energy and water vapor from one place to another utilizes moving air as its major agent. Because horizontal air movement, or wind, is caused primarily by horizontal differences in atmospheric pressure, the former will be considered first. Despite many short-term fluctuations in pressure and the characteristics of air motion, regional consistencies may be observed about the earth surface and within the troposphere above. We shall note some of these patterns and indicate the changes that take place within them. The oceanic circulation system will also be briefly treated. While the global seas lack the powerful heat-exchange mechanisms of water vaporization and condensation, there is a direct transportation of both heat and mechanical energy from place to place within the great volume of ocean water. Further, water temperatures at various places within the surface oceanic circulation pattern have an important influence on the streams of air that flow over them.

Within the streams of horizontal airflow, or winds, at the surface, it is possible to distinguish several more or less homogeneous air masses whose properties are related largely to the broad areas from which they originated and to trace certain changes in these air masses as they pass into areas that are warmer, colder, wetter, or drier than their source regions. Finally, we shall consider the sudden paroxysms of the atmosphere—the hurricanes, tornadoes, and thunderstorms that periodically unleash their violence on man and have led him to learn the places where they are most frequent and how to live with them.

Atmospheric Pressure

Unlike fluctuations in temperature and precipitation, small differences in air pressure have little direct relationship to human affairs. Their indirect importance, however, is indicated by the presence of barometers in many homes and business establishments. Man has long realized that pressure differentials are associated with weather changes and that alterations in pressure often precede drastic changes in temperature, precipitation, and winds. Air, a highly compressible and fluid substance, is extremely sensitive to forces directed against it. As an integral part of the energy distribution system in the world, however, pressure patterns constitute significant indicators of various general weather and climatic patterns.

6-1 TERMINOLOGY AND MEASURE-MENT The atmospheric pressure at any given point on the earth surface and at any given time is a measurement of the total mass weight, or gravitational force, of a column of air extending upward from the surface to the outer limits of the earth gravitational attraction. At sea level, the mean atmospheric pressure is approximately 14.7 pounds per square inch, which means that there are 14.7 pounds of air above a single square inch of sea-level surface. Small differences in pressure are important in weather and climatic patterns; therefore, a finer scale than pounds per square inch is used. For many years, pressure was expressed in inches or millimeters of mercury, or the height to which mercury rises in the tube of a mercurial barometer.

Mean sea-level pressure, or the equivalent of 14.7 pounds per square inch, is 29.92 inches, or 760 millimeters of mercury. More recently, in the interest of international standardization, the U.S. Weather Bureau adopted the millibar as the basic pressure unit on its weather maps. A millibar is roughly equal to the weight of 1 g/cm², or 0.014 lb/in.². For a more precise definition, see footnote 1 in Chapter 5. Mean sea-level pressure expressed in millibars is 1013.2. Fig-

ure 6-1 shows a simple conversion scale which relates millibars to inches and millimeters of mercury. Most home barometers are calibrated in inches. Two types of barometers, the mercurial and aneroid barometers, are used to measure atmospheric pressure. Their operating principles are demonstrated in Figure 6-2. While the mercurial barometer is simpler and generally more accurate, the aneroid is much more portable. It is an important instrument in recording pressure variations during balloon and rocket flights, and can be used in surface surveying to record selected elevations where extreme precision is not required.

Figure 6-1 Graphic conversion scale of atmospheric pressure; millibars to inches and millimeters of mercury.

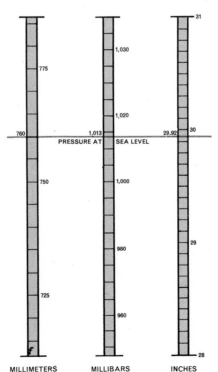

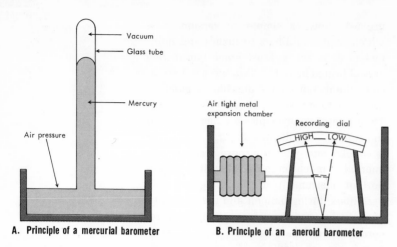

A. Principle of a mercurial barometer   B. Principle of an aneroid barometer

Figure 6-2   Principle of the mercurial and aneroid barometers.

Pressure patterns on maps usually are shown by *isobars,* or lines connecting points of equal barometric pressure. It is customary for pressure maps to have all readings reduced to sea level in order to eliminate the factor of pressure decrease with elevation. Pressure maps are similar to contour maps. High points and low points are found, having different shapes and separated by different slopes or gradations of pressure. Low atmospheric pressure areas are known as *cyclones, troughs,* or simply *lows,* whereas the areas of high pressure are termed *anticyclones, ridges,* or *highs.* The direction of pressure change always is at right angles to the isobars, and the pressure slope is determined by the spacing of the isobars. A steep pressure gradient is indicated by closely spaced isobars. When presenting pressure differences at elevations that are several thousand feet above the surface, it is customary for upper-air charts to show the contours (lines of equal elevation above sea level) of an isobaric surface, or the surface upon which all points have the same air pressure. The 700-, 500-, 300-, and 100-millibar

surfaces, which lie roughly at about 10,000, 17,000, 25,000, and 50,000 feet, respectively, above sea level are those most frequently used (see Figures 6-5 and 6-6). The upper-air charts are similar in appearance to surface weather maps that use isobars to indicate horizontal variations in pressure. Both types of pressure maps reveal the configuration of horizontal pressure differentials.

6-2   CAUSES FOR CHANGES IN ATMOS-PHERIC PRESSURE   Changes in atmospheric pressure result from changes in mass within the column of air that lies above a particular place. These changes involve movement of mass and hence require the transfer of energy into or out of the column of air. The causes for alteration in air pressure can be grouped into two types: *thermal* and *dynamic.* The former involves a transfer of heat energy, the latter mass motion or mechanical energy.

Thermal causes for pressure changes. The thermal causes for pressure changes are related to the lateral transfer of mass resulting directly from increases or decreases in the content of sensible heat within the air column. As we learned earlier (see Section 4-10), when air above a heated surface has sensible heat imparted to it, it expands. Some of the air mass is transferred laterally, and thus the column contains less mass than it did previously; therefore, its weight or pressure drops. Conversely, when air loses heat energy (as by radiation), it shrinks or contracts, and because air moves into the column, the total mass becomes greater and the pressure rises. Sensible heat may be added to the air column by direct conduction, by the absorption of direct or indirect radiation, or by the release of latent heat following condensation. The first two take place mainly near the earth surface. Thermal "lows" at the ground surface are likely to be extremely shallow, however, and are rarely more than a few thousand feet in depth. The reason for this is that convection, turbulence, and eddy diffusion are confined largely to lower levels because their generation requires interaction between the air and the ground surface. Excessive cooling because of contact with a cold, radiating ground surface, as in snow-covered polar areas, may cause surface air to lose heat and to become more dense. Thermal high-pressure areas at the surface, like thermal lows, tend to be shallow, since cold air is stable, hugs the ground, and is much less likely to mix vertically than warm air. Measurements of air pressure over Greenland and Antarctica, for example, indicate sharp drops in pressure and increases in temperature within 100 feet of the surface during calm, cold periods.

Air is heated several thousand feet above the surface mainly by the release of latent heat of condensation and by the absorption of terrestrial radiation. The former may be at extremely high elevations in the tropics or much nearer the surface at higher latitudes. The latter decreases rapidly with elevation. Air cooling through the loss of outbound radiation is an important factor in removing heat anywhere in the troposphere, but it is especially effective immediately above the cloud zone. The upper troposphere is an effective *heat sink,* so called because it drains away by radiation the heat energy that is brought to it from the earth below. In Section 4-5 it was noted that the upper air consistently emits more radiation than it receives. This balances the net positive radiation that prevails at the surface. The outgoing loss of radiation from the upper air is much greater at low latitudes than it is at high latitudes because more heat energy is transferred upward from the surface in the tropics. The greater loss of energy at high

elevations but low latitudes means that pressure in the upper troposphere tends to be highest near the equator and to decrease toward the pole (see Figures 6-5 and 6-6), especially during the winter season, when the greatest latitudinal energy contrasts are observed.

Dynamic causes for pressure changes. The dynamic causes for horizontal pressure differences are complex, and we shall not concern ourselves with the details except to note that the major ones involve such factors as earth rotation, friction and angular momentum. They are by far the most important causes for the surface patterns of pressure differences. Evidence of this is the large number of pressure centers at the surface that obviously are unrelated to temperature conditions. For example, one of the lowest pressure centers on earth is located in winter over the Aleutian Islands in the North Pacific. Few people would consider visiting these bleak islands during the winter for swimming or sunbathing!

The pressure patterns at the surface are exceedingly complex because of the many factors that influence dynamic pressure changes, as well as the thermal factor, shallow though its effect might be. The local pressure changes that take place from day to day do not appear on the maps of monthly averages and do not especially concern us in our global survey, although they are of great concern to the forecaster of daily weather changes. Keeping in mind that the seasonal maps of pressure changes mask an infinite variety of short-term changes, let us now turn our attention to the major patterns of pressure differences that exist around the world during the two seasonal extremes of the year. We shall use July and January as our index months because they represent the seasonal extremes in energy accumulations.

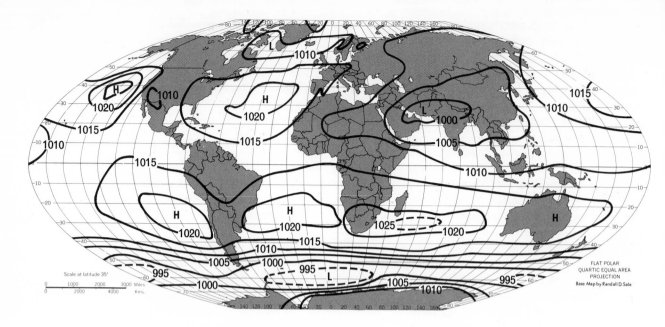

6-3 GLOBAL SURFACE PRESSURE PATTERNS FOR JULY

Figure 6-3 shows the mean pressure conditions at the surface for July and indicates the patterns for the northern hemisphere summer season and the southern hemisphere winter. All isobars are related to sea-level pressures. The major regional pressure zones are discussed in the following subsections:

Equatorial low-pressure trough. The axis of a low-pressure trough in the low latitudes is indicated in Figure 6-3. It lies entirely north of the equator during July, because this is the summer hemisphere. The trough is by no means uniform in width, depth, or position. It is deepest and most pronounced where it lies farthest from the equator, such as over an area extending from the Persian Gulf eastward to northwestern India, and also over northern Mexico and the southwestern United States. It is least distinct over the western and central Pacific. The extreme heating that the dry, subtropical land areas

in the northern hemisphere undergo at this time of year is influential in decreasing pressures at the surface. As was indicated earlier, however, such continental, thermal lows are relatively shallow and have an influence only on surface wind patterns. Temperatures within the equatorial low-pressure trough, such as those encountered in the Amazon Basin or over the Pacific often are not as high as over many of the land areas that lie north of them and that have much higher pressures. Convergence of air streams moving toward each other from opposite hemispheres, and the lifting and expansion of air associated with convectional precipitation in the tropics, can result in dramatic lowering of surface pressures.

The equatorial trough should not be considered as permanently structured as shown in Figure 6-3 at this season. The map was made from mean data which are not nearly so complete as for some of the mid-latitudes. Weather stations set up in the low latitudes during World War II revealed that considerable daily pressure changes occur within the trough and that at times the trough pinches out entirely, being replaced by separate low-pressure cells, separated by low ridges or waves of higher pressure.

Subtropical anticyclones or highs. Two large cells of high pressure, elongated slightly east-west, are located in the northern hemisphere during July, with their centers in the eastern third of both the Atlantic and Pacific, roughly between lat 30° and 40° N. In the southern hemisphere, an elongated ridge extends around the world, centered approximately at lat 25° S. This high-pressure ridge is not of uniform height, and low crests occur on it, most of which are located over the ocean areas. Central Australia and the Kalahari Desert of South Africa have secondary centers of high pressure that are more pronounced than the general belt, although they are not indicated on the map. The position of the subtropical highs, which at this season is somewhat farther poleward north of the equator than south of it, is related to the seasonal shifting of the pressure belts, such highs being farthest north during the northern hemisphere summer.

The subtropical highs are developed largely by dynamic, rather than thermal, causes. Subsiding air from high levels is largely responsible for their greater pressures. Variations in their strength and form are among the more important features of the global energy balance, since the air that diverges from them comprises a large part of the entire air circulation system on the surface. The number of high-pressure cells within the subtropical zone is closely related to the demands for energy flow. The summer hemisphere shows only the large oceanic high-pressure centers. The winter hemisphere has a greater number of anticyclonic highs and the most active day-by-day pressure changes along the general subtropical pressure zone. For the most part, the day-by-day changes are produced by wavelike undulations that ripple along the flanks of the pressure cells, mainly from east to west (*easterly waves*) on the equatorward sides of the cells and from west to east (*westerly waves*) on their poleward flanks.

The Antarctic low-pressure trough. One of the deepest and most persistent low-pressure troughs in the world borders the Antarctic Continent, centering roughly at lat 55° S during July. It is present, however, at all seasons, shifting slightly southward in January. This low-pressure trough marks the zone of energy transfer between warm and cold air. It is a strong frontal zone, and the dynamics of air interchange, implemented by the upward displacement of the warmer air, the high condensation, great angular momentum, rotational deflection, and other factors produce an almost continuous succession of deep cyclones that move around the world through this zone. It is low in pressure statistically because individual cyclonic storms follow one another throughout the zone in rapid succession, particularly in July, or the winter season, when the contrast between the cold air of Antarctica and the warm waters of the South Pacific and other oceans is greatest. Anticyclones, or high-pressure cells with accompanying descending, diverging air, sometimes occupy positions within the trough and at rare intervals may persist for several days. The predominant low-pressure centers that spin through the zone with steep pressure gradients and strong winds justifiably give rise to the various epithets of mariners who refer to these latitudes as the "Roaring Forties," "Furious Fifties," and "Slashing Sixties."

The subpolar lows of the northern hemisphere. One might expect to find a strong frontal zone between lat 60 and 70° N, with accompanying cyclonic lows, similar to conditions in the Antarctic low-pressure trough. The heating of the continents, however, even at high latitudes in the northern hemisphere, restricts significant chilling to the ice cover of the central Arctic Ocean and the Greenland icecap. This cooling is not sufficient to generate cold air masses that can move far out of their source areas and produce dynamic low-pressure centers following interaction with warmer winds from the south. Thus, since the frontal activity between cold and warm air in these latitudes is weak during the summer season, strong subpolar lows do not develop. The weak low-pressure zone bordering southern Greenland and extending across the North Atlantic to Norway is related to shallow summer cyclonic depressions that pass eastward through the region.

The polar highs. Until recently, the lack of suitable weather data for most of the polar regions had been a serious handicap in generalizing the pressure conditions at high latitudes. This is why the global maps of air pressure (Figures 6-3 and 6-4) do not include the Arctic and Antarctic regions. Data from Antarctica, collected during the International Geophysical Year program, now suggest that, contrary to earlier belief, the continent is not at the heart of a pronounced high-pressure center that is thermally induced and present throughout the year. Both high- and low-pressure centers are present, which change in intensity and shape seasonally, with the low (located near the continental margin) dominant during most of the year but less

strongly developed during the winter months. A high-pressure ridge tends to occupy the highest portion of the continent but is extremely shallow and not well developed. This appears to be similar to the shallow polar high that is indicated in central Greenland.

6-4 GLOBAL SURFACE PRESSURE PATTERNS FOR JANUARY The pressure patterns for January are shown in Figure 6-4. The general zones of pressure that tend to develop at this season are the same as those indicated for July, including the equatorial trough, the subtropical highs, the subpolar lows, and the polar highs. Considerable changes take place in these pressure zones, however, during the course of the 6 months between July and January. There is a shift southward in the general location of the pressure centers. The equatorial low-pressure trough in January, for example, lies just north of the equator over most of the oceans but bends far southward and deepens in the interior of South America and Africa, following the areas of maximum heating. These are typically shallow thermal lows, similar to those that developed over the hot, continental, subtropical deserts north of the equator in July.

The subtropical highs in the southern hemisphere still are observed over the oceans, but low gaps appear between them over the continents. The subtropical high-pressure zone at this season is much less interrupted than that in the northern hemisphere during July. The January subtropical highs north of the equator broaden in an east-west direction to include some of the adjacent continental surfaces. Although not

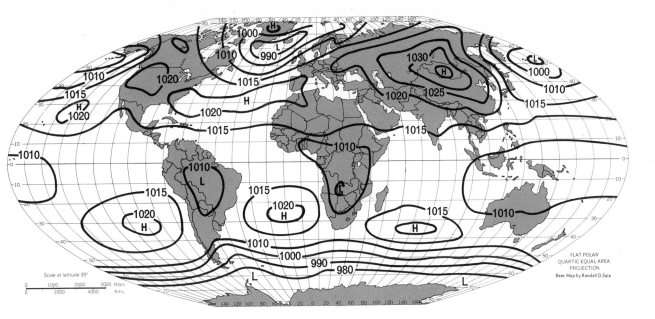

Scale at latitude 35°

1000 2000 3000 Miles
2000 4000 Kms.

FLAT POLAR
QUARTIC EQUAL AREA
PROJECTION
Base Map by Randall D. Sale

clearly indicated on the map, localized continental high-pressure centers in the subtropics north of the equator tend to develop to a greater degree over highlands than over low plains in the winter season. Examples include the high plains of the southwestern United States, the Sahara Desert in northern Africa, and the Tibetan and Yunnan Plateaus of Asia. Another characteristic of the subtropical high-pressure centers, also not clearly indicated on the mean pressure maps, is that there are more individual high-pressure cells within this general latitudinal zone in winter than in summer. This can be noted clearly in daily sequences of weather maps. The Pacific subtropical cell north of the equator tends to split into two distinct cells, one centered northeast of Hawaii and the other northeast of the Mariana Islands.

The subpolar low-pressure trough bordering Antarctica remains essentially unchanged, except for a slight shift southward. In the northern hemisphere, two strong subpolar low-pressure areas develop in the winter season: one over the North Pacific, termed the *Aleutian low,* and the other over the North Atlantic, termed the *Iceland low.* These are typical areas featuring a high frequency of deep, low-pressure centers that follow each other in rapid succession along the fronts established between the warm water of the northern oceans and the extremely cold air masses from off the northern continents. These two winter lows are similar to the Antarctic low-pressure trough that prevails throughout the year.

A huge high-pressure area is present over eastern Asia during January, roughly centering on the vicinity of Lake Baikal in the eastern Soviet Union. This is believed to be influenced mainly by the rapid radiational cooling, indicated by the extremely low surface temperatures that characterize this area during the winter season. Pressure conditions prevailing directly above in the upper troposphere support the hypothesis of a thermal origin for this strong anticyclonic center. Regardless of its origin, it is relatively shallow compared with the subtropical highs. A somewhat weaker high-pressure center also develops over northwestern Canada during this season. Although supporting data are few, it is believed that a low ridge of moderately high pressure extends over the North Pole, connecting the Asiatic and Canadian highs during the winter.

6-5 PRESSURE CONDITIONS ABOVE THE EARTH SURFACE Atmospheric pressure everywhere decreases with increasing elevation. This is caused by the compressibility of air and the cumulative effect of the weight of overlying air. This cumulative effect is responsible for the decrease in the *rate* of pressure drop with greater elevation. The vertical pressure gradient is shown graphically along with the thermal gradient in Figure 4-18. Note that at the top of the stratosphere, or about 30 miles above the surface, the pressure averages only 1 millibar, or about one thousandth of that at sea level. The 500-millibar level is found at about 18,000 feet, which means that half the total mass of the atmosphere lies within the lowest $3\frac{1}{2}$ miles. The cumulative weight effect signifies rapid changes near the surface. This has many implications with respect to man's requirements of oxygen in metabolic rates as he attempts to perform normal activities at various elevations. There appears to be a definite upper limit to human habitation based on physiological reasons at an elevation of about 17,000 feet. Beyond that limit man can, through special acclimatization, exist and even perform heavy manual labor for short periods without recourse to supplementary oxygen, but sooner or later he will be forced to replenish a systemic deficiency. Above 17,000 feet, the air does not contain sufficient oxygen to maintain the normal demands of living for an indefinite period.

Horizontal pressure differences occur at all levels in the troposphere and even well within the stratosphere, but the general arrangement of such differences is quite unlike those at the surface. Figures 6-5 and 6-6 show pressure conditions at sea level and at two upper levels for January and July, 1964. Analyses of these average conditions for two particular months are highly revealing. First, the sea-level pressure in Figure 6-5 should be compared with that of Figure 6-4, which is a long-term average. Minor differences between the two can be found, but the Aleutian and Iceland lows, the eastern Siberian high, and the subtropical (Azores) high in the North Atlantic are clearly indicated on both charts. The major difference consists of a pronounced high-pressure cell in central Europe during January, 1964 (Figure 6-5), indicating a winter in central Europe that probably was drier and colder than normal. The two July surface maps are almost identical, except that the shallow Iceland low of Figure 6-3 has enlarged to encompass the entire north polar area in 1964. This is not surprising, since long-term records for the Arctic were not available to fill in details on Figure 6-3.

The 500- and 300-millibar charts for January, 1964, indicate a pronounced change from the surface patterns. First, the separate high-pressure and low-pressure cells that characterized the surface pattern have disappeared, except for a slight tendency for shallow ones to develop at low latitudes (note the one over India). Instead, the basic upper-air winter pattern shows a pronounced vortex of low pressure near the pole with troughs extending out from it. Pressure tends to increase toward the low latitudes. At even higher elevations, such as at 100 millibars, well within the stratosphere, the vortex for this month shows up as deeper, smaller, and more symmetrical than at the 300-millibar level. The arrows of airflow indicate that much

Figure 6-5 Average pressure conditions in the northern hemisphere at sea level, 17,000 feet, and 30,000 feet for January, 1964.   Note that at the surface the pressure is broken into distinct cells, whereas at high elevations the pattern consists of a high-pressure ridge near the equator and a low-pressure vortex near the pole. Winds, shown by arrows, cross the isobars at an angle near the surface but parallel the isobars in the upper air. Pressure changes in the upper air consist largely of waves and troughs that ripple around the polar vortex. [Adapted from *Weather and Climate* by R. C. Sutcliffe, F.R.S., by permission of W. W. Norton and Co., Inc. © by R. C. Sutcliffe, 1966]

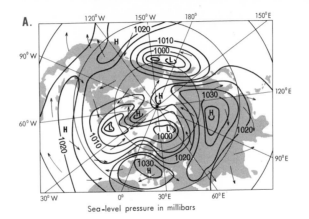

Sea-level pressure in millibars

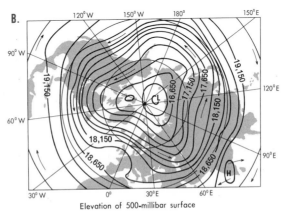

Elevation of 500-millibar surface

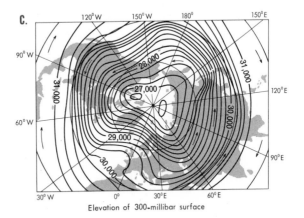

Elevation of 300-millibar surface

different wind directions prevail in the upper troposphere than at the surface. In the latter, the winds generally cross the isobars at an angle, whereas at the upper levels, they consistently parallel the isobaric slope and are westerly almost throughout the entire hemisphere during this northern-hemisphere winter season.

The 500- and 300-millibar charts for July, shown in Figure 6-6, indicate somewhat the same tendencies. A polar vortex again is present and becomes deeper near the top of the tropopause (300 millibars near the pole), and the winds tend to blow parallel to the isobaric surfaces. There are distinct differences, however, in the upper-air pressure patterns between winter and summer. First, the individual shallow cells of high pressure that were barely noticeable in the winter at low latitudes on the 500-millibar surface are clearly indicated during the summer on both the 500- and 300-millibar charts between 20 and 30°. A pronounced reversal of the normal wind direction, or a flow from east to west, appears on the equatorward side of these subtropical cells at both the 500- and 300-millibar levels. Second, the general pressure gradient between the equatorial regions and the pole is not nearly so great or so steep

as during the winter season. The latitudinal variation in height of the 300-millibar surface in January was about 4,500 feet; in July, it

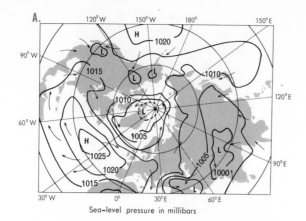

A.

Sea-level pressure in millibars

Figure 6-6 Average pressure conditions in the northern hemisphere at sea level, 17,000 feet, and 30,000 feet for July, 1964. The seasonal differences in global pressure conditions are seen by comparing these diagrams with those in Figure 6-5. In winter, the pressure gradients at all levels are steeper and the pressure differentials greater, reflecting the much greater latitudinal difference in energy balances during the winter season. [Adapted from *Weather and Climate* by R. C. Sutcliffe, F.R.S., by permission of W. W. Norton and Co., Inc. © by R. C. Sutcliffe, 1966]

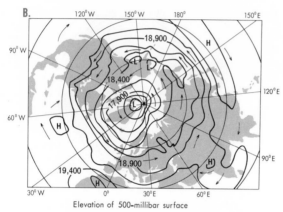

B.

Elevation of 500-millibar surface

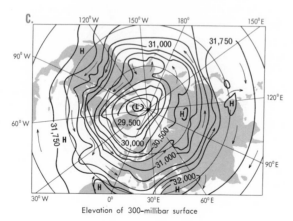

C.

Elevation of 300-millibar surface

collapse of the polar vortex, a much more irregular isobaric surface throughout the hemisphere, and generally weak pressure gradients, indicating little latitudinal temperature differences.

It was noted in Chapter 4 that the greatest contrast in surface temperatures and in the net radiation balance between the low and high latitudes takes place during the winter season. The strong surpluses of energy over the low-latitude oceans contrast greatly with the net deficiencies of radiation poleward of 40° at the surface in winter. Despite the manner in which these energy imbalances are adjusted by energy flow both horizontally and vertically, the strong winter latitudinal contrast continues to mark the pressure patterns in the upper troposphere. The relationship between pressure and winds both at and above the surface will be treated in the next division of this chapter.

Daily upper-air charts reveal the following fluctuations that are not indicated on the mean monthly charts (Figures 6-5 and 6-6):

1. Major pressure troughs and ridges move slowly eastward with the polar vortex as their center; the troughs taper equatorward, while the ridges taper poleward.

2. Minor ripples or waves also move along with the major waves.

was only slightly above 3,000 feet. If the 100-millibar chart had been included for July, it would have shown an almost complete

3. The pressure gradient varies at any one place and in a roughly cyclical pattern.

4. The most changeable daily pressure conditions occur during the winter season and within a zone roughly between 35° and 45° N.

Winds and the Global Air Circulation System

Wind is air in horizontal motion. The principal cause of wind is the equalization of horizontal differences in air pressure. Gravity seeks to create an equalization of air pressure at a given elevation and will cause lateral air motion toward this end. Since wind occurs to relieve horizontal differences in pressure, the flow is horizontal. It is true, of course, that upward and downward air movements take place, but in comparison to the total amount of horizontal flow, these vertical components are small indeed. There is appreciable upward flow only where quantities of air are mechanically displaced upward, such as with rapid condensation and along the boundaries that separate air masses of contrasting temperatures. Upward or downward air movements approach the strength of horizontal winds only under highly localized conditions, such as in a severe thunderstorm. One of the most common misconceptions about global air circulation is that strong, steady trade winds descend at a rapid rate from the upper air, rush toward the equator, and rise vertically as great upward flows of air in the rainy tropics. The total vertical displacement is not comparable to horizontal flow anywhere on earth. This does not mean that there are no strong winds in the upper air. On the contrary, winds consistently are stronger at the upper levels than they are at the surface. They are, however, horizontal winds, set in motion by horizontal pressure gradients. The term *wind* customarily refers to horizontal air movement, whereas upward or downward movements are called *air currents*.

Since wind is air movement in the process of equalizing horizontal differences in air pressure, it would seem likely that air would move directly down a pressure slope at a speed proportional to the steepness of the pressure gradient, in much the same way as water flows down a slope on the earth surface. Observations indicate, however, that such direct flow is exceedingly rare. Factors other than differences in air pressure influence the speed and direction of winds. Let us now consider the important variables in wind patterns.

6-6 THE APPARENT DEFLECTIVE FORCE OF THE EARTH'S ROTATION A French scientist, Foucault, discovered long ago that, if a heavy weight is suspended by a wire to form a long pendulum and is set into motion along a straight line, the arc described by the pendulum will, after a few hours, no longer follow the straight line but instead will veer away from it. The earth, because of its rotation on its axis, turns beneath the swing of the pendulum. This apparent deflective effect later was found to influence all freely moving objects on the earth surface. It is absent at the equator and reaches a maximum at the poles, and the apparent deflection is *toward the right of the direction of motion in the northern hemisphere and toward the left of the direction of motion in the southern hemisphere.* At times, this apparent deflective effect has been somewhat inappropriately termed the *Coriolis force.* It is not a true force, since it does not involve any gain or loss of energy (see Figure 6-7). In fact, if an observer were situated in space beyond the earth, the trajectory of movement would not appear to be a deflection at all, but a con-

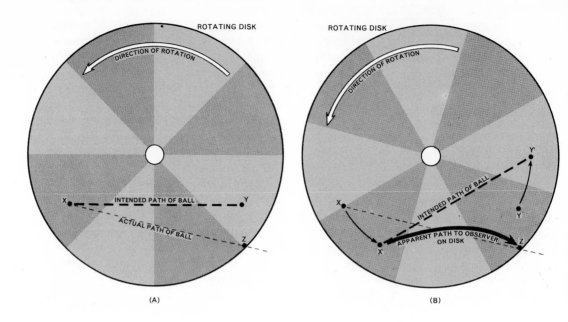

(A)   (B)

tinuance in the direction toward which it had originally been set in motion. It is only from a vantage point on the rotating earth that the direction appears to be deflected.

A simple demonstration of the Coriolis effect can be made by placing a paper disk on the turntable of a phonograph. If a pencil line is drawn on the rotating disk by moving one's hand in a straight line, the resulting line will always be a curve. Furthermore, the arc of the curve will have a smaller radius the nearer the line is drawn to the center of rotation. The deflection will be to the left of the direction of hand motion if the table is rotating clockwise and to the right with a counterclockwise rotation. The direction of hand motion makes no difference in the amount or the direction of deflection.

If we consider the hand moving the pencil on the turntable as comparable to the motion of air down a pressure gradient and let the counterclockwise rotation of the turntable represent the rotation of the earth north of the equator (which is counterclockwise), a

useful analogy can be presented. All moving air, regardless of direction, is deflected toward the right in the northern hemisphere and toward the left in the southern hemisphere. The deflection is zero at the equator and increases with both latitude and velocity.

6-7 GEOSTROPHIC WINDS   Although the Coriolis effect may be only an apparent deflection, it produces some significant effects on the observed behavior of wind movement. In terms of observed wind direction, air does *not* flow down a pressure slope but curves away from it. If the deflection were to continue, in time it would appear to circle back toward its origin. Opposing this return motion is, of course, the pressure gradient (see Figure 6-8). A balance between (1) the deflective component of the earth's rotation, which tends to turn wind back toward its high-pressure source, and (2) the pressure gradient, which tends to move air directly away from the high-pressure source, produces a resultant direction of air movement that *parallels*

Figure 6-7 A graphic analogy to the Coriolis effect. Imagine a rotating disk, like a merry-go-round, large enough to carry two persons, one of whom (x) throws a ball to the other (y). Between A and B, x moves to x′ and y moves to y′. The ball itself follows a straight line relative to the surface of the disk, leaving it at z by the time the disk has moved to B. To both x and y, however, the ball has not followed its intended path and has described an apparent arc in reaching z.

the isobars, with little exchange of air taking place between the high- and low-pressure areas. Such an airflow, representing a balance between the pressure gradient and the Coriolis effect, is known as a *geostrophic wind.* It is the characteristic wind-flow pattern well above the surface and where frictional influences are negligible. Near the equator, at the surface or above it, geostrophic influences are weak because of the low Coriolis deflection.

The velocity of geostrophic winds depends on the following three variables:

1. *The steepness of the pressure gradient.* Other conditions remaining constant, wind velocities increase as the isobars become closer together or as the steepness of the pressure gradient increases.

2. *Latitude.* A given pressure gradient produces decreasing wind velocities with increasing latitude. The most rapid change takes place in low latitudes. The Coriolis effect increases with latitude. Thus, at higher latitudes, the wind curves back against the pressure gradient in a tighter circle, weakening the effect on velocity produced by the pressure gradient. The quantitative expression of this factor decreases somewhat near the surface because of friction.

3. *Density.* Other conditions remaining constant, the velocity of geostrophic winds decreases with increasing air density. Differ-

ences in temperature and atmospheric pressure, which are variables in air density, thus affect the speed of winds. One of the main results of the density factor is an increase in wind velocities at higher elevations and with equivalent horizontal pressure gradients.

It was noted in Section 6-5 that at high elevations the horizontal pressure pattern consists of a high-pressure belt at low latitudes and a low-pressure area near the poles. The isobars encircle the globe roughly in an east-west direction. The tendency to develop geostrophic winds in the upper air thus leads to a great dominance of westerly winds except at low latitudes, with much less latitudinal exchange of air than at the surface. The wavelike irregularities on the upper-air pressure charts, however, indicate that some such transfer across parallels takes place. Wind as an effective means of transporting energy directly from the lower to higher latitudes,

Figure 6-8 The cause of geostrophic winds. In *A,* the pressure factor directs air at right angles to the isobars, or straight down the pressure gradient. The Coriolis effect (*B*), however, leads to a curvature toward the right (northern hemisphere), which eventually leads back against the pressure gradient. The effect of these two directions of motion, shown in *C,* is an air movement paralleling the isobars, termed a *geostrophic* wind.

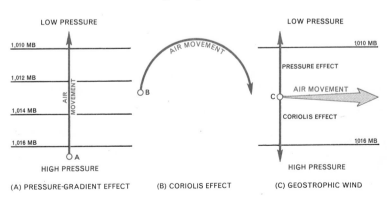

then, is much more important near the earth surface than in the upper air. Exceptions will be noted later, in the section on the behavior of the jet streams (see Section 6-12).

6-8 FRICTION AND ITS EFFECTS ON WIND Earlier it was noted that friction tends to decrease air velocities. Air that moves over the earth surface has some of its mechanical energy transformed into other forms of energy because of friction, and thus its velocity decreases. Friction with the surface, since it decreases wind velocity, also tends to decrease the amount of Coriolis deflection, which is proportional to straight-line velocity. This decrease in Coriolis deflection, in turn, tends to alter the resultant wind direction with respect to the pressure gradient (see Figure 6-9). The wind direction no longer parallels the isobars, as in geostrophic flow, but crosses them somewhat in the direction of the pressure gradient. The angle of intersection with the isobars is roughly proportional to the amount of friction encountered at the earth surface.

Figure 6-9 The effect of friction on wind deflection. Since friction tends to slow down velocity, it decreases the Coriolis effect and favors the influence of the pressure gradient.

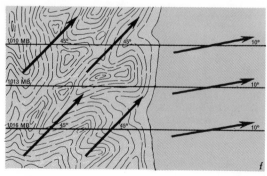

(A) ROUGH LAND SURFACE   (B) WATER SURFACE

Surface winds usually cross isobars at about a 10° angle over oceans and may reach angles of up to 45° over rough land surfaces. Although the loss of energy by friction takes place only at the surface, the "braking" effect may be transmitted upward into the higher levels of horizontal wind flow. Usually, however, it is contained within 1,500 to 2,000 feet of the surface.

6-9 MOVEMENT OF WINDS AROUND PRESSURE CENTERS The net effect of the pressure gradient, the Coriolis deflection, and friction induces a rotational motion in air diverging from a high-pressure center, or anticyclone, and also in air converging toward a low, or cyclone. In the northern hemisphere, the deflection toward the right produces an outward, clockwise rotation for winds around anticyclones and an inward, counterclockwise rotation for those circulating around cyclones (see Figure 6-10). In the southern hemisphere, since the deflection is toward the left, anticyclones rotate counterclockwise and cyclones clockwise. Each high- and low-pressure area on earth thus can be thought of as generating a spiraling wind pattern about itself. The dominant global highs and lows are not only centers of pressure but also centers of air rotation. The airflow around some of the great permanent anticyclonic high-pressure centers forms some of our major planetary winds at the surface.

If the great pressure centers are elongated in an east-west direction, as sometimes happens, a zonal (east-west or west-east) flow tends to predominate. However, when several separate cells of high or low pressure develop, there is greater airflow across the parallels, both east and west of the pressure centers. The traditional concept of fairly uniform latitudinal wind belts encircling the globe is generally not valid at the earth surface, although it appears to be generally true with respect to the airflow at high elevations. As was indicated earlier, contrasts in temperature conditions between oceans and continental areas tend to be influential in the development of the more or less permanent cellular pressure areas. In general, the northern hemisphere, with its greater area of land, has more interrupted latitudinal pressure belts than the southern hemisphere, and consequently a greater amount of meridional (north-south or south-north) airflow. In relation to the global energy balance, this indicates that the differential between high and low latitudes with respect to the net accumulation of solar energy is greater in the northern hemisphere than in the southern hemisphere; hence, there is a greater need for the transfer of energy across the parallels. Assisting in this is the greater development of cellular pressure centers north of the equator for equivalent seasons. For similar reasons, there is a greater number of cellular pressure areas in winter than in summer in both hemispheres. This is because latitudinal energy differentials (as indicated by temperature contrasts) are greater during the winter season, and thus the demand for meridional airflow across the parallels is greater. It should be recognized, however, that changes in the amount and intensity of pressure centers occur throughout the year and are largely responsible for the day-to-day fluctuations in weather conditions.

6-10 GENERALIZED AIRFLOW PATTERN AT THE EARTH SURFACE A generalized pattern of airflow at the surface for January and July is presented in Figures 6-11 and 6-12. The most noticeable feature of both maps is that near the surface, the airflow consists of a series of pressure cells around which the air streams circulate. There is some air movement across the isobars because of friction,

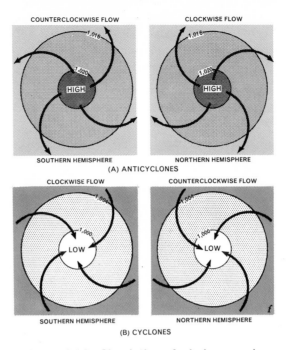

Figure 6-10 Circulation of winds around cyclones and anticyclones in the northern and southern hemispheres.

however, that produces divergence around the high-pressure cells and convergence toward the low-pressure centers. Especially noticeable as source areas for the surface wind patterns are the oceanic subtropical highs or anticyclones, each of which forms a center for diverging and circulating air. The steadiest winds, indicated by the longer wind arrows, are found in the eastern and equatorward-facing quadrants of these oceanic highs. These steady winds with easterly components, which blow day after day throughout the year, are known as *trade winds*. The trade wind zone generally is an area of much sunshine and relatively little precipitation, mainly because the air has recently subsided from upper levels and has been heated adiabatically. Upper-air tem-

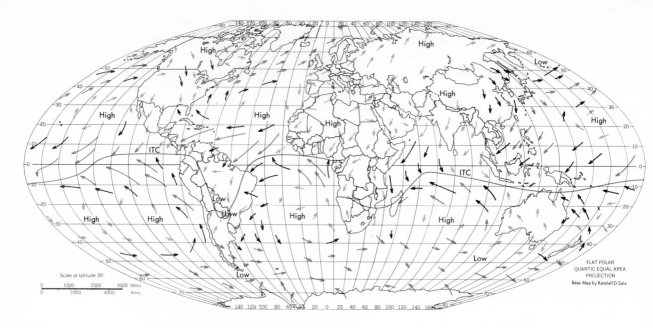

Figure 6-11 Idealized airflow pattern for January. The general arrangement of the surface winds into great whorls or eddies is illustrated. Note that they turn clockwise around the high-pressure centers north of the equator and counterclockwise south of it. Winds near their centers are light and irregular. The steadiest winds are always in the trade wind zones, but these are not always the strongest. The heavy gale winds north of Antarctica are not steady, and wind directions there shift rapidly, owing to the influence of cyclonic eddies that pass in succession through this stormy part of the world. Note also that strong gale winds are rare over land areas, where friction decreases their velocity. [After Garbell and Gentilli]

Force of wind

→ Light

→ Moderate

→ Heavy

Steadiness of wind is proportional to length of arrow

perature inversions resulting from subsidence are normally present. The amount of water vapor pressure at the surface depends on how long the air has moved over the water surface since its descent from high in the troposphere. Evaporation is exceedingly rapid because of the low relative humidity when it first reaches the surface. Ship captains in the trade wind zone during the days of sailing vessels knew these winds well, often detouring far from their course in order to benefit from their constancy. Trade winds from the subtropical oceanic highs of opposite hemispheres tend to converge along a line or zone known as the *intertropical convergence zone,* or *ITC.* It is marked on Figures 6-11 and 6-12. This convergence zone will be discussed in Section 6-14.

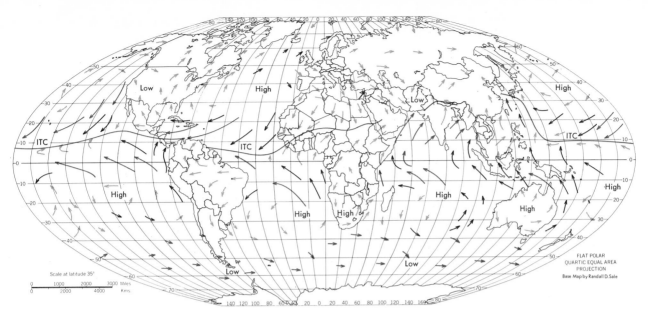

Figure 6-12 Idealized surface airflow pattern for July. Several differences can be noted between this map and that in Figure 6-11. The monsoon circulation in the Indian Ocean has reversed direction, the southern hemisphere trades have crossed the equator at several points, and the intertropical convergence zone (ITC) has shifted into the northern hemisphere over the continents. The gale winds of the North Pacific and North Atlantic have subsided, and in general the southern hemisphere experiences a decrease in wind velocities. [After Garbell and Gentilli]

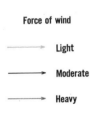

Force of wind

———→ Light

———→ Moderate

———→ Heavy

Steadiness of wind is proportional to length of arrow

The poleward side of the subtropical oceanic highs has a predominance of winds with westerly[1] components, often referred to as *prevailing westerlies*. The seasonal wind maps (Figures 6-11 and 6-12) at the surface indicate that the surface westerlies are by no means as steady as the trades but are likely to be much stronger, particularly in the oceanic areas between 40° and 60° and during the winter season. These strong but variable westerly winds frequently are associated with the small migrating pressure cells or eddies that mark the contact between cold and warm air. Each small cell, consisting of both migrating high-pressure, divergent cells and low-pressure, convergent ones, has its own surface circulation system (see Figure 6-10) that does not appear on the average monthly airflow patterns. The short, wide

[1] In describing wind directions, a *westerly* wind is one that blows from west to east. Wind directions normally are given according to the direction from which they come. Occasionally, however, the suffix *ward* is used, such as a *northward* wind. In this case, it signifies the direction toward which the wind is blowing.

arrows on Figures 6-11 and 6-12 over the southern oceans off Antarctica throughout the year and in the North Atlantic and North Pacific Oceans during the winter indicate the conditions that meant trouble to the masters of sailing vessels in these waters: first, the winds of gale force and mountainous sea waves that are generated, and second, the sudden shifts in wind direction. The prevailing westerly zone has winds from every compass direction. When the migrating cells subside, however, and zonal flow predominates, as is more common during the summer months, southwesterly winds in the northern hemisphere and northwesterly winds south of the equator become much more frequent, as is indicated by the increased arrow lengths on Figure 6-12 during this season.

The high-latitude circulation, poleward of the belt of prevailing westerlies, is not clearly marked on the two seasonal surface wind maps. For many years it was believed that a zone of polar easterlies prevailed in a wide band at these higher latitudes. The convergence of air into the deep, strong, migrating lows that concentrate near the Aleutians, Iceland, and off Antarctica, especially during the winter season, provides for a flow with easterly components, but it is strongly seasonal in the northern hemisphere and highly variable in occurrence. During the summer months there seems to be some outflow of air from the frozen surfaces of Greenland and the frozen arctic ice pack, but it is extremely shallow and disappears soon after leaving its source areas.

The atmospheric circulation at the surface might be likened to a complicated system of "big wheels" (the major high-pressure and low-pressure centers, ridges, and troughs) and "little wheels" (the ripples along the edges of the major pressure cells), each of which generates its own miniature circulation system. Laboratory experiments, using

a rotating "dishpan" and differentially heated from below, have clearly demonstrated that an exchange of heat or kinetic energy between cold and warm sectors of a fluid medium, in which the entire mass is undergoing rotation, will yield surface flow patterns of large and small eddies that bear a close resemblance to the air circulation system on earth. Pressure and energy inequalities that involve horizontal flow at the surface must assume roughly circular paths. It is also beginning to be clear that the most important corrective, dynamic air motions involved in energy exchanges are not the large cells, but the small ones—the little wheels that are related to changes in energy imbalances not only at the surface of the earth but also far above, in the upper troposphere. Modern theory now indicates that the migrating cyclones and anticyclones impart some of their kinetic energy to the major pressure cells, changing the latters' shape and strength, and directing the amount of meridional versus zonal flow originating from these major centers. A useful generalization is that the greater the number of wheels within the subtropical zone, the greater is the amount of meridional flow (across the parallels of latitude), whereas few wheels with long east-west axes signify a greater proportion of zonal flow (paralleling latitude). There are many wheels in winter and relatively few in summer, because the greatest demands for air and energy transport north and south take place during winter, when the hemisphere's radiation surpluses and deficiencies are greatest.

Wind patterns at the surface are complex, because the pressure patterns are cellular and many. We shall treat the latitudinal sectors of the surface section of the global circulation again later, but because upper-air conditions are also intimately related to the global air circulation system, we shall now examine the pattern of airflow within the troposphere, ranging from the surface to the tropopause.

6-11 AIR MOVEMENT IN THE TROPOSPHERE

A clear picture of all the patterns of air motion above the earth surface is not yet possible, because of the scarcity of upper-air observations compared with those at the surface. Enough of them have been taken, however, to establish several important facts. The upper air is neither a homogeneous medium nor one in which change is related only to decreased air pressure and temperature with altitude. Figures 6-5, 6-6, 6-13, and 6-14 contain several important characteristics that may be summarized as follows:

1. The prevailing direction of wind throughout most of the troposphere above 10,000 feet is westerly, except for a zone about 30 to 35 degrees wide at low latitudes.

2. The zone of tropical easterlies[2] shifts northward and southward with the seasons and tends to narrow with increasing elevation. It has a broad base at the surface of about 60° in width.

3. The highest wind velocities are found in narrow core zones within the westerlies during the winter hemisphere and located immediately below interruptions in the tropopauses (see Figure 4-19). Velocities within these core zones increase from west to east across the continents. These core zones of strong upper winds are the so-called *jet streams*, which will be discussed in Section 6-12.

[2]The entire belt of winds with easterly components, including the trade winds at the surface, is generally termed the *tropical easterlies*. As can be noted in the vertical cross sections of Figures 6-13 and 6-14, portions of the tropical easterlies at the surface may be overlain by westerly winds. These sometimes are referred to as *anti-trades*, because they move opposite to the direction of the surface winds.

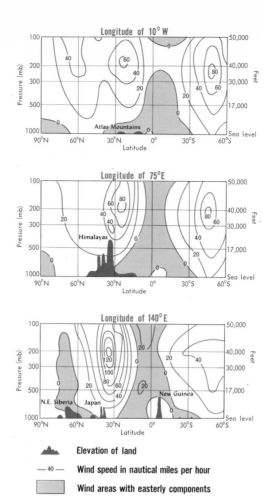

Figure 6-13 Idealized meridional cross sections of the troposphere for January. Note how much stronger the westerly winds are aloft than the easterlies. The diagram also indicates that the jet stream core is larger and much faster near the eastern part of Eurasia than in the western portion. A semblance of a double easterly jet appears near the equator at longitude 140° E. Note also that the southern hemisphere jet is weaker toward the east. [Adapted from *Weather and Climate* by R. C. Sutcliffe, F.R.S., by permission of W. W. Norton and Co., Inc. © by R. C. Sutcliffe, 1966]

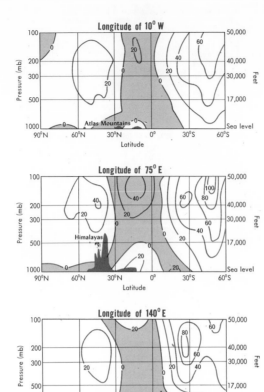

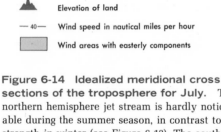

Figure 6-14 Idealized meridional cross sections of the troposphere for July. The northern hemisphere jet stream is hardly noticeable during the summer season, in contrast to its strength in winter (see Figure 6-13). The southern hemisphere jet and the easterly equatorial jet are more pronounced, however, because July is the winter season south of the equator and latitudinal energy contrasts are greater than north of the equator. Note the monsoonal influence in India (75° E), which results in a westerly airflow at the surface within a zone that normally has an easterly one. [Adapted from *Weather and Climate* by R. C. Sutcliffe, F. R. S., by permission of W. W. Norton and Co., Inc. © by R. C. Sutcliffe, 1966]

4. An easterly *equatorial jet stream* of much less velocity than the westerly jets develops within the easterly wind zone above 40,000 feet; it is strongest during July.

5. The upper-air, westerly flow poleward of 60° lies near the surface. Since the tropopause lies within 25,000 to 30,000 feet of the surface at these latitudes, much of the air motion above is typically weak and variable. Occasionally a westerly arctic jet appears during the winter season above the Iceland and Aleutian subpolar lows.

Nearly all of the airflow above 5,000 to 7,000 feet is geostrophic, that is, it tends to flow parallel to the isobars or at right angles to the pressure gradient because of the Coriolis effect. This is indicated clearly in Figures 6-5, 6-6, 6-13, and 6-14. The major exception is near the equator, within the belt of tropical easterlies. Weak cells of high pressure at various levels within this zone (see Figure 6-6) tend to move air poleward toward the equatorial edges of the mid-latitude jets. This conforms with theory, because the deflective force of rotation is least at or near the equator.

The most dynamic portion of the upper-air circulation system lies in the changing characteristics of the jet streams, and the next section will be devoted to the characteristics, causes, and effects.

6-12 THE JET STREAMS The upper-air jet streams in the general westerly flow of the upper troposphere are remarkable, túbular-shaped streams of air that encircle the globe, with their cores lying near the top of the troposphere, especially near a break in the tropopause. They were first discovered by U.S. bomber pilots during World War II when they climbed to high altitudes before or after their bombing runs over Japan in order to avoid antiaircraft fire or pursuit craft. As noted in Figures 6-13 and 6-14, the jets are found in both hemispheres and are much stronger during the winter season when they shift equatorward.'

The most consistent westerly jet streams are found at elevations of from 30,000 to 45,000 feet and between latitudes 30° and 40° N and S. They have been termed the *mid-latitude jets*. A much more irregular jet, also with a westerly flow, appears when a strong polar front lies immediately below a break in the tropopause. It occurs near latitude 60° and at a considerably lower elevation than the mid-latitude jet. It rarely is present during the summer and sometimes is only weakly developed in the winter. It has been termed the *arctic* or *polar jet* to distinguish it from the more prevalent mid-latitude variety.

The westerly jets vary widely in velocity during their course, as indicated in Figure 6-15. They seem to gather velocity as they cross the continents from west to east, and they have their lowest speeds toward the eastern parts of oceans or nearest the oceanic subtropical high-pressure centers at the surface. Velocities range from 30 to 300 miles per hour.

The jets vary in other ways. At times, they fluctuate in elevation and may divide into separate branches. Flow within them is not laminar, and there often are filaments of variable speeds within the jet core. Planes riding the jets encounter considerable turbulence at times. The trajectory of the jets may be directly eastward, but at other times, following the irregularities in the upper-air pressure surfaces, it may be likened to the course of a winding stream, twisting and turning through a broad zone. Like meandering streams, the jets twist sufficiently at times to cut back on themselves, intercepting loops to break off segments from the main stream to be incorporated as twisting masses of air

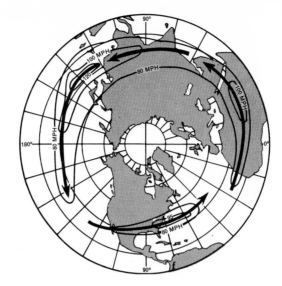

Figure 6-15   The northern hemisphere jet stream in January.   Note the maximum velocities off the continental east coasts and the minimum velocities off the west coasts. These velocities vary considerably cyclically. [After Namias]

within the pressure zones on either side (see Figure 6-16). In general, the amount of twisting is proportional to the velocity of the jets, which in turn is related to the steepness of the pressure gradient. An analogy might be made here between the jet behavior and that of a garden hose, which lies straight on the ground when discharging water at low velocity but which will twist and kink when the discharge increases beyond a certain point. There is also a rough cyclical nature to this feature: following a period of high jet velocity and kinking, the pressure gradient weakens, the jets decrease in velocity and straighten their courses, and a new cycle begins.

The jet streams appear to be a major mechanism for the transfer of both energy and mass latitudinally and vertically in the

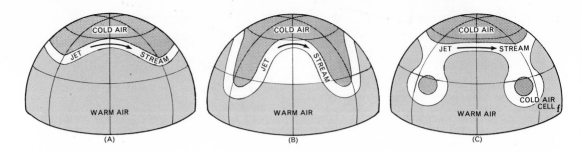

upper air. They often lie along the border of sharply contrasting temperatures below, with a cold side toward the pole and a warm side toward the equator (see Figure 6-16). Energy and mass differentials between high and low latitudes are strongest in the winter, but with geostrophic flow and little friction in the upper air, there can be little airflow across the isobars to relieve the pressure gradients. If the gradients become great enough, however, major and minor waves and irregularities begin to appear along the isobars and air is forced into the flanks of the jets to accelerate their speed. If sufficiently warped in their course, the jets inject warm air masses into the colder zone of higher latitudes by means of centrifugal action or upward into the stratosphere via breaks in the tropopause. On the opposite side of the jets, cold air may be forced equatorward into the warmer zone of the tropical easterlies. Some also may be forced downward, feeding into the most receptive areas for subsidence: the high-pressure subtropical cells, or, at higher latitudes, the tops of migrating anticyclones.

The paths taken by the migrating high- and low-pressure cells at the surface bear a close relationship to the configuration of the major waves that appear on the upper-air pressure surfaces and the trajectories of the jet streams. The locational correlation between surface storm centers and the areas of strongest jet activity is pronounced. Surface storms involve air-mass interaction, precipi-

tation, and energy exchange. Continents produce the steepest energy imbalances, especially during winter, and their eastern sectors are the stormiest and have the greatest precipitation. This is especially true north of the equator, where the continents are larger. Figures 6-13 and 6-15 show the greater strength of the jet in this sector.

The much weaker easterly equatorial jet is far less known, and its role is less clear than that of the westerly subtropical jet. Lying as it does at a much higher elevation than the latter, it may have more affinities with the stratosphere than with the surface. Tropical and subtropical forecasters are beginning to pay more attention to it, if for no other reason than that high-altitude plane flights heading west find it useful in saving fuel.

6-13 THE TROPICAL VERTICAL CIRCU-LATION CELLS The vertical circulation within the troposphere takes place in a triple series of cells, shown in Figure 6-17. These include the *tropical vertical cells,* the *mid-latitude vertical cells,* and the *polar vertical cells.* The arrangement still is hypothetical, especially as to the arrangement of flow at the upper sector of the cells and the relationship between the troposphere and stratosphere as to air interchange. Some of the latter is known to take place through breaks in the tropopause, but details are largely lacking.

Figure 6-16 Jet-stream oscillations. Increased meanderings of the jet, probably caused by an increase in velocity, may produce a dynamic transfer of air latitudinally and a reestablishment of stable zonal flow. [After Namias]

The low-latitude segment of the tropical vertical cells consists of local updrafts of air, associated with the formation of rain. The total volume of air displaced upward is not large, but the amount of energy transported in the form of latent heat is great. The height of the rising air columns varies with the amount of water vapor, whose latent heat exchange runs the lifting mechanism. At or near the summit of the air columns, the air begins to move laterally. The weakness of the Coriolis effect near the equator makes it possible for air to move down the pressure gradients directly to relieve horizontal pressure differentials. Airflow in this sector probably is weak, irregular, and variable in direction. Some of it feeds into the easterly equatorial jet, while some of it moves poleward at various levels, gradually feeding into the westerly flow as the influence of rotation becomes more effective.

The downward component of the tropical cells subsides into the subtropical anticyclones, thence into the general surface trades

Figure 6-17 Hypothetical vertical circulation cells in the troposphere. The arrows that indicate an upward or downward movement of air should not be considered as a steady flow. This is a cross section of the troposphere, and the predominant flow is zonal, that is, either toward the reader or away from him (westerly or easterly flow). Note that the mid-latitude jet stream is shown in cross section. Most vertical arrows are associated with precipitation updrafts. Downward subsidence feeds into surface high-pressure cells. [After Palmen et al.]

or tropical easterlies, thus completing the vertical circulation. What proportion of the subsidence in the subtropical high-pressure cells originates in the jets is not yet clear. It is known, however, that the subsidence slowly spirals downward from high elevations and that the axes of the columns are not vertical but slant from west to east.

The tropical, vertical circulation takes place not within a single cell, but within a series of cells aligned roughly in the same latitudinal zone.

As was noted in Section 6-11, there is a strong, steady, horizontal flow at the surface around the equatorward flanks of the subtropical anticyclones that greatly exceeds the movement of air in the vertical circulation. Equatorward winds are dominant near the eastern part of the cells, and a poleward flow prevails near the western boundaries (see Figures 6-11 and 6-12). Subsidence from the upper air is strongest in the eastern portion of the anticyclones. Thus, the trades in this sector are unusually stable, dry, and frequently have temperature inversions aloft to

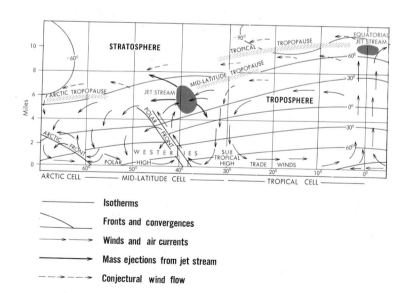

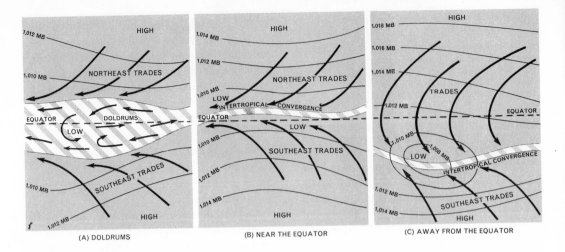

(A) DOLDRUMS	(B) NEAR THE EQUATOR	(C) AWAY FROM THE EQUATOR

check the upward passage of air currents. The centers of the subtropical anticyclones are regions of calms, or light, variable winds. These calm, cloudless regions, dreaded in the days of sailing vessels, were termed *horse latitudes*. An old but disputed explanation for this term is that animals often had to be thrown overboard because of the shortage of drinking water on becalmed vessels in these dry zones.

The trade winds within the tropical circulation cells are not necessarily confined to a single hemisphere but often cross the equator into the opposite hemisphere under the influence of a strong thermal low or some other type of convergence. Once across the equator, the Coriolis deflection in the opposite hemisphere begins to take effect, and the trade winds gradually shift their trajectory. A northeast trade wind crossing into the southern hemisphere, for example, is changed into a northwesterly wind. Conversely, a southeast trade wind crossing into the northern hemisphere is turned toward the right and becomes a southwesterly wind.

6-14 THE INTERTROPICAL CONVERGENCE ZONE Surface trade wind air, whose direction represents the equatorward

component at the surface within the tropical vertical circulation cells, moves obliquely toward the equatorial low-pressure trough from the subtropical highs of both hemispheres. A zone of air convergence, the ITC, thus is established in the low-pressure trough. The converging air streams do not have strong contrasts in temperature. Therefore, sharp discontinuity zones, or fronts, do not often develop—at least, not fronts comparable to those along the boundary between tropical air and the air from polar highs. The seasonal location of the ITC is indicated on Figures 6-11 and 6-12.

Figure 6-18 shows three types of convergences that are characteristic of the general convergence zone between the trades of opposite hemispheres. In the type of convergence shown in *A*, the trades weaken and lose momentum as they approach the equator, eventually developing into a slow westward drift. No real convergence boundary or front develops, and the air movement within this zone is weak, fitful, and variable in direction. The air is highly unstable up to great heights, and convectional rain occurs in frequent showers. This zone, often termed the *doldrums*, may be considered a broad, transitional zone between the trade winds of op-

Figure 6-18 Types of intertropical convergence zones. [After Trewartha et al.]

posite hemispheres. It is not always present, however, because the trade winds from opposite hemispheres may meet more directly along a sharper convergent zone, as shown in B and C of Figure 6-18. Some parts of the equatorial zone are more likely to develop a broad, intertrade, or doldrum, zone than others—for example, the area straddling the equator in the Indian Ocean in the summer, the western Pacific, and also for a short distance off the west coasts of Central America and Sierra Leone in western Africa. The doldrums are not usually found over continental areas.

In Figure 6-18, type B indicates a narrow zone of convergence between the trades of opposite hemispheres. Strong fronts such as those in mid-latitudes do not occur in this zone, because the air masses involved do not show marked temperature contrasts. It is believed by some, however, that weak frontal activity can take place in such a zone. In any event, the convergence creates favorable conditions for instability and precipitation. Shallow low-pressure centers, with inward- and upward-spiraling air, move slowly from east to west along the narrow convergence zones, bringing extended periods of precipitation and overcast skies (see Section 6-22).

The type of convergence shown in C of Figure 6-18 develops more frequently at or following the solstices. Thermal, or "heat," lows developing over a continent in the subtropics tend to result in a steep pressure gradient, which, combined with friction, produces an airflow from the equatorial zone. Note that the Coriolis deflection, which elsewhere works against the pressure gradient, is absent at the equator. Air moves across the equator and gradually increases in accel-

eration toward the east. Such transequatorial trades are exceptions to the more predominant easterly flow in low latitudes. Various interactions between the easterly and westerly winds may produce a series of local convergences, with accompanying precipitation and local wind shifts. This type of convergence is best developed over continental areas.

All types of convergence that take place in the intertropical zone are favorable to the ascent of air, and if the involved air is conditionally unstable, rain usually results. This carries energy to upper levels, where it can be lost through radiation. Weather changes take place from day to day in rapid succession in the convergent zones of low latitudes. Temperature is about the only weather element that remains relatively unchanged. The traditional concept of monotonous weather in the rainy tropics does not conform to the observed facts in the convergent zones.

6-15 MONSOONS It has been known for centuries that in some parts of the low latitudes there is a summer and winter reversal of winds accompanied by a distinct seasonal change in rainfall—the wind from one direction being humid, unstable, and conducive to precipitation, and the other being dry and stable. The term monsoon is derived from the Arabic word mausim, which means "season." India has the classic example of a monsoon, or seasonal wind reversal, and it is notable for the suddenness with which the onset of the rainy season takes place. Usually the wind shift from dry to wet season occurs shortly before the summer solstice. The appearance of the dry winter monsoon in southern Asia is not so sudden, and it lacks the drama and contrast of the opposite seasonal shift. A temperature change is hardly noticeable, and except for a different "feel," owing to the dryness of the wind, a gradual widening in the interval between rains, and the shift in wind direction, this change from wet to dry season is scarcely noticed.

The term *monsoon* has been variously defined, but it typically refers to wind systems that have a pronounced and fairly sudden seasonal reversal of direction, in which one season has a humid, landward wind and the other a dry, seaward wind. Seasonal wind shifts are quite common in low latitudes and occur wherever transequatorial trade winds are found. Such conditions are most usual in areas where subtropical low-pressure centers develop over the interior of continents, some owing to thermal reasons, others perhaps to dynamic causes. Such seasonal reversals merely signify the dominance of different trade winds, and they may or may not be accompanied by seasonal changes in precipitation. Equatorward-facing coasts between lat 5° and 25° are good places for seasonal reversals of land and ocean winds, and such conditions may be observed not only in India but also along the Guinea coast of Africa and the northern coast of Australia. Monsoon tendencies, although not so pronounced, can be found along the Gulf coast of the United States and the east coast of Asia.

The suddenness of the shift in the monsoon of southern and southeastern Asia has puzzled climatologists for a long time. An explanation that has been proposed recently relates the breaking of the monsoon to the shifting of the mid-latitude jet stream. Subsidence of air and anticyclonic conditions normally are found on the earth surface a short distance equatorward of the position immediately below the jet. It has been suggested that the jet alternately swings around the northern and southern flanks of the high mountain core of central Asia, being north of it in summer and south of it in winter (see Figures 6-13 and 6-14). During the winter season, subsidence along its equatorward

Figure 6-19 The southeast Asia monsoon system. [After Garbell]

side induces a subtropical anticyclonic cell over northern Central India and another over the high plains of southwestern China. Also at this season, a strong polar anticyclone develops over Mongolia and eastern Siberia. Divergences from these centers cause the dry, oceanward winter monsoons of India and eastern Asia. It should be noted here, however, that little horizontal airflow passes across the high mountain barriers that border India. The monsoons of India and eastern Asia, although they may be similarly caused, are not directly connected in their airflow. As the winter season recedes, the jet stream slowly moves poleward west of the high Pamir ranges but must continue to skirt the southern ranges until it suddenly "turns the corner" and flows around the northern cordilleras, such as the Tien Shan and Nan Shan. Following this topographic diversion of the jet stream course, subsidence is reduced in northern India, and the clear skies, the heated surface, and the accompanying convergence result in a humid, unstable airflow from the equatorial areas of the Indian Ocean and the western Pacific. As this summer air stream moves northward away from the equator, it is deflected to the right and becomes the southwest monsoon. The seasonal movement of the southern Asiatic monsoon is illustrated in Figure 6-19. Climatologists still are not agreed about the causes of monsoons, but the old hypothesis of seasonal continental heating and cooling apparently does not explain some of the observed features of monsoons satisfactorily.

6-16 THE MID-LATITUDE VERTICAL CIRCULATION CELLS A vertical cross section of one of the mid-latitude circulation cells

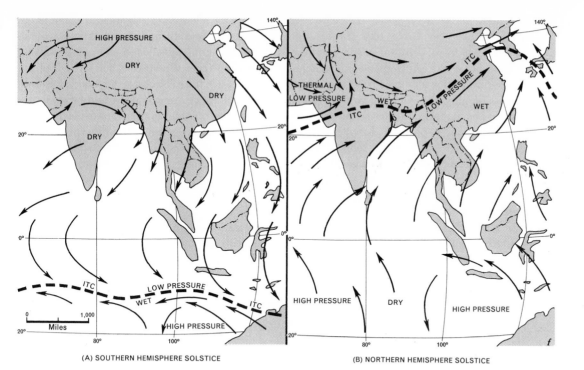

(A) SOUTHERN HEMISPHERE SOLSTICE (B) NORTHERN HEMISPHERE SOLSTICE

is shown in Figure 6-17. At the surface, warm horizontal airflow from the poleward flanks of the diverging subtropical anticyclones moves poleward and meets much colder air, whereupon it rises along a frontal zone formed by the juncture of cold and warm air masses. Aiding in the ascent is the addition of latent heat and lowering of the wet adiabatic rate of cooling associated with precipitation along such fronts. Again, the disposition of the upwardly displaced air is conjectural, but undoubtedly some of it subsides poleward into the cold polar air masses following radiational cooling in the upper air. Some of it undoubtedly becomes incorporated into the jet stream, which is a prominent feature of the mid-latitude vertical circulation system. As indicated in Section 6-12, air is drawn into this jet stream gradually from all sides, each addition adding its own kinetic energy to the jet stream flow. Eventually some of this air is ejected suddenly, following a critical rise in velocity and corresponding kinking of the jet stream. Some air feeds downward into the subtropical anticyclones, some may be forced laterally into the cold, polar air masses, and some may even be thrown vertically through the tropopause gap into the stratosphere.

The fluctuations of the frontal system that are a principal feature of the mid-latitude cells will be discussed in greater length in Section 6-21, because weather conditions throughout the mid-latitudes are closely related to them. The location of the *polar frontal zone,* or the poleward margin of this mid-latitude circulation cell, corresponds closely to that of the transitional zone between the net positive radiation balance of the earth-atmosphere energy system at low latitudes and the net negative deficiencies at high latitudes (see Table 4-3). The location of the polar front is by no means as stable in position as the subtropical anticyclones. During the winter the polar front may swing widely, roughly following the loops in the jet stream above, from the edges of the subtropics near 30°, to positions between 50° and 60°. Each great equatorward loop in the jet stream or elongated pressure wave on the barometric surface usually is accompanied by an equatorward extension of the polar front toward the tropics. These intrusions of cold air in the United States may threaten the citrus fruit growers of Florida and discourage the sun lovers of New York who go south to escape the rigors of northern winters. At other times during the winter, a poleward swing of the jet and the accompanying frontal system below can bring a "January thaw" that imports warm, moist air all the way from the Gulf of Mexico into the northeastern United States.

6-17 THE POLAR VERTICAL CIRCULATION CELLS The vertical upward and downward circulation of air at high latitudes is much less pronounced and occurs at relatively low elevations above the surface. The so-called *arctic front* is a separation plane between cold or cool air and much colder air. The latter is generally found only in the winter. Convergence and a certain amount of lifting occur at the frontal surface, but the air is relatively stable and contains little moisture for conditional instability; precipitation, if any, is likely to be in the form of light snow. There is a strong negative radiation balance at these latitudes, and surface air in winter and upper air throughout the year both lose more heat by outbound radiation than they receive from the sun. This helps to explain some of the subsidence within the polar cell. Were it not for the occasional injection of warmer air poleward from the jet stream, a slight addition of latent heat released during snowfalls, and a possible dynamic subsidence and compressional heating from the stratosphere which is not far above the surface, the high latitudes both at and above the surface would become progressively colder. Much more needs to be learned of the mechanics of the polar circulation cell. Even the curious arctic jet that suddenly appears and disappears is a newly discovered phenomenon about which little is known. The deep, low-pressure vortex that lies over the poles during the winter and extends far into the stratosphere is another feature of this part of the vertical cell. Perhaps the arctic jet is a mechanism like that in the mid-latitudes for relieving the steep pressure gradients that surround the polar vortex during the winter, even within the stratosphere.

6-18 AIR MASSES An air mass is a large quantity of air that has more or less uniform properties horizontally. Such uniformity may be considered to be an approach to equilibrium with respect to general atmospheric environmental factors. Some areas are much more conducive to the formation of homogeneous air masses than others and may be considered as *air-mass source areas*. One type of source area is that in which air stagnates prior to an outward movement. The areas of calms associated with the centers of large, stationary anticyclones are illustrative. A second type of source area includes extensive surface regions of general climatic uniformity, such as the broad expanses of warm oceans in the low latitudes and the snow-covered surfaces of continents in the high latitudes during the winter season. Such areas are large enough to produce a certain degree of uniformity within the air streams that pass over them.

The characteristics of air masses are derived, first, from the source areas where they

Table 6-1 Characteristics of Major Air Masses

Name of air mass	Source areas	Typical characteristics	Major modifications	Associated weather
Arctic (A)	Ice- and snow-covered surfaces in Greenland, Antarctica, Hudson Bay, Mackenzie Basin, Mongolia–Lake Baikal region, Scandinavia	Low water vapor content; temperature inversion above surface cooling zone (2,000–3,000 ft); subsidence above; generally stable	Rapid rise in humidity in lower levels when passing over water, creating instability; inversion aloft destroyed by passage over rough terrain	Clear skies, except when modified; low temperatures at night; strong winter storms (blizzards) along leading edge (arctic front)
Continental polar (cP)	In winter, the cold continental surfaces mostly free of ice and snow; in summer, cool surfaces in Canada and U.S.S.R.; rare in summer south of U.S.S.R.	Found only in northern hemisphere; often modified cA air; clear skies at night; scattered cumulus clouds during the day; some subsidence above	Generally becomes highly unstable away from source areas; persistent overcast skies with passage over rough terrain	Clear air; frost hazard in late spring and early fall; summer convectional showers near water bodies; involved in most frontal activity in mid-latitudes
Maritime polar (mP)	Ocean areas lat 40–60° N and S	In winter, mainly modified cA and cP air brought into ocean areas, resulting in instability (mPk); surface temperatures rarely subfreezing; summer air generally stable (mPw)	mPk air on lee side of mountains (as in the Rockies) dry, clear, and warm; passage over cold surfaces produces stability at lower levels	Fog and low overcast skies when passing over a cold surface; cyclonic and orographic conditions produce heavy precipitation; fronts with mPw air produce drizzly rain
Maritime tropical (mT)	Warm tropical and subtropical ocean areas beneath subtropical anticyclones	Usually subsidence and low lapse rate above; generally fairly stable; some low-level instability; often an inversion aloft	Rising humidity and increasing instability away from source areas; passage over cold land in winter increases stability (mTw); in summer, air sometimes unstable over warm land	Hazy air; scattered cumulus clouds; passage over cold land results in broad overcasts; produces most continental snowfall in mid-latitudes; convectional rain with mTk in summer
Continental tropical (cT)	High plains with subtropical anticyclones (mainly in winter); Sahara, S. Africa, central Australia, S.W. China, N. India, Andean Plains	Extremely low humidity; hot, dusty, turbulent air; subsidence of warm air aloft	Becomes highly unstable in lower levels when passing over water (as over the Mediterranean Sea)	Hot, dry weather; much haze; sometimes associated aloft with tornadoes

The code system above consists of three parts. The first lower-case letter refers to the continental (c) or maritime (m) source of the air mass. The second letter, which is capitalized, refers to the general latitudinal location of the source area: i.e, A = Arctic; P = Polar; T = Tropical. The third, lower-case letter indicates whether the air mass is generally colder than the surface over which it is passing (k = kalt = cold) or warmer (w = warm). This third letter is an indication of general stability or instability.

originate and, second, from the areas they cross once they leave their source areas. The modifications en route may take many forms, but generally the lower portions of the invading air masses are the first to be altered. The great air streams that form the circulation systems at the earth surface thus are continually undergoing some alteration outside their source areas. The extent of modification frequently can be observed by examining the vertical profiles of air masses. Nearly all modified air masses exhibit distinctly different properties in their upper and lower portions, and some remarkably sharp boundaries can be found separating the lower modified portion from the upper. One of the most frequent modifications is formed by changes in temperature. For this reason, a sharp change in the lapse rate of temperature often marks the upper boundary of the modified portion. Temperature inversions in air masses are far more common than was formerly believed, and they were not recognized until measurements of vertical temperature changes could be regularly and easily taken. The principal modifications that take place in surface air masses may be summarized as follows:

Thermodynamic modifications
1. Heating at the earth surface.
2. Cooling at the earth surface.
3. Cooling above the earth surface by loss of radiation.
4. Addition of water vapor at the surface.
Mechanical modifications
1. Mixing by passage over rough terrain.
2. Orographic, convergent, and convectional lifting.
3. Subsidence from above.

Table 6-1 presents the essential characteristics of the major global air masses, and Figures 6-20 and 6-21 show the distribution

Figure 6-20 Typical world air-mass distribution for July. [After Haurwitz and Austin]

——— Average frontal positions

1000 Pressure in millibars

T Tropical

m Maritime

c Continental

P Polar

s Stable aloft

u Unstable aloft

W Warmer than underlying surface

K Cooler than underlying surface

Figure 6-21 Typical world air-mass distribution for January. [After Haurwitz and Austin].

——— Average frontal positions

1000 Pressure in millibars

T Tropical

m Maritime

c Continental

P Polar

s Stable aloft

u Unstable aloft

W Warmer than underlying surface

K Cooler than underlying surface

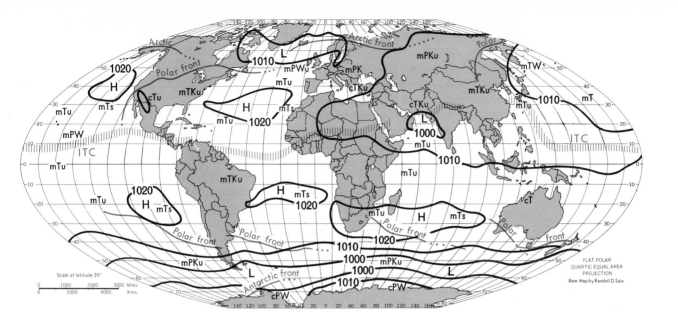

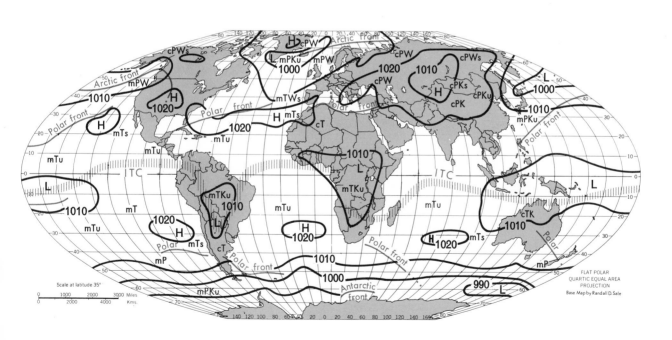

219

of the principal source areas. Each air mass has certain identifying characteristics which are most clearly revealed in its vertical profile. This is why, centuries ago, keen observers could make reasonable weather predictions by studying cloud patterns at different levels. Cloud associations reflect differences in the upper air that are symptomatic of air-mass stratification, modification, or interplay. The classification of air-mass types given in Table 6-1 is similar to the classification of other natural forms and phenomena in that there is a gradual change from one type to another. No two continental, polar air masses are exactly alike, and each of them is subject to dynamic change. Nevertheless, certain broad characteristics enable the classification system to have useful validity.

6-19   SUMMARY OF THE GLOBAL AIR CIRCULATION SYSTEM   The air circulation system of the earth exhibits a highly complex pattern of circular air movements that have both vertical and horizontal components. The entire system represents an accommodation of the fluid atmosphere to remove energy surpluses in those areas where the net radiational balance is positive and to supply those sectors of the earth-atmosphere system where a net radiation deficit prevails, and surrounding a rotating planet. The greatest positive balance is found at the surface in the low latitudes, especially in the oceanic areas. The greatest negative balance lies in the same latitudes, but at high elevations. A vertical airflow, mainly in the form of many conditionally unstable air columns, provides the initial start of the circulation system. Mass and energy are both carried upward in the towering cumulus clouds of the tropics. Much of the latent heat released by condensation at various levels is to be radiated, some of it into space, some back to earth. The air columns moving upward

Figure 6-22   Diagram of a typical chinook wind.   On the windward side, the temperature of the rising air at 4,000 feet would be 33.5°F. The depressed adiabatic rate above the point, caused by the release of latent heat, results in the rising air being cooled to 15.5° at 10,000. On the lee side of the mountain, the air will be warmed at the dry adiabatic rate throughout its descent, and will reach 65° at 1,000 feet, or 15° higher than the 1,000-foot level at which it began its rise.

eventually develop density differentials horizontally that favor movement toward the poles.

If the earth were not rotating and had a homogeneous surface, a simple convection cell probably would have evolved, with the air at various elevations above the low latitudes moving slowly toward the poles and gradually descending. Near the surface the air would move as a slow, irregular flow toward the equator. The effect of rotation, however, is to deflect airflow away from the pressure gradients. This deflection increases away from the equator, and the poleward-moving air high above the low-latitude surface is channeled into a predominantly westerly flow. Accumulations of mass and energy within the westerly air streams form the jets that cyclically discharge both mass and energy away from the jet cores and across the isobars. A major portion descends in great spiral vortices to reach the surface in the subtropical high-pressure cells. Some is discharged poleward, whereupon it may again be involved into a westerly stream and a second mid-latitude jet, but at a lower elevation. Another series of downward mass discharges from the "polar" jet feeds the polar highs at the surface. Downward-spiraling vortices of air thus are supplies of mass for the various high-pressure centers on the surface, both stationary and migrating. Diverging air masses from these anticyclones

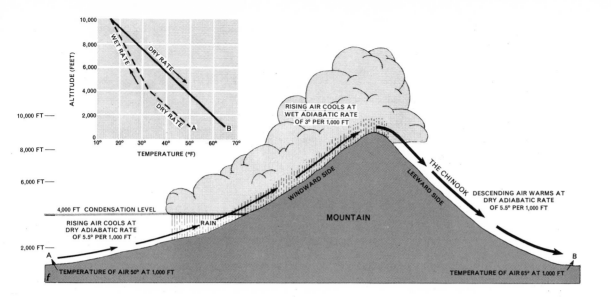

RISING AIR COOLS AT
WET ADIABATIC RATE
OF 3° PER 1,000 FT

THE CHINOOK

DESCENDING AIR WARMS AT
DRY ADIABATIC RATE
OF 5.5° PER 1,000 FT

WINDWARD SIDE

LEEWARD SIDE

MOUNTAIN

4,000 FT CONDENSATION LEVEL

RISING AIR COOLS AT
DRY ADIABATIC RATE
OF 5.5° PER 1,000 FT

RAIN

TEMPERATURE OF AIR 50° AT 1,000 FT

TEMPERATURE OF AIR 65° AT 1,000 FT

are deflected by the earth rotation. Curved, converging air streams originating from these pressure centers and having contrasting temperatures promote the equalization of energy differentials.

Despite the net annual positive radiation balance on the earth surface from the equator to the pole, the winter season in each hemisphere features a negative radiation balance poleward of about 30°. The resultant sharp change from a high positive net radiation balance near the equator to negative balances in the mid-latitudes is reflected in the steep temperature gradients across the parallels of latitude during this season. This sudden change promotes energy transfers latitudinally to eliminate the energy differentials. The convergence of air from both polar highs and subtropical highs aids in this transfer. Some of this energy transfer is in the form of latent heat produced by convergent or frontal precipitation; some of it is by advection, and some by mechanical energy seen in severe storms. Various methods thus stimulate the transport of energy into the higher

latitudes to prevent the net radiation deficits from becoming greater. These are most effective in winter. Precipitation at the higher latitudes involves vertical circulation cells, just as in the tropics, except that the former are shallower and the zones of energy transfer are sharper.

6-20 LOCAL WINDS Local conditions provide exceptions to the global patterns of air circulation, as they do with the generalized patterns of all the other elements in the environment. Many parts of the world have distinctive winds that result from local terrain conditions and that have been given local names. Many of these have been noted for some unpleasant feature, usually temperature. Among such local winds is the *chinook* (snow-eater), an unusually warm wind that is heated adiabatically by compression as it descends the eastern slopes of the Rocky Mountains (see Figure 6-22). Another reason for the unusually high temperatures of the chinook winds is that the air contains the heat of condensation acquired during its as-

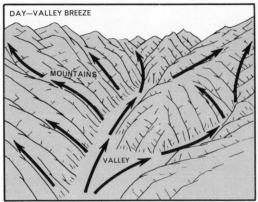
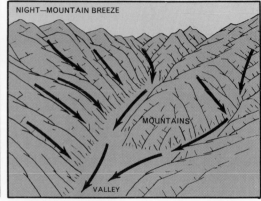

(A) MOUNTAIN AND VALLEY BREEZES

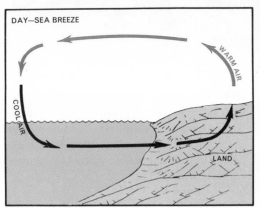
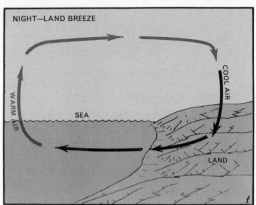

(B) LAND AND SEA BREEZES

cent up the windward side of the mountains following precipitation.

The Mediterranean region is the home of hot, searing, dust-laden winds off the deserts that lie to the south and are known in various countries as *sirocco, khamsin, ghibli, leveche,* or *samiel.* These desert winds occur most frequently in February, March, and April, when low-pressure centers within the Mediterranean Basin still are deep enough to attract the first heated desert air in the late spring. They are not unknown at other times of the year. Rises in temperature of as much as 12° within 15 minutes may mark their onset. Wind velocities reach gale force and

dust carried for hundreds of miles may limit surface visibility to less than 100 yards. The meteorological features associated with three khamsins in March, 1961, are shown in Figure 8-5.

Two other examples of local winds that are thermally induced include: (1) *land-and-sea breezes,* which affect narrow coastal strips in areas where land temperatures are high during the day, and (2) *mountain-and-valley breezes,* which involve air drainage at night and convectional movements upslope during the day. These two types of local winds are diagramed in Figure 6-23.

A close scutiny of air movement reveals

Figure 6-23 Diagram of two types of local winds.

that every building corner or land-surface irregularity produces swirls and eddies in the winds that stream by. Frontal conditions between conflicting air masses can result in sudden wind shifts through all quarters of the compass. Rare indeed are those areas where the planetary air currents are regular and dependable. The steadiest winds are found mainly over the broad, open expanses of oceans in the trade wind zone, where no topographic obstacles or unequal heating surfaces occur and where only an occasional convergence or easterly wave creates diversions from the normal airflow. Combine all the regional variations in wind patterns with the changes that occur from day to day, year to year, and century to century, and it becomes apparent that the geographic generalization of wind patterns is subject to the same limitations that are found in the geography of all other natural elements.

The Secondary Air Circulation

Mention has been made in this chapter of traveling cyclones and anticyclones. These temporary and moving pressure disturbances in the atmosphere are secondary features of the air circulation pattern. We now turn our attention to these "lesser wheels within the larger wheels" of the air circulation system. As with all the other weather and climatic phenomena, their behavior is linked with the distribution of energy received from the sun and the necessity of evening out the local surpluses and deficiencies that develop at various places about the earth surface and within the envelope of air that surrounds it.

The details of the origin and operational aspects of cyclones and anticyclones will not concern us here; the interested student is referred to standard textbooks in meteorology for such material. Like the larger permanent or quasi-permanent pressure centers on earth, the traveling highs and lows feature the subsidence of air associated with anticyclones and the rising of air that is typical of cyclones. Disturbances like these involve the atmosphere to great heights, and most well-developed pressure centers are related to waves in the streamlines of airflow throughout the troposphere.

6-21 EXTRATROPICAL CYCLONES The cyclones of the middle and upper latitudes nearly always develop along frontal boundaries of air masses that differ in their basic properties, particularly temperature. Essentially they are eddies or spiral vortices of converging and rising air along a front. These low-pressure cells vary considerably in shape, size, intensity, speed, and duration. A typical mid-latitude cyclone of the northern hemisphere is shown in Figure 6-24. Note that it has roughly circular isobars that are slightly kinked at the boundaries of the air masses. Wind directions are angled across the isobars and down the pressure gradient, giving the entire cyclone a convergent, counterclockwise rotation in the northern hemisphere and an opposite rotation south of the equator. Wind velocities depend on the steepness of the pressure gradient and on the latitude. Two separate fronts can be identified: a *warm front* in the southeastern quadrant and a *cold front* in the southwestern quadrant. Rising air along the frontal zones produces cloudiness and precipitation. The size of extratropical cyclones varies, but most of them are between 600 and 1,000 miles in diameter.

Extratropical cyclones have different de-

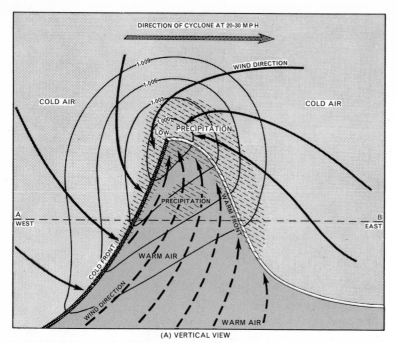

DIRECTION OF CYCLONE AT 20-30 M P H

WIND DIRECTION

COLD AIR

COLD AIR

1.009

1.006

1.003

1.000

LOW

PRECIPITATION

PRECIPITATION

WARM FRONT

A
WEST

B
EAST

COLD FRONT

WARM AIR

WIND DIRECTION

WARM AIR

(A) VERTICAL VIEW

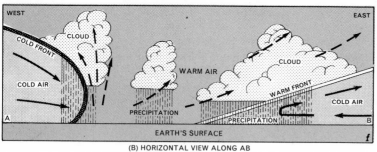

WEST

EAST

CLOUD

COLD FRONT

COLD AIR

WARM AIR

CLOUD

WARM FRONT

COLD AIR

PRECIPITATION

PRECIPITATION

A

B

EARTH'S SURFACE

f

(B) HORIZONTAL VIEW ALONG AB

Figure 6-24 Model of a typical mid-latitude cyclone in the northern hemisphere. [After Bjerknes]

Figure 6-25 The development of a cyclone, showing stages in the development and occlusion of an extratropical cyclone. The smaller diagrams (A', B', C', D', E', F') represent vertical cross sections of the air masses along the line x-y. [After Garbell]

gent lifting and the various thermodynamic processes associated with this. Their distribution is related to surface contrasts in energy availability. Figures 6-20 and 6-21 show the mean location of the polar front at the solstice seasons. It is along this front that the extratropical cyclones are most active, forming, occluding, and regenerating. The winter season marks the period of maximum activity, because the latitudinal contrasts in the net radiation balance are most pronounced at this time. The North Atlantic, North Pacific, and the waters off Antarctica during the winter have exceptionally great cyclonic activity, because this is where the greatest contrasts between net radiation balances occur: the high positive balance over warm tropical water and the high radiation deficits over the frozen land areas nearby. The general direction of movement varies somewhat, but usually there is an eastward component because of their location within the general flow of the westerly winds.

A study of the cyclonic model shown in Figure 6-24 will reveal the weather changes associated with the passage of a cyclone, including cloud types, temperature and pressure changes, wind direction, and precipitation. The sequence of these features varies, depending on whether the cyclonic center passes to the north, to the south, or directly through the point of observation. Some are elongated rather than circular, with the long axes usually extending from northeast to southwest in the northern hemisphere, and have unusually sharp shifts in wind direction along the cold-front sector.

velopmental sequences. Some are highly dynamic, with rapid changes in form. Others, once formed, remain relatively unchanged for long periods. Mid-latitude cyclones are destroyed mainly by the process of *occlusion,* shown diagrammatically in Figure 6-25. Their main function is to provide a means of transferring energy quickly by inducing conver-

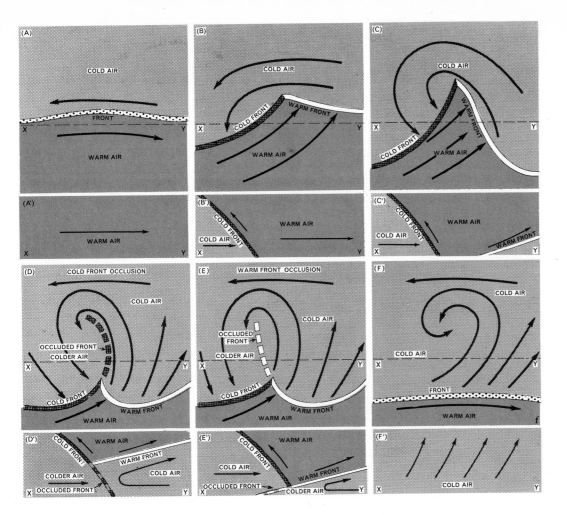

6-22 TROPICAL CYCLONES Tropical cyclones typically are broad, shallow, low-pressure centers that are found mainly within the intertropical convergence zone between the trade winds. The absence of strong fronts between air masses in this zone, due to the small differences in temperature, prevents the development of cyclones like those of higher latitudes. The tropical cyclone rarely has a pressure range of more than about 3 to 5 millibars and is much larger in size than the extratropical type, often reaching 1,500 to 2,500 miles in diameter. They rarely develop occlusions and generally slowly weaken and disappear. Their direction is from east to west within the general easterly flow of low latitudes. Movement is slow, and they frequently remain motionless for several days. Weather changes accompanying them are little more than "spells of weather," with periods of overcast skies and fairly steady rains that may last for several days. They appear to be more frequent near the equinoctial periods but may occur in any season.

An interesting feature of some tropical cyclones is that they appear within the streamlines of the trade-wind flow and thus do not require an intertropical convergence. Such cyclones are associated with wavelike undulations in the trade wind flow, termed *easterly waves*. This type of tropical cyclone tends to be located somewhat farther from the equator than the intertropical convergence and is of special interest because it sometimes develops into severe storms, the dreaded tropical hurricanes. In the easterly

waves (see Figure 6-26), the bending of the airflow at the axis of the wave (roughly from east-north-east in front of the wave to east-southeast in its rear) results in a local convergence in the rear of the axis of the wave. The dynamic lifting here destroys the normal trade wind inversion in the upper air, thus increasing the lapse rate, making the upper air more unstable, and stimulating condensation and precipitation. Subsidence generally is stronger in front of the easterly waves, resulting in exceptionally clear, dry weather there. The easterly wave cyclones are somewhat smaller than other tropical cyclones. Usually the waves are more pronounced in the upper atmosphere than they are at the surface and may well have been generated as the result of energy patterns in the upper zone of the tropical easterlies. It has been estimated that fewer than 10 percent of them develop into hurricanes.

Figure 6-26 Isobaric patterns associated with weak tropical lows and easterly waves in the northern hemisphere.

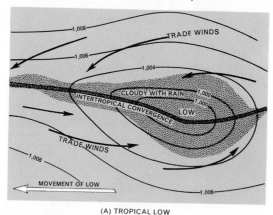

(A) TROPICAL LOW

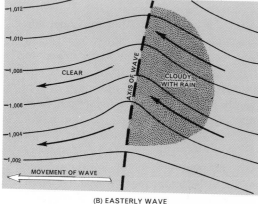

(B) EASTERLY WAVE

6-23 TRAVELING ANTICYCLONES *Traveling anticyclones* are restricted largely to the mid-latitudes and are of two types: the cold, polar anticyclones and the warm, subtropical type. The former are synonymous with the outbreaks of arctic and polar air from their respective source areas. The warm variety represents offshoots from the subtropical, permanent, or quasi-permanent, anticyclonic cells. Both types intrude into the general westerly circulation, but from opposite directions, and tend toward the east.

The polar traveling anticyclones are shallow, rarely over a mile or so in depth. Subsidence within them thus takes place over only a short vertical distance, in contrast to the high columns of rising air typical of cyclonic centers in the subtropics. Wind patterns are typically divergent, with a deflection that produces a rough, clockwise rotation in the northern hemisphere and a counterclockwise rotation south of the equator. Unusually steep gradients behind a cold front during the winter sometimes generate cold gale winds. Winds in the center of anticyclones are light and variable with intermittent calms. Cold anticyclones generally do not travel as fast as cyclones and are more likely to remain stationary, especially over rough, hilly terrain. Weather conditions associated with polar anticyclones are characterized by clear, cool weather. Once into the westerly airflow pattern, especially over warm surfaces, the traveling polar anticyclones tend to become shallower and broader and may disappear because of progressive warming from below, since they are maintained thermally by conductive and radiational cooling. A few of them may penetrate into the subtropics and become incorporated into the subtropical anticyclones. Traces of these old polar anticyclones are preserved for long periods in the upper air. The traveling polar anticyclones form equatorward lobes in the polar front.

Traveling subtropical anticyclones are not so well defined as the cold, or polar, type, and they have a much greater tendency to stagnate for long periods. They have much warmer air and often are responsible for spells of clear skies and unseasonably warm weather in the mid-latitudes.

The Oceanic Circulation

Not all of the surpluses of energy that accumulate at the surface in low latitudes are led away toward the heat sinks at high altitudes and high latitudes by vertical and horizontal airflow. Some of them also are transferred to higher latitudes by the kinetic mass transfer of water within the oceans and by the loss of heat through conduction and diffusion to the cooler waters of high latitudes. Were it not for the rotation of the earth, a relatively simple convectional circulation system probably would have developed, with surface water moving toward the poles, progressively losing its heat en route. Near the poles, where it would have become more dense because of the loss of heat, it would have sunk toward the bottom and moved equatorward, where it would have eventually arisen to replace the surface water being displaced poleward. This is similar to the convectional circulation system in a home which is being heated by radiators and hot water. The effect of the Coriolis deflection, however, works upon the oceanic circulation much as it does that of the atmosphere, except that being more dense than air, sea water is not as susceptible to deflection. Ocean water moving poleward cannot move in a straight line, yet is not deflected in circles as tight as those of air. Other factors also interfere with the direct exchange of warm tropical water and that from the polar seas. These include: (1) the surface push of prevailing winds at the surface which are roughly circular in their trajectories; (2) the configurations of the continents and the ocean bottom that channel the movement of both surface and bottom water into complex patterns in places; (3) tidal motions, which, when restricted into narrow passageways, may be a powerful impelling factor in water motion; and (4) variations in density, owing to differences in both temperature and salinity. Unlike air, where the composition has little influence on density or pressure, ocean water may vary greatly in the proportion of water to contained soluble salts. Ocean areas of low rainfall and high evaporation tend to have dense, highly saline surface waters, hence are likely to result in subsidences. Figure 6-28 shows a local current that enters the Atlantic from the Mediterranean and subsides to deep water because of its greater salinity and density.

Figure 6-28 Subsurface circulation (north-south) in the Atlantic Ocean. [Karl K. Turekian, *Oceans,* copyright © 1968, Prentice-Hall, Inc., Englewood Cliffs, N.J.]

currents is only part of the total system, and, like the airflow pattern near the surface, is highly complex. Speed and steadiness vary in different parts of the ocean system, and there are many interesting analogues between the behavior of the two fluid mediums, including such features as fronts, jet streams (the Gulf Stream), convergences, divergences, and eddy diffusion. There is one important difference, however: The ocean waters do not have an efficient energy transfer system equivalent to the evaporation and condensation of water vapor. Evaporation and condensation influence oceanic circulations only in their interactions with the atmosphere. The nearest equivalence is a small amount of freezing and thawing of sea water at high latitudes and the seasonal melting of icebergs in mid-latitude waters. Speeds also are incomparably slower in the oceanic flow, ranging from a maximum of about $5\frac{1}{2}$ miles per hour (excluding local tidal currents) to an almost imperceptible drift. The slowest movement generally is along the bottom, and the most rapid is located at the surface in the western parts of oceans. Probably every molecule of water in every ocean has made the round trip between the tropics and the polar seas and has alternately visited the surface and the ocean bottom. Scientists now estimate that this round trip takes from 300 to 600 years.

The general circulation of ocean currents is shown in Figures 6-27 and 6-28. The influence of the earth rotation in developing gyres, or giant circular eddies, is clearly shown in Figure 6-27. The configuration of the ocean basins also affects the flow patterns. Note the distribution of cold and warm currents in Figure 6-27. One general area of cold water

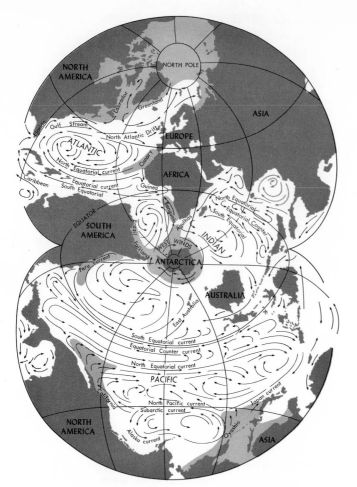

Figure 6-27 World map of ocean currents. The circular gyres into which the global ocean currents arrange themselves at the surface is clearly seen. [From "The Circulation of the Oceans," by Walter Munk. Copyright © 1955 by *Scientific American,* Inc. All rights reserved.]

Like the atmospheric circulation system, therefore, the movement of ocean water eventually transfers energy to high latitudes, but the route taken is devious. Comparably, the flow patterns of both air and water are more or less circular or ovate, and are three-dimensional. The surface pattern of ocean

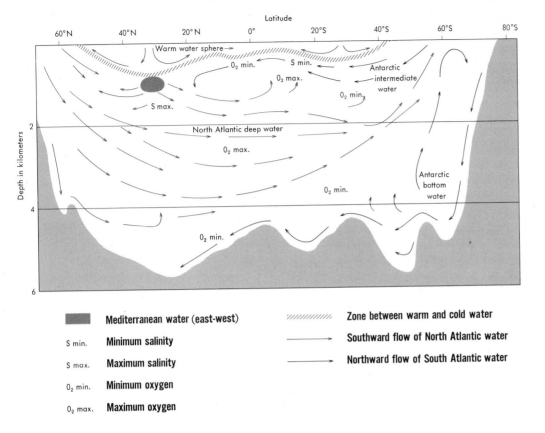

Latitude

| | | | | | | | |
| 60°N | 40°N | 20°N | 0° | 20°S | 40°S | 60°S | 80°S |

Warm water sphere →

O_2 min.   S min.

O_2 max.

S min.

Antarctic intermediate water

O_2 min.

S max.

North Atlantic deep water

O_2 max.

O_2 min.

Antarctic bottom water

O_2 min.

Depth in kilometers

Mediterranean water (east-west)	
S min.	Minimum salinity
S max.	Maximum salinity
O_2 min.	Minimum oxygen
O_2 max.	Maximum oxygen

//////////	Zone between warm and cold water
———→	Southward flow of North Atlantic water
———→	Northward flow of South Atlantic water

should be particularly noted and explained. This lies off the west coast of continents roughly between 10° and 30°. It is cold mainly because of an upwelling of bottom water brought about by the offshore movement of surface water heading westward into the westward-moving equatorial currents.

Cold ocean water, wherever it is found, is favorable to marine life. This is because cold water contains much more dissolved carbon dioxide and oxygen than warm water, and which it obtains by contact with the atmosphere. The oxygen-rich arctic and antarctic waters are shown in Figure 6-28. The lowest content of these gases is found in the warm waters in the center of the subtropical gyres. Cold, gas-rich sea water generally is green

because of its rich microlife forms, whereas the warm, gas-poor waters are deep blue and sterile. There is a striking contrast between the blue, warm water of the Gulf Stream off the eastern coast of the United States and the green, cold water of the Labrador Current off New England and the Grand Banks. Where the cold water has resulted from the upwelling of bottom water, as off the subtropical west coasts of continents and in high-latitude waters, there is an upward movement of important chemical nutrients derived from the slow decomposition of organic life on the bottom of the sea. These nitrogenous and phosphatic materials, combined with the gas-rich, cold, surface water, give rise to some of the richest concentrations of marine life on earth. They are focal points for the channeling of some of the net radiational surpluses into the biotic world and comprise the most valuable areas of commercial fishing on earth.

Storms

Storms may be defined as any sudden, severe increase in air motion. They result either from the sudden transformation of potential energy into the kinetic energy of air motion (as when cold air subsides and warm air rises) along fronts, or from the sudden release of latent heat following condensation in warm, humid air masses and the transformation of this heat energy into the kinetic energy of air motion (as in a thunderstorm). Among the many different types of storms, the most important in terms of potential loss of life and property are thunderstorms, tornadoes, and hurricanes (typhoons), each of which has a distinctive pattern of occurrence and characteristics. From an economic standpoint, the geographic patterns of storms are a significant factor in world climates, and they are briefly described and partially explained in the following sections.

6-24 THUNDERSTORMS Thunderstorms are by far the most frequent convulsions of the atmosphere. When viewed on a global scale, they are tiny phenomena, but their beautiful but awesome short-lived forms, their strength, and their virtually worldwide distribution have been responsible for impressing humans everywhere with the potent forces that lie hidden within the atmosphere. By definition, *thunderstorms* are rainstorms that are accompanied by thunder and lightning. The rain clouds, in order to have these features, must be high—usually at least 20,000 feet—and some may rise as high as the tropopause, 6 or 7 miles above the surface.

Causes for thunderstorm heights. There can be only one reason for such great vertical ascents of air—a high content of latent heat, associated with air that has a high vapor pressure. Warm air is essential, because only such air can contain the water vapor in quantities sufficient to produce lifting to such great heights. The initial triggering of the rise may be caused by different reasons (see Section 5-12), but once condensation begins, the lowered wet adiabatic rate keeps the rising air warmer than its surroundings. The ascent continues until the rising air begins to run out of water vapor to condense. When this happens the wet adiabatic rate of cooling rises, and once the air is cooled to the level of the surrounding air, its ascent stops and it diverges laterally. The broad, flat tops to most cumulonimbus clouds, which often give the storms an anvillike appearance, are the result of this high-altitude divergence.

Distribution of thunderstorms. The distribution of thunderstorms on a global scale is shown graphically in Figure 6-29. Compare this pattern with the one for vapor pressure (see Figure 5-3). The greater frequency over the continents is related to the fact that more mechanisms exist for the initial lifting of air over land than over the oceans. The orographic effect is obvious. Continents also are likely to have more convergent lifting because of frontal and pressure-wave activity associated with continental temperature contrasts, and convection is likely to be more active. The world's record for thunderstorm frequency is held by Buitenzorg, a mountain station in western Java, which averages 322 per year. The map of thunderstorm frequency in the United States (Figure 6-30) clearly indicates several reasons for their distribution. The lowest frequency, and a seeming paradox, is along the entire west coast of the country. Even the Olympic Penin-

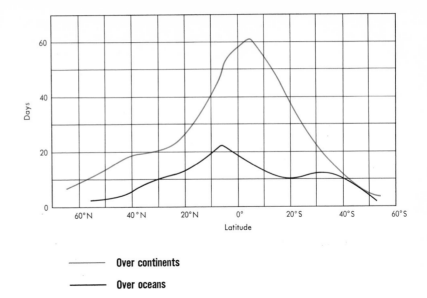

Figure 6-29 World distribution of thunderstorms by latitudinal zones. The much greater incidence of thunderstorms over land than sea is caused by the higher temperatures over land. The warmer the air, the greater is its capacity to hold moisture for generating the thermodynamic mechanisms of thunderstorms. [After Brooks]

——— Over continents

——— Over oceans

sula, which has the heaviest rainfall in the country, rarely hears thunder. Cool air prevails along this coast, and despite the maritime position, the vapor pressure of its polar maritime air is relatively low, certainly too low to generate great cloud heights. The high rainfall amounts on the coastal or near-coastal mountain slopes occur in almost steady drizzles, day after day throughout much of the year. The northern edge of the United States also has few thunderstorms, because warm, moist air rarely reaches these latitudes, even during the short summer season. Lightning is likely to be especially active, however, during the summer rains of this north country. The high incidence of thunderstorms in Florida and along the Gulf Coast is related to the predominance of maritime tropical air that is consistently warm and moist. Frontal disturbances may produce thunderstorms at any season. The orographic effect of the southern Appalachians and the central Rocky Mountains is also illustrated. Thunderstorms are likely to be more common in the mountain areas than is indicated on the map, owing to a much lower frequency of weather stations. Not all of the many thunderstorms in mountain areas should be attributed to orographic lifting, however. Thunderstorms are defined by the presence of lightning and thunder, and, as will be indicated in more detail later, the opportunities for a steeper electrical potential gradient between the surface and the ionosphere above are greater in mountain areas. Lightning is a real hazard for starting fires in

Figure 6-30 Distribution of thunderstorms in the United States. [U.S. Weather Bureau]

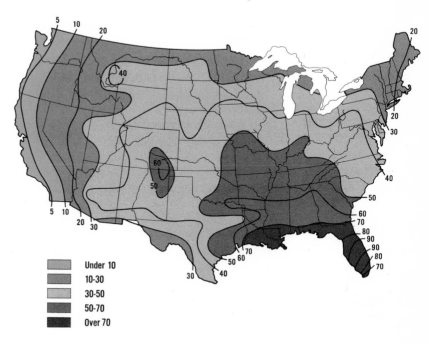

Under 10

10-30

30-50

50-70

Over 70

Colorado, New Mexico, and Wyoming. It is less of a hazard, but still exists in the Cascades and Sierra Nevada Mountains.

Morphological features of thunderstorms. Figure 6-31 shows a typical cross section of a mature thunderstorm, or rather, a single thunderstorm cell. Each thunderstorm is a mechanism that is run thermodynamically within itself, usually has a short life, and is associated with clusters of other cells in various stages of development and decadence. Each storm has distinct vertical columns of ascending and descending air, both of which may reach speeds of 10 to 12 feet per second on their way up or down. The updrafts are generated by the release of latent heat. The downdrafts, which generally increase as the storm continues, are caused mainly by (1) friction of the rapidly falling drops that drag some of the cold air of high levels down with them, and (2) evaporation of some of the

water droplets that, having come from the supersaturated and supercooled zone far above, have unusually high vapor pressures at their surfaces, in comparison to the surrounding lower environmental air, and hence evaporate rapidly on the way down. This is a cooling process, and the downdrafts are the reason for the decay and disruption of the thunderstorm cell. This is because the cooling effect stabilizes the surface air, leading to a blockage of the initial updrafts that began the storm. Nearby, however, are the incipient updrafts that may grow to full storm height. While most cells have a breadth of only a few miles and often last only for an hour or two, giant cells 30 to 40 miles across have been studied and tracked for 100 miles or more. Thunderstorms are especially dangerous to aircraft, not only because of the rapid ice formation that may be found within the upper portions of these clouds but also because of the sharp boundaries between rising and falling air currents. These currents may subject plane wings to severe shearing stresses. Lightning also may fuse the electrical circuits in the plane. Today pilots of high-flying jet planes much prefer to fly over such weather disturbances, and with pressurized cabins and powerful engines, flights within the lower stratosphere at 40,000 to 50,000 feet are commonplace.

Lightning and thunder. The causes for the evolution, structure, and distribution of thunderstorms are well known. The relationship with lightning and its associated thunder is somewhat less clear. It now is believed that the major cause for lightning lies in the role of thunderstorms as equalizers in the electrical field that surrounds the earth, where the positive "pole" is found in the ionosphere and the negative "pole" is at the surface of the earth. Normally, during fair weather, there is a regular leakage of elec-

Figure 6-31 Model of a typical thunderstorm.

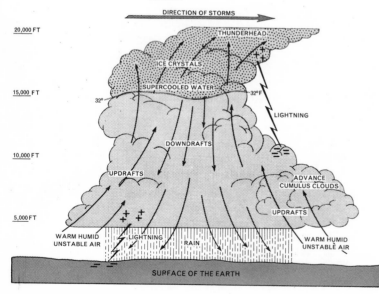

trons downward from the upper atmosphere toward the earth surface, although the mechanism of transport is not definite. The earth has a strong attraction for these tiny, negatively charged particles. The fairly regular downflow of electrons is similar to a weak electrical current. The constant growth of the negative charge at the earth surface is matched by an increasing positive charge within the ionosphere, thus producing an electric potential that increases steadily. The only way to reduce this potential is to reverse the flow of electrons into the upper air. Apparently, a steady upward flow similar to the downflow is not possible. The towering clouds of warm-weather rainstorms, however, form a possible linkage and mechanism for this return flow. The number of thunderstorms occurring around the world, estimated at about 3,000 per day, is sufficient to maintain such a flow.

6-25 TORNADOES *Tornadoes* are the smallest yet most violent major storm type. They are rotational storms, or miniature cyclonic eddies, often much less than a few hundred yards in diameter. Their funnellike form, hanging downward from a large thunderstorm cloud, is clearly visible, owing to the extremely rapid condensation within the upward-spiraling air and the dust and debris within it. Hail often precedes or follows tornadoes, indicating the strength and height of the updrafts. The atmospheric pressure in the center of tornadoes is extremely low, so low, in fact, that buildings in their path may explode outward because of the sudden drop in air pressure.[3] No accurate measurements of wind velocities ever have been taken in a tornado, instruments having been destroyed

before they could record maximum strengths. Estimates place these velocities at between 100 and 600 miles per hour. The speed of the storm cell is relatively slow, seldom more than 20 to 40 miles per hour. The funnel-shaped mass of destructive energy also is somewhat erratic in its course, sometimes bounding upward without touching the ground or winding irregularly in hilly country. Tornadoes rarely travel far, because their violence makes equilibrium precarious. The average distance covered by an individual tornado is only about $1\frac{1}{2}$ miles, although exceptional ones have been known to follow paths over 100 miles long. Courses are somewhat longer over flat, open country.

The global distribution of tornadoes is not well known because of their small size and local occurrence. They have been reported throughout the world within middle and lower latitudes. Their highest frequency appears to be in the central continental plains east of the dry climates. The United States and Australia have by far the largest reported number, averaging about 140 per year. There has been a notable increase in number in the United States during the past 15 years. Kansas, Arkansas, Iowa, Oklahoma, and Nebraska have the largest number among the states.

Tornadoes are almost always associated with frontal conditions and with a deep accompanying cyclonic disturbance, indicating sharply contrasting air masses with respect to energy potentials. Like all storms, they represent one of the safety valves that redistribute the imbalances in the global energy system.

Two minor storms that are similar to tornadoes but smaller and much less violent are *dust devils* and *waterspouts*. The former vary from 50 to 600 feet in height and are about 50 feet in diameter. Their winds are not particularly destructive. Dust devils are formed

[3]A sudden drop of 34 millibars, or 1 inch of mercury, exerts a force equal to approximately .49 lb/in.[2]. Applied to a single 30 × 8 foot wall of a frame building, such a pressure would approximate $8\frac{1}{2}$ tons. This is why houses blow apart during such storms.

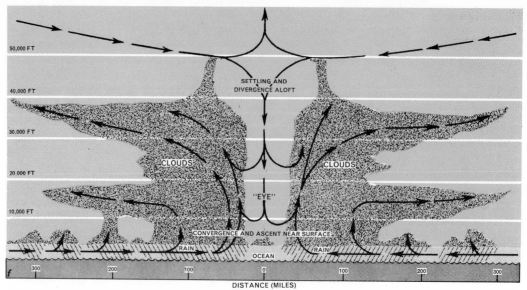

(A) CROSS SECTION OF A HURRICANE

Figure 6-32   Typical vertical and horizontal structures of tropical cyclonic storms (hurricanes and typhoons).   [After Phelps and Pollard]

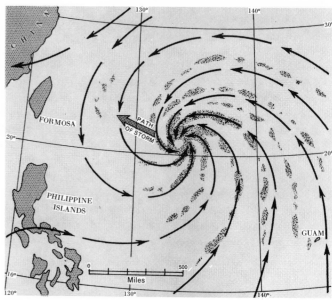

(B) A TYPHOON (HURRICANE) OFF SOUTH EAST ASIA

mainly in hot, dry areas where the extreme heating of the bare surface establishes mechanical instability in the lower air. The resultant rapid overturning produces the local vertical air columns. They have no accompanying condensation. Waterspouts are tornadoes over water, but they have much less velocity than true tornadoes and are related mainly to eddy diffusion. The characteristic pendant column consists largely of water droplets formed as the result of condensation. Only the bottom of the column is water-lifted directly from the surface. They generally travel in groups and are found mainly in the tropics and subtropics.

Much work has been done in the attempt to forecast tornadoes because of their destructiveness. As yet, however, forecasters can give warning only a few hours in advance of general conditions favorable to tornado development. They have not been able yet to predict the occurrence and trajectory of individual "twisters." It is hoped that tracking by mobile radar units flown into predicted danger zones is aiding greatly in plotting the course of individual storms, because they can be recognized on radar screens.

6-26 HURRICANES *Hurricanes* are large, violent tropical storms with convergent winds rotating about a low-pressure center. They are the most destructive storms on earth because of their great size—usually about 100 miles in diameter—although they do not have the concentrated violence of tornadoes. By definition, a tropical storm becomes a hurricane when its wind velocity averages more than 75 miles per hour. Associated features (see Figure 6-32) include the following:

1. A counterclockwise rotation north of the equator with winds spiraling inward and upward.
2. Heavy condensation and precipitation, mainly in the quadrants preceding the storm center.
3. A central "eye," with cloudless skies and warm, calm air.
4. A sudden barometric drop, sometimes as much as 1 inch of mercury within a half hour.
5. Isobars that have a symmetrically circular shape, principally during the storm climax.
6. Spiraling bands of clouds and thunderstorm centers radiating out from the center like the arms of a pinwheel (see Figure 6-34).
7. A surface tongue of unusually warm air preceding the storm.

The heavy precipitation accompanying hurricanes may result in disastrous floods, such as those associated with Hurricane Diane in August, 1955, which devastated valley sections in northeastern Pennsylvania, southeastern New York, and western Connecticut. Nearly 200 persons lost their lives in the floods, despite the fact that the storm winds by this time had diminished to only light gale force (30 to 40 miles per hour).

Hurricanes are found principally in the western parts of oceans between lat 10° and 30°. Some can be found off the west coast of Central America, where they are related to Caribbean atmospheric conditions, and several have been reported in the Bay of Bengal, southeast of India. As indicated in Figure 6-33, they are absent in the South Atlantic. This is because the equatorial zone of converging trade winds seldom lies south of the equator there. The two areas where hurricanes are most common are the Caribbean Sea and the Philippines-Formosa-South Japan region in the western Pacific. In the latter area they are generally known as *typhoons* (Chinese: *tai-fung* = big wind). Locally in the Philippines, they are termed *baguios*, and in northern Australia, *willy-willies*. Al-

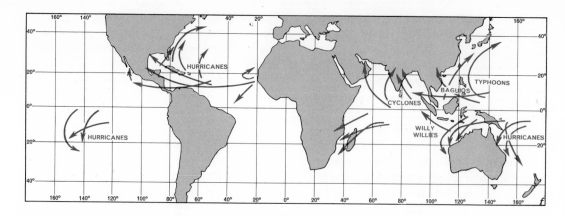

Figure 6-33 Typical paths of tropical cyclonic storms. [After Garbell]

though hurricanes may form during most times of the year, a large percentage of them occur during the late summer and fall months. In the Caribbean Sea-North Atlantic region, they generally occur from late August to the middle of October. In the western Pacific, north of the equator, they are somewhat more frequent during other seasons, but the maximum occurrence is in October and November. South of the equator, March to May is generally the hurricane season.

Although hurricanes develop from tropical cyclones, they are basically the result of conditions in the upper troposphere, in the more or less turbulent upper zone of the tropical easterlies. In this zone there are migrating cyclones and anticyclones similar to those at the earth surface, although much less strongly developed. If one of the anticyclonic centers in the upper air happens to coincide in position with the summit of one of the tropical cyclones at the surface, especially those associated with easterly waves, conditions are favorable for the lateral removal of air aloft and subsidence in the center of the cyclonic vortex. These upper-air waves apparently also play some role in steering the

course of a hurricane once it is developed, for a low-pressure center usually is observed in the upper air well in front of the storm center.

The hurricane is similar to a pump that removes warm, conditionally unstable air from near the ocean surface and forces it to spiral upward, condense, and liberate heat. An extremely warm column of descending air in the center of the hurricane helps to feed warm air into the rising spiral and check the inflow of cold air from the surrounding air that destroys the equilibrium of hurricanes. Both the strong divergence aloft and the central warm eye appear to be critical in maintaining hurricane equilibrium. The temperature of air in the eye is usually about 15°F higher than that of the air outside the rising column and may be as much as 25° to 30° higher. The potential energy of adjacent air masses with such contrasts in temperature is enormous, and this explains why such violent winds can be generated. Little air feeds into the center of a hurricane except at the bottom and the top. The latter generally is about 40,000 to 50,000 feet above sea level. Hurricanes that move away from the equator and around the flanks of subtropical anticyclones, as many of them do, tend to shrink in size and grow in intensity. This is

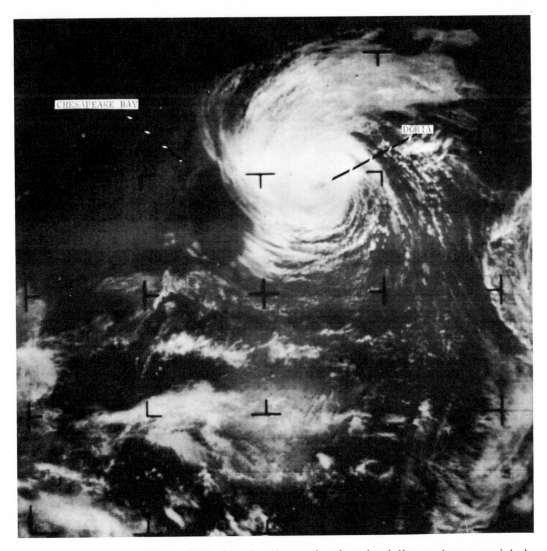

Figure 6-34 Cloud patterns showing circulation system associated with a hurricane. Hurricane Doria (September 15, 1967) is shown here as viewed by NASA's Nimbus II weather satellite off the Delaware coast. The east-west diameter of the storm is about 300 miles. Note the counterclockwise circulation system and the small, dark "eye" in the center of the hurricane. Doria was accompanied by a pair of other tropical storms, Beulah and Chloe, shown together in Figure 5-6. [NASA]

due to increased deflection owing to the earth rotation, a smaller vortex, and increased angular momentum. The smaller and more violent the hurricane becomes, the more vulnerable it is to destruction by an interruption in its equilibrium.

The paths of hurricanes, though always uncertain, tend toward normative patterns—in particular, a curved course poleward and toward the east after leaving an earlier westward direction. When hurricanes intercept continental coastlines, they soon lose much of their internal generating strength because of lack of moisture and friction with the surface. Since these severe tropical storms are found in the western portions of oceans and tend to curve eastward and poleward, many of them constitute no serious threat to continental coastlines. A depression of lower pressure over the eastern part of the continents between two stationary anticyclones, however, may lead a hurricane in this direction. Such a condition was responsible for the highly destructive hurricanes that passed northward across the eastern United States during the autumn of 1954 and the late summer of 1956. The adjacent anticyclones usually extend to high elevations and thus are able to bar the passage of the entire hurricane column.

Hurricanes have been studied perhaps more than any other type of storm, and considerable success and knowledge have been achieved in following their forward progression by radar plotting and direct observation by planes. Adequate warning, several hours in advance of these storms, has generally been available in the United States. Theoretical projections of paths well in advance of these storms, however, are not yet possible.

Study Questions

1. Explain why the rate of pressure drop with increasing altitude is not fairly constant, as it is with temperature.
2. a. Explain why the pressure patterns at an elevation of 30,000 feet (see Figures 6-5 and 6-6) are so much more regular than at the surface.
   b. On the 17,000-foot charts in Figures 6-5 and 6-6, where are the wind directions the most irregular? Why?
3. a. Explain the statement that the subpolar lows in Figure 6-4 are "statistical."
   b. Which major surface anticyclones are permanent, and which are seasonal?
4. a. Explain the short, wide arrows in the South Pacific Ocean (see Figures 6-11 and 6-12) and the long, medium-width arrows near Hawaii.
   b. In which kinds of areas are the winds the most variable and weakest? Why?
5. a. In Figure 6-13, explain the difference in the velocity of the jet streams on opposite sides of the equator.
   b. Compare summer upper-air wind velocities in the southern hemisphere with the summer velocities north of the equator. Explain.
6. Describe the cyclical changes in the major jet streams. What functions do they perform with respect to the changes in surface air movements?
7. Explain how the changes in the subtropical anticyclones influence the exchange of mass and energy across the parallels of latitude.

8. Which of the following air masses would be least likely to produce heavy, convectional rains: *mTk* or *mTw*? Why?

9. a. Distinguish between zonal and meridional airflow.

   b. What factors influence the proportion of each, and what are the mechanisms for accommodating each on the surface and in the upper air?

   c. Does most of the meridional flow at the surface occur over land or over the oceans? (see Figures 6-11 and 6-12.) Explain.

10. a. In Figure 6-18, which of the three ITC conditions would be most favorable for heavy, fairly continuous rain? Explain.

    b. At what season would this be most likely to take place? Explain.

    c. What would be the typical weather patterns within the doldrum belt in (A)?

    d. In which direction would the low-pressure center in (C) tend to move?

11. Interpret the weather changes shown in Figure 4-13 in terms of the cyclonic model shown in Figures 6-24 and 6-25.

12. Explain why California leads all states in the United States in terms of the value of its commercial fisheries.

13. Explain in terms of air mass analysis why the people in Seattle rarely hear thunder.

14. Contrast and explain the general pattern of ocean currents in the high latitudes of the northern and southern hemispheres.

15. Explain the origin and operation of a hurricane.

CHAPTER 7
THE HUMID TROPICS AND DRY CLIMATES

This chapter and Chapter 8 are concerned with portraying the general long-run condition of the atmosphere from place to place on the earth surface. One of the problems of generalizing the conditions of the earth surface on a global scale is the availability of quantitative data on which to base valid comparisons. Fortunately, in the case of climatic data, the importance of recording weather changes from day to day is well recognized. Nearly all nations today maintain a network of weather stations that have reasonably comparable observation standards and that forward their data to a central depository for dissemination. The World Meteorological Organization, an agency of the United Nations, is the main coordinating body. In 1963 it designated the U.S. Weather Bureau as the world depository agency at the Fourth WMO Congress in Geneva. The U.S. Weather Bureau's publication, *Monthly Climatic Data for the World,* also was selected to be the official world data source. Published summaries[1] of climatic data for the world usually include the mean monthly pressure (mb), temperature (°C), and precipitation (mm). The British data cited in the footnote include relative humidity but not pressure, whereas the U.S. source contains pressure data but not relative humidity. The British data also include average and absolute maximum and minimum temperatures, which are in British measurement units. Examples of long-term averages of monthly and annual temperature and precipitation for scattered representative weather stations are given in Appendix C.

The Koeppen-Geiger Climatic Classification

The climatic classification used in this book (see Appendix A for the code definitions in tabular

[1] The two most widely used current sources are: U.S. Department of Commerce, Environmental Science Service Administration: *World Weather Records,* 1951–1960, 5 vols., Superintendent of Documents, Washington, D.C., 1962–1968; and Great Britain, Meteorological Office: *Tables of Temperature, Relative Humidity and Precipitation for the World,* 6 vols., H.M. Stationery Office, London, 1962–. All of the climatic statistics in this text were taken from one or the other of the above sources.

form) was developed by W. Koeppen, a German botanist, who recognized that vegetation patterns usually reveal major gradations of climate. Geiger, a German climatologist, later elaborated the system. Koeppen placed the climatic boundaries at places where he observed major vegetation changes and defined these empirically in terms of mean monthly temperatures and precipitation. Most of his field work was done in Africa and Europe; hence, the correlation there between climate and vegetation is fairly close. The agreement between these two is not so close elsewhere, although on a global scale the exceptions are not serious. The rear endpaper of this volume reproduces the distribution of the Koeppen-Geiger climates and is based on the two sources listed in footnote 1.

As with any regional classification, the Koeppen-Geiger system is not universally applicable. It utilizes, for example, only the data on mean monthly temperature and precipitation. There is no provision for variations in the strength or constancy of winds, temperature extremes, precipitation intensity and range, amount of cloud cover, or the net radiation balance. Its greatest inadequacies perhaps lie in its application to humid and dry boundaries, and it should not be considered for land management planning purposes, where more precise and varied factors should be utilized. The Thornthwaite or Penman formulas for determining potential evapotranspiration should be consulted for such purposes (see Section 5-20). Despite these and other disadvantages, the Koeppen-Geiger classification has been used here mainly because of four reasons that have special value for classroom teaching purposes: (1) It has precise definitions that can be applied easily to standardized data that are available for locations throughout the world. (2) There is a reasonable correlation globally with major vegetation regions. (3) It

requires a minimum amount of calculation. (4) It is widely used in educational circles throughout the world.

The Koeppen-Geiger system utilizes an alphabetical code (with capital and lower-case letters) for generalizing climatic data. Some of the smaller subdivisions are omitted here, but most of the details of the code are presented in Appendix A. The student should become familiar with it before proceeding with Chapters 7 and 8. The five major climatic divisions of the system, designated by the first five capital letters in the alphabet, indicate the sequence of climates that proceed from the equator to the poles. The letter A refers to the humid climates that have no winter; the B climates are dry; C signifies the presence of a mild winter; the D climates have a severe winter; and finally, the E or polar climates have no true summer.

The humid tropics (A) and the dry climates (B) treated in this chapter are not among the most hospitable sections of the earth for humans. Yet they are far from uninhabitable, and both have nourished some of the greatest achievements of early man. Both have one important feature—the accumulation of large net surpluses of radiation. This can be troublesome and annoying to man's comfort as he goes about his daily toil, but it also provides enormous sources of potential energy. Where water is easily available, the biotic elements are able to tap the energy surpluses, as is testified by the lushness of tropical forests and desert oases. We are only beginning to realize the potential of tropical agriculture for food production. The problem is largely one of marshaling the right combinations of materials. The dry lands are deficient in water, but the high incidence of sunshine gives them the highest intake of solar energy on earth. Despite the loss of solar energy by reflection and convection, the desert usually blossoms profusely where water is available.

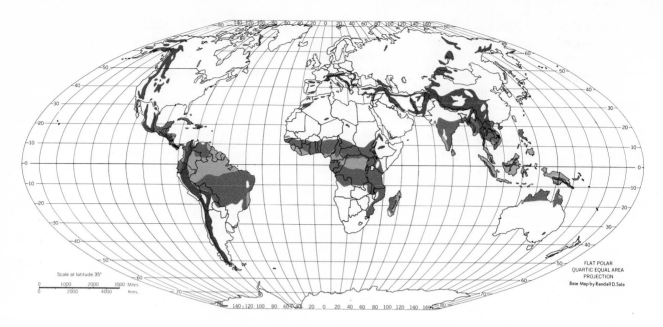

When man develops techniques for tapping the incoming radiation directly and diverting it into controlled uses, the low-latitude deserts may well contain the greatest industrial and urban concentrations on earth. There is water in the hot deserts, but it now is locked in vapor form. Cheap solar power some day may make this water available through refrigeration or air compression, or it may supply fresh water by extraction from adjacent seas. The latter method already is being used in some urban areas.

The A and B climates of the world can be thought of as areas of real challenge to modern man, because their energy potentials are greater than elsewhere on earth. For this reason, special attention will be given to the implications of these climates for human living, their advantages and their hindrances.

The A Climates (humid tropical)

7-1 GENERAL CHARACTERISTICS AND DISTRIBUTION OF THE A CLIMATES The A climates, or humid tropics, are located in a zone that straddles the equator and tends to be somewhat wider on the eastern side of continents than on the western side. Average monthly temperatures do not fall below 64.4°F (18°C), and the rainfall is sufficient to support tree growth. The major subtypes of the A climates are distinguished by differences in the amount and distribution of rainfall, ranging from continuously moist (Af) through a short compensated[2] dry season (Am) to a pronounced uncompensated dry period (Aw). The A climates have no real winter season as far as temperatures are

[2]A compensated dry period implies that sufficient rain falls during the wet season to provide moisture for normal tree growth during the dry period. Most trees in the Am climates thus do not have to shed their leaves or develop other types of special mechanisms for resisting drought, although some of them may do so. An uncompensated dry season requires such adaptive mechanisms. Koeppen's precise but empirical definition for the Am/Aw boundary is $3.94 - (r/25)$ for the driest month, where r is the mean annual precipitation. The values are in inches. The metric equivalent is $100 - (r/25)$, where the quantities are in millimeters. To illustrate, a station having a total of 75 inches of rain per year would be an Aw climate if its driest month were less than .94 inch $(3.94 - \frac{75}{25})$. It would be an Am climate if its driest month were between .94 and 2.4 inches. Table 1 in Appendix A calculates this boundary for various annual amounts.

Figure 7-1 Distribution of the A climates.

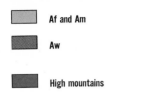

▦	Af and Am
▦	Aw
▦	High mountains

concerned because of their position in the low latitudes. Daily variations in temperature regularly exceed seasonal changes, although some of the *Aw* climates have unusually warm seasons near the end of the dry period. Representative statistical data for various types of *A* climates throughout the tropics are given in Appendix C and in Tables 7-1 and 7-2. Seasonal graphs of temperature and precipitation are given for each of the three major types of *A* climates in Figure 7-2. The characteristic daily and seasonal march of temperatures for two contrasting *Af* and *Aw* stations are presented in Figure 7-4. Their characteristics will be noted later.

The *A* climates in general extend from about lat 5° to 10° S to lat 15° N on the western side of continents and to about the Tropics of Cancer and Capricorn on the eastern sides (see Figure 7-1). The moist *Af* and *Am* subtypes are located mainly within an inner zone (nearer the equator), with the *Aw* climate occupying an outer or poleward zone, although the *Af-Am* region may extend to the margins of the *C* climates along the eastern parts of continents without the *Aw* transition. This poleward extension along continental east coasts is illustrated by the *A* climates of eastern Brazil, southern Florida, eastern Madagascar, Hainan Island off the coast of southern China, and part of Queensland in Australia.

7-2 THE RAINY TROPICS (*Af-Am*) The *Af* and *Am* climates have been included together on the world climatic map (see Figure 7-1). There is frequently little difference between the two as to major impacts on vegetation. Both are forest climates, and while a short

Table 7-1 Typical Climatic Regimes in the Rainy Tropics

	Jan.	Feb.	Mar.	Apr.	May	June	July	Aug.	Sept.	Oct.	Nov.	Dec.	Yr.
Victoria Point, Burma (Amgi) lat 09°58′ N; long 98°35′ E; elev. 154 ft (Monsoonal)													
Temp. (°F)	80.4	81.9	83.8	84.7	82.6	79.2	80.0	79.9	79.2	80.1	80.2	79.5	81.0
Precip. (in.)	1.1	0.4	1.6	4.0	16.8	27.5	25.9	27.6	28.4	15.2	6.2	1.4	156.1
Ponape I., Caroline Is. (Afi) lat 07°00′ N; long 158°14′ E; elev. 128 ft (W. Oceanic)													
Temp. (°F)	80.0	80.5	80.5	80.5	80.0	79.5	79.0	79.5	79.0	79.5	79.5	79.5	79.5
Precip. (in.)	12.3	9.1	13.0	17.4	20.6	17.9	16.8	15.6	16.1	16.1	15.9	20.5	191.3
Buta, Congo (Amgi) lat 02°47′ N; long 24°47′ E; elev. 1,436 ft (Continental Equatorial)													
Temp. (°F)	74.1	76.5	77.0	76.3	76.5	75.4	74.5	74.5	74.8	75.4	75.2	75.0	75.6
Precip. (in.)	1.1	2.5	5.1	7.0	6.4	4.6	6.1	6.0	7.5	8.8	5.2	1.9	62.2
Kisangani, Congo (Afi) lat 00°31′ N; long 25°11′ E; elev. 1,361 ft (Continental Equatorial)													
Temp. (°F)	76.8	77.0	77.4	77.2	76.8	75.9	74.7	74.7	75.6	76.1	76.1	76.1	76.3
Precip. (in.)	3.3	4.0	5.6	6.8	7.1	4.4	4.6	7.0	7.0	8.5	6.3	4.6	69.2
Cairns, Queensland, Australia (Ami); lat 16°55 S; long 145°47′ E; elev. 16 ft (E. Coastal)													
Temp. (°F)	82.0	81.5	80.0	77.5	73.5	71.5	69.5	71.0	73.2	77.0	79.0	81.5	76.5
Precip. (in.)	16.6	15.7	18.1	11.3	4.4	2.9	1.6	1.7	1.7	2.1	3.9	8.7	88.7

Figure 7-2 Climatic charts for selected A climates (Colombo, Manaus, and Caracas). Note the flatness of the temperature curves, owing to lack of seasonal variation, and the wide fluctuations in precipitation. May is an important month in Ceylon because of the water needs for growing rice. In the Colombo graph, note that even within a 10-year period, rainfall totals for May varied between 1 and 26 inches. The possible low amounts during the summer in monsoon Asia are critical to subsistence agriculturalists, who depend on the rice harvest for their lives. The end of the dry periods at Manaus and Caracas is marked by a slight rise in temperature. The cloud cover during the rainy season helps to keep temperatures down, as does the removal of latent heat by evaporation.

Colombo, Ceylon, 6°54'N.; 79°52'E., Elev. 24'

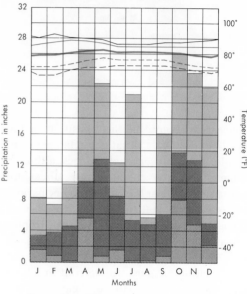

Koeppen Class: Afgi
Mean annual temperature: 80.4°F.
Mean annual precipitation: 99.4 inches

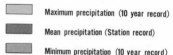

Maximum precipitation (10 year record)
Mean precipitation (Station record)
Minimum precipitation (10 year record)
Average monthly maximum temperature
Average daily maximum temperature
Mean monthly temperature
Average daily minimum temperature
Average monthly minimum temperature

dry period exists in the *Am,* the dryness is compensated by heavy rains the rest of the year. Most tropical areas have a short dry period, especially when the ITC shifts farthest away during the low-sun period; hence, the *Am* areas generally exceed those of *Af.* The rainy tropics are located near the equator, because this is the zone of most frequent air stream convergence, through which the ITC passes on its way north and south. The conversion of radiant energy to latent heat is great in this zone, and ascendance of air containing water vapor is the main process for removing surplus energy from the surface. These humid climates are not in a continuous belt along the equator, however, and continental and topographic conditions can be important in determining the details of rainfall distribution. As observed in Figure 7-1, the eastern portion of the equatorial regions of both South America and Africa illustrates this lack of continuity in the rainy equatorial belt. Extremely wide latitudinal swings in the ITC, drawn poleward by thermal, low-pressure centers (see Figure 7-6), are responsible for dry periods in these eastern, continental, equatorial regions. The highlands of eastern Africa also prevent the shallow, moist tropical easterlies from penetrating far inland on this continent. Oceans that straddle the equator, on the other hand, tend to have *Af* climates throughout their equatorial zones.

There are noticeable differences in the rainfall patterns or regimes within the broad definitions of the *Af-Am* climates. There may be, for example, single or double high- or low-rainfall periods, and the driest or rainiest times of year may occur in any month. Generally, however, rainfall tends to "follow the

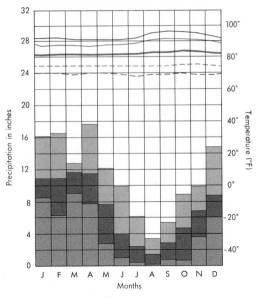

Manaus, Brazil, 3°08'S.; 60°01'W., Elev. 144'

Koeppen Class: Amgi

Mean annual temperature: 80.0°F.

Mean annual precipitation: 82.5 inches

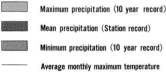

- Maximum precipitation (10 year record)
- Mean precipitation (Station record)
- Minimum precipitation (10 year record)
- —— Average monthly maximum temperature
- —— Average daily maximum temperature
- —— Mean monthly temperature
- --- Average daily minimum temperature
- --- Average monthly minimum temperature

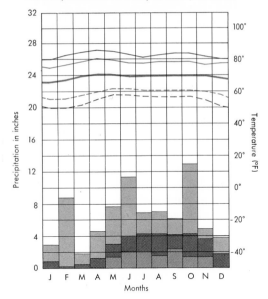

Caracas, Venezuela, 10°30'N.; 66°56'W., Elev. 3418'

Koeppen Class: Awi

Mean annual temperature: 69°F.

Mean annual precipitation: 32.8 inches

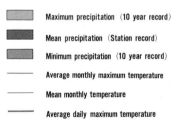

- Maximum precipitation (10 year record)
- Mean precipitation (Station record)
- Minimum precipitation (10 year record)
- —— Average monthly maximum temperature
- —— Mean monthly temperature
- —— Average daily maximum temperature
- --- Average daily minimum temperature
- --- Average monthly minimum temperature

sun." That is, the highest rainfall period tends to occur in the months when the sun stands highest in the heavens. This is related not so much to solar heating as to the seasonal shift in the ITC, which in turn tends to move northward and southward with the seasons. The air stream convergences in the tropics are treated separately in Section 7-5. The humid tropics have several exceptions to the generalization that rainfall increases with directness of solar radiation. Frequently the period of maximum rainfall is accompanied by a drop in temperature that is related to an increase in cloud cover with the onset of the rainy season. Koeppen recognized this characteristic by designating the letter g to indicate tropical climates whose maximum temperature occurs before the summer solstice, or high-sun period. Monsoons almost always exhibit this feature. This is indicated in the climatic data for Victoria Point, Burma, shown in Table 7-1. Manaus, a city on the

Amazon River, also exhibits a slight pre-solstice temperature maximum (see Figure 7-2) owing to a wind shift, but because of its continental position, the seasonal precipitation contrasts are not so great as in the true southern Asiatic monsoons. The data for Cairns, Australia (see Table 7-1), indicate that a monsoon climate does not always feature the drop in temperature with the onset of the rainy season (g). The highest monthly temperature here occurs at the height of the rainy season. The double maximum precipitation at Buta and Kisangani, both of which are in the Congo Basin near the equator, represents the swing in the ITC across this zone twice each year at or near the two equinoxes. No significant drop in temperature accompanies these rainy periods.

It has sometimes been asserted that the rainy tropics do not have weather, they have only climate. This may have some validity with respect to temperatures, since both daily and seasonal variations normally are slight, but this is by no means true of precipitation, winds, or humidity. The climatic graphs for Colombo (Af) and Manaus (Am) in Figure 7-2 both illustrate the wide fluctuations in monthly precipitation even during the relatively short 10-year period (1950–1960) which was used for the maximum and minimum figures. There also are considerable variations in precipitation locally from day to day, month to month, and year to year. Weather stations only a few miles apart may show wide differences in amounts, although over a long period of time the differences become less noticeable in the mean figures. The principal reason for the variability locally lies in the type of precipitation, which commonly is in the form of rainstorms of short duration. Nearly all tropical rain involves conditional instability, and local thunderstorms are common. The amount of rain in any one shower, however, depends on a number of variables, including the initial vapor pressure and slight variations in the lapse rate aloft. Wind conditions aloft also may interfere with the stability of thunderstorm columns. Upper-air subsidence and temperature inversions aloft, however, are exceedingly rare throughout the rainy tropics. Since thunderstorms generally are highly localized and short in duration, they are responsible for the high variability of tropical precipitation.

Nonfrontal convergences and orographic conditions are the principal initial lifting mechanisms, although in the interior of continents, convectional updrafts may induce local rainstorms. The latter now are believed to be much less important in tropical rains than was formerly thought. There is a remarkable uniformity in the general characteristics of air masses in the rainy tropics, regardless of their source areas. Some generalizations of weather differences in the rainy Af and Am climates can be stated and summarized as follows:

1. Inland positions tend to have most of their rain from mid-afternoon to late evening, while marine locations experience their maximum precipitation from early to late morning hours (see "Convectional Lifting," p. 161). Coastal positions sometimes will have a double daily maximum, derived from both continental and marine regimes.

2. Seasonal maximum rainfall periods occur during convergences of planetary air streams, especially the trades, and are marked by a lack of any daily rhythm in precipitation.

3. Wind velocities throughout most of the rainy tropics are low and irregular except during storm squalls. Atmospheric pressure differences are slight, and isobars on weather maps usually are far apart. Major exceptions include the hurricanes near the borders of the rainy tropics and in the western part of

Figure 7-3 Cumulonimbus clouds over Viti Levu, Fiji. Such thunderstorm clouds are a common sight during the afternoon over mountainous islands in the humid tropics. The orographic effect, plus conditional instability and steep lapse rates, are largely responsible. [Van Riper]

oceans, generally between 8° and 25° N and S lat. Daily land and sea breezes are a feature along most coasts in the tropics.

4. Radiation fog and low overcasts are common to inland positions from about 4 A.M., usually not disappearing until 8 to 10 A.M. This is especially true in areas of irregular surface relief. Marine fogs or low overcasts seldom occur during either night or day.

5. Overcast skies and intermittent showers that lack diurnal maximum and minimum patterns for extended periods are rare, and are associated with broad convergences, such as the ITC and easterly waves (see Section 6-22).

6. Tornadoes and waterspouts are not unknown in the rainy tropics, but are not common. It is believed that strong local convectional updrafts below a towering cumulonimbus cloud are an important requisite.

7. Temperatures are remarkably constant, and usually there is a greater diurnal than seasonal range. Figure 7-4, which shows

Figure 7-4 Isothermographs in °F for Belem and Poona. These diagrams of two localities in the humid tropics indicate the march of temperatures daily and seasonally. Note that at Belem, near the mouth of the Amazon River, which has an Af climate, there is an extremely low variation in temperature, in which the daily range exceeds the seasonal range. The outdoor temperature never gets within the comfortable range (below 70°). At Poona, India, which is a typical monsoonal climate (Awg), the extremely high temperatures that occur toward the end of the dry period and shortly before the summer solstice are indicated. December is the most comfortable month at Poona. [After Troll]

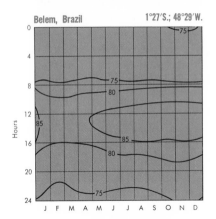

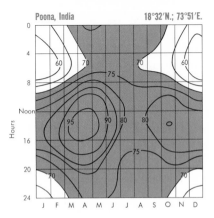

both the diurnal and seasonal fluctuations for Belem, Brazil, is representative. Low daily temperatures occur shortly before sunrise, and the highest are between 1 and 2 P.M.

7-3 TROPICAL CLIMATES WITH PRONOUNCED DRY SEASON (Aw and As)

This is sometimes termed a savanna climate, because the *savannas*, or tropical grasslands, often with scattered trees that shed their leaves during the dry period, are common in this climatic zone. Many ecologists believe, however, that most of the savannas have been culturally induced by seasonal burning, and that the dominant plants were deciduous woodland trees at one time and would be again if fire were eliminated. The small letter *w* or *s* signifies the time of year in which the dry season occurs and indicates winter or summer.[3] The *Aw*, or winter-dry, tropical climate greatly predominates over the *As*, or summer-dry, climate. The latter is the result of a unique combination of coastline configurations and orographic influences. One such area is found along part of the southeast coast of India and the northeast coast of Ceylon (see the climatic data for Trincomalee, Ceylon, in Table 7-2). Here, because of the Coriolis effect, the northwest winter monsoon in northeast India recurves to the right and becomes a northeast wind. Picking up moisture rapidly as it crosses the Bay of Bengal, it produces heavy winter rains as it meets the exposed northeast coasts in this area.

The predominance of the *Aw* over the *As* climates in the tropics is the result of two major factors: (1) an intermediate position between the humid tropics (*Af-Am*) and the major low-latitude deserts; and (2) monsoon influences. The seasonal latitudinal shift of

the sun produces a corresponding shift in the entire planetary air circulation system. During the summer period, equatorial rainy conditions prevail, with their characteristic convergent airflow, high relative humidity, and conditional instability. During the winter, on the other hand, when the sun is lower, the effect of the subtropical high-pressure cells becomes dominant in this intermediate zone. Sometimes this takes the form of typical trade wind flow, air that is warm and dry, having recently subsided from levels high in the troposphere. Upper-air inversions are common, produced either by upper-air subsidence or by the advection of lateral sheets of warm, dry air at upper levels derived from the thunderstorm cells at lower latitudes. The effect of monsoons on producing rainy summers and dry winters was discussed in Section 6-15.

The *Aw* climates generally have a somewhat greater seasonal variation in temperature than the *Af-Am* types, but this still is much less than in higher latitudes. Temperatures are related almost entirely to the transformation into sensible heat rather than into latent heat. The highest temperatures occur near the end of the dry period, when the sun at noon is higher than at the onset of "winter" but before the cloudiness and rain of the wet season begin. This tendency is accentuated in the true monsoons that experience sudden and dramatic shifts from the dry, continental winds of winter to the moist sea winds of summer. The daily and seasonal march of temperatures in a typical monsoonal climate is shown in the chart for Poona (see Figure 7-4).

There are noticeable variations in the arrangement of both temperature and precipitation within the *Aw* and *As* climates which are illustrated in Table 7-2. One such variation is a pronounced maximum of rainfall during the autumn, a special circumstance

[3]The terms *winter* and *summer* as applied to tropical regions refer to the low-sun or high-sun periods, respectively, instead of to a temperature regime.

Table 7-2 Climatic Data for Selected Aw and As Locations

	Jan.	Feb.	Mar.	Apr.	May	June	July	Aug.	Sept.	Oct.	Nov.	Dec.	Yr.

Trincomalee, Ceylon (Asi); lat 08°35′ N; long 81°15′ E; elev. 23 ft

	Jan.	Feb.	Mar.	Apr.	May	June	July	Aug.	Sept.	Oct.	Nov.	Dec.	Yr.
Temp. (°F)	78.6	79.2	81.1	83.7	85.6	85.8	85.5	84.9	84.7	82.0	79.3	78.3	82.4
Precip. (in.)	8.3	3.7	1.9	3.0	2.7	.7	2.1	4.1	3.5	9.3	14.0	14.7	68.0

Balboa Heights, Canal Zone (Aww'i); lat 08°57′ N; long 79°33′ W; elev. 118 ft

	Jan.	Feb.	Mar.	Apr.	May	June	July	Aug.	Sept.	Oct.	Nov.	Dec.	Yr.
Temp. (°F)	80.2	80.8	82.0	82.9	81.0	80.2	80.2	80.2	79.7	79.0	79.3	79.9	80.4
Precip. (in.)	1.9	1.0	.3	2.5	9.4	7.1	7.9	8.5	7.9	12.0	10.5	5.9	74.9

San Fernando, Venezuela (Awgi); lat 07°54′ N; long 67°25′ W; elev. 250 ft

	Jan.	Feb.	Mar.	Apr.	May	June	July	Aug.	Sept.	Oct.	Nov.	Dec.	Yr.
Temp. (°F)	80.1	81.7	83.8	84.2	81.1	78.6	78.1	79.2	80.6	81.0	81.0	80.4	80.8
Precip. (in.)	0	0	.2	3.4	6.7	10.8	12.9	11.7	6.9	4.7	2.0	.8	60.1

Caetite, Brazil (Awi); lat 14°05′ S; long 42°37′ W; elev. 2,880 ft

	Jan.	Feb.	Mar.	Apr.	May	June	July	Aug.	Sept.	Oct.	Nov.	Dec.	Yr.
Temp. (°F)	71.6	72.1	72.3	71.0	68.7	66.7	65.7	67.3	70.9	73.0	72.0	71.8	70.2
Precip. (in.)	4.8	3.6	4.2	2.7	.7	.4	.4	.3	.5	2.3	6.8	7.0	33.7

Bangalore, India (Aww'gi); lat 15°58′ N; long 77°35′ E; elev. 3,021 ft

	Jan.	Feb.	Mar.	Apr.	May	June	July	Aug.	Sept.	Oct.	Nov.	Dec.	Yr.
Temp. (°F)	69.6	73.6	78.3	81.1	80.4	75.7	73.8	73.9	73.9	73.9	71.1	68.9	74.5
Precip. (in.)	.1	.4	.2	1.8	4.6	3.1	4.6	5.8	5.6	7.3	2.1	.6	36.2

which Koeppen recognized by a code symbol *w′*. In most cases this is related to the heavy rainfall associated with hurricanes, which are most common at this time of year. The data for Balboa Heights in Table 7-2 illustrates this. In other cases, as in Bangalore, India, the autumn maximum is related to the passage of the ITC. The importance of the seasonal shift in this convergence zone with respect to low-latitude climates is treated in Section 7-4.

The dry winter season of the *Aw* climates is a gaunt, barren time of year when streams and water holes become dry and trees shed their leaves and enter a semidormancy. The abundant tall grasses wither, becoming brown, brittle, and crackling underfoot, and an acrid smoke haze hides distant landscapes. This is the season for burning the savannas, a practice that is followed throughout much of the world where such tropical climates occur (see Figure 7-5). The

lowest temperature inversion level is clearly revealed, because the smoke haze extends upward only to this level, where it spreads laterally in a great horizontal sheet from horizon to horizon. The appearance of this smoke layer as observed from above is remarkably similar to that seen when flying over radiational-cooling or advection inversions above mid-latitude urban areas, except that it usually lies at a higher elevation. Not all smog zones are urban in origin! In the latter part of this dry season, the sun rises and sets like a huge red ball, only its red rays penetrating the murky path near the horizon. During the forenoon the temperature rises swiftly, and in early afternoon nearly everyone seeks a shady spot or the coolness within mud walls. The horizon shimmers, dust devils play across fallow fields, and flies seem to be everywhere. The early part of the dry season is the most pleasant time of year. Temperatures are the lowest and, combined

Figure 7-5 Burning the savanna, or tropical grassland, in Colombia. This scene, taken on the *llanos* of Colombia, could be duplicated in the savannas of many other parts of the world during the dry season. The savanna in this photo is somewhat unusual in that it lacks trees of any kind. [Standard Oil Company of N.J.]

with low relative humidity, produce the most favorable comfort index of the year. Note in Figure 7-4, for example, that at Poona, India, January is a pleasant month, with temperatures rising only slightly above 80° at the warmest time of day and the early morning temperatures dropping slightly below 60° to aid in good sleeping. The month of May, on the other hand, is blisteringly hot, with afternoon temperatures over 100° and night-

time readings in the low 80s. Compare this with the even temperatures typical of the rainy tropics as shown by the isothermograph for Belem, Brazil.

The rainy season in the *Aw* climates features the typical weather patterns of the rainy tropics: variable intense showers, frequently accompanied by thunder and lightning, good visibility, high relative humidity, and low diurnal range in temperature. The

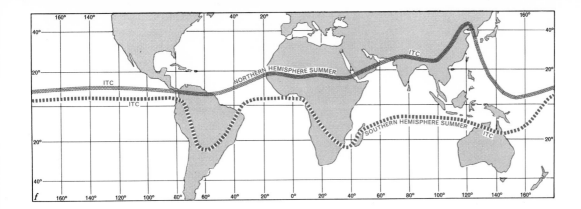

Figure 7-6 Seasonal migration of the intertropical convergence zone. Note that the ITC shifts much farther seasonally over the continents than over the oceans. [After Garbell]

onset of the rainy season always is welcome because of the long drought that precedes it, yet the comfort index is unfavorable because of the combination of high heat and humidity. The landscape changes color rapidly from the browns and yellows of the dry season. New green growth appears suddenly within a few days, and the sudden onset of life is reminiscent of the bursting of spring in the upper mid-latitudes. This is the planting and washing season, and both animals and humans seem to revel in cleansing themselves of the dust and grime accumulated during the long dry period. The times of seasonal change vary from year to year.

7-4 VARIATIONS IN THE INTERTROPICAL CONVERGENCE ZONE (ITC)

Additional material on variations in the behavior of the ITC is given here to supplement the discussion in Section 6-14, because these variations are extremely important in explaining the seasonal pattern of precipitation from place to place in the tropics. The ITC, which represents the convergence between the trades, may produce heavy convergent lifting along a strip from 1 to 50 miles wide and over 2,000 miles long. Large cyclonic eddies passing along this zone, however, may spread precipitation over a belt 500 to 600 miles wide. In some parts of the world the ITC is more strongly developed and may remain stationary for long periods during a particular season. Its location is influenced largely by the location and strength of the subtropical anticyclones and by the location of the heat lows that tend to develop over the interior of continents. Figure 7-6 shows the shifts in the location of the ITC between January and July. Note that the greatest latitudinal shift occurs over continental areas and in the western parts of oceans. Nearness to the subtropical anticyclones in the eastern parts of oceans south of the equator prevents the ITC from passing south of the equator in this area. Heat lows determine the location of the ITC in the subtropical continental areas. When the convergence is far from the coast and involves fairly dry trade wind air, it may produce little rain. An ITC usually is found in the Sahara during July, but little precipitation results. In contrast to this, the ITC that is found south of the Amazon Basin during January brings copious rains to the Chaco area. Convergence aids in precipitation only if the air is conditionally unstable.

When the ITC is farthest poleward, usually just after the solstice, the conflicting air masses are likely to present the greatest contrasts in temperature and humidity. Unusually strong and violent thunderstorm activity is present along the ITC at this season, provided that conditionally unstable air is involved. Cyclones are less likely to develop into tropical hurricanes at this time. Near the autumn equinoxes, however, when the ITC lies nearer the equator, the zone is likely to be broader and less violent; cyclonic vortexes are larger, more frequent, and more likely later to develop into hurricanes. The rainfall pattern usually is broader and more continuous. Thus, the ITC, along with its accompanying cyclonic disturbances, produces definite interruptions in the weather pattern of the tropics as it swings back and forth, bending and straightening and shifting at variable rates of speed. Intense *line squalls,* periods of overcast skies and steady drizzles, and scattered convectional thunderstorms all are associated with frontal disturbances. Once the ITC passes, normally at the beginning and end of the rainy season, weather conditions become more regular.

The passage of the ITC is clearly shown by the rainfall regimes of six weather stations in Africa, presented in Table 7-3. Their locations extend through the rainy tropics from 10° N to 25° S. At Addis Ababa there is a single rainfall maximum that occurs in August, when the ITC reaches its northernmost position. From here southward there is a double maximum period representing the passage of the ITC on its way northward and

[4]A *line squall* is a sudden violent windstorm of short duration associated with a sudden shift in wind direction along the leading edge of a severe thunderstorm.

Table 7-3 Rainfall Regimes for Representative African Stations Illustrating the Change in Periods of Maximum Rainfall from North to South*

Jan.	Feb.	Mar.	Apr.	May	June	July	Aug.	Sept.	Oct.	Nov.	Dec.	Yr.
				Addis Ababa, Ethiopia; lat 09°02′ N; long 38°45′ E; elev. 7,900 ft								
.9	1.0	2.6	3.7	2.1	4.1	9.4	_10.5_	6.8	1.7	.1	.7	43.6
				Yaunde, Cameroon; lat 03°52′ N; long 11°32′ E; elev. 2,493 ft								
.9	2.5	5.7	7.2	_8.0_	5.9	2.2	2.9	7.9	_11.8_	5.0	.8	60.8
				Coquilhatville, Congo; lat 00°03′ N; long 18°16′ E; elev. 70 ft								
3.2	4.0	_6.1_	5.5	5.2	4.7	3.9	4.3	8.1	_8.4_	7.7	4.8	65.9
				Luluabourg, Congo; lat 05°53′ S; long 22°25′ E; elev. 2,204 ft								
4.7	4.5	_7.3_	6.1	3.2	.5	.7	2.0	4.6	5.7	_9.2_	8.1	56.6
				Zomba, Malawi; lat 15°23′ S; long 35°19′ E; elev. 3,140 ft								
10.8	_11.4_	7.8	3.0	1.1	.5	.2	.3	.7	.7	5.3	_11.0_	52.8
				Lourenco Marques, Mozambique; lat 25°58′ S; long 32°36′ E; elev. 177 ft								
_5.7_	5.4	2.4	2.5	1.0	.7	.5	.7	1.8	2.0	2.6	4.7	30.0

*Data are in inches, and maximum months are underlined.

Table 7-4   Precipitation at Lourenco Marques*

	Jan.	Feb.	Mar.	Apr.	May	June	July	Aug.	Sept.	Oct.	Nov.	Dec.	Yr.
1951	1.1	.5	<u>6.2</u>	3.9	1.9	.4	.4	1.7	1.7	<u>3.1</u>	.5	1.4	22.8
1953	4.6	<u>14.7</u>	2.2	2.2	.6	0	0	0	1.9	2.2	<u>3.5</u>	2.0	33.9
1955	<u>14.5</u>	5.9	4.8	2.2	2.0	1.7	.3	0	0	3.1	<u>7.1</u>	2.9	44.5
1957	2.5	<u>7.8</u>	2.8	2.7	.5	.3	3.4	1.8	4.4	3.6	.8	4.6	35.2
1959	1.9	1.6	1.6	.5	<u>5.0</u>	.1	.5	.2	<u>4.1</u>	2.0	1.2	2.7	21.4

*Data are in inches, and maximum months are underlined.

southward. Note that at the equator (Coquil-hatville), the maximum rainfall period tends to occur almost exactly at the time of the equinoxes. Farther south, the two maxima again tend to converge, but in the southern hemisphere summer, reaching a single maximum in January at Lourenco Marques. Again, it should be understood that the statistics represent long-term averages, and the swinging of the ITC through its latitudinal course is far from regular, being influenced each year by the uneven distribution of energy surpluses and deficiencies on a global scale. To illustrate, Table 7-4 shows the precipitation data for five years at Lourenco Marques, Mozambique. The maximum rainfall months show a noticeable variation from the mean pattern indicated in Table 7-3.

7-5   EXCEPTIONS TO THE CLIMATIC GENERALIZATIONS IN THE RAINY TROPICS   The previous section indicated in several places that most weather stations in the rainy tropics experience significant fluctuations in precipitation from the statistical averages. Similarly, there are several areas in the low latitudes that have precipitation norms considerably different from those expected. Some of the principal anomalies are given in the following subsections.

The north coast of Venezuela.   Although the north coast of Venezuela (between 10° and 12° N lat) is well within the latitudinal range of the humid tropics, much of it is semiarid to arid. The prevailing wind direction here throughout the year is typically from the east, paralleling the coast or at a slight angle to it. The fingerlike eastern extensions of the Andes Mountains, jutting obliquely into the Caribbean Sea, form bold headlands and intercept the trade wind flow, thus producing copious precipitation on their windward flanks. The intervening coastal lowlands, however, lying in the lee of these ridges, normally have little rainfall. The lowland bordering Lake Maracaibo, for example, lies in the rain shadow of the high Cordillera de Merida and is a true desert (see the climatic data for Maracaibo in Appendix C). The narrowness of the continent in Central America is not conducive to the northward swing of the ITC, which usually does not pass north of the Guiana highlands. For this reason, the trade wind flow in the southern Caribbean rarely is interrupted.

Northeastern Brazil.   Northeastern Brazil is located along the eastern side of the continent and within 10° of the equator, yet it has an Aw climate, with small areas of BS, instead of the Af or Am that might be expected because of its position. This condition is largely the result of the dominance of trade wind air that is exceptionally stable, having a low lapse rate aloft due to subsidence (the

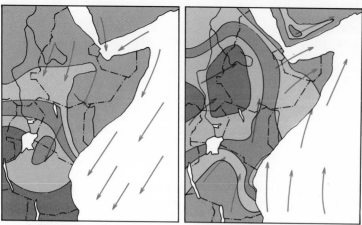

A. May to November   B. November to May

Precipitation in inches

Under 5
5 - 10
10 - 20
20 - 30
Over 30

←—— Prevailing wind direction

Figure 7-7 Wind and rainfall patterns in eastern Africa. Note the influence of the Indian Ocean monsoon on wind direction and rainfall along the coast of Somalia. Because the winds parallel this coast at both seasons, there is little convergence or orographic influence.

reason for this is that the monsoon winds of the western Indian Ocean parallel this coast during both monsoon seasons; hence, no importation of moist air into the interior is possible. The normal tropical easterlies thus are not present as they are south of the equator. The uplands of Kenya and Tanzania intercept more rain than the dry Somalia coast, but not enough to make them continuously humid. During the winter season, wind directions are from the desert areas to the north. The high elevation of the East African interior plains, ranging from 4,000 to 7,000 feet, results in temperatures somewhat below the minimum for the *A* climates. Thus, they are mostly *Cw* climates, but without the seasonal temperature contrasts usually found within the *C* group. These highland tropical climates are discussed at greater length in Chapter 8.

Mountainous tropical islands. High, rugged islands in the humid tropics generally produce a wide variety of rainfall and temperature patterns, depending on exposure to prevailing winds and elevation. The monsoonal reversal of winds that occurs throughout much of Indonesia produces an added complicating factor. Weather stations located on opposite sides of a high mountain range that crosscuts the monsoonal airflow may have pronounced wet and dry periods at opposite times of the year at the two sites. The high islands of the western Pacific, Indonesia, and the West Indies exhibit especially sharp contrasts in rainfall regimes, in both total amount of rainfall and seasonal variation.

South Atlantic subtropical anticyclone being nearby). The shape of the coast in this sector also is a factor, acting as a wedge to divert the air streams away from the interior. A third factor is that the ITC rarely enters this portion of Brazil, usually leaving the continent somewhere between the mouth of the Amazon and Fortaleza during the summer season (see Figure 7-6). Farther south, the steep highlands back of the coast result in greater convergence and higher rainfall for the coastal zone.

Eastern Africa. A glance at a rainfall map of East Africa (see Figure 7-7) clearly indicates the unexpected feature of a dry coastal zone northward from the equator. The main

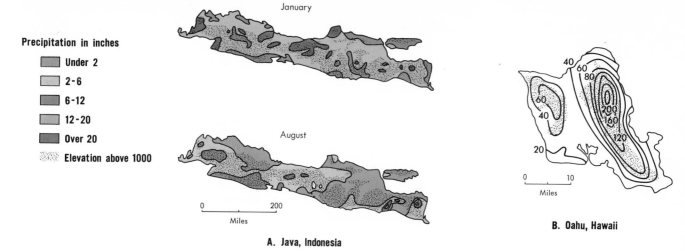

Precipitation in inches

- Under 2
- 2-6
- 6-12
- 12-20
- Over 20
- Elevation above 1000

January

August

0 — 200
Miles

A. Java, Indonesia

0 — 10
Miles

B. Oahu, Hawaii

Climatic generalizations are highly unreliable in such areas. Figure 7-8 illustrates some of these contrasts.

7-6 PROBLEMS OF HUMAN ADAPTABILITY IN THE HUMID TROPICS Although much more is known now than before regarding the physiological adaptability of man in the humid tropics, many of the long-range climatic effects still are inadequately understood. It has not been easy to eliminate psychological and social factors in studies of physiological reactions. Experiments in acclimatization during and following World War II, however, have led to a number of reliable conclusions about human physiology which include the following:

1. Combinations of high heat and humidity can produce actual cell deterioration in the human body, but such combinations are exceedingly rare on earth.

2. Given a reasonable adjustment period,

Figure 7-8 Rainfall patterns on mountainous tropical islands. The seasonal precipitation maps of Java indicate the typically complex rainfall patterns on such islands. The Oahu map indicates that mountains in the trade-wind zone may produce extreme contrasts in rainfall within short distances.

the human body is remarkably capable of adapting itself to increases in skin temperature—through changes in metabolic rate (utilization of oxygen), cooling mechanisms, and various other systemic adjustments.

3. Clothing and diet are highly important in maintaining body comfort in high heat and humidity. Clothing should be loose, absorbent, and kept to a minimum. Constrictive articles of clothing, such as belts, neckties, and garters, should not be worn. They not only restrict capillary blood circulation but also check air circulation and provide sites for fungus infections. Alcohol and foods high in fats and sugars should be avoided as much

as possible. The more slowly but easily digested starchy foods, fruits, reasonable quantities of proteins, and the various protective foods, such as vitamins, are advised.

4. Regular physical exercise appears to be an important requirement in maintaining body "tone" under conditions of high heat and humidity.

5. Racial differences apparently are of little significance in general physical adaptability to climate, although skin color is important in the selected absorption of direct solar radiation.

The psychological effects of high heat and humidity are important throughout the tropics. The most important effect is the reduction of the desire for physical activity, and this general physical inertia also affects mental activity. Several tropical countries move their government administrative offices to summer capitals at higher elevations to escape the more unpleasant time of the year, and hill stations are becoming more and more popular sites for middle- and upper-class residential purposes. The rainy tropics unquestionably can be discomforting but need not be limiting in terms of human effort or achievement.

Climate often has been blamed for the supposedly high incidence of disease in the tropics, yet this statement is highly debatable. There undoubtedly are many diseases, such as yellow fever, yaws, dengue, and leprosy, that occur mainly in the humid tropics, but there also are many others that are endemic to much cooler climates. Medical science has been much more rewarding in controlling mid-latitude diseases, such as measles, smallpox, or diphtheria, than many of the tropical disorders, mainly because of the relatively greater research efforts in the mid-latitudes.

The reservoir of small primates in the tropical forest areas may well be an especially troublesome factor in the control of many diseases that affect humans in the tropics. Malaria, for example, has almost been eliminated as a serious problem from the mid-latitudes, but its control in the tropical forest areas presents exceptional problems because monkeys can act as intermediate hosts for the malarial protozoa that are transmitted by mosquitoes. The tropical forests, with their primate populations, also are believed by some authorities to be latent breeding grounds for some of the virus infections that periodically develop highly prolific mutant strains, "explode" out of the tropics, and bring about some of the influenza epidemics of the mid-latitudes, often by way of the dense populations in southeast Asia.

Poverty undoubtedly has been a prime factor in giving the humid tropics a bad name as regards disease incidence. Living standards, while improving, still are relatively low for a large part of the tropical population. With poverty appear such side effects as inadequate diet, improper sanitation, and a general insufficiency of medical care and disease prevention. Dietary deficiencies, particularly an inadequate protein intake, produce serious body disorders, such as *kwashiorkor,* that formerly were incorrectly blamed on unknown viruses. Fungus disorders of the skin and various parasitic worms and protozoa are frequently met with in the tropics, but they usually can be avoided by proper precautions in clothing, footwear, and personal hygiene.

The variability in the length of the wet season near the outer margins of the rainy tropics is perhaps more significant in human activities than either the direct impact of climate on body functioning or its indirect effects in fostering certain diseases. The *Aw* climates, or those with a pronounced dry season, are notorious as disaster areas. Three such areas stand out in the world: (1) part of the savanna country that stretches across the continent of Africa south of the Sahara; (2) central and eastern India; and (3) northeastern Brazil. In each of these areas, populations are moderate to dense, standards of living are low, self-sufficient economies predominate, and famines occur periodically. The people of these regions subsist largely by agriculture, and crop planting and growth depend greatly on the arrival of the wet season within fairly narrow time limits. The change from the hot, dry period to the wet season in most *Aw* climates comes suddenly with the passage of the ITC. The movement of this zone, then, is extremely critical. It shifts roughly with the sun, but with by no means the same regularity, since its more immediate regulators are the changes in the position and intensity of the subtropical anticyclones on both sides of the equator. Meteorologists are only beginning to perceive the three-dimensional operations of these great anticyclonic cells and the jet streams that feed into them from above.

Until long-range forecasting and its dissemination are possible in the tropics or until the respective countries can absorb the economic and social consequences of variations in the movements of the intertropical convergence, dense populations in *Aw* climates will continue to experience difficulties in planning their agricultural activities, and large losses of life due to droughts will continue to occur where living standards are depressed.

Floods may also cause much damage within these marginal tropical climates. Near the dry margins, stream courses may not be able to hold the abnormally high rainfalls that accompany occasional tropical cyclones or stagnated convergences. Hurricanes are a seasonal hazard in the east coast areas between lat 10° and 30°. Throughout the humid tropics, much heavier downpours of rain occur than are usually experienced in midlatitudes. The large quantities of water absorbed by hillside soils during typical thunderstorms lead to landslides and mass slumping, especially where the forest vegetation has been destroyed or greatly reduced.

The greater number and activity of decay-producing bacteria in the tropics, as compared with the higher latitudes, result in an additional problem: preserving dead organic materials, particularly food, wood, paper, and leather. Food preserving must be done immediately. Wooden structures, such as frame houses, fence posts, and railroad ties, last for much shorter periods than they do in drier or cooler climates. Certain types of wood are much more resistant to decay than others, and impregnation with decay-inhibiting chemicals such as creosote is frequently an effective preventive measure. Galvanized-iron or sheet-aluminum materials are useful for the outer surfaces of more durable buildings but are expensive. A major benefit received from air conditioning is the preservation of materials within building interiors.

The *B* Climates
(arid and semiarid)

7-7 GENERAL DISTRIBUTION AND DEFINITION OF THE DRY CLIMATES The dry climates of the world are found characteristically over the eastern parts of oceans, roughly between lat 15° to 30° N and S, extending onto the continents and thence inland and poleward to approximately lat 45°. They are relatively rare along the eastern margins of continents, although South America (Patagonia) and Africa (Somalia) both have dry east coast regions. For the most part, deserts are areas where a subsidence

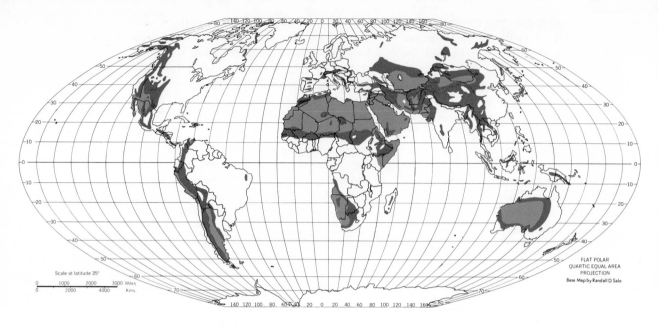

FLAT POLAR
QUARTIC EQUAL AREA
PROJECTION
Base Map by Randall D. Sale

Scale at latitude 35°

0   1000   2000   3000 Miles
0   2000   4000 Kms.

of air associated with subtropical anticyclones prevails over an ascendance, where
lapse rates and relative humidities are low,
and where there is relatively little interaction
between contrasting air masses.

Orographic dryness, or that caused by a
position in the lee of high mountains, also
is responsible for some of the dry lands. A
special feature of the Patagonian desert of
Argentina is that the driest section does not
lie close against the eastern slopes of the
Andes, but some distance to the east. This
probably is related to the subsidence of air
in a wavelike form and is produced by turbulence in the airflow across the mountain
divides. Eddying on the lee slope may even
produce a weak upflow of air along the eastern slopes. Aiding in the dryness of this unusual east coast desert is the cold Falkland
Island Current offshore, which stabilizes any
air that might intrude onto the continent
from the east. The orographic influence of the
mountain chains of the western United States
is responsible for the westward displacement

of the interior dry lands close against the
eastern mountain slopes in Colorado.

A decrease in evapotranspiration normally
checks the poleward extent of deserts in the
interior of continents at about 40°. In Asia,
however, the dry lands extend to about 50°,
owing to the large size of the continent, the
towering mountain barriers along its eastern
and southern flanks, the long reach from
Atlantic maritime sources, and a powerful
winter high-pressure zone of subsidence.

Dryness is a relative term, and definitions
of what constitute arid, semiarid, and subhumid conditions are subject to various interpretations. Dryness is related not only to
the amount of precipitation but also in large
degree to evapotranspiration, as was indicated in Section 5-20. This, in turn, has several controlling variables, the most important
by far being the net radiational surplus, or
energy available to supply latent heat. The
interrelationship between precipitation, the
net radiation balance, and the availability
of moisture for plant growth is obvious. Ten

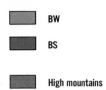

Figure 7-9 Distribution of the *B* climates.

- BW
- BS
- High mountains

Table 7-5 Formulas for the Determination of the Koeppen-Geiger *B* Boundary

	English system $(R = \text{in.}; T = °F)$	Metric system $(R = \text{mm}; T = °C)$
Summer dry season	$R = .44T - 14.0$	$R = 20T$
Even distribution	$R = .44T - 8.5$	$R = 20(T + 7)$
Winter dry season	$R = .44T - 3.0$	$R = 20(T + 14)$

inches of precipitation in Alberta or northern Finland is sufficient to support a forest growth, because much of this moisture remains in the ground, little of it being evaporated in the cool air where the relative humidity always remains high and where the net radiation surplus is low. Ten inches in Arizona or in Iraq, on the other hand, will support only a sparse covering of desert shrubs, because the hot, dry air of summer pulls the moisture out of the ground like a suction pump. The term *rainfall effectiveness* sometimes is used to designate the degree to which precipitation is made available for plant growth. Among the many variables involved in it, other than the amount and seasonal distribution of precipitation and temperature, are the intensity of precipitation, wind velocity, relative humidity, soil permeability, rate of runoff of surface water, cloud cover, slope, and type of vegetation cover. A general system of climatic classification for global use including all of these variables would be impossible, since insufficient data are available.

The dry-humid boundary, or the border of the *B* climates, according to the Koeppen-Geiger classification, is established according to empirical mathematical formulas using three variables: average annual rainfall,

average annual temperature, and the seasonal distribution of the rainfall. These formulas are given in Table 7-5. The definition of the threefold division based on seasonal rainfall distribution is as follows:

1. A summer is termed dry if the rainiest month of winter has *three* times as much rain as the driest month of summer.

2. A winter is termed dry if the rainiest month of summer has *ten* times as much rain as the driest month of winter.

3. An even distribution prevails when neither of the above conditions prevail.

The values corresponding to the boundary between semiarid and arid climates (*BS/BW*) are half those given for the *B*/humid boundary as shown in Table 7-5. Table A-2 (Appendix A) gives the numerical values for the *B*/humid boundary using English measures; they are derived from the formulas in Table 7-5. The *BS/BW* values may be obtained by dividing the figures in the Appendix by two. To illustrate the procedure in identifying *B* climates, the following example is given:

Problem: To determine whether or not the following climatic data conforms to a *B* climate, and if so, whether it is an arid (*BW*) or a semiarid (*BS*) climate.

	Jan.	Feb.	Mar.	Apr.	May	June	July	Aug.	Sept.	Oct.	Nov.	Dec.	Yr.
Temp. (°F)	86	85	85	83	76	71	70	72	77	81	85	86	80
Precip. (in.)	5.0	6.4	3.8	1.4	0.4	1.2	0.3	0	0.1	0	0.9	3.5	23.0

Procedure: Note first that this is a southern hemisphere station, with its summer season between November and March. The distribution of rain is almost entirely during the summer. The rainiest month of summer (6.4 inches) has well over ten times the driest month of winter (0 inches). The appropriate formula thus is $R = .44T - 3.0$. Substituting 80 (average annual temperature) for T in the formula yields $R = 35.2 - 3.0 = 32.2$ inches. Since the annual rainfall at the station is 23 inches, the station is within the B climates. Also, since 23 inches is more than half of the 32.2-inch figure, the station is semiarid rather than arid, and hence bears a *BS* classification in the Koeppen system.

Besides the division of dry climates into arid (*BW*) and semiarid (*BS*), the system further distinguishes the hot, low-latitude B climates with the letter h ($h = heiss = $ hot) and the cooler mid-latitude types with k ($k = kalt = $ cold). The isotherm separating the two is based on the average annual temperature of 64.4°F (18°C). Unbalanced seasonal distribution of precipitation is symbolized by the use of s when the rainiest month of winter has three times the precipitation of the driest summer month and w when the rainiest month of summer has ten times the driest month of winter. The absence of an s or w indicates the lack of a pronounced seasonal imbalance. A *BSks* climate, for example, is a semiarid variety with cold winters and a predominance of rain during the winter season. Several varieties of B climates are indicated in Table 7-6.

A special symbol n ($n = neben = $ fog) is used to designate those areas that have an unusually high prevalence of fog. They are especially common where the B climates are found along continental margins, especially

Table 7-6 Climatic Summaries of Selected Types of *B* Climates

	Jan.	Feb.	Mar.	Apr.	May	June	July	Aug.	Sept.	Oct.	Nov.	Dec.	Yr.
Kamloops, B.C., Canada; lat 50/43 N, long 120/25 W; elev. 1,131 ft — Koeppen Classification: BSk													
Temp. (°F)	22.6	27.7	38.7	49.6	58.5	64.2	70.0	67.6	59.4	47.5	35.2	28.2	47.5
Precip. (in.)	1.1	.7	.4	.4	.8	1.5	.9	1.0	.7	.8	.7	1.0	10.0
Karachi, W. Pakistan; lat 24/55 N, long 67/09 E; elev. 72 ft — Koeppen Classification: BWhw													
Temp. (°F)	63.9	68.4	76.1	82.8	86.9	88.5	86.2	83.8	83.5	81.7	75.0	67.3	78.6
Precip. (in.)	.3	.5	.2	.1	0	.3	4.1	1.8	1.1	.1	.1	.2	8.8
Gabes, Tunisia; lat 33/53 N, long 10/06 E; elev. 16 ft — Koeppen Classification: BWhs													
Temp. (°F)	52.2	54.9	59.2	63.3	69.1	75.9	79.7	81.1	78.1	71.2	62.1	54.5	66.7
Precip. (in.)	.7	.7	.7	.7	.3	.1	0	0	.5	1.6	1.2	.7	7.2
Port Augusta, S. Australia; lat 32/29 S, long 137/45 E; elev. 18 ft — Koeppen Classification: BWh													
Temp. (°F)	78	78	73	67	60	55	53	56	60	67	72	76	66
Precip. (in.)	.6	.5	.7	.7	1.1	1.1	.7	.9	.9	.9	.7	.6	9.4

Locality	Av. ann. precip. (in.)	Locality	Av. ann. precip. (in.)
Antofagasta, Chile	.02	Yuma, Ariz., U.S.	3.0
Asswan, U.A.R.	.08	Mossâmedes, Angola	3.15
Kufra, Libya	.08	Dalbandin, W. Pakistan	3.3
Wadi Halfa, Sudan	.12	Krasnovodsk, U.S.S.R.	3.6
Aoulet, Algeria	.60	Nasiriya, Iraq	4.8
Aden	1.5	William Creek, S. Australia	5.0
Jedda, Saudi Arabia	2.2	Abadan, Iran	5.0

the west coasts between lat 15° and 30°. These fogs are related to the upwelling of cold water offshore and the resultant cooling of the air above.

The *B* climates differ considerably in the amount of precipitation. Table A-2 indicates that under high temperatures and where the rainfall occurs mainly in the hot season, annual totals of over 30 inches are insufficient to produce a humid climate. At the other extreme, in the cool climates of high latitudes, evaporation is so low that *B* climates cannot occur even when the precipitation falls below 5 inches per year. The exceptionally dry deserts are not extensive. Some rain falls even in the driest areas, although many years may go by without any. The driest deserts in the world include the Atacama Desert of southern Peru and northern Chile (unique in having such dry conditions close to the coast), the central Sahara, and southern Arabia. Examples of annual averages for some of the driest localities in the *B* climates are given in Table 7-7. The annual averages shown in this table are somewhat misleading in that with such small quantities, short, heavy rains may influence averages for many years. To illustrate, Wadi Halfa, ranking fourth in the table, had experienced only one measurable amount of rain (1 millimeter in August, 1952) in 20 years, yet in 1960, 14

millimeters fell in October and 6 in December. This is an extreme case, but irregularity is a feature of most desert rainfall. To illustrate this point further, using a desert station that is not extremely dry, see Table 7-8, which clearly illustrates the limitations of using averages when dealing with desert stations.

7-8   CHARACTERISTIC TEMPERATURE PATTERNS IN THE *B* CLIMATES   The previous section indicated wide irregularities in the precipitation of dry climates. Variability in temperature also characterizes these climates. The high luminosity and low cloud cover of most dry lands tend to produce a high daily range of temperatures. The desert surfaces heat rapidly after daybreak during the summer season (see Figure 7-10), and nearly all of the solar energy reaches the surface. While it is true that the large amount of bare ground reflects more insolation than the surface of humid areas, enough is absorbed to raise ground surface temperatures to high levels. Lapse rates during the day are exceedingly high within a short distance above the surface. Convectional currents produced by mechanical instability (a lapse rate greater that 19°F per 1,000 feet) are active throughout the day but rarely lift high enough to produce clouds. The difference between sun and shade temperatures is no-

Table 7-8   Precipitation Record for Yuma, Arizona in mm* (1951–1960)

	Jan.	Feb.	Mar.	Apr.	May	June	July	Aug.	Sept.	Oct.	Nov.	Dec.	Yr.
1951	10	T†	T	10	T	0	14	29	T	7	3	3	76
1952	9	7	16	9	0	T	1	0	T	0	9	25	76
1953	T	3	5	T	0	T	T	1	0	T	T	T	9
1954	1	1	6	0	T	T	3	1	T	3	1	T	16
1955	33	T	0	0	0	0	9	33	0	T	T	T	75
1956	1	2	0	T	0	T	5	0	T	T	0	0	8
1957	17	1	7	3	0	0	T	26	0	68	T	1	123
1958	T	46	9	T	1	T	27	2	T	14	2	0	101
1959	1	6	T	T	1	0	3	3	1	1	T	27	43
1960	19	2	1	T	T	1	T	T	11	T	T	3	37
10-yr. mean	9.1	6.8	4.4	2.2	0.2	0.1	6.2	9.5	1.2	9.3	1.5	5.9	56.4
Long-range mean	10	9	6	2	T	T	6	13	10	10	3	8	77

*For a simple conversion into inches, use 25 mm = 1 in.   †T = Trace

tably high, and any shade is welcome. Single gaunt acacia trees in the Sahara sometimes are preempted as shade shelter by nomadic families, and in the southwestern United States gasoline stations lure customers with the sign "FREE SHADE" next to a small clump of cottonwoods.

Immediately following sundown the surface temperature drops suddenly, as does the lapse rate. Heating of the atmosphere by the absorption of terrestrial radiation, however, continues throughout the night and is influenced greatly by the content of water vapor. Some dry lands have unusually high vapor

Table 7-9   Seasonal Temperature Ranges at Selected Localities (°F)

Location	Latitude	Av. ann. precip. (in.)	Koeppen class.	Av. ann. max. temp.	Av. ann. min. temp.	Range
B CLIMATES						
Lima, Peru	12°05 S	0.7	BWh	89°	51°	38°
Maracaibo, Venez.	10°39 N	15.2	BWhw	99	69	30
Salala, Oman	17°03′ N	3.2	BWh	102	57	45
Bilma, Niger	18°41′ N	.9	BWhw	108	45	63
Yuma, Ariz., U.S.	32°45′ N	3.0	BWh	113	31	82
Peshawar, Pakistan	34°91′ N	13.5	BShw	116	31	85
Kashgar, U.S.S.R.	39°24′ N	3.2	BWh	102	0	102
Reno, Nev., U.S.	39°30′ N	7.1	BSks	98	−1	99
Kazalinsk, U.S.S.R.	45°46′ N	4.9	BWk	103	−20	123
HUMID CLIMATES						
Baker, Ore., U.S.	45°46′ N	11.8	Cfb	96°	−10°	106°
Minneapolis, Minn., U.S.	44°53′ N	28.6	Dfa	99	−21	120
Eastport, Maine, U.S.	44°54′ N	38.8	Dfb	86	−12	98
Bordeaux, France	44°50′ N	35.5	Cfa	96	19	77
Belgrade, Yug.	44°48′ N	24.6	Cfa	99	4	95

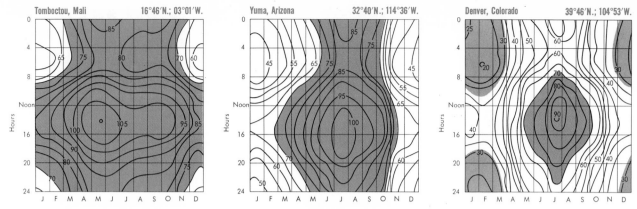

Figure 7-10 Isothermographs for Tomboctou (Timbuktoo), Yuma, and Denver.
Temperatures in the subtropical deserts, as at Tomboctou or Yuma, feature high daily ranges throughout the year, and wide seasonal contrasts as well. Denver is typical of the higher mid-latitude dry lands in the continental interior. Note the coolness of the nights in summer and freezing during the winter, related to Denver's high elevation. Despite the long period of high temperatures at Tomboctou, in the southern Sahara Desert, the nights are cool during December and January. [Tomboctou diagram after Trewartha]

pressures, such as those near warm seas[5] or extensive marshlands, and experience extremely uncomfortable nights because the temperatures remain fairly high while the relative humidity climbs to nearly 100 percent.

The highest daily ranges of temperature generally are related more to the height to which the daytime temperature rises than to the level to which the nighttime temperature drops. For this reason, the low-latitude deserts in the interior of continents have the greatest diurnal ranges on earth. A combination of a large landmass and a subtropical

position produces extremely high diurnal ranges in the Sahara Desert. A range of temperature within a single 24-hour period has been known to reach 60° in this area, although this is exceptional. Individual daily ranges of 30° to 40°, however, are common.

An additional feature of the desert temperature patterns is that the daily maximum temperatures during the hot season consistently reach near the absolute maximum, and there are few breaks in this heat. At Riyadh, in the interior of Saudi Arabia, for example, the average daily temperature for June, July, and August remains at 107°, while the absolute maxima recorded for these months are 113°, 113°, and 112°, respectively. Even at Yuma, Arizona, which is within reach of occasional importations of cooler air masses, the average daily maximum for July is 106°, while the absolute maximum recorded was 120°, and the average monthly maximum for this month is 113°. To illustrate further the continuity of summer desert heating, a desert oasis in southern Algeria (in Salah) experi-

[5]Muscat, near the entrance to the Persian Gulf, has the dubious distinction of having one of the highest average combinations of relative humidity and temperature on earth. In August, when the daytime temperatures average 92° and the nights 84°, the relative humidity averages 82 percent at 8 A.M. and 80 percent at 4 P.M. Unlike most parts of the world, the relative humidity is lowest in winter and highest in summer. This is related to a sea breeze which brings to the coastal areas additional moisture during the day. The Persian Gulf and the Red Sea regions both are infamous for the debilitating effect of their summer heat.

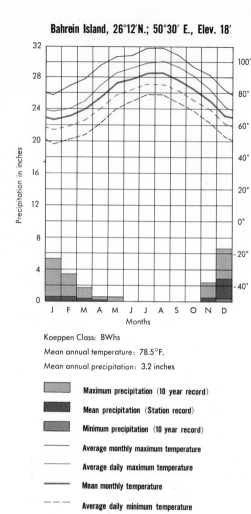

Figure 7-11 Climatic charts for Bahrein Island, Santiago del Estero, and Kazalinsk.

Bahrein Island, 26°12′N.; 50°30′ E., Elev. 18′

Koeppen Class: BWhs

Mean annual temperature: 78.5°F.

Mean annual precipitation: 3.2 inches

- ▢ Maximum precipitation (10 year record)
- ▢ Mean precipitation (Station record)
- ▢ Minimum precipitation (10 year record)
- — Average monthly maximum temperature
- — Average daily maximum temperature
- — Mean monthly temperature
- – – Average daily minimum temperature
- – – Average monthly minimum temperature

enced daily temperatures during 1931 that reached 118° or more for 45 continuous days!

The seasonal range in temperatures in dry lands is largely a function of latitude. Table 7-9 lists a series of weather stations with *B* climates that are arranged according to latitude. The seasonal range of temperatures is related closely to the length of day, to the height of the sun at midday, and to the degree of continentality. The influence of the latter is shown in Table 7-9 by the variation in seasonal temperatures near the parallel of 45°. Note that Minneapolis has a greater seasonal range than Baker, Oregon, and one almost as great as that at Kazalinsk, despite having a much higher rainfall. Bordeaux, France, on the other hand, has a relatively low seasonal range for its latitude because of its position on the windward side of the continent. The consistent increase in seasonal range of temperature with latitude in the *B* climates, as indicated in Table 7-9, holds true for all continents.

The climatic charts shown in Figure 7-11 indicate long-range temperature averages for two subtropical desert stations and one mid-latitude desert. The much higher seasonal range of temperature at Kazalinsk is illustrated. Note that there is a pronounced difference between the average daily range of temperature at Bahrein Island and Santiago del Estero. The much lower diurnal range of the former is related to its oceanic position and also to an unusually high relative humidity which varies only slightly during the 24-hour period (see Table 7-10).

7-9 WINDS Deserts generally are windy and gusty places, especially during the day, when steep lapse rates cause convectional overturning of air near the ground. Bare rock, soil, and sandy surfaces differ greatly in their reflectivity of solar energy, and the conductivity of these materials also influences their surface temperatures during the daylight hours. Surface temperatures may vary as much as 15 to 30° within a few feet, and the resultant local temperature and pressure gradients result in strong but localized air movements. The sparseness of desert vegeta-

Santiago del Estero, Argentina, 27°47′S.; 64°18′W., Elev. 653′

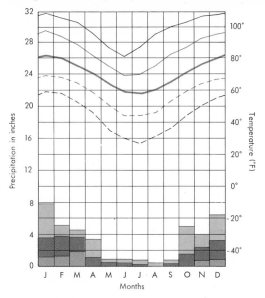

Koeppen Class: BShw

Mean annual temperature: 71°F.

Mean annual precipitation: 20.4 inches

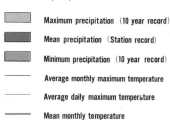

Maximum precipitation (10 year record)

Mean precipitation (Station record)

Minimum precipitation (10 year record)

—————— Average monthly maximum temperature

·············· Average daily maximum temperature

—————— Mean monthly temperature

— — — Average daily minimum temperature

— — — Average monthly minimum temperature

Kazalinsk, U.S.S.R., 45°46′N.; 62°06′E., Elev. 207′

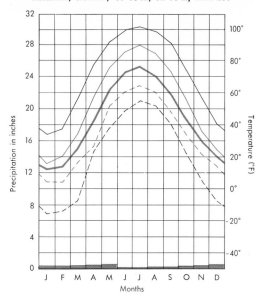

Koeppen Class: BWk

Mean annual temperature: 45°F.

Mean annual precipitation: 4.9 inches

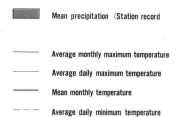

Mean precipitation (Station record)

—————— Average monthly maximum temperature

·············· Average daily maximum temperature

—————— Mean monthly temperature

— — — Average daily minimum temperature

— — — Average monthly minimum temperature

tion results in a relatively small frictional loss of energy by desert winds, and the sand-blasting action of wind equipped with fine rock fragments decreases surface friction further. The importance of heating on air movement is most noticeable toward evening, when the customary drop in temperature immediately brings with it a pronounced drop in wind velocity. Desert nights are as noteworthy for their stillness as the days are for their gustiness. Gone at night are the sudden blasts of hot air that seem to have originated in a furnace. Gone, too, are the swirling dust·devils that play across areas where the soil is loose and powdery. And gone are the shimmering and dancing images near the horizon that produce the heat haze that·blurs distant views during the day. The stars stand close and clear, and once the night is set, the lapse rate begins to reverse itself

The Humid Tropics and Dry Climates

near ground level, and relatively cooler surface air begins to flow gently down dry ravines toward some distant basin.

The dreaded sandstorms and dust storms that occur occasionally in most deserts of the world are not of local origin, nor are they related to surface heating. Instead, they are usually associated with wavelike disturbances or fronts that move along the flanks of the continental subtropical high-pressure centers creating cyclonic eddies that result in unusually high-pressure gradients extending for hundreds of miles. Winds of gale force sweep across the desert plains, laden in their upper reaches with dust and carrying a stinging load of sand below. This is when the dunes "smoke" and the sand sings as it rasps across the dry surfaces. Visibility may be lowered to a few feet, and all living things stop and seek shelter. Such storms may last for 2 to 3 days or be over within a few hours. Some of them leave the deserts and penetrate far into adjoining climates. The Sahara Desert of northern Africa is a notorious breeder of regional desert storms, and their hot breath is felt along all parts of the western, northern, and northeastern borders of the continent. Late winter and spring seem to be the most frequent time for them to occur, although they are encountered during the summer. The *harmattan* of the Mauritania coast, the sirocco that descends to the Mediterranean after pouring through the passes of the Atlas Mountains, and the khamsin of Egypt and the Levant are important weather phenomena in this part of the world. These desert winds are exceptionally warm along the continental edge of Africa, because they have descended several thousand feet from their interior source areas and have been warmed by adiabatic compression during their descent.

Gale winds of winter accompanying polar fronts sometimes visit the mid-latitude dry lands in the interior of continents. In North America and the east-central part of the Soviet Union, drifting snow is a serious problem in land surface depressions, despite the low quantities of winter precipitation in these areas. Winter is the dust season of northwestern China, and storm after storm sweeps out of the Gobi Desert, thickening the dust layers that lie over most of northern China. The Yellow River and the Yellow Sea are appropriately named.

7-10 MIRAGES The perception of objects in places where they do not exist, or *mirages,* are among the most intriguing desert phenomena noted by visitors. Apparent sheets of water whose surfaces are rippled as though disturbed by wave action are the most common types, although clumps of trees, hills, and mountains have been observed. They are most frequently observed during the warmest times of day and are seen at or near the horizon. Far from being the hallucinations of heat-crazed minds, they have a sound basis for their existence. There are several types of mirages, although the type most frequently observed is related to the refraction of light at the boundary between very warm air near the surface and denser, cooler air above. Light passing through the density interface, or from more dense air, is bent slightly, thus projecting the image of the sky slightly below the normal horizon line. The interface boundary itself appears as a faint line that may be near the position of the real horizon. The constant shimmering of this density boundary under the play of local convection currents gives the impression of wavelets on a water surface. This phenomenon is closely similar to the common illusion observed by motorists on a hot day when the highway appears to have a wet surface wherever it approaches the horizon.

Magnification also may take place so that clumps of trees or hills lying on the distant horizon may seem to be only a short distance away. An unusually large setting sun is a characteristic sight and is the result of this magnification effect. Apparent wave-rippled lakes rimmed by date palms can be real visions and have enticed unwary strangers who have lost their way in deserts during the hot season.

7-11 DESERT DEW A high diurnal range of temperature can be responsible for heavy nocturnal dew in dry areas that contain unusually high humidities. Many deserts exist not because of a shortage of water vapor in the air but because of low lapse rates in the upper air. The deserts that adjoin seacoasts in the tropics and subtropics are especially noted for their heavy dews. Aiding in this are land and sea breezes that import sea air onto the land during the warmest time of day. Desert dew, however, is not confined to coastal areas. Contrasting humidity readings for coastal and interior desert locations are presented in Table 7-10. The deserts located at the interior positions show low relative humidities during the afternoon hours and remain far below the saturation point even in the early morning. The coastal positions all show relative humidities in the early morning hours exceeding 75 percent, and even the afternoon readings are high. The only coastal position adjacent to cold water shown in the table is Lima, Peru, as is indicated by its temperature of 82° during the warmest time of day. With a relative humidity averaging 93 percent during the early morning hours at Lima, dew is more likely to occur there than at any of the other positions, but the amount is not likely to be as great as in the Persian Gulf area, where the quantity of water in the atmosphere is large. To illustrate, the vapor pressure saturation during the afternoon at Muscat is about twice that at Lima for the same time of day.

Table 7-10 Relative Humidity and Temperature Readings at Selected Coastal and Interior B Climatic Locations for July*

Location	Observation hours and relative humidity		Av. daily max. (°F)	Av. daily min. (°F)
			Temperatures	
	INTERIOR POSITIONS			
Baghdad, Iraq	6:00 A.M.–32%	3:00 P.M.–12%	110°	76°
Tashkent, U.S.S.R.	7:00 A.M.–63	1:00 P.M.–33	92	84
Isfahan, Iran	7:30 A.M.–41	3:30 P.M.–15	98	67
Yuma, Ariz., U.S.	5:30 A.M.–63	noon–27	106	77
	COASTAL POSITIONS			
Bahrein Island	7:30 A.M.–69%	3:30 P.M.–67%	99°	85°
Bushire, Iran	7:30 A.M.–75	3:30 P.M.–73	95	84
Karachi, Pakistan	8:00 A.M.–88	4:00 P.M.–73	91	81
Muscat, Muscat	8:00 A.M.–82	4:00 P.M.–72	97	87
La Paz, Mexico	7:30 A.M.–80	2:30 P.M.–67	95	75
Lima, Peru*	5:30 A.M.–93	1:00 P.M.–69	82	66

*The data for Lima, Peru, are for the month of January.

Nocturnal dew is an important source of water in the coastal deserts of the world; it may amount to as much as 10 inches of precipitation. A sparse vegetation growth that maintains a meager grazing for animals is often found in the coastal deserts, a growth that is far greater than that which the average precipitation seems to warrant. Nocturnal dew also greatly reduces the amount of irrigation water that is required in deserts for the successful growth of vegetables and fruits that grow close to the ground, such as squash, tomatoes, cucumbers, and melons. For the same reason, some semiarid regions are able to incorporate such crops in their nonirrigated cultivation patterns.

The possibility of extracting water from the air in coastal deserts has been the object of much research recently, and cheap power seems to have much promise through refrigeration or pressure processing.

7-12 MAN IN THE DRY LANDS The dry lands of the world, despite their limiting factors of high summer temperatures and scarcity of water, have nurtured man since he first appeared on earth. Ruins of past cultures can be found in nearly all of the dry lands, preserved in the dry air and belonging to a wide range of historical periods. The desert, though harsh and demanding, can support man if certain conditions are met and a local water supply is available. Adaptability includes manual labor in full sun in the hottest areas on earth. There are unique advantages that accrue to desert inhabitants who are willing to pay the price, as will be seen in Section 7-13. The cost of desert living, however, is high to all of its inhabitants, from the aboriginal inhabitants of Australia and Africa, and the Bedouin Arab, to the suburban dweller in desert oases in Arizona.

The extremely high temperatures of low-latitude deserts place a strain on the human body when man seeks to live and work in the open sun. Such conditions can be tolerated, however, if reasonable precautions are taken. The most important requirement is a high intake of water, because evaporation from body surfaces (perspiration) and transpiration during exhalation are the two major methods of removing surplus internal heat from the body. The quantity of water needed in the heat-exchange mechanism varies with outside temperatures and with the amount of energy consumed in the work performed by muscles and internal metabolism. Figure 7-12 indicates the results of some experiments carried out by the Quartermasters Corps of the U.S. Army in connection with environmental conditioning. With temperatures over 100°F, manual labor requires about a quart of water per hour to maintain the body at normal temperatures. Further, the loss may be so rapid as to precede the feeling of thirst. Water intake amounts should thus be planned accordingly and not be based on individual desires. With insufficient water for heat exchange, and in the high daytime temperatures of low-latitude deserts, body temperatures rise. This results in a series of physiological reactions, including the withdrawal of water from internal tissues, which lowers the basic body reserves, a reduction in kidney functioning to retain fluids, a dilation of blood capillaries through stages of surface rash (prickly heat) to surface blood hemorrhaging and a withdrawal of blood from extremities that can cause fainting (heat stroke), cell deterioration, and eventual death. Without water and with high air temperatures, body deterioration and death following coma may take place within a few hours. The hot deserts can be death traps to unprepared, untrained strangers and should command the respect and fear of all who enter them.

Profuse sweating may lead to difficulties

Figure 7-12 Daily water requirements of humans at three levels of activity. Following laboratory tests, the Quartermasters Corps recommends the amounts of water shown in relation to varying air temperatures and at three normal daily activity patterns. [U.S. Army Quartermasters Corps]

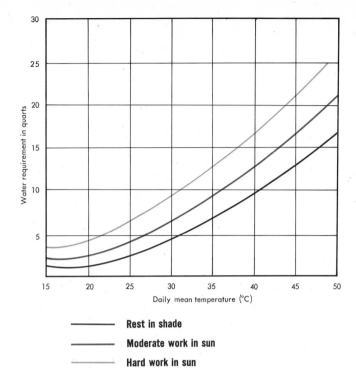

——— Rest in shade

——— Moderate work in sun

············ Hard work in sun

even when a sufficient water supply exists. The triggering mechanism for sweat glands involves the sodium chloride (salt) balance within the vital fluids of the body. Because this salt is secreted along with water on the skin surface, body deficiencies well below normal not only interfere with the process of sweating but can also lead to agonizing body cramps. Four to five grains of sodium chloride daily above normal intakes are recommended when sweating is continuous during the day. Experience with day laborers in the oil fields of the Sahara has led to the conclusion that sweat glands have a certain fatigue point after which they do not function as well. Experience suggests that hard manual labor cannot be continued indefinitely, and periodic rest periods are required. An efficient cycle appears to be a 2-week rest period following each 6 weeks of day labor.

There is no doubt that a certain degree of acclimatization greatly assists in the human body's adaptation to hot desert climates. Among the conditioning techniques are the following:

1. Darkening the skin by careful tanning is an important factor, because sunburn greatly interferes with the normal functioning of the sweat glands.

2. A reduction in body volume through a decrease in general cell size is helpful in decreasing both surface area and the demands for temperature control. Even without deliberate dieting, a rapid loss of weight seems to be a common experience of nearly all active newcomers to desert life.

3. A general reduction takes place in the volume of urine excretion and an increase in urea concentration with acclimatization. This is related to the general rationing of the body fluids, which is signaled by the demands of the body heat-exchange system.

4. Some reduction in the body metabolism rate takes place with conditioning, in which the internal heat production is reduced through less efficient "combustion" of sugars, fats, and starches.

Of all the elements in biological conditioning, however, the most important is learning how to conserve muscular energy. To be a true son of the desert requires moving slowly, deliberately, without wasted motion, and with an equanimity of mind to match the slowness of body movements. Strangers never cease to be amazed at the small amount of water needed by a Bedouin nomad when on a desert trek. This is not an inherited genetic characteristic but is partly a state of mind, partly a long period of conditioning, and partly a matter of suitable dress and food.

Proper clothing is another important means of reducing the inclemency of desert heat. Loose-fitting and rather heavy, loose-woven clothing is recommended. The flowing, heavy robes of the desert Arab are ideal because they permit free circulation of air about the body, aiding in evaporation. The head and back of the neck should be shaded by means of a loose, insulated hat such as a pith helmet, combined with a cloth back drop. Again, the Bedouin's shrouded and heavy head *khaffiya*, or shawl, is an extremely functional garment for protection against both excessive heat during the day and the chill of a desert night. Sandals are ideal for those whose feet are conditioned to tolerate high temperatures, but they are not advised for desert use by the newcomer. Heavy, open-meshed, woolen

socks worn within heavy-soled shoes will provide insulation and yet accommodate needed perspiration.

7-13 THE RESOURCE SIGNIFICANCES OF DRY CLIMATES The dry climates, though harsh and demanding in many ways, have attracted people from earliest times and for many reasons, not all of which can be mentioned here. One of the greatest advantages of deserts, or rather desert oases, is their isolation. Except for wandering nomads and certain aboriginal and mining settlements, desert people have been concentrated into compact groupings near water sources. Most of the desert area by far consists of relatively empty land, incapable of supporting any but sparse, wandering groups of pastoralists or aborigines, and where cross-country travel is difficult and dangerous. This confers an insulating quality on desert oases against most potential invaders that has been sought by many people to whom isolation was important. Such folk included religious, political, or cultural rebels who were willing to pay at times the unbelievably high cost of desert living. The early Mormons of Utah (see Figure 7-13), the Mzabites of the Algerian Sahara, the Druzes of the Middle East, and the Touaregs or enshrouded "Blue Men" of the central Sahara are among many who have sought oasis havens protected by long stretches without water or roads. A few scattered groups of aboriginal peoples still exist in the

Figure 7-13 Irrigation works developed by the Mormons along the flanks of the Wasatch Mountains. The narrow irrigation ditch on the right, still in use after more than 90 years of service, diverts water from within Logan Canyon and distributes it over the slope of an alluvial fan beyond the mouth of the canyon. The concrete retaining wall was added later. [U.S. Bureau of Reclamation]

empty barrens of northwestern Australia and the Kalahari Desert of South Africa, far from oases and maintained by a highly specialized and demanding way of life. They remain as cultural anomalies only because their land has little value for anyone else. Like the smaller forms of desert life, they have learned how to extract a bare existence from one of the most restrictive environments on earth.

Desert soils can be highly productive for agriculture when adequately watered and drained. Unleached of their soluble plant nutrients by rainfall, such soils yield astonishingly well when irrigation water plus nitrogenous materials are added. Of the soluble plant foods, only nitrogen is likely to be in short supply, because it is derived largely from organic sources. *Exotic rivers,* or those that carry large amounts of water and topsoil from distant humid climates into or through the deserts, have been choice sites for settlement. Many of the early great civilizations of the Old World were developed as the result of significant surpluses of food that accumulated along the flanks of such rivers once the principles of cultivation, irrigation, and drainage were understood. Fungus and bacterial diseases in general are much less common in arid lands than in humid regions because of dryness, and the great amount of sunshine provides exceptional conditions for photosynthesis and plant growth. Desert fruits are always sweeter and command premium prices in world markets. The date palm of the hot desert oases is a remarkable tree, providing a fruit of extremely concentrated food value without human processing, with high yields per acre, and admirable preservative qualities for desert travel. Sun-dried fruits are a specialty of desert oases, where nature does the preserving, with little risk of loss.

A new asset has begun to be realized in recent decades with the advent of higher living standards and new technologies. Air conditioning has greatly decreased the impact of high daytime temperatures on people residing or working indoors. Machines utilizing inanimate energy are performing more and more of the work of basic production that formerly required human labor outdoors. Today more people are making a living in cities and towns by performing personal services to others, and there is no reason why deserts should not be prime locations for urban concentrations, provided water and food can be brought to them. The spectacular growth of cities in the southwestern United States indicates that cheap land, much sunshine, lack of severe winters, a healthy climate, and exotic landscapes can be powerful attractions to residential and industrial development (see Figure 7-14). Furthermore, urban growth in the open desert is not competing for the land that could be used for agriculture; instead, it is competing for the use of water and can afford to pay high prices for it.

The semiarid lands differ distinctly from the true deserts in that they support a much larger population and one that is far less restricted to scattered water points. Until recently, an important problem of agricultural settlement in these marginal areas has been the irregularity of rainfall, which has led to alternate expansion and contraction of settlement accompanied by serious economic and social dislocations. The application of mass-production techniques in agriculture, however, has turned what once was a hindrance into an asset. The hazardous position of semiarid lands for normal nonirrigated field agriculture renders such land inexpensive—a basic reason for their general use as grazing lands. Many of these areas also are in open plains in the interior of continents, which makes them suitable for large-scale machine agriculture. Mass production

Figure 7-14 Desert suburban development, Tucson, Arizona. The typical desert-bush vegetation of mesquite, creosote bush, and cactus is shown in the dotted texture in the foreground. These cities of the American southwest are expanding into the desert at an extremely rapid rate. [Courtesy Fred Wehrman]

in agriculture, as well as in mining or manufacturing, is most efficient under conditions of low rent, cheap raw materials, high labor costs, and available capital. Crop yields in this type of agriculture are not high. They do not have to be, but large-scale production and low unit costs can greatly reduce the impact of low yields per acre.

Large farm machinery units also can perform rapidly certain special cultivation practices that provide an important part of *dry farming,* the general term for various

techniques in agriculture that are designed to retain moisture in the cultivated soils. One such technique where machinery can be highly effective will serve to illustrate. The use of alternate cultivated and fallow strips, termed *strip cropping* (see Figure 7-15), has long been found to increase productivity near the humid-dry boundary. Its success well within the *BS* climates, however, often was precarious because of the susceptibility of the fallow strips to erosion by wind when temperature, wind, and surface-soil moisture conditions reached certain critical levels. It was learned, however, that this could be greatly reduced and virtually eliminated by a shallow disking or harrowing of the top few inches of soil, replacing the powdery surface soil with moisture-knit clods from below. Without large tractor-drawn units that rapidly harrow a square mile of land in a few hours, such a technique would be worthless when conditions become critical. Where soil and terrain conditions permit, some hitherto uncultivable semiarid lands have been made highly productive—and without the deterioration that accompanied earlier attempts to cultivate them.

7-14 DRY CLIMATES AND THEIR HINDRANCES TO HUMAN USE Water short-age is the most obvious hindrance to human settlement in the dry lands and will continue to explain the relatively low population in these areas around the world. To illustrate, the B climates make up about 36 percent of the total land surface of the earth, yet they contain only about 13 percent of the population. Of this, almost three-fourths is located in the semiarid BS climates. The true desert (BW) climates make up about a fifth of the world area and have slightly less than 4 percent of the total population. In the deserts, human settlement is concentrated mostly in small and widely scattered oases where water is available. The hot desert climates have the highest evapotranspiration rates on earth; hence, water needs are greatest where the supply is least and most expensive. Pres-

Figure 7-15 Strip cropping, a type of dry farming in which part of the area is kept bare in order to conserve moisture in the soil. The fields here are cultivated in wheat. This scene is near Cut Bank, Montana, in the American Great Plains. [Standard Oil Company of N.J.]

Country	Total area	Cultivated annually	Area irrigated	% of irr. land to total area	% of irr. land to cult. area
China	2,405,837	335,000	77,275	3.2	23.0
India	810,777	296,400	59,057	7.3	19.9
U.S.	1,934,256	340,998	26,253	1.4	7.7
Pakistan	230,000	45,000	21,310	9.1	47.4
U.S.S.R.	5,502,917	?	16,062	.3	?
Egypt	247,100	7,000	7,000	2.8	100.0
Mexico	486,787	57,700	5,330	1.1	9.2

SOURCE: N. D. Gulhati, "Irrigation in the World: A Global Review," International Commission on Irrigation and Drainage, Delhi, India, 1955.

sures for water use in the dry lands are high, because the rewards are great. Some of the advantages of desert living were discussed in the previous section. Figure 5-29 illustrates that irrigation in the United States accounts for almost half of the total water diverted from the hydrologic cycle for human use —and approximately 95 percent of the consumed loss of this water—yet the total area of irrigated land is only about 1.4 percent of the total land area and 7.7 percent of the cultivated land in the country. These figures, moreover, include some irrigated land within the humid eastern parts of the country, such as the rice lands of Arkansas and Louisiana. Table 7-11 shows that the percentage of irrigated land to total area is small even within countries that contain much desert land. Less than 3 percent of Egypt, for example, is under cultivation, yet all of the cropland is irrigated.

The wide seasonal and daily ranges of temperature constitute problems of human adjustment, and the maintenance of body comfort is expensive. The price may be paid in many ways, including the high energy consumption for air conditioning, shortened daily work periods through daily rest periods during the hot hours, or frequent work rest periods on a weekly basis. Water costs are high because of the high demands on this scarce resource.

Dry climates require and reward irrigation, yet this is not an unmixed blessing. Irrigated land values and taxation are likely to be high and require continued and intensive soil exploitation. Salt accumulations in the soil resulting from poorly planned and administered irrigation projects have ruined highly productive desert acres in many parts of the world. Several diseases and parasites, including waterborne organisms such as the parasitic worm Schistosoma, that use certain freshwater snails as their intermediate hosts and whose infestation in humans is known as bilharzia, various bacterial and protozoan infections, and hookworms seem to follow unerringly the introduction of irrigation into newly developing areas. It is extremely difficult to enforce sanitation in areas of open irrigation canals and ditches in the desert without a long period of education and discipline. Irrigation, wherever found, requires careful administration, a high degree of technical competence, and constant maintenance of a complex water distribution system. There must be continual guarding against the silting of reservoirs and canals, the crumbling of aqueducts and conduits, fluctuating water tables, salinization, poor sanitation, and military vulnerability. The deserts of the world exhibit many ruins representing the dry bones of civilizations that reached remarkable heights of technological and cultural achievements but relaxed the vigilance needed for permanence.

Highway construction and maintenance operations pose special problems in dry lands. The lack of distinct channels in desert watercourses clogged with fragmental debris and sudden runoff of rare torrential rains hinders the selection of good sites for bridging streams, and desert bridges, once built, usually are highly vulnerable to damage. Road washouts are a common occurrence in dry lands everywhere. Blowing sand may be troublesome in some of the sandy deserts, and numerous attempts have been made to stabilize dunes adjacent to surfaced highways. Such airborne sand and dust can also be a serious problem in semiarid cultivated areas. The dust storms of the middle 1930s in the United States were tragedies to settlers in the areas where erosion took place and on the fertile fields nearby, where crops were buried in a few days by wind deposition. Pitted, frosted automobile headlights and windshields, and fine sand in bearings are expected annoyances to people who do much motoring in sandy deserts.

Another serious problem in areas that depend on groundwater is the maintenance of stable reserves in the face of increased economic pressure and the growth of population. Like petroleum, underground water is a "fugitive" resource that is able to move from place to place below the surface and remain hidden from view. In the United States legal regulations have generally held that a person is entitled to the use of water on his own land. Hence, there is little basis (under English law) for preventing free competition in the use of subsurface water supplies or restricting the use of such water in order to

maintain water levels in wells. Some areas that depend on this type of water supply are threatened with disaster unless the trend toward depletion can be reversed. New and more efficient pumps have added to the problem.

Probably the most serious hindrance to efficient land use in the dry lands occurs in underdeveloped, semiarid parts of the world that lack capital and cheap inanimate energy sources. Irregular periods of above-normal rainfall are responsible for speculative expansions of crop cultivation into areas where risks are high. Large-scale machinery has been able to reduce some of these risks, but such risks can be serious in areas where dense populations living close to subsistence levels press tight against the dry limits of cultivation. Some of the worst areas of famine incidence in the world are located here.

Insects are a constant annoyance in many dry lands, and in some areas they may cause serious problems. Insects are well adapted to dry climates. They are small in size, require minute quantitites of water that may easily be supplied from nocturnal dew, have excellent mobility and eyesight for night foraging, and can use small soil openings in the bare ground for their pupa or larval stages. Except in localities where insecticides have been used to destroy them, desert oasis settlements are notoriously attractive to flies, fleas, cockroaches, mosquitoes, and various types of arthropods, including spiders and scorpions. All of these add to the problems of maintaining sanitation and isolating infectious diseases. Locusts are the worst insect problem associated with human settlement in the dry lands of Africa and Asia (see Figure 7-16). Periodically breeding in enormous numbers, clouds of them great enough to darken the sky can strip every green leaf or stem for dozens of miles within a day or two. Their worst ravages are in the open grain fields of

Figure 7-16   A flight of locusts in the Somali Republic, eastern Africa.   The voraciousness of these insect swarms must be seen to be believed. The tall spire on the right is a *termitorium,* or giant anthill. Insects can be a scourge of the low-latitude dry lands. [Courtesy, Desert Locust Control Organization for Eastern Africa]

semiarid lands. Their most vulnerable period is when they first begin to shed their pupal shells and emerge ready to eat any green thing. At this stage they usually are highly concentrated into a restricted area, and food poisoning can greatly reduce their numbers. Finding these areas and rushing poison by planes and helicopters is expensive and requires cooperation on regional and international levels, because once in flight, not much can be done to stop destruction.

There are no Edens on earth, and while some well-watered desert oases may seem like miniature paradises, they pose problems for the humans who seek to settle there and to compete for the use of water.

Study Questions

1. Give two reasons why the *A* climates extend farther poleward along the east coasts of continents than along the west coasts.
2. Explain why the west coast of Madagascar is so much drier than the east coast.
3. Why does the ITC swing so much farther over the continents than over the oceans?
4. Locate Bombay, Hyderabad, and Madras on a map of India. From your knowledge of the Indian monsoons, grade these in order of their mean annual precipitation, and explain.
5. Explain why the *As* climates are rarely found in the tropics.
6. A weather station has 75 percent of its precipitation in the 6 summer months, with a mean annual total of 30 inches. If its mean annual temperature is 60°F, determine whether the locality has a humid, semiarid, or arid climate, using the Koeppen formulas.
7. List the factors that influence rainfall effectiveness, and explain each.
8. Would a *BShs* climate be more likely to be found to the north or to the south of the Sahara Desert? Explain.
9. Explain why vacation periods are so vital to manual workers in dry lands.
10. Why are there so many flies in desert oases?
11. List several advantages and disadvantages of living in the *A* and *B* climates.
12. Explain why October is the rainiest month in Miami.

CHAPTER 8
THE MIDDLE AND HIGH-LATITUDE AND HIGH-ELEVATION CLIMATES

The *C*, *D*, and *E* climates in the Koeppen-Geiger climatic classification system include the humid climates of the middle and high latitudes and the climates at high elevations at any latitude. They may be generally described as humid climates with mild winters, with severe winters, and with no summers, respectively.

The Mediterranean Climates (*Csa* and *Csb*)

8-1 CHARACTERISTICS AND DISTRIBUTION OF THE *Cs* CLIMATES The *Cs*, or Mediterranean, climates are characterized by mild, humid winters and arid summers. The letter *C* indicates a climate whose coldest month is above 26.6°F (−3°C) but below 64.4° (18°C). The letter *s* indicates that at least three times as much rain falls in the wettest month of winter as in the driest month of summer. The two subtypes, *Csa* and *Csb*, are distinguished by the temperature of the summer season, the *a* designating hot summers, whose warmest month is above 71.6°F (22°C), and the *b* indicating cool summers, whose warmest month is below 71.6° but above 50°. If all months have more than 1.2 inch of rain, the symbol *f* also is used.

Quantitative features of the *Cs* climates are presented in Table 8-1 and in Figures 8-2, 8-3, and 8-5. The seasonal contrast in precipitation between winter and summer is sharp and clear. The water budget for Beirut (Figure 8-6) indicates a water surplus during the winter with stream runoff from December until May. Precipitation falls off quickly in early summer, and June is almost rainless. A water deficit exists throughout the summer months, and normal field agriculture at this season is impossible without irrigation. This is typical of the *Cs* climates. The temperature pattern of both the *Csa* and *Csb* climates is illustrated by the isothermographs for San Francisco and Sacramento in Figure 8-3. Cool to cold water offshore, as in California and Portugal, tends to keep summer temperatures moderate where the *Cs* climates lie next to the open

Table 8-1 Climatic Data for Selected Stations in the Cs Climates

	Jan.	Feb.	Mar.	Apr.	May	June	July	Aug.	Sept.	Oct.	Nov.	Dec.	Yr.
		Eureka, Calif.; lat 40°38′ N long 124°10′ W; elev. 79 ft (Csbi)											
Temp. (°F)	47.5	48.2	48.7	50.4	53.1	55.6	56.3	56.7	56.5	54.3	51.3	48.9	52.3
Precip. (in.)	6.7	5.5	5.2	2.7	2.2	.7	.1	.1	.6	3.2	4.6	6.7	38.3
		Boise, Idaho; lat 43°34′ N long 116°13′ W; elev. 2,887 ft (Csa)											
Temp. (°F)	28.6	34.0	41.2	49.8	57.7	64.8	74.7	72.1	63.1	52.5	39.0	32.2	50.7
Precip. (in.)	1.3	1.3	1.3	1.1	1.3	.9	.2	.2	.4	.8	1.2	1.3	11.3
		Adana, Turkey; lat 36°59′ N long 35°18′ E; elev. 66 ft (Csa)											
Temp. (°F)	48.4	50.4	54.9	62.4	70.2	77.0	81.7	82.4	77.4	69.4	59.9	51.6	65.5
Precip. (in.)	4.4	3.7	2.6	1.8	1.8	.7	.2	.2	.7	1.6	2.4	4.0	24.1
		Konya, Turkey; lat 37°52′ N long 32°30′ E; elev. 3,366 ft (Csa)											
Temp. (°F)	31.6	34.9	41.0	51.8	60.6	67.6	73.6	73.2	64.8	54.5	43.7	34.9	52.7
Precip. (in.)	1.6	1.3	1.2	1.2	1.5	1.0	.2	.2	.4	1.1	1.2	1.5	12.4

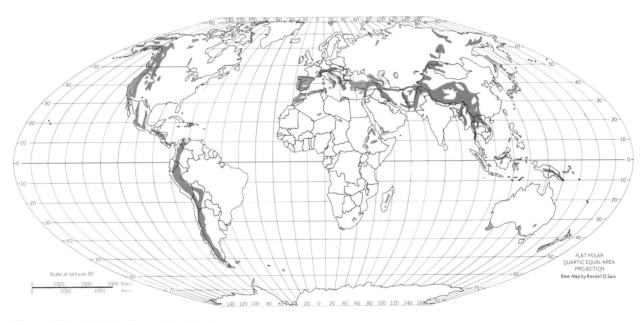

Figure 8-1 Distribution of *Cs* climates.

 **Cs**   **High mountains**

The Middle and High-Latitude and High-Elevation Climates   *279*

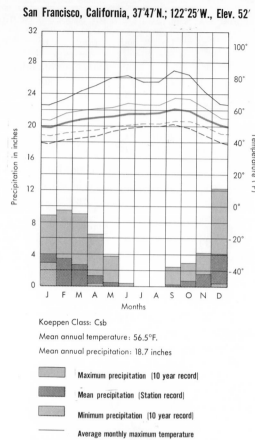

Izmir, Turkey, 38°27'N.; 27°15'E., Elev. 92'

Koeppen Class: Csa
Mean annual temperature: 63.5°F.
Mean annual precipitation: 27.3 inches

	Maximum precipitation (10 year record)
	Mean precipitation (Station record)
	Minimum precipitation (10 year record)
———	Average monthly maximum temperature
———	Average daily maximum temperature
———	Mean monthly temperature
– – –	Average daily minimum temperature
— — —	Average monthly minimum temperature

San Francisco, California, 37°47'N.; 122°25'W., Elev. 52'

Koeppen Class: Csb
Mean annual temperature: 56.5°F.
Mean annual precipitation: 18.7 inches

	Maximum precipitation (10 year record)
	Mean precipitation (Station record)
	Minimum precipitation (10 year record)
———	Average monthly maximum temperature
———	Average daily maximum temperature
———	Mean monthly temperature
– – –	Average daily minimum temperature
— — —	Average monthly minimum temperature

Figure 8-2 Climatic charts for San Francisco and Izmir, two representative *Cs* stations. Note that while both of these stations are coastal, Izmir (*Csa*) has a much higher daily and seasonal range of temperature than San Francisco. The extremely warm surface water in the eastern Mediterranean helps to maintain high summer temperatures, whereas the water off San Francisco in California is cold.

ocean. Summer temperatures are high and uncomfortable inland, or around the Mediterranean Sea, which has unusually warm surface water during the summer. Frost may occur in nearly all parts of the Cs climates, although its incidence decreases toward the coasts and equatorward. Frost rarely occurs in the coastal districts of southern California or along the eastern and southern shores of the Mediterranean, but is common farther

Figure 8-3 Isothermographs of average hourly temperatures, by months, in 1964 (°F) for San Francisco (*Csb*) and Sacramento (*Csa*). These two stations, less than 100 miles apart, contrast markedly as to daytime temperatures during the summer season. Cool water offshore, with an onshore daily sea breeze in the summer, keeps San Francisco cool at this season. This cooling does not reach Sacramento, in the Central Valley of California. Note, however, that the rapid gain and loss of radiation at the interior location produce a radical diurnal change in temperature. Nights are cool and comfortable in Sacramento, as they are in San Francisco.

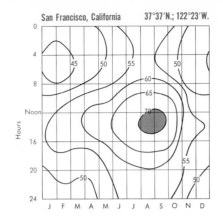

San Francisco, California   37°37′N.; 122°23′W.

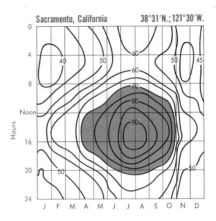

Sacramento, California   38°31′N.; 121°30′W.

north and at interior positions. Table 8-1 indicates, for example, that both Boise and Konya have mean monthly temperatures below freezing. Usually a mean monthly temperature of 50° or higher for the coldest month indicates that frost is relatively uncommon.

The general distribution of the Cs climates is between latitudes 30° and 43° on the western side of continents (see Figure 8-1). The poleward extent in North America is greater than this, however, reaching a point slightly north of Vancouver, British Columbia, or almost to 50°. Each of the five major continents has an area in this climatic zone, although the only large extent is found bordering the Mediterranean Sea. The Cs climates in North and South America are restricted mainly to a relatively narrow zone between the western mountain cordilleras and the sea, although a small tongue crosses southern Idaho and extends southward along the western flanks of the Wasatch Mountains of Utah. In South Africa and Australia, the continental landmasses terminate within the latitudinal zone of the Cs climates; hence, only the southern extremities have Cs climates. Australia has two separate areas: a western sector, including the southwestern corner of the continent near Perth, and an

eastern one along the west-facing coast near Adelaide. The largest extent of Cs climates is in the vicinity of the Mediterranean Sea, where these climates reach eastward from the Atlantic far into central Asia and border both sides of the Mediterranean Sea. The high plains throughout central Turkey and the mountain slopes and high basins within parts of central Asia exhibit characteristic humid winters and arid summers. High elevations in these eastern reaches reduce temperatures and evaporation sufficiently to produce a *Csb* climate. Lowlands in the same area tend to have a *BShs* climate. Note, for example, in Table 8-1 the contrast in rainfall between Konya, on the high Anatolian plains

The Middle and High-Latitude and High-Elevation Climates   281

of Turkey, and Adana, only about 150 miles to the south and on a small plain along the coast. Were it not for Konya's much lower average temperatures, owing to its elevation of more than 3,300 feet, the city would have a *B* climate. This same condition is duplicated in many other parts of Turkey, Iraq, Iran, and parts of the southern Soviet Union. Boise, Idaho (see Table 8-1), also would have a *B* climate if not for its elevation.

The distribution of the two subtypes of Mediterranean climates (*Csa* and *Csb*) is irregular and difficult to generalize, and is influenced greatly by latitude, land-surface features, and offshore water temperatures. In the United States, the *Csb* climates are widest in the north, extending several hundred miles inland from the coast, across Washington and Oregon into Idaho. The area of cool summers narrows southward into California and is generally restricted to the coastal zone and intermediate elevations in the mountains. Near its southernmost extent, in the Los Angeles lowland, the *Csb* subtype is restricted to a narrow coastal strip only 5 to 10 miles wide corresponding to the zone which experiences the daily summer sea breeze. The cool water offshore in this region has a significant role in making this coastal area pleasant during the summer months. The interior lowlands of California have either *Csa* or *B* climates. The two isothermographs shown in Figure 8-3 indicate the sharp increase in summer daytime temperatures as one leaves the San Francisco Bay area and passes inland into the Great Valley of California (Sacramento).

All of the *Cs* climates in the southern hemisphere are of the cool summer type (*Csb*) except for Australia, where there are narrow strips of *Csa* climates bordering the inland sides of the cool-summer subtype. The Mediterranean region including Europe, the Middle East, and North Africa, shows an over-

whelming predominance of the hot summer subtype (*Csa*). The only spot of low-elevation *Csb* in this entire region is found along the northwestern coast of Portugal, from Cape Finisterre south to Oporto. The main reason for the dominance of the hot summer subtype is that the Mediterranean Sea becomes an unusually warm body of water during the summer months, lacking the importation of cooler water from higher latitudes or the upwelling of cold water from great depths that feature the western margins of continents at these latitudes. The surface water in the Mediterranean Sea reaches its highest temperatures along the Levantine or eastern coast, from Egypt north to Turkey, and at times may reach over 90°F. This warm water is responsible for the unusually high relative humidity of the sea breeze and the resultant low comfort index during the summer afternoon hours in this general area.

8-2 INTERPRETATION OF THE SEASONAL REVERSAL OF PRECIPITATION IN THE *Cs* CLIMATES The seasonal reversal of the precipitation regime in the *Cs* climates, in which the winters are humid and the summers dry, is based on the seasonal latitudinal shift of the major global air circulation system. The *Cs* climates of both the northern and southern hemispheres are bordered on their equatorial sides by the dry or *B* climates, and on their poleward sides by humid climates. The *Cs* climates thus are in an intermediate position. With the seasonal shifting of the planetary circulation system that is related to the latitudinal shift in the zone of direct solar radiation, this intermediate zone comes under the dominance of the *B* climates during the summer season and the humid *C* climates during the winter season.

The principal mechanism influencing the seasonal change in precipitation is not a change in direction of wind but rather the alteration of the lapse rate. During the summer season, the subsidence of air from the equatorial flanks of the mid-latitude jet streams feeding into the great subtropical high-pressure centers at or near the surface shifts poleward and overlies the surface air strata of the Cs climates. High pressure normally is not observed near the surface because of high solar radiation, surface heating, and resultant rising convectional currents. The low lapse rates a few thousand feet above the surface, however, are a powerful deterrent to rain-making, because updrafts of air are quickly cooled adiabatically to the temperature of the surrounding air. All of the other requirements for precipitation may be present—high vapor pressure, onshore and upslope winds, frontal convergences, etc.—but unless the upper-air subsidence and its accompanied low lapse rates disappear, no precipitation can take place.

A further feature that may influence upper-air lapse rates during the summer is horizontal intrusions (advections) of warm air strata into the upper-air space. Not much is known about them, but a possible explanation is that they originate in the heat sinks of the tropical thunderstorms in low latitudes and extend slowly poleward between the subsiding cells of the subtropical highs. Radiosonde temperature soundings at high elevations indicate these intrusions as sharply defined temperature boundaries. Two or three separate such strata may be present at elevations of from 10,000 to 20,000 feet. Regardless of their origin, they effectively check thermal updrafts and may prevent precipitation during the spring and fall even when the normal regional upper-air subsidence has disappeared.

As the global air circulation belts slowly shift equatorward during the autumn months, the upper air undergoes a distinct modification. Sometimes the subsidence is present, and sometimes it is not. Lapse rates are highly irregular. If a frontal system moves in at a time when the subsidence is weak or absent and the lapse rate is high (see Figure 8-5), favoring conditional instability, heavy instability showers will result. The frequency of such rains increases into the winter season, usually with December and January as the rainiest months. Table 4-6 shows the average monthly variation in lapse rates in Lebanon, which has a typical Csa climate. By comparing this table with the monthly rainfall averages for Beirut in Appendix C or in Figure 8-6, the correlation can be easily noted.

8-3 VARIABILITY WITHIN THE Cs CLIMATES Mountains are landscape features that accompany nearly all of the areas of Cs climates, the only exception being southwestern Australia. The varying height and direction of the individual mountain crest lines result in sharp differences in rainfall, winds, and temperatures within short distances. As a general rule, the average annual totals of precipitation tend to increase with latitude, altitude, and exposure to marine winds, but there are many exceptions. The local differences in weather and climate within the general areas of the Cs climates in California are well known to residents of that state and help to create the diversity of agricultural specializations for which the state is famous. The same variability can be observed throughout the Mediterranean Basin. Figures 8-4A and B show great contrasts of winter precipitation within short distances. Note the steep precipitation changes in Turkey and the Balkans. Wind directions are not responsible for the seasonal variations in precipitation, but they have a great influence on the amounts that fall during the rainy season.

Mountains within the Cs climates not only are significant in channeling winds and in-

A. April - September

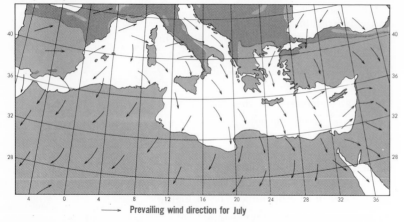

→ Prevailing wind direction for July

B. October - March

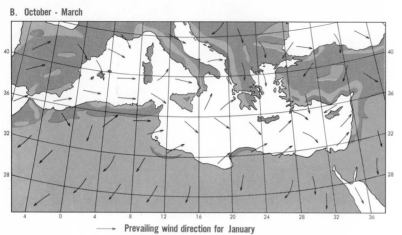

→ Prevailing wind direction for January

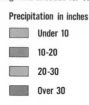

Precipitation in inches

- Under 10
- 10-20
- 20-30
- Over 30

Figure 8-4 *A* and *B* Precipitation and wind patterns in the Mediterranean Sea region during summer and winter. Note the seasonal latitudinal shift in precipitation. There is no corresponding strong seasonal shift in wind directions, however. Exposure of mountain slopes to prevailing winds during the winter season results in strong local contrasts in precipitation.

fluencing precipitation and temperature patterns; they also may be sufficiently high to receive most of the winter precipitation in the form of snow. This is an important asset in that the precipitation is retained on the upper slopes well within the spring and early summer months, to be released gradually with melting into streams at a time when it is needed most for irrigation on the lowlands. Newspapers throughout southern California publish current measurements of snow accumulations at selected points within the upper watersheds of the Sierra Nevada Mountains. These reports are regularly noted by horticulturalists who must plan their water requirements for the next summer.

8-4   WEATHER VARIATIONS WITHIN THE Cs CLIMATES Pronounced day-to-day changes in weather often occur in the Cs climates because of the transitional nature of these climates between predominantly arid and humid regions, between the high energy surpluses of low latitudes and the winter deficiencies of higher latitudes, and because of the coastal position by which they share features of both marine and continental climates. While the Cs winters may be considered mild compared with the continental climates that lie farther east, they occasionally feel the icy touch of polar continental air masses. Continental tropical air masses, with their arid, searing winds often laden with dust, also visit the Cs lands. Mention already has been made (see Section 6-20) of the siroccos or khamsins in the area of the Mediterranean Basin that are weather phenomena born in the arid lands to the south, but which produce sudden disturbances in human lives far beyond the air-mass source areas. Similar winds from the Arizona desert visit the coastal lowlands of southern California.

Figure 8-5 shows the variety of air masses

Figure 8-5 Weather conditions during March, 1962, at Beirut, Lebanon. The three peaks of temperature represent three khamsins, or importations of continental tropical air (*cTw*), from Arabia. Each of the three rains of the month was associated with a cold front at the end of a khamsin. Note the correlation between the steepness of the lapse rate, precipitation, and air masses. The temperature at 18,000 feet indicates the great thickness of the khamsins and the tendency for these upper levels to arrive slightly prior to the lower levels, which are held back by friction. The *cPk* intrusion that destroyed the first khamsin was somewhat unusual in originating on the Russian steppes and arriving via the Anatolian plains of Turkey. It brought unusually cold air for March to this subtropical country. ["Climat du Liban," *Bulletin Statistique Mensuel,* Ksara Meteorological Observatory, Ksara par Zahle, Lebanon, March, 1962]

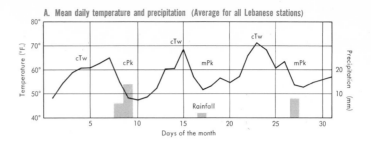

A. Mean daily temperature and precipitation (Average for all Lebanese stations)

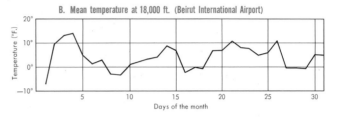

B. Mean temperature at 18,000 ft. (Beirut International Airport)

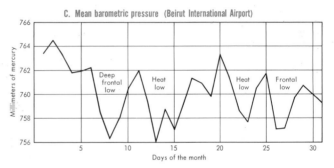

C. Mean barometric pressure (Beirut International Airport)

D. Lapse rate (°F. per 1000 ft.) sea level to 5000 ft. (Beirut International Airport)

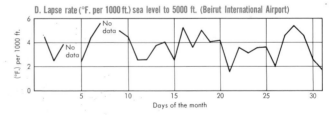

that passed across Beirut, Lebanon during March, 1962. The usual dominant air mass at this time of year is the modified polar maritime air (*mPk*) that passes over northwestern Europe before reaching Beirut. The three khamsins were prominent weather features and brought with them high temperatures and dust-laden air. The onset of the second, on March 14, was especially sudden. In the 15-minute period following 9 A.M., the temperature rose from 51°F to 63°, and by 3 P.M. it had stabilized at 71°. The small rain that ended this desert wind was so laden with dust that it plastered all exposed surfaces with a layer of fine, yellowish mud. Each air mass visiting the Cs climates has its distinctive properties. The March, 1962, pattern at Beirut, shown in Figure 8-5, indicates the different properties of pressure, temperature, and lapse rates that are associated with different air masses. It is interesting to note that during this month, a lapse rate of about 4°F per 1,000 feet of elevation seemed to be critical for the production of rain. The variety of weather changes shown

in Figure 8-5 is not always so great in Beirut, and even the month shown is not necessarily typical. The long-range precipitation average at the Beirut International Airport, for example, is 77 millimeters (3 inches) for the month of March. In 1960, 4.9 inches of rain fell, and the khamsins of 1962 reduced the monthly total to only 1.1 inches, nearly all of which fell during 3 days (see Figure 8-5).

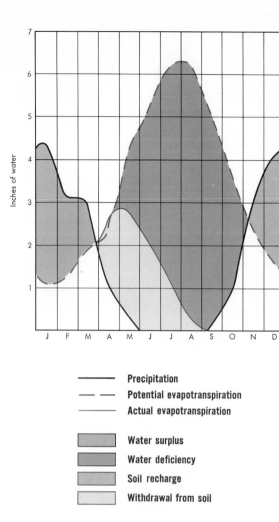

Precipitation
—— —— Potential evapotranspiration
———— Actual evapotranspiration

Water surplus
Water deficiency
Soil recharge
Withdrawal from soil

Figure 8-6 Water budget diagram for Beirut, Lebanon. For an explanation of the water budget and evapotranspiration, see Section 5-20. In this diagram of the water budget for Beirut, typical of many Cs climates, note the existence of a water surplus during the late winter and a pronounced water deficiency that lasts all summer until early autumn. The amount of water deficiency for any month is a rough measure of the amount that should be applied by irrigation to supply plants with the maximum amount that they could use.

8-5 HUMAN IMPLICATIONS OF THE Cs CLIMATES The areas of Cs climates are attracting people from other climatic areas. Other factors than climate are important, but there are many indications that the Cs climates are attractive, especially when employment location requirements are not critical. Much of the extremely rapid population growth in California and the Pacific Northwest has been by people seeking a residential location without ties to a specific occupation. California has a large number of retired people who have earned their income elsewhere, hence are free to choose their retirement locations. Mediterranean Europe has long been an area sought by the wealthy from northwestern Europe who did not have to live at home. This is especially true of the Csb or adjoining areas, as in Portugal and the French and Italian Rivera, where a mild winter coupled with a reasonably moderate summer produces a climate close to body comfort for nearly all of the year. Other attractive features are the low incidence of cloud cover and long periods of sunshine. The rainy winter period has a relatively high incidence of sunshine compared with the humid climates farther poleward, because most of the rains are intense and of short duration. Coastal locations adjoining cold, ocean water in the Csb climates result in regular morning fogs brought in by daily sea

Variability of rainfall also is indicated in the climatic graphs for Izmir and San Francisco (see Figure 8-2). Note that both stations experienced at least one rainless February during the 10-year period of 1951–1960, yet the long-term average for this month is almost 4 inches for both stations. The most consistent weather season at these stations, as with most others near the equatorward margins of the Cs climates, is the aridity of the summer months.

breezes that may detract from the advantages of even, mild, summer temperatures. The coast of California north of San Francisco has an extremely low population, partly owing to its rugged, cliffed shoreline, but the dense sea fogs and cold water (see Figure 5-7) inhibit almost any recreational or residential development.

All five of the Cs areas in the world are noted for their horticultural specialties, especially fruits and vegetables. The advantages of desert agriculture where a water supply is available were indicated earlier in Section 7-13. Summer agriculture in the Cs climates has all of these advantages, combined with an unusual accessibility to irrigation water stored in the snowfields of adjacent mountains during the humid winter season.

The smog of the Los Angeles lowland has been the subject of many jokes, but it is a source of irritation, literally and figuratively, to residents of the area. It is perhaps well to examine if the incidence of smog is peculiar to Los Angeles alone, or if it has implications within other areas of Cs climates. The factors involved in serious air pollution are complex but generally involve two major elements: (1) an atmospheric trap in which the pollutants are retained within a limited layer near the ground; and (2) a major source of pollutants. The Los Angeles atmospheric trap is produced by a low-level temperature inversion combined with a topographic barrier that intersects it. The inversion is associated with the advection of cool air chilled by contact with cool water offshore that wells up from great depths. Subsidence within the subtropical high-pressure cell that extends into this sector during the late summer also tends to accentuate the inversion by decreasing the lapse rate in the upper air. The level of the temperature inversion gen-

Table 8-2 Suspended Particulate Matter		
.0001 gm/m^3 (1961) Mean		90th percentile
Chicago	190	254
New York	173	241
Philadelphia	173	302
Milwaukee	146	234
Baltimore	142	180
St. Louis	141	214
Pittsburgh	137	214
Cleveland	136	214
Washington	128	214
Detroit	118	173

SOURCE: U.S. Dept. of Health, Education and Welfare, Public Health Service, *Air Pollution Measurements of the National Air Sampling Network*, Table 2.3, U.S. Government Printing Office, Washington, D.C., 1966.

erally is between 2,000 and 4,000 feet in elevation. Since air cannot pass across the inversion boundary from below, the urban pollutants are trapped above the city. The nightly land breeze helps blow some of the smog out to sea, but its influence is not great.

The city of Los Angeles is a somewhat unusual pollutant source in that its air is much freer of dust and smoke particles than many other U.S. cities. Table 8-2 indicates that it is not among the 10 leading cities in the country as to the amount of suspended particles in its air. Its air pollutants are of a different type and are much more difficult to overcome. Particle pollutants, as measured in Table 8-2, are derived mainly from the smoke resulting from space heating, especially by coal furnaces. The pollutants in the Los Angeles area are largely gaseous molecules that are particularly related to gasoline or diesel oil combustion. With strong solar radiation under clear skies, some complex oxidation reactions take place that may be summarized as follows:

The relatively large amounts of nitrogen oxides and various hydrocarbons released by auto exhausts and other sources coupled with ample sunshine for photochemical reactions provide the essential ingredients. One important chain of events can be summarized in the following fashion. Nitrogen dioxide (which has a brownish color and is a good absorber of solar radiation) is readily dissociated by sunlight into nitric oxide and oxygen atoms. Some of these oxygen atoms react with oxygen molecules to form ozone. These oxidants attack organic material, including the unburned hydrocarbons from automobile exhausts. The result is the formation of more ozone and various other oxidation products. Some of these products cause eye irritation, odors, and plant and material damage. . . . Thus all of the smog manifestations: eye irritation, reduced visibility, and plant damage can be attributed to the photochemical reaction products of nitrogen dioxide and hydrocarbons.[1]

The metropolitan area of Los Angeles is unusually large for its population, and automobiles are a necessity to reach almost any location in the area. Hence, the number of cars, busses, and trucks per unit of area is higher than it is in most U.S. cities. The oil refineries near Long Beach do not help the situation. The cars, the sunlight, the mountains, and the strong regional low-elevation temperature inversion combine to produce the critical Los Angeles air pollution problem.

Temperature inversions resulting from upper-air subsidence and reflected in low lapse rates are a common feature of the equatorward portions of the Cs climates throughout the world. Most of these localities also have mountains near coastlines; hence, any large metropolitan areas on a coastal lowland backed by mountains have this problem. Beirut, Lebanon, is beginning to experience smog, and its government banned all diesel motors in trucks, cars, and busses

[1]Reproduced by permission from Resource Paper #2, "Air Pollution," published by the Commission on College Geography of the Association of American Geographers.

Figure 8-7 Distribution of the *Cf* and *Cw* climates.

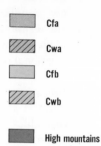

Cfa

Cwa

Cfb

Cwb

High mountains

in 1961 for that reason. The warm water of the Mediterranean Sea helps to make the inversion contrast aloft less strong than in Los Angeles, but the almost constant blue haze throughout the Mediterranean Sea area during the summer and fall months is an indication that pollutants have little opportunity of escaping vertically anywhere in this region. With an increase in population, especially urban population, the problem of air pollution can be expected to become more serious in many Cs areas.

The Humid Subtropical Climates (*Cfa* and *Cwa*)

8-6 GENERAL DESCRIPTION AND LOCATION OF THE CFA AND CWA CLIMATES The Ca climates have mild winters (coldest month below 64.4° but above 26.6°) and hot summers (warmest month above 71.6°). The two subtypes are distinguished by the continuously humid conditions (at least 1.2 inches in each month) in the *Cfa* and the dry winters (ten times as much rain in the rainiest month of the summer 6 months as in the driest month of winter) of the *Cwa* subtype. Because frost is to be expected at some time and in varying degrees of frequency during the winter season, plant growth and agricul-

Climate

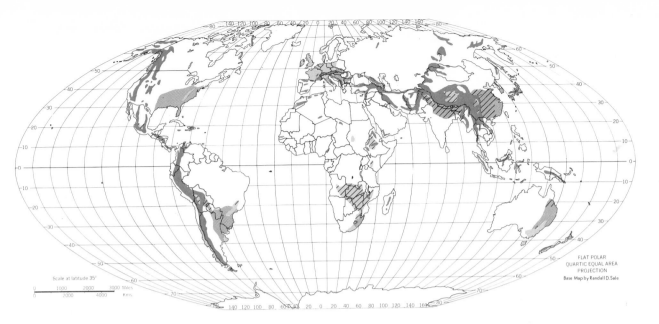

FLAT POLAR
QUARTIC EQUAL AREA
PROJECTION
Base Map by Randall D. Sale

Scale at latitude 35°

ture have definite restrictions. The dryness of winter in the *Cwa* climate thus becomes progressively less meaningful to land use as one moves poleward through the areas of its occurrence, since freezing temperatures become more and more restrictive, regardless of the amount of precipitation that falls during the winter.

The *Ca* climates typically are located in the eastern parts of continents, immediately poleward of the humid tropics and east of the *B* climates that occupy the continental interiors (see Figure 8-7). They extend along the east coasts of continents from about 25° to 40° N and S lat. Small areas of *Cfa* climates also are located in the lowland areas of southern Europe immediately north of the *Cs* climates, as in northeastern Spain, the Po Valley of northern Italy, in Yugoslavia, and on the Hungarian and Romanian plains.

The *Cwa* climates form a logical poleward extension of the dry winters of the *Aw* climates east of the dry lands. The largest areas of them are found in northeastern India and in China, where the winter is dominated by continental air masses. The powerful winter high-pressure center in Mongolia and the eastern Soviet Union sends a continuous series of cold, dry air masses over much of China, preventing the influx of moisture-bearing marine tropical air from the Pacific, except in southeastern China. A separate high-pressure cell during this season lying over the high plateaus of southwestern China also helps to maintain the winter dry period. The dry winters of India are the result of the Indian monsoonal circulation. Northeastern India is barely poleward enough to have its winter temperatures drop below those of the *A* climates. In Brazil and in a small section of northeastern Australia, the *Cwa* climates also adjoin *Aw* climates, and their winter temperatures are only slightly lower than the latter. In South America, most of the *Cwa* climates lie at intermediate elevations that depress the average winter monthly temperatures below 64.4° but are not sufficiently high to lower the summer

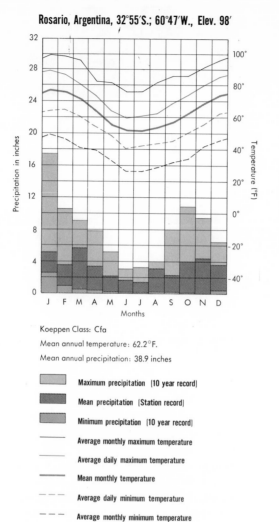

Rosario, Argentina, 32°55′S.; 60°47′W., Elev. 98′

Koeppen Class: Cfa

Mean annual temperature: 62.2°F.

Mean annual precipitation: 38.9 inches

Maximum precipitation (10 year record)

Mean precipitation (Station record)

Minimum precipitation (10 year record)

Average monthly maximum temperature

Average daily maximum temperature

Mean monthly temperature

Average daily minimum temperature

Average monthly minimum temperature

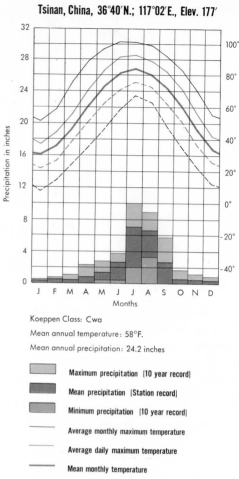

Tsinan, China, 36°40′N.; 117°02′E., Elev. 177′

Koeppen Class: Cwa

Mean annual temperature: 58°F.

Mean annual precipitation: 24.2 inches

Maximum precipitation (10 year record)

Mean precipitation (Station record)

Minimum precipitation (10 year record)

Average monthly maximum temperature

Average daily maximum temperature

Mean monthly temperature

Average daily minimum temperature

Average monthly minimum temperature

Figure 8-8   Climatic charts for Rosario, Argentina (*Cfa*), and Tsinan, China (*Cwa*).

temperatures below 71.6°. No *Cwa* climates are found in North America, primarily because the Gulf of Mexico provides a source for moisture-bearing winds that can penetrate into the heart of the continent even during the winter season.

8-7   MAJOR FEATURES OF THE C*FA* AND C*WA* CLIMATES   The most noteworthy feature of these climates is their hot, humid summers. A comparison of Figures 8-9 and 4-22 will indicate that, according to Terjung's comfort index, June, July, and August in Charleston are hot and uncomfortable throughout both day and night. Daytime

Figure 8-9 Average hourly temperature (°F) and relative humidity (%) by month for Charleston, South Carolina, in 1964. The *Cfa* climates near the continental margins, of which this is an example, have unusually high combinations of temperature and humidity. These two diagrams illustrate why the Middle Atlantic states in the United States have such uncomfortable summer nights. Unlike most localities, where relative humidity is higher in winter because of lower temperatures, here the relative humidity rises in summer because of the advection of marine tropical air. The diurnal changes are caused mainly by solar heating.

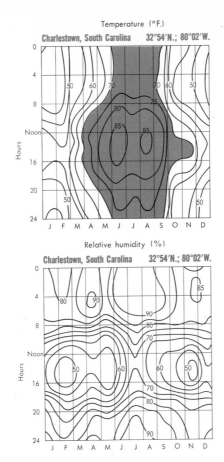

temperatures frequently are higher than they are in the humid tropics, because direct rays of solar radiation near the summer tropic ($23\frac{1}{2}°$ of lat) are combined with longer hours of sunlight than near the equator. Summer rains commonly are of the convectional type, consisting of short, heavy showers resulting from the conditional instability of the *mTk* (unstable, tropical marine) air masses that pass over the heating land surfaces. The cost of maintaining body comfort is high, because air conditioning is desirable in both summer and winter. Values for total *heating and cooling degree-days*[2] are among the highest in the world, being exceeded only by the *Da* climates, where hot summers are combined with severe winters.

Snow rarely is a significant obstacle to transportation, although an unusually heavy fall of several inches may block highways for short periods. Although conditions vary somewhat, the *Cfa* climates rarely have a

snow cover for more than a couple of weeks, most of it melting soon after it reaches the ground. Sleet and glaze are much more troublesome, especially near the poleward margins of these regions, since the freezing line associated with the winter polar front normally shifts back and forth across this climatic zone during the winter months. The east coast areas within this climatic group are likely to experience hurricanes, especially in the autumn months, and in some cases, they influence the average monthly precipitation records (note the September and October rainfall for Jacksonville in Table 8-3).

[2]A *degree-day* is a useful measurement of temperature excesses or deficiencies. It is based on the number of degrees of variation in the mean daily temperature from 65°F, taken as the base point for body comfort. A single day with a mean temperature of 46°, for example, is the equivalent of 19 *heating degree-days*, whereas a day's mean temperature of 80° represents 15 *cooling degree-days*. The total number of heating and cooling degree-days for a year is a convenient indication of the requirements for air conditioning (both heating and cooling) in buildings.

Table 8-3 Climatic Data for Representative *Cfa* and *Cwa* Stations

	Jan.	Feb.	Mar.	Apr.	May	June	July	Aug.	Sept.	Oct.	Nov.	Dec.	Yr.
Jacksonville, Fla.; lat 30°25′ N long 81°39′ W; elev. 38 ft; Koeppen class: Cfa													
Temp. (°F)	55.9	57.6	62.2	68.7	75.7	80.8	82.6	82.2	79.3	71.1	61.7	56.1	69.4
Precip. (in.)	2.4	2.9	3.5	3.5	3.5	6.3	7.7	6.8	7.6	5.2	1.7	2.2	53.3
Allahabad, India; lat 25°27′ N long 81°44′ E; elev. 321 ft; Koeppen class: Cwag													
Temp. (°F)	61.5	66.4	77.2	87.3	94.5	93.7	86.2	84.4	84.2	79.9	70.0	62.8	79.0
Precip. (in.)	.8	.9	.6	.2	.3	3.9	11.2	13.1	7.7	1.6	.2	.2	40.7
Barcelona, Spain; lat 41°24′ N long 02°09′ E; elev. 305 ft; Koeppen class: Cfa													
Temp. (°F)	48.9	49.8	54.1	58.3	63.9	70.9	75.9	75.6	71.1	63.5	56.3	50.4	61.5
Precip. (in.)	1.2	1.6	2.1	1.8	2.1	1.6	1.2	1.8	3.1	3.0	2.1	1.9	23.5
Nagasaki, Japan; lat 32°44′ N long 129°52′ E; elev. 88 ft; Koeppen class: Cfa													
Temp. (°F)	43.5	45.7	50.4	58.5	65.3	71.4	79.3	81.3	75.0	65.5	56.8	48.4	61.7
Precip. (in.)	2.8	3.8	4.6	7.2	8.1	11.6	11.3	7.4	10.0	4.2	3.3	3.4	77.7

The *Cwa* climates of northern India are different from those in most parts of the world in that they have extremely high temperatures during the spring months that are related to the typical monsoonal feature of a presolstice temperature maximum. As is illustrated in the data for Allahabad in Table 8-3, these climates barely miss being included in the tropical monsoonal climates (*Aw*).

8-8 HUMAN IMPLICATIONS OF THE *Cfa* AND *Cwa* CLIMATES These climates are not highly restrictive to human settlement, despite their uncomfortable summer weather conditions, and in some places they support exceedingly dense populations, as in China and northern India. Perhaps the most limiting factor is the frost hazard of winter, but even this does not prevent the winter cultivation of some of the grains such as wheat or barley, or some of the vegetables that tolerate light frosts such as peas, carrots, lettuce, etc. Except for about 2 months following the winter solstice, frosts are not likely to interfere with agriculture; hence, double cropping is possi-

ble throughout most of the region. The maintenance of body comfort is a problem but is not restrictive. Air conditioning, while desirable during both winter and summer, is not essential. Summer rainfall in these climates, although subject to considerable fluctuation, rarely drops below the quantities necessary for normal agricultural demands; hence, crop failures are relatively rare because of insufficient rainfall in summer.

Hurricanes and tornadoes are the major weather hazards, although the former mainly affect coastal areas, and the latter are found principally in the western margin of the *Cfa* climates.

The Marine West Coast Climates (*Cfb-Cfc*)

8-9 GENERAL CHARACTERISTICS AND DISTRIBUTION OF THE *Cfb-Cfc* CLIMATES The *Cfb-Cfc* climates by definition have mild winters, cool summers, and are without a pronounced dry season. The coldest month averages more than 26.6°F (-3°C), and the warmest month is below 71.6° (22°C). The only difference between the two sub-types is that the *Cfb* climate has more than 4 months with average temperatures above 50°F (10°C), and the *Cfc* has between 1 and 4 months with average temperatures above 50°. The mild winters and cool summers of these climates indicate a low seasonal range of temperature, a condition that is brought about by the prevalence of maritime air masses. The *Cfc* subtype occurs in extremely small areas, mainly islands, that are found near the poleward limit of the climatic zone. For that reason, they cannot be shown separately from the *Cfb* group on the world map of climates (see back endpaper and Figure 8-7).

The general location of the *Cfb-Cfc* climates is on the western side of continents poleward of the *Cs* group, or from about 43° to 65° N and S lat. In North America, this climatic zone begins just within the United States border and extends as a narrow and irregular belt along the broken coastline of British Columbia and into southern Alaska. The high mountain slopes of the Canadian coastal range and the Alaskan range prevent these marine climates from penetrating inland more than a few miles. A similar situation occurs in South America, where these climates typically extend southward along the Chilean coast south of 43° to the tip of the continent. The climate of the broken, fjord coast of Chile thus matches its counterpart north of the equator.

Africa and Australia have a small area of *Cfb* climate along their southeastern sides, caused mainly by the cooling effect of Antarctic sea water which barely intercepts the continental margins before plunging below the warmer waters of the South Indian and South Pacific Oceans that are headed poleward along the east coasts. New Zealand extends between latitudes 38° and 50° S, hence is well within the typical latitudinal range of the marine west coast climates (*Cfb-Cfc*).

The largest area of *Cfb* climate is found in Europe. Here there is no mountain barrier to bar the incursion of cool maritime air from the North Atlantic far into the interior, except in Norway, where this climatic zone is restricted to the southern fjord zone along the coast. *Cfb* climates predominate over all of northwestern Europe from Scandinavia to the Pyrenees and Alps and from the Atlantic coast to central Poland.

Small areas of *Cfb* climate are located on the easternmost extensions of Newfoundland, Nova Scotia, and Cape Cod. The *Cfb* climates also are important in the highland areas of low latitudes and will be discussed later in Section 8-19.

The wide latitudinal sweep of the *Cfb-Cfc* climates along the west coasts is a noteworthy feature. Despite broad differences in the length of day and directness of insolation along these coasts, temperatures remain remarkably even. The main reason for this is the advection of air from maritime regions that exhibit strong positive temperature anomalies (see Figures 4-20 and 6-27), especially in winter. The predominant air masses are *mPk* or *mPw*.

8-10 WEATHER AND CLIMATIC CONDITIONS WITHIN THE MARINE WEST COAST CLIMATES Characteristic temperature and precipitation regimes for several *Cfb-Cfc* stations are given in Table 8-4, Figures 8-10 and 8-11, and Appendix C. While some variations in the averages are indicated, all of the stations show the two dominant features of these climates: a low seasonal range of temperatures, and consistently humid conditions. In most of these climates sweaters or topcoats are welcome wear for

Bergen, Norway, 60°24'N.; 05°19'E., Elev. 141'

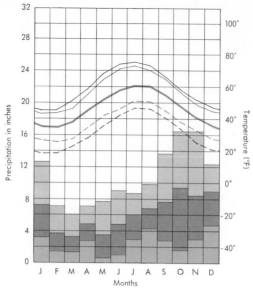

Koeppen Class: Cfb

Mean annual temperature: 46.0°F.

Mean annual precipitation: 77.1 inches

- Maximum precipitation (10 year record)
- Mean precipitation (Station record)
- Minimum precipitation (10 year record)
- Average monthly maximum temperature
- Average daily maximum temperature
- Mean monthly temperature
- Average daily minimum temperature
- Average monthly minimum temperature

Valdivia, Chile, 39°48'S.; 73°14'W., Elev. 42'

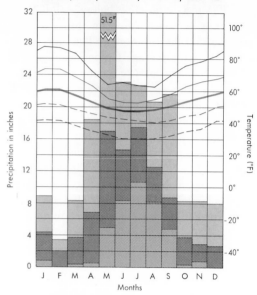

Koeppen Class: Cfb

Mean annual temperature: 53.4°F.

Mean annual precipitation: 97.9 inches

- Maximum precipitation (10 year record)
- Mean precipitation (Station record)
- Minimum precipitation (10 year record)
- Average monthly maximum temperature
- Mean monthly temperature
- Average daily minimum temperature
- Average daily maximum temperature
- Average monthly minimum temperature

comfort throughout the year. The monthly averages are not indicative of extremes. Yet compared with most of the other mid-latitude climates, even the extremes are confined within a relatively narrow range, especially as to temperature. At Bergen, Norway (see Figure 8.10), for example, the average monthly temperature for January is 35°F, yet the average of the lowest temperatures for that month is 19°. To illustrate a contrast using a typical mid-latitude continental climate, Louisville, Kentucky, at lat 38° N has an average temperature for January that is identical to that of Bergen, but the average of the lowest temperatures for January is 4°F, or much lower than that for Bergen. Polar continental air masses usually do not move westward within the general belt of westerly winds. This means that the polar continental air masses (cPk) of North Amer-

Kodiak, Alaska, 57°45'N.; 152°30'W., Elev. 104'

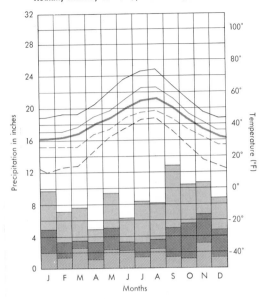

Koeppen Class: Cfc

Mean annual temperature: 41.5°F.

Mean annual precipitation: 60.4 inches

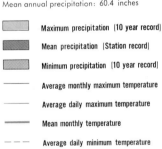

Maximum precipitation (10 year record)

Mean precipitation (Station record)

Minimum precipitation (10 year record)

Average monthly maximum temperature

Average daily maximum temperature

Mean monthly temperature

Average daily minimum temperature

Average monthly minimum temperature

Figure 8-10 Climatic charts for the marine west coast climates, *Cfb-Cfc* **(Bergen, Valdivia, and Kodiak).** Note that for the Valdivia record, the Koeppen classification calls for both an *f* (at least 1.2 inches per month) and an *s* (three times as much rain in the rainiest month of winter as in the driest month of summer). This unusual combination also is found along the British Columbia and Alaskan coasts.

in Europe and may be sufficient to freeze the canals and estuaries in Britain and the Low Countries. Nightly frosts are common but not persistent throughout most of the *Cfb* climates, and temperatures in winter generally climb above freezing during the day. Figure 8-11 indicates that at Oxford, England, temperatures average above freezing for any hour during the winter. The average minimum temperatures recorded at Oxford for both January and February, however, are 23° and 24°, respectively, and a reading as low as 3° has been recorded.

Precipitation varies widely in the *Cfb* climates, depending largely on terrain conditions and exposure to marine air streams. Sharp local contrasts are observed on opposite sides of mountain ranges, such as along the rugged coastal regions of British Columbia, southern Chile, New Zealand, and Norway. The windward side of Vancouver Island, British Columbia, for example, regularly has more than 100 inches of rain, whereas the lee side receives only about 20 to 25 inches on the average. The high mountainous coasts within the *Cfb* climate regularly receive much more precipitation than the flat lands of northwestern Europe, but even in the latter, prolonged droughts are exceedingly rare, and usually there is sufficient rainfall for normal plant growth at any time of the year. The monthly minimum precipitation figures for the 10-year period shown for Bergen and Valdivia in Figure 8-10

ica and Eurasia rarely extend out to the west coasts in the upper mid-latitudes. South of the equator, Antarctica is too far away to bring cold winter weather to the southern hemisphere continents. The west coast of British Columbia and the coast of southern Alaska are especially sheltered from cold Canadian air by the high mountains that border these coasts. Cold waves from continental sources are somewhat more frequent

Table 8-4   Climatic Data for Selected *Cfb* and *Cfc* Stations

	Jan.	Feb.	Mar.	Apr.	May	June	July	Aug.	Sept.	Oct.	Nov.	Dec.	Yr.
Dublin, Ireland; lat 53°26' N long 06°15' W; elev. 264 ft; Koeppen class: Cfb													
Temp. (°F)	40.3	40.5	43.7	47.1	51.1	56.1	59.2	58.6	55.8	50.9	45.1	42.4	49.3
Precip. (in.)	2.8	2.0	2.0	1.6	2.5	2.2	2.6	3.3	3.1	2.5	2.6	3.1	30.3
Port Hardy, Vancouver Is.; lat 50°41' N long 127°22' W; elev. 75 ft; Koeppen class: Cfbs													
Temp. (°F)	36.3	38.3	38.8	43.9	49.5	53.4	56.7	57.0	53.4	48.0	41.7	39.0	46.2
Precip. (in.)	7.1	6.3	5.2	3.7	2.4	2.9	1.6	2.6	4.8	8.1	9.0	10.7	64.4
Hobart, Tasmania; lat 42°53' S long 147°20' E; elev. 177 ft; Koeppen class: Cfb													
Temp. (°F)	62.0	62.0	59.5	55.5	51.0	47.0	46.0	47.5	51.0	54.5	57.0	60.0	54.5
Precip. (in.)	1.9	1.5	1.8	1.9	1.8	2.2	2.1	1.9	2.1	2.3	2.4	2.1	24.0
Reykjavik, Iceland; lat 64°00' N long 20°56' W; elev. 92 ft; Koeppen class: Cfc													
Temp. (°F)	32.0	32.5	34.5	38.5	44.5	49.5	53.0	52.0	46.5	40.0	35.5	34.0	41.5
Precip. (in.)	4.0	3.1	3.0	2.1	1.6	1.7	2.0	2.6	3.1	3.4	3.6	3.7	33.9

indicate that the growing season has plenty of precipitation for normal agriculture and tree growth.

The much higher rainfall of the coastal regions of British Columbia and southern Chile, as compared with that over most of northwestern Europe, is not entirely the result of orographic lifting. This is indicated by the continuation of the high precipitation well out to sea. Much of this precipitation is associated with cyclonic disturbances that move along fronts that are pocketed along the coast, being prevented from moving eastward by the high mountain ranges. The highest amounts of precipitation along the marine west coasts occur during the late fall and winter seasons, because this is the time of maximum frontal activity.

Torrential rains are rare throughout most of the *Cfb-Cfc* climates, although New Zealand has some heavy, intermittent rain showers. Steady precipitation, ranging from a light drizzle to a continuous rain, may continue for days and even weeks, as many American tourists discover to their dismay when touring northwestern Europe, especially in the fall and winter months. Thunderstorms are rare occurrences along the northwest coast of North America (see Figure 6-30) and in southern Chile. The coolness of the marine air masses that prevail does not permit the air to hold the large quantities of water vapor necessary to generate the thermodynamic processes of thunderstorms. Heavy precipitation totals exceed 100 inches. Even amounts of up to 200 inches per year are not unknown. They are found along exposed coastal mountain slopes, where precipitation is almost continuous throughout the year. The heaviest rainfall recorded in the United States, for example, is located on the windward slopes of the Olympic Mountains in northwestern Washington and exceeds 100 inches per year.

Not much of the total precipitation falls as snow on the low-lying areas, although poleward of about 60°, temperatures at sea level drop low enough to bring heavy winter snows. Mountainous regions in these climates, on the other hand, have exceedingly heavy snowfalls, with total amounts occasionally exceeding 500 to 600 inches per year.

One of the more unpleasant features of the marine west coast climates is the general low frequency of sunny days, which is perhaps the lowest of any climatic region on earth. Less than a third of the days are sunny over many parts of the *Cfb* regions, and a cloud cover may persist for more than 80 percent of the time on some of the exposed mountain slopes. New Zealand is exceptionally favored in this climate by having an unusually high percentage of sunny skies. Most of the air masses that reach New Zealand are composed of highly modified Antarctic air that is warmed and humidified in its lower reaches. The lapse rate is steep, and the air is unstable and overturns easily. Short, sudden showers are much more frequent in New Zealand than the steady drizzles and rains of most *Cfb* regions. Stable fronts also are rare in New Zealand, and, while fronts are common, they pass relatively rapidly across the islands.

Gale winds associated with cyclonic storms occur frequently during the winter season along the *Cfb* west coasts. Pressure gradients usually tend to become steeper with increasing latitude, because the increasing Coriolis effect reduces the tendency for airflow to cross isobars and thus reduce pressure differentials. Angular momentum of winds also is greater at higher latitudes. The winter gale winds of the North Atlantic, North Pacific, and off the Antarctic continent are well known, and many of these storms lash the west coasts poleward of 50°. Friction with the land soon decreases the velocity of these storms, and their principal fury is expended

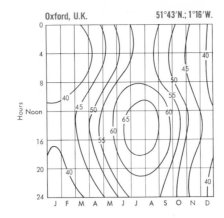

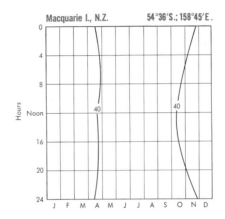

Figure 8-11 Isothermographs of average hourly temperatures (°F), by month, for Oxford, United Kingdom (*Cfb*), and Macquarie Island, New Zealand (*ET*). Oxford illustrates a typical temperature pattern for the *Cfb* climates, in which frosts are as rare as hot days. Both occur at Oxford, as do temperatures near zero, but they are relatively unusual, as the averages indicate. Macquarie Island, located south of New Zealand, exhibits an extreme case of marine influence at high latitudes. It is unique in having a polar climate in which frosts are almost unknown. The lowest temperature recorded here over a 7-year span was 25°F, and the highest a chilly 50°.

on the immediate coastal areas. Winter wind stress often prevents the development of forest cover along the coastal margins, and the predominant vegetation usually is a low bush growth of such plants as laurel or gorse, streamlined to resist wind friction. Many of the immediate outer coastal zones in Ireland, Scotland, Norway, British Columbia, and Chile have a bleak, barren appearance.

Fog is another common feature of the marine west coast climates, although conditions vary widely in different sectors. The areas bordering the North Sea and the Norwegian coast have an unusually high incidence of fogs, ranging from 30 to 60 days per year. They also are frequent in southern Chile. The British Columbia coast experiences fog on about 10 percent of the days in the year. The coastal fogs tend to be most common in spring and early summer, when intrusions of *mTw* air become chilled by passage over water that still retains some of the coolness of winter. Winter fogs are especially common over interior positions and may occur either as the result of radiational cooling at night or by the advection of marine air over a land surface that has been dusted with snow or chilled by an occasional *cPk* air mass. Air pollution over the industrial cities of Britain adds to the intensity and frequency of their winter fogs.

8-11 HUMAN IMPLICATIONS OF THE *Cfb-Cfc* CLIMATES The marine west coast climates vary widely in terms of general body comfort. Perhaps the most consistently miserable climate on earth in this respect is the *Cfc* type, found near the poleward margins of this general climatic zone. Temperatures here are near freezing throughout the year, and the relative humidity is near saturation at all times. Much of the precipitation is in the boundary zone between rain, sleet, or snow. At the same time, this climate has no physiological barriers to human life. Charles Darwin noted with amazement that the early human inhabitants of Tierra del Fuego were able to live without clothing and with simple sod shelters. Apparently the human body is capable of becoming acclimatized to temperatures that are persistently near the freezing point. This cannot be accomplished, however, without an unusual cultural motivation, and there may possibly be some biological conditioning as to the heat sensory and distributive mechanisms following centuries of exposure to these climatic conditions. Agriculture is extremely limited in the *Cfc* climates, not because of frost hazard but because most plant reproductive processes and growth begin not at the freezing point, but at temperatures of around 43°. A dairy economy based on the cultivation of grasses and root tubers seems to be the most feasible agricultural system, but for the most part the inhabitants of the *Cfc* climate consist largely of fishermen, since the offshore waters are unusually prolific as to sea life, especially during the summers.

At the other extreme, the equatorial margins of the *Cfb* climates are among the more pleasant climates of the world for human body comfort. Table 8-5 presents a few of the more pleasant areas of *Cfb* climate.

The cost of air conditioning to maintain indoor temperatures near body comfort level is lower than for any of the low-elevation climates on earth, at least for sedentary whites. Temperatures during the summer are close to the ideal of 68°F, and winter heating

Table 8-5 Average Temperature of the Coldest and Warmest Months at Some of the Warmer *Cfb* Stations (°F)

Station	Warmest month		Coldest month	
Bordeaux, France	July	(67.3°)	January	(41.4°)
Vancouver, B.C.	July	(64.0°)	January	(36.5°)
Mar del Plata, Arg.	January	(68.2°)	July	(46.2°)
Canberra, Australia	January	(68.5°)	July	(42.5°)
Port Elizabeth, S. Afr.	February	(69.3°)	January	(55.9°)

costs are low. The heating degree-days for the extreme northwestern tip of the United States are between 5,000 and 6,000, or roughly the same as those in St. Louis, Columbus, Philadelphia, and New York City.

The frost-free season in most *Cfb* climates lasts roughly from 9 to 10 months, and winter temperatures are not low enough to interfere with the growth of winter grains such as wheat, oats, or barley. Some of the highest wheat yields in the world are found in the *Cfb* climates of Europe. Summer grains such as corn (maize), flaxseed, rice, and the grain sorghums do not thrive because of the high humidity during the late summer and early fall maturing season. Fungus infections rather than temperatures are the limiting factor. Grasses, root tubers, and acid fruits find ideal conditions in the cool, damp summers. Drying hay and forage crops is a problem because of cloudy, wet conditions, but special techniques of drying and the widespread conversion of green plants into ensilage have helped to develop the dairy industry of the *Cfb* areas into one of the most productive agricultural systems in the world.

The relatively mild winters, combined with grass growth throughout the year, favor all types of livestock raising. Expensive, large barns are not required for winter shelter for animals, and pastures can be used throughout the year. Nearly all of the *Cfb* areas, with the possible exception of coastal British Columbia, have developed sheep raising for both wool and meat.

As though to compensate for the drabness of the leaden skies, fog, and drizzly weather, the *Cfb* climates favor the growth of flowers of brilliant colors, and these seem to thrive almost throughout the year. The relative ease of growing flowers, especially roses, azaleas, rhododendrons, and the bulb plants such as tulips, narcissus, lilies, and hyacinths, is an inducement that lures residential development by retired people from other climates.

The *D* Climates (humid continental with severe winters)

8-12 THE *D* CLIMATES: GENERAL The *D* climates include the humid parts of the world that have a summer but that also have a long, severe winter. By definition, this type of climate is one in which the mean temperature for the coldest month is below 26.6°F (-3°C) and that of the warmest month is above 50°F. It is thus characterized by wide seasonal ranges of temperature which are the result of continental heating and cooling. Except for a few small spots of *D* climates in the mountains of New Zealand and in the Andes of South America, this group is restricted to the continents of the northern hemisphere. There is insufficient land area at the higher middle latitudes south of the equator to produce wide temperature ranges seasonally.

The members of the *D* group are differentiated in the Koeppen classification by the degree of summer warmth (*a*, *b*, or *c*) and by the presence or absence of a winter dry season (*w* or *f*, respectively). The *Dw* climates are found only in eastern Asia, where the Siberian polar anticyclone prevents the importation of moist air during the winter season. The generalized distribution of the *D* climates is from lat 60° to 70° on the western sectors of the northern hemisphere continents to lat 40° to 60° on the eastern sectors. The distribution of the various *D* climates is shown in Figure 8-12.

8-13 THE *Dfa* AND *Dwa* CLIMATES (HUMID CONTINENTAL WITH SEVERE WINTERS AND HOT SUMMERS) The *Dfa* and *Dwa* climates, which are distinguished by the latter's having a pronounced dry season in winter, are characterized by a wide seasonal range of temperatures, with hot summers similar to those in the subtropics (warmest month above 71.6°), and cold win-

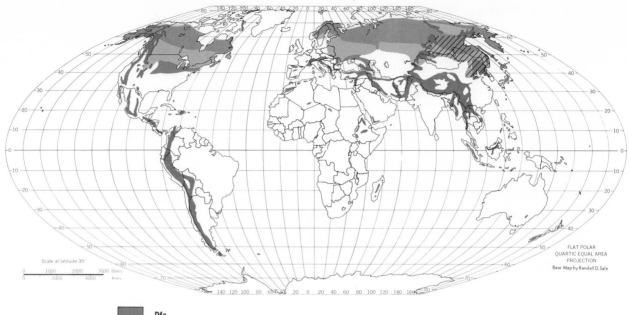

■	Dfa
▨	Dwa
▨	Dfb
▨	Dwb
■	Dfc
▨	Dwc
▨	Dfd
■	High mountains

Figure 8-12   Distribution of the *D* Climates.

ters (coldest month below 26.6°). Examples are given in Table 8-6 and Figure 8-13. The average hourly range of temperatures by months for Omaha, Nebraska, a typical *Dfa* station, is given in Figure 8-14.

The *Dfa* and *Dwa* climates are found immediately north of the *Cfa* or *Cwa* climates in both eastern North America and eastern Asia. They also lie just east of the

B climates that occupy the interiors of these continents. The southern margin of these climates is located at about lat 35° N in China and about lat 40° N in the United States. In the latter area, the *Dfa* region is widest in the west near the 100th meridian, where it extends from southern Nebraska northward to the center of South Dakota. It narrows eastward and becomes discontinuous east of northern Ohio. Portions of the upper Hudson Valley and central New York have a *Dfa* climate. The *Dwa* subtype predominates in eastern Asia, occupying most of northeastern China north of the Shantung Peninsula, southern Manchuria, and northern Korea. The drier winter season in Asia is the result of the greater dominance of *cP* air, which originates in the great winter polar anticyclone of eastern Siberia and Mongolia. The smaller size of North America, the deep indentation of the subtropical Gulf of Mexico, and the lesser strength of the Canadian polar highs result in a fairly humid winter

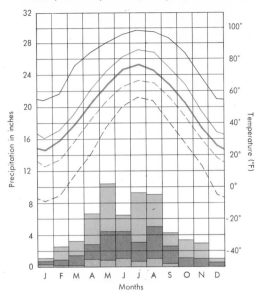

Omaha, Nebraska, 41°18′N.; 95°54′W., Elev. 992′

Koeppen Class: Dfa

Mean annual temperature: 51.4°F.

Mean annual precipitation: 27.6 inches

☐ Maximum precipitation (10 year record)

▨ Mean precipitation (Station record)

▨ Minimum precipitation (10 year record)

—— Average monthly maximum temperature

----- Average daily maximum temperature

········ Mean monthly temperature

– – – Average daily minimum temperature

— — Average monthly minimum temperature

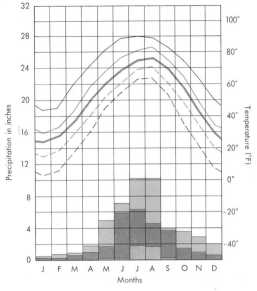

Talien (Dairen), China, 38°55′N.; 121°38′E., Elev. 316′

Koeppen Class: Dwa

Mean annual temperature: 51.0°F.

Mean annual precipitation: 23.9 inches

▨ Maximum precipitation (10 year record)

▨ Mean precipitation (Station record)

▨ Minimum precipitation (10 year record)

—— Average monthly maximum temperature

----- Average daily maximum temperature

········ Mean monthly temperature

– – – Average daily minimum temperature

— — Average monthly minimum temperature

Figure 8-13 Representative climatic charts for the *Dfa-Dwa* climates (Omaha, Nebraska, and Talien [Dairen], China).

season in the upper Mississippi drainage basin.

The distinction between *Dfa* and *Dwa* climates has little significance with respect to vegetation and agriculture. Since the winter season is too cold for active vegetative growth, most plants curtail their life processes then, so that it makes little difference whether the precipitation of the winter season is low or high. The difference in amount of snow cover, however, is of some importance. In northern China, the small amount of winter snow leaves cultivated fields exposed to winter wind erosion. The winter season in Peking, as elsewhere on the North China Plain, is characterized by an almost

Table 8-6 Representative Statistical Averages for the *Dfa-Dwa* Climates

	Jan.	Feb.	Mar.	Apr.	May	June	July	Aug.	Sept.	Oct.	Nov.	Dec.	Yr.
Chicago, Ill.; lat 41°47′ N long 87°45′ W; elev. 613 ft; Koeppen class.: Dfa													
Temp. (°F)	26.1	27.9	36.3	49.1	60.1	70.7	75.7	74.5	66.4	55.4	39.9	29.1	50.9
Precip. (in.)	1.8	1.6	2.8	3.0	3.7	4.1	3.4	3.1	2.7	2.8	2.2	1.9	33.1
Huron, S. Dak.; lat 44°23′ N long 98°13′ W; elev. 1,288 ft; Koeppen class.: Dfa													
Temp. (°F)	12.6	16.5	28.8	45.0	57.6	67.6	75.0	72.0	61.9	48.7	31.3	19.0	44.8
Precip. (in.)	.4	.6	1.1	1.8	2.4	3.2	1.8	2.1	1.6	1.1	.7	.5	17.3
Mukden, China; lat 41°48′ N long 123°23′ E; elev. 141 ft; Koeppen class.: Dwa													
Temp. (°F)	11.5	17.0	32.0	49.0	62.0	72.5	78.0	76.0	62.5	49.5	31.5	15.0	46.5
Precip. (in.)	0.3	0.3	0.7	1.1	2.7	3.3	7.2	6.7	2.5	1.4	1.1	0.6	27.9

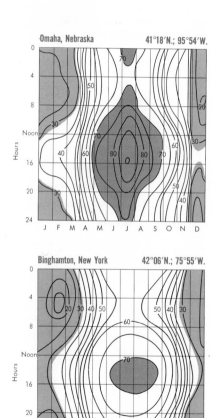

continuous dust haze, some of which is blown in from the Gobi Desert to the northwest, but some of it also is from the exposed fields throughout this area. Although occasional heavy snows occur along winter fronts in the American *Dfa* region, snow does not remain on the ground for more than a month or so. Occasional importations of warm *mTw* air from the Gulf of Mexico during the winter often will remove the snow cover rapidly. As in the northern portions of the *Cfa* region, sleet and glaze are troublesome winter hazards in the lower Great Lakes region. Even in the United States, however, precipitation is greater in summer than in winter, and the discrepancy increases toward the west. A comparison of the rainfall statistics for Chicago, Illinois, and Huron, South Dakota, both of which have a *Dfa* climate, is illus-

Figure 8-14 Isothermographs of average hourly temperature (°F), by month, for Omaha, Nebraska (*Dfa*), and Binghamton, New York (*Dfb*), in 1964. Note that continentality at Omaha is greater than at Binghamton at roughly similar latitudes; that there is a rapid change in temperature in the spring and autumn; and that the summer temperatures at Omaha are comparable to those of the subtropical deserts.

trative (see Table 8-6). The increased winter dryness of the *Dfa* region west of the Mississippi River is aided by the increasing importance of masses of modified *mPw* air from the Pacific Ocean which lose much water vapor during their passage across the mountains and become warm and dry upon their descent into the western Great Plains.

The *Dfa* and *Dwa* climates not only regularly exhibit wide seasonal ranges of temperature and precipitation, but there are wide departures from mean figures for temperature and precipitation in both the summer and winter seasons. This is illustrated by the climatic charts in Figure 8-13. The position of Omaha in the interior of the country, unlike Talien, which lies at the end of the Liaotung Peninsula in southern Manchuria, is responsible for the greater continentality and more irregular rainfall of the former. Compare the data for Talien with that of Mukden (see Table 8-6), which is in central Manchuria, and note how rapidly the continentality in eastern Asia becomes effective away from the coast. The major reason for the wide fluctuations from the statistical norms in most *Da* stations is that the eastern interior sections of the northern hemisphere continents are visited by air masses of many

types. In the United States, for example, which is the more variable of the two sections, the *Dfa* region may be under the influence of most of the major air masses, including *mP*, *cP*, *cT*, and *mT*, which show varying degrees and kinds of modifications.

8-14 THE *Dfb* AND *Dwb* CLIMATES (HUMID CONTINENTAL WITH COOL SUMMERS) The mild-summer variants of the *D* climates (*Dfb* and *Dwb*) are much more extensive than the *Da* types and are found in a broad belt across southern Canada and northern United States, and also in Eurasia from southern Sweden and central Poland eastward across the Soviet Union to northern Manchuria, northern Korea, the Soviet Far East, and the northern Japanese island of Hokkaido. The *Dwb* subgroup is found only in eastern Asia, southeast of Lake Baikal (see Figure 8-12).

Winters in the *Dfb* and *Dwb* climates are more severe than in the *Da* types. Minimum temperatures drop to −20 or −30°F nearly every winter, but general daytime temperatures range from 0° to 30°. Snow usually remains on the ground during the months (2 to 5) when the mean average is below 32°, depending also on the amount of snowfall.

Table 8-7 Representative Climatic Data for Selected *Dfb-Dwb* Stations

	Jan.	Feb.	Mar.	Apr.	May	June	July	Aug.	Sept.	Oct.	Nov.	Dec.	Yr.
Montreal, Canada; lat 45°28′ N long 73°45′ W; elev. 117 ft; Koeppen class.: Dfb													
Temp. (°)	14.7	17.1	28.0	42.6	55.6	65.5	70.5	68.5	59.7	48.9	36.5	20.5	44.1
Precip. (in.)	3.3	3.2	3.1	2.8	2.8	3.3	3.5	3.0	3.2	3.1	3.3	3.5	38.1
Leningrad, U.S.S.R.; lat 59°51′ N long 30°18′ E; elev. 12 ft; Koeppen class.: Dfb													
Temp. (°)	18.3	17.8	24.3	37.9	49.8	59.7	65.1	62.2	52.2	41.2	31.6	24.1	40.3
Precip. (in.)	1.4	1.3	1.0	1.3	1.6	2.1	2.7	3.0	2.3	2.0	1.8	1.4	21.9
Khabarovsk, U.S.S.R.; lat 48°31′ N long 135°10′ E; elev. 235 ft; Koeppen class.: Dwb													
Temp. (°)	−7.6	0.7	16.9	37.0	52.0	63.9	70.3	68.5	57.0	41.2	17.2	−1.1	34.7
Precip. (in.)	0.3	0.3	0.5	1.3	2.2	2.9	4.0	4.5	3.4	1.5	0.6	0.4	21.9

Semipalatinsk, U.S.S.R., 50°24′ N.; 80°13′ E., Elev. 709′

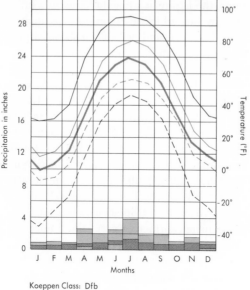

Koeppen Class: Dfb

Mean annual temperature: 38.1°F.

Mean annual precipitation: 10.3 inches

	Maximum precipitation (10 year record)
	Mean precipitation (Station record)
	Minimum precipitation (10 year record)
——— Average monthly maximum temperature
——— Average daily maximum temperature
——— Mean monthly temperature
- - - Average daily minimum temperature
- - - Average monthly minimum temperature

Duluth, Minnesota, 46°50′ N.; 92°11′ W., Elev. 1415′

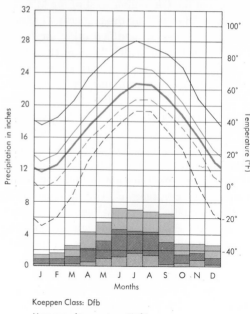

Koeppen Class: Dfb

Mean annual temperature: 38.1°F.

Mean annual precipitation: 29.0 inches

	Maximum precipitation (10 year record)
	Mean precipitation (Station record)
	Minimum precipitation (10 year record)
——— Average monthly maximum temperature
——— Average daily maximum temperature
——— Mean monthly temperature
- - - Average daily minimum temperature
- - - Average monthly minimum temperature

Figure 8-15 Representative climatic charts for the *Dfb-Dwb* climates (Duluth, Semipalatinsk, and Khabarovsk). Duluth has a *Dfb* climate, while the latter two stations are *Dwb*. Note the continentality of Semipalatinsk.

Summers are mostly pleasant and cool, with daytime temperatures in the low 70s, but occasional spells of warm weather may bring temperates into the 90s. The isothermographs shown in Figures 8-14 and 8-16 indi-cate that there is considerable variation in the degree of continentality within the *Db* climates. This is also indicated by the data in Table 8-7.

Summer air masses include mainly polar continental (*cP*) and highly modified marine tropical (*mT*) types. The cool polar air is highly unstable, because it is being warmed below and is also picking up moisture in its lower layers during its passage over a damp

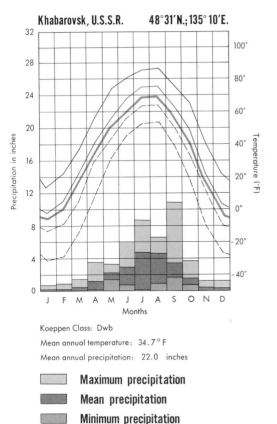

Khabarovsk, U.S.S.R.   48°31′N.; 135°10′E.

Koeppen Class: Dwb

Mean annual temperature: 34.7°F

Mean annual precipitation: 22.0 inches

Maximum precipitation

Mean precipitation

Minimum precipitation

—— Average monthly maximum temperature

—— Average daily maximum temperature

—— Mean monthly temperature

--- Average daily minimum temperature

--- Average monthly minimum temperature

boundary with the cooler *cP* air results in heavy summer rains near the eastern margins of the continent. Inland, rainfall decreases rapidly, and only the cool temperatures and lower evaporation keep large areas from having a *B* climate (see the climatic chart for Semipalatinsk in Figure 8-15). The summer

Figure 8-16 Isothermographs of average hourly temperature (°F), by month, for Duluth, Minnesota (1964), and Irkutsk, Soviet Union. The general configuration of these two charts is similar, but the length of the summer season is greater in Duluth, and the seasonal range is much smaller. Note how the isotherms in these charts tend to run vertically in the high latitudes and horizontally in the low latitudes. [Irkutsk data after Troll]

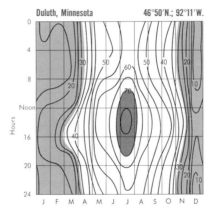

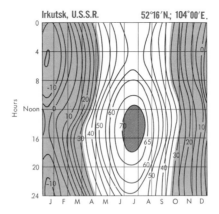

land surface. Summer rainfall alternates between heavy instability showers and long periods of light rain. The Great Lakes and the many small lakes of the Canadian Shield region in southern Canada afford evaporation surfaces, and summer rainfalls are somewhat greater than might be expected south of the lakes. There is a strong inflow of tropical Pacific air (*mPw*) into eastern Asia, and almost continuous frontal activity along the

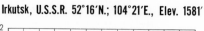

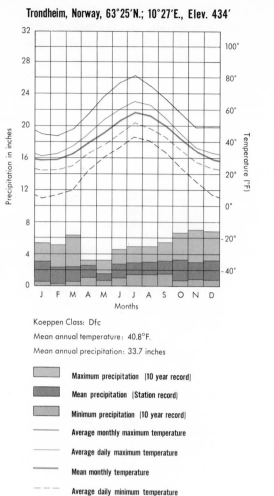

Trondheim, Norway, 63°25′N.; 10°27′E., Elev. 434′

Koeppen Class: Dfc

Mean annual temperature: 40.8°F.

Mean annual precipitation: 33.7 inches

	Maximum precipitation (10 year record)
	Mean precipitation (Station record)
	Minimum precipitation (10 year record)
——	Average monthly maximum temperature
——	Average daily maximum temperature
······	Mean monthly temperature
– – –	Average daily minimum temperature
— — —	Average monthly minimum temperature

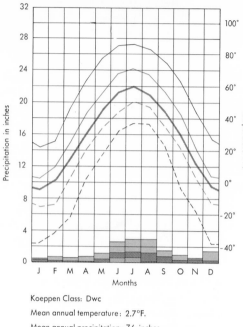

Irkutsk, U.S.S.R. 52°16′N.; 104°21′E., Elev. 1581′

Koeppen Class: Dwc

Mean annual temperature: 2.7°F.

Mean annual precipitation: 7.6 inches

	Maximum precipitation (10 year record)
	Mean precipitation (Station record)
	Minimum precipitation (10 year record)
——	Average monthly maximum temperature
——	Average daily maximum temperature
······	Mean monthly temperature
– – –	Average daily minimum temperature
— — —	Average monthly minimum temperature

Figure 8-17 Representative climatic charts for the *Dfc*, *Dwc*, and *Dwd* climates (Trondheim, Irkutsk, and Verkhoyansk).

cyclonic lows of the *Db* regions are not nearly so deep and intense as they are during the winter.

Winters are dominated by cold, polar continental or arctic air masses, and precipi-tation is less than in the summer season. The unfrozen portions of the Great Lakes, how-ever, provide an important source of winter moisture, and this is released as heavy oro-graphic snowfall on slopes along the lee side of the lakes. Heavy orographic snowfall also occurs on the western slopes of the Ural Mountains in the central Soviet Union, in northern Japan, and on Sakhalin Island.

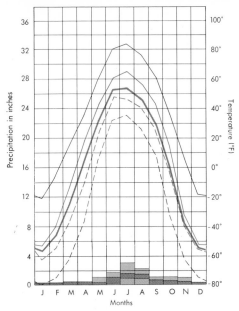

Verkhoyansk, U.S.S.R., 69°33'N.; 133°23'E., Elev. 446'

Koeppen Class: Dwd
Mean annual temperature: 4.6°F.
Mean annual precipitation: 6.1 inches

▩	Maximum precipitation (10 year record)
▩	Mean precipitation (Station record)
▩	Minimum precipitation (10 year record)
——	Average monthly maximum temperature
——	Average daily maximum temperature
——	Mean monthly temperature
- - -	Average daily minimum temperature
- - -	Average monthly minimum temperature

8-15 THE Dfc, Dwc, AND Dwd CLIMATES (SUBARCTIC) The Dfc-Dwc-Dwd[3] group of the D climates is the most extensive of all the groups, covering broad, uninterrupted expanses from Alaska to Labrador in North America and from Norway to Kamchatka

[3]Koeppen devised the special symbol d to categorize the unusually low temperatures of eastern Siberia. This letter is used when the average mean temperature of the coldest month drops below −36.4°F (−38°C).

along the Pacific coast of the Soviet Union. It thus extends farther poleward along the west coasts (roughly from 60° to 70° N lat) than in the east, where it lies between 50° and 65° N lat. This climatic group is sometimes referred to as the *taiga* climate, since it closely corresponds to the region of the northern coniferous forest, that seemingly endless sea of spruce, fir, larch, and birch trees. It also has been aptly described as a country with "10 months of winter and 2 months of bad sleighing." Although all such areas have at least 1 month with an average mean temperature above 50°, the summers are short (less than 4 months above 50°) and cool. Frost may occur at any time during the summer, although it is neither common nor persistent. Rare warm spells may bring soaring temperatures into the high 80s.

Precipitation in these northern regions usually is light, but the low temperatures checking evaporation keep the ground well supplied with moisture. Despite the long hours of daylight during the summer, the sun never rises very high in the sky and solar radiation is not intense. The snowfall is not so heavy as in the Db climates in the south, but since less of it disappears through melting or direct evaporation during the winter, a snow cover 2 to 3 feet deep generally persists for 5 to 6 months.

Winter minimum temperatures in this climatic group are among the lowest ever recorded outside the permanent ice fields of Antarctica and Greenland. A series of mountain-girt basins in eastern Siberia, northeast of Lake Baikal, combine perfect conditions for air drainage on long, clear, calm, winter nights with their continental interior position to develop a general region of extremely low winter averages (Dwd) and wide seasonal ranges of temperatures. An unofficial reading of −108°F was recorded at Oymyakon, a

Table 8-8 Representative Climatic Data for Selected *Dfc, Dwc,* and *Dwd* Stations

	Jan.	Feb.	Mar.	Apr.	May	June	July	Aug.	Sept.	Oct.	Nov.	Dec.	Yr.
	Fort Simpson, Canada; lat 61°52′ N long 121°21′ W; elev. 420 ft; Koeppen class.: Dfc												
Temp. (°F)	−15.9	−9.9	4.8	25.9	46.2	57.2	62.7	57.9	46.2	29.8	6.3	−10.8	25.0
Precip. (in.)	0.7	0.5	0.6	0.7	1.1	1.4	2.1	1.9	1.3	0.8	0.9	0.9	12.7
	Chita, U.S.S.R.; lat 52°01′ N long 113°20′ E; elev. 2,233 ft; Koeppen class.: Dwc												
Temp. (°F)	14.8	−6.5	11.3	32.4	47.5	60.8	64.8	59.9	45.5	30.4	5.7	−10.1	27.1
Precip. (in.)	0.1	0.1	0.2	0.3	0.8	2.0	3.7	3.5	2.0	0.5	0.3	0.1	13.6
	Oymyakon, U.S.S.R.; lat 63°16′ N long 143°09′ E; elev. 2,367 ft; Koeppen class.: Dwd												
Temp. (°F)	−53.0	−45.0	−29.6	4.3	34.5	52.9	58.6	51.6	34.9	2.8	−32.6	−47.2	2.6
Precip. (in.)	0.3	0.2	0.2	0.2	0.4	1.3	1.6	1.4	0.8	0.5	0.4	0.3	7.6

small village east of the Lena River in this area. At this village, the 1952 record indicates an average temperature during January of that year of −59.4° and one for July of 62.2°, or an annual range of 121.6°! Over a 23-year period at Verkhoyansk, the average monthly minimum was −80° during January. Winter temperatures in this area are much lower than any observed in northern Canada, where monthly averages do not drop below −25°F and where the seasonal range of monthly averages does not exceed 82°.[4] Apart from those in the cold basins of northeastern Siberia, temperatures elsewhere in the Soviet Union are not appreciably lower than those in Canada. As a general rule, free air temperatures below −30° are the result of air drainage, where air chilled by radiational cooling slips downslope and accumulates in topographic depressions. Sharp differences of as much as 20° can be observed within a few miles. As noted in Section 5-6, at air temperatures below −40°, spontaneous sublimation of atmospheric moisture takes place, and the air becomes hazy with tiny ice crystals. As soon as the temperature rises above this level, the air becomes remarkably clear. Clouds are rare during the long winter nights, and bright starlight and the displays of northern lights (aurora borealis) make the winter night far less dark than at lower latitudes.

[4]It should be remembered that the monthly averages represent the average for the mean daily temperatures during the month, although at these high latitudes there is not much difference during any one 24-hour period.

Table 8-9 Range of Mean Monthly Temperatures at a Series of *Dwd* Stations

Station	Jan. mean temp. (°F)	July mean temp. (°F)	Range
Oymyakon	−53.0°	+58.6°	111.6°
Verkhoyansk	−52.2	+60.1	112.3
Ust-Maya	−46.1	+64.0	110.1
Yakutsk	−44.9	+60.0	104.9
Shelagontsky	−41.6	+58.8	100.4

8-16 HUMAN DISADVANTAGES OF THE *D* CLIMATES The principal impact of the *D* climates on human living is the coldness of the winter season. Temperatures below freezing require that energy be expended in some way by men and animals in order to maintain internal temperatures above freezing. This is an energy input into the human physiological or cultural system that is compulsory and over and above the amount needed in the *A, B,* and *C* climates. It takes

Figure 8-18 Heavy snow blower at work in Antrim County, Michigan. Maintaining highway traffic in the *Dfb-Dfc* regions during the winter season requires specialized equipment. [Courtesy, Michigan State Highway Dept.]

several forms—increased expenditures for clothing, shelter, and heating fuel—and even the food caloric intake must be somewhat greater for outdoor workers in the winter. Air-conditioning costs are especially high in the *Da* climates, since even the summers require expenditures for cooling to maintain comfortable room temperatures. Below-zero weather, which is common throughout the *D* climates, necessitates added costs for buildings, because insulating materials are needed to reduce heating costs. Commercial warehousing presents special problems of heating large open spaces.

Snow removal to maintain surface transportation routes is an additional cost of living

and may constitute the major item in highway maintenance costs. The *Dfb* areas are especially bothered by heavy snows, and many of the cities within or near this zone are occasionally made temporarily helpless by 2- or 3-day storms that may drop 10 to 20 inches of snow on city streets. Technology has made amazing developments in solving the snow-removal problem (see Figure 8-18), and the coordinated efforts of highway departments demand highly sophisticated logistical services in marshaling both equipment and men following early warnings by the weather services.

The *D* climates have definite limiting effects on agricultural land-use patterns.

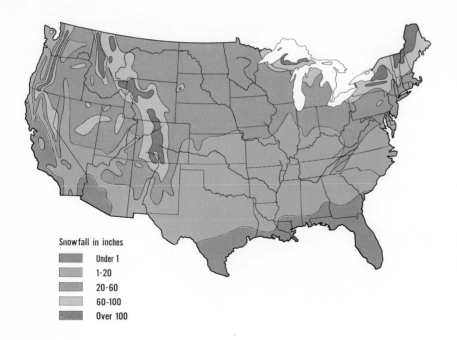

Figure 8-19 Snowfall in the United States. Terrain has a great influence on the distribution of snowfall, and a much larger map would indicate a greater regional variation and correlation with topographic conditions than is indicated here. Note the influence of the Great Lakes in producing instability in the *cP* air masses that sweep across the lakes from northwest to southeast and cause heavy snowfalls along the southeastern side of the lakes. [U.S. Dept. of Agriculture]

Snowfall in inches

- Under 1
- 1-20
- 20-60
- 60-100
- Over 100

Within the scope of these climates, the frost-free period varies widely, ranging from 90 days or less to 200–250, but several major food crops generally are not possible. Low winter temperatures do not permit the cultivation of winter grains—such as winter wheat or barley—that are planted in the fall to remain in the ground during the winter to obtain an early start for early summer maturity. This means that double-cropping generally is not possible in the *D* climates, and thus total food yields per acre tend to be less. The subtropical crops such as rice and cotton require longer growing periods than are found in these climates, although some exceptionally fast-maturing rice hybrids have been developed for use in the southern margins of the *D* climates in Asia. Maize and soy beans are well adapted to the *Da* climates, but the cool, moist summers of the *Db* and *Dc* group are not favorable for their growth as grains, although they are grown for ensilage. Except for the *Da* sub-type, the *D* climates are not favorable climates for the food grains and tend to favor forage crops (grasses and forage legumes), root crops, and acid fruits (berries, apples, cherries, etc.). Agriculture is extremely hazardous in the *Dc-Dd* group because of the short growing season, and livestock must be kept indoors and fed with stored hay or ensilage for at least half the year.

The extremely low temperatures in the subarctic climates (*Dc-Dd*) present an additional set of problems. *Permafrost,* or permanently frozen subsoil, begins to appear near the poleward margins of this group (see Figure 16-6). It presents many serious problems to human living, especially with respect to transportation. Mechanized equipment of any kind requires special lubricants, since the customary ones tend to become gummy at −30° to −40°. Also, special clothing is necessary for outdoor wear, largely because the accumulation of any moisture will rapidly destroy the insulating properties of ordinary outer clothing. This can present serious dangers of freezing. Clothing that provides an insulating layer of air between inner and outer coverings is recommended. The tensile strength of metals, plastics, and many other materials rapidly decreases at low temperatures, and breakage of structural materials under stress is a constant winter problem.

Snow cover for several months, combined

with bright sunny periods, produces much eye irritation, and "pink eye" (conjunctivitis) is an annoying regional malady in late winter. Another annoyance of the northern continental areas is the host of blood-sucking insects during the summer, particularly mosquitoes and black gnats. The extensive marshy areas, lakes, and rivers provide ideal incubating areas for winged insects that pass through an aquatic larval stage, and the relative scarcity of animals makes the blood-sucking insects particularly voracious. Fortunately, few of them are carriers of human diseases.

8-17 RESOURCE SIGNIFICANCES OF THE D CLIMATES The previous section indicated a bleak picture of the D climates as places in which to live, yet the world population distribution indicates that the Da and Db groups are supporting large numbers of people both within and outside urban centers. Almost all of the population of Canada and the Soviet Union lives within the D climates, and parts of the D climates of China and Japan include dense rural populations living off the land.

Some writers have indicated that the mid-latitude climates, with their fluctuating seasons and variable weather conditions, are especially stimulating to both mental and physical human activity. If this were true, the southern portions of the D climates would represent especially favorable environments for human activity. The difficulty in weighing the many factors responsible for human incentives makes it practically impossible to demonstrate the validity of such a statement. The fact remains, however, that populations press close against some kind of climatic limitation within the D climates, a limit beyond which human living is not easy, and possible only through special cultural and technological adaptations to a harsh environment.

The Da climates, within the permissiveness of their growing-season length, are highly productive agriculturally, combining long hours of daylight with reasonably direct insolation. The winter season inhibits soil leaching, or the washing away of soluble plant foods by the percolation of rainfall, and expenditures for fertilizers to maintain fertility generally are less than in the warmer, moist climates. Much of the northern portion of the Corn Belt in the United States has a Dfa climate, and this region is one of the most productive farming areas on earth. Perhaps the greatest agricultural resource of the D climates, especially the Db group, is their suitability for the cultivation of forage and root crops. This is the center of the world's dairy and poultry industries, both of which are ideally suited as a rural adjunct to the growth of urbanization. Of all the protein products, milk, milk products, poultry, and eggs are the most efficiently produced in terms of conversion of feed into food.

The potato is another important agricultural product of the D lands. Consisting of little more than starch and water, the potato crop does not produce a high nutrient drain on soils. The suitability of these climates for potato cultivation lies in the long hours of daylight, when photosynthesis may continue to form starches for 14 to 16 consecutive hours. Acid soils at these latitudes also tend to prevent fungus blights that decrease yields and marketability at lower latitudes. Crop yields of as much as 500 to 750 bushels per acre have been reported on potato farms along the northern border of the United States. This crop has achieved a special significance in the agricultural system of the Soviet Union, in eastern Germany, and in Poland, not only for a wide range of food uses but also for industrial uses, such as the production of alcohols and related products.

The summer season of the *Db* climates is nearly ideal for outdoor body comfort, and consequently the region has a great attraction at this time of year for people on vacation seeking a climatic retreat from the extremely hot, humid regions immediately to the south. For many parts of the *Dfb* climate in the United States and Canada, especially in the upper Great Lakes region, summer recreation is the most important source of regional income, far exceeding agriculture and industry. The heavy snows of this same area are attracting growing numbers of winter sports enthusiasts. Recreation is an ideal source of regional income, because it does not require the consumption of nonrenewable resources.

The *Db* regions lie at the periphery of the major urban areas of the world. Many people in North America and Eurasia, with a choice of locating their homes or places of business, would prefer more comfortable climates, such as the *C* group, but except for the *Csb* and *Cfb* climates, a choice must still be made between hot summers and cold winters, in which case the choice frequently is the cool, rather than the hot, summers. Developments in air conditioning have tended to reduce the effect of climate on urban living. Climate is only one of many factors that influence the location, size, and frequency of cities, but the evidence seems conclusive that somewhere within the *D* climates, there is an effective boundary beyond which the severity of winter has effectively hindered not only urban living but also rural habitation.

Population is sparse in the *Dc* and *Dd* areas, and it is not likely to increase, except possibly in scattered mining, hunting, and lumbering regions. The greatest pressure for settlement here has occurred in the Soviet Union, where there is a continuing effort to force the limit of agriculture northward into this climatic region. Progress has been slow,

and for the most part the taiga, or northern coniferous forest, remains a wilderness. People are willing to undergo the rigors of this climate only where their culture knows no other environment, or where there are unusual economic opportunities, particularly in mining or recreation, to compensate for the difficulties in maintaining a comfortable way of life. Areas of the latter are for the most part spotty and temporary.

The *E* Climates (polar)

8-18 DESCRIPTION AND DISTRIBUTION OF THE *E* CLIMATES The *E* climates may be considered to have no true summer. Specifically, in the Koeppen classification they are defined as climates whose mean monthly temperatures do not exceed 50°F (10°C). The two subtypes include the *ET*, or tundra, climate, where the mean temperature of the warmest month is between 32° and 50°, and the *EF* regions, having permanent ice and snow, where the warmest month does not exceed 32°. The distribution of the *E* climates is indicated in Figure 8-20, and selected representative stations are presented in Table 8-10.

Except for mountain locations and an exceptional marine position at high latitudes, the *ET* climates are found only in the northern hemisphere. A strip of *ET* climate rims

Figure 8-20 Distribution of the *E* and high-elevation climates.

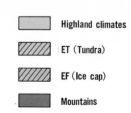

Highland climates

ET (Tundra)

EF (Ice cap)

Mountains

Table 8-10 Representative Climatic Data for Selected ET–EF Stations

	Jan.	Feb.	Mar.	Apr.	May	June	July	Aug.	Sept.	Oct.	Nov.	Dec.	Yr.
Chesterfield Inlet, Canada; lat 63°20′ N long 90°43′ W; elev. 13 ft; Koeppen class.: ET													
Temp. (°F)	−24.9	−26.0	−13.7	2.5	21.0	36.7	47.8	47.3	37.0	21.7	0.0	−15.3	11.3
Precip. (in.)	0.3	0.3	0.4	0.7	0.5	1.0	1.8	1.6	1.5	1.2	0.8	0.6	10.7
Macquarie Is., N.Z.; lat 54°30′ S long 158°57′ E; elev. 20 ft; Koeppen class.: ET													
Temp. (°F)	43.5	43.0	42.0	40.0	39.5	37.0	37.0	37.0	37.5	38.0	40.0	42.0	40.0
Precip. (in.)	4.0	3.5	4.1	3.8	3.3	2.9	3.2	3.2	3.8	3.3	2.8	3.0	41.8
Little America, Ant.; lat 78°34′ S; long 163°56′ W; elev. 30 ft; Koeppen class.: EF													
Temp. (°F)	20.0	3.0	−6.5	−19.5	−24.5	−16.5	−36.0	−33.0	−40.0	−15.0	1.5	18.5	−12.5
Eismitte, Greenland; lat 70°53′ N long 40°42′ W; elev. 9,843 ft; Koeppen class.: EF													
Temp. (°F)	−43.0	−53.0	−40.0	−25.5	−12.0	2.0	10.0	12.0	12.0	−32.5	−45.0	−37.0	−23.0
Precip. (in.)	0.6	0.2	0.3	0.2	0.1	0.1	0.1	0.4	0.3	0.5	0.5	1.0	4.3

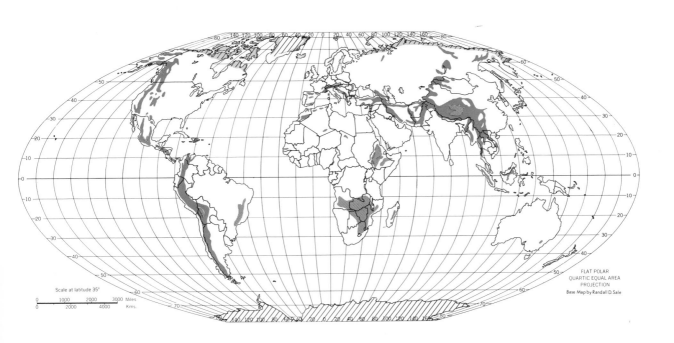

the northern borders of the North American and Eurasian continents, extending southward along the eastern continental margins to about 55°. The most significant feature of the ET climates is the contrast in length of day between winter and summer. This determines more than anything else the changes of temperature that occur. As indicated by the isothermograph for Sagastyr (see Figure 8-21), there is almost no diurnal variation, and temperatures slowly change with the height of the sun above and below the horizon, the pattern broken occasionally by short-lived periods of unusual warmth and cold. Two or three times during the summers, importations of warm air from lower latitudes find their

way into these high latitudes and bring short periods of temperatures that may be as high as 60° to 80°. For the most part, however, temperatures in the high-sun period are in the 40s. Winter temperatures are consistently below zero but do not usually reach the low figures of the continental basins of eastern Siberia or northern Canada. The influence of the Arctic Ocean, despite being frozen during the winter, is felt along the continental margins. Snowfall usually is low (1 to 4 inches), there being little moisture possible for condensation in the cool to cold air. Wind patterns are likely to be at the extremes, either with long periods of dead calm or spells of gale strength. In the latter case, the air is filled with snow particles blown from the surface, and visibility can be reduced to only a few feet.

The ET climates of the tundra may not have a true summer in comparison with climates farther south, but the beginning of the high-sun period produces sharp changes in the lives of the few Arctic human inhabitants. Most of the snow disappears, the tundra becomes a riot of flowering plants, the coastal waters and great northern rivers break free of their winter ice cover, and fish provides a readily accessible source of food. Teeming flocks of birds of many types return from the

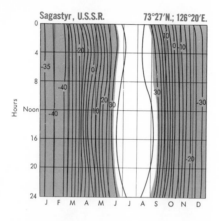

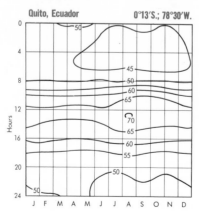

Figure 8-21 Isothermographs of average hourly temperature (°F), by month, for Sagastyr, Soviet Union (*ET*), and Quito, Ecuador (*Cfbi*). The Sagastyr chart shows that at extremely high latitudes, temperature patterns are entirely seasonal, without diurnal variation. This is because day and night are almost continuous at opposite times of the year. The Quito chart shows almost the opposite tendency: the virtual absence of seasonal variation near the equator. Note that a high elevation makes little difference in the daily or seasonal ranges of temperature. A lowland station near Quito would have a similar pattern except that the values would be higher. [After Troll]

south to nest on the tundra, relatively free of predators, and the air hums with swarms of insects. As soon as the pack ice moves away from shore, the open reaches of water presage the arrival of supply steamers at the coastal villages. The dog sleds are replaced by boats often equipped with outboard motors, and houses are opened and aired for the first time in months. Cross-country travel on the tundra is much more difficult in summer than in winter because of marshy ground, broad rivers, and ponds and lakes. Water dominates the lowland landscapes despite extremely low precipitation.

A special type of *ET* climate is found on a few isolated islands in the high latitudes. These represent a continuation of the exceptionally marine conditions of the *Cfc* climates into higher latitudes. Macquarie Island, located south of New Zealand, has such a climate, in which, despite the lack of a summer, the temperature rarely goes below freezing. Table 8-10 and Figure 8-11 present climatic data for this unusual location. A similar *ET* climate occurs in some of the islands near the tip of South America.

The *EF* climate is a cold desert where the maintenance of human life is extremely difficult, especially during the long nights of winter. Not much was known about these polar climates until the Wegener Expedition to Greenland in 1931 and the first U.S. Navy Expedition to Antarctica under the command of Admiral Byrd in 1947. Since then, several nations have cooperated in both Arctic and Antarctic exploration with accompanying meteorological measurements, and some fifty stations in Antarctica have been manned for different lengths of time since the start of the International Geophysical Year in July, 1957. Collected data indicate not only that cold weather persists throughout the year, but that the temperature minima are far below those recorded elsewhere. The lowest temperatures ever recorded under standard observational standards were taken by the Russian scientists at Vostok, in the east-central portion of Antarctica in 1957–1958. The year's record is given in Table 8-11. During this year of joint research, there was some friendly rivalry between the Russian and American meteorologists.

Records taken by the American stations, including one at the South Pole, although consistently low, were not nearly so low as those at Vostok. The Russians kiddingly referred to the American base at Little America as "Little Florida." Throughout Antarctica, however, winter temperatures average below −40°. Precipitation is mostly in the form of tiny ice crystals, being sublimated water vapor. Some of them are too small to be clearly visible and at times may be so plentiful as to give the air a weird translucent quality, without shadows even in full sunlight. True measurements of the amount of precipitation are extremely difficult, if not impossible, because the tiny particles are

Table 8-11	Temperature Data for 1958 at Vostok, lat 78°27′ S–106°52′ E												
	Jan.	Feb.	Mar.	Apr.	May	June	July	Aug.	Sept.	Oct.	Nov.	Dec.	Yr.
Mean temp. (°F)	−24	−48	−69	−83	−81	−90	−86	−97	−87	−74	−46	−28	−68
Max. temp.	−9	−20	−53	−45	−45	−53	−47	−67	−49	−42	−29	−15	−9
Min. temp.	−42	−76	−89	−99	−108	−113	−114	−125	−116	−96	−76	−47	−125

SOURCE: G. D. Cartwright and M. J. Rubin, "Inside Antarctica No. 6—Meteorology at Mirny," *Weatherwise*, Vol. 14, June, 1961, p. 114.

borne by almost imperceptible air movements. Some estimates have been made by measuring yearly accumulations of snow and ice. Amounts vary, but converted to liquid equivalents, the totals rarely are more than 2 to 6 inches. The great ice masses of Greenland and Antarctica represent accumulations of thousands of years.

Wind is almost constant and tends to be stronger near the margins of the subcontinents. Cyclonic gales are almost continuous in some sectors in Antarctica, such as Adelie Land. One weather station there reported an average annual wind speed of almost 45 miles per hour, and winds of 80 to 100 miles per hour are common. The slower but much more constant wind is mainly a gravity wind, or air drainage on a large scale. The air is chilled by radiational loss of heat, by contact with a cold, frozen surface, and also by loss of latent heat removed from it by the evaporation of ice and snow. Air slips down the concave slopes of the great ice masses and is concentrated along the margins onto the moving tongues of glacial ice that occupy topographic depressions near the surrounding sea. Pronounced temperature inversions are common throughout the year, with the warmest layers varying from 1,000 to 5,000 feet above the surface, and highest during midwinter.

The Highland Climates

8-19 GENERAL DESCRIPTION AND LOCATION OF THE HIGHLAND CLIMATES The lapse rate, or vertical temperature gradient, assures that areas at high elevations have climates that differ considerably from those at low elevations. Mountain slopes that rise 10,000 to 20,000 feet above their valley floors may have a sequence of climatic zones ranging from A, through C and D, to the frozen

E climates of snow-capped summits. No attempt has been made to chart all of these sequences on the world map of climates (see endpaper), and the high mountain areas are grouped together into a single category to indicate the areas where vertical zonation prevails.

A few broad areas of upland plains occur, however, that are sufficiently large to be shown on a global map of textbook size and that fall within a Koeppen classification that differs from the lowland types within the same latitudinal zone. They are indicated on Figure 8-20, along with the E climates.

Besides the mountain areas, five major areas of highland climates occur: (1) the high plains of central Mexico; (2) the eastern portion of the Brazilian uplands; (3) a large area in southeastern Africa; (4) Ethiopia, and; (5) the intermontane plains of Tibet. The first four have mostly a *Cwbi* climate, and the last is high enough to be in a transition between *Cwb* and the E climates. Smaller areas of *Cb* climate occur over parts of the Columbia Plateau in eastern Washington, central Turkey, southwestern China, and in the pocket basins and plains contained within the sweep of the Andes Cordillera of South America.

From a human resource point of view, the significant highland climates are the *Cfb*, *Cwb*, and *Csb* of the tropics and subtropics. At higher latitudes the highland climates become too cold for much human habitation. The D climates of low latitudes are so high that the problems of rarified air are added to those of low temperature to discourage permanent settlement.

The temperature pattern of the high-elevation *Cb* climates of low latitudes differs from that of the *Cb* climates in middle latitudes in having a lower seasonal range in temperature. Many of the former, in fact, have less than 10° difference between winter and summer, thus meriting the use of the Koeppen symbol *i* (compare the data for Bogota in Table 8-12 with the *Cfb* stations in Table 8-4). This indicates that seasonal range of temperature is related to variations in the length of seasons and directness of solar radiation, and is not significantly correlated with elevation. Another illustration of this is shown on the isothermograph for Quito, Ecuador (see Figure 8-21), which is a high-elevation station almost on the equator. Except for the values of the isotherms, the Quito graph pattern does not differ much from that of Belem (see Figure 7-4), which is a coastal station near the equator.

Mountain slopes at high elevations tend to have higher rainfalls than at adjacent lower elevations because of their orographic influence in forcing air to rise. This factor does not hold, however, for high plains, and often the latter will have lower precipitation than lowlands nearby, because the orographic effect has removed considerable moisture in reaching the high elevations. The orographic influence of mountains on precipitation is illustrated by the data for Mount Washington

in Table 8-12, which indicates an annual precipitation of about twice that of the general New England area. Lower temperatures at high elevations will frequently decrease evaporation and make the same amount of rainfall more effective than at nearby low elevations. Attention earlier was given to the situation in the Cs climates (see the data for Konya and Boise in Table 8-1), where low temperatures associated with elevation were able to maintain a Cs climate despite low rainfalls. Many of the mountain areas adjacent to deserts have humid climates, not only because of somewhat higher precipitation than the areas below but also because of decreased temperatures and evaporation.

8-20 HUMAN IMPLICATIONS OF THE HIGHLAND CLIMATES From a human standpoint, the most receptive highland climates are the *Cfbi* locations at low latitudes. Here solar radiation is just as favorable as that in the *A* climates below for agricultural purposes, and temperatures are nearly ideal for body comfort. Precipitation is sufficient

Table 8-12   Representative Climatic Data for Selected Highland Stations

	Jan.	Feb.	Mar.	Apr.	May	June	July	Aug.	Sept.	Oct.	Nov.	Dec.	Yr.
Lilongwe, Malawa; lat 15°58′ S long 33°42′ E; elev. 3,697 ft.; Koeppen class.: Cwb													
Temp. (°F)	70.0	69.6	69.3	67.6	63.7	60.0	58.8	62.2	67.2	73.2	73.9	71.2	67.3
Precip. (in.)	8.0	8.3	5.3	1.5	0.2	0	0	0	0.2	0.2	3.0	6.7	33.4
Bogota, Colombia; lat 4°38′ N long 74°05′ W; elev. 8,333 ft.; Koeppen class.: Cfbi													
Temp. (°F)	55.8	56.5	57.7	57.7	57.4	56.7	55.8	55.9	55.9	56.3	56.5	56.3	56.5
Precip. (in.)	1.5	1.9	2.6	3.8	4.1	2.4	1.9	1.5	2.1	6.2	5.5	3.7	37.2
Lhasa, Tibet; lat 29°40′ N long 91°07′ E; elev. 12,013 ft.; Koeppen class.: Cwb													
Temp. (°F)	33.4	36.7	41.5	47.3	55.6	62.6	63.0	63.1	59.5	52.7	42.3	32.5	48.9
Precip. (in.)	0	0	0	0	0.7	2.8	6.2	6.0	2.7	0.2	0	0	18.6
Mount Washington, N.H.; lat 44°16′ N long 71°18′ W; elev. 6,262 ft.; Koeppen class.: Dfc													
Temp. (°F)	10.0	10.5	12.0	23.0	34.0	45.0	50.0	48.5	40.5	32.0	21.0	11.5	28.5
Precip. (in.)	4.3	4.2	5.9	5.9	6.2	9.1	10.3	8.6	8.4	7.5	6.6	5.2	82.2

for maximum plant use, and the range of permissiveness for different crops is extremely high, especially in those regions where frost does not occur. The major disadvantage of such areas is that nearly all of them are found on mountain slopes or in small basins, and thus terrain is a serious hindrance to widespread human settlement. The broad upland plains with temperatures similar to the *Cfbi* mountain slopes are likely to be much drier or to have at least a seasonal shortage of water (*Cw* or *Cs*), and thus are not as favorable. The *Cfbi* locations are becoming increasingly popular throughout the world, despite their mountain position. Some of them are being used as *hill stations,* a term for residential communities that are planned as climatic retreats by people who work in adjacent tropical lowlands and who can afford the cost of commuting daily, weekly, or seasonally. Several of the tropical capitals of the world have hill-station summer retreats to which many of the governmental services are transferred during the most unpleasant season. Examples include Bogor, the summer capital for Djakarta, Indonesia; Baguio, the summer retreat for Manila, Philippine Islands; and Simla, the hill station for New Delhi, India. As an illustration of the contrast in comfort between lowland climates and nearby hill stations, see Table 8-13. Newara Eliya can be reached from Colombo by automobile in about four hours. Kandy, which is at an intermediate level and reached within two hours, is used more as a residential station, and Newara Eliya as a recreational station. The suitability of these hill-station climates is splendidly illustrated by the enormous variety of flowering plants that can be observed in the botanical gardens at Peredeniya, near Kandy, and also at Hakgale, at the edge of Newara Eliya.

It is interesting to note that most of the broader areas of highland climate have gained fairly dense populations only during the present century and mostly by European settlement. This is true of the *Cfb* eastern edge of the Brazilian uplands (see Sao Paolo in Appendix C) and some of the high plains in Kenya or Zambia. Exceptions include some of the high plains in Mexico and within the Andes, where early Aztec and Inca civilizations developed a preference for these cool, highland climates. Even the high-elevation *D* and *E* climates are populated to a limited degree, and the permanent inhabitants of such harsh environments probably sought them as a place of refuge, as in Tibet or the Andes. The Indians of Peru, for example, in attempting to escape the subjugation

Table 8-13 Contrasting Temperature Data for Colombo, Ceylon, and an Interior Hill Station, Newara Eliya

	Jan.	Feb.	Mar.	Apr.	May	June	July	Aug.	Sept.	Oct.	Nov.	Dec.	Yr.
Colombo, Ceylon; lat 06°54′ N long 79°53′ E; elev. 25 ft.; Koeppen class.: Afgi													
Mean monthly temp.	79.2	79.5	81.0	81.9	82.4	81.3	80.8	81.0	81.0	79.9	79.2	79.0	80.4
Av. daily max.	86	87	88	88	87	85	85	85	85	85	85	85	86
Av. daily min.	72	72	74	76	78	77	77	77	77	75	73	72	75
Newara Eliya; lat 06°46′ N long 80°46′ E; elev. 6,168 ft.; Koeppen class.: Cfbi													
Mean monthly temp.	58.5	58.6	59.0	61.2	62.4	61.2	60.4	60.6	60.6	60.6	60.1	59.2	60.3
Av. daily max.	67	70	71	71	70	66	65	67	67	68	68	68	69
Av. daily min.	47	44	46	49	53	55	55	54	53	52	51	48	51

of their Inca and Spanish masters, sought the upper part of the mountain slopes just below the snow line and, in so doing, gave to the world the "Irish" potato. The tiny, bitter tubers of this mountain plant provided sustenance in a land that no one else wanted. Barley, the hardiest of all grains since it may mature within 60 days, likewise supplies a precarious food supply to permanent dwellers of high-elevation slopes and plains, and may well have been developed from one of the mountain grasses in the Middle East. Today it is the staple food for the inhabitants of Tibet and adjacent mountain lands. The highest permanent settlement consists of a few scattered Indian huts at an elevation of about 17,000 feet in the Andes. Beyond such heights, rarified oxygen in the atmosphere seems to be as limiting as low temperatures.

The vertical zonation of climates in mountain lands everywhere is responsible for vertical zonation of land use patterns, ranging from the multiple zones of the low-latitude mountains to the simple forest-alpine pasture zones of mid-latitudes. The early Spanish recognized the multiple zonation within the Andes, and their terms for such zones are in common use today, including: *tierra caliente* (the hot lands), or *A* climates at elevations generally below 2,000 feet; *tierra templada* (the temperate lands), or *C* climates, occupying the intermediate slopes up to about 8,000–9,000 feet; *tierra fría* (the cold lands), or *D* climates that extend up to 13,000–15,000 feet; and finally, the *paramos* (the summits), which is the zone of barrens and snow fields, included in the *E* climates of the mountains. Figure 8-22 indicates the zonation of climates and corresponding land use patterns in the

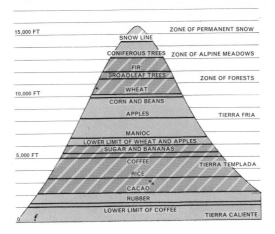

Figure 8-22 Diagram of land use and altitudinal zones in the Andes. [After Sapper]

northern part of Venezuela. The elevation of these vertical zones tends to lower with increasing latitude until the *E* climates appear at sea level in the polar regions.

The high-elevation climates within the mid-latitudes are generally within the *D* or *E* range, and the climatic advantages are somewhat more limited than at the lower latitudes. Distinct regional advantages, however, accrue in their use for both winter and summer resorts; in forestry, principally for the exploitation of mountain conifers (cone-bearing trees) that thrive within the upper *C* and lower *D* elevational range; and for livestock raising, mainly for summer alpine pasturage. Even the areas of snow and ice, high above the timber line, have their resource values as scenery, for winter sports, and as reservoirs for water supply needed during the summers at lower elevations.

Study Questions

1. Explain the extreme dryness of the summers in the Cs climates.
2. Explain why the Cs climates penetrate so far into central Asia and so little into South America.
3. Explain why there is so much Csb and so little Csa in the United States, and why there is so much Csa and so little Csb in Europe.
4. If there were no Gulf of Mexico and the two Americas constituted a large, single continent, how would the climates of the United States likely be different from those at present?
5. Name South American cities that have climates roughly comparable to the following:
    a. Los Angeles, Calif.   c. Denver, Colo.
    b. Baltimore, Md.   d. Bordeaux, France
6. Explain the fog that is such a common winter phenomenon in London.
7. What climate in the United States has the lowest total requirement for air conditioning, measured in terms of both heating and cooling degree-days? What climate has the highest? Discuss the correlation with density of population.
8. Why is the yield of potatoes so much higher in the northern than in the southern United States?
9. If a retired British civil servant wished to emigrate to the part of Canada where the climate would be most like home, where would he find it?
10. What difference is there between the climate above the tree line in the Rocky Mountains of Colorado and that above the tree line in the Andes of Ecuador? Explain.
11. Explain the difference in the two isothermographs of Duluth and Irkutsk shown in Figure 8-16.
12. Study the two charts in Figure 8-9 and describe the general correlation between temperature and relative humidity in Charleston on a daily and seasonal basis.

Part 2:   References

Battan, L. J., *The Nature of Violent Storms,* Doubleday & Company, Inc., New York, 1961.

Blair, Thomas A., and Robert C. Fite, *Weather Elements; A Text in Elementary Meteorology,* 5th ed., Prentice-Hall, Inc., Englewood Cliffs, N.J., 1965.

Blumenstock, David I., *The Ocean of Air,* Rutgers University Press, New Brunswick, N.J., 1959.

Budyko, I. M., *The Heat Balance of the Earth's Surface,* trans. by N. A. Stepanova, U.S. Government Printing Office, Washington, D.C., 1958.

Cotter, C. H., *The Physical Geography of the Oceans*, Hollis and Carter, Ltd., London, 1965.

Critchfield, Howard J., *General Climatology*, 2d ed., Prentice-Hall, Inc., Englewood Cliffs, N.J., 1966.

Fairbridge, Rhodes W., (ed.), *The Encyclopedia of Atmospheric Sciences and Astrogeology*, Reinhold Publishing Corporation, New York, 1967.

Flora, S. D., *Tornadoes of the United States*, University of Oklahoma Press, Norman, Okla., 1953.

Garbell, Maurice A., *Tropical and Equatorial Meteorology*, Pitman Publishing Corporation, New York, 1947.

Geiger, R., *The Climate Near the Ground*, Harvard University Press, Cambridge, Mass., 1965.

Gentilli, Joseph, *A Geography of Climate*, 2d ed., University of Western Australia Press, Nedlands, W.A., 1958.

Griffiths, John F., *Applied Climatology*, Oxford University Press, New York, 1965.

Hare, F. K., *The Restless Atmosphere*, Harper & Row, Publishers, Incorporated, New York, 1963.

Haurwitz, Bernard, and James M. Austin, *Climatology*, McGraw-Hill Book Company, Inc., New York, 1944.

Kendrew, W. G., *Climates of the Continents*, 5th ed., The Clarendon Press, Oxford, England, 1961.

King, C. A. M., *Introduction to Oceanography*, McGraw-Hill Book Company, Inc., New York, 1963.

Kondrat'yev, K. V., *Radiative Heat Exchange in the Atmosphere*, Pergamon Press, New York, 1965.

Landsberg, Helmut, *Physical Climatology*, Gray Printing Co., Inc., DuBois, Pa., 1958.

Penman, H. L., "Vegetation and Hydrology," *Technical Communication No. 53*, Commonwealth Bureau of Soils, Harpendun: Commonwealth Agricultural Bureaux, Farnham Royal, Bucks, England, 1963.

Petterssen, Sverre, *Introduction to Meteorology*, 2d ed., McGraw-Hill Book Company, Inc., New York, 1953.

Riehl, Herbert, *Introduction to the Atmosphere*, McGraw-Hill Book Company, Inc., New York, 1965.

Rumney, George R., *Climatology and the World's Climates*, The Macmillan Company, New York, 1968.

Sutcliffe, R. C., *Weather and Climate*, W. W. Norton & Company, Inc., New York, 1966.

Thornthwaite, C. W., "An Approach Toward a Rational Classification of Climate," *Geographical Review*, vol. 38, no. 1, pp. 55–94, 1948.

Thornthwaite, C. W., and J. R. Mather, "Instructions and Tables for Computing Potential Evapotranspiration and the Water Balance," *Publications in Climatology*, vol. X, no. 3, Drexel Institute of Technology, Laboratory of Climatology, Centerton, N.J., 1937.

Trewartha, Glenn T., *An Introduction to Climate,* 4th ed., McGraw-Hill Book Company, Inc., New York, 1968.

Watts, I. M. E., *Equatorial Weather,* Pitman Publishing Corporation, New York, 1955.

Wolman, Abel, "Water Resources, A Report to the Committee on Natural Resources of the National Academy of Science; National Research Council," *Publication 1000-B,* National Academy of Science, National Research Council, Washington, D.C., 1962.

Recommended articles in *Scientific American* available as offprints from W. H. Freeman and Company, San Francisco, Calif.

Emiliani, Cesare, "Ancient Temperatures"   Feb., 1958
Landsberg, Helmut E., "The Origin of the Atmosphere"   Aug., 1953
Malkus, Joanne S., "The Origin of Hurricanes"   Aug., 1957
McDonald, James E., "The Coriolis Effect"   May, 1952
Munk, Walter, "The Circulation of the Oceans"   Sept., 1955
Myers, Joel N., "Fog"   Dec., 1968
Opik, Ernst J., "Climate and the Changing Sun"   June, 1958
Plass, Gilbert N., "Carbon Dioxide and Climate"   July, 1959
Revelle, Roger, "Water"   Sept., 1963
Roberts, Walter O., "Sun Clouds and Rain Clouds"   Apr., 1957
Rubin, Morton J., "The Antarctic and the Weather"   Sept., 1962
Starr, Victor P., "The General Circulation of the Atmosphere" Dec., 1956
Stewart, R. W., "The Atmosphere and the Ocean"   Sept., 1969
Tepper, Morris, "Tornadoes"   May, 1958
Woodcock, A. H., "Salt and Rain"   Oct., 1957
Wright, Sir Charles, "The Antarctic and the
     Upper Atmosphere"   Sept., 1962

PART 3

LANDFORMS AND SURFACE CONFIGURATION

The objective of this division is to examine the solid outer surface of our planet, to note its variable composition from place to place, the surface forms which help give character to areas both large and small, and the processes that are at work on the surface, changing it to accommodate the sweep of energy that swirls across the face of the earth. At the outset, two major observations should be made concerning the outer surface of our planet: (1) it is remarkably different from that of either Mars or the moon, the only other celestial bodies whose surfaces have been observed by man; (2) it is in a constant state of alteration by processes that cause surface irregularities and then remove them. These latter processes of change have been operating on the earth surface since early in the planet's history, over a span of some 2 to 4 billion years, as determined by reliable geophysical rock-dating techniques.

The first of these observations has been verified only since the development of space exploration techniques by earth-based probes. The second was suggested in the eighteenth century by James Hutton, a Scottish geologist, who denied the cataclysmic creation of the major earth lineaments. He suggested instead that the history of the evolution of the earth surface could be obtained from a study of the processes of change at work upon the surface today. At first regarded with religious skepticism, this theory of *uniformitarianism* stands firm today under the testing and observation of many generations of geologic investigators. Both of these observations pose some fundamental questions that will be presented in this introduction, along with suggested answers.

The clear photographs obtained by near-contact space probes investigating the surface of the moon and Mars revealed, first, that neither body has continental upland blocks or pronounced basins with or without water. These primary surface units of the earth, then, are unique. Second, nearly all of the surface irregularities of both Mars and the moon appear to be extremely old and related mainly to meteoric impacts. The clear evidence of some past volcanic activity on the moon suggests it was left from an extremely hot, formative period, or, as has been suggested by Van Dorn, was a secondary result of huge meteoric impacts. The circular scars of impact craters that are so prevalent on Mars and the moon are rarely encountered on the earth surface. Since there is no reason why the earth should have been spared such visitations, the conclusion must be reached that such irregularities have long been removed on earth by the processes of *weathering and erosion*,[1] and likewise, that similar processes are not operative on these celestial neighbors of ours, or certainly not at rates comparable to those observed on earth. The more subdued topography of Mars, however, and its relative lack of the stark, jagged, cliff-edged crater rims so characteristic of much of the lunar surface show that slight weathering and transport of material may occur there.

If we examine the weathering, erosional, and depositional processes that remove surface irregularities on earth by attacking the high places and filling in the low ones, we find all of them dependent on incoming solar radiation for their energy supply. As we learned in the earlier chapters, the work of rivers, waves, winds, and glaciers is derived from solar energy. Neither the moon nor Mars, however, is deprived of receiving solar energy. In fact, the amount of energy received per unit area on the sunlit side of the moon is greater than that reaching the earth surface, because there are no clouds to reflect the incoming radiation or atmospheric gases to intercept it. As a result, two questions come to mind. First, why does solar energy do so little to alter the face of the moon and Mars? Second, if the leveling forces generated on earth by solar radiation have been operative for 2 to 4 billion years, why have not all of the surface irregularities been removed, including the continents and ocean basins?

The answer to the first question lies in the absence of an atmosphere. Without water vapor, oxygen, and carbon dioxide, solar energy on earth would do little more than heat the surface. There would be no wind, no rain, no rivers, no ocean currents or glaciers to perform work on the surface. The only means of transporting energy would be by molecular conduction. We shall see in the next chapter that the work of sculpturing the earth surface is intimately tied to the global energy flow system.

The second question has no such easy solution, but a large amount of geologic evidence, derived from reading the rock record, indicates that the crust of the earth has never been still since it was formed and that opposing the work of surface beveling are forces that produce surface irregularities other than meteor-impact craters. Evidence also indicates that the continental blocks and ocean basins, if not always in their present arrangement, have been major surface irregularities for at least 2 billion years. As will be indicated at greater detail in Chapter 9, there is believed to be a causal connection between the continued maintenance of continental uplands through crustal warping and the removal of these uplands by erosion and weathering and the subsequent deposition of this material into depressions both within and along the margins of continents.

The term *geologic cycle* has been proposed for the removal of surface material on earth from the uplands, its deposition and subsidence into the crust, and its alteration back into rock with a shift of position, to emerge later on upland rock surfaces. In this cycle, there has been no removal or addition of material. Instead, there has been work per-

[1]*Weathering* refers to the processes by which solid rock is altered into loose, fragmental material through atmospheric agents (see Sections 9-5 and 9-6). *Erosion* involves the removal of material by transporting agents.

formed to alter the form, composition, and location of this material. The energy supply to perform this work is a detour in the global energy supply system derived from solar radiation. There is a similarity in the operation of the geologic cycle and the hydrologic cycle. Energy flows into both from incoming solar radiation. It performs work within the cycle by altering material, moving it from place to place, operating within fixed limits and obeying certain signals and rules. Eventually it leaves the system in the same form which it had when it entered—as radiant energy, unchanged except for its longer wavelengths and its decreased capacity to perform work.

The earth surface, like the atmosphere, is dynamically active, a surface that fulfills all of the conditions expressed in Chapter 1 relative to the principles of natural distributions as follows:

1. No two spots on earth are exactly alike as to surface configuration because of the application of energy by means of many different processes upon various kinds of materials, and at different rates of intensity. There are, however, areas of greater or lesser stability and areas where similar processes, operating on somewhat similar materials, make possible a certain degree of classification.

2. The differences in landforms and surface configuration from place to place exhibit different gradients of change. Some processes lead to extremely narrow terrain boundaries, while others induce gradations that are almost imperceptible.

3. The tempo of change in surface forms is not uniform in area or in time, but eventual change is inevitable. Some processes of change produce short, sudden bursts of activity separated by long periods of stability, while others are steady and continuous. Some portions of the earth surface are more susceptible to the dynamic processes of change than others. In the long sequence of geologic time, evidence points to periods of intense crustal instability and mountain-building, separated by long intervals when the beveling processes of weathering, erosion, and deposition prevail.

4. Whatever the processes of change that influence a particular portion of the surface, there is a tendency for an alteration toward short-term and long-term equilibria, in which all of the dynamic elements in the environment tend to adjust to each other. Such equilibria never are permanent, and some of the longer ones may never be attained. The surface patterns of such equilibria can be expressed in certain geometric attributes of the surface configuration. Equilibrium slopes, for example, can be observed in many places, such as in the longitudinal profiles of streams, the angle of rest of fragmental material on the back slope of a sand dune, a forested slope in the tropical rain forest, along beach gradients, and in many other places.

This part of the text is concerned with four major aspects of the configuration of the earth surface, and each will be treated in a separate chapter. The first will deal with the processes of change resulting from the application of energy. The second will discuss the types of rocks and minerals that occur at the earth surface, their different reactions to weathering and erosion processes, and some of the results of the interaction of processes and materials. The third chapter will treat the major or global features of the planetary surface with respect to characteristics, arrangements, and significances of continents, ocean basins, mountains, plains, and hills. The last of the four chapters alters the scale of observation and treats some of the more common topographic or local landforms that can be seen at various places on the surface and that represent the more visual results of the processes that perform work on the surface.

The substance of these chapters represents only a glimpse of the complexity of the earth surface. It also represents a short overview of the results obtained in the field of *geomorphology*, a specialized investigative discipline that seeks to understand the configuration and dynamics of the earth surface. It is hoped that the interested reader will explore some of the excellent references that are suggested and that summarize achievements in this field more adequately than the abbreviated glimpse that is presented in these chapters.

CHAPTER 9
ENERGY SOURCES AND THE PROCESSES OF CRUSTAL CHANGE

This chapter seeks to define briefly the major processes by which energy is applied to the earth surface and to note the changes that take place within the geologic cycle.

The Sources of Energy for Surface Modifications

9-1 ENDOGENIC ENERGY The energy that is derived from sources within the interior of the earth is termed *endogenic energy*, in contrast to *exogenic energy*, which originates outside the earth, such as solar energy. Two sources provide energy from within the earth interior. The first is *gravitational energy* which was derived from the compaction of mass and the addition of angular momentum during the formation of the planet. The second is *radiogenic* or *atomic energy*, which is derived from the breakdown of atomic nuclei in certain radiogenic earth materials, such as uranium, radium, or thorium.

Neither of these energy sources provides an energy flow to the surface that is important in the day-to-day processes taking place at the surface. There is no doubt that an outflow from the interior takes place, but it is extremely small. To illustrate, the heat flow from the interior for an average continental position amounts to only about one-tenth of 1 percent of the intake of solar radiation at the equator at noon. For this reason, the heat flow from the interior usually is disregarded in calculations of the earth-atmosphere energy budget. Its amount is slightly higher on the ocean floors and is least in the interior of the continents. Accumulations of radiogenic (atomic) energy probably are significant in initiating some volcanic activity below the surface, but it is highly improbable that this is the only, or even a major, cause of earth vulcanism. Adding heat energy is not the only way of inducing a phase change from solid rock to molten rock. If rock temperatures are high enough to induce melting or fusion but are prevented from doing so by rock pressure, a shift in overlying material that

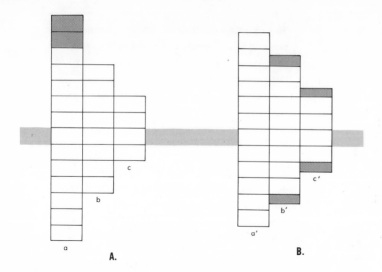

A.

B.

Figure 9-1 Principle of isostatic balance.
The diagram indicates a series of blocks, loosely fastened together and suspended in a fluid whose density is twice that of the blocks. Each column of blocks has half of its mass above the fluid surface and half beneath. If the two shaded blocks at the top of column *a* in (*A*) are transferred to columns *b* and *c*, the new arrangement will be as shown in (*B*). Note that there is a readjustment of the level of each column to fit the alteration in mass. It has been suggested that the removal of material by erosion from mountains and its deposition in sedimentary basins result in such isostatic adjustments of the earth crust.

would relieve some of the compression could produce fusion even at depths within 10 to 20 miles of the earth surface. The thermal gradient for rock depths has not been tested for more than a few miles, but within this range it amounts to an increase in temperature of about 50 to 90°F per mile of depth. This rate probably varies greatly with depth.

9-2 ENDOGENIC PROCESSES ATTRIBUTABLE TO EXOGENIC ENERGY SOURCES The earth crust is sensitive to vertical changes in pressure exerted upon it. Deformation of the crust daily by lunar tides indicates that it is not a completely rigid surface. The evidence of postglacial land uplift following the retreat of continental ice sheets indicates that the crust can be warped downward or upward for hundreds of feet by the addition or subtraction of mass at the surface. The readjustment of the earth radius to accommodate such changes in mass is known as *isostasy*. The principle of isostatic balance is diagrammed in Figure 9-1. While it is not entirely correct to consider the continental masses as floating on a liquid

base, as illustrated in the diagram, the material beneath the crust can be considered a plastic solid with some of the properties of liquids, including fluidity. Isostatic depression of the surface resulting from deposition and uplift following erosion were first suggested by Jamieson, a Scottish geologist, in 1908. Vening Meinesz, a Dutch geophysicist, has worked out a noteworthy hypothesis for relating compressive stresses in the crust to lateral compensatory stresses following the downward buckling of the crust caused by sedimentation.[1] According to his hypothesis, the major areas of deposition are likely sometime in the future to be centers of compressional buckling and regional uplift.

Later in this chapter (see Section 9-3) it will be noted that, given the relative permanence of continental masses and ocean basins throughout most of geologic time and the relatively rapid rate of *denudation* (Latin: *denudare* = to remove), or the removal of material from uplands, some sort of subsurface transfer of material must have taken place below the surface, or the continental masses would not have persisted. There appears to be no doubt that most of the stresses operating on the crustal rocks are isostatic

[1]F. A. Vening Meinesz, "Indonesian Archipelago: A Geophysical Survey," *Bulletin, Geological Society of America*, vol. 65, 1954, pp. 143–164.

compensations for the processes of denudation and deposition, and hence are derived
from exogenic sources.

In comparing the earth surface again with
that of the moon and Mars, neither of the
latter shows any signs of crustal compressional stresses or buckling, probably because
the means for transporting material from one
place to another does not exist as it does on
earth. The rock material found on the lunar
surface crystallized from a molten state approximately 3 to 4 billion years ago. This
suggests that the moon at one time was extremely hot and probably molten, and has
changed little since it cooled and solidified.

Another combination of an exogenic energy source influencing endogenic processes
is the possible gravitational influence of lunar
tides. While the influence of lunar tides on
the earth crust is slight, it could act as a
trigger to set off crustal dislocations whose
stresses had accumulated for other reasons.
There is some evidence that earthquakes and
volcanic eruptions are more frequent during
periods of high tides.

Regardless of the sources of energy, endogenic processes keep the earth surface from
being beveled to a smooth plain. Great regional uplifts, the buckling of the surface into
folds, the sliding of ruptured blocks upward,
downward, and sideways, the slow sag of
basins, the outpourings of lava upon the surface, and the opening of huge fissures—all
create obstacles to the smooth flow of energy
upon the surface. Eliminating them diverts
solar energy into the geologic cycle—a wheel
that may take millions of years to rotate, but
whose turning keeps the surface of the earth
in a state of flux. Figure 9-2 is a diagram-
matic representation of this cycle.

This figure contains the term *tectonic,*
which has not yet been defined. It is a useful
adjective that distinguishes all landforms and
processes that are related to crustal move-

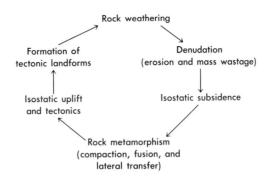

Figure 9-2 The geologic cycle.

ments. It might be considered to represent
the converse side of the geologic cycle from
denudation and deposition. Some of the
major tectonic landforms are described in
Chapter 12.

9-3 EXOGENIC ENERGY SOURCES Solar
radiation comprises nearly all of the energy
that does the work of denudation and deposition on the surface of the earth, and as was
noted in the previous section, it probably is
involved in most tectonic processes as well.
The total quantity of energy that passes
through the global energy system is enormous. Some idea of the quantities involved
are indicated in Table 4-3. Chapter 4 indicated the many diverse paths taken by this
energy within the atmosphere. Not much was
indicated there, however, of its work in
modifying the earth surface. Every stream or
every storm wave that pounds a coastal cliff
receives its energy supply from solar radiation. The disposition of this energy is highly
irregular, and its capacity to perform work
in altering the surface varies from place to
place and from time to time. The relationships between the processes of surface
alteration (which involves work or energy
applications), climate, and materials will be
discussed later.

9-4 THE BEVELING ROLE OF SOLAR EN-
ERGY AT THE EARTH SURFACE In the
treatment of the global energy balance in
Chapter 4, it was noted that the existence of
net radiation surpluses, as in the low-latitude
oceans, and the areas of net radiation deficits,
as at the heat sinks of high latitudes and high
altitudes, are responsible for a highly complex flow of energy to reduce these imbalances. At any point along the course of the
energy flow, work may be performed. This
may take many forms: the evaporation of
water, the growth of plants, the flow of rivers,
the oxidation of metals, a flash of lightning,
a severe storm, or the melting of ice and
snow. The entire air circulation system, the
system of oceanic circulation, the hydrologic
cycle, and the activities of all living things
are involved in this flow. It also was noted
earlier that the flow is irregular, that short-
term and long-term imbalances sometimes
occur that interfere with the smooth operation of the system. A characteristic of this,
as in all systems, is the tendency to maintain
an equilibrium, and whenever imbalances
occur, the signals change, and work patterns
are altered to correct the imbalances.

Surface irregularities are obstacles to the
smooth flow of energy within the global system. The higher precipitation in mountain
areas and the resultant high potential energy
of both water and ice at high elevations is
but one illustration of the accumulation of
energy at the sites of these irregularities.
Denudation can be considered as the application of work to remove obstacles to the free
flow of energy. The factors that influence the
configuration of the surface, therefore, are
numerous; they involve not only the composition and arrangement of surface material
but also the entire climatic pattern on earth.
The kind and rate of work done to remove
mountains at high latitudes is different than
that in the rainy tropics, because energy

availability is different in the two places.

It is impossible to combine all of these
factors into a relatively simple system by
which regional landforms can be classified
genetically, because the major factors are
unrelated except within the vastnesses of the
geologic cycle. There is no direct, immediate
connection between climate, materials, and
structure (the attitude or arrangement or rock
material). It is possible, however, to separate
some of the variables in a general way so as
to indicate the relative importance of different types of erosional processes in different
climates.

The remainder of this chapter will be concerned with the major processes of denudation and deposition, or the ways by which
solar energy and its derivatives are directed
against irregularities on the surface of the
earth. It also will discuss the relationships
between these processes and the global climatic variables of temperature and precipitation.

Denudation and Deposition

Before treating the denudational processes,
it is appropriate to present a general idea of
the total scope of denudation on earth. An
accurate measurement of the total amount of
material removed from the continents and
deposited into the oceans each year would
be impossible to obtain. Yet, through sampling
and by deductive reasoning, students of the
subject have arrived at various estimates of
this rate. Table 9-1 presents estimates of the
total erosion rates for three different areas
totaling 10 percent of the global land area.
The table has been adjusted to eliminate the
factor of human intervention.

Judson concludes that man's present activities have accelerated the natural erosion
rate by all erosive processes about $2\frac{1}{2}$ times.

Table 9-1   Rates of Erosion for the Amazon River Basin, the United States, and the Congo River Basin

Drainage region	Drainage area $(km^2 \times 10^6)$	Load, (tons $\times 10^6$/yr) Dissolved	Solid	Total	Tons/ km^2/hr	Erosion/ cm/yr $\times 10^3$
Amazon River (Gibbs)	6.3	232	548	780	124	4.7
U.S. (Judson & Ritter)	6.8	292	248	540	78	3.0
Congo R. Basin (Spronck)	2.5	99	34	133	53	2.0
	15.6	623	830	1453	93(av)	3.6 (av.)

*SOURCE: Sheldon Judson, "Erosion of the Land, or What's Happening to Our Continents," *American Scientist*, vol. 56, no. 4, 1968, p. 367.

His conservative estimate of the total mass of material deposited annually in the oceans, excluding the influence of man, is approximately 10^{10} metric tons. His conclusions are worth repeating:

> Whether we use the rate of erosion prevailing before or after man's advent, our figures pose the problem of why our continents have survived. If we accept the rate of sediment production as 10^{10} metric tons per year (the pre-human intervention figure) then the continents are being lowered at the rate of 2.4 cm per 1000 years. At this rate the ocean basins, with a volume of $1.37 \times 10^{18}m^3$, would be filled in 340 million years. The geologic record indicates that this has never happened in the past, and there is no reason to believe it will happen in the geologically foreseeable future. Furthermore, at the present rate of erosion, the continents, which now average 875 m in elevation, would be reduced to close to sea level in about 34 million years. But the geologic record shows a continuous sedimentary history, and hence a continuous source of sediments. So we reason that the continents have always been high enough to supply sediments to the oceans.[2]

It is evident from the above data that the material being removed from the continents somehow is returned. Hundreds of millions of years may be required, but somehow the geologic cycle is completed. Like the global energy flow, imbalances in the geologic cycle

[2]Sheldon Judson, "Erosion of the Land, or What's Happening to Our Continents," *American Scientist*, vol. 56, no. 4, 1968, pp. 371–372.

or in the flow of material away from or toward the continents contain or induce the means for their eventual correction. Solid rock is altered into fragmental or soluble material and eventually finds its way back into rock again. The total supply of rock material, like the supply of water in the hydrologic cycle, does not change appreciably. The surface material, however, is changed in form and position to accommodate the flow of energy that sweeps across the face of the earth. It is more than coincidental that some of the highest mountain areas on earth present evidence of once having been areas of excessive deposition, and that the major areas of crustal instability are near the continental margins.

The earlier chapters on climate revealed that winds, clouds, water vapor, snow, ocean currents, and many other materials related to the atmosphere were bound together in an interrelated system designed to regulate the flow of solar energy toward the earth, in various directions upon the earth, and finally its transport away from the earth. Now we shall demonstrate that the forms and composition of materials that make up the solid surface of the earth, and events that occur several miles below the surface, likewise constitute a subsystem that is involved in the global energy flow.

9-5 MECHANICAL WEATHERING *Weathering* refers to processes by which atmospheric agents alter rock material in place. *Mechanical weathering,* sometimes also referred to as *rock disintegration,* consists of "making little ones out of big ones," or a physical fragmentation of rock without necessarily involving any chemical alteration of the material, although the two frequently work together in the weathering process.

The earth surface, or the interface between the atmosphere, the lithosphere (rock sphere), hydrosphere, and biosphere, is not a smooth plane surface at equilibrium, but a zone of admixture. A fragmental soil mantle covering solid rock and gradually merging with it is a truer expression of the interfacial balance—the result of mutual interactions between solid, gases, and liquids. Mechanical fragmentation of rock or other surface material is only a part, but an essential part, of the process of creating this balanced interfacial zone.

Fragmentation of rock material requires the application of energy, whether the process is directed by man or by nature. By far

the most effective agent of rock weathering is frost action. We learned earlier that phase changes of water were among the most effective methods of heat transfer in the global energy balance, and it is not surprising that water should be involved in energy manifestations of this type. The molecular growth of ice crystals exerts a pressure when confined up to 2,000 lb/in². Alternate diurnal freezing and thawing of water contained within rock openings of various sizes provide ideal conditions for rock fragmentation. The incidence of frost action depends on several factors, including frequency of water-phase changes (related to the position of the freezing boundary), the quantity of water available, and the tendency of the rock to provide openings for water penetration. Ideal conditions are found in *periglacial* regions.[3]

Daily variations in the expansion and contraction of mineral particles in rock, caused by solar heating and cooling, once were believed to be important in fragmenting rocks, especially in hot, arid regions where rock temperatures may rise to near the boiling point of water. Careful investigations indicate, however, that such temperature changes have little direct effect but may have some influence on the rate of chemical weathering.

Minor forms of mechanical weathering

[3]*Periglacial* refers to climates that are peripheral to glacial conditions where rapid changes occur between freezing and thawing. Subpolar areas and high elevations (see Figure 9-3) well above the timberline but below the areas of permanent snow fields constitute the major periglacial regions.

Figure 9-3 Mechanical weathering in the periglacial environments of high elevations. This scene, at the Ljubljana Pass in the Alps along the Austrian-Yugoslavia border, shows how the fragmental material weathered by frost action in the periglacial zone collects in a steep depositional slope at the foot of cliffs. Such debris is termed *talus* or *scree.* It is a characteristic feature of high mountain landscapes. [Van Riper]

result from forest or grass fires and from the expansive force of salt crystallization. The latter results from the entrapment of water from rare rains in desert rock crevices and its subsequent evaporation.

The general comparative efficiency of mechanical weathering under different global temperature and precipitation conditions is diagramed in Figure 9-4. It will be noted that mechanical fragmentation is most effective in the periglacial climates of upper mid-latitudes, where annual temperatures range between 0 and 40°F. The only area where mechanical weathering is important at high temperatures is the arid regions. Figure 9-4 clearly indicates the importance of frost action. Mechanical weathering is least effective in the humid tropics. This is because of the dominance of chemical weathering and the cover of vegetation, both of which tend to preserve the soil mantle and prevent direct exposure of the bed rock. Then, too, frost action does not take place in these climates.

9-6   CHEMICAL WEATHERING   The chemical action of atmospheric gases and water derived from precipitation is the most effective process within humid regions for reducing rock to fragmental materials. This is sometimes referred to as rock *decomposition* or *corrosion.* Rock corrosion by chemical means involves chemical reactions of many kinds, but the most important by far involves *solution* by weak, highly dilute acids, particularly carbonic acid, which is formed by a combination of water and carbon dioxide. The latter is universally found within the lower atmosphere; it is released as an end product of decaying organic material and the respiratory function of animals. Other acid solutions found in nature include a wide range of humic acids, complex organic compounds that are formed during intermediate stages of decay. *Hydration,* or the direct sol-

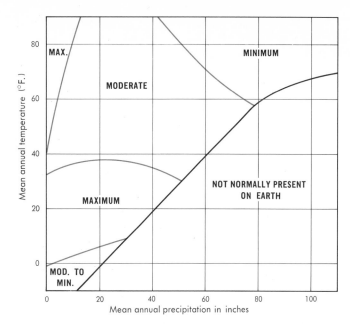

Figure 9-4   Relationship between mechanical weathering and climate.   [After Wilson]

vent action of water, can be important in the breakdown of rocks containing soluble salts, and it is an effective means of opening the crystal lattices of certain types of minerals. Alkaline solutions, most frequently found in arid regions, have a particular affinity for removing silica (SiO_2) from mineral compounds. This molecule is an important building block in most continental crustal rocks. The quantity of alkaline solutions, however, is far less than that of the more acid types. *Oxidation,* or the union of oxygen with various elements and compounds, is another form of rock weathering. The rusting of iron is illustrative.

Only the simplest chemical reactions involved in weathering have been indicated here. Many others are exceedingly complex. The processes of chemical weathering in general indicate that the chemical elements in rocks were arranged under conditions of high heat and pressure far below the surface and were not at balance with the environment existing at the surface; hence, the changes. Since chemical weathering always involves some transfer of energy (usually heat to chemical energy, or vice versa), the reaction speed is related to the quantity of available heat energy. Later, in the chapter on soils (Chapters 15 and 16), much more attention will be paid to the complex processes by which the balanced interface between rock, gas, and liquid is established. As might be expected, such a balance differs with climate, with parent rock material, and with time.

Figure 9-5 presents a rough picture of the relationship between chemical weathering, temperature, and precipitation. The major variable is precipitation, because of its solvent function once it reaches the surface. Temperature also is important because it is

an index of the energy supply that is required for chemical reactions, especially oxidation. Chemical weathering thus tends to be most effective within the warm, humid regions and least effective in the polar and dry lands. Aiding in the dominance within the humid tropics is the covering of vegetation. This tends to reduce the runoff of water, diverting more of it into the soil for chemical solution. The inevitable decay of organic material also aids in supplying acidic solvents. A lush vegetation cover tends to reduce soil erosion, and while a soil covering reduces the opportunities for mechanical weathering, it aids in the incidence of chemical weathering by retaining soil solutions and increasing their potential for chemical reactions by adding various solutes from the soil column.

Rock composition can be another extremely important factor in the distribution and intensity of chemical weathering. By far the greatest amount of chemical weathering and subsequent removal is accomplished within areas underlain by massive deposits of carbonate rocks, mainly limestones. Such rocks are unusually susceptible to solvent action by weak acids, which may completely remove the rock material via groundwater solution, leaving behind only a thin residue of insoluble impurities (see Figure 9-6). In this case, chemical solution is not only a weathering agent; it performs erosion by removing the original rock material. In many areas bordering the Mediterranean Sea, where limestone is especially common, landforms resulting from limestone solution constitute the most common surface features. Such landforms are described in Chapter 12. Not all limestone areas, however, are highly weathered and feature caves, caverns, underground drainage, and fluted rock columns. Some have a high content of impurities, such as clays or sand, and hence make the rock much more resistant to solution than pure,

Figure 9-5   Relationship between chemical weathering and climate.   [After Wilson]

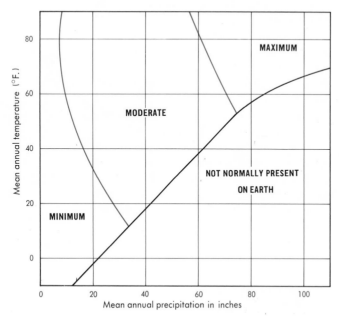

thick beds of limestone. Chemical weathering is minimal in areas of extreme dryness and coldness (see Figure 9-5). In such climates, limestones stand out in bold relief, because its smooth surfaces do not yield as many openings for frost action as many other rocks.

The results of chemical weathering vary, depending on the climate, the composition and structure of rock, and the time involved. Weathering always works inward from the surface. Hence, its first effect usually is a spalling-off of the outside layers of rock in concentric layers. This is sometimes termed *exfoliation* (Latin: *folium* = leaf). Where the rock is highly fractured, weathering takes place inward from the cracks as well. Weathered blocks of rock usually show curved surfaces, because sharp corners invariably are removed by the weathering processes. The depth of weathering may be only an inch or it may reach over 100 feet, as has been observed in tropical areas. Normally weathering proceeds irregularly, because of the varied arrangement of minerals in rocks, each of which may have a different susceptibility to weathering, and also because the work of weathering penetrates along cracks and fissures. Thick layers of weathered material often contain rounded unweathered rock, the cores of blocks that have not yet been decomposed.

9-7 THE WORK OF RUNNING WATER

Running water is of extreme importance in the work of denudation and deposition. The results of its work are evident even from views taken from high altitudes. Whereas circular scars of meteoric craters form the characteristic feature of the lunar surface, the acute-angle branching or tree-branch forms (*dendritic*) assumed by most types of stream networks characterize most land surfaces on earth. Figure 9-7 shows a view of southern Arabia taken by one of the first manned

Figure 9-6 Erosion of limestone by rainwater solution. Rain that picks up a small amount of carbon dioxide during its descent through the atmosphere is able to sculpture pure, massive beds of limestone where they are exposed at the surface. The highly irregular surfaces and fluted columns are typical. Scene near Bikfaya, Lebanon. [Van Riper]

space satellites. This is one of the driest areas on earth, yet the familiar dendritic stream network is the principal surface pattern seen. Stream valleys are taken for granted in humid regions, but desert landscapes are a reminder that here, too, the major work of denudation is performed by stream erosion and deposition.

Stream erosion. The great importance of water in motion as an agent of attack on surface irregularities should be expected, because, as was noted in Chapter 5, water is the most important heat-transfer mechanism in the global energy system. Almost half of the solar energy that is absorbed at the surface is transferred to the atmosphere by the evaporation of water. The lifting of this water above the surface confers a certain

Figure 9-7 Dendritic drainage pattern in an arid region. This photo, taken from the manned Gemini VII satellite in December, 1965, is a view of the southern part of the Arabian peninsula, one of the driest regions on earth. The body of water is the Arabian Sea, and the major watercourse is the Wadi Hadramaut. Running water is the major sculpturing agent in arid as well as humid regions. Much of the dissection shown here probably was accomplished during the much more humid Pleistocene period, although some work still is performed by runoff following rare rains. The watercourses now are clogged with detritus washed or blown from the adjacent uplands. [NASA]

2. Hydraulic compression within crevices and resultant prying and lifting.

3. Chemical solution of material from rocks exposed to the stream by solvents carried by the stream.

4. Abrasion (*corrasion*) of the channel floor and sides by the impact of the *bedload* (the material that is too large to be carried in suspension).

5. Transportion of material supplied by hydraulic action, solution, and abrasion.

6. Reduction in size of the transported detrital material.[4]

Several types of energy transfer can be noted in this list, including the production of heat, mechanical energy, and chemical energy.

Despite the many different ways by which energy may perform work along a river course, the primary objectives of running

amount of potential energy to each molecule of water that is lifted. Sooner or later this water finds its way back to the sea, where its potential energy again becomes zero. On its way back to sea level, the potential energy is not lost but is transferred into the kinetic energy of motion. It is this kinetic energy that performs work of various kinds, including the beveling of the earth surface. The amount of work that can be done by water is proportional to the amount of potential energy which it has and which is a function of mass and height of fall.

The most significant ways by which the potential energy of water contained in streams is changed into kinetic energy include:

1. Production of friction through eddying and turbulent flow.

[4]There is a lower mass limit of the suspended detrital particles in a stream beyond which no abrasion takes place, either on these particles or by them. This is because of the cushioning effect of the water molecules that cling to the surfaces. This lower limit is smaller with more dense material, but for the more common rock debris, it lies within the general textural size of fine sand, or from .1 to .2 millimeters in diameter. The extremely fine-textured material carried by streams, which does no abrasive work, is largely the result of other types of weathering or erosion agents. Despite this limitation of particle size, abrasion is an important part of stream erosive work, as is testified by the rounded shapes of bounders, cobbles, gravel, and sand contained in stream beds.

water are fourfold: (1) to set up a geometric network of drainage ways that will most efficiently remove the water surpluses within an upland area; (2) to remove the soluble and insoluble material brought to the stream edges by other agents of erosion; (3) to transfer the potential energy of water in the upper drainage system to kinetic energy during its passage to the mouth; and (4) to arrange the work performed by this kinetic energy so that it acts most efficiently, given the variety of conditions encountered along the stream course. This last objective warrants further explanation. The priorities for expenditure of work of various kinds are set by such conditions as stream gradient, volume, shape and configuration of the channel, the type of material found in the stream bed, and the amount of load being carried downstream, either in suspension or by being shifted along the stream bed. To illustrate how these various conditions operate to determine the kind of work done by a stream, let us observe two typical river sections: one a canyon section within mountainous terrain near the headwaters, and the other a section near the mouth.

The typical mountain stream is relatively clear of sediment and rushes with abandon down its steep course. Its channel is highly irregular, is confined in some places to a narrow notch between rock walls, and spread out in others onto a boulder-clogged bed. Some abrasive work is being done, as witnessed by the rounded nature of many of the boulders, and here and there in the relatively slack water behind an unusually large stone mass, one can find a small strip of sand or a gravel bar whose fragments bear the unmistakable roundness of stream wear. The water boils and churns, bouncing from one boulder to another, plunging downward, swirling in violent eddies, and even breaking into spray and tossing into the air at the foot of small falls or cascades. Despite the evidence of stream abrasion, not much of it is usually visible. As with most small streams, most of the erosive work is done during extremely short periods of high water. The cracking noise of boulders being smashed into each other during highwater periods on a mountain stream is an impressive and unforgettable sound. Most of the time, however, the only abrasive work is done by a small amount of small pebbles and sand moving irregularly with the current. However, a swift mountain stream, even during normal water periods, still has a large amount of potential energy to put to work. If it lacks tools to abrade, it dissipates this energy largely through eddying and turbulent flow, in which the energy is altered mainly into frictional heat. Erosion and load transport thus compose a relatively minor part of the work performed by kinetic energy along mountain streams. Farther dowstream, where the stream channel becomes deeper and the bed smoother, where the gradient is less and less of the kinetic energy is lost from the stream by friction, much more of the stream's work can be done in both abrasion and in load transport (see Figure 9-8).

Now let us examine the work done along a stretch of river near its mouth, using the Mississippi as an example. The river water here is brown and turgid because of its high silt content derived from topsoils throughout the drainage basin. The river flows slowly and smoothly within its channel, and one must look closely during normal water periods to see eddies other than those produced by an occasional tree trunk protruding from the bank. At flood time the eddies become more pronounced. The coarsest material found in the river bottom is fine sand that appears in streaks and lenses within the swifter portions of the channel. The river swings in great bends, and the resultant vari-

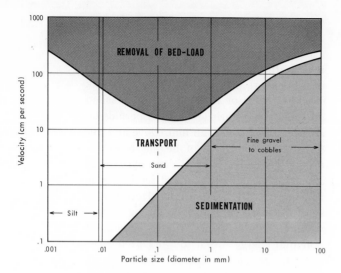

Figure 9-8 Velocity-load ratio in the work performance of a stream on its bed load. Close study of this graph reveals many things about the work of streams. Note that streams are able to remove sand particles easier than fine silts or clays. This is because of the co-hesiveness of the smaller particles. Once carried, however, the finest particles are not deposited unless the velocity drops close to zero. Note also the extremely narrow range between removal and deposition of the coarsest material. A stream's load generally is much greater in its lower course than in its upper, steeper sections. [After Hjulstrom]

characteristics. In the uppermost reaches,[6] volume is relatively small and irregular, the speed is great, turbulence, eddying, and hy-draulic work are at a maximum, the sus-pended load is slight, and the bed load is extremely coarse. Downward abrasion of the channel is rapid but takes place mainly dur-ing short periods of high water. Downstream, the volume increases and becomes more reg-ular, the slope of the channel and speed decreases, the bed load becomes finer, more material is carried in suspension, and down-ward erosion is reduced. This is reflected in the reduced stream gradient. Beyond the point at grade, the land surface is modified mainly through either deposition or lateral cutting at the channel sides.

The amount of potential energy expended at any point along the stream is a function of its volume and velocity. How that energy is put to work depends on many factors, as was illustrated in the previous paragraphs of this section, but the many applications of energy must add up to the total amount available. An alteration in any one of the

ations in speed on the outside and inside of the bends produce lateral cutting of the banks along the former and deposition of silt along the latter. Virtually no abrasion of any kind occurs. The major part of the kinetic energy is used to transport the stream load. Nearly all of the material constituting the stream load is being carried in suspension. Only a small amount of energy is lost through friction.

As indicated in the examples given in the previous paragraphs, different portions of a stream have different attributes of slope, channel, load, and volume. Because of some normal progressions in these attributes along the length of streams, most of the streams that reach *grade*[5] tend to develop what might be termed a *curve of equilibrium,* of a bal-anced longitudinal profile. An idealized dia-gram of such a curve is shown in Figure 9-9. Different sections of the profile have different

[5]A stream whose gradient represents a balance between the work capacity and load transport is said to be *at grade* and the stream designated as a *graded stream.* Work capacity here is assumed to be the kinetic energy that remains after frictional losses to the bottom and sides of the channel.

[6]Many streams that originate, not along steep, mountain, crest lines, but on as yet undissected uplands, will show a gently sloping crest on their longitudinal profiles. This is the zone of rill collection, where the potential energy and the capacity for work are extremely low because of the small volume or mass. The curve does not tilt sharply downward until enough water collects in the stream.

variables of energy application will influence all of the others, and changes are always taking place, if one looks closely enough. This is the meaning of *dynamic equilibrium.*

Deposition by streams. Attention has been paid in the previous paragraphs largely to the erosive work of streams. Deposition is also important in smoothing the land surface, so let us look closer at the work of streams in this capacity as they lose potential energy and release the load that they are carrying. The ultimate *base level* toward which a stream flows is the point at which it can do no more work, since it has lost all of its potential energy. This may be at sea level or at some local base level, such as a lake or an inland basin without an outlet. When a stream reaches its mouth, its velocity gradually slows until all of its flow ceases. During this reduction in velocity, the load carried by a stream is deposited and is gradually sorted in the process, with the heavier fragments deposited first and the lightest last. The deposits laid down in water as the result of the halting of stream velocity at base level are known as *deltas.* Their characteristics will be described more fully in Chapter 12. The lightest clay particles carried by streams are so fine that they may remain in suspension within quiet water for months; they are carried far beyond the borders of deltas.

As the deltas are enlarged and extended farther into the body of water at base level, the stream lengthens. With added length, the average gradient of the lower course is reduced, and some load is deposited within the channel upstream until grade is again reestablished. This process can be a major cause of floods along the lower course of great rivers. With delta growth and the raising of the channel bed, the stream banks are less capable of containing flood waters, and the river seeks a new and lower channel across its

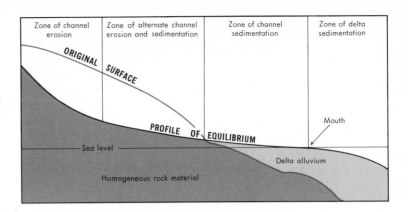

Figure 9-9   Profile of equilibrium at a graded stream.

lower plain. Thus, under natural conditions, the growth of a delta tends to build up not only the stream bed but also the entire river plain on which the stream is located. Since the depositional plains of a river often are prized for agricultural use, people try to contain the lower courses of streams within their channels by constructing dikes or levees. The full load of river sediment then is directed toward delta growth, the river gradually is raised higher and higher above the level of its plain, and the potential flood hazard grows. Along the lower course of the Hwang Ho (Yellow River) of northern China, where delta growth has been particularly rapid, the bed of the river has been elevated as much as 20 feet above the level of the surrounding plain. Only extremely high artificial dikes prevent the river from flooding large areas.

The idealized profile of equilibrium, as diagramed in Figure 9-9, represents a smooth flow of the energy of running water. If the profile is considered as extending from the main stream up each tributary to its uppermost end, it might be thought of as the controlling base for an arterial energy distribution network which includes the entire three-dimensional surface of the drainage basin. It grows in length at both ends—the upper end to tap the unused potential energy of water at high elevations to reduce the slopes of the entire drainage basin surface—and it lengthens at its mouth to fill in a depression.

The idealized profile is rarely met in nature because of many local variables that may exist during the growth of a drainage system. Variation in the erodibility of underlying material, in the arrangement of the rock structure, in the degree of crustal warping, and in the amount of rainfall at different places within the drainage system are among the more important factors that influence the longitudinal profile of streams. Many streams have profiles that alternate between stretches at grade and those where active cutting or active deposition is taking place. The major rivers of Africa, for example, afford excellent examples of rivers whose longitudinal profiles are far different from that of the idealized profile (see Figure 10-16).

Minor variations in the longitudinal profile can be found along nearly all streams. Variations in the underlying material and in the random behavior of stream currents produce alternate stretches of swift and slow water, stretches of active channel-cutting separated by depositional sections, and interruptions in the smoothness of the profile. Small depositional flats can be found even at the bottom of the Grand Canyon in Arizona, in the midst of an overwhelming predominance of erosive work. None of these exceptions, however, negates the concept of the goal of stream dynamic equilibrium. Work is performed in beveling the surface, and as the drainage system grows in length, in width, and in the number and order of tributaries, the longitudinal profile of equilibrium tends to be established throughout the drainage basin.

The development of a stream drainage system may be likened to the growth of a tree as its branches extend in length and divide into further branch ends, or to the extension of the arterial and vein systems in animals as they grow to maturity. These regularities stimulate certain questions. Why do small streams tend to assume forms similar to large ones? Why are branching tributaries usually at similar angles? Is there a blueprint of growth that directs the hearts or lungs or legs of children to exhibit the same proportions during growth as those of adults? All of these regularities in growth patterns seem to represent adjustments of materials to energy applications, adjustments that are called for by a complex informational signal system that all organic and inorganic materials possess. For example, what "information" induces a stream to begin lateral meandering back and forth across its valley when it reaches grade? Without energy, such signals would have no significance; there would be no change, no movement. The regularities that appear in nature are not pure chance, but rather the result of informational signals built into materials, signals by which materials react to variations in energy receipts. These regularities often can be expressed in mathematical terms. The geometries of specific patterns resulting from the work of running water, such as the longitudinal profile of equilibrium in Figure 9-9, never are ideally met in nature any more than are the shapes of individual trees, human hearts, or volcanoes, but each exhibits recognizable regularities that conform to the "blueprints" that are set for them.

The work of running water, although widely distributed over the face of the earth, varies greatly in intensity. A general diagram of the relationship between it and the major climatic variables is presented in Figure 9-10. The maximum activity does not show a direct correlation with precipitation, as one might believe. Areas having precipitation of more than 60 inches show only moderate erosive work by streams, first, because of the stabilization of slopes under the influence of a thick soil mantle and a dense vegetation cover, and second, because of the increased proportion of the stream transport devoted to removing solutes (see Table 9-1). The solution load of tropical streams utilizes energy just like their load of silt or sand. The former, however, does no abrasive work on the stream bed.

Figure 9-10 indicates that the work of running water is least in areas of driest and coldest climates. Even in deserts, however, a large part of the total work of denudation and deposition is performed by streams, during the infrequent periods of intense rainfall. The maximum activity, as shown in Figure 9-10, lies mainly in the areas of semiarid to subhumid climates, or those areas having between 10 and 30 inches of rain. Rainfall in this zone, although irregular and of short duration, is likely to be intense. Runoff is rapid, and the supply of abrasive tools is likely to be great because of the dominance of mechanical weathering.

Climate is not the only variable influencing the intensity of erosion by running water. Rock resistance is an independent variable everywhere. The *stage* in the development of the profile of equilibrium is a time factor of relative length. It was noted earlier that different sections of the profile underwent different intensities and types of erosive processes (see the last division of this chapter). Crustal stability is still another variable influencing the intensity of stream activity.

9-8 MASS WASTAGE The term *mass wastage* refers to the general movement of surface material downslope under the control of gravity alone. No transporting medium such as water is required. A large variety of individual processes are involved, and the

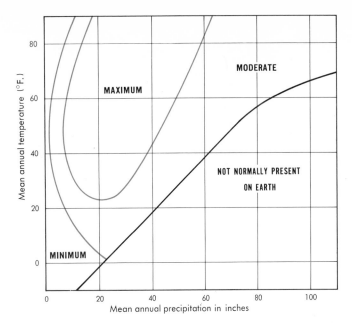

Figure 9-10 Relationship between the work of running water and climate. [After Wilson]

movement of material may be massive and sudden, as in landslides, or extremely slow and localized, as in the progressive creep of a large boulder that is half buried within the soil of a forested slope. The total volume of material moved by mass wastage is great, although much of it is too slow to be observed directly. The total volume eroded probably exceeds that of running water, except that the latter generally removes the material brought to it by mass wastage. The two types of erosion supplement each other. Stream valleys, with certain exceptions caused by lateral stream erosion, may be considered as having their depths excavated by stream abrasion and their widths excavated largely by mass wastage.

The evidences of mass wastage can be observed almost everywhere on sloping ground. Figure 9-11 illustrates only a few of them. There are many human implications for this type of process, ranging from the

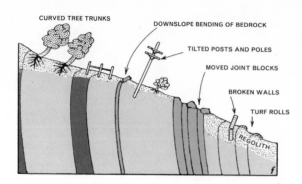

Figure 9-11 Evidences of mass wastage on a slope covered with soil material. [After Sharpe]

serious threat to life and property caused by landslides and avalanches to the subtle shifting of slopes causing cracked basements of houses, broken sewer and water lines, the heaving of road pavements and dislocated fence lines. Many factors influence the amount and rate of work done by mass wastage. Some of the more important variables include steepness of slope, thickness and texture of the underlying unconsolidated material, the weight of the detritus, its water content, vegetation cover, susceptibility to freezing and thawing, and the structural resistivity of the underlying material to shearing, plasticity, and other forms of deformation. One of the best places to observe mass wastage at work is on the excavated slopes in unconsolidated material along a newly built highway, where slumping often produces large scars on the newly graded slopes.

The position of the mass wastage processes in the energy flow system is related to the potential energy imparted to the rock material when it is raised to an elevated position through tectonic forces, generally triggered by isostatic readjustments. Just as water is given potential energy by being raised into the atmosphere by evaporation, so too is energy imparted to rocks and rock material

by their being raised to higher elevations. The rigidity of rock material prevents gravity alone from deforming it by flowage to produce a smooth, even slope. When this rigidity is gradually destroyed by weathering processes, however, a point is reached when gravity can translate the potential energy into the kinetic energy of motion through deformation of the mass, and the latter moves down slope in various ways and at various speeds. Weathering on extremely steep slopes, such as a cliff face, is sometimes followed immediately by the fall of fragments to the base of the cliff, where they collect and assume an angle of rest that varies with the coarseness of the fragmental material. In most cases involving gentler slopes, the process is much slower, and weathering must proceed deeply into the rock material before mass deformation can take place. Some of the forms related to mass wastage are described in Chapter 12. The general collective term for material moved by mass wastage is *colluvium*. It differs from *alluvium* (material deposited by running water) in having no distinct textural stratification.

Figure 9-12 indicates the impact of the major long-range climatic variables on the incidence of mass wastage. The maximum activity is revealed as taking place in the humid tropics and also within subpolar regions. The high incidence in the humid tropics is associated partly with the thickness of the weathered soil mantle under the effective work of chemical weathering and partly because of the readily available water supply.

The capacity of the soil material to absorb and hold water increases as the constituent rock particles in the *regolith* (fragmental rock mantle) become smaller with continued weathering in the humid tropics. This increases the total mass, which in turn increases the deformational stresses. A further aid to mass wastage in the rainy tropics is the constant flow of the streams toward which the material tends to move. Accumulations of colluvial material are less likely to interfere with further mass wastage when they are relatively rapidly removed by running water. A noteworthy feature associated with mass wastage in the humid tropics is the unusual steepness of weathered slopes. The thick mantle of vegetation is largely responsible for this, binding the soil mass together by root growth and delaying the mass wastage process. The deformational stresses accumulate for a longer period, but when the deformation takes place, it is likely to be massive and sudden, frequently in the form of landslides. These sudden ruptures tend to produce steeper slopes on valley sides (see Figure 9-13) than the more gradual movement of material where the binding action of vegetation is not so great, and they may be entirely independent of the type of underlying rock material.

The maximum zone of mass wastage shown in Figure 9-12 as occurring in the subpolar regions is related to the influence of alternate freezing and thawing, to the occurrence of permafrost (permanently frozen subsoils), and to the low evaporation rates and availability of water despite low precipitation rates during the summer season. Freezing of groundwater causes an expansion of soil volume on any slope, due to the expansive force of water crystallization. The major direction of expansion is vertical to the slope surface. After thawing, contraction takes place under the influence of gravity and the direction of this is toward the center of the earth, not straight back toward the slope surface. Every thaw thus displaces the material slightly downslope, as diagrammed in Figure 9-14. In subpolar areas, where alternate freezing and thawing occurs almost every day during the summer period, this

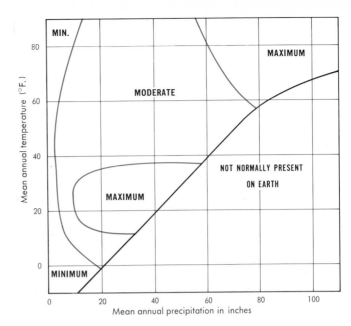

Figure 9-12 Relationship between mass wastage and climate. [After Wilson]

Figure 9-13 Characteristic steep hill slopes in the humid tropics. Note how chemical weathering has rounded off the summit of this hill in central Ceylon. Fragmental debris soon becomes weathered, and vegetation maintains the typical steep slopes. A tea plantation is at lower left. [Van Riper]

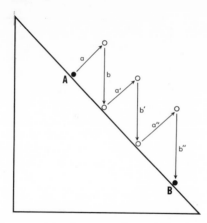

Figure 9-14 Movement of material down-slope by alternate freezing and thawing.
A pebble at position *A* is moved along path *a* by freezing. This is at right angles to the slope and represents the path of least resistance against the expansion of freezing water in the soil. Upon thawing, gravity moves it along path *b*. Alternate freezing and thawing thus could shift the pebble downslope to position *B*.

process is almost continuous. The role of permafrost is mainly one of preventing the downward percolation of groundwater, thus concentrating surplus water within a narrow surface zone. Under almost continuous saturation by water, the fine-textured topsoils on slopes frequently are susceptible to plastic or even to liquid flowage. *Solifluction* (soil flowage) is a term given to this unique type of mass wastage, found on slopes underlain by permafrost. Some of the other effects of permafrost on soil deformations will be discussed in Chapters 11 and 16.

9-9 THE WORK OF THE WIND Moving air contains kinetic energy that is capable of smoothing the surface in ways that resemble those of water, but, being much less dense and buoyant, its capacity for transporting material is far less. As with water, the erosive work of wind consists of prying and lifting by compression within cracks and crevices, and abrasion by rock fragments being carried or rolled along the surface. The former is far less powerful than the hydraulic compression of water, but it suffices to remove small particles from the fragmental mantle. Generally the maximum size that can be moved by wind is about 1 millimeter in diameter. There is also a lower limit of about .01 millimeter below which molecular adhesion tends to hold particles together and to prevent their being removed by wind. The process of removing fine particles from their resting places at the surface is termed *deflation*. The factors favoring wind deflation and abrasion are clearly related to dry land environments, as indicated in Figure 9-15. The particle load to be removed must be fine but not too fine, a condition that is fulfilled by the processes of mechanical weathering that prevail in dry regions. The material must be dry, because the cohesiveness of water molecules helps bind small particles together. Water also decreases the size of pore openings in the soil for air compressibility. Dry environments also tend to be windy because of the prevalence of convectional air currents and reduced friction, owing to the sparseness of vegetation. The vertical components of convectional air currents resulting from steep lapse rates near the ground also help to lift small particles of silt well above the surface, where stronger wind velocities may carry them hundreds of miles from their place of origin.

Particles are moved by being carried in suspension, by *saltation* (jumping), and by being rolled or slid along the surface. Wind abrasion is not a particularly effective sculpturing process, mainly because of the low-mass limitation of particles that are being carried. Fine sand is the most important textural class for wind abrasion, but these particles rarely are carried more than a foot or two above the surface under normal wind conditions. The principal work of wind abrasion is to smooth the surface by sandblasting and by the removal of minor surface irregularities. Wind deflation, on the other hand, is capable of removing a large volume of fragmental material from dry surfaces.

Material carried by wind, as by water, is deposited when the carrying capacity is reduced by a drop in velocity. The load carried by wind rarely is great enough to absorb the supply of kinetic energy and thus cause deposition. Since wind velocity is not primarily a function of surface slope as in water velocity, wind deposition is much less related to surface configuration. Deserts are the major supply points for wind deflation. Hence, depositional areas occur mainly along the desert margins or in the semiarid to subhumid climates in the direction toward which the prevailing winds are moving and where the velocities normally decrease. The deposits consist of two types: *sand dunes* and *loess* deposits. Loess is a fine-textured, windblown silt with particles ranging from about .02 to .07 millimeters in diameter. Large areas of loess deposits are found in various parts of the world, some of them far from dry lands (see Figure 10-12).

The work of the wind is not exclusively restricted to dry lands. Sand dunes are commonly found along most sandy sea coasts. Storm waves here toss beach sand into ridges sufficiently high to be dried and removed later by wind action before vegetation has had an opportunity to stabilize the sand. Glacial debris is particularly susceptible to wind deflation in front of glacial ice because of the large amount of finely pulverized rock material that results from glacial abrasion.

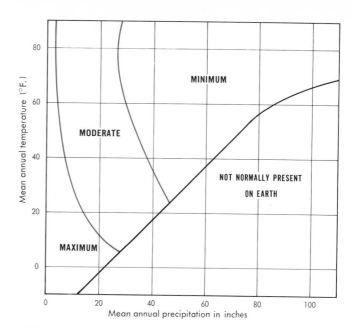

Figure 9-15 Relationship between wind erosion and climate. [After Wilson]

Figure 9-16 Exposure of loess along a road cut near Vicksburg, Mississippi. Note the columnar structure in this material. Initials carved in the vertical face of the loess columns may remain for years (see lower right); yet where a little water trickles over the side of the cliff, the loess erodes easily. [U.S. Geological Survey]

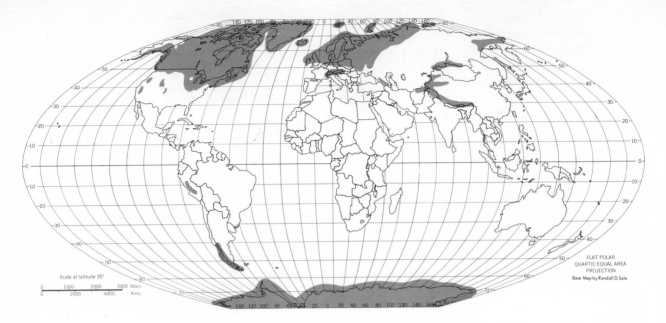

The principal work of wind deflation and erosion—the work that tends to smooth the surface and to create fewer obstacles to air movement—for the most part is slow and is found mainly in arid and semiarid regions, not favored places for human settlement except where water is available. Man's continued attempts to cultivate soils in these areas frequently have stimulated the work of wind by destroying the natural vegetation cover. The "Dust Bowl" conditions of the 1930s were a frightening experience to the marginal farmers in the semiarid western plains of the United States. Techniques of soil stabilization have improved since then, but to a farmer who has seen his entire topsoil removed during the course of a 2- or 3-day dust storm—or to another, whose fertile fields have been buried beneath a foot or two of fine sand—the work of the wind can be catastrophic.

9-10 THE WORK OF GLACIATION The present work done by glacial action is con-

fined to a small part of the earth surface, largely uninhabited and located in high latitudes or high altitudes. Features associated with glaciation, however, have been found in areas far from any present glaciation and are associated with dated sedimentary material that belongs to various periods of geologic time. A cyclical series of alternating warm and cold periods, with average annual temperature fluctuations of about 18 to 25°F, apparently characterized the major glacial periods of the Cambrian, Permian, and Pleistocene (see Table 2-2). The last of these periods, the Pleistocene, began around 1.6 to 2 million years ago and featured the spread of continental ice masses a mile or more in thickness over a large part of North America and Eurasia (see Figure 9-17). During this period, several major and minor advances and retreats of ice took place, with ice movements originating at several centers. The earliest clear evidence of Pleistocene glaciation in North America is dated at only 500,000 years ago, but European evidence in the Alps

Figure 9-17 Maximum extent of Pleistocene glaciation. Not all of the northern part of North America and Eurasia were covered with ice. Much of central Alaska and Siberia never were glaciated because of insufficient precipitation. [After Lobeck]

indicates cool periods comparable to those of later glacial advances extending almost to 2 million years ago. Table 9-2 presents the sequence of major glacial fluctuations.

The last major ice advance, which exceeded previous Pleistocene limits in most places, reached its maximum extent approximately 18,000 years ago and covered an area equivalent to about 30 percent of the total land surface of the earth. The total amount of ice involved has been estimated at about 19 million cubic miles, or an amount of water sufficient to lower sea level by about 600 feet. This much ice caused many side effects, such as the isostatic deformation of the crust, a significant alteration in the global energy balance (mainly through an increase in the albedo), and widespread alteration of global climatic patterns. Current theory suggests an increase in precipitation in the desert areas of the world and along the ice margins and a significant decrease in the humid tropics. There was much less solar energy available for evaporation over the warm, tropical oceans. Shifts in the center of earth gravity and a wobbling of rotation with consequent changes in the arrangement of material beneath the mantle also have been proposed.

The melting of such large masses of ice requires enormous drains from the earth energy budget, and readjustments in the total energy flow must be spread over a long period of time. Undoubtedly there is a significant lag of thousands of years between the removal of the long-range causes of glaciation and the establishment of a normal ice-free world. The rate of removal also will decrease as the ice makes its last stand in the polar regions. This is why the Antarctic and Greenland ice sheets still exist, and why it is believed that they existed during the interglacial periods. Time is still much too short to indicate if we are in the early stages of an interglacial period or if the entire glacial

Table 9-2 Tentative Dating and Correlation of the Quaternary Glacial and Interglacial Stages in North America and the Alps

Dating (before present)	North America	Alps
0–67,000	Wisconsin-Iowan Glacial	Wurm Glacial
67,000–128,000	Sangamon Interglacial	Uzuach Interglacial
128,000–180,000	Illinoian Glacial	Riss I and II Glacials
180,000–230,000	Yarmouth Interglacial	Hotting Interglacial
230,000–300,000	Kansan Glacial	Mindel I and II Glacials
300,000–330,000	Aftonian Interglacial	—
330,000–470,000	Nebraskan Glacial	Gunz I and II Glacials
538,000–548,000	—	Donau I Glacial
585,000–600,000	—	Donau II Glacial
600,000–2,000,000	—	Villafranchian Preglacial

SOURCE: Fairbridge, Rhodes W., *The Encyclopedia of Geomorphology*, Reinhold Book Corporation, New York, 1968, p. 923.

period is waning. The estimated duration of previous glaciations during the Cambrian and Permian periods and the length of the previous Pleistocene interglacial stages suggest the former.

Although continental glaciation produces many side effects, including crustal warping, it modifies the face of the earth mainly by erosion and deposition. Its work continues in high mountain regions throughout the world today and is the predominant agency for removing rock material from above the snow line, a line that corresponds rather closely to the average temperature of 50° for the warmest month. The erosional and depositional processes of glaciation will be treated in the next two subsections.

The erosive work of glacial ice. Moving masses of ice, nourished by ice fields that accumulate more frozen water than melts, have potential energy that varies with mass and elevation, similar to that of water at high elevations. This energy is transformed into kinetic energy in three major ways by which surface irregularities are removed: plucking, abrasion, and transportation. Other losses, such as by friction or by evaporation and melting also take place. *Plucking* occurs when ice is forced by compression into cracks by the weight of overlying material, widening them and prying blocks of rock loose. This is somewhat similar to the hydraulic prying action of streams. Glacial melt-water, flowing into rock crevices during warm days and later freezing and being incorporated into the bottom or sides of the glacial mass prior to movement, can also aid in plucking.

Like wind or running water, glacial ice itself can do little abrasive work on rock surfaces. But equipped with rock fragments frozen into the bottom of the ice, it may act as a gigantic rasp, gouging deeply into soft

sedimentary rocks such as shale or limestone, and even grinding away at extremely hard crystalline rocks. Unlike running water, which performs most of its abrasion by using relatively small rock tools rolled along the stream bed, glacial ice may carry huge boulders within its mass to be used as powerful tools. Where the material clogging the bottom of the ice mass is fine-textured, the abrasive work consists largely of rock polishing. Some of the eroded hard-rock surfaces of the Canadian and Scandinavian uplands exhibit highly polished surfaces. Despite the great amount of kinetic energy applied by the weight of ice and its forward progression, it is doubted that the continental ice sheets did much to lower general upland surfaces. Even in the northern Appalachians, where some of the best conditions for glacial abrasion can be found, glacial erosion was largely confined to deepening and widening valleys that roughly paralleled the direction of ice flow. The bottom of Seneca Lake, one of the scenic Finger Lakes in central New York State, has been cut to a depth of 200 feet below sea level, due to unusually soft shales in the valley, but adjacent hills still rise several hundred feet above it. In many places there is evidence of glacial overriding of older glacial deposits, and interglacial soils complete with humus layers have been discovered, buried beneath the debris of later glacial advances. This indicates that continental glacial abrasion at times can be negligible. Overloading of ice with rock material probably is responsible for such weak abrasion. This is not unlike the condition of a stream at grade.

Probably the most effective work accomplished by continental glacial erosion was the transportation of the regolith or mantle of weathered soil and rock fragments that existed prior to glaciation. Preglacial soils covered by debris are exceedingly uncommon in the areas covered by the Pleistocene ice sheets, and the total amount of debris removed must have been enormous. Not all of this material was pushed ahead like snow in front of a plow. Glacial ice, being a plastic solid, develops internal turbulent flow patterns induced by surface irregularities. These movements shift glacial debris throughout the ice mass. The incorporation of material originating from mass wastage along the flanks of a valley glacier into the center can be observed in Figure 9-18. The transportation of material in a continental ice sheet is largely independent of the underlying terrain, but always is in the general direction of ice movement. This movement generally radiates from the major centers of origin, but near the outer edges of the ice mass it tends to become more irregular, due to the lobate nature of the outer ice margins. No simple analogy can be drawn between the transportation of rock material by ice and by water. To illustrate, whereas the nourishment of energy in a stream is derived from potential energy mainly in the headwater areas the principal nourishment in a continental ice sheet is likely to be by the accumulation of snow relatively near the ice margins. The relative independence of ice movement with respect to the slope of the land surface beneath is another major difference. Mountain and valley glaciers, on the other hand, are miniature energy subsystems that operate much more like river systems.

Deposition by glaciers. Glaciers deposit their load, like streams, when their kinetic energy becomes insufficient to transport the load. In most cases this is because melting exceeds the supply of fresh ice derived from the snow fields. As indicated earlier, however, deposition may also take place along with forward progression when the ice becomes overloaded with debris.

Figure 9-18 A view down the Tasman Glacier, South Island, New Zealand. Note how the fragmental debris from frost weathering accumulates along the sides of the glacier. Rock debris is so abundant near the terminus of the glacier that it hides the ice surface. Active mountain glaciers such as this have an abundance of tools for abrasive use. [Van Riper]

Several unique features may be noted concerning glacial deposition. First, the deposition is not related to the slope or elevation of the land surface. This is especially true of continental ice sheets, where the ice mass may be a mile or more in thickness. Some of the depositional plains in central Illinois and Indiana completely mask an underlying terrain having a local relief of 100 to 200 feet. In the rougher terrain to the east, in the northern Appalachians, the glacial debris is likely to be thicker in the valleys than on the uplands. Second, glacial deposits that result from material dropped from suspension in ice are unassorted and unstratified. Large glacial boulders may be mixed amid extremely fine clays or coarse sand. This material, termed glacial *till,* is likely to have a wide textural range. Third, most till deposits, even those associated with continental ice sheets, show a close correlation in composition and texture to that of the underlying rock. Owing to the lobate feature of continental ice sheets, in which the outer edge is in a series of lobes or loops, especially in rough terrain, portions of the ice may become broken away from the main mass during the retreat of the ice front and stagnate. A large part of the depositional material in this case will be water-worked and sorted as to size. Lastly, glacial deposition always involves extensive supplemental transportation and deposition by running water. Regardless of their location, glaciers always have a melting outer edge, and glacial melt-water is an important adjunct to glacial deposition.

Unless stagnated, a condition that is localized, an ice front appears to retreat because the rate of melting exceeds its forward progression. The ice continues to move forward during the melting process, although the front of the ice may be stationary, retreat, or slowly press forward, depending on the ratio between melting and ice movement. When an ice margin remains constant for a long period of time, indicating a close balance between melting and movement, a large amount of material may accumulate at the ice margins, forming thick deposits of glacial till termed *moraines*. More detail of these forms will be presented in Chapter 12.

9-11 WAVES AND CURRENTS Surface water in the ocean and in lakes has energy imparted to it by wind friction. This may assume the form of surface waves or below-surface currents. Tides also may result in water motion where the ocean surface deformation is channeled by bottom configuration into frictional and mechanical energy. Waves along shorelines cause erosion mainly by chemical weathering, hydraulic compression, and abrasion. Seawater is an effective corrosive agent, and the constant saturation of rocks within the reach of tides or waves tends to etch out the less resistant rock minerals. Cracks and fissures widened by chemical action provide lines of attack for concentrated hydraulic compression during the pounding of storm waves and the resultant loosening and removal of material.

The erosive work of waves is concentrated within *surf depth*, or the depth within which surf breakers tend to form. This depth varies with the length and height of storm waves. Usually it is less than 40 feet in the case of sea coasts and from 3 to 10 feet in inland lakes. Under exceptional conditions, surf action may begin to be effective in 60 to 70 feet of water. Within surf depth, the work of waves tends to produce an equilibrium slope along the bottom, in which the kinetic energy of the waves can be transformed gradually into friction (see Figure 9-19). If the slope is too steep, erosion by hydraulic compression, abrasion by stones shifted in the breaker zone, and the movement of fine material out from shore will adjust the bottom slope to grade. Both erosion and deposition work to develop this equilibrium curve. Where deep water lies close to shore, and where the full kinetic force of storm waves can be directed directly against shore cliffs, the erosive action is powerful and rapid, especially where the material is unconsolidated or highly fractured. As in the case of stream erosion, most of the erosive work is performed during short periods of exceptional activity during storm periods. Again the association of denudation with the global energy balance is seen. Local imbalances of energy are dissipated in a variety of ways.

The active zone of coastal wave erosion, since it must be contained within about 40 feet of water, is not likely to be wide along any shoreline unless there are successive changes in sea level. It was commonly believed for many years that the continental shelf zone, or the gradual slope of the continental masses to depths of about 600 feet, was the result of a fairly stable balance between marine erosion and sedimentation, a kind of continental curve of equilibrium. Now, however, it is believed that benches and terraces produced by erosion or sedimentation and associated with a particular shoreline position are likely to be narrow and localized.

Successive changes of sea level caused by *eustatic*[7] fluctuations may result in a series of separate wave-cut and wave-built terraces. The volume of water in the oceans may fluctuate widely, as, for example, during periods of continental glacial advances and retreats. The erosive work of waves during such times is likely to be exceedingly active throughout the world. The remnants of old shorelines, produced during the Pleistocene period, with their characteristic forms resulting from marine erosion and deposition, can be found widely distributed around the world, both above and below sea level. Many of them can be traced for thousands of miles and are much too continuous to be caused by tectonic forces.

Ocean currents in deep water do little to transport material, but wind-driven waves and tides moving obliquely along a coast may cause a strong flow of water parallel to the shore that can distribute detrital sediments for hundreds of miles and help smooth both the coastline and the equilibrium slope within surf depth. Material brought to the sea at the ends of deltas may be distributed along the coast in the same way. Beyond surf depth, however, while some wave motion still can take place, it generally is too weak to move more than extremely fine material. Sands, gravels, and cobbles belong largely to the active surf zone. Fine silts and clays, on the

---

[7]*Eustatic* changes in sea level are those caused by changes in the total volume of ocean water, in contrast to local deformations of sea level caused by tectonic processes.

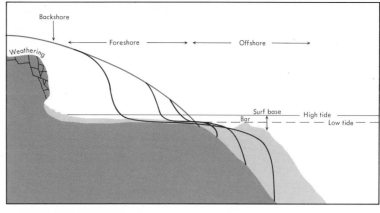

Figure 9-19 Development of equilibrium curves in the erosion of a cliffed shoreline. Beveling a steep shoreline involves subaerial weathering, as well as wave abrasion and deposition. In the early stages, much of the energy is expended in hydraulic impact directly against the cliff. In later stages, much of the wave energy is used in surf action, gradually moving detrital material back and forth along the wave-cut terrace, abrading it and shifting the finer particles out to the wave-built terrace. Weathering becomes more and more important in adding debris to the back shore in the later stages. The splash of storm waves helps in the chemical weathering process.

other hand, are distributed over much greater depths.

The predominance of fine-textured sediments in deep water produces some interesting features connected with submarine erosion. The edges of the continental shelf, for example, frequently are indented by deep canyons, usually opposite the mouths of rivers (see Figure 9-20). These also usually terminate on small, flat plains on the ocean floor. In profile and in size, submarine gorges resemble the great canyons that entrench the

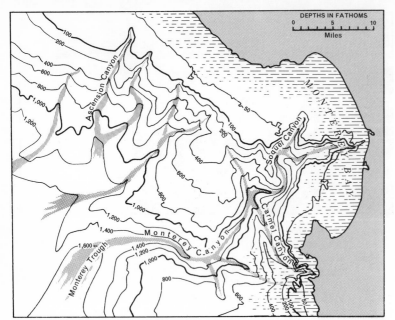

Figure 9-20 **Bathymetric map of the submarine canyons off the coast of California.** Such canyons are found in several places along the outer margin of the continental shelf bordering the continents. [After Sverdrup]

high plains of continents, such as the Grand Canyon of Arizona. Such characteristics led investigators to believe at first that the canyons resulted from stream erosion at a time when the entire continental slope was above water. There is little evidence to support such an extreme shift in sea level. The most accepted hypothesis today is that they are the result of scouring by *turbidity currents,* a type of submarine mass-wastage consisting of sudden avalanchelike shifts of silty material that is brought to the edge of the continental slope by currents and later shaken loose down the steep, outer slope of the continental edge by earth tremors. Hundreds of miles long in places, and often over a mile deep, the submarine gorges are major gashes on the continental flanks. Knowing the location of these canyons is important in choosing suitable routes for submarine cables. Mass slumping of fragmental material in the canyons off the Colombia coast in the Caribbean Sea broke the same cable seventeen times between 1930 and 1960.

Study Questions

1. What is the *geologic cycle?* Show how it operates as a subsystem for regulating the flow of solar energy across the face of the earth.
2. Indicate how volcanic activity might have been caused by a diversion of solar radiation input.
3. Explain and illustrate the principle of isostatic balance with respect to the earth crust. What kind of evidence is there for assuming that it has operated for billions of years?
4. Relate the speed of surface beveling to the distribution of the net radiation balance on the earth surface. What other variable besides the net radiation surplus is important in this overall process?
5. Name some variables other than temperature and precipitation that might be important in the distribution of the erosional processes. See Figures 9-4, 9-5, 9-10, 9-12, and 9-15.

6. Relate Figure 9-8 to the normal activities of a stream at different parts of its course. Why is there so little fine clay in stream alluvium?

7. Give and explain some exceptions to the normal rules that river volumes increase downstream, that velocities decrease downstream, and that the gradient of a stream is greatest near its source.

8. What is mass wastage, and what is its role in the formation of valleys? Give some evidences for its presence.

9. Explain the chart in Figure 9-12.

10. Examine Table 9-2 and relate the global temperature fluctuations to the operation of the earth-atmosphere energy system, with its multiple controls for regulating changes in the input-output ratio. Why should multiple climatic changes be expected rather than a single fluctuation in global temperatures?

11. Discuss the similarities and differences between the process of erosion by running water and that of mountain glaciation.

12. Explain this statement: "Ocean waves and currents cannot cut deeply into continents without successive changes in sea level." Would you expect the Pleistocene period to have been a period of exceptional coastal erosion and deposition? Why?

13. Chemical weathering always is involved in the erosion of coastlines by waves and currents. Explain why and where.

CHAPTER 10

MATERIALS OF THE EARTH CRUST AND THE RELATIVITY OF TIME

The varied nature of the lithosphere, either solid rock or fragmental mantle, is apparent almost everywhere. The previous chapter emphasized the processes by which energy alters the surface through denudation and deposition. Now the physical and chemical nature of the lithosphere will be examined to see how it varies from place to place and from time to time, and to discover what role it plays in the utilization or transformation of energy. First, the general composition of the outer crust will be described. Then the major rocks and materials of the continental masses will be classified and discussed as to their general adaptability to modification by weathering and erosion. A few important conclusions summarized in advance should be kept in mind:

1. The composition of the continental rocks remains remarkably constant despite eons of weathering, erosion, and tectonic disturbances.

2. Relatively few chemical elements and a few closely related mineral groups make up the major portion of the continental rocks and rock materials.

3. There is a movement of continental material from rocks to sediments and back again into rocks within the geologic cycle that is roughly comparable to the movement of water within the hydrologic cycle.

4. The transformation of rock material from one form to another involves the transfer of energy and constitutes a subsystem of the entire global energy system.

Sima and Sial

10-1 THE GLOBAL ARRANGEMENT OF ROCK MATERIAL Figure 2-7 reveals that the material within the earth is arranged in a series of density zones, with the lightest rock materials located in the outermost layer, or *crust*. The boundaries between the zones vary somewhat in sharpness, but some of them are remarkably well defined. They have been plotted by noting variations in the speed of earthquake shock waves as the latter pass through the earth. Our focus is upon the surface of the earth. Hence, we shall be mainly concerned with the crust. The latter generally is considered to be that portion lying above a pronounced and sharp boundary known as the *Mohorovicic Discontinuity Line,* or more simply as the *Moho.* The speed of earthquake waves rises suddenly from about 260 to about 300 miles per minute immediately below the Moho, suggesting either a sudden change in composition or some phase change in the rock material that produces a sudden increase in density. The depth of the Moho below the surface varies, but it tends to be deepest (about 30 miles) beneath the stable central portions of the continents and shallowest beneath the ocean floors, where it reaches within 5 miles of the surface. Beneath the continents, it tends to be somewhat higher under areas of mountains and high plains and lowest beneath the main continental platforms and plains (see Figure 10-1).

There is another discontinuity zone lying above the Moho, or within the crust, that separates two zones of different densities. It is found, however, only beneath the continental masses and is much less clearly defined than the Moho. The material in the upper, or lighter-density, material is termed *sial* (*si* = silica; *al* = aluminum), and the denser material below is termed *sima* (*si* = silica; *ma* = magnesium). The sima is continuous around the earth and lies immediately above the Moho. The arrangement of these two zones, the Moho, and the plastic mantle below is shown diagrammatically in Figure 10-1. It indicates clearly that the sialic rock materials form blocks of lighter material supported upon a denser sima. The diagram also indicates that a much larger amount of the sial lies below the general level of the upper sima surface than above it, much as an iceberg shows only about a tenth of its mass above the ocean in which it floats.

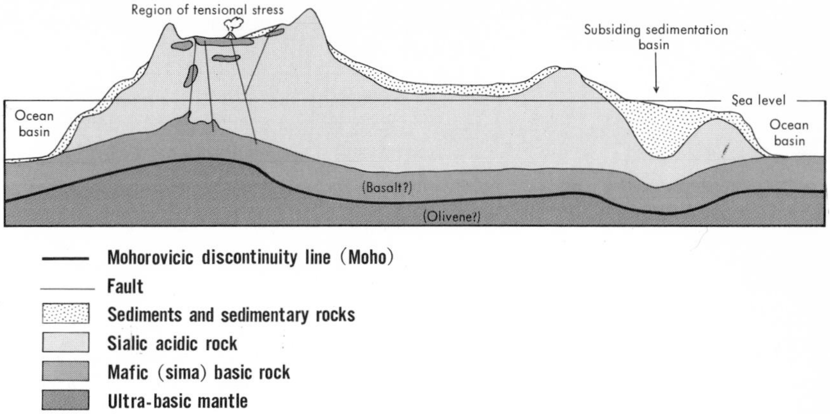
Region of tensional stress
Subsiding sedimentation basin
Sea level
Ocean basin
Ocean basin
(Basalt?)
(Olivene?)
Mohorovicic discontinuity line (Moho)
Fault
Sediments and sedimentary rocks
Sialic acidic rock
Mafic (sima) basic rock
Ultra-basic mantle

Figure 10-1 Schematic representation of the earth crustal zones. Note that the continental rocks (*sial*) appear to be floating on a denser substratum, the *sima.* Relate this diagram to that in Figure 9-1.

The continental, sialic rocks as a group are relatively lightweight, with a specific gravity of about 2.65. Their colors are predominately pink, light gray, and white. Their constituent minerals have a high proportion of silicon, oxygen, aluminum, potassium, and sodium. The sima rocks, on the other hand, are dark, with shades ranging from gray, green, to black. Their specific gravities are slightly higher (2.67–2.7), and their minerals show a lower content of silicon than the sialic rocks. The bases of magnesium, iron, and calcium substitute for the potassium and sodium of the sial. Aluminum and oxygen appear to be important elements in both types of rocks. Table 10-1 presents the proportional distribution of elements within a varied assortment of rock samples taken from different continental locations. Presumably most of the samples used were sialic rocks. Despite the fact that over 1,500 different minerals have

Table 10-1 Composition of the Earth Crust

Element	Percentage
oxygen	47.3
silicon	27.7
aluminum	7.9
iron	4.5
calcium	3.5
sodium	2.5
potassium	2.5
magnesium	2.2

SOURCE: F. W. Clarke, "The Data of Geochemistry," *U.S. Geological Survey Bulletin 770*, 1924.

been identified in crustal rocks, only a few elements make up the mass of rock material, as indicated in Table 10-1. Further, a quantitative measurement of some 700 igneous rock samples taken from various locations showed that most of the minerals in the rocks that crystallized from a molten state, and that constitute the bulk of both sial and sima, comprise a few closely related mineral families, as shown in Table 10-2. The mineral families indicated in Table 10-2 represent a grouping of minerals that are closely similar in chemical composition, both within and between the families. All of the first four, for

Table 10-2 Mean Mineralogical Composition of 700 Igneous Rock Samples

Mineral family	Percentage
feldspars	59.5
hornblende and pyroxene	16.8
quartz	12.0
micas	3.8
all others	7.9
	100.0

SOURCE: F. W. Clarke, "The Data of Geochemistry," *U.S. Geological Survey Bulletin 770*, 1924.

example, or about 92 percent of the total, are silicates; that is, they possess a basic silicon-oxygen combination within their crystal structures. Similarly, the first two and the fourth are aluminum silicates. The quartz family includes various forms of *silica*.[1]

The predominance of the four leading elements in the above mineral families and in all of the rocks of the earth crust (see Table 10-1) apparently is related to a commonality that exists throughout the universe. All of them were present in the rock samples returned from the first trip to the moon, and many meteorites reaching the earth from outer space are aluminum silicates. There is a group of meteorites, however, that are metallic and composed of pure iron, or combinations of metallic iron and nickel. The latter is why the dense interior of the earth is believed to be made up of an iron-nickel mixture.

It should not be surprising that the most common elements at the exterior of the earth are oxygen, silicon, and aluminum. The first is the most active chemically of all of the gases in the lower part of the atmosphere. The next two elements are relatively light-weight materials that in some way were segregated into the outer crust at an early period in earth history. Iron lies in the top four because it is a common element in space for some reason. Although much denser than any of the other elements listed in the table, some of it finds its way into combinations with other elements, especially within the sima rocks.

The second group of four elements listed in Table 10-1, including calcium, sodium,

[1]The term *silica* should not be confused with *silicon*. The latter is a chemical element with a definite atomic structure. *Silica* is a general term for the common molecule SiO_2, or silicon dioxide. The latter may take many different forms, including the pure crystalline *quartz*, or different amorphous forms such as *agate, flint, chert, jasper,* and *chalcedony*.

potassium, and magnesium, all are light-weight metals that are extremely active chemically and tend to form *salts* upon exposure to acids. Collectively they are referred to as *bases,* and rocks that are rich in them are termed *basic rocks.* Since silicon is chemically active in producing silicic acid, rocks that have high percentages of silicon are termed *acidic rocks.* The four common bases are exceedingly interchangeable, being easily transferred from one chemical compound to another, especially when they are exposed to different types of solutions. Most soil solutions on earth contain base ions, and the oceans constitute an important reservoir for their accumulation. Most of the difference between the chemical composition of the common rocks and minerals of the lithosphere is related to the varying proportions of these bases. An additional feature is that their atomic weight is related to the kind of rocks in which they are found. Potassium and sodium, for example, are smaller and lighter elements than calcium and magnesium. The former thus are more common in sialic minerals, whereas the latter appear more frequently in the darker, heavier, sima minerals. The group of bases plays an important role in the global energy exchange process, but within the subsystem of the geologic cycle. A primary feature of the chemical weathering of igneous rocks (rocks that once were molten) is the disassociation of bases from the aluminum-silicate crystal lattices. The role of calcium, the most abundant base element in the group, in the formation of calcium carbonate in the oceans and in the chemical weathering on land will be discussed more fully under the section on sedimentary rocks (see "Limestones," p. 365).

The chemical composition of the great mass of crustal rocks, then, differs mainly in the arrangement of the leading constituent elements. The same elements and minerals appear over and over again. This relative homogeneity is not obvious at the surface, which usually is masked by detrital material resulting from weathering, erosion, or deposition. Furthermore, rocks closely related chemically may differ widely in form and appearance. There is no great difference in chemical composition between granite and ordinary clay or shale. There is, however, a notable difference in the physical arrangement of the elements in each. The relative homogeneity of the sial is best found in the ancient eroded roots of old mountains or in places where the unweathered sial is exposed over wide areas. A motor trip around the northern end of Lake Huron or Lake Superior will reveal rock outcrops in sialic granites for hundreds of miles. The pink mass of *orthoclase* (potash) feldspar, interspersed with gray, glassy quartz, is seen in nearly every road cut, occasionally intersected by veins of darker rocks, representing intrusions of the dark sima rocks from below.

Weathering, erosion, deposition, and crustal deformation all are ways through which the same chemical elements are rearranged in somewhat different proportions and in different physical forms. In the subsequent sections in this part of the chapter, the principal materials that make up the earth crust will be described, along with their general role in the geologic cycle. They are divided into four main groups: igneous, sedimentary, and metamorphic rocks, and unconsolidated or fragmental rock material.

10-2 IGNEOUS ROCKS *Igneous* rocks are those whose minerals have crystallized from a liquid or near-liquid state. They form the bulk of both sial and sima. The classification used here and shown in Table 10-3 is based on the proportion of sial or sima material and on the rate and size of crystal growth during the crystallization process. Continental rocks

Table 10-3 Classification of Igneous Rocks

Textural class	Sialic (acidic)	Intermediate	Sima or mafic (basic)
	Maximum ← ———— Silica content ———— → Minimum		
	Minimum ← ———— Base content ———— → Maximum		
	Potassium Sodium	Calcium	Magnesium Iron
	Light ← ———— Color ———— → Dark		
	Low ← ———— Melting point ———— → High		
	Maximum ← ———— Feldspar content ———— → Min.		
	Minimum ← — Content of hornblende, olivene, pyroxene, and amphiboles — → Maximum		
Granitic	GRANITE (with quartz) SYENITE (without quartz)	DIORITE	GABBRO; AUGITE; PERIDOTITE
Aphanitic	RHYOLITE (with quartz) TRACHYTE (without quartz)	ANDESITE	DIABASE BASALT
Vesicular (with gas holes)	PUMICE		SCORIA

are not exclusively sialic in composition. Tensional stresses at times result in deep fissures that weaken the compressive bonds within the sima, give it fluidity, and provide lines of weakness for the dark, heavy sima material to work its way toward the surface in volcanic areas (see Figure 10-1).

The textural variation of the included minerals, or the vertical differentiation shown in Table 10-3, is based on the conditions prevailing during the time of crystal growth. Under some conditions of chemical composition and rapid cooling, molten lava may solidify into an amorphous, glassy rock such as *obsidian,* a kind of volcanic glass. This commonly is dark green to black, because only a small amount of ferromagnesium material is needed to color the entire mass. Igneous rocks that have cooled rapidly at or near the surface with corresponding small crystals are sometimes referred to as *extrusive* rocks. Where the cooling is much slower, as under conditions of high pressure far

below the surface, the mineral crystals grow slowly to a much larger size to form a coarsely granular or *granitic*[2] texture. Such rocks are termed *intrusive,* in contrast to the extrusive surface lavas. An even, fine-textured, but visibly crystalline texture is termed *aphanitic.* A mixture of large crystals set in a matrix of finer ones is termed a *porphyry.* A *pegmatite* is an extremely coarse-textured intrusive rock in which the mineral crystals may range from an inch in diameter to several feet. Terms combining the mineralogical and textural class frequently are used to designate a particular rock, such as a *diorite porphyry* or a *granite pegmatite.* The distinction in Table 10-3 between gabbro, augite, and peridotite is based largely on the proportion of specific ferromagnesium minerals. *Gabbro* is a general mixture of the more basic feldspars (calcium), dark micas, horn-

[2]The term *granitic* does not imply any chemical relationship with the rock *granite.* Instead, it signifies a granitelike or coarsely crystalline texture.

blende, and pyroxene. *Augite* consists largely of the dark ferromagnesium silicate pyroxene, and *peridotite* is composed principally of representatives of the pyroxene and amphibole families without any of the basic feldspars. *Olivene,* a green, glassy mineral, is the heaviest and most basic of all the ferromagnesium minerals. It has a specific gravity of 3.3, which is well above that of the more common rock minerals in both the sial and sima. It is a common constituent of the stony meteorites, and some geologists believe that it is the main constituent of the dense material within the plastic mantle that lies below the Moho.

Granites, syenites, and diorites make up the bulk of the continental, sialic, intrusive rocks. Large masses of basic intrusives, however, are sometimes found that possibly represent old magmatic reservoirs, where some of the sima worked its way toward the surface and later was exposed by erosion. The large nickel deposits of Sudbury in Ontario, Canada, are associated with such a mass. The extrusive, or volcanic, rocks include both basic and acidic types. The more liquid types that poured out onto the surface and spread into broad sheets are likely to be composed of basalt or diabase. The latter differs from basalt mainly in being slightly coarser in texture. The acidic extrusive rocks are found principally within the unstable parts of the surface, especially within the major mountain cordilleras, and are associated with the explosive types of volcanic eruptions.

The minerals found within igneous rocks generally are hard and resistant to abrasion. They are highly susceptible, however, to chemical weathering, because the feldspars and many of the ferromagnesium minerals have crystal structures that permit the relatively easy ingress of water molecules. Of the major families of minerals found within the igneous rocks of the sial and sima (see Table 10-2), only quartz and the micas resist chemical weathering. Quartz is resistant because its molecular crystal lattice is among the tightest and most stable found on the earth surface. The micas are chemically resistant because their crystal lattices already are hydrated (contain water molecules), and solvents find entry difficult.

The process of chemical weathering as it alters the feldspars and ferromagnesium silicates first involves hydration. This loosens the crystal lattice bonds and permits the solvents contained in soil solutions to remove the bases, leaving behind aluminum and iron silicates in various degrees of hydration. These are the clays that are found in soils. Although they bear little resemblance to the minerals and rocks from which they were derived, they differ chemically only in having a higher content of water molecules and a much lower content of bases. Chemical weathering in the tropics tends to further the decomposition of clays by removing some of the silica "building blocks" in the microscopic clay particles, leaving behind a residue composed mainly of oxides of aluminum and iron. It is likely that most of our commercial ores of aluminum and iron originated as the result of enrichment by weathering processes over millions of years. The large amount of energy that man must apply to separate the metallic aluminum and iron from their oxides is a reminder that such ore deposits may well represent stored chemical energy derived from the sun, similar in principle to the deposits of the fossil fuels petroleum, natural gas, and coal.

Figure 10-2 shows the distribution of surface materials that underlie soils throughout the world. The igneous rocks are included in two groups: Volcanic (extrusive) and nonvolcanic (intrusive). The map indicates an irregular distribution of these two groups. The intrusives comprise the main mass of

Figure 10-2 World distribution of surface materials.

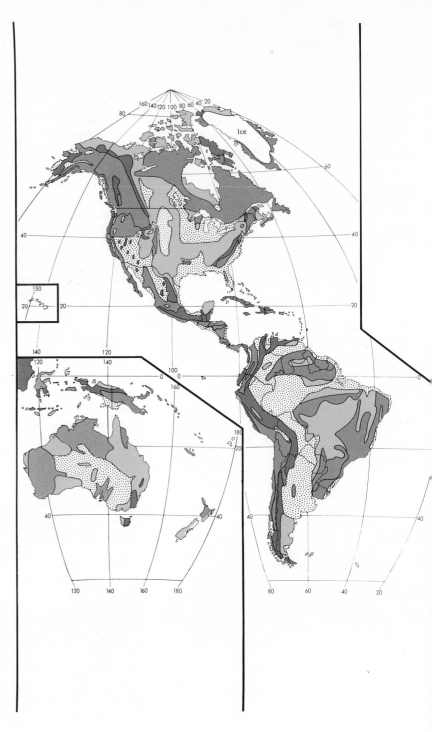

Unconsolidated sediments

Crystalline rocks (Non-volcanic)

Highly faulted or folded sedimentary rocks

Volcanic rocks

Flat or gently dipping sedimentary rocks

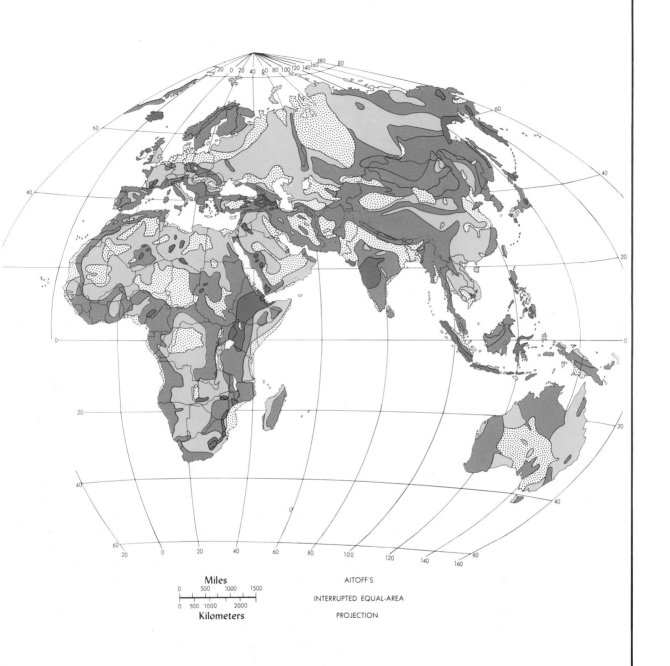

Miles

| 0 | 500 | 1000 | 1500 |

| 0 | 500 | 1000 | | 2000 | |

Kilometers

AITOFF'S

INTERRUPTED EQUAL-AREA

PROJECTION

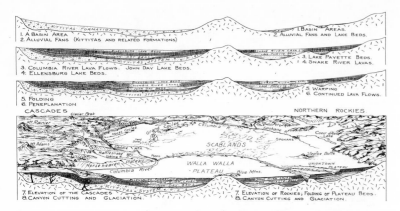

Figure 10-3 Stages in the development of the Columbia Plateau. These basaltic lavas, laid down during the Miocene, form individual sheets 10 to 200+ feet thick. Total accumulations are more than 4,000 feet in places. Not all of the volcanic material was fluid. Thin layers of volcanic ash and coarser fragmental material were interspersed between some of the basaltic flows. [A. K. Lobeck, *Atlas of American Geology,* Sheet 75. Map courtesy of Hammond, Inc.]

exposed sialic rocks. These underlie all of the continental blocks but are mantled here and there by relatively thin deposits of sedimentary rocks or fragmental materials. The location of the volcanic or extrusive rocks is related closely to areas of crustal instability, and especially to areas that have been uplifted to high elevations relatively recently, but without having been buckled into great folds. In such areas, following crustal warping, relaxation of the compressional stresses and uplift often result in tensional stresses and deep fracturing, which in turn may lead to volcanic activity. Figure 10-1 indicates that the sima tends to bend upward in these upwarped areas. If the sima becomes involved in volcanic activity, the basaltic lavas that result are highly fluid and may cover wide areas at the surface. The Columbia Plateau of eastern Washington and Oregon, for example, is underlain by thousands of feet of

basalts that resulted from highly fluid basic lavas welling up to the surface and spreading outward in all directions (see Figure 10-3). Other such areas include Bolivia, Ethiopia, southern Brazil, northwestern Arabia, and west-central India. Not all extrusive flows are basic in composition, however. Thick flows of acidic rhyolites constitute high plains in the vicinity of Yellowstone National Park.

The intrusive, sialic igneous rocks generally are not among the most favorable surface materials for human use. Soils derived from granites tend to be sandy and to have a high content of quartz, a substance that has little chemical influence in fertility. The feldspars decompose into clays, but their base content is low and lacks much variety. This is important because several of the soluble bases are among the principal nutrients for plants. Granites and other intrusive rocks have been quarried for centuries for use as building and paving material. Although their hardness makes them difficult to cut and shape, it also makes the stonework durable except for surface weathering in the warm, moist climates. Their advantages for tombstones and other monument markers are well known. Associated metallic minerals, occurring either in veins within intrusive rocks or disseminated through large masses of intrusive rocks, provide many of the world's metallic resources.

The volcanic extrusives, which usually contain holes or gas pockets, do not have the durability of the intrusive igneous rocks, even outside the tropics; hence, they are seldom quarried except for use as crushed stone. Unlike the intrusive rocks, they are not likely to contain important metallic mineral veins, although some of our most precious gemstones, such as diamonds, rubies, and emeralds, are found in old volcanic rocks. The holes in ancient porous lavas have sometimes been filled with molten metals, as exemplified by the copper ores of Northern Michigan and the gold ores of South Africa. Sulfur is extracted from active or recently active volcanic craters. Sulfur dioxide is one of the common gases discharged during explosive eruptions. Italy and Japan both are self-sufficient in sulfur because of their position in volcanic areas.

The basaltic lava flows of the world, almost without exception, have a unique functional significance to man: they are the parent material for exceptionally fertile soils. There are two reasons for this. First, their vesicular structure, or porosity due to gas pockets, makes them decompose unusually rapidly into deep soils. Second, they tend to have unusually high proportions of the two bases calcium and magnesium, which act as soil conditioners, giving soils derived from basalts unusual capacities for retaining water and plant foods. The cotton of India and the coffee of Brazil owe much to the inherent fertility of soils weathered from basalt.

10-3 SEDIMENTARY ROCKS *Sedimentary rocks* are those that have resulted from the cementation, compaction, or hardening of depositional sediments. The last may include fragmental detritus such as sand or gravel, fine-textured materials such as clay or silt, or even chemical precipitates such as lime oozes or silica gels. For this reason, sedimentary rocks usually occur in distinct beds, or *strata*. The more common types of sedimentary rocks are shown in Table 10-4 and are briefly described in the following subsections. In the formation of sedimentary rocks, the cementing process usually involves the precipitation of chemical compounds from solutions contained in the openings between the grains of coarse-textured rocks. Among the most common cements are lime ($CaCO_3$), iron oxide (Fe_2O_3), and silica (SiO_2). The first is by far the most common and usually is associated with marine environ-

Table 10-4 Types of Sedimentary Rocks and Their Constituent Materials

Sedimentary rock	Corresponding sediment
conglomerates, breccias	gravels, pebbles, cobbles
sandstones	sands
shales, siltstones, graywackes	clays, muds, and silts
limestones	lime oozes, marine shells, chalk, marl, etc.
chert and flint	precipitated silica gels
bituminous coal	peat, lignite

ments. Compaction is brought about by the weight of overlying material, which may be enormous when the sediments in subsiding basins accumulate to thicknesses of several thousand feet. Hardening can result from dehydration, from chemical reactions, or from heat. The last can be encountered in sedimentary basins deep within the crust.

Conglomerates. These are a mixture of coarse and fine fragmental particles cemented together. The most common cemented sediment is gravel from river or beach deposits. Conglomerates composed of sharp, angular fragments are known as *breccias*. Their angular fragments may result from mechanical weathering, the shattering of a volcanic explosion, or rock fracture during earthquakes. Conglomerates are not usually widely distributed because the deposition of coarse fragmental material requires rather specialized and local conditions.

Sandstones. Sandstones are simply sand that has been turned into stone, principally by cementation. They are much more widely distributed than conglomerates, because sand is a more widely distributed sediment than gravel. The most extensive beds of sandstones are of marine origin, representing the progression of beaches and wave-built terraces as the marginal seas alternately invade or recede from the continental interiors. Most marine or alluvial sand has a high percentage of the durable mineral *quartz;* hence, this mineral is especially common in most sandstones. There are exceptions, such as the *arkose sandstones,* which are composed mainly of feldspar fragments resulting from mechanical weathering in arid regions. The properties of sandstones are largely derived from the natural cement that holds the grains together. *Calcareous (lime-cemented) sand-*

stones are by far the most widely distributed, lack strength, and tend to weather rapidly in warm, humid regions. *Ferruginous (iron-cemented) sandstones,* usually marked by their reddish color, and *siliceous (silica-cemented) sandstones* tend to resist weathering in moist climates, because their cementing materials are stable chemical compounds. These two types of sandstones are important cliff-makers in many parts of the world, and their strata stand out boldly when they are in association with other sedimentary rocks. The ferruginous and siliceous sandstones are prized for building stone because of their durability. Their distribution, however, is much more restricted than that of the calcareous types.

Figure 10-4 Black shale with associated joint planes near Utica, New York. Regional stresses tend to produce a geometric pattern of smooth fracture planes or *joints* in these soft rocks. Such shales crumble easily between the fingers and are easily eroded by streams bearing rock debris for tools. [Van Riper]

Shales and siltstones. Compaction and dehydration are the principal means of altering clays, muds, or silts into rock. *Shales* are compacted clay or mud, while silt becomes siltstone or graywacke. As the weight of overlying material becomes greater, the volume of the original sediments is greatly reduced. In contrast to granites, these rocks are highly resistant to chemical weathering but erode easily. Fine-textured shales frequently have vertical cracks or *joints* that are geometrically arranged with respect to regional crustal stresses. An example is shown in Figure 10-4. Although shales usually are grayish in color, some of them are red because of a high content of ferric oxide, some are green because of ferrous oxide, some are black owing to organic compounds with some sulfides, and those with a high lime content may be almost white. Shales often are used as a source of clay for pottery, brick, and tile manufacture. Unless protected by more durable rock strata above, shales erode quickly and tend to lie at the surface within lowlands and plains.

Limestones. Limestones are amorphous masses of calcium carbonate ($CaCO_3$) that have been turned to stone. Limestones are relatively soft compared with most rocks. They can easily be scratched with a knife, are usually gray in color, and effervesce with applications of acids. Lime, another term for calcium carbonate, is formed in two principal ways: first, by being precipitated from solution (mainly from seawater), and second, by being synthesized by marine life that extracts calcium ions from solution and unites them with carbon dioxide (CO_2) and oxygen to build lime shells or skeletons. Corals, shell fish, and some of the protozoa are among the most effective lime-secreting marine organisms. The calcium in all limestone probably was derived originally from basic feldspars and other lime-rich minerals within igneous crustal rocks, mainly of the sima types. Although limestones are found in sedimentary rocks throughout the world, they are being formed today mainly in relatively warm, clear, tropical and subtropical seas.

The role of limestone in the operation of the global energy balance is extremely important, and special attention will be given here to its role in the adjustment of the carbon dioxide balance. As was noted in Chapter 4, carbon dioxide is one of the principal warming agents in the atmosphere, and fluctuations in its content there influence the amount of radiation that is absorbed by the atmosphere, especially the long-wave radiation from the earth (see "Carbon dioxide," p. 104). This gas is the basic source of the carbonate radical in all carbonate rocks, including limestone. The quantities involved are prodigious. It has been estimated, for example, that only a little more than 2 cubic miles of pure limestone contains an amount of chemically bound CO_2 that is equivalent to that of the entire atmosphere. The oceans also contain dissolved CO_2 in amounts that average roughly sixty times that contained within the atmosphere.

There is a close balance maintained in ocean water between the amount of calcium and magnesium ions in solution and the amount of dissolved CO_2. (The latter usually is in the carbonate radical HCO_3, derived from carbonic acid.) Several factors related to the global energy balance influence this latter quantity. For example, the oceanic content of CO_2 can be reduced by an increase in temperature, an increase in evaporation, and an increase in the proportion of marine flora to fauna. Microfloral organisms such as algae frequently change rapidly in numbers. When CO_2 is released from solution, for example, following an increase in temperature, it unites with various hydroxyl (OH) ions and bases to form carbonate salts.[3] In the choice of bases to form carbonate salts, the least soluble salts are the first to be formed; hence, calcium and magnesium are especially important in this process. When CO_2 is increased by a reverse of the conditions mentioned above, the calcium and magnesium carbonates are again brought into solution via the bicarbonate form to maintain the proper ionic balance in the seawater.

Another factor that can influence the CO_2 content in the ocean is the relationship to the

[3]The chemical reactions in the processes of solution and precipitation of calcium carbonate are outlined as follows:

1. *Formation of solvent for chemical weathering (solution):* water (H_2O) + carbon dioxide (CO_2) = carbonic acid $[H_2^{+2}(CO_3)^{-2}]$

2. *Solution of calcium carbonate ($CaCO_3$):* carbonic acid (H_2CO_3) + lime ($CaCO_3$) = calcium bicarbonate $Ca^{+2} + [2(HCO_3)^{-1}]$

Note: calcium bicarbonate is soluble in water up to a given ionic concentration that is set by temperature and the percentage of other soluble ions. The concentration in seawater is close to the saturation point at all times.

3. *Precipitation of calcium carbonate from seawater:* calcium bicarbonate − CO_2 = calcium hydroxide $[Ca^{+2}2(OH)^{-1}] + CO_2$; calcium hydroxide + CO_2 = $CaCO_3 + H_2O$

Note: Heat is necessary to remove the CO_2 from the bicarbonate. This last process thus is reversible with changes in the heat budget.

amount of CO₂ in the atmosphere. An unusually high content of CO_2 will result in its
being absorbed directly by seawater at the
ocean surface (a partial-pressure function),
thus increasing the solubility of carbonates
and stimulating marine plant life. An unusually low content in the atmosphere will have
a reverse effect by withdrawing CO_2 from
seawater and thus inducing the precipitation
of limestones or other carbonates and checking marine plant life to maintain the ionic
balance in the sea.

Lime in any form thus is an important
storage mechanism for surplus CO_2 and calcium ions in the oceans. It might be thought
of as a bank where withdrawals or deposits
can be made at any time when needed. The
narrowness of the ionic balance in the oceans
has been verified by careful observers who
have noted that in the tropics, small amounts

of lime can be alternately precipitated and
dissolved in shallow water during the day
and night, respectively, associated with relatively minor temperature changes. Limestones and carbon dioxide help regulate the
major global energy balance in the same way,
but on a much larger scale.

The process of regulating the global energy
balance by means of oceanic CO_2 and lime
is traced in the following paragraphs. Figure
10-5 is helpful in tracing the complex interrelationships of carbon. The latter is contained within a wide variety of compounds
while in storage, but it is mobile mainly in
the form of the gas carbon dioxide or the
liquid carbonic or humic acids.

An abnormal increase in atmospheric CO_2
for some reason, such as an accelerated oxidation of hydrocarbons or a decrease in vegetation, can result in several adjustments.
Temperature and precipitation will increase,
and plant growth will be stimulated. The
increased rainfall will remove more CO_2 than
before from the atmosphere by direct solution, and the increased vegetation signifies
not only an increased assimilation of CO_2
through photosynthesis but also a greater
storage of carbon in various organic compounds. Chemical weathering also will increase because of the increase in temperature, rainfall, and the amount of carbonic and
humic acids added to soil solutions. The last
is aided by increased organic decay. The
increase in weathering will remove more
bases from rocks in the form of soluble
bicarbonates and carry them either to the sea
or into desert basins. Since the sea maintains
a narrow ionic balance, the added bicarbonates and the increase in heat energy will
result in the production of insoluble carbonates in the reservoir of carbonate rocks and
sediments, especially those of lime.

An abnormal decrease in atmospheric CO_2,
on the other hand, would result in reverse

Figure 10-5 The carbon cycle. The flow
of carbon from one position to another is exceedingly complex and varied in form, ranging
from sugars and starches to CO_2 and bicarbonates. The most common form in transit probably
is CO_2. The carbon cycle might be considered a
subsystem of the global energy flow. Its role with
oxygen in the biotic world is well known.

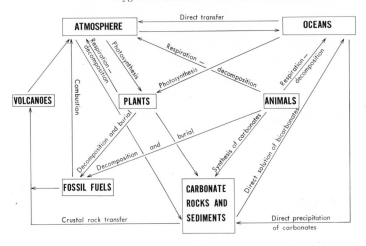

reactions. Decreases would take place in each of the following: temperatures, photosynthesis, vegetation volume, the rate of chemical weathering, and the amount of bicarbonates and carbonic and humic acids carried into the sea. The capacity of the sea to retain CO_2 in solution would increase because of the decreased temperature (note an analogy with a cold bottle of soda water), and the marine carbonate rocks and sediments would begin to decompose, adding base ions and CO_2 to the sea water. Part of the increased quantity of CO_2 in the seawater would be absorbed directly into the atmosphere, thus adding to the quantity in the air. In short, an atmospheric shortage of CO_2 would indirectly release both CO_2 and calcium ions from the marine reservoirs of lime material.

A logical question arises at this point. If the atmosphere now contains so little CO_2 relative to the quantity locked within the great strata of carbonate rocks and sediments, within the total content of life forms, and in the deposits of fossil fuels—from where was all of this CO_2 derived? Another question follows: If calcium and magnesium are relatively minor elements in sialic rocks, how did such concentrations take place in what now are among the more common sedimentary rocks?

One proposed hypothesis for both suggests that the earth atmosphere once had vast quantities of CO_2 and water vapor, somewhat similar to the condition of the planet Venus today. Being farther from the sun, atmospheric temperatures were not so high on earth, but still were too high originally to condense water vapor into the liquid phase. Following some cooling, however, torrential rains began to remove enormous quantities of water from the atmosphere and to fill the ocean basins. Large quantities of CO_2 were washed from the atmosphere during these deluges to form carbonic acid. There must have been exceedingly active chemical reactions between this carbonic acid and the sima rocks that were exposed on the ocean floor. Large quantities of bases, especially calcium and magnesium, were released from their silicates and transformed into soluble bicarbonates.[4] As soon as the oceans reached their saturation point for these bicarbonates, additional CO_2 and base ions would form insoluble carbonate deposits. The formation of these carbonate deposits, mainly lime, was concentrated around the ocean margins, where heat energy sources were greater. This led to a progressive decrease in the CO_2 content of the atmosphere. This decrease was later aided by the development of plant life both within and outside the oceans. Once plant life developed, oxygen was added to the atmosphere and oceans by plant photosynthesis and transpiration. Finally, the amount of CO_2 in the atmosphere dropped to a level where the temperatures could be maintained indefinitely within certain limits by the present regulatory mechanisms in the ocean, on land, and in the air. The oceans are the vital sector in this regulatory apparatus. They operate by removing or adding CO_2 within relatively narrow limits set by temperature, and by relying on their abundant supply of lime and CO_2 for storage and supply functions. The weathering and synthesis of limestone, therefore, which represent two sides of the geologic cycle, might be likened to a thermostat regulating the flow of energy through the system.

Limestone and lime sediments also can be produced in freshwater. Freshwater algae, living in waters heavily charged with soluble calcium bicarbonates, will cause lime pre-

[4]The possible sources of the two most important ions in seawater, sodium and chlorine, have not yet been wholly resolved. Sodium is a minor element in some of the more acidic feldspars of sialic rocks, and chlorine is exceedingly rare in both sima and sial. Chlorine probably was one of the original planetary gases during the formative period.

cipitates, such as *marl,* a white lime mud. This is done simply by removing CO_2 from solution. The release of CO_2 by evaporation at the sites of thermal springs is another example of lime deposits formed in freshwater (see Figure 10-6).

Chalk is a special form of limestone composed largely of the calcareous shells or exoskeletons (*testa*) of microorganisms. Most of them belong to the *Foraminifera,* an order of marine protozoa. Chalk strata hundreds of feet thick were deposited in many parts of the world, especially during the *Cretaceous* period (Lat.: *creta* = chalk). Figure 10-7 shows one of the thick beds of chalk that can be found in the Mediterranean region. The well-known chalk deposits of the English downs and the cliffs at Dover are other examples.

Limestones are among the most useful rocks to man. Relatively abundant, easily quarried and crushed, they form the bulk of crushed stone for construction purposes. They also are the principal source of cement, and the various calcium salts derived from limestone are important raw materials in the chemical industry. Since limestone is easily cut and shaped, the stone is by far the most important one used for exterior ornamental purposes and for general building purposes.

Chert and flint. These are amorphous forms of silica. Chert is whitish to grayish, while flint is dark gray to black. Both are extremely hard, brittle, and resistant to acids. Like all members of the quartz family, they exhibit a characteristic curved, *conchoidal* fracture similar to the broken edge of glass. Chert and flint occur mainly as nodules or lumps within beds of limestone or chalk; they represent the precipitation of colloidal silica gels. Colloidal silica is carried only in neutral to alkaline solutions, hence their relationship to lime deposits. The sharp edges of broken flint served as an important source of early hand tools for man, and later refinements led to the manufacture of spearheads and arrowheads from this relatively common material. Highly resistant to chemical weathering by acid solutions, chert and flint nodules accumulate in soil that results from the weathering of limestones and chalks. Where these durable members of the quartz family are particularly

Figure 10-6 Rock terraces of calcium carbonate (travertine) deposited by hot springs at Pamukkale, Turkey. These hot springs were the site of the ancient Greek city of Hierapolis, a resort center, some 2,000 years ago. [Turkish Ministry of Tourism and Information]

Figure 10-7 Massive accumulations of chalk deposited during the Cretaceous period near Chekka, Lebanon. These soft chalks erode easily, and rapid absorption of water makes it difficult to establish a woodland cover. This, in turn, aids the erosion process. An important cement industry at Chekka utilizes the chalk as its basic raw material. [Van Riper]

abundant, they are a farmer's bane because of their hardness, which shortens the effective life of tools, shoes, and tires.

The arrangement of sedimentary rocks. Sedimentary rocks occur in stratified rock series that grade vertically and horizontally from one type to another. This is related to the processes of sedimentation, especially within quiet water, in which the coarsest materials are found near shore, with the finest farther from land. Several changes in sea level, resulting in a fluctuating shoreline, may result in a complex pattern of sedimentary strata, as shown in Figure 10-8.

Figure 10-2 indicates the location of sedimentary rocks that are level to gently dipping. Note that they are not exclusively found along the margins of continents but are scattered in broad sheets, many of which are in shallow depressions mantling the sialic, continental platforms. A large part of the interior plains of continents lies below 500 feet in elevation above mean sea level. Relatively small rises in sea level or subsidences of the continental shorelines have often occurred

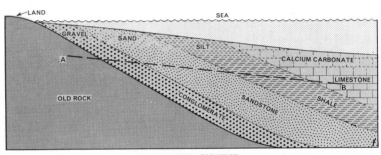

(A) HORIZONTAL SEQUENCE

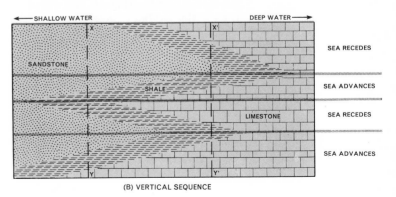

(B) VERTICAL SEQUENCE

Figure 10-8 Typical stratigraphic sections in sedimentary rocks. The horizontal sequence (*A*) from coarse to fine material with increasing depth of water is typical of relatively stable coastlines. Rapid fluctuations in sea level, however, may produce complex vertical rock strata sequences (*B*). Two well records of rock sequences (x-y and x'-y') relatively near each other may vary widely in rock sequences.

throughout geologic time, resulting in broad seaways invading continental interiors. Hudson Bay is an example of such a continental penetration by marginal seas. These shallow interior seas have been good places for the accumulations of fine-textured clays and lime oozes in past geologic periods. The sedimentary rocks that occur in geosynclines, or rapidly subsiding basins, tend to be located near the margins of the continents, where the major isostatic crustal readjustments take place. The strata here are highly variable in composition, ranging from coarse conglomerates to pure, massive limestones and dolomites. Siltstones, however, probably tend to be more common than other rock types in these basins.

10-4   METAMORPHIC ROCKS   Metamorphic rocks are those that have been altered in form when in a solid state, principally as the result of pressure, heat, and/or chemical alteration. The most frequent cause of rock metamorphism is the compressional stresses that are associated with tectonic forces in mountain-building (*orogenies*). There are different degrees of metamorphism, ranging from slight physical and chemical changes to severe alterations in which the rock material is recrystallized into minerals that are scarcely distinguishable from those in igneous intrusive rocks. Rock metamorphism may be considered to lie on the opposite side of the geologic cycle from weathering and erosion. The results of rock metamorphism present a broad spectrum of physical characteristics that differ with the process of rock alteration, the kinds of rock that are being altered, and the intensity and duration of the metamorphism. Table 10-5 lists several rock types and their metamorphic equivalents.

The properties of the original rock change noticeably under severe metamorphism. In *quartzite,* for example, the sand grains of the original sedimentary sandstone are fused together so that there is no longer any cement, and the constituent grains disappear. Quartzites are extremely resistant rocks that weather and erode slowly. When limestone or dolomite is subjected to heat or pressure, the amorphous carbonates begin to crystallize, and the resulting *marble* is an agglomeration of calcite or dolomite crystals. *Calcite* is the pure, crystalline form of calcium carbonate. Impurities tend to become segregated into streaks or bands, giving some marble its prized patterns for ornamental purposes. Shale, when metamorphosed by pressure into *slate,* tends to harden, develop a tendency to break into parallel sheets (*cleavage*), and rearrange the aluminum silicates of the original clays into new crystal lattices. Slate represents a relatively early stage in the metamorphism of shale. Its ability to split cleanly into smooth sheets makes it useful for roofing and for schoolroom blackboards. *Phyllite* is a somewhat more crystalline form of slate. *Schists* are laminated metamorphic rocks in which the minerals have been under such extreme rock pressure that their molecules are arranged into thin, sheetlike, or foliated forms. There are many types of schists, subnamed according to the predominant mineral:

Table 10-5   Selected Rock Types and Their Metamorphic Equivalents

Original rock	Metamorphic equivalent
sandstone	quartzite
limestone	marble
shale	slate, phyllite, schist
bituminous coal	anthracite
granite, slate, shale	gneiss, schist

Figure 10-9 Gneiss, a banded crystalline rock that is folded and contorted under heat and pressure. [U.S. Geological Survey; W. C. Alden]

i.e., *mica schist, chlorite schist,* and *talc schist.* The flaky crystals of many schistose minerals are in various degrees of hydration, in which water molecules are inserted by pressure into the crystal lattices. Such minerals have chemical compositions that are similar to the clays in one direction of the geologic cycle and to the most common igneous intrusive rocks on the other. Another interesting characteristic of many schists is that they often contain some types of non-flattened mineral crystals that require extreme pressure metamorphism for their growth. The gem minerals *garnet* and *tourmaline* are of this type.

When granites and other igneous intrusive rocks are subjected to extreme metamorphism, there is a tendency for the different mineral crystals to arrange themselves in parallel bands, as in bands of quartz, feldspar, mica, or occasional hornblende or pyroxene. These banded rocks, with their alternate layers bent and twisted to conform to the enormous pressure alignments, are

known as *gneisses* (see Figure 10-9). Gneisses and schists are usually found in the old roots of eroded mountain ranges.

An example of degrees of progression along the geologic or rock cycle is afforded by examining what happens to carbon after it enters the cycle. Our illustration starts with the synthesis of carbon in the cellulose of plant tissues by converting it from atmospheric carbon, oxygen, and water. Under restricted decay, resulting from burial in a marsh or swamp, it first becomes *peat,* a mass of brown, fibrous, partly decayed plant remains. With compression and induration, it loses some of its water and volatile material to become *lignite,* or brown coal, and finally, *bituminous coal.* It now is a sedimentary rock. Continued pressure and possible rock folding alter the bituminous coal to its metamorphic equivalent, *anthracite,* sometimes termed "hard coal." Extreme rock pressure, such as that occurring at the innermost bends of folded seams of anthracite, will further change anthracite to *graphite,* a pure, but still amorphous form of carbon. Graphite schists have been found that may represent some original organic life that has been altered by extreme metamorphism. The crystalline form of carbon, the diamond, is apparently not formed in rock metamorphism, but since it appears in the lava conduits of ancient volcanoes, it may have originated as a bit of early life or as some of the carbonate rock material that has completed the geologic cycle back to the igneous crust.

The gradation of sedimentary and metamorphic rock types indicates that almost any rock or rock material can be altered physically and chemically to become adjusted to any environment, ranging from the warm, humid climate of a tropical hillside to the immense heat and pressure deep within the crust and subject to strong deformations on a continental scale. Material must pass from one environment to another and adjust its chemical and physical form in each. Only through a rock cycle such as this could the continents have survived billions of years of denudation and deposition.

The map of world surface materials (Figure 10-2) includes the metamorphic rocks as highly folded or *faulted*[5] rocks. Some of them also are included within the intrusive crystalline rocks, because there is a gradual transition from metamorphic to igneous rocks and because many metamorphics are crystalline.

The resource significance of the folded rocks is generally not so great as that of the less disturbed sedimentary rocks, despite the similarity in chemical composition. Marble and slate are desired for specific purposes, but the local variations in physical structure of contorted rocks are handicaps to large-scale quarrying. Anthracite, particularly desirable for heating, is found only in folded strata, but the amount is extremely small on

a global scale as compared with the vast deposits of lignite or bituminous coal.

10-5   UNCONSOLIDATED OR FRAGMENTAL SURFACE MATERIALS   A large part of the land surface beneath the soil is underlain by loose, unconsolidated material (see Figure 10-2). This rock debris was deposited after having been carried by transporting agents such as rivers, wind, ice, and ocean and lake currents. Great variety exists in this material, since it was derived from rocks of all types, worked on by various weathering and erosive processes for different lengths of time, and deposited under different conditions. The world map indicates a broad but scattered distribution. Comparing this distribution with the world map of mountains, hills, and plains (Figure 11-14), note that the fragmental materials are mainly, but not exclusively, found on plains. This is mainly because many of the global plains are depositional rather than erosional surfaces. Not all fragmental material deposited relatively recently and not yet altered into rock is found on plains, however. The thick deposits of volcanic ash that are ejected from high volcanic peaks in mountainous areas comes to mind, and the glacial rock debris deposited over hills and valleys in northern North America and Eurasia is another example. The following subsections present a series of generalizations concerning the nature and meaning of different types of fragmental surface materials, classified according to their method of transport.

Alluvium.   Alluvium is one of the most common types of fragmental surface materials. It consists of the debris that is carried and deposited by streams. It is found mainly along the floodplains of rivers, on delta plains, and on the gentle slopes that often mark the junction of mountains and plains. The allu-

[5]A *fault* is a fracture in the earth crust along which some rock displacement has taken place.

vium of wide river plains and deltas generally is admirably adapted for agricultural use. Its texture, usually fine sand or silt (see Figure 9-8), is not too fine, yet is open enough for easy cultivation and drainage. It is composed of a wide variety of small rock fragments recently derived from the entire drainage basin and thoroughly mixed by the turbulent river waters. Hence, it is not likely to be deficient in any of the essential mineral nutrients for plants. It is commonly dark, owing to the organic material derived from topsoils upstream or formed in place, and the high organic content improves the ability of the alluvium to hold plant foods and moisture. The flat, low position of most alluvium, however, is a detriment because of the liability to flooding and the difficulty of removing excess water from the surface and from within the soil.

Streams that suddenly debouch from mountain canyons onto adjacent plains drop suddenly in energy as the streams reach grade. The sudden decrease in velocity forces them to deposit material to build a graded slope. The sudden break in the stream profiles as the rivers leave the mountain canyons results in much coarser material than in a stream that has a more gradual slope and where the bed load can be gradually reduced in size downstream (see Figure 9-8). The coarseness of alluvium is roughly proportional to stream velocity. The *piedmont alluvial plains,* or *bajadas* (see Figure 12-19), are convex, with their steep upper slopes consisting of coarse cobbles and gravels. The lower, gentler slopes have finer clays and silts. Individual deposits, because of their shape, are termed *alluvial fans* or *alluvial cones* (see Figure 10-10). Piedmont alluvial plains are choice areas for agriculture in dry regions, since their sloping surfaces provide ideal conditions for irrigation. Their permeable texture also drains off surplus water,

Figure 10-10 Alluvial cone south of Skopje, Yugoslavia. Here a tributary of the Vardar River, descending from its mountainous source area, has deposited a steep, coarse alluvial cone where it enters the main Vardar Valley, a structural, downwarped trench. The Vardar Valley long has been an important gateway into central Europe from Greece. [Van Riper]

thus helping to prevent the accumulation of toxic salts caused by poor drainage and evaporation.

Colluvium. This is the fragmental debris that is shifted directly by mass wastage from upslope positions. It is characterized by a lack of stratification and a mixture of sizes, ranging from large stones to fine clays. It is localized in occurrence, the main deposits being found along the lower slopes of hills and mountains. It is not so satisfactory as alluvium for agricultural use because of its tendency to be poorly drained. The heavy clays that usually are found in colluvium are highly retentive of water. Colluvial slopes also are not graded by surface runoff, hence tend to have a hummocky surface that retains surface water. Large stones sometimes interfere with cultivation. Colluvial material in the humid tropics is likely to be more fertile than upland soils, however, because it is less leached of its soluble plant foods by rainfall. Containing topsoils from upslope, it also tends to have a somewhat higher humus content.

Lacustrine and marine sediments. Most of the rock debris carried by rivers is finally deposited in the ocean or in lakes. Alluvium deposited in lakes and then redistributed by waves and currents is termed *lacustrine* material. It is similar to *diluvium,* which is alluvium that is deposited and reworked by marine processes. Lacustrine deposits are likely to be much more local than those of marine origin. Lacustrine sand deposits, for example, tend to be in narrow linear strips representing former lake beach or dune deposits. Silty lake delta deposits also tend to be small.

The coastal plains along the southeastern Atlantic and the Gulf of Mexico coasts of the United States are good examples of the zonal arrangement of marine sediments (see Figure 10-8). Strong contrasts between heavy clays, lime muds, and coarse sands are characteristic. Soils derived from the coarser sandy

to gravelly deposits are among the poorest soils in the world in terms of inherent fertility. The sparse pine barrens of eastern New Jersey are illustrative. Such soils may, however, become highly productive with proper soil conditioning. Diluvial clays, silts, and marls have a higher natural fertility than the sands but frequently suffer from poor drainage unless they occur in elevated benches or terraces.

Lacustrine material is likely to be much more suited than diluvium for agriculture, and the flat lake terraces of former glacial lakes in the United States, Canada, and northern Europe are generally prized agricultural land after they have been properly drained. Even the sandy beach ridges around the edges of the glacial lake terraces are useful for orchard fruits, which do not tolerate water-saturated soils even for short periods.

Wind-blown sediments. The deposits resulting from wind activity include *sand dunes* and *loess.* There are small areas of sand dunes scattered widely around the world, especially along the sandy beach coasts of oceans and lakes. Wide expanses of them, however, are found in many of the major deserts of the world, especially in the Sahara, Saudi Arabia, northwestern India, central Tarim Basin in central Asia, and west-central Australia. Narrow dune belts also can be found bordering former glacial lakes. Most dunes, especially those in humid regions, are composed of quartz sand, which is chemically stable and resists further abrasion in waterborne sediments. Even most desert sand has been partially worked over by streams and brought down into desert basins by occasional runoff. The high percentage of quartz in dunes, and the susceptibility of the sand to blow when the vegetation cover is removed, reduce their value for agriculture. Some desert dune sands, however, have

Figure 10-11 **Air view of an oasis in the midst of an erg or dune desert near Ouargla, Algeria.** Groundwater lies near the surface in this desert basin, and the date palms are able to tap it directly without irrigation. Cultivation consists mainly of preventing the moving sand from burying the trees. The small squares toward the left are the mud walls of family compounds. Note the longitudinal dunes in the background, which are arranged at right angles to the prevailing wind direction. [Institut de Recherches Sahariennes]

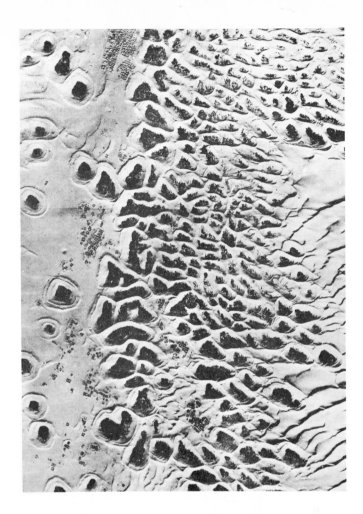

much higher percentages of other mineral sand grains which can release plant nutrients when water is brought to them. Some of the finest date gardens in the world are located within interdune depressions in the Sahara (see Figure 10-11).

Loess consists of remarkably even-textured, yellowish silts whose particles tend to fall mainly within the range of .01 to .05 millimeter in diameter. Relatively recent, unweathered deposits of typical loess exhibit a vertical, columnar structure. The tendency to split into vertical columns makes it possible for loess to retain steep slopes for long periods of time without slumping (see Figure 9-16). When the structure is destroyed by cultivation or weathering, however, loess erodes extremely rapidly. Controversy still continues over the origin of loess, but there is a general agreement that at least some wind deposition has been involved at one time or another, although many deposits may have been reworked or even originated as stream or glacial silts. The global distribution of loess, shown in Figure 10-12, indicates broad areas of deposition, some of which are far from arid or semiarid regions. The four major areas in the world include the Argentine pampas, the American Midwest, central Europe, and northwestern China.

The largest deposits are found in northwestern China, in the vicinity of the great bend in the Yellow River (Hwang Ho), where loess has accumulated up to thicknesses of 300 feet. Much of this loess has been blown out of the dry desert plains of central Asia, aided by the strong winter gales that originate in this area. Streams have eroded a large part of the loess and have deposited it in the lower courses and deltas of the rivers. Such river silts, exposed during the dry winters, are again subject to wind deflation, and almost all North China is mantled with the fine yellowish dust. The Yellow River and the Yellow Sea are appropriately named.

The loess deposits of southern Illinois, Iowa, southern Indiana, and Europe clearly are related directly or indirectly to continental glaciation. Much of it had its main source of material in the drainage channels choked with detritus from the melting ice. During winter seasons in the waning stages of glaciation, the river floodplains were dry and barren. Wind swirled great clouds of dust out of the valleys onto the adjacent interstream areas. Even the uplands between the ice margins and the encroaching vegetation must have been dusty places, littered as they were with the fine rock flour produced from glacial abrasion.

Separate deposits of loess correspond to

each of the interglacial periods, and each one differs from the others because of the differential period of weathering since it was formed. The deposits are thickest (60 to 100 feet) near the major river valleys (especially on the eastern valley sides) and gradually thin to depths of 3 to 10 feet on the general interstream areas. The older loess, such as was laid down during the Aftonian interglacial stage, about 300,000 years ago, has lost most of its columnar structure, has weathered to depths of 8 to 10 feet, and is a rather poor soil for agriculture in southern Illinois.

Fragmental volcanic detritus. Molten magma that works its way out to the surface as lava differs in chemical composition, as was indicated earlier. The basic lavas, such as basalt, form sheets of rock. The acidic types of sialic origin, on the other hand, tend to form fragmental material, such as ash and cinders—the kind of material that results from explosive eruptions. The acidic lavas are unusually high in silica and included gases.

Figure 10-12 World distribution of loess deposits. The loess deposits on a world scale appear to surround the dry lands. The major exceptions are those associated with glaciation in the American Midwest and in Europe. [After Lobeck]

�*(dark)* Areas of major loess deposits

▪*(light)* Areas of irregular loess deposits

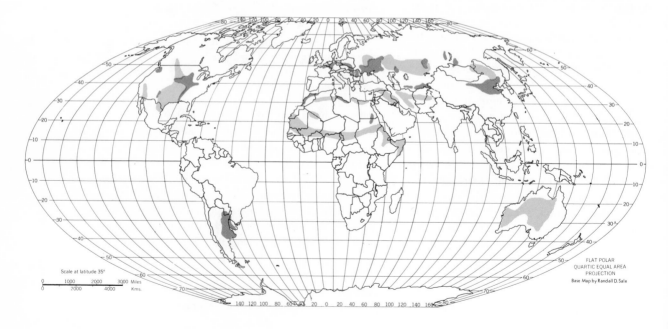

Landforms and Surface Configuration

Figure 10-13 Erosional cones in what is left of a once ash-filled basin near Goreme, Turkey. Thick deposits of volcanic ash nearly filled this large basin southeast of Ankara. Rainwash subsequently has sculptured the soft ash into these unusual, cone-shaped spires. Some of them were excavated for dwellings and churches and served as refuges for remnants of the Byzantine followers after the conquest by the Seljuk Turks. [Turkish Ministry of Tourism and Information]

The rapid release of these gases, which have been under great rock pressure while in the volcanic conduit leading below, ejects liquid particles high into the air. There they are cooled soon to form the somewhat glassy, cindery, or frothy debris that accumulates near the vent to form coneshaped peaks. Unusually powerful volcanic explosions may spread ash material for miles and to great thicknesses. Small ash plateaus and ash-filled basins adjoin cone-shaped explosive volcanoes in parts of the world, especially along the major mountain belts (see Figure 10-13).

There is a considerable range of chemical composition, even among the fragmental volcanic types. The variation in the content of bases makes a great difference in suitability for the development of agricultural soils. Those that are glassy and contain little more than silica have small value. The Japanese volcanoes, for the most part, are of this type and form some of the poorest soils in that country. The darker, more basic material, on the other hand, generally contains sufficient potassium, phosphorus, calcium, and magnesium to support high productivity, especially when young. While the base content of these fragmental materials is not nearly so great as in the basaltic lavas, their rapid availability makes them useful soon after they are deposited. The rice terraces that climb for thousands of feet up the ash slopes of Java indicate the fertility of such material. In several other parts of the world, as in Italy, Central America, and the Philippines, farmers continue to cultivate the

slopes of active volcanoes despite the constant threat of renewed activity. Fresh falls of ash might be considered to be a natural form of fertilization.

Fragmental glacial material. Glacial detritus constitutes the parent material for soils throughout the regions where glaciers now are active and also in areas that were subject to continental glaciation during the Pleistocene period (see Figure 9-17). Fragmental glacial material comprises two types of material: *glacial till,* an unassorted mixture of boulders, finely ground rock, sand, and preglacial soil material that was included within the ice mass and deposited as the ice melted (see Figure 10-14); and *fluvioglacial outwash,* which is waterborne and sorted material derived entirely from glacial detritus. The relative amounts of these two types of material differ widely in various regions. Where the ice has stagnated and had little forward movement, fluvioglacial material makes up most of the surface deposits. Glacial till tends to predominate where the ice was active.

Glacial till varies greatly in composition, depending largely on the kind of local underlying material, despite the ability of continental ice sheets to transport rock debris for hundreds of miles. The major part of the till rarely is carried more than 10 to 20 miles, even in continental ice sheets. Glacial tills in areas underlain by hard, igneous, intrusive rocks, such as those in the Canadian shield area north of the Great Lakes and in the crystalline uplands of central Sweden and in Finland, are extremely sandy and stony, with only small quantities of clay (see Figure 12-33). The tills in the area south of the Great Lakes and in parts of the Baltic states in Europe, where shales and limestones form much of the underlying rock, are heavy with included clay and have only a small percentage of sand and stones. Glacial boul-

Figure 10-14 Glacial till, Syracuse, New York. Note the mixture of fine clay and stones of various sizes, and the unassorted characteristics of the material. [Van Riper]

ders are so rare in the heavy clay tills of central Illinois that they are considered local oddities.

The quality of glacial tills for agriculture likewise differs greatly. By far the best tills are those where a variety of different sedimentary rock strata occur, such as sandstones, shales, and limestones. Limestone that is ground to a fine powder by glacial abrasion and thoroughly mixed with sand and shale produces exceptionally fine agricultural soil. Most of the fertile soils of the Corn Belt in the American Midwest owe much of their fertility to the calcareous tills from which they were derived. Glacial tills in areas of sialic granites, on the other hand, are coarse, sandy, and relatively infertile, largely because they lack the fine-textured constituents needed for the retention of water and plant foods.

Fluvioglacial material is not nearly so variable as glacial till. The sorting action of water soon separates the fine clays and silts from the coarser fractions through fluctuations in velocity (see Figure 9-8). The clays are the last to be dropped from suspension and usually end up at base level as lacustrine or marine deposits. Sands and gravels form the bulk of fluvioglacial material. The quality of these as parent materials for agricultural soils is far from uniform. Some have a high percentage of shale and limestone pebbles. The latter, relatively soft, rarely survive stream abrasion for more than 10 to 20 miles unless they were unusually large stones to begin with. Such gravels, despite their limited extent, form unusually good soils. Glacial outwash in areas of granites or other sialic igneous rocks, like the till in these areas, have an unusually high percentage of quartz sand and pebbles that weather slowly in the upper mid-latitudes. Many of the outwash plains in the northern part of the United States and southern Canada are composed almost entirely of quartz sand and are almost worthless for agriculture without expensive soil conditioning. Originally covered with vast stands of pine, many of these plains have been cut over and burned so often that they now cannot even support a forest growth and are covered with low shrubs.

The Relativity of Time in Geological Processes

The processes of shaping the earth surface should always be viewed within an appropriate time perspective. Such a perspective is more obvious with respect to the changes that go on within the atmosphere, and quantitative data are available throughout the world to place the current fluctuations of atmospheric processes within the proper time relationship. On the other hand, the rates of change of the energy flow involved in geological processes are much more difficult to measure and usually must be obtained from indirect evidence. Such rates of change are extremely important, because it is becoming evident that they are linked to the overall flow of exogenic energy, just like day-by-day fluctuations of the weather. Questions such as, "How much more time is required to weather a granite tombstone in Florida than one cut from limestone?", "At what rate is the Mississippi delta subsiding?", or "Are the continents still moving laterally, and if so, at what rate?" are not easy to answer, because such rates of change have many variables. Figure 10-15 illustrates this point.

Many new tools and methods have been developed to measure the time factor in geological processes (both short-term and long-term). Some include: (1) techniques of rock dating by measuring the amount of decomposition of various radiogenic materials

Figure 10-15 Detail of the Treasury Columns, Petra, Jordan. Weathering in the desert proceeds extremely slowly, revealing the relativity of time in the task of shaping the earth. These remarkable columns were carved in place from the sandstone cliffs by the Sabateans about the time of Christ. Protected from scattered rains by the overhanging cliff, the sculpturing detail has hardly been touched by weathering in almost 20 centuries. [Van Riper]

predictions of dangerous earthquakes comparable to those of atmospheric storms, much progress has been made.

Developments in the fields of hydrology, geomorphology, and soil mechanics have been highly rewarding in measuring and understanding the dynamic aspects of such geologic processes as the formation of hydrologic equilibrium curves, the systems analysis of drainage areas, and glacial flow. There is a growing awareness that many different processes are interrelated, provided they can be placed in their proper time sequences. These include major global temperature fluctuations, variations in volcanic activity, eustatic changes in sea level, unusual fluctuations in limestone deposits, and changes in the configuration of ocean basins and the pattern of ocean currents.

Throughout this volume, in order to understand the influence of *process* (the application of energy) upon *materials* to produce *forms,* these three must be considered as segments of open energy systems (see Section 1-6). The systems that concern us here are segments or subsystems of the main supersystem, that of the global energy flow. In the chapters on climate, the atmospheric circulation system was described as a part of the global system by which a major part of the energy input from the sun is led from areas of surpluses to areas of deficiencies and finally out of the earth-atmosphere environment into outer space. The hydrologic cycle

ranging from carbon 14 to uranium-lead or potassium-argon; (2) precision geodetic mapping of positions and elevations using satellite control; (3) *palynology*, a subscience that specializes in pollen analyses to reconstruct vegetation and climatic patterns for thousands of years; and (4) new developments in seismology. Internal rock stresses are now being measured directly, and while the science of seismology has not yet made the

was considered another kind of subsystem, in which only part of the atmospheric energy flow was circulated and regulated. In the present chapter it was noted that even the changes in rocks through tectonic, denudational, and depositional processes constituted subsystems into which energy flows from the sun, performs work, modifies forms, and passes on, but not without producing a continuous modification of the earth surface.

It is not necessary to assume fluctuations in the solar constant (solar output) to produce irregular changes in the energy application at local points within the system. What may appear to be a major imbalance in the energy flow system may well be a necessary change in form or materials within a long-range movement toward equilibrium. Mountains today may be a result of stresses brought about by hundreds of thousands of years of deposition. The time required for a mass of loose soil balanced precariously on a steep mountain slope in the humid tropics to react to the addition of water from a single rain may be measured in seconds. The time taken for sedimentation to accumulate so as to produce a readjustment of rock material within the crust, and subsequent tectonic stresses on a continental scale, may be measured in hundreds of millions of years.

The time involved in maintaining equilibrium in the global balance is relative to many factors, including *scale* (a change in river course versus continental warping), *process* (solution in Florida versus mechanical weathering in the Arctic), *material* (limestone versus granite), and *form* (the longitudinal profile of a stream versus the configuration of an explosive volcano).

The main heading of this portion of the chapter indicates the importance of understanding the relativity of time when studying geologic processes. The same could be said for all of the other processes involved in the arrangement of differences and similarities on the earth surface. Without a systems approach by which all of the forms and processes on the surface are interrelated in space and time, the study of any single process, or any single area, or any particular time, is incomplete. The complexity is enormous, but the logic is simple—and perfect.

Recent reports indicate that the major features of the lunar surface probably are billions of years old, or as old as the earth crust. Compared with the dead, unchanging lunar orb, the crust of the earth exhibits a kaleidoscope of change. Despite this continual change, our planet has not been much hotter or colder, nor has the total environment changed much since the first rains fell, the oceans filled with water, the first life evolved, and the atmosphere attained roughly its present composition. This is what is meant by dynamic equilibrium. Without it, the earth would have become another star or another "dead" mass of rock material like the moon.

10-6 THE NILE RIVER: A CASE STUDY

The Nile River will be examined and analyzed with respect to the kinds of work being done along its varied course. This is done in order to illustrate the variations of process within a particular hydrologic system, one that accomplishes only a small fraction of the work of sculpturing the earth surface. The reader is referred to Figure 10-16, which shows the longitudinal profile of the river. Instead of the smooth, even profile of a stream that is at grade throughout, four distinct divisions of this river can be recognized:

1. *The lower Nile,* below Aswan, where the river is at grade and where the principal work being done (until the construction of the Aswan High Dam) was depositional, in extending the delta into the Mediterranean Sea, and in raising the channel bed corre-

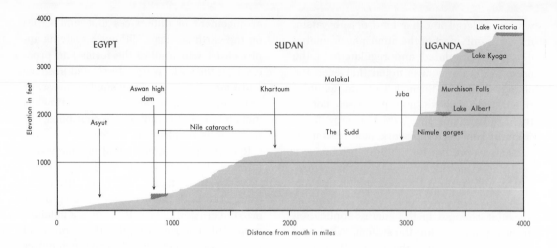

spondingly above the mouth to preserve grade.

2. *The cataract section,* a series of discontinuous falls and rapids that extend upstream from the head of the Aswan High Dam reservoir to within about 100 miles of Khartoum in the Sudan.

3. *The graded section from near Khartoum to Juba,* which includes the vast marshlands of the *Sudd,* where the river is below grade and where deposition of detritus obtained upstream is slowly raising the slope of the stream channel as well as that of the entire Sudd basin.

4. *The headwater section,* from Juba to Lake Victoria, where the Nile exhibits wide extremes in its gradient, including several magnificent falls and cascades, as well as placid navigational stretches at grade, such as immediately below Lake Albert. Abrasion predominates throughout most of this upper course but is extremely slow, owing to the shortage of tools as well as to the durability of the crystalline rocks in the area. Deposition occurs within Lake Kyoga and Lake Albert but likewise is extremely slow.

The major tributary of the Nile, the Blue Nile, which rises in Ethiopia's Lake Tana and joins the White Nile[6] at Khartoum to form the main river, has a steep but fairly smooth profile, with a graded stretch along its lower course above the junction with the White Nile. Lake Tana lies at an elevation of about 6,000 feet above sea level, but the length of the Blue Nile is only about half that of the White Nile. The average gradient of the Blue Nile thus is much greater than that of the White Nile. It also has an entirely different regime of flow, as indicated by the flow chart given in Figure 10-17. Ethiopia has typical *Aw* and *Cw* climates with a pronounced dry season between November and April. The headwaters of the White Nile, on the other hand, lie far to the south in regions of *Af* or

[6]The upper course of the main Nile is known by several names. The *White Nile* refers to the section of the river between Khartoum and the Uganda border. The section between Lake Albert and the Uganda-Sudanese border is termed the *Albert Nile,* and the *Victoria Nile* includes the remainder of the river. Lake Victoria usually is considered to be the source of the Nile, although streams that flow into Lake Victoria carry the ultimate source to slightly higher elevations. The entire river course thus has a length of slightly more than 4,000 miles, the longest of any river on earth.

Landforms and Surface Configuration

Figure 10-16 Longitudinal profile of the Nile River. This profile is highly irregular and far unlike a profile of equilibrium. Note the contrast between the depositional sector of the Sudd and the erosional sectors below and above.

Am climates, which have a much more even distribution of rainfall. The lakes and the Sudd marshes along the White Nile act as storage reservoirs to regulate the flow of the river and to collect sediment. Hence, the White Nile at Khartoum has a remarkably even and clear flow, virtually without load. The Blue Nile, on the other hand, is swollen, brown, and turbid during the summer season, laden with silt eroded from the steep slopes carved in the easily weathered basalts that underlie much of Ethiopia to depths of thousands of feet. The outlet of Lake Tana is a small stream. Hence, most of the Blue Nile water is obtained from direct surface runoff, plus additions from groundwater.

Complicating the factors operating within the hydraulic system of the Nile are variations in the underlying rock. From Aswan downstream, the rock floor below the alluvial fill in the valley was cut into relatively weak, sedimentary rock strata. At Aswan, the sialic igneous rocks of the continental block lie at the surface of the stream. While the gray Aswan granite might be highly susceptible to chemical weathering in the humid tropics, its hard quartz and feldspar resist strongly the abrasion of stream particles and the extremely slow weathering processes of this desert land. These durable rocks extend far into the interior of the continent, and they are mantled here and there by thin sheets of sedimentary rock that long ago were removed along the course of the river and now form cliffs high above the stream. Upstream from Juba, crystalline igneous rocks again make river abrasion slow and difficult, especially

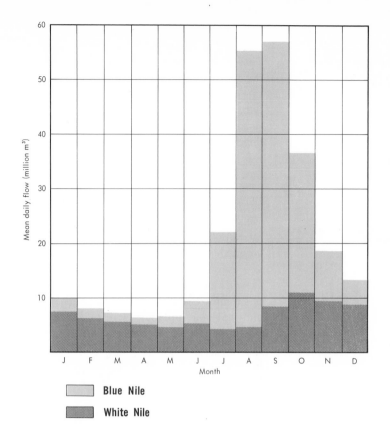

Figure 10-17 Mean daily flow of the Nile River at Khartoum. The data are given in millions of cubic meters for the 1912–1952 period. Note how much more even is the flow of the White Nile, which comes out of the Sudd, than the Blue Nile, which plunges down from high Ethiopia. [After Hurst]

below the lake outlets where the water is silt-free.

Climate is a factor that influences the amount and distribution of river volume, partly through precipitation variability and partly through the effect of evaporation. It has been estimated, for example, that almost a third of the total river intake is lost through evaporation along its course. Downstream, the dry Sahara air withdraws water from the river like a pump.

Tectonic forces have played another major role in the operation of the river system. These forces were responsible for warping the continental block upward at differential rates. The presence of Cretaceous and Tertiary sediments lying above the sialic crustal rocks along much of the valley side within the cataract stretch indicates that this region was not nearly so high during these periods (see Table 2-2). The great sedimentary basin of the Sudd marshes has been interpreted as an area of crustal subsidence during post-Pleistocene time that has almost been filled by both sedimentation and rapid organic growth. It is entirely possible that the basin still may be subject to irregular subsidence as additional loads of stream debris from upstream are washed into it. The fault blocks and volcanic uplands that make up the structure of the East African lake country, the headwater region for the Nile, clearly have their major surface features controlled by rock structure as developed by tectonic faulting and warping. The active volcano Mount Elgon, near the Uganda-Kenya border, indicates that internal tectonic stresses still are active in the region.

Despite the many variables that influence the operation of this complex hydrologic system, each individual stretch of the river has its own dynamic equilibrium based on river velocity, volume, load, slope, configuration of the channel floor, and resistivity of the underlying rock. A change in any one of these factors at any one place along the stream would result in an immediate readjustment of form to reestablish equilibrium. The time required for the adjustment is relative to the scale of the change, whether it is crustal warping measured in inches per thousand years or channel alteration during the flood season, measured in inches per minute. Within the cataract stretch, for example, there are many sections where the river is at grade, each graded stretch related to a local obstruction downstream caused by a resistant rock outcrop. Each year the river fluctuates widely in volume corresponding to the Blue Nile flow. No two years are exactly alike with respect to this flow because of the climatic fluctuations caused by changes in the atmospheric reaction to the global energy flow. Deposition on the small, discontinuous floodplains along the gentler gradients of the cataract section varies from day to day and from year to year, and the bed of the channel is continually shifting with changes in velocity, volume, and load.

Assuming no further crustal warping, the Nile would continue to erode its cataract barriers, seeking to establish grade from the mouth upstream through the hard-rock section. The excavated material would lengthen the delta (assuming no subsidence), and eventually a smooth profile of equilibrium for the entire river would be reached. The cataracts would disappear, as would the Sudd and the lakes upstream. The ultimate goal of sculpturing this curve through the sialic granites and other igneous rocks to establish this profile free of irregularities probably would require millions of years, and it is doubtful that the continental surface would remain stable for such a long period of time. A gradual long-range shift of the planetary air circulation system might also take place during such a long period of time, and if the present Sahara Desert were to become humid and the tropical headwater areas arid, the entire hydrologic system would be altered. There is good evidence that this occurred during the Pleistocene period.

The construction of the Aswan High Dam provides an interesting example of an entirely new factor introduced into the hydrologic system of the Nile. Man has intervened by applying energy to rearrange materials in order to produce a local base level at the foot of the cataract section. The effect along the lower Nile below the dam is considerable. The volume of the river in its lower course is now regulated entirely by human decision to serve human needs for irrigation and navigation. Sedimentation has almost ceased below the dam. Floods no longer will occur and thus will constitute neither a boon nor a threat, as they have throughout recorded history. Water is available in uniform quantities throughout the year, and when the distribution system is complete, it will add a third more cropland to Egypt's crowded acres. The lower Nile is an artificially controlled water supply system, free from the vagaries of natural energy flow—at least temporarily. This was accomplished at a price that can be measured either in monetary units or in total energy diversion, including human labor, diesel fuel, electricity, and even chemical energy.

The High Dam, however, has a definite life-span, which is but a fleeting moment as compared with the total age of this mighty river. The reservoir behind the dam, slightly more than 300 miles long and almost 350 feet deep immediately behind the dam, is the local base level for the entire cataract stretch. The full load of sediment, derived mainly from the Blue Nile in Ethiopia, is discharged into the lake. Compared with the time required to cut into the hard granites of the cataract section, the process of filling the reservoir with stream alluvium, or even the weathering of the concrete dam itself, is much faster. Yet, in relation to the span of human needs, the dam has been judged to be worth the effort and cost. Sedimentation rates are not yet accurately known, but 500 years has been given as a rough estimate of the time required for filling the reservoir with silt. Unless man builds a new dam upstream or raises the height of the present one, the Nile flood eventually will pour over the top of the reservoir and resume its uninterrupted flow. The turbulence of the fall from the top of the dam probably would find the dam structure a great deal easier to remove than the hard Aswan granite located upstream.

The entire river system may be modified at any time and place by a change in any of the variables that influence the flow of its energy, derived entirely from the rain that fell thousands of miles to the south. Through constant yet compensated change, the river system adjusts its kinetic energy flow to a dynamic equilibrium at all points. The time factor in all of its many alterations is entirely relative to the processes and work to be done. From the human point of view, someone decided that the temporary interruption caused by the High Dam was sufficiently long to warrant the local energy application necessary to control the local flow. In contrast, within the relative long-term goal of earth sculpture by running water, the High Dam is but a temporary obstruction to the smooth flow of energy—a tiny ripple in the gigantic task of denuding the continent of Africa.

A variety of changes operating within the context of dynamic equilibrium can be observed along almost every stream and throughout all the processes of shaping the earth surface. They exist in the work of waves on shorelines, in the operation of mass wastage in fixing the angle of slope on hills and mountains, in the filling of lakes and depressions by sedimentation, and even in the slow, inexorable turning of the geologic cycle.

Study Questions

1. Consult an appropriate reference book and determine how the study of earthquake waves reveals the structure of the earth interior.
2. Would earthquake waves tend to travel faster between Honolulu and San Francisco than between San Francisco and New York? Explain.
3. Contrast the properties of rocks that belong to the sial and sima. Explain.
4. Of the principal elements found in the earth crust, which ones tend to be most mobile within the processes of weathering? Of these, which single element is most abundant?
5. Explain why granites are so common among the continental rocks of the earth crust.
6. Distinguish between igneous, metamorphic, and sedimentary rocks. Are they interchangeable? Explain and illustrate.
7. Describe the important role of CO_2 and calcium as major regulators of the global energy balance.
8. Explain why basalts frequently tend to produce fertile agricultural soils. Give some examples.
9. Explain the high fertility of most alluvium. Give an example of alluvium that is not likely to be fertile, and explain.
10. Why does fluvioglacial outwash contain so little clay? Refer to Figure 9-8.
11. Give some examples of the relativity of time in geologic processes.
12. Draw an analogy between the energy exchange function of water and that of rock material.

CHAPTER 11

GLOBAL LANDFORMS

Having had a glimpse of the processes that shape the earth surface and the kinds of materials which react to the flow of energy about the surface, we shall now examine the major global landforms on earth, their distribution, characteristics, and significance to man.

As we shall observe in more detail later, the arrangement of continents, ocean basins, mountains, hills, and plains is not haphazard. There is a design—a pattern that, though not exhibiting perfect symmetry, still is sufficiently clear to be noticed and to stimulate earth scientists in seeking logical, consistent explanations.

Continents and Ocean Basins

The arrangement of continents and ocean basins reveals a number of characteristic features that may be summarized as follows:

1. *Antipodal position.* The landmasses, with few exceptions, are antipodal (directly opposite) to ocean areas (see Figure 11-1).

2. *Hypsographic curve.* The hypsographic curve of the earth surface, or the relationship between area and distance above or below sea level, indicates almost an opposite correlation between continents and ocean basins (see Figure 11-2). Stated another way, whereas most of the continental areas lie a short distance above sea level, most of the ocean basins lie in the ocean deeps (excluding the narrow chasms of the oceanic trenches). Seawater occupies almost 70 percent of the total surface of the earth, and most of this is deep.

3. *Continental tapering.* The continental masses, with the exception of Antarctica and Australia, tend to widen toward the north and taper toward the south. Thus, there is much more land north of the equator than south of it. Also, many more peninsulas extend southward from the continents than in any other direction. Such peninsulas are especially noticeable on the southern sides of the northern hemisphere continents.

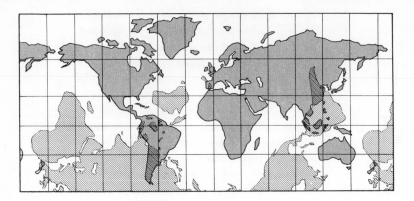

Figure 11-1 Antipodal arrangement of continents and oceans.

4. *Continental "fit."* The major outlines of the continental blocks appear to fit together like pieces of a jigsaw puzzle (see Figure 11-5).

5. *Continental structural pattern.* There are many continental and oceanic structural lines aligned in a northeast-southwest or northwest-southeast direction. This can be observed in the trend lines of coasts and in the major mountain cordilleras. The trend lines of deep ocean trenches and submarine ridges show similar alignments. A minor set of trend lines are oriented north-south and east-west.

6. *Flooding of the continents.* The distribution of marine sedimentary strata indicates that shallow seas have invaded the interior of continents several times during the last few hundred million years, but there is no good evidence that the ocean deeps ever existed within the continental blocks.[1] The hypsographic curve of the earth surface in-

[1]The Mediterranean Sea appears to be an interesting exception, and its great depth distinguishes it from the other marginal or adjacent seas that invade the continents. The Mediterranean either represents a major foundering of a continental block within the crust or is a gap that remained when a northern and southern continental block drifted apart.

Figure 11-2 Hypsographic curve of the earth surface. This diagram indicates the bimodal nature of the relationship between area and elevation. The two curves intersect at the top of the continental shelf. Whereas the area of continents becomes greater near sea level, the areas of ocean basins becomes greater near the ocean floor. The greatest variations from sea level have only a small areal expression. [After Kossena]

dicates that relatively slight changes in sea level could greatly alter the area above sea level.

7. *Continental shelves.* The continents are bordered by a *continental shelf zone of variable width which slopes gradually* downward to depths of 300 to 600 feet, beyond which there is a steep descent to depths of 10,000 to 17,000 feet (see Figure 11-3).

8. *The mid-oceanic ridge.* The largest single relief feature on the surface of the earth is the *mid-oceanic ridge,* a long, sinuous, submarine mountain range of variable height (3,000 to 10,000 feet) and width (1,200 to 2,500 miles). It extends for over 40,000 miles across the Arctic Ocean southward through the middle of the North Atlantic, the South Atlantic, eastward across the Indian Ocean and the South Pacific, turning northward, and finally ending at the peninsula of Lower (Baja) California. A branch in the Indian Ocean terminates near the southwestern part of the Arabian peninsula (see Figure 11-4).

11-1 THE PERMANENCE OF THE CONTINENTS AND OCEAN BASINS The continental sialic rocks are exceedingly old. Igneous rocks can be dated, using methods that measure the rate of radioactive decomposition of selected elements in these rocks. Such methods indicate that igneous, intrusive, sialic rocks from each of the present continents solidified at least 2 to 3 billion years ago. These are ancient rocks even within the

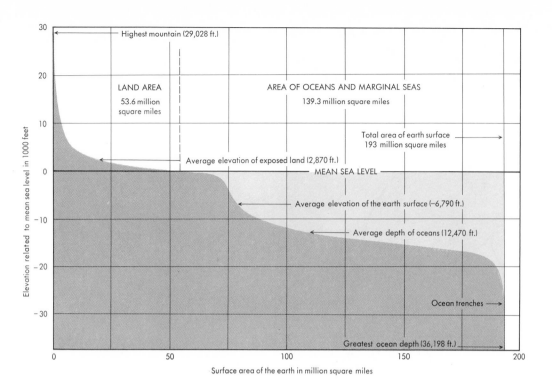

Elevation related to mean sea level in 1000 feet

Highest mountain (29,028 ft.)

LAND AREA
53.6 million
square miles

AREA OF OCEANS AND MARGINAL SEAS
139.3 million square miles

Total area of earth surface
193 million square miles

Average elevation of exposed land (2,870 ft.)

MEAN SEA LEVEL

Average elevation of the earth surface (−6,790 ft.)

Average depth of oceans (12,470 ft.)

Ocean trenches →

Greatest ocean depth (36,198 ft.)

Surface area of the earth in million square miles

vastness of geologic time (see Table 2-2). In contrast, the rock samples taken from the ocean floor and analyzed are much younger—generally less than 200 million years in age.

There is much evidence for a relatively recent drifting apart of the present continental masses to their present position. The "fit" of the present continents has already been noted. The similarity of rock types, structure, and fossil life forms in the older rocks, the parallel magnetic alignment of minerals, the correspondence of old erosion

Figure 11-3 Continental margin divisions off the northeastern United States. The continental rise in this sector is smoother than elsewhere, owing to sedimentation and the work of unusually strong bottom currents. Note the absence of any abyssal trenches and the steepness of the continental slope. [After Heezen, Tharp, and Ewing]

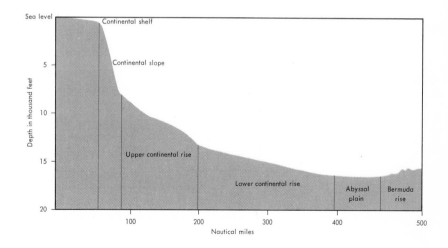

Sea level

Continental shelf

Continental slope

Depth in thousand feet

Upper continental rise

Lower continental rise

Abyssal plain

Bermuda rise

Nautical miles

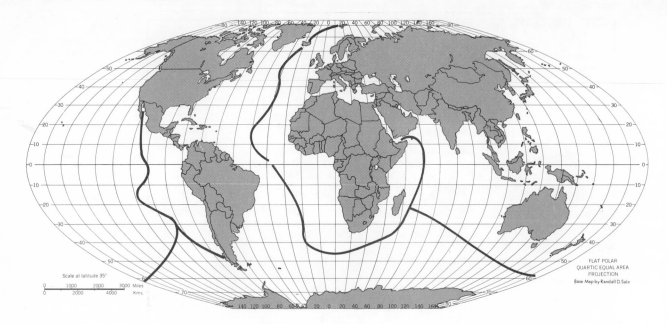

FLAT POLAR
QUARTIC EQUAL AREA
PROJECTION
Base Map by Randall D. Sale

Scale at latitude 35°

levels, and the distinct differences that are found in the younger rock strata along the continental margins on opposite sides of the oceanic basins all support the continental "drift" hypothesis. Estimates of the time of the event place it as beginning in the Mesozoic era, or approximately 200 million years ago, and continuing into the Tertiary (see Table 2-2). This time closely agrees with the calculated age of most of the rocks from the ocean floors. It is further believed that the mid-oceanic ridge represents the original juncture of the continental blocks. Some investigators believe that during earlier periods than the Mesozoic there may have been a northern and a southern continental block, the southern including Africa, Arabia, peninsular India, Australia, Antarctica, and South America. The differences in fossil life forms between the southern and northern continents had been noted by early paleontologists, who named the old, southern, hypothetical continent *Gondwanaland* (see Figure 11-5). A northern double-continent has

also been suggested, with a western Euro–North American and an eastern Asiatic block. The separation of North America from Europe supposedly took place at the same time as the Gondwanaland separation.

Earth scientists, while now convinced of a lateral movement of the continental blocks and also of the approximate time of split, do not agree as to the cause. One hypothesis, supported by Urey, utilizes the principle of convection movement of material within the plastic mantle below the crust. This movement supposedly was set up by convection within the fluid core, the number of convective cells depending upon the size of the liquid core. With a small core, only a single upward and downward vortex could be formed. Lighter material would rise toward the surface within the rising vortex and then drift laterally, finally accumulating above the downward vortex. The sialic material would be brought to the surface because it was lighter, and having solidified near the surface, would not be drawn into the downward

Figure 11-4 Location of the mid-oceanic ridge.

——— Mid-oceanic ridge

vortex. Thus, one large continental block would be formed. Urey further suggested that no continental blocks exist on the moon and Mars because neither body was large enough to develop a liquid core for convectional circulation and the subsequent separation of lighter constituents. The recent analysis of surface lunar rocks as having cooled from a molten state over 3 billion years ago seems, however, to contradict this hypothesis—or at least the idea of a "cold" moon evolution.

If the liquid core of the earth were increased in size, either as the result of increased total mass by the gathering of fragments from space or by a possible increase in radiogenic heat, multiple convection cells would evolve according to the convectional hypothesis. Internal stresses in the mantle, set up by the convection, would rupture the original block of sialic material and shift the individual blocks to positions above the new downward vortices. Such a rifting of the larger block would take place suddenly after the core reached a critical size. Successive increases in the number of convectional vortices could lead to continental rupture and drifting more than once.

One objection to the convectional hypothesis is the unusually high endogenic heat flow and outflow of material from the earth interior continuously throughout most of the length of the mid-oceanic ridge, or along the line where the continents may have been joined prior to the Mesozoic. Such a long, continuous line of upwelling of heat and material does not appear to be consistent with a series of separate convection cells. The sudden merging of the mid-oceanic ridge with the continents in Lower California and southwestern Arabia is also puzzling. Other

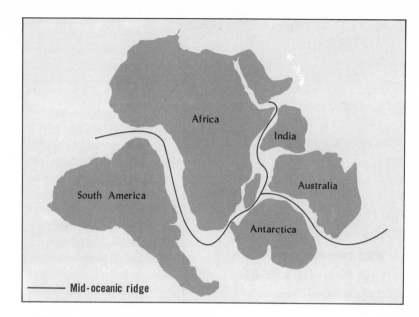

Figure 11-5 Suggested split of the ancient continent of Gondwanaland. There is much supporting evidence for the hypothesis suggesting a split of the ancient continent of Gondwanaland and the subsequent drifting apart of the constituent segments. The event probably took place during the late Mesozoic era.

geophysicists, such as Hsu, favor the theory of continental rupture and drift due to an internal shift of material because of variations in earth rotation and a migration of the rotational axis.

Whatever may be the cause, it now seems clear that the continental sialic rocks separated from heavier material below at an early stage in earth history, and that the present continents have drifted apart during the last 200 million years of earth history.

11-2 THE MAJOR STRUCTURAL DIVISIONS OF THE CONTINENTS Each of the continental blocks has a series of similar structural divisions, although the area of each division varies with different continents. These divisions are shown in Figure 11-6.

Figure 11-6 Structural divisions of the continents.

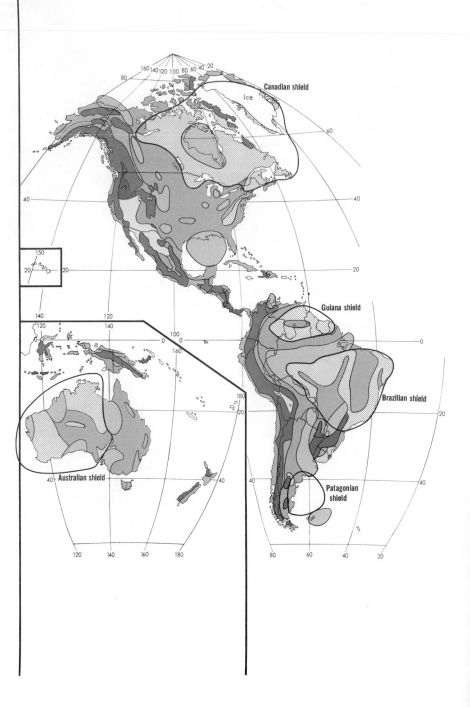

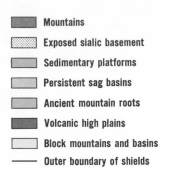

Mountains

Exposed sialic basement

Sedimentary platforms

Persistent sag basins

Ancient mountain roots

Volcanic high plains

Block mountains and basins

Outer boundary of shields

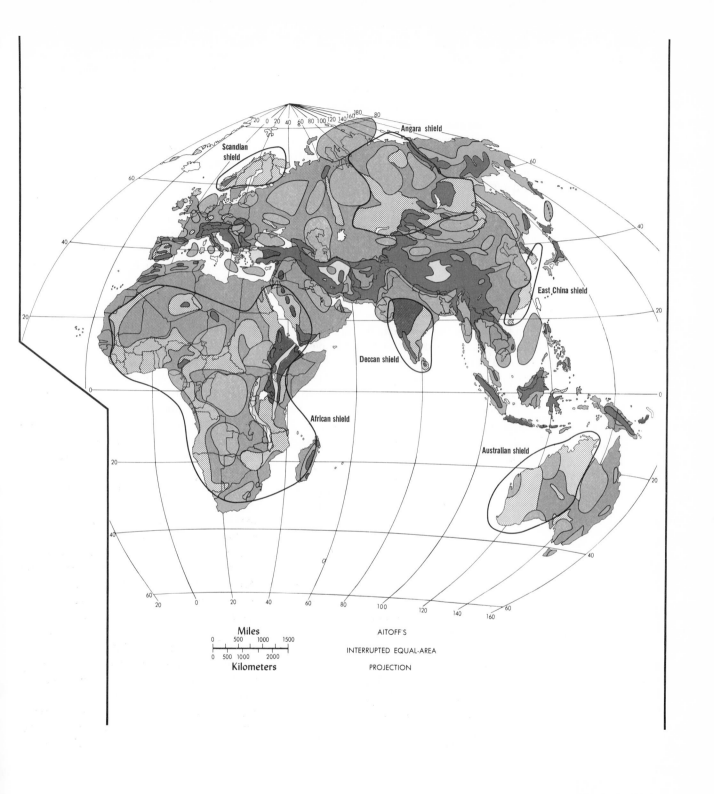

Scandian shield

Angara shield

East China shield

Deccan shield

African shield

Australian shield

Miles

0   500   1000   1500

0   500   1000   2000

Kilometers

AITOFF'S

INTERRUPTED EQUAL-AREA

PROJECTION

Sialic shields. These are broad, stable areas of typical sialic, igneous intrusive rocks and represent portions of the continental sial that are exposed at the surface. They are termed *shields* because typically their surface is broadly convex in general profile, like that of a shield. The granites and syenites that compose them are extremely old. The continental shields have been areas of consistent but extremely slow uplift since early geologic time. Most of them have been above sea level and subject to erosion throughout most of the last third of earth history, or since the beginning of the Paleozoic. Most of the great shields are hilly areas lying less than 2,000 feet in elevation, but vast areas of interior plains in central Africa have been beveled out of the great continental shield that makes up the bulk of the continent, despite the fact that they now stand over 2,000 feet above sea level. The stability of the shields is indicated by the fact that they are almost free of earthquakes and volcanic activity. They are subject at times to broad regional arching, as the present surface of the African shield indicates, but on the whole they tend to resist compressional stresses.

The sedimentary platforms. These are sections of the shields that have been mantled with a relatively thin veneer of sedimentary rock, representing areas where the shield rocks were beveled, lowered slightly, and invaded by shallow incursions of the marginal seas into the continental interiors. Because of the stability of the underlying shield rocks, such sedimentary strata are likely to be nearly horizontal and relatively undisturbed.

The geosynclinal[2] basins. These are closed continental basins into which weathered and erosional material has been or is being de-

[2]A *geosyncline* is a large "sag" basin whose strata generally dip and thicken toward the center.

Figure 11-7 Cross section of the Barton (Gulf Coast) geosyncline. Deposition has been continuous in this geosyncline for about 300 million years. The salt domes are formed by the load of overlying material forcing the salt upward. Pure salt deposits are able to flow as a plastic solid under pressures much below those required by most ocean sediments. Continual subsidence of the earth crust beneath this geosyncline cannot continue forever. [After Carsay]

posited. Their sedimentary deposits differ mainly from those of the platform strata in being much thicker. Accumulations of sediments up to 50,000 feet occur in some of the larger ones. As the sediments accumulate in the basins, the added weight tends to force them downward into the crust, much like a shovelful of gravel sinks as it is tossed into a mudhole. Geosynclinal basins characteristically undergo periodic subsidence. Figure 11-7 shows a cross section of the geosyncline at the mouth of the Mississippi River. Not all deltas are geosynclines, however, and not all geosynclines occur at the continental margins. The enormous thickness and extent of sedimentary rock strata found in the Appalachian region of the eastern United States, for example, indicates that a huge geosyncline once existed during the Paleozoic period and extended for thousands of miles from near the mouth of the St. Lawrence River southwestward to Alabama. The poorly drained sedimentary plains that lie east of the Andes Mountains in the Chaco area of Paraguay, Bolivia, and Brazil are an example of a currently active interior basin of this type. Most geosynclines are, or have been, formed along the continental margins or near the great mountain cordilleras.

Umbgrove, Meinesz, and others believe that the geosynclines eventually become compressed and elevated. The hypothesis suggests that when the detrital material is forced deep into the crust, perhaps even warping the mantle downward, gravity

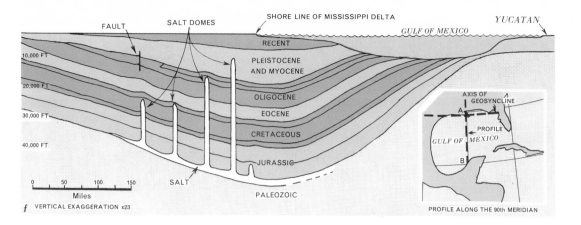

SHORE LINE OF MISSISSIPPI DELTA

FAULT   SALT DOMES   YUCATAN
                                    GULF OF MEXICO

10,000 FT   RECENT

                                   PLEISTOCENE
20,000 FT   AND MYOCENE

                                   OLIGOCENE

30,000 FT   EOCENE

                                   CRETACEOUS
40,000 FT

0   50   100   150   JURASSIC
                          SALT
Miles   PALEOZOIC
f   VERTICAL EXAGGERATION x23

AXIS OF
GEOSYNCLINE

A   PROFILE
GULF OF MEXICO

B

PROFILE ALONG THE 90th MERIDIAN

eventually seeks a readjustment through iso-static balance, resulting in compression from the sides and vertical arching of the entire basin or trough from below. Many of the high, alpine mountain ranges of the world are carved from thousands of feet of sedimentary rock strata (see Figure 11-8).

The global geosynclines are especially important to man because of their deposits of the fossil fuels coal, petroleum, and natural gas, and for extensive salt deposits, especially those of sodium and potassium.

The mountain cordilleras. Great mountain ranges are located on each continent and are arranged in a series of long, narrow belts, most of which are near the borders of continents. One belt of *alpine*[3] ranges sweeps in a series of great festoonlike arcs around the margins of the Pacific Ocean, and another

[3]The term *alpine* is reserved for mountains that have the jagged, irregular crest lines and peaks characteristic of mountain glaciation.

Figure 11-8 Mount Assiniboine in the Canadian Rockies, near Banff. The sharp crest lines produced by glacial erosion are typical of high mountains throughout the world. Note the horizontal rock strata, indicating that this region once was in a sedimentary basin. [Courtesy Royal Canadian Air Force]

runs transversely from west to east from the Straits of Gibraltar across southern Europe and southern Asia, joining the circum-Pacific belt in southeast Asia. Evidences of ancient orogenies (mountain-building periods) also exist in areas now eroded into hills and plains. In the eastern United States, for example, the contorted gneisses and schists of the Great Smoky Mountains, the Virginia Blue Ridge, the rolling Piedmont section, and

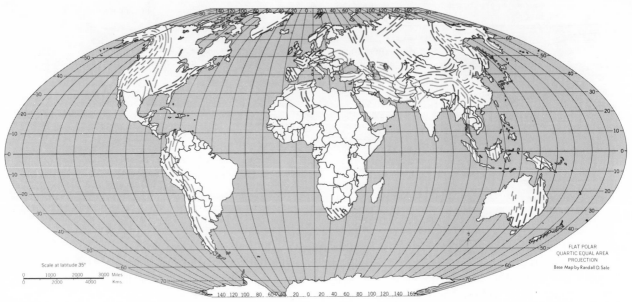

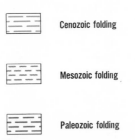

Cenozoic folding

Mesozoic folding

Paleozoic folding

Figure 11-9 World map of crustal folding and mountain-building during the last three major eras of geologic time. These three eras cover approximately the last 600 million years of earth history. Note that with few exceptions, the zones of crustal instability were located around the borders of the major continental blocks. This would become even more evident if the southern continents were united, as in Figure 11-5. [After Umbgrove]

much of western New England represent the beveled roots of a towering mountain range that was raised at the close of the Paleozoic era. Figure 11-9 shows the location of the major orogenic activity and crustal buckling during the last three eras of geologic history. If South America, Africa, and Australia were to be joined, the continental marginality of the orogenic zones would be even more striking.

The main global cordilleras exhibit such common characteristics as the following:

1. Compressional folding and faulting with associated landforms prevail, especially along the mountain flanks.

2. There is a predominance of sialic rocks and highly contorted metamorphics, even in the eroded cores of the main mountain areas.

3. There are frequent explosive-type volcanic eruptions and earthquakes (see Figure 11-10), especially in the circum-Pacific belt.

4. Thick series of sedimentary strata are exposed by erosion, especially on the mountain flanks.

5. The main mountain cordilleras bifurcate into festooned arcs, often enclosing areas of high plains, as those in the western United States, central Mexico, Bolivia, Turkey, Iran, and Tibet.

6. There are strong contrasts in gravimetric readings along the cordilleras. Figure 11-11

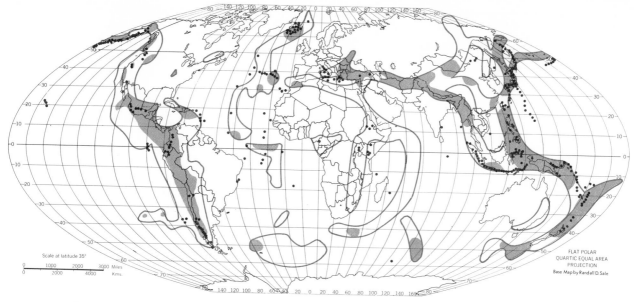

•   One or more active volcanoes (last 100 years)

▨   Intensive earthquake activity

⬭   Moderate earthquake activity

Figure 11-10 World distribution of active volcanoes and seismic (earthquake) activity. This map clearly indicates that the present crust is most unstable along three lines: (1) the circum-Pacific zone of island arcs and continental folded cordilleras; (2) the transverse mountain cordillera zone across southern Eurasia; and (3) the mid-oceanic ridge.

indicates some of the positive and negative gravimetric anomalies[4] within the Indonesian archipelago.

The high mountains today are relatively young geologically, having been the result of orogenies during the Miocene and Pliocene epochs, or between 2 and 25 million years ago. They represent the most unstable portions of the earth surface, and it is not surprising that they tend to be located at or near the continental margins, where the isostatic stresses are likely to be greatest (see Figure 9-1).

Block mountain and basin areas. These regions feature landforms on a grand scale that are associated with tensional rupturing of the crust. Mountain blocks that have extremely abrupt sides and generally are short in length and separated by debris-filled basins are particularly characteristic (see Figure 11-12). Block mountain and basin regions usually are located mostly in the areas of high plains contained within the bifurcating arcs of the major mountain cordilleras. The Basin-Range region of the southwestern United States is a classic example. Other

[4]A positive gravimetric anomaly indicates that the gravity pull of the earth is greater than normal and that the crust below is denser than normal. This, in turn, could be caused by the intrusion of mafic (sima) material, such as basalt, into the upper crust. A negative anomaly indicates a lower than normal gravity pull and could be related to a subsiding sedimentary basin, where relatively light weight, fragmental material is forced deep within the crust.

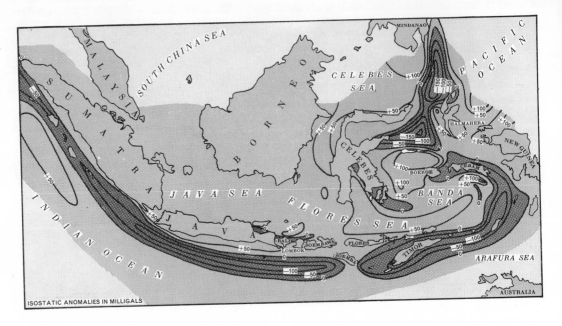

Figure 11-12   Block mountains south of Teheran, Iran.   These short, jagged, mountain blocks rise abruptly out of the Iranian plains. Note the steep, truncated edges of the mountain spurs and the gentle slopes of the alluvial *bajadas* that sweep away from the mountain flanks. [Van Riper]

similar regions are found in parts of Turkey, Iran, and Tibet. The deep trenches of eastern Africa represent a similar structure but were uniquely developed within one of the greatest shields on earth.

Many of the continental margins today that rise steeply from great ocean depths are undergoing vertical rock displacements along faults and may be associated with isostatic adjustments of the continental margins. The active fault lines along or near the coasts of California, Peru, and the Mediterranean sea are of this type. Another frequent feature of the block-mountain regions is the occurrence of *strike-slip faults* (see Figure 12-11), where the rock displacement is parallel to the rupture line.

Volcanic high plains.   Broad, upland plains composed of extrusive lavas, mainly basalts, occur on nearly all of the continents. These regions generally occur near the areas of block faulting described in the previous paragraph, where the sima or lower crust for

Figure 11-11 Gravimetric anomalies in southeast Asia and Indonesia. A gravimetric anomaly is a local gravity pull that is greater or less than average for the earth. Note the sudden change from positive to negative anomalies on opposite sides of the Indonesian archipelago, indicative of a sharp change in the density of the underlying rock. Positive anomalies indicate dense, heavy rock near the surface. [After Vening Meinesz]

some reason lies near the surface and where upward crustal stresses prevail. The principal area in North America is the Columbia Plateau of eastern Washington and Oregon (see Figure 10-3). The basaltic high plains of southeastern Brazil, the highlands of Yemen in the Arabian peninsula, the high plains of Ethiopia and Kenya in Africa, and the lava flows that cap the Deccan Plateau in India are the major global examples.

11-3 MAJOR FEATURES OF THE SURFACE BELOW SEA LEVEL The floor of the oceans differs markedly from the exposed surfaces of the continents. Except for a small amount of erosive work done by waves and currents along the landward margins of the continental shelf and island margins, a limited amount of scouring by turbidity currents, and some mass wasting along steep slopes, nearly all of the processes that change the configuration of the sea bottom involve tectonic forces or sedimentation. Even the latter is restricted mainly to the submarine continental margins. Within the ocean deeps, sedimentation rarely exceeds more than a few hundred feet, and often it is entirely missing. The deep-sea sediments generally are fine-textured clays and oozes, the only types that can be moved by the slow bottom currents. Sedimentation by the accumulation of organic detritus such as calcareous and siliceous skeletons takes place, but again, most of it occurs within the continental shelf zone where light is available.

As a result of the relative unimportance of denudational processes, mainly owing to the unavailability of powerful exogenic energy sources as on land, the sea bottom has a much more irregular surface than that of the continents (see Figure 11-13). There is a little erosion by currents and some deposition of fine sediments far from land, but the processes are exceedingly slow. The dominant features are great chasms, sharp edges of fault escarpments where rock slippage has taken place, submarine volcanoes of many shapes and sizes, and huge mountain ranges such as lie along the mid-oceanic ridge. Broad, flat plains and true plateaus can be found on the ocean floor, but they make up only a small part of it and tend to be concentrated near the foot of the continental rise. Another contrast with the continents is that the tectonic stresses present are mainly tensional, suggesting a tendency for crustal stretching. The reverse is generally true on continents. The greatest single relief features of the ocean basin floors are deep fault rifts produced by tension. The greatest relief features on the continents are mountain ranges produced by compression.

Three major divisions of the ocean floor can be recognized and will be described briefly in the following subsections. These include the continental margins, the basin floors, and the mid-oceanic ridge.

The continental margins. The continents, as indicated earlier, do not end abruptly at sea level. A typical sequence of transition between the coastal margins and the main ocean basin floor is shown in Figure 11-3. All or none of the typical slope divisions may be present, and their widths differ greatly. The *continental shelf zone,* or shallow coastal rim, for example, is virtually absent along the coast of Peru and northern Chile, while it extends for hundreds of miles out

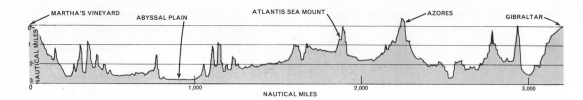

from shore east of Malaya, China, and Newfoundland. The shelf zone differs also in the
degree to which it has been dissected. Mention already has been made of the curious
submarine canyons (see Figure 9-20) that
have been cut into the outer flanks of the
continental slope, the steep outer edge of the
shelf zone. Fault rifts are fairly common
along the lower part of the *continental rise,*
which is the general transition toward the
ocean basin floors. The surface of this continental rise exhibits a wide range of irregularities. Some portions are remarkably
smooth because of sedimentation. Such areas
are more common along the western margins
of the ocean basins, especially in the vicinity
of the major deltas of the world. Note how
few major rivers empty into the eastern parts
of oceans. The high frequency of block faulting along the lower margins of the continental rise is not surprising, since this area
forms the principal boundary zone between
continents and ocean basins and where the
continental sialic blocks pinch out. If the
continents are to be kept high and the ocean
basins low, despite eons of erosion and deposition, the compensating stresses should be
greatest in this sector.

Great festoons, or arcs of islands, rise abruptly from the foot of the continental rise
in some areas and mark its position at sea
level. The entire western margin of the
Pacific Ocean, for example, is fringed with
island arcs, beginning with the Aleutians in
the north, southward through the Kuriles, the
Japanese archipelago, the Ryukyus, the Phil-

ippines, and the great island chains of the
southwest Pacific. The West Indies in the
Atlantic and the Maldives and Andaman
Islands in the Indian Ocean afford other examples. The arcs of these islands are remarkably similar to the arcs of the high
mountain cordilleras on the continents, and
in general where these arcuate cordilleras
occur on land, the island arcs are absent
offshore. Note, for example, the contrast between the eastern and western sides of the
Pacific. In the eastern Pacific, the coastlines
are steep, the continental shelf zone is either
absent or extremely narrow, and there are
no island arcs. Mountain arcs, however, extend along the entire length of North and
South America. The opposite is true of the
western Pacific. Wherever the island arcs
intersect, seismic and volcanic activity is
especially active, but tectonic forces operate
along the entire length of the arcs (see Figure
11-10). The volcanic material of most of these
island arcs is more the acidic, andesite type
than the basaltic material that forms so much
of the volcanic activity in other parts of the
ocean basins, as in the Hawaiian Islands. The
acidic lavas indicate a tie to sialic crustal
material and tend to produce explosive volcanoes. The island arcs throughout the world,
similar to the high, continental, alpine ranges
of the continents, are dotted with impressive
symmetrical cones built by successive ash,
cinder, and fluid ejecta. Sharp changes from
negative to positive gravimetric anomalies
indicate, however, that the sima lies close by
(see Figure 11-11).

Figure 11-13 A surface profile across the North Atlantic. This is a generalized profile, in which many of the minor irregularities have been eliminated. The Azores represents the crest of the mid-oceanic ridge. Broad, flat plains are rare in most of the ocean basins. Erosion of tectonic landforms is exceedingly slow. [After Heezen]

The underlying material of the continental blocks that lie beneath the sea can only be inferred. The much slower speed of earthquake waves as they pass through this zone indicates that these continental marginal areas are mainly composed of sialic, less dense material, thus justifying their inclusion within the continents. Except for localized areas of geosynclines contained within the shelf zone, or sedimentation in some of the rift depressions, the sedimentary components are probably relatively shallow. The evidence is against attributing a sedimentary origin to the continental shelf zone.

The potential significance of the submarine continental margins to man should be neither exaggerated nor underestimated. The successful exploitation of petroleum and natural gas from off-shore wells in the Gulf of Mexico, the Persian Gulf, and the Caspian Sea is no indication that the continental shelf zone represents a continuous potential area for successful drilling operations. There are a few other promising shelf areas in this respect, but they are not extensive. Such petroleum deposits are restricted mostly to thick sedimentary rock strata, a relatively rare feature of most of the continental shelf zones. The potential human utilization of the shelf zone has only begun to be investigated, and some aspects appear to be promising, including uses for waste disposal, controlled marine food production, exploitation of certain chemical elements that are common in marine sediments but rare on the continents (manganese, titanium, phosphates, rare earths, etc.), and even for shelter, if air pollution continues toward dangerous levels.

The ocean basin floors. The ocean basin floors lie generally between 12,000 and 17,000 feet below sea level. On the whole, the surface is rough (see Figure 11-13), and large, flat plains are exceedingly rare. Whereas most of the surface irregularities on the continental rise are depressions, such as rifts or canyons, most of the irregularities on the basin floors are lines of hills, frequently of ancient basaltic volcanic origin. The linear pattern of these submarine hills tends to follow the general structural lines in the ocean basin, paralleling the mid-oceanic ridge, the island arcs, or the lines of mid-oceanic volcanic islands. The North Pacific Ocean, lacking a mid-oceanic ridge, appears to have a somewhat different surface configuration in which lines of volcanic hills run transverse to the general trend lines of the margins and tend to divide the basin floor into a series of minor basins. Great fracture lines also are found, extending for hundreds of miles. They are especially common in the eastern part of the North Pacific, moving from north-northeast to south-southwest. Tabular, plateaulike rises, such as the platform near Bermuda, sometimes rise abruptly from the basin floor and are capped by thin sedimentary strata.

The greatest single relief features of the ocean basins, however, are the *abyssal deeps,* or *ocean trenches.* These are long, narrow clefts that frequently reach depths of from 28,000 to 35,000 feet below sea level. Most of them are located immediately along the oceanic side of the island arcs, hence might be considered the junction of the ocean basins and the continental blocks. All of them are immense fault basins or rifts of tectonic origin, are active seismically, and represent major global lines of weakness. The deepest is the Marianas trench in the western Pacific, which reaches a maximum depth of 36,198 feet and thus ranks as the greatest local irregularity on earth. A short trench in the Atlantic (the Romanche Gap) is unusual in being located in the center of the constriction between the North and South Atlantic and cutting directly across the mid-oceanic ridge. An interesting feature of the ocean trenches is their relatively low heat flow from the earth interior, which is about half the normal amount as observed around the world. This indicates a possible downward shift of material within the mantle. Despite the tectonic activity in the vicinity of the trenches and the many evidences of old volcanic activity throughout the basin floors, most of the basin floor area is relatively stable. The principal action takes place along the edges of the basins or along mid-oceanic ridge.

Secondary but striking relief features of the basin floors are sea mounts and guyots. *Sea mounts* are isolated volcanic peaks whose summits are usually from 2,000 to 5,000 feet below sea level. *Guyots* are sea mounts high enough to have had their summits beveled by wave or current erosion. They are frequently capped with coral growth in the tropics and subtropics and are especially common in the Pacific basin. Many of the volcanic islands in the Pacific, as well as the coral *atolls* (circular coral islands), are former guyots that now extend above sea level owing to regional uplift. Good examples are found in the Tuamotus, the Tongas, and the Cook Islands. These sharp volcanic peaks that rise from the floor of the ocean frequently are aligned in rows. In the Pacific they seem to be somewhat more frequent in the vicinity of the ocean trenches, although they occur throughout the ocean basins.

The mid-oceanic ridge. This enormous submarine mountain range has already been discussed (see Figure 11-4). It is not only the

Figure 11-14 Distribution of mountains, hills, and plains.

Mountains (local relief more than 2000 feet)
Hills (local relief more than 300 feet)
Plains (local relief less than 300 feet)

most massive and longest single relief feature on earth but also one of the most active seismic and volcanic areas. It varies considerably in width and height. In a few places, as in the Azores Islands, its summits rise above sea level. It has an exceedingly rough surface, being broken not only by great fault rifts but also by lines of volcanic peaks. An unusually large, prominent rift depression lies near its summit along a large part of its length. Unlike the deep oceanic trenches that border the peripheral island arcs, the rifts associated with the mid-oceanic ridge have an unusually high heat flow from the interior, indicating an upward movement of heat and material from below. Heezen suggests that sialic material is still being introduced from the mantle beneath the crust and that it gradually drifts away from the ridge, much as the continents themselves did during the Mesozoic. He cites as evidence the increasing age of rocks and the decreasing amount of *surface relief*[5] away from the ridge. Seismic evidence indicates that the ridge is composed of much lighter rock material than that of the basalts that predominate over most of the ocean basin floors.

We have described the distributional pattern of the major geomorphological divisions of continents and ocean basins and related these to the general global forces that shape

[5]*Surface relief* is a common geographical term and refers to the relative amount of irregularity of the land surface. *Local relief*, for example, signifies the difference in elevation between the bottom of the local valleys and the tops of the adjacent summits. *Relief* should not be confused with *elevation*, which usually signifies a measurement from a prescribed datum plane, such as mean sea level.

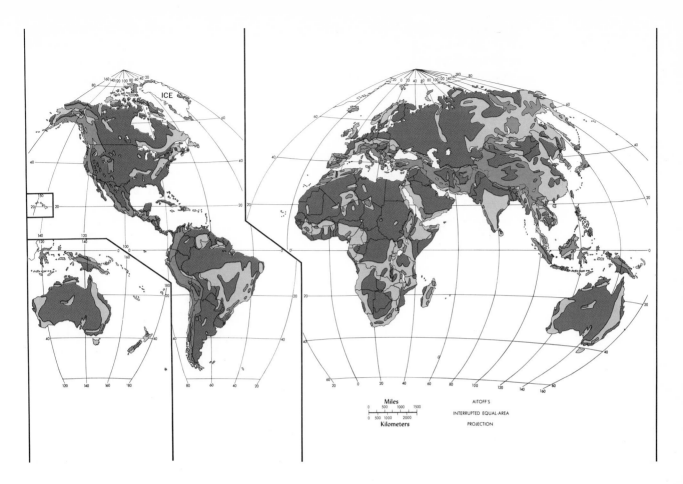

ICE

Miles
0   500   1000   1500

0 500 1000 2000
Kilometers

AITOFF'S

INTERRUPTED EQUAL-AREA

PROJECTION

the earth surface. Now we shall look more closely at a threefold division of the continental surface relief into mountains, plains, and hills (see Figure 11-14), paying special attention to their surface forms and their significances to man.

Mountains

What is a mountain? There is no universally accepted definition, except for the general acknowledgment that the term implies a prominent relief feature. The word *prominent*, however, is a relatively descriptive term.

No one would deny that the Himalayas, the Alps, or the Rockies are mountains. The difficulty arises in determining the point at which mountains become hills. Since our task is to delimit the mountain areas of the world, some arbitrary limit must be used. The definition selected is that *mountainous areas are those having a local relief greater than 2,000 feet and having narrow summits.* Local relief, as used here, signifies the difference in elevation between the summits of the peaks or ridges and the bottoms of the included or adjacent valleys and basins. The figure "2,000 feet" has no special significance, except that this difference in elevation is usually sufficient to produce a distinct vertical zonation of vegetation.

As indicated earlier, the great mountain ranges of the world are but diminutive wrinkles on the surface when compared with the entire size of the planet, but they exert a major influence on the flow of exogenic energy along the surface and on the distribution of life forms, including man. Their general distribution, as described in Section 11-2, is along the margins of the continents, especially so if the ancient Gondwanaland blocks are considered together (see Figures 11-6 and 11-9). Geologically, they are rather temporary global landforms. Mountain-building periods occur in irregular cycles, and such periods are relatively short and separated by long periods of crustal stability. We are living now in a geological period immediately following an exceptionally active period of crustal unrest that began in the latter part of the Tertiary era (see Table 2-2). Compared with the usual condition of the earth surface throughout most of the last third of earth history, the mountains are high and abundant, the continental blocks have risen to unusual heights, the shallow interior seas have retreated to the continental margins, climatic conditions show great latitudinal contrasts, and more of the energy flow across the face of the earth is being expended in the work of denudation. Mountains are geologically temporary features because they represent major obstructions to the transfer of energy. Mountains are the best places to observe weathering and erosional processes at work. Wind, rain, roaring mountain streams, freezing and thawing, mass wastage, glacial abrasion, solution—all operate intensely to destroy high elevations.

11-4 TYPES OF MOUNTAINS Mountains assume a wide variety of forms whose features, like those of all other landforms, are determined in part by the processes of formation and alteration, by the structure or arrangement of the rock material from which they have been formed, and by the relative stage in the general work of denudation. It is not possible to develop a simple generic or morphological classification suitable for all mountains because of the many variables that influence their form and substance. Each mountain has its own history, and no two mountains are exactly alike. There are, however, a number of general types that are sufficiently common to be easily recognized in various parts of the world. These are described in the following subsections.

Volcanic mountains. Volcanoes are perhaps the most spectacular and distinctive mountains on earth. This is especially true of the great cone-shaped peaks that were built by the accumulations of fragmental ejecta or pyroclastic (Gr.: *pyros* = fire; *klastos* = broken) material. Towering high above the surrounding region and often snow-capped throughout the year (see Figure 11-15), the immense cones are awesome sights, and it is not surprising that many of them—Fujiyama in Japan, Ararat in eastern Turkey, Popocatepetl and Orizaba in Mexico, and Kilimanjaro in east Africa—have had religious connotations from early history.

The different types of volcanic eruptions and the landforms that subsequently result are related largely to the composition of the constituent lavas. Basic lavas, or those that contain a high percentage of iron, magnesium, and calcium, have a high melting point, are highly fluid, and tend to lose their included gases (steam, carbon dioxide, sulfur dioxide, chlorine, etc.) rather quickly unless confined under high pressure. Such lavas, because of their fluidity and low gas content, flow out of fissures at the surface to form the sheetlike basaltic strata of many high plains.

Figure 11-15 Mount Hood, a typical explosive volcanic peak, seen from above The Dalles, Oregon. Note the eroded edges of basaltic lava flows in the foreground. Such flows form a large part of the Cascade Range, as well as the Columbia Plateau to the east. [U.S. Air Force (MATS)]

At the other extreme is the acidic, sialic material of the andesitic or felsitic lavas that tend to be highly viscous. Such lavas retain their gas content, cool quickly, and tend to form the explosive type of eruption and general conical shape of volcanic peaks. Many intermediate types of lavas can be found, each reacting differently to cooling and accumulation.

One intermediate type of lava produces the *dome volcanoes,* so named because of their broad, convex shapes. The Hawaiian volcanoes are of this type. Their rounded summits cap enormous volcanic structures, because they have been built up, layer upon layer, from the floor of the ocean at about 18,000 feet below sea level to heights of 13,000 feet above the sea. Their sides are scarred by the twisted, corded surfaces of numerous lava flows that typically break out on the sides of the mountains and flow down toward the sea (see Figure 11-16).

Some volcanic mountains have been formed from a mixture of lavas derived from the same *magma* (molten rock beneath the surface) reservoir. Mount Vesuvius and Mount Etna in Italy, for example, not only eject cinders and ash from their craters but also emit streams of liquid lava on their sides to help build up the mountain flanks. Composite cones, with multiple-crater rims, and craters of various sizes and shapes add to the complexity of form. The construction of the ash cones can be remarkably rapid, as in the case of Paricutin, an ash cone in west-central Mexico, which appeared suddenly on February 20, 1943, in a farmer's cornfield and grew into a mountain cone 6,000 feet high within a few weeks.

The volcanic mountains built by the accumulation of pyroclastic material may also be destroyed by volcanic activity. Gigantic explosions caused by the gradual accumulation of unrelieved gas pressure within the volcanic conduit may destroy the top of the mountain in a single blast. Foundering of the volcano within its own magma reservoir following a long period of activity is also possible. Crater Lake, a former explosion cone in southwestern Oregon (see Figure 12-2), is believed to have lost its summit in precisely this manner.

Block mountains. Block mountains, or *fault-block mountains,* are formed as the result of crustal rupture or faulting by tectonic forces seeking to stretch sections of the earth surface. Figure 11-17 illustrates two ways by which block mountains may be formed: (1) stretching and block foundering, leaving upland blocks to be eroded later into mountains; and (2) vertical thrusting from below. As mentioned earlier, areas contained within bifurcated segments of the great continental alpine chains of the world often show the effects of regional tensional stresses following the relaxation of tension. The form taken by individual block mountains may vary widely, as indicated in Figure 11-17, but generally they have a rough outer rectilinear form, occur in parallel, linear patterns, and have steep, truncated spurs along their flanks. The faults that border them usually are steep and almost vertical. Symmetrical block mountains that have not been tilted or folded are known as *horsts*. Block mountains often still retain some of the tectonic activity that

Figure 11-16 Lanai Island, Wahapuu, Hawaii. This island shows the characteristic domed surface of the Hawaiian *dome* volcanoes. Deep ravines have begun to score the sides of the volcanic island. [U.S. Air Force (MATS)]

Figure 11-17 Block faulting by regional tension and irregular vertical thrusting. In (A), broad regional arching by slow compression is followed by tensional stresses when the compression ceases. This is the most common origin of the block mountain-basin regions found between bifurcating segments of the great alpine cordilleras, as in the southwestern United States. In (B), the region is domed upward by irregular vertical thrusting from below.

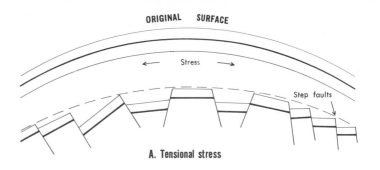

A. Tensional stress

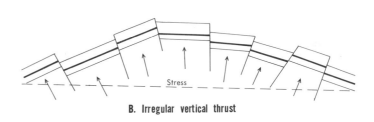

B. Irregular vertical thrust

formed them; hence, earthquakes are fairly common in areas where such mountains occur.

Figure 11-6 indicates that large areas of block mountains occur in many parts of the world. Some individual fault blocks are unusually large and form some of the major world mountain ranges. The Sierra Nevada Range of California, for example, extending from north to south for a distance of over 700 miles, is a gigantic, tilted fault block of ancient metamorphic and igneous rocks that has a precipitous fault escarpment along its eastern face. Slippage along its eastern flanks still takes place periodically. The much gentler western side of the range represents an arched erosion surface. Extending from the eastern face of the Sierra Nevada for hundreds of miles is the Basin Range or Great Basin region, which contains dozens of block mountains of various shapes and arrangements. This region continues far to the south into central Mexico. Parts of interior Iran are almost mirror images of the block mountain region of Arizona (see Figure 11-12).

Folded mountains. Compressional stresses are the main tectonic processes that operate along most of the great mountain arcs of the world, buckling and arching the crustal rocks upward. The reaction of different rock materials to compressional stresses varies. Some of the softer and thinner sedimentary strata fold and crumple easily into small, tight folds.

Thicker and more durable strata tend to produce folds of great width and height (see Figure 11-18).

Mountains whose present form is directly related to compressional folding are rare, mainly because erosion tends to destroy the complete folds rather quickly. There are a few exceptions, however, such as some of the great anticlinal mountain ridges in Lebanon and Algeria. Many ranges still preserve the internal structure of their original folding, and the erosional forces that are destroying them are guided by variations in rock resistance.

The complexity of folds in the structure of many mountain ranges may be great. This is especially true of the Alps in southern Europe. The sedimentary strata there were twisted and warped into shapes that chal-

Figure 11-18 A large Pliocene (late Tertiary) fold in massive limestones near Kousba, Lebanon. The vertical face of this eroded anticline (upfold) is approximately 2,000 feet. The small white dot in the center of the cliff face is a small Maronite monastery. Massive beds such as these do not buckle easily and tend to form great arches under regional compression. [Van Riper]

lenged the research of many geologists who for many years tried to decipher the sequence of events that produced the present complex mountain landscape.

Erosional mountains. Most of the mountains in the world probably owe their forms largely to erosional processes rather than to tectonic forces. The latter are always responsible for the earth surface's attaining elevations far above sea level, but the shapes and forms assumed in mountainous terrain are largely the result of weathering and erosional processes acting upon different kinds of rock material arranged in different ways. Despite the infinite differences between mountains, each type of weathering or erosional process leaves its distinctive mark on the mountain landscape. To illustrate, many mountains are sufficiently high to have permanent snowfields and glaciers. The jagged crest lines, rounded passes, sharp divides, and amphitheaterlike depressions of glaciated mountains give them a characteristic appearance everywhere on earth (see Figures 11-8 and 11-19). The distinctive cliff forms and lack of linear valleys associated with solution (see Figure 11-27) and the unique features associated with mass wastage in the periglacial mountain zones (see Figure 9-3) are also illustrative.

Remnants of ancient plains often can be found in some of the mountain cordilleras, indicating how far erosion has proceeded in the beveling of upwarped surfaces. Figure 11-20 shows the remnants of a concordant

Figure 11-19 Typical glaciated alpine landscape in the New Zealand Alps. The characteristic sharp, jagged edges of the crest lines are clearly evident. Note the rilling by snow avalanches and the lateral cracks (*bergschrunds*) that result from the slipping of ice and snow out of the amphitheaterlike *cirques*. [Van Riper]

Figure 11-20 Remnants of an ancient peneplain or erosional surface in the Beartooth Mountains, Montana. The flat summits of these mountains represent an ancient erosional surface uplifted far above sea level and now being dissected by erosion. [U.S. Air Force (MATS)]

level at the highest summits of one of the Rocky Mountain ranges. The photograph illustrates the enormous amount of excavation that has taken place since the regional upwarping of the Tertiary. These clearly are erosional mountains.

11-5 THE HUMAN SIGNIFICANCES OF MOUNTAINS The significance to man of mountains, like the meaning of all the other elements in his physical environment, cannot be treated apart from local cultural or technological developments. We usually consider mountains as obstacles to human aspirations. Yet for some people, mountain lands have provided welcome havens against war-minded neighbors or have furnished valuable resources of minerals, lumber, water, and scenery (see Figure 11-21). The Swiss and the Austrians would stoutly deny that the Alps were a hindrance to their national goals. Ifugao tribesmen, in the interior of Luzon, shunned the relatively empty humid coastal plains 50 miles away to select some of the steepest mountain terrain on the island to construct their remarkable rice terraces. To them, the centuries of hand labor required

Figure 11-21 Terraced fields and mountain villages related to rock type in Lebanon. Relatively soft, friable sandstones and shales lying between durable limestone beds can be easily excavated into terraces for cultivation. A frequent site for villages is the "spring" line, or contact between permeable sandstones and less permeable shales. Springs also frequently are found near the base of the massive limestone beds that form cliffs in this rough country. [Van Riper]

to create these incredible feats of hydraulic engineering were entirely justified by the isolation of their tribal territory and the relative ease of defending it.

Mountains, of course, can be obstacles even in the modern world of advanced technology, but the specific meaning of such resistance changes with the times. The high passes in the Alps have entirely different meanings to the tourist who speeds through the Simplon tunnel in a comfortable railroad car than they had to Hannibal, who struggled through them with his elephants on his historic campaign against the Romans. Yet, even today, the cost of overcoming the mountain terrain must be paid for in some way. The leading source of income today in Colorado is not agriculture or mining, but recreation, not only because of the presence and accessibility of its mountains but also the leisure time and surplus wealth created by our production system. Mountains are simultaneous assets and hindrances. The proportion of each on our cultural balance sheets cannot be measured solely in terms of slope or elevation.

An elaboration of the meaning of mountains on a global scale would be both misleading and unfruitful. The geographer or economist should perform this task on the local or topographic level. We may, however, list some of the more important resource significances and obstacles that mountain lands have presented to human beings at different times and at different places:

Resource significances of mountainlands

1. Protection
2. Isolation
3. Minerals
4. Timber
5. Water power
   a. regulation
   b. storage

6. Recreational
   attractions
   a. scenery
   b. camping,
      fishing, etc.
   c. winter sports
7. Agricultural
   attractions
   a. volcanic soils
   b. crop variety
      owing to vertical climatic
      zones
   c. irrigation

Obstacles of mountain lands

1. Long, steep slopes
2. Avalanches
3. High cost of maintaining transportation lines
   a. landslides
   b. washouts
   c. deep snow
4. Earthquakes and volcanoes
5. Forest fires
6. Erodable, thin, rocky soils
7. Rarefied air at high elevations
8. Wind damage
9. High rainfall
10. High incidence of lightning

Plains

Plains are areas of low local relief. In mapping the global distribution of plains, however, a more specific definition is needed.

As here considered, plains include *areas where the common local relief is less than 200 feet and where slopes of less than 5 percent predominate.* Isolated hills or even mountains may rise more than 200 feet above the surrounding plain without affecting the estimate of common local relief. Moreover, plains are not restricted to low elevations. In Bolivia and Tibet, for example, they may occur at elevations well over 12,000 feet above sea level.

11-6 THE GLOBAL PATTERN OF PLAINS

Plains comprise slightly more than half of all of the surface of the continents. This is not surprising, because as we have previously noted, the flow of exogenic energy tends to reduce surface irregularities, and the tectonic processes that produce them are relatively localized, being found mainly at or near the margins of continents. Even the present distribution and height of mountains appear to be unusually great compared with the more normal condition over long periods of geologic time. The predominance of plains also is reflected in the shape of the hypsographic curve of the earth surface (see Figure 11-2), which shows the predominance of land surfaces below 600 feet in elevation. The main platforms of the continental blocks are relatively stable, and stability signifies a plain surface, some of which is erosional and the rest depositional.

For our purposes, the global plains may be classified into four main groups: (1) interior plains, (2) alluvial trough and delta plains, (3) high plains, and (4) coastal plains. Each of these major types is discussed in subsequent sections, and their location is shown in Figure 11-22.

11-7 INTERIOR PLAINS

Broad plains extend throughout much of the interior of the continents, mostly at elevations less than 600 feet above sea level. With only a few local exceptions, the great continental interior plains have been without appreciable relief since the early Paleozoic, or the last 600 million years of earth history. During this time they often were invaded by shallow seas, as their thin covering of sedimentary strata of marine origin testifies. The existence of persistent shallow basins, troughs, arches, and domes in the rock structure of the interior plains indicates local areas of uplift or depression. Such movements generally have been gentle and slow, and since deposition and erosion have been able to compensate for them, a subdued relief has been maintained.

The interior plains of the world differ widely in detail of form, depending on the processes that have shaped their low relief. Plains resulting from erosion by running water and the accompanying mass wastage are likely to have not a flat but an undulating-to-rolling land surface, with occasional local areas of steep slopes. Much of the central drainage basin of the Mississippi River south of the glaciated region is an example, as is much of the broad expanse of plains in the central portion of the Soviet Union. Preferred areas for human settlement on these plains include the strips of alluvium that follow the drainage paths through these regions.

Deposition has leveled the surfaces of other sections of the interior plains. Continental glacial deposition, for example, smoothed the old erosional surfaces of much of the American Midwest and parts of the northern European plains. Deposition of alluvium in great shallow basins formed the surfaces of other sections of the interior plains. Parts of the Amazon Basin (especially the western portion), the Congo Basin, the Ob River basin in the Soviet Union, and the Chaco area of Paraguay are good examples. Lake plains, sand- and ash-filled basins, and loessal plains make up other types of depositional surfaces.

Regardless of the individual process of

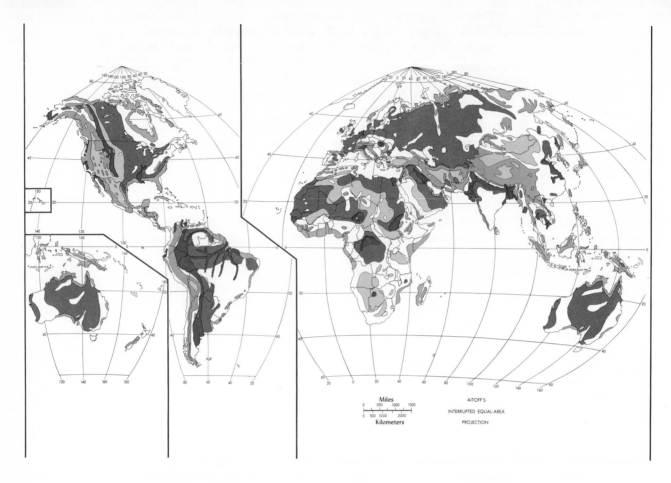

Miles
0   500   1000   1500

0   500 1000   2000
Kilometers

AITOFF'S

INTERRUPTED EQUAL-AREA

PROJECTION

gradation or deposition, the interior plains have a basically low relief because they have been relatively stable portions of the continental masses for long periods of geologic time. In general, their elevations have fluctuated only occasionally from slightly below to slightly above sea level.

11-8 ALLUVIAL TROUGH AND DELTA PLAINS The alluvial trough and delta plains are underlain by thick beds of sediment resulting from long and continuous deposition. Many are in geosynclines that are subsiding slowly into the crust under the weight of accumulated sediments (see Figure

11-7). Some are delta plains that may or may not be located on active geosynclines. This group of plains is generally located within the unstable portions of the earth, such as around the continental margins and near high mountains. They are the major collection places for the material that is removed from high elevations. Eventually they may become highlands themselves, as the result of isostatic readjustments. The most active area by far in the United States is near the mouth of the Mississippi. Others include the sedimentary basins along the west coast, such as the Willamette Valley-Puget Sound trough, the Central Valley of California, and the

Figure 11-22   World distribution of types of plains.

Coastal Plains

Alluvial Troughs and Delta Plains

High Plains (+2,000' elev.)

Interior Plains

Mountains

Imperial Valley at the head of the Gulf of Lower (Baja) California. In late Tertiary times, enormous depositional troughs occurred along the eastern margins of the Rocky Mountains and in the basins that separated the individual ranges. Not much deposition is taking place in these areas today, and they are classified as high plains. Erosion working its way westward along the tributaries of the Mississippi and Missouri Rivers is beginning to dissect some of the great alluvial strata out of which many of these plains were formed.

Examples abroad include the flat, sedimentary plains immediately east of the Andes Mountains from Venezuela to northern Argentina; the Po Valley of northern Italy; the delta of the Rhine River in northwestern Europe; and the Hungarian and Romanian plains. The wide marshes and poor drainage of the Tigris-Euphrates plain in Iraq, south of Baghdad, indicate a slow subsidence of this large plain. Other such plains include the large fertile deltas of the great rivers in eastern and southeastern Asia, such as the Indus, Ganges, Mekong, Yangtze, and Hwang Ho. Some of the interior basins of central Africa, such as the Sudd in the Sudan, are slowly subsiding, probably under the weight of sedimentation.

One major feature of these plains is their extremely low relief. Another is the unconsolidated sedimentary material that underlies the surface everywhere. In humid areas, such plains are subject to flood hazards because of their low slopes and their nearness to great rivers.

11-9   HIGH PLAINS   High plains are those that lie at elevations of more than 2,000 feet. Many of them have been termed *plateaus,* but this term is avoided as much as possible in this text, mainly because it has been used to mean different things in the past. Presumably the term signifies a fairly level tableland whose upland surface is higher than much of its surroundings. Most of the so-called plateaus of the world do not fit this definition; in fact, true plateaus that do are relatively rare. The best examples are the plains of interior Africa and the Colorado Plateau of the southwestern United States. For this reason, the term is used here only where it is part of a proper name—for example, the Appalachian Plateau or the Bolivian Plateau.

High plains differ from other plains principally in elevation. Although this feature does not influence the topographic definition of plains, it produces at least two characteristics that are common to most high plains. First, such plains are likely to be drier than other plains, mainly because moisture is removed from air masses as they ascend abruptly to elevations above 2,000 feet or as they pass over the much higher mountains that often adjoin high plains. Second, many (but not all) such plains have steep-sided, deeply entrenched river valleys, representing early stages in the processes of erosion by running water. These deep gorges greatly influence the factor of accessibility and constitute primary terrain obstacles to surface transportation routes. Gorges of more than a few hundred feet or so in depth and width are generally more difficult to cross than mountain ranges.

The general distribution of high plains in the world is principally within the zones of crustal instability. They usually are bordered by great mountain cordilleras. The highest and best-known high plains are located between bifurcating segments of the continental cordilleras, such as those in Tibet, Bolivia, Mexico, Turkey, Iran, Spain, the western United States, and western Canada.

Although the continent of Africa is not located within the major global areas of crustal instability, it contains a large area of high plains. The interior of most of this continent was elevated relatively recently in geologic time, but not to heights comparable with those of the cordillera zones. Most of the African high plains lie at elevations between 2,000 and 3,000 feet above sea level. Exceptions include the volcanic lava surfaces of Ethiopia and Kenya, which rise far higher.

The initial cause for many high plains is broad regional arching, in which the arches may be several hundred miles across. In the United States and Canada, for example, regional arching along a north-south axis elevated much of the western portion of these two countries. The present Middle Rockies of the United States represent durable, resistant rock areas that remained after stripping of the less resistant materials near the higher parts of the regional arch. Slightly to the east of the Rockies, the surface of the Great Plains represents the gentle eastern slope of the regional arch, and there is a gradual transition between the high plains and the interior plains to the east. The amount of dissection of the regional arch, which often determines whether the surface becomes mountains, hills, or plains, is related to the amount of runoff of surface water. The drier portions of the regional arches in the world have thus been preserved as high plains. This is perhaps the major reason for the low relief of much of the Colorado Plateau.

The statements made in the previous paragraph on the origin of the high plains should be regarded as broad generalizations. There are several other ways, supplementing broad regional uplift, in which high plains may be formed. Some such plains are of volcanic origin, consisting of thousands of feet of successive lava flows. The Columbia Plateau east of the Cascades and the high plains of Kenya and Ethiopia are examples. Other high plains are produced by deposition of sediments in dry interior basins, supplemented by the development of rock plains in the erosion of adjacent mountain ranges. The original surface may have had a much greater relief. The high plains of much of Iran, Anatolia, the Basin Range region of the southwestern United States, and parts of the Gobi Desert in eastern Asia are examples. In eastern Mongolia and northwestern China, some of the high plains are made up of basins that have been partially filled by loess.

11-10 COASTAL PLAINS Coastal plains are the shelving edges of the land portion of continents, plains formed as the result of deposition or erosion below sea level and later raised above sea level. By definition, then, coastal plains are of marine origin and do not necessarily include all plains that lie along the coast. Some are rock plains that represent ancient erosional surfaces, such as those that border the Arctic Sea. Others represent areas of coastal marine sedimentation, such as the Atlantic–Gulf of Mexico coastal plain of the United States and eastern Mexico.

There are only two extensive areas of coastal plains in the world: the Atlantic–Gulf of Mexico plain and the low plains that border parts of the Arctic Sea in Canadá, Alaska, and the Soviet Union. Smaller coastal plains, however, may be found bordering most of the continents. Examples include the narrow plains along parts of eastern Brazil, Mozambique, Nicaragua, and the Guinea coast of Africa.

The characteristics of coastal plains depend partly on the amount of uplift and erosion that has taken place since they emerged from the sea and partly on the composition of the underlying material. In plains that have emerged relatively recently, many of the original depressions have collected water and have become lakes or swamps. The marshes of the Florida Everglades and the Lake Okeechobee region are illustrative. Sometimes the terrain is so flat that it is difficult for water to drain from the surface, and extensive swamps or marshes may result, such as the Dismal Swamp of southeastern Virginia and northeastern North Carolina. Other poorly drained areas are former lagoons back of beach ridges. Some of the coastal swamps of eastern Brazil and the Guianas are of this type.

Not all coastal plains are flat. Sections of the coastal plain in the eastern United States have been uplifted and eroded to form a rolling terrain with a local relief of between 50 and 100 feet. When depositional coastal plains are uplifted, the soft, unconsolidated material is usually rapidly dissected by streams.

The rock coastal plains of the world show a variety of surface characteristics, and their specific landforms, like those of the depositional plains, depend on the degree of uplift, the composition of the underlying material, and the erosional processes that now are working on them. The surface features of the Arctic coastal plains are related largely to

recent glacial erosion and deposition and to the unique features of mass wastage in this periglacial region. Limestone underlies some of the rock coastal plains, as in the Yucatan Peninsula in Mexico, in parts of Florida, and on some of the narrow tropical coastal plains which represent former coral reefs that have been raised above sea level. Solution by underground water generally has modified such surfaces.

11-11   THE HUMAN SIGNIFICANCES OF PLAINS   The outstanding significance of plains to human beings everywhere is that surface irregularity generally becomes a relatively minor consideration in the physical environment. Exceptions may be noted, however, in the deep canyons of some of the

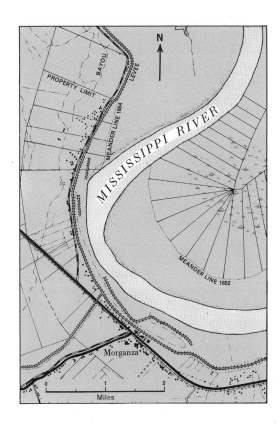

Figure 11-23 Cultural patterns developed on a typical floodplain.   Note how the buildings and main roads tend to follow the arc of the natural levee next to the river. The radiating lines are drainage ditches that drain into the back marshes away from the river. They also tend to follow the ancient property lines set by the early French settlement, which traditionally ran at right angles to the river frontage. [From AMS sheet accompanying H. N. Fisk Report on Geological Investigation of the Mississippi River alluvial valley]

Figure 11-24 Flat, glacial till plain near Towanda, Illinois.
The levelness of this plain is caused by glacial deposition. The combination of level land, black, fertile, and stoneless soils, and skillful farming makes this some of the most productive land in the world. [Standard Oil Company (N.J.)]

high plains and in areas where minor topographic irregularities have a special significance because of their drainage or land-use features—for example, natural levees on alluvial plains and deltas (see Figure 11-23), and beach ridges on lake plains. In general, however, slope considerations on plains have mainly local significances. It is appropriate to note here that, although slope rarely is an important independent variable in the settlement of plains, dense populations generally do not occur except where many of the environmental factors are favorable, *including low relief.* There are fertile plains and sterile plains and plains whose significances to man have changed with the times. As with the other environmental factors, the significances of plains cannot be treated apart from the associated factors of climate, soil, drainage, vegetation, mineral wealth, and human drives and abilities. Some generalizations about the resources and obstacles of the major types of global plains are presented in the following paragraphs.

The interior continental plains, despite their low relief, have wide areas of low-to-moderate population density. In the rainy tropics, such areas are generally characterized by infertile soils and dense forests. In the central Asiatic plains and in central Australia, the sparsely inhabited sections are dry regions, while in the plains of northern Canada and the Soviet Union, they are areas of excessive winter cold. By far the most suitable for agriculture are the humid interior plains of the mid-latitudes, especially where soils are underlain by limestones or material that contains much lime (see Figure 11-24). Examples include the glacial till or loessal plains of the American Midwest, the limestone plains of Kentucky, Tennessee, and southern Indiana, the limestone plains of central France, and the loessal plains of southeastern Germany and the Ukraine.

The alluvial troughs and delta plains, where they are adequately watered and have been occupied for a long time, include some of the most favored agriculture areas on earth. Their soils are mixtures of topsoils from extensive areas of the drainage basins and are largely stoneless and easily cultivated. The rice plains of eastern and southern Asia are examples. The quality of alluvial soils depends partly on their age, and some of the older alluvial deposits, such as those along the middle course of the Ganges, are relatively poor because of leaching (the washing out of soluble plant foods). In some of the structural troughs and basins adjacent to mountains, the sloping plains are well adapted to the use of irrigation. Floods are a recurrent threat on many of these plains, and as the clearing and cultivation of watershed areas continue, this problem becomes more and more difficult to control. Not all the alluvial troughs and delta plains are favored places for human settlement. The delta of the Amazon has relatively few inhabitants, and agriculture is only beginning in that area. The alluvial basins included within the mountain ramparts of Asia, such as the Tarim Basin, are too dry to support more than a sparse population subsisting by means of irrigation and grazing. The great alluvial trough that lies immediately east of the Andes has been handicapped by inaccessibility, dense tropical forests, and alternate floods and droughts.

The high plains in general are inhospitable to man. Windswept, usually arid, generally far from the main centers of world population, and generally hindered by terrain obstacles that separate them from more suitable plains at lower elevations, the high plains generally have only a sparse population. Some of them support clusters of oasis settlements where irrigation water from adjacent mountains is available, and others support a sparse pastoral population that depends on grazing for subsistence, as in interior Asia and Africa. Exceptions always should be noted, and in the United States and Canada, the great granaries of the western Great Plains and portions of the Columbia Plateau constitute valuable segments of our productive terrain. Inhabitants of Mexico City, Nairobi, Ankara, and La Paz could also rightfully claim that their environs had opportunities which they would not exchange for the advantages of many lowland plains.

Coastal plains are not among the most preferred sites for human settlement, although there are some exceptions to this statement. Marine shallow-water sediments frequently are composed of quartz sands or gravels and are deficient in minerals that release plant nutrients following rock decay. Limestones produce the best soils for agriculture, but solution by underground water and rapid removal of water via underground streams frequently result in deficiencies of water in the soils for plant use. The dry, rocky soils of Yucatan are illustrative. Poor drainage is another major handicap on coastal plains. Their rivers, although relatively far apart, are likely to be broad, deep, and with low banks. Suitable sites for bridges are few. Coastwise travel by land requires unusually large expenditures for drainage and bridges.

Hills

Hill lands are areas where local relief is between 200 and 2,000 feet and where slopes of more than 5 percent predominate. Hill lands thus include the continental surfaces midway in surface irregularity between plains and mountains (see Figure 11-14).

11-12 THE GLOBAL PATTERN OF HILL LANDS The global distribution of hills is related to the geological processes involved in their formation. Hilly terrain usually results from one of two basic causes: (1) regional uplift to elevations at which subsequent erosion produces a rough terrain with a local relief of between 200 and 2,000 feet; or (2) the advanced stages of erosion of former mountainous areas, in which the hills represent the remaining "roots" of the earlier mountains. Most mountain areas have a transitional zone of hills, sometimes termed *foothills,* between the mountains and the adjacent plains. Such intervening hill zones result either from intermediate heights of crustal warping or from the progression of erosion from the borders of the adjacent plains. Perhaps the most extensive hill lands in the world are associated with the great platform blocks of sialic rocks that are known as *shields* (see "Sialic Shields," page 394). Some broad areas of these continental platforms apparently undergo periodic slow uplift that maintains some of the platform blocks at elevations where subsequent erosion produces hilly terrain.

There are two major areas of hills in North America. The first is the Canadian shield area of Canada (including a small portion of the northern United States), and the second is the general Appalachian region. The latter is an excellent example of a former sedimentary basin that later was uplifted, highly folded by compression on the east, beveled by erosion to a plain during the Cretaceous period, then arched upward again, and is now in various stages of dissection by streams. Only a small part of it has sufficient relief to be termed mountainous.

The major area of hills in South America includes the Brazilian and Guiana highlands, both of them shield areas with segments of sedimentary rock strata lying upon them. Figure 11-25 shows the edge of some of the sedimentary strata that lie on the underlying crystalline base. The hydroelectric potential of the Amazon tributaries as they descend from the crystalline uplands is unusually large.

There are many small areas of hill lands in Europe separated by both plains and mountains. The Scandinavian countries as a group comprise a shield zone that is tilted upward irregularly. Hill land prevails throughout most of Norway, Sweden, and Finland except on the southwest, where the shield block was raised the highest and has been dissected into mountainous terrain. Several old eroded highlands occur in continental Europe that once were mountainous. Among these are the highlands of Scotland and Wales, Brittany in France, the Vosges and Black Forest regions bordering the central Rhine River, and the other patches of hilly terrain in central and southern Germany, such as the Bohemian Forest (Bohmerwald), the Hartz Mountains, the Thuringia Forest, and the Erz Gebirge.

The Ural Mountains of the Soviet Union, which traditionally separate Europe from Asia, are inappropriately named, because only a small part has a local relief sufficiently great to be termed mountainous. The Urals are a good example of hilly country that represents the ancient eroded remnants of what once was a high mountain range. The shield areas of Asia include the major areas of hills: (1) the elevated shield that borders the sedimentary basin of the Lena River; (2) the southeast China-South Korea shield; and (3) the crystalline shield block of peninsular India that is tilted upward on the west and may well have been joined to Africa in early Mesozoic times. A large part of China is hill country, and west of the sialic shield the country is underlain with thick sedimentary rock strata deeply eroded by streams.

Africa has a large percentage of its area in hilly terrain. The continent constitutes a vast shield and is only masked in places by a veneer of sedimentary rocks. Large sections of the Atlas Mountains are hilly rather than mountainous and are comparable to the highly folded and eroded strata of the eastern Appalachians.

Australia, similar to Africa and India, to

Landforms and Surface Configuration

Figure 11-25   Typical tablelands in southeastern Venezuela.
The high, cliffed edges of these upland surfaces make them almost
inaccessible. Erosion of these uplands consists mainly of the recession of the cliff faces. Note the sharp boundary between grassland
and forest, a good indication of the influence of clearing and the use
of fire in removing the original forest vegetation. [U.S. Air Force
(MATS)]

Global Landforms   419

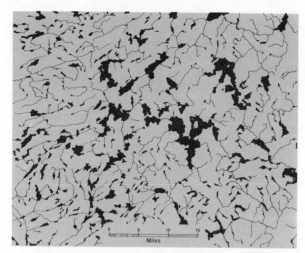

Figure 11-26 Typical drainage pattern in the Canadian shield. The exceedingly complex drainage pattern illustrates how a rough, rock-knob terrain and irregular deposition of glacial detritus can combine to interfere with normal stream processes. The pattern also illustrates why it is so easy for strangers to lose their way in such country. [Toronto-Ottawa Sheet No. 31 SW., National Topographic Series, Canada, Dept. of Mines and Resources]

which it once probably was joined, comprises a crystalline shield block with an irregular mantle of sedimentary rocks. Streams are just beginning their work of dissecting the main shield, and the major hill lands mark the areas where erosion is most active, a condition that is largely related to rainfall. This is especially true in Africa around the outer margins of the Congo Basin and along the slopes north of the Guinea coast.

Several different types of hilly terrain may be distinguished and are described separately in the following sections. In general, however, the texture, or grain, of hilly terrain is more even than that in plains or mountain areas; that is, the areal spacing of valleys and summits is likely to be more uniform regionally. Slope is usually the dominant environ-

mental factor, and sudden areal changes in slope are reflected in complex local patterns of soils, vegetation, and land use. These variations, however, are topographic in scale. At broader scales of observation, hilly areas exhibit a regional uniformity which frequently contrasts with the irregularities that are discernible in adjacent plains and mountains.

11-13 THE HILL LANDS OF THE SHIELDS
Figure 11-6 shows the location of the principal sialic shields in the world. By comparing this map with that in Figure 11-14, which shows the distribution of hill lands, it will be noted that both plains and hills make up nearly all of the shield areas. Continental glaciation during the Pleistocene period modified the terrain of the Canadian and Fennoscandian shields in several ways to produce a highly distinctive terrain. The ice sheets, on their initial advances, removed practically all of the soil and broken rock mantle. Despite an abundance of durable rock tools for glacial erosion, the hard shield rocks of the hills resisted glacial scouring remarkably well. The hills were rounded and polished, but not removed. It is doubted that glacial erosion ever removed much of the unweathered, unfractured bedrock. Valley sides were widened and scoured and became grooves for the passage of the ice. With the retreat of the ice fronts, a mass of glacial detritus was let down upon the eroded surfaces, and both till and fluvioglacial material were irregularly distributed, thickest in the valleys and basins and thinnest on the rounded hilltops. Erosion and deposition destroyed almost all traces of the preglacial drainage system. Today lakes, swamps, and irregular stream patterns dot the land surface within irregular depressions and between rounded, rocky knobs (see Figure 11-26).

The amount of local relief differs widely

within these shields. In many parts of northern Minnesota and Wisconsin, the knobs rise only 50 to 200 feet, so that the shield area is only a rough plain. In others, such as in the Adirondacks of New York State or the White Mountains of New Hampshire, the rounded rock knobs are sufficiently high to be mountains. Only the topmost peak of Mount Washington (elevation 6,288 feet) in the White Mountains protruded as an island above the ice. The northern shield hill lands, including those in Scandinavia, are extremely submarginal for agriculture, and forestry and recreation have proved to be much more stable sources of income than farming. The stony, acid, and droughty soils of these regions are among the poorest in the world. Occasional patches of heavier soils, such as those found on some of the glacial lake plains, are better, but the short growing season is an additional handicap to the farmer.

Tropical and subtropical shields have been uplifted so that denudation is dissecting them. These processes result in hilly terrain that is entirely different in appearance from that in the shield areas of the upper mid-latitudes. Intrusive igneous rocks weather rapidly in the humid tropics and soon are mantled with a thick soil cover. Stream erosion, on the other hand, cannot cut downward into these hard rocks much faster here than anywhere else, and dissection proceeds extremely slowly. As a result, the broad areas of homogeneous rocks are etched into a fine network of ravines, which are relatively steep-sided because of the soil-retaining capacity of the forest cover. The heavy rainfall is efficiently removed by a completely integrated drainage network, which slowly etches its way downward into the upland platforms. This is rough country, with few easy cross-country routes. The predominance of weathering and mass wastage over stream abrasion in these humid climates, then,

produces a much different type of terrain from where streams find downward cutting relatively easy.

11-14 ERODED SEDIMENTARY ROCK UPLANDS Extensive hilly areas also are located around the borders of some of the interior continental plains and apart from the crystalline shields. Examples include the Appalachian uplands, the hill lands of southern Europe, and the southwestern Soviet Union. These are mainly areas of former depositional plains that have been warped upward and are in the process of being eroded. For the most part, the topographic landforms of these hill lands are related to the processes of erosion operating on various types of rock structure and to the stages of denudation. There are various stages of gradation, from flat plains to extremely rough hill country, throughout the continental interiors.

11-15 KARST HILL COUNTRY An irregular land surface that has resulted from the solution of limestone or dolomite is termed *karst terrain*. The name was derived from the Kars district of Yugoslavia, where some of the classic studies of this type of erosion first were made. Not all karst terrain is hilly. Gently undulating plains, such as the Highland Rim country of Kentucky, some of the Paris Basin, and the limestone country of central Florida and the Yucatan Peninsula, have resulted from this kind of erosion. Occasionally some of the limestone strata may be so thick and massive that solution produces mountainous terrain. The extremely rugged country of southwestern Yugoslavia and northern Albania is of this type (see Figure 11-27).

The largest areas of karst terrain, however, are hill lands. Such terrain occurs in many parts of the world, and in general bordering

Figure 11-27 View of mountainous karst terrain near Cetinje, Yugoslavia. Nearly all of the irregularity in the land surface observed in this photograph has been produced by chemical weathering in the massive underlying limestones. [Van Riper]

the shields and the interior plains. It is especially prevalent in two general areas: bordering the Mediterranean Sea and encircling the southern portion of the crystalline shield in southeast China. The latter comprises thousands of square miles in south-central China and in northern Vietnam and northeastern Thailand. Smaller areas of karst hills are found in southern France (the Causses district), some parts of the West Indies (Cuba, Jamaica, and Haiti), western New Guinea, Venezuela, and the Philippines.

Probably the most unique feature of karst hilly terrain is the infrequent occurrence of surface streams. Drainage usually is underground, following fractures and bedding planes that are enlarged to form solution channels. The underground drainage sometimes appears at the base of a limestone cliff as large springs, as shown in Figure 11-28. The surface is pitted with depressions ranging from small hollows to large basins bordered by steep limestone cliffs. Caves and caverns are abundant. Cross-country travel is extremely difficult, because there are no continuous upland or lowland routes. The rock exposures typically are rilled with solution grooves (see Figure 9-6) and have an extremely rough microsurface.

Despite a residual soil that is generally good for cultivation, karst hill country rarely constitutes important agricultural land. Rock outcrops preclude the use of machinery, and only pockets of soil are available for cultivation in the bottoms of the solution depressions. Accessibility also is likely to be poor. Karst areas are extremely difficult for modern offensive military operations, and the rocky cliffs honeycombed with caves and caverns with multiple outlets afford ideal protection for determined defenders.

11-16 PARALLEL RIDGE AND VALLEY HILL LANDS Parallel ridges and valleys form another distinctive type of hilly terrain. The parallel alignment of the included landforms is caused by the differential erosion of highly folded sedimentary rocks of varying degrees of resistance. The best example is the folded Appalachian area of the eastern United States, a belt 10 to 100 miles wide extending for a distance of about 800 miles from New York State southward into Alabama. Other examples can be found in Arkansas, in western Burma, in North Africa within the general structural lines of the Atlas Mountains, along the eastern flanks of the Andes, and near the southern tip of Africa. Local ridges may reach mountainous relief in some of these areas.

11-17 OTHER TYPES OF HILLY TERRAIN Other types of hilly land surfaces are interspersed within the major plains of the world, but they do not have sufficient areal extent to be distinguished on a global scale.

Examples include continental glacial deposits, *badlands* (highly sculptured areas of soft unconsolidated sediments), sand dunes, dissected lava flows, and elevated coral reefs.

11-18 THE HUMAN SIGNIFICANCES OF HILL LANDS The major hill lands of the world generally are neither particularly disadvantageous nor especially attractive for human settlement. It is true that some of the hill country of the northern high latitudes is essentially unoccupied and that the narrow stream valleys included within the hilly terrain of India and China contain dense populations, but, on the whole, the above generalization is valid. As noted earlier, hilly terrain is characterized by sharp and frequent areal changes in slope conditions. This variability tends to decrease both the obstacle effects of steep slopes and the advantages of level or gently sloping terrain, for neither of these features has local continuity. Any given gradient in hill lands is likely to be less of a hindrance than the same gradient in mountain lands, because the slopes are not so long. The areas of flat land in hill country are likely to be narrow and discontinuous, despite the fact that there are many of them. Population concentrations everywhere usually tend to be cumulative, since the opportunities for earning a living increase where populations are continuous. A single block of favorable land 50 square miles in size offers more advantages to humans than the same amount of comparable land that is fragmented and scattered over a total area of 5,000 square miles. Thus, it is possible that some mountain lands may be more densely populated than hilly areas, since the included plains are likely to be larger in individual size, if not in total area.

Many hill lands throughout the world have been designated as problem areas with respect to increasing or maintaining standards

Figure 11-28 Source of the Nahr Ibrahim (Abraham's River) near Afka, Lebanon. The river appears suddenly at the foot of a limestone cliff 3,000 feet high. The height of the cavern opening is 50 to 60 feet. [Van Riper]

of living. This is closely linked with their susceptibility to deforestation and soil erosion. Hill lands are much more accessible than mountain lands; hence, hill forests are likely to be favored areas for forest exploitation. The narrow stream valleys of hill lands are often as favorable for agriculture as adjacent plains, but the restricted extent of the small strips of superior land may force an expanding population to extend agriculture onto adjacent slopes.

Hill lands, however, are not necessarily submarginal for human occupancy. An understanding of the erosion forces and the limitations of different land practices will indicate the extent to which cultivation or grazing may fit into the natural equilibrium of hillside environments. The barren, eroded slopes of Syria and Lebanon, which once supported magnificent stands of timber, the gullied, brush-covered slopes of Korea, and the scarred, grassy hillsides of India testify to a lack of human foresight in the management of hill areas. Even in the United States, where so much has been done to advertise the dangers of soil erosion, where technical assistance is available to every farmer and sheepherder, where so much surplus production can be invested in corrective and preventive measures, and where available machinery can duplicate the work of many hands, harmful crop and grazing practices still are lowering the carrying capacity of many hillside soils. Fortunately, such practices are becoming much less common, and the smooth, rounded patterns of fields under contour cultivation and terracing are becoming a characteristic feature of hill landscapes in many parts of the country. It is probably true that increased alternatives for making a living and easy mobility in employment have more to do with decreasing the ecological abuse of hillside environments than any increased knowledge of the suitable practices to ensure slope equilibrium. Stable land use practices have other parameters than the physical conditions of the terrain itself.

The hilly shield areas of crystalline rock contain local areas of exceptional importance because of their mineral wealth. The major iron deposits of the world—including the Lake Superior district of the United States, the Swedish and Brazilian deposits, and the newly discovered iron ores of the Knob Lake district in northern Quebec and western Australia—all are found in hilly shield areas. The many ore-mining districts of Rhodesia, the Republic of the Congo, Canada, and the eastern Soviet Union, most of which are included within typical shield country, contain many valuable metals, including uranium, copper, nickel, precious metals, manganese, chromium, and cobalt.

The hilly shields, apart from their mineral wealth, have decided disadvantages for human settlement. The Canadian shield of North America, the Fennoscandian shield of northern Europe, and part of the Angara shield in Siberia were covered by sheets of glacial ice about 10,000 to 18,000 years ago. This ice removed the accumulated soil from the hard-rock hills and swept it away, leaving only fresh, unweathered sandy and stony till and outwash in its place. Deposition in the valleys interrupted the normal drainage pattern and resulted in a succession of rocky knobs, lakes, and swamps. Constructing highways and railroads through such terrain is extremely expensive, frequently more so than in mountain country. Swamps must be dredged of their organic accumulations and filled with rock ballast, lakes must be skirted, and roads must be cut into a succession of hard, durable knobs—a difficult, costly, and slow process. Problems in the construction of the new Trans-Canada Highway through the hilly terrain north of Lake Superior caused it to be one of the last segments of this transcontinental highway to be completed. Until the middle 1960s, the Canadian motorist seeking to cross his own country was forced to detour into the United States. Recently, the mineral exploitation of the Canadian shield has forced new innovations in solving transportation problems, especially the use of cargo planes and tractor trains. The latter are particularly suitable for winter travel, since they need no roads and can utilize the frozen surfaces of lakes. The hundreds of thousands of lakes of the Canadian shield, once a major obstacle to overland travel, are becoming one of the region's major assets, not only for winter use by tractor trains and ski planes but also for summer recreational use.

The shield areas of low latitudes, while not easy for cross-country travel, do not present the barriers of the high-latitude shields. The thick soil weathering means relatively easy grading for roads, and although gullying by streams tends to create a fine network of slopes, road construction is relatively easy with suitable mechanical equipment. Swamps and lakes are notably absent.

Dissected upland hill country, such as the Appalachian region, affords both advantages and disadvantages for human settlement. Soils are likely to be stony and poor except in some of the limestone valleys or along narrow alluvial *bottoms* bordering rivers. Cheapness of land has attracted many of the urban poor who, with rural backgrounds, prefer being poor in the hills to being poor in the urban slums. Erosion has deteriorated many of the small, cleared hillside farms, and in general, the Appalachian and Ozark hills have the lowest living standards of any rural area in the United States. At the same time, this hilly country has significant assets, including: (1) sources of inanimate energy derived from deposits of the fossil fuels and hydroelectric power; (2) forest wealth, increasing as man learns how to manage this renewable resource properly; and (3) recreational attractions, including fishing, camping, touring, and simply spaces left in their natural state for human contemplation and enjoyment. Hill lands are neither destined to become poverty areas nor blessed for human living by virtue of their physical characteristics. Their role as assets or hindrances to man is determined by man himself.

Study Questions

1. Describe the hypsographic curve of the earth surface. If the earth were reduced to a circle 3 feet in diameter, what would be the distance between: (a) the maximum height of land and maximum depth of the sea; and (b) the mean height of land and the mean depth of the oceans?
2. How do we know the age of the rocks that make up the bulk of the continents? If the surface rocks of the moon are about the same age as our continental rocks on earth, can you think of a reason why the moon does not have pronounced upland blocks or lowland basins?
3. Has the segregation of the sialic material been completed for the earth? What evidence can you give for your answer?
4. What two parts of the ocean basins are the least stable? Give evidence and explain.
5. Contrast and explain the surface configuration or general terrain features of the Canadian shield and the shield region of the Rhodesias in Africa.

6. Relate the general location of block mountains and basin terrain, which is mainly formed by tensional stresses, to the great compressional folds of the global mountain cordilleras. Explain.

7. Define mountains, hills, and plains as used in this book. What justification can you present for these definitions? Why are plateaus not included among these major global landforms?

8. What is the generalized pattern for the location of major mountain ranges in the world? What exceptions are there to this pattern? Explain why Africa is such an unusual continent structurally.

9. Examine Figure 11-7. As this geosyncline subsides deeper, what is likely to be happening to the Moho beneath? Is the material that comprises the geosyncline related more to sialic or to sima material? Show diagrammatically and explain why both horizontal compressional stresses and vertical pressure from below might eventually arch this geosyncline far above sea level.

10. Explain why it is more likely for mountain lands to contain greater concentrations of population than hill lands. Use Japan as an illustration. Is China an exception?

11. Contrast the barrier effect of the Pyrenees and the Alps from a physical, economic, and political standpoint, using a basic atlas for your data. What factors contribute to these contrasts?

12. Compare the world map of plains with a world map of population density. From the map evidence, would you say that most plains were hospitable or unhospitable for human settlement? Explain. Discuss the correlation between dense populations and plains and explain.

CHAPTER 12
TOPOGRAPHIC
LANDFORMS

The Problem of Landform Classification

Landforms are surface features having characteristic shapes and compositions. They have been formed by various processes working on different materials through a variable length of time. Some of them, such as lakes, are of relatively short duration. Others, such as the continents and ocean basins, are almost as old as the planet itself. In Chapter 11, we considered the major global landforms of the earth. The present chapter focuses on the local level and summarizes briefly some of the more common topographic landforms that give distinctiveness to local landscapes. The continuous process of change resulting from the interplay of tectonic forces, on one hand, and the denudational processes, on the other, combined with a complex arrangement of surface materials, help to explain the great differences in landforms from place to place. As the first principle of natural distributions Section 1-7 stated, no two forms or areas on earth are exactly alike. The gradual evolutionary process within the biological world, or the structural linkage of present forms with their predecessors, has made possible a genetic classification of living things. All humans, for example, have distinct, recognizable characteristics because of the "blueprints" of growth built into the gene structure by millions of years of evolutionary processes. This genetic form of classification is impossible to use with landforms, because consistencies of developmental "signals" are absent and because of the large number of variables that determine form. Geomorphologists have been trying to develop a consistent genetic classification of landforms for over a hundred years, but without much success. Today landforms are studied not so much for comparative purposes as for the clues they yield in the understanding of natural processes.

Topographic landforms will be grouped in this chapter under the basic processes by which their principal characteristics were formed, but the reader should be warned that, while a cone shape appears to characterize many explosive volcanoes with highly acidic lavas, there is a wide range of differences in both composition and shape. This chapter presents a sampling of forms that appear more or less frequently across the face of the earth. It is hoped that the perceptive student or reader may compare some of the forms that he sees around him with the relatively few selected samples presented here, and through such comparisons, work out the processes whereby energy applications have altered materials through time to produce the distinctive forms which he sees before him.

One of the most influential attempts to erect a genetic classification of landforms was made by William Morris Davis, an eminent American geographer and physiographer, near the beginning of the present century. Focusing on denudational landforms, Davis used three basic variables in his classification: (1) *process,* (2) *structure,* or the kind and arrangement of material out of which the forms are shaped, and (3) *stage,* or the degree to which the processes have completed their work. His *cycles of erosion* assumed an initial elevation of land by tectonic processes. Following this, different erosional processes beveled the upland surface in successive stages to produce a graded lowland near base level. The stages within the cycle were represented in a continuum ranging from youth, through maturity, to old age, the presumed end result of the beveling process—a plain at or near base level. Figure 12-1 shows a typical series of forms associated with stages in the erosion of an upland area by running water, in a humid climate, and assuming a homogeneous underlying material or structure. The Davisian classification system was a useful pedagogical device, but it fell far short in its application to reality, principally in the use of his sequential stages. Some of the more serious limitations are summarized as follows:

Figure 12-1 Stages in the cycle of erosion by running water. [After Strahler]

1. Forms associated with old age are rarely encountered over wide areas, primarily because the earth crust, at least during the last 50 or 60 million years, has rarely remained stable enough to enable erosional processes to develop broad lowland plains. Erosional surfaces are much more frequently found on uplands than on lowlands.

2. The variation in underlying material creates so many differentials in the rate of erosion that the ideal sequential forms have little reality. Similarly, differential rates of earth crustal movements also tend to produce inconsistencies in the application of process sequences to specific areal interpretations. This was exemplified in the case study of the longitudinal profile of the Nile (see Section 10-6).

3. Many low, broad plains are the result of sedimentation, not erosion. The lower Mississippi Valley often has been cited by teachers using the Davisian classification as typical of old age in the cycle of erosion by running water in humid regions. Instead, it is an alluvial plain deposited upon the surface of an ancient delta. The breadth of the Mississippi flood plain has not been created by lateral erosion. Its rock floor lies deep below the surface.

4. Assumptions were made by Davis in the processes themselves that have been seriously questioned by later students. For example, Davis assumed that as a humid region was dissected by streams, the slopes of the valley sides gradually decreased, and that such slope angles could be used as indicators of the stages in the cycle of erosion. Penck, a German geomorphologist, disagreed, arguing that valley-side slopes normally change from convex to concave and then gradually reduce

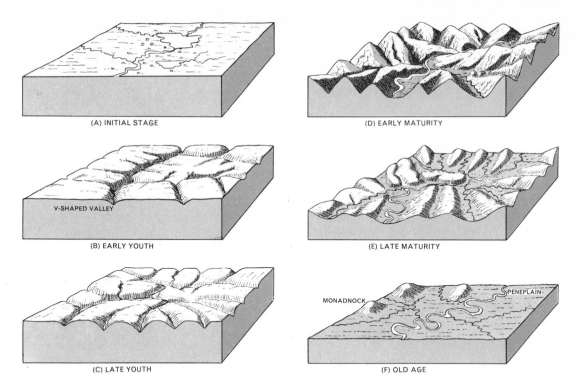

(A) INITIAL STAGE

V-SHAPED VALLEY

(B) EARLY YOUTH

(C) LATE YOUTH

(D) EARLY MATURITY

(E) LATE MATURITY

MONADNOCK   PENEPLAIN

(F) OLD AGE

the uplands by parallel recession of the slopes, eventually producing a gently sloping erosional surface, or *pediment,* terminating at steep-faced erosional remnants at or near the drainage divides. More recently, it has been recognized that valley slopes may retreat in different ways, depending upon several factors, including underlying material and climate. Figure 12-22 indicates a variety of slope forms that can be found along valley sides. All or a few of the segments may be present in any one place. Many factors influence the dominance of any one slope sequence, and each transverse profile across a valley represents an adjustment of many variables involved in the movement of material downhill by mass wastage. Just as variations occur within the longitudinal slope of a valley to adjust to variable factors influencing energy flow, side slopes also develop slope se-

quences in response to the specific controls in the mass wastage process.

It is becoming increasingly clear that all landforms, including erosional surfaces, are highly variable in process, structure, and stage and that simplified classifications are misleading and unrealistic. With this in mind, the reader will not be misled by an assumption that the landforms described in this chapter are types that accurately portray genetic groupings. Instead, they are merely a sampling of some of the more common forms, loosely grouped according to the dominant process of formation, or to the materials out of which they were shaped. Paraphrasing a statement that originally referred to cultures, all mountains are like *all* other mountains, *some* other mountains, or *no* other mountains.

Tectonic Landforms

12-1 LANDFORMS ASSOCIATED WITH VULCANISM Mention has been made of the relationship between the composition of *magma* (parent molten lava) and the type of eruption (see "Volcanic Mountains," pp. 404–406). Some of the common landforms associated with different types of eruptions are described in the following subsections. Again, it should be kept in mind that the classification of eruption types is arbitrary. There are many examples of composite and transitional types, based on the mixtures of magmatic materials.

Landforms of explosive volcanoes. These are types of eruption forms most frequently associated with acidic lavas. The most characteristic landform types are *explosion cones*, which are conical mounds of pyroclastic (Gr.: *pyros* = fire; *klastos* = broken) material that form around the explosion vents. Their flanks usually are concave, partly because the material near the vent tends to be coarser and partly because of the wash of fine material downslope and subsequent spreading out at the base. The finest dust material may be carried for hundreds of miles by winds. Cross sections of large cones indicate that the type of ejecta may differ with successive eruptions. Many of the largest explosion cones have liquid flows as well as fragmental material. The pyroclastic material varies from extremely fine dust to blocks several feet in diameter. The most common particle size seems to be in the textural range of coarse sand to fine gravel. This granular material sometimes accumulates to depths of 100 feet or more, masking underlying erosional surfaces to form *ash plains* or *ash-filled basins*. A large part of the eastern Anatolian plains in central Turkey (see Figure 10-13) is composed of volcanic ash. The Fort de Kock

district of northern Sumatra and parts of northern Kyushu, the southernmost island of the Japanese archipelago, are other examples.

The explosion cones range in size from small cinder cones between 10 and 20 feet high to huge cone-shaped mountains (see Figure 11-15). Perfectly symmetrical, cone-shaped volcanic peaks such as Mount Fujiyama in Japan are rare. Much more common are those with multiple vents and asymmetrical sides. Most large volcanic cones have smaller cones, termed *parasitic cones*, grouped in clusters or placed irregularly on their flanks. Groupings of pyroclastic vents sometimes are referred to as *volcanic fields*. These are especially common at the intersections of the island arcs along the western borders of the Pacific Ocean.

Most explosion cones have depressions or *craters* at or near their summits. Large, more or less circular craters are termed *calderas*. Most of these probably are the result of the downward foundering of the mountain summit into the void left below in the magmatic reservoir by the extremely rapid removal of large quantities of material. A good example is the large caldera at Crater Lake National Park in Oregon (see Figure 12-2). Lake Rotorua, on North Island, New Zealand, is another immense subsidence caldera, measuring about 10 miles in diameter.

The loose, unconsolidated material of explosion cones is highly susceptible to erosion by rainwash and stream erosion, and the cones are quickly rilled and grooved. As the ash and other fragmental material is removed by erosion, the central lava conduit tube may be left behind as an erosional remnant, because the molten material cooled more slowly there and became solid rather than fragmental. Such erosional remnants are termed volcanic *necks* or *plugs* (see Figure 12-3). Devil's Tower National Monument in northeastern Wyoming is a fine example.

Figure 12-2 Crater Lake, Oregon, a large caldera. It is believed that this huge volcanic crater resulted from an internal collapse or foundering of the mountain summit into the magmatic reservoir below, following a long period of eruption. [U.S. Dept. of the Interior; National Park Service]

Landforms of quiet, or liquid-flow, volcanoes. Basaltic lavas, or those containing relatively high percentages of calcium, magnesium, and iron, because of their high melting point tend to lose their included gases quickly as they near the surface; thus, they tend to remain fluid longer. Eruptions consist mainly of upwellings along faults or fissures. Clusters of vents often are found along such lines. The fissure-flow eruptions of highly basic lavas generally occur only once from the same vent. Once the vent is plugged by solidified rock, the lava tends to work its way out onto the surface from a nearby position along the same fracture line, hence the linear arrangement. Massive regional reservoirs of basaltic magma, however, may result in many vents scattered irregularly over a large territory. Small *spatter cones* 20 to 40 feet high (see Figure 12-4) composed of coarse, dark scoria are found frequently near the vent outlets. These are formed by jets of gas that force some of the liquid basalt into the air, where it cools unusually fast to form the clinkerlike scoria.

Dome volcanoes were mentioned in the subsection dealing with volcanic mountains. They often feature a large central crater such as Kilauea on the side of Mauna Loa (see

Figure 12-3 Volcanic neck north of Rotorua, North Island, New Zealand. The protruding spire represents the solidified lava that cooled in the central conduit of a cinder cone. Rainwash has removed the loose pyroclastic material from around it. [Van Riper]

Figure 12-4 Spatter cone at a volcanic vent, Craters of the Moon National Monument, Idaho. Note the shiny, black, corded surface of the congealed basaltic lava at the base of the cone. This cone is about 30 feet high. [Van Riper]

Figure 12-5). This crater is over 2 miles in diameter. These volcanoes in the Pacific rank among the largest landforms on earth. Rising from the sea floor some 18,000 feet below sea level, they climb to heights of 10,000 to 12,000 feet above sea level. They do not differ greatly in composition from the fissure-flow basalts of the Columbia Plateau, and they are distinctive mainly in their continual venting from individual lava conduits. The eruptions usually consist of a rise of molten lava in the crater and the emission of large quantities of smoke and gases. The rise of lava within the crater usually signals an outbreak of fluid lava someplace along the mountain flanks. The lava rarely rises to the brink of the crater. There is a tendency for the dome volcanoes to increase their viscosity and thus become more explosive toward the end of their period of activity. Dome volcanoes are mostly found within the ocean basins and are relatively rare on the continents. Besides the

Hawaiian group, others include Samoa and Tahiti in the Pacific, the Azores in the Atlantic, and Reunion and Mauritius in the Indian Ocean. They undoubtedly are fed from the sima that lies close to the surface beneath the ocean floor.

Landforms resulting from the surface exposure of intrusive rocks. Lavas may cool and solidify into rock below the surface, either within their conduits or within their magmatic reservoirs. Later, as a result of erosion, they may become surface forms. Such intrusives, in contrast to the extrusive surface forms, exhibit a much wider range of chemical composition and mineral crystallization. When the liquid magma is injected between layers of sedimentary rock to become a flat sheet of igneous rock, it is termed a *sill* (see Figure 12-6). Upon cooling, basaltic sills often develop a structure of rough hexagonal columns whose vertical axes lie at right angles to the cooling surfaces. The well-known Palisades along the Hudson River near New York City represent the rough columnar structure of a large sill injected between parallel layers of Triassic sandstone and now exposed at the surface. The most famous example is the Giant's Causeway (see Figure 12-7) on the north coast of Ireland, where the hexagonal columns are especially well developed. Small domelike masses of magma, intruded between sediments, are termed *laccoliths*.

Batholiths are huge magmatic reservoirs that have solidified well below the surface, but, because of uplift and subsequent erosion, have been exposed at the surface. They are on a much smaller scale than the continental shields but are similar in being large masses of relatively homogeneous intrusive

Figure 12-6 Landforms associated with extrusive and intrusive vulcanism.

Figure 12-5 Boiling lava in Halemaumau pit within Kilauea crater, Hawaii, during an eruption. This eruption took place on November 5, 1967. Some of the lava fountains rose to a height of 250 feet within the pit. [Hawaii Visitors Bureau]

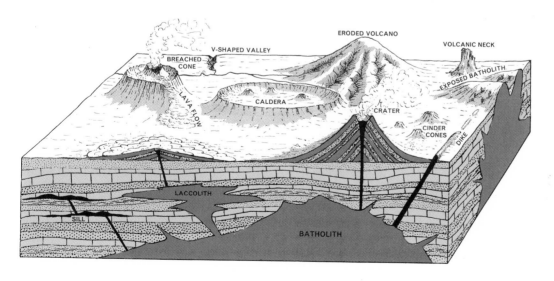

Figure 12-7 The Giant's Causeway, County Antrim, Northern Ireland. This is a famous tourist attraction. The unusual red and yellow hexagonal and pentagonal stone columns resulted from contraction following cooling of a basaltic lava sheet. [Courtesy Bord Failte Eureann, Dublin]

rocks, such as granites. A large part of the northern Rocky Mountains in central Idaho is a large batholith that was uplifted and has been dissected into mountainous terrain. As batholiths slowly cool beneath the surface, there is some segregation of mineral constituents within the magma, and often some of the metallic ores tend to arrange themselves in concentric zones within the batholith. For this reason, batholiths often are important areas for metallic ores. The Coeur d'Alene mining district of Idaho and the Hartz Mountains of east-central Germany are examples.

12-2 STRUCTURAL LANDFORMS ASSOCIATED WITH FOLDING Compressional stresses are especially associated with mountain building. Rocks react in various ways to such stresses, depending on the composition and thickness of the beds and the intensity of the stresses. The simplest reaction is the buckling of the rock strata into elongated upfolds and downfolds, much as a pad of paper will buckle when subjected to lateral compressions. The elongated upfolds and downfolds are known as *anticlines* and *synclines,* respectively. The axes of such folds usually are not horizontal, in which case the folds are said to be *pitching.* Note the effect of pitching on the arrangement of ridges and valleys following the erosion of alternate hard and soft rocks, as shown in Figure 12-8. The zigzag pattern of ridges and valleys in the folded Appalachians in Pennsylvania and in the Zagros Ranges of western Iran well illustrates the influence of pitching structures on surface erosional landforms. Also, as shown in Figure 12-8, not all upfolds result in ridges or synclines in depressions. Depending on the arrangement

of the more durable layers, former high places may become low terrain, and vice versa. *Anticlinal valleys* and *synclinal ridges* are examples of this structural control, which is sometimes referred to as *relief inversion.*

Extremely intense folding may produce *overturned folds,* where the limbs are roughly parallel and sloping at a steep angle from the vertical.

The orogenic stresses are not always horizontal. Sometimes they may press upward from below, in which case a *structural dome* may be formed. Conversely, a centralized sag in a sedimentary basin may warp the strata downward to produce a *structural basin.* Examples of both are shown in Figure 12-9. Unusually large basins of this type are termed *geosynclines.*

The differential erosion of sedimentary rock strata that have been warped by compressional stresses produces many distinctive types of landforms. Among the most common are *cuestas* (see Figure 12-9), which are asymmetrical ridges that have a steep slope on one side and a gentle one on the other,

Figure 12-8 Landforms associated with rock folding.

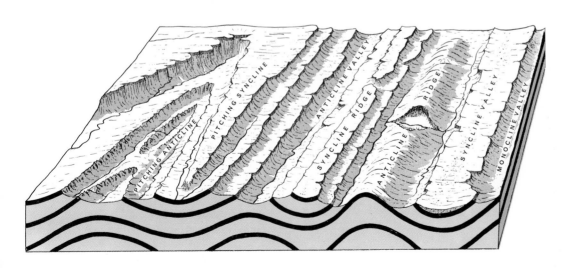

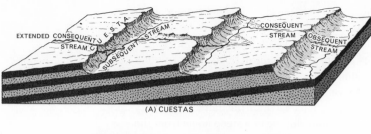

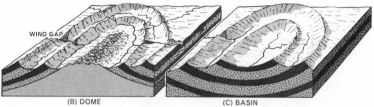

Figure 12-9 Cuestas, domes, structural basins, and associated landforms. [*B* after Lobeck; *C* after Trewartha, Robinson, and Hammond]

Figure 12-10 Dendritic and trellis drainage patterns.

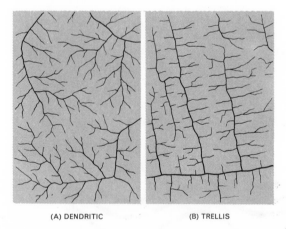

(A) DENDRITIC   (B) TRELLIS

the latter conforming to the slope or dip of the rock strata. In the United States, the cuesta of the Niagara limestone, a particularly massive rock stratum that forms the crest of Niagara Falls, can be traced from New York State northward across Ontario into the northern peninsula of Michigan, thence southward through Wisconsin, recurving to the northwest into Minnesota and North Dakota, a total distance of over 1,000 miles. Cuesta escarpments with cliffed edges comprise major surface landforms in many parts of the world.

Drainage is influenced markedly by rock structure in folded regions. Parallel ridges and valleys, such as are found in the folded Appalachians, tend to produce a rectangular drainage pattern, termed *trellis drainage,* which is in sharp contrast to the normal *dendritic* or treelike type (see Figure 12-10). The relationship between *consequent streams,* which follow the original structural slope, *subsequent streams,* which follow the general direction of the weaker beds after the latter are excavated by erosion, and *obsequent streams,* which cut back into the cuesta faces, is diagramed in Figure 12-9. A *radial* drainage pattern generally results from the erosion of domes or basins containing rock strata having different resistances to erosion.

12-3 LANDFORMS ASSOCIATED WITH FAULTING Several distinctive landforms are related to *faulting,* or the slipping of blocks of the earth crust along great fractures. Faults can be grouped into three main classes: normal faults, thrust or reverse faults, and strike-slip faults.

Normal faults. The most common faults are those that result from tensional stresses or an attempt at crustal lengthening. They frequently occur following the relaxation of compression after regional arching, as shown in Figure 11-17, or at the summit of an arch during lateral compression. Such faults are termed *normal* or *gravity faults.* Their properties are diagramed in Figure 12-11A. It will be noted that the overhanging side of the fault, termed the *hanging wall,* is down with reference to the opposite, or *foot-wall,* side. Normal faults usually are high-angle faults; that is, they are not far from the vertical.

Thrust and reverse faults. These are the result of compressional stresses and often occur in connection with folding. In both of these faults, the hanging-wall side of the fault has been thrust over the foot wall, as in Figure 12-11B. The only distinction between the two is that the thrust faults have extremely low angles that are not far from the horizontal. Figure 12-12 is a low-angle thrust fault.

Strike-slip faults. These feature a pronounced horizontal displacement along the faults rather than a vertical movement. They are so named because the displacement is in the direction of the *strike,* or compass direction of the fault line at the surface. They are somewhat less common than the other faults but are sometimes encountered in areas of tensional stress. They are diagramed in Figure 12-11C.

Geologists distinguish between *fault escarpments* and *fault scarps.* Both of these are steep slopes that result from faulting, but in the fault escarpment the slope represents the displacement surface, while in the fault scarp the abrupt slope results from differential erosion in rocks of unequal resistance on opposite sides of the fault. Displacements along faults differ greatly. The movement during a single earthquake rarely exceeds a few feet, but major faults may experience displacements irregularly for many centuries until the total vertical movement or hori-

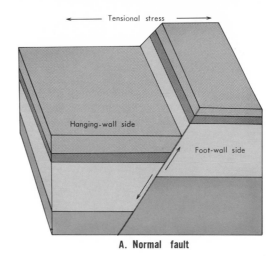

A. Normal fault

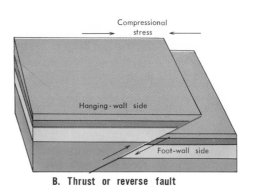

B. Thrust or reverse fault

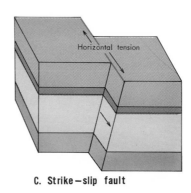

C. Strike—slip fault

Figure 12-11 Normal, thrust, and strike-slip faults.

Figure 12-12   A thrust fault. The fault may be seen as an almost horizontal line across the photo. Above the fault, the sedimentary beds have moved from right to left and have been folded as a result of frictional drag along the fault. [U.S. Geological Survey; M. R. Mudge]

zontal thrust may be measured in thousands of feet. In some thrust faults, the horizontal displacement may be several miles. The Lewis Overthrust, for example, located at the eastern base of the Rocky Mountains in Glacier National Park, Montana, was the site of an enormous horizontal displacement, in

which thousands of feet of early Paleozoic sedimentary rocks were thrust eastward for 10 miles or more over young, unconsolidated sediments of Cretaceous age.

Horsts and grabens.   These are local landforms often found in areas of normal faulting and diagramed in Figure 12-13. A *horst* is a more or less elongated block that has been elevated with respect to the areas on either side and without appreciable warping or folding. Not all horsts are highlands, because the erosion of a horst exposing weaker beds below could result in a horst depression, another example of inverted relief. *Grabens* (Ger.: *graben*=grave) are the opposite of horsts, as shown in Figure 12-13. They

Figure 12-13   Landforms associated with normal or gravity faulting.   [After Finch et al.]

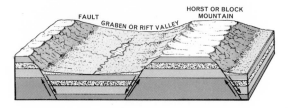

Landforms and Surface Configuration

are fault depressions with normal faults on either side. The classic example of a graben is the upper Rhine Valley between the upland block of the Vosges Mountains on the west and the Black Forest (Schwartzwald) highlands of Germany on the east.

Unusually large and elongated grabens, such as the great trenches in east Africa, as well as those in the ocean basins, are sometimes termed *rift valleys,* or simply *rifts.* The fault escarpments along the east African rifts are 2,000 to 3,000 feet high in places. Lakes Albert, Tanganyika, and Nyasa are among the larger bodies of fresh water that filled these great rift depressions up to their lowest outlets. As indicated earlier, volcanic activity often is associated with areas of strong tensional faulting. Both explosive and fissure-type eruptions, for example, adjoin the rift depressions in eastern Africa.

Landforms Associated with the Work of Running Water

12-4 EROSIONAL LANDFORMS ASSOCI-ATED WITH RUNNING WATER The work of running water is largely linear and serves three main functions: (1) to remove surpluses of water that accumulate on the surface; (2) to lower the gradient of the drainage network or system so as to set up a smooth runoff curve that is in dynamic equilibrium with velocity, load, bed resistance, and other variables; and (3) to remove the load brought to the streams by other eroding agents, such as mass wastage. The erosive work of a stream is confined mainly to the channel bed and sides. Thus, the associated landforms are related largely to linear drainage patterns.

The form of drainage basins. The arrangement of streams in a drainage system is not so haphazard as might appear. As Horton

discovered in his research,[1] there are distinct mathematical ratios between stream length, number, and order. By *stream order* is meant the order in which tributaries are gathered. A first-order stream consists of the small, unbranched upper-stream ends. A second-order stream unites at least two first-order streams, as shown in Figure 12-14A. After examining Figure 12-14, it should not be inferred that all drainage basins have the same numerical ratios. For example, several factors could influence the *bifurcation ratio* or the ratio between the number of stream tributaries in one order with those in the next and within the same drainage basin e.g., climate, underlying material, or rock structure.

Other mathematical relationships can be found between volume, channel width, channel depth, gradient, and velocity. Proceeding downstream from the headwaters of most streams, generally there is a progressive increase in volume, an increase in depth, and an increase in the number of tributaries. The river channel becomes wider and more regular, and the bed load becomes finer in texture. No single mathematical formula has yet been devised that can relate all of these forms within the basin to a set of variables, but certain regularities appear over and over again and reflect relationships between form and the performance of work as the potential energy of streams is converted into kinetic energy within the system.

Each drainage basin can be considered an open system into and through which energy flows. In the early stages of river basin development, there is a highly disordered state, and the forms and materials within the system are subject to rapid change. With the passage of time, the system becomes more

[1]R. E. Horton, "Erosional Development of Streams and Their Drainage Basins: A Hydrophysical Approach to Quantitative Morphology," *Bulletin of the Geological Society of America,* vol. 56, pp. 275–370, 1945.

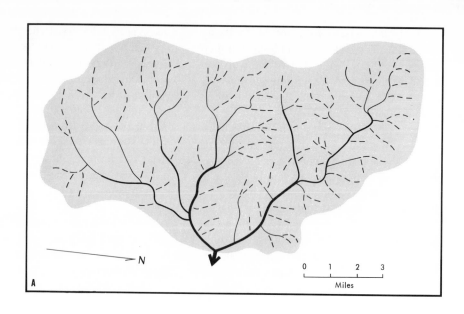

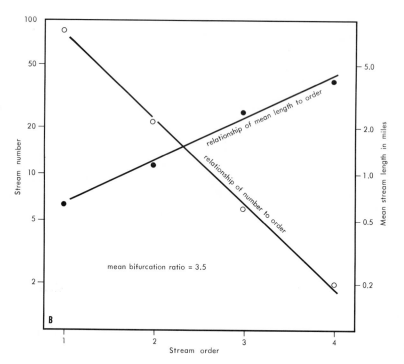

Figure 12-14 *A* and *B* **Mathematical relationships between stream order, number, and length.** The principle of stream ordering is illustrated by *A*, a small drainage basin in Georgia. The mathematical relationship between order, number, and length is given in *B*. These illustrate the definite orderliness in the development of a drainage basin that is related to the flow of energy through the system. [After Horton]

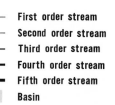

- – – –   First order stream
- ——   Second order stream
- ——   Third order stream
- ——   Fourth order stream
- ——   Fifth order stream
-       Basin

Table 12-1 Characteristics in the Evolution of Drainage Basins

Forms	Early instability	Late stability
Longitudinal valley profile	Gentle in headwater collection zone; steep middle section, with irregular gradients; curve beginning to flatten toward the mouth. Increasing length by headward extension and delta growth.	Relatively smooth curve; unstable only in upper reaches; increase in length by delta growth, meandering, or braiding (see "The Course or Trajectory of Streams," pp. 442–444).
Channel features	Unstable along most of length. Rapid alteration between deposition and bed erosion except near mouth. Much dissipation of energy in friction and eddy diffusion. Highly variable bed-load size.	Slight channel alteration except lateral erosion and deposition in sections at grade. Bed load normally granular (fine sand or silt); more transverse shift of bed load than downstream when at grade.
Relief ratio	Increasing. Local relief greatest along middle portion of drainage basin.	Stable, with slight decrease as divides are lowered.
Divides (interstream areas)	Broad and poorly drained. Haphazard runoff; beginning of cross-grading and rilling.*	Narrow; irregular in height and trajectory.
Transverse valley profiles	Downward cutting faster than reduction of slopes by mass wastage. Result: frequent convexity of profiles.	Downward cutting minimal along most of the stream course; rate less than reduction of slopes by mass wastage. Result: frequent concavity of profiles.

Cross-grading is the early development of stream order by the coalescence of rills (see Figure 12-15). *Rilling* is the tendency for surplus water to run off on an initial, even slope in parallel grooves.

orderly and the flow of energy smoother and more even. The frenzied rush of a mountain stream in its upper reaches is symptomatic of the disordered, unstable state of the energy flow. Once at grade, however, an equilibrium state is reached in which the energy flow is much smoother and expended largely in load transport, modification of channel sides, and internal friction.

There are changes in form within the drainage basin as it passes from an unstable to a stable state. The changes represent work done to expedite the flow of energy through the system. The various features associated with instability and stability in a drainage basin are summarized in Table 12-1. As noted

earlier, it would be rare to have all of a drainage basin at grade. Usually, with the passage of time, assuming no changes in base level or climate, the stable sector of a drainage basin slowly moves upstream. The evolutionary sequence toward stability may also, of course, be interrupted at any time by tectonic processes that could alter the energy flow pattern. For example, a subsidence of the lower part of a drainage basin would increase the potential energy within the system, while conversely, an uplift of the lower portion would decrease the potential energy. Another important interruption in the evolutionary sequence could result from significant climatic changes. A change from a humid to

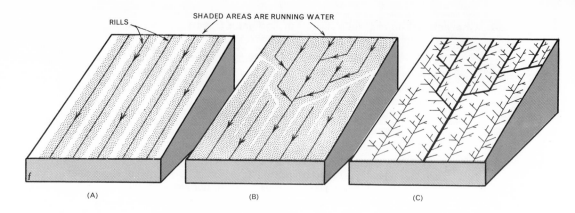

RILLS

SHADED AREAS ARE RUNNING WATER

(A)

(B)

(C)

an arid climate or vice versa would tend to decrease or increase greatly the potential energy within the system. The erosion-deposition ratio along stream courses can be altered greatly because of this. This is clearly illustrated by Figure 9-7, which shows a well-integrated drainage system, typical of those developed in humid regions, but which now has insufficient potential energy because

of aridity to remove the detritus brought to the streams by other erosional agents. Thus, the level of grade has moved far upstream, and deposition has proceeded to a point where it is beginning to bury some of the lower interstream divides.

Rock structure also may influence the evolution of drainage basins. Drainage patterns associated with the erosion of highly folded sedimentary beds of unequal resistance may have forms distinctly different from those of homogeneous materials. Figure 12-9 illustrates this point. *Trellis drainage* is a rectangular pattern that contrasts sharply with the more typical dendritic patterns. Faulting or joint development may direct drainage patterns into a trellis form. Stream bifurcation ratios in structurally disturbed areas are likely to be different from ratios in areas where the underlying material is uniform.

The course or trajectory of streams. The trajectory of a stream not at grade is usually irregular and subject to many local factors. It may be conditioned by random choice or chance, or by some directive factor, such as resistance of the underlying rock or the alignment of fracture lines. Straight, ungraded streams are rare and generally are related to structural weakness lines, such as faults or

Figure 12-16 Typical landforms on a river floodplain. [D. Johnson; *Geographical Review*]

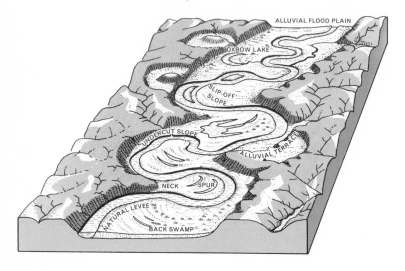

Figure 12-15 Development of stream order from rills. The initial runoff on a uniform slope is in parallel rills (*A*). Minor irregularities, located at random, divert some of the water diagonally to adjacent rills, producing the beginning of a dendritic pattern, as in *B*. Later, as in *C*, the normal first-, second-, and third-order tributaries develop. [After Horton]

Figure 12-17 Braided streams southwest of Christchurch, New Zealand. These streams have their sources in the glaciers of the New Zealand Alps. They are at grade, and their load of granular fluvioglacial debris causes them to develop a typical braided pattern within their broad channels. [Van Riper]

joints. When a stream reaches grade, however, its trajectory becomes closely related to the characteristics of the bed load and is largely independent of the underlying structure. Figure 9-8 indicates the relationship between particle size and velocity in the processes of stream erosion, deposition, and transportation. The importance of granular material in transit is indicated. Coarse material has an extremely narrow range in which it is transported. Fine silts and clays (not shown) can be transported easily but are rarely involved in stream bed deposition. The colloidal or finely textured nature of clay enables it to remain suspended even in quiet water for long periods of time. Its binding qualities, however, make it less liable to be subjected to stream bed erosion, as indicated in the diagram. Granular materials, ranging from silt through fine sand, constitute the major bed load of streams at grade. When the channel bed of a stream is made up of granular material such as silt and fine sand, the stream trajectory takes one of two forms, depending on the presence or absence of binding material such as clay or organic material (humus). If such material is present to produce a plastering or matting effect on the bottom and sides of the channel, the stream tends to retain a single channel and to develop a winding course across its graded floodplain. Such loops are termed *meanders* (see Figures 1-2 and 12-16), and the streams are termed *meandering streams.* Where fine-textured binding material is absent, the river channel tends to break and divide into multiple channels across a broad section of the flood plain. Such a stream is termed a *braided stream* (see Figure 12-17) because of the braided appearance of the channel course. Humid regions are more likely to contain meandering streams, because mass wastage is likely to bring to the streams fine-textured products of rock and organic decomposition from the valley slopes. Braided streams, on the other hand, are more common in semiarid to arid regions, where the fine material is much less likely to be present. Such streams can sometimes be found in humid regions where only coarse material is found, as on young glacial outwash plains in crystalline rock areas, where the till is likely to be coarse and without fine clay fractions (see Figure 12-17).

Streams at grade occasionally may have a channel bed on solid rock or on clays or soft shales, in which case the trajectory will be neither meandering nor braided. Here the trajectory is likely to be either straight or irregular, without any distinct pattern.

A meandering stream continues to utilize energy to perform work, despite being at grade. The looping course causes a change in the flow of water, with a curved surge upward, outward, and down-valley toward the concave portion of the meander and a downward motion accompanied by a decreased velocity inward toward the convex part of the loop. In this way, lateral erosion takes place on the outside or undercut portion of the bend, and deposition of material occurs on the *slipoff slope,* or convex part of the meander. Meanders thus tend to migrate slowly down-valley and to intersect each other. The result is a general horizontal distribution of bed load across the valley. The movement of material downstream is slowed, and to compensate, the course is lengthened by the growth of the meanders. The kinetic energy of the stream thus is distributed laterally across the floodplain, thus widening the equilibrium plane beyond the immediate channel. There is a limit to the width of the meander loops, and generally this is related to the width of the channel bed in a ratio of between 12 and 20 to 1. Meanders that intersect the valley sides may widen the valleys somewhat (see Figure 12-16).

Meandering streams apparently represent a condition of stability with respect to the flow of energy, as do braided streams. Given the right type of bed load and the energy equilibrium of a graded condition, the kinetic energy of streams tends to be directed away from the channel bed and toward the sides, and to perform some lateral erosion. The processes of meandering and braiding hinder the use of streams for navigation or the floodplains for agriculture. Flood hazards are increased by both processes because of the tendency of the streams to escape from their existing channels. Attempts to maintain straight channels by dredging only create temporary relief, and, unless constantly maintained, are soon overcome by deposition or channel-side attacks.

The elevation of a meandering stream by crustal upwarping sometimes will reestablish a new grade with increased potential energy and result in a rejuvenation of the stream gradient. Resultant downward cutting within the channel may produce *entrenched meanders,* in which the stream no longer is at grade. Its trajectory is inherited from a previous graded condition.

12-5 DEPOSITIONAL LANDFORMS RESULTING FROM THE WORK OF RUNNING WATER   The characteristic processes and materials of stream deposition were discussed in earlier chapters. We shall now be concerned with some of the characteristic landforms that result from these processes.

Floodplains.   These are the depositional surfaces formed once a stream reaches grade, and across which a stream migrates as it expends some of its energy laterally instead of vertically. As indicated in the previous section, the depositing streams generally have either meandering or braided courses, depending on the kind of sediment in transit. Both types construct floodplains, and there are similarities and differences in the surface forms associated with each. Meandering rivers, for example, are likely to have well developed *natural levees,* or low embankments, that are highest next to the river and slope gradually away from it. They are caused by differential rates of sedimentation as water leaves the channel at times of flood.

The maximum deposition occurs when the velocity is first checked, or immediately next to the channel. Braided streams have no distinct channel. Hence, there is a large amount of random movement of water and depositional material back and forth across the floodplains (see Figure 12-17). Unless man creates artificial dikes or levees to contain braided streams within prescribed channel widths, there is little surface irregularity on braided stream plains.

Old abandoned channels and crescentic levees are prominent features of the depositional floodplains where the rivers are meandering. The ridges make good sites for roads and settlements (see Figure 11-23) because they lie somewhat above the level of the plain and are less likely to be inundated during flood periods. Artificial embankments often are constructed upon the natural levees to add additional flood protection. As a general rule, the floodplains of meandering streams are more suitable for agricultural use than those of braided streams, because the alluvium has a finer texture and hence retains more moisture and plant nutrients. For this reason, there is a tendency for the artificial levees to be higher and closer together on such streams.

Alluvial terraces, which are flat benches of alluvium situated on floodplains, mark former levels of deposition below which the streams have since cut their present graded plain. Such terraces may result from a lowering of base level or from a sudden decrease in load. Many of the tributaries of the Mississippi and Ohio Rivers were overloaded with fluvioglacial debris during the waning stages of Pleistocene glaciation. Isostatic uplift of the area following the removal of the ice in irregular movements, as well as a rapid decline in load, resulted in a series of new base levels for the region and the development of a series of alluvial terraces along these rivers and their tributaries. The terraces locally are referred to as first, second, or third *bottoms.* The first, or lowest, bottom is the preferred site for cultivation since it contains the youngest and most fertile alluvium.

Deltas. These are alluvial plains built by streams at their mouths, that is, in ponded water. The term was derived from the Greek letter *delta,* which resembles the general shape of the Nile delta. Not all deltas take this shape, however, as indicated by the representative examples given in Figure 12-18. Being floodplains, they have many of the features associated with the lower valleys of large rivers, including natural levees and marshes. The delta depressions lie so near the base level of the stream that floodwaters have difficulty being removed. Shallow lakes often are found on deltas. Rivers here usually divide into branches or *distributaries* upon reaching their deltas. The natural levees of distributaries that intersect each other may coalesce and enclose broad, shallow depressions or lakes. Silting at the mouths of delta distributaries makes such waterways relatively unsuited to the passage of deep-water vessels unless dredging is used to keep the channels clear. Deep water and swift currents may prevent the development of deltas or limit their growth. Neither the Hudson nor the Columbia River has an appreciable delta for this reason. The St. Lawrence River lacks a delta because the river contains only a small amount of alluvium, owing to its origin in Lake Ontario, which acts as a settling basin.

Delta plains generally are favorable areas for agricultural land use because of their fine-textured alluvial soils, but flood danger, poor drainage, and salinity in swales and depressions can be local problems. The delta plains in eastern and southeastern Asia have some of the densest rural populations in the

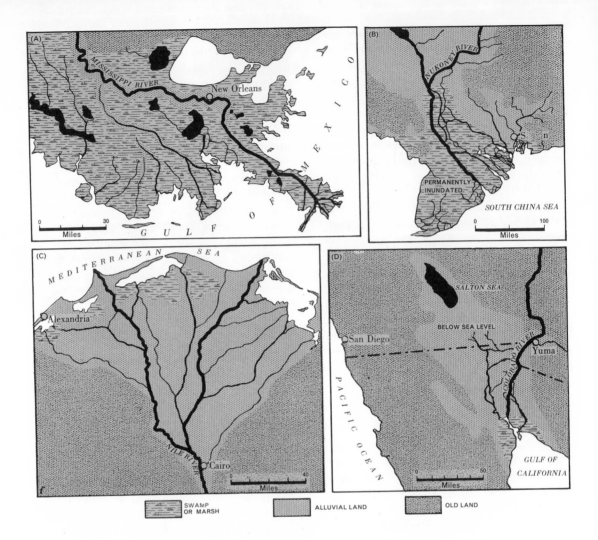

| SWAMP OR MARSH | ALLUVIAL LAND | OLD LAND |

world, based principally on the cultivation of rice. The availability and suitability of the river water are fully as important to this cultivation as the quality of the underlying soils. Other intensively cultivated delta plains include those of the Mississippi, Nile, Po, Rhine, Tigris-Euphrates, and Colorado Rivers. The broad delta of the Amazon is unique in that it is largely unused and supports only a sparse pastoral population.

Some coastal plains are produced through the joining of adjacent deltas. The plains along the eastern side of Sumatra, the northern part of Java, the coast of the Gulf of Guinea in western Africa, and parts of the Guiana coast of northern South America are of this type. Many alluvial plains that border the ocean in such tropical areas as these are almost always bordered by dense tangles of mangroves and swamp palms that hinder the reclamation of the seaward margins of deltas.

Figure 12-18 Types of deltas.

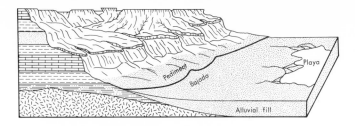

Figure 12-19 Diagram of a pediment and bajada. A pediment is an erosion surface, while a bajada is a depositional surface produced by coalescing alluvial fans. The pediment surface usually has a thin covering of alluvial material; hence, the contact line between pediment and bajada often is not visible.

Stream depositional landforms in dry lands. *Alluvial fans* are among the most common depositional landforms in dry lands, although they are not restricted to such climates. These are fan-shaped deposits of alluvial material that result from the sudden drop in velocity as upland streams debouch onto plains. Many of the streams that form such fans empty into desert basins that have no drainage outlet. Unusually steep fans are termed *alluvial cones* (see Figure 10-10).

The longitudinal profile of alluvial fan surfaces is concave. That is, the depositional surface is somewhat steeper upslope, gradually flattening below. This is related largely to the texture of the alluvial material, which is coarsest upstream. The lateral profile of the fans has a tendency to be convex, or the central portion somewhat higher than the sides. The fans are graded by successive lateral diversions of both flood waters and included sediments during high-water periods. The debris carried by these occasional floods clogs the old channels and sweeps back and forth across the fan surface, maintaining an equilibrium surface as the fan grows longer and thicker. The temporary stream channels that diverge into lesser distributaries and gradually lose themselves within the fan debris are termed *wadis,* the general name for all temporary dry-land drainage channels. Characteristically they have vertical sides and flat bottoms.

The union of fans along the foot of mountain slopes produces a sloping plain known as a *bajada* or *piedmont alluvial plain* (see Figure 12-19). This should not be confused with a *pediment,* which is an erosional surface produced by the retreat of escarpments or cliffed slopes. Bajadas and pediments frequently are found together, especially in areas of block mountain and basin topography. Since pediments often may be covered with a thin veneer of rock waste in transit, the transition between them and bajadas may be imperceptible unless revealed by erosion in a deep ravine or gully.

Interior basins or *bolsons* that are bordered by bajadas and have no outlets often have temporary lakes, or *playas,* in their lowest portions. These collect water draining into the basins through seepage from the base of bajadas, or more rarely by the direct discharge from wadis. This water always contains quantities of soluble salts dissolved from rock minerals, and when it is evaporated in the playa basins, these accumulate at the surface, much as a kitchen teakettle accumulates a crust of lime. The salt- or alkali-encrusted flat that results from repeated evaporation from a playa is termed a *salina* in the western hemisphere (see Figure 12-20) and a *sebkha* in the Arab world, which includes so much of the desert land in the eastern hemisphere.

Similar saline mud flats may be formed along coastal regions, as along the western side of the Persian Gulf or the Rann of Cutch in northwestern India. Salina or sebkha sur-

Figure 12-20 A salina and bajada near Las Vegas, Nevada. The white strip in the center is the salt-encrusted salina, too toxic for normal plant growth. Note the pediments and bajada sloping toward the salina. [Van Riper]

faces may be extremely treacherous for wheeled vehicles and even for men or animals. The indurated salt crust at the surface may seem deceptively solid, but it may also contain thin, weak sections that break to expose a wet, highly plastic, saturated salt clay that could mire humans as well as vehicles. Unusually large playas are found in the basins that adjoin the Atlas Mountains in Algeria and Tunisia, where they are known as *shotts.* Some playas or salinas have been important sources of highly soluble salts such as nitrates, borates and potash salts. Probably the best-known of these are the nitrate deposits of northern Chile, which are composed largely of sodium nitrate and which have been concentrated into a shallow, hardpan soil layer by evaporation. Sometimes, however, the surface of desert basins is formed of cracked, sun-baked clay without noticeable salt encrustations.

The alluvial fans and bajadas in arid regions form characteristic sites for desert oases. The fans form groundwater reservoirs that can be tapped by underground wells; the sloping surfaces are well suited to conduct floodwaters directly to cultivated fields; and the permeable underlying material prevents the accumulation of surplus water that may lead to salinity problems. Agricultural land use on bajadas often is in zones corresponding to variations in the texture of the fan material. Coarse gravels and stone usu-

Figure 12-21 A mud slide near Akoura, Lebanon. The source of this mud slide is a basaltic sill high on the drainage divide. The basalt has weathered to a heavy, plastic clay which is water-soaked by melting snow each spring. The slide moves only at irregular intervals several years apart. Note the entrenchment of the slide, indicating its potential for erosion, and the lack of terracing or any kind of land use on the slide surface. [Van Riper]

ally lie at the apexes of the fans where the velocity of flood torrents first is checked. The gradation from coarse sands to fine silts in the descent down the bajada surface often is accompanied by a gradation in the types of crops grown. Tree crops normally do well on intermediate portions of the slopes, since their roots can draw on water from subsurface supplies. In areas where frost may be hazardous, the intermediate positions also are favored locations for frost-sensitive crops, partly because they are above the cold air that accumulates in the basins on clear nights and partly because air movement associated with air drainage helps to retard freezing. Shallow-rooted crops generally are located farther downslope, where the water table begins to approach the surface. These lower sites, however, are more susceptible to toxic conditions resulting from salinity, especially where irrigation is practiced.

Landforms Associated with Mass Wastage

12-6 EROSIONAL LANDFORMS ASSOCIATED WITH MASS WASTAGE Mass wastage includes the processes in which the movement of weathered material is controlled by gravity alone. These processes were described in Section 9-9. The principal work done by mass wastage is to reduce subaerial slopes and to move weathered material toward streams where it can be removed by running water. Unlike stream erosion, whose work is linear, mass wastage involves the shift of material areally. The major types of mass wastage are indicated in Table 12-2, which is adapted from a classification system developed by Savage.

Slope profiles along the sides of hills and valleys everywhere are related to the work of mass wastage. The particular kind of work

---

Table 12-2   Classification of Mass Wastage Types on Land Surfaces: Description, Location, and Subtypes (after Savage)

Free fall	The fall of rock, soil, or debris from slopes greater than 60°. The accumulated material at the foot of the slope is known as *talus* or *scree*.
Landslide	A rapid movement of rock, soil, or debris on slopes below 60°. A movement of surface material parallel to the surface and along shear planes is termed a *slide* and is most common on talus or scree slopes. A rotational movement produces *slump slides* and generally requires fine-textured clays or silts. *Avalanches* involve unusually sudden movements of snow, ice, and other included material.
Landflow	A slow, irregular movement of rock, soil, and debris. Slopes are usually 50° to 10°. Water lubrication is usually but not always present. Most common forms are *mud flows* (see Figure 12-21) and *solifluction* (soil flowage). The latter is especially prevalent in periglacial climates, where permafrost is an important factor.
Landcreep	An imperceptible movement of rock, soil, or debris downslope. Associated forms are *rock glaciers, stone stripes, stone* and *soil polygons, soil creep,* and *terracettes* (see Glossary for definitions). These forms are especially prevalent on all slopes in periglacial climates. Soil creep, however, is found on low slopes anywhere.
Landslump	Vertical subsidence, resulting in surface depressions; not to be confused with *slump-slides*. It is found mainly in periglacial regions but may occur elsewhere where material is removed from below, as in limestone and salt solution, mining, water or oil pumping, etc. Its dominance in periglacial regions results from the melting of subsurface ice.

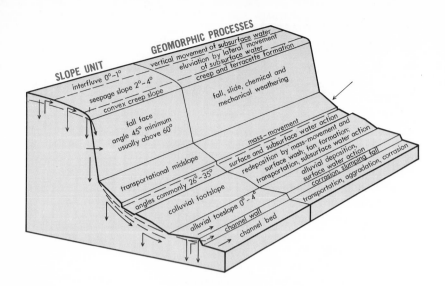

Figure 12-22 Slope process model. This diagram illustrates the relationship between form and process along valley slopes. Not all of these segments need be present. In some cases, for example, the fall face (cliff face) may be absent. The arrows indicate the direction of groundwater and fine soil particle movement. [Adapted from Dalrymple, Blong, and Conacher]

being done is closely related to position with respect to the slope profile, as shown in Figure 12-22. The predominance of particular segments of the slope profile, in turn, is related to climate and underlying material. This profile can be considered a miniature energy system somewhat comparable to that along the longitudinal profile of a stream. Rock material at the summit of the hill has potential energy that cannot be released until the cohesion of the rock is destroyed. When weathering breaks the rock apart and the fragmental regolith and soil develop the capacity for movement, the material begins to move downslope, aided by the potential energy of absorbed water. At first, the movement is slow and can be compared with the formation of first-order tributaries of a drainage system or largely at random. Volume and gradients are low. Soon, however, the slope profile becomes steep and highly unstable. Rapid changes in form take place, and the slope is likely to be irregular, with alternate erosional and depositional segments. Whether this slope is a vertical cliff face or a lesser slope depends on the relative importance of mechanical and chemical weathering, on the vegetation cover, and on the type of material. There are several parallels that can be drawn between the fall-face slope of a valley side and the steep, unstable upper portion of a longitudinal stream profile.

The *fall face,* or most unstable and steepest section of the transverse slope profile, is a particularly dominant feature of mass wastage slopes in arid, semiarid, and periglacial regions, where mechanical weathering is relatively more significant than chemical weathering. The frequent occurrence of *mesas* and

Figure 12-23 Typical landforms developed on nearly horizontal sedimentary rocks in dry regions. [After Strahler]

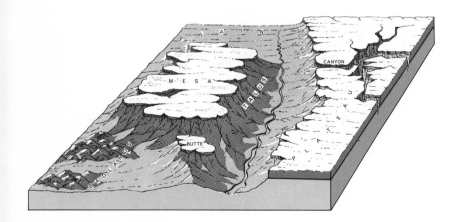

Figure 12-24 View of the Grand Canyon of the Colorado River, Arizona. The upper portion of the canyon walls, cut in sedimentary rock strata, is a succession of fall faces, talus slopes, and rock terraces caused by the retreat of the fall faces. Note the even skyline, indicating the evenness of the high plains into which the canyon is cut. [U.S. Dept. of the Interior; National Park Service; M. W. Williams]

buttes, which typically are small, flat table-lands with pronounced fall faces and associated talus slopes, is evidence of this (see Figure 12-23). The slow recession of fall faces is an extremely important process in the beveling of dry regions and high elevations. An excellent example of fall-face recession is the Grand Canyon of the Colorado River in Arizona, where alternate fall faces and talus slopes reflect variable resistances among the thick series of sedimentary rocks (see Figure 12-24). Fall-face recession also can be found in warm, humid regions but tends to be restricted to particular structures, such as thick, massive beds of siliceous or ferruginous sandstones.

The *transportational midslope* section of the profile in Figure 12-22 represents a surface of mass movement where the material derived from the fall face is moving relatively rapidly downslope, mainly by *slides* or *slump slides.* The angle of the transportational midslope is usually determined by the coarseness of material, with the coarsest rock fragments developing the steepest slopes, as in *talus cones* (see Figure 9-3). The transportational midslope thus is both a depositional and an erosional surface that is roughly convex in profile and in a constant state of dynamic equilibrium. Material is continually being added by fall-face weathering above, and weathering continues to alter the size of the fragmental debris along the slope. Movements of various kinds alter

this section of the slope profile to adjust the general slope to equilibrium. The continuous mass movement of material down the slope surface tends to abrade the surface beneath, especially in the downslope portion, where the material is smaller and thinner. The erosional surfaces thus produced are termed pedislopes (*pedi* = feet), and when coalesced produce *pediments* or *pediplains.* These erosional surfaces are best developed in dry climates. Many of the broad interior plains of South Africa and Australia are believed to be pediments, testifying to the stability of these continental shield areas over long periods of geologic time. The frequent occurrence of fall-face slopes in these areas is partially explained by the extensive occurrence of unusually durable cemented layers[2] in ancient soils which, after erosion, are exposed and become the cap rock or fall face for low tablelands. Some of these cemented layers are believed to predate the continental split of Gondwanaland and the drift southward out of the humid tropics in late Mesozoic times. *Pediplanation,* or the parallel retreat of scarps by mass wastage, can occur in warm, humid regions, but its effects are masked by deep chemical weathering. Miniature pediments frequently can be found in humid areas along valley sides in sedimentary rocks of varying resistance, producing what sometimes are termed *rock terraces,* or gently sloping benches lying above durable beds (see Figure 11-21).

The steep concave sector of the generalized slope profile of Figure 12-22 lying immediately below the transportational midslope may or may not be present, as is the case with any particular segment of the profile. In miniature, or on a local scale, it can be represented by the back slope of landslides or slump slides. In general, this break in slope represents the change from the erosional part of the profile to that of deposition. Below this point, fragmental material accumulates, awaiting its turn to be removed by stream transport below.

After mass wastage reduces the slopes of interstream areas, remnants of the former upland level sometimes remain behind as

[2]One of the most common types is *plinthite,* formerly termed *laterite,* composed mainly of iron oxides. Once formed, these undergo practically no chemical weathering and may persist as a resistant topographic feature for millions of years on flat plains.

Figure 12-25 Stages in the cycle of erosion by underground solution in a limestone region. [After Lobeck]

isolated knobs or hills rising above an undulating erosional plain. Such knobs are termed *monadnocks.* Usually they occur in areas of crystalline rock. Stone Mountain in Georgia and Mount Monadnock in New Hampshire are examples. Much more common but somewhat related forms are *inselbergs* (Ger.: *insel* = island; *berg* = mountain), which have characteristic rounded tops and steep sides rising abruptly in fall-face slopes from adjacent erosional plains. These are abundant in parts of Africa, especially along the low watershed divides to the north and northeast of the Congo Basin.

12-7 DEPOSITIONAL LANDFORMS ASSOCIATED WITH MASS WASTAGE *Colluvial slopes,* or depositional slopes formed of colluvium, are found almost everyplace where mass wastage is in operation. It has already been noted that talus or scree accumulates at the base of fall-face slopes, and nearly all valley sides in humid regions contain colluvium within part of their slope profiles. In Figure 12-22, the major sector of the lower part of the profile is formed by the *colluvial footslope.* This is composed of a heterogeneous mixture of coarse and fine material, often convex in surface profile and exhibiting characteristics related to poor drainage. The reason for the latter is that subsurface drainage of groundwater from the upper slopes keeps this material saturated. Temporary hillside springs often are found here. Where the groundwater is charged with solutes, such as lime or iron, these may be precipitated within this zone as the water evaporates at or near the surface. In tropical regions, these colluvial slopes are sites for the development of *plinthites,* the ironstone hardpans mentioned in footnote 2.

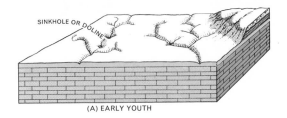

(A) EARLY YOUTH

(C) MATURITY

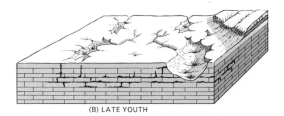

(B) LATE YOUTH

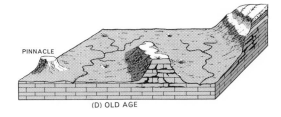

(D) OLD AGE

Landforms Associated with the Work of Solution

12-8 EROSIONAL LANDFORMS ASSOCIATED WITH SOLUTION The landforms resulting from solution are mainly found in limestone areas under humid climatic conditions. Figure 12-25 shows the gradual dissection in stages of an upland limestone surface by underground solution. At first, the principal work is below the surface and beneath the water table, where groundwater fills all cracks and crevices. Cracks and drainage channels under such conditions are rapidly widened and enlarged, producing large underground caves and caverns. As the cracks near the surface enlarge, soil material collapses into the void below, forming funnel-shaped surface pits or depressions that are termed *sinkholes* or *dolines*. Soil material slipping into the center may plug up the drainage tube, and water will collect to form small ponds. With time the sinkholes become widened and coalesce to form steep-sided depressions without surface outlets known as *uvalas*. Unusually large uvalas are re-

ferred to as *poljes*, some of which may be several miles long, with broad, flat floors and cliffed sides not unlike fault depressions or grabens. As indicated in the discussion of weathering processes in Section 9-7, rainwater plus CO_2 is able to rill or groove the surfaces of exposed limestone if the latter is sufficiently pure. Such microrelief forms are common features of areas containing limestones without many impurities (see Figure 9-6).

The remnants in the late stages of erosion of limestone areas by solution are generally much steeper than in areas dominated by stream erosion and mass wastage. Cliffs are characteristic features of limestone-solution areas, and they tend to persist into the later stages of erosion, even in humid tropical areas where solution is rapid. Isolated steep-sided limestone pinnacles rising from a flat plain are termed *mogotes*. The classic regions for such pinnacle forms are in southern China and northeastern North Vietnam (see Figure 12-26), although such terrain appears in many parts of the humid tropics, from New Guinea to Jamaica, Venezuela, and Cuba.

Figure 12-26 Limestone pinnacles near Kwei-yang, China.
The steep-sided hills in the background are typical of a vast area of
karst topography that is found in southern China and North Vietnam.
[Courtesy Emerson Comstock]

Sometimes the rapidity of solution in the tropics will produce a dense association of mogotes to form a rough, hummocky surface appropriately termed *haystack topography.*

12-9 DEPOSITIONAL FEATURES ASSOCIATED WITH SOLUTION Not all of the work of solution is erosional. Deposition from solution occurs for several reasons, including (1) evaporation, (2) organic precipitation, and (3) chemical precipitation. Distinctive landforms can be related to these processes. Of the materials involved in deposition from solution, by far the most important is calcium carbonate (see "Limestones," pp. 365–368) which becomes marl or limestone following deposition.

Once underground water is drained from caves or caverns, deposition of *dripstone* begins to take place, resulting from the evaporation of lime-charged water dripping into the empty underground openings. *Stalactites* are the icicle-like pendant forms of precipitated lime that hang from the ceiling of many caves, while *stalagmites* are the protuberances of dripstone that grow upward from the cave floors.

The terraces of hot springs (see Figure 10-6) represent other landforms caused by the evaporation of lime-charged water. In this case, evaporation is accelerated by the high temperature of the groundwater that was heated by contact with hot volcanic rocks below the surface. Such landforms are found in only a few places on earth but are spectacular features.

Coral reefs are perhaps the most widespread type of landform associated with solution. They are massive deposits of calcium carbonate built by colonial sea animals that are able to extract lime from seawater to build protective exoskeletons. The habitat of corals is restricted to tropical and subtropical coastal waters which do not cool below 65°F and are normally less than about 150 feet in depth. With slow subsidence, coral deposits several hundred feet in depth may be constructed. Once within tidal or surf range, the reefs are attacked by erosion, and the debris is tossed over to the lee side of the reef. The most active living reefs, however, are found in the surf zones, where the water is constantly aerated and where tiny microscopic algae and other small food particles are carried within reach of the organisms. Among the more common reef structures is the *fringing* type (see Figure 12-27), which is found where waves beat directly against the shore, and the *barrier reefs,* which are found in the surf zone within shallow water some distance from shore. A gradual subsidence of low tropical islands may be matched by the upward growth of coral reefs, where the coral colonies extend upward in order to reach the light and aerated zone near the surface. In this way, circular reefs, termed *atolls,* are formed. Low islets built of coral sand often surmount the atoll reefs. Narrow passageways through the reefs are kept open by strong tidal currents. The lagoons, or sheltered water between the reefs and the shore or within the center of atolls, afford protective anchorages for ships. Navigation of the narrow tidal inlets that are rimmed by jagged coral edges, however, may be hazardous. Seen in detail, coral reefs are wonders of complexity and beauty, alive with a profusion of marine life forms, all living together

Figure 12-27 Fringing coral reef, Kusiae Island, Caroline Islands. Note how the waves break against the outer margins of the reef, indicated in white. Kusiae is typical of the high, volcanic islands that rise abruptly from the floor of the Pacific Ocean. [*Geographical Review*]

in a symbiotic, or mutually beneficial, environment. The combination of seafood and coconut, breadfruit, and other land plants on the coral islands has nurtured human inhabitants for thousands of years.

Landforms Associated with Glaciation

12-10 EROSIONAL FEATURES ASSOCIATED WITH GLACIATION The processes of erosion and deposition by glaciers were described in Section 9-11. While the processes of mountain glaciers closely parallel those of continental glaciation, the forms resulting from the two are different and are distinguished in the following subsections.

Erosional landforms of continental glaciation. The distribution of continental glaciation during the Pleistocene period is shown in Figure 9-17. Evidence obtained by quantitative analysis of radioactive carbon in organic material indicates that the last ice sheet in North America reached its maximum extent about 18,000 to 20,000 years ago and was present over the northern Great Lakes area as recently as 10,000 years ago. Greenland and Antarctica are the remnants of these waning ice sheets. The erosional work of continental glaciation thus is unusually recent within the range of most geological processes.

Erosional landforms resulting from continental glaciation prevail today mainly in those parts of the glaciated regions that have hard, crystalline rocks from which a shallow cover of soil and rock rubble was quickly removed to expose bedrock. This condition is best found within the crystalline shield areas of Canada and Scandinavia. Frost weathering at high latitudes tends to form a mantle of broken rock rubble relatively rapidly. Prior to glaciation, during a long cool period with periglacial conditions, there must have been a plentiful supply of rock fragments lying on the surface. Such fragments were incorporated into the ice masses as they moved out from their source areas. Thus equipped, they became gigantic rasps that scoured underlying rock surfaces. The hills were rounded and polished by the abrasive action of the glaciers, producing the smooth, polished surfaces known as *glacial pavements.* Scratches on these surfaces, termed *glacial striae,* have been used to determine the direction of ice flow. The plucking or quarrying action of ice in areas of highly fractured rock often resulted in steep cliffs, especially on the lee sides of hills, that is, the side opposite the direction from which the ice came. These rock cliffs remain today

as prominent features of the hilly shields and add to the general difficulty of cross-country travel.

Wherever the ice masses were channeled into narrow valleys, oriented in the direction of ice motion, typical *U-shaped valleys* were formed, with steep sides and rounded bottoms. Such forms are often encountered in areas of mountain glaciation but were also produced by continental ice sheets. The broad, north-south valleys in the northern part of the Appalachians, such as in the Finger Lakes area of New York State, are good examples of such valleys.

Patterns of stream drainage were extensively modified by a combination of glacial erosion and deposition. In the northern Appalachian uplands of southern New York State, for example, evidence indicates that the preglacial drainage was mainly toward the southwest. The ice sheets, by scouring deep trenches in a north-south direction and depositing glacial till in the previous northeast-southwest drainage ways, altered the entire regional drainage pattern.

Other erosional features resulting from glaciation are glacial spillways and cross channels. *Glacial spillways* are large valleys eroded by meltwater from the ice. During the waning stages of continental glaciation, large quantities of water from glacial melting flowed from the ice margins. Where the terrain was underlain largely by unconsolidated glacial debris, the streams of meltwater that contained many abrasive rock fragments could carve valleys quickly. Sometimes also, the glacial waters, ponded behind a terrain obstacle, sought an outlet laterally across uplands to produce lateral drainage ways across the normal grain of the land surface. Such lateral valleys are termed *cross channels* (see Figure 12-28). Spillways and cross channels are often found in glaciated terrain that is rolling to hilly, as in parts of Wiscon-

Figure 12-28 An abandoned plunge basin in a glacial cross channel; Clark State Reservation near Syracuse, New York. A waterfall rivaling Niagara Falls in volume and height once poured over this cliff from the left. When a lower channel to the north was uncovered by the continental ice retreat, the waterfall and cross channel were abandoned, and only the deep plunge basin remains as a small pond. [Van Riper]

sin, Southern Michigan, eastern Ohio, and central New York. The east-west cross channels in northeastern Germany have provided excellent routes for canals that connect the north-south rivers and thus provide a network of inland waterways.

Upon leaving the hard-rock areas of southern Canada and Scandinavia and continuing southward, the ice masses suddenly encountered soft sedimentary rocks, including shales, limestones, and sandstones. With the weight of thousands of feet of ice and with durable boulders of hard rock as abrasive tools, the ice masses gouged deeply at first and were channeled into major lobes or tongues by irregularities in the margins of the crystalline rock uplands. Thus were excavated the large basins that later filled with water to become the Great Lakes in North America and the Baltic and North Seas in Europe.

The ice mass in the United States was diverted into the basins of western Lake Superior and Lake Michigan by an "island" of hard granites in northern Wisconsin and the Northern Peninsula of Michigan. The lee side of this upland region, located in southwestern Wisconsin and extending into parts of Iowa and Minnesota, was never covered by any of the four ice advances, although it was surrounded by ice on all sides. This unglaciated area is known as the *Driftless Area*. The landforms of this area, clearly related to stream erosion instead of glacia-

tion, contrast sharply with those of the surrounding areas that were glaciated.

Once out of the basins adjoining the crystalline uplands, the continental ice sheets rapidly lost much of their erosive power. The old crystalline tools were increasingly worn away, the newly incorporated rock debris lacked the hardness necessary for good abrasive action, and unconsolidated material began to clog the lower portions of the ice. Ice energy was expended more in transporting material than in performing erosive work. In many parts of the American Midwest, unconsolidated glacial deposits have been found buried under later glacial material,

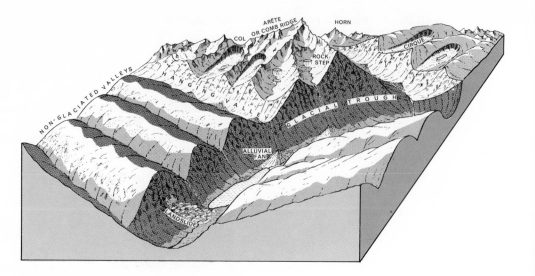

indicating that the subsequent ice advances were not able to remove previous loose material. In a few places, interglacial topsoils have been found, complete with surface humus layers. Except for the northern Appalachians and the northern British Isles, erosional landforms produced by the ice sheets are rarely encountered much beyond the margins of the crystalline shields.

Erosional landforms in areas of mountain glaciation. Glaciation today is producing characteristic features of erosion wherever mountains are high enough to have mean temperatures during the warmest month of less than 50°F, and where precipitation is sufficient to nourish snowfields. During the Pleistocene, when temperatures were lower over the entire earth, the snow line descended to much lower elevations than at present. Thus, the distribution of glacial erosional features in mountainous regions is much more widespread than the extent of glaciation today.

The major landforms of mountain glacial erosion (see Figure 12-29) include the sharp, steep-sided mountain spires termed *horns* (such as the famous Matterhorn of Switzerland), the jagged crest lines termed *arêtes* or *comb ridges,* the amphitheaterlike depressions or *cirques* that have been gouged out as if by some gigantic ice cream scoop, U-shaped passes termed *cols* through the crest lines with their knife-edged summits, *U-shaped valleys* leading down to adjacent lowlands, and tributary valleys that enter high on the side slopes, termed *hanging valleys.* A series of cirques occupying the upper portion of a glacial valley may rise in a series of rock steps toward the back wall of the valley head.

Erosion by glacial scour is found mainly within the valleys below the snow line. As in the case of continental glaciers, moving streams of glacial ice containing stones and boulders deepened and widened the valleys through which they flowed. The amount of scour is related to the volume of the valley ice and to the resistance of the rock material beneath. Main valleys that receive much more ice than their tributaries tend to be scoured more deeply than the latter, resulting in the characteristic hanging valleys.

Above the snow line, and within the

Figure 12-29 Erosional landforms associated with mountain glaciation. [After Lobeck]

snowfields of the cirques where the glaciers are born, the dominant erosive work is performed by the freezing of water in cracks and crevices and by the plucking action of the accumulated ice as it pulls away from the back wall of the cirques. This outward and downward movement of cirque ice usually leaves a noticeable crack near the back wall that is termed a *bergschrund* (see Figure 11-19). Accumulation of meltwater within the bergschrunds plays an important role in the sapping action of mechanical weathering in cirque enlargement. Fragments of rock from the upper-rock fall faces fall into the bergschrund, to be incorporated into the cirque ice and act later as a rasp when the ice moves out of the cirque. Cirques formed during the Pleistocene, and now abandoned by glacial ice, frequently feature gouged depressions that lie lower than the outer lip of the cirques, thus collecting water to form *cirque lakes* or *tarns*.

High, slender waterfalls are often encountered in glaciated mountain landscapes, marking the entry point of hanging valleys into the main valleys. Bridal Veil Falls, in Yosemite National Park, is a famous example. Hanging valleys containing lakes with appreciable outflow are choice sites for hydroelectric power. The lakes help to regulate the flow. There is a large hydrostatic "head" because of the long fall to the valley below, and there is no need for expensive storage dam structures. Many such power sites have contributed to the cheap electric power of such countries as Switzerland and Norway.

The U-shaped valleys of mountain glaciers are impressive when they open directly to the sea. These long, narrow inlets, scoured at times to depths which today are well below sea level, are termed *fjords* (see Figure 12-30).

Figure 12-30 A narrow fjord south of Alta, Norway. Note the steep sides of this drowned glacial valley. A small end moraine extends almost across the fjord. [Van Riper]

Some of them, such as Sogne Fjord in Norway, form superb natural harbors and extend far inland for 100 to 150 miles. Plunging steeply into the heads of the fjords are the glacial valleys that lead toward the highlands. Fjord coasts are found along the entire west coast of North America north of the Canadian border, in Labrador, along the Norwegian coast, southern Chile, South Island in New Zealand, and bordering Greenland and Baffin Island.

12-11 DEPOSITIONAL LANDFORMS ASSOCIATED WITH GLACIATION Deposition by the continental ice sheets during the Pleistocene prevailed over erosion in determining the major features of the present land surface throughout most of the glaciated areas lying south of the Canadian and Scandinavian shields. Most of the depositional material was dropped at the margins of the ice sheets during the retreat of the ice fronts.

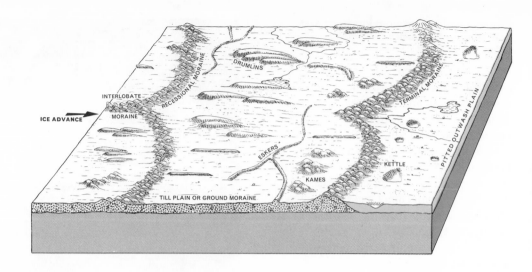

Figure 12-32 Depositional landforms in the Great Lakes region. Note how the pattern of moraines marks the various lobes of the ice advance in this area. [After Graetz and Thwaites]

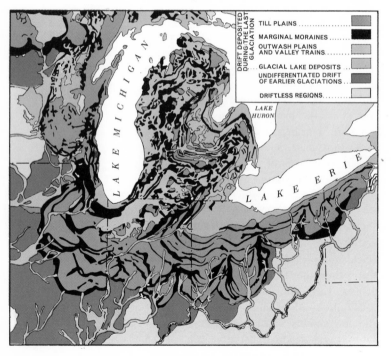

Landforms composed of glacial till are collectively known as *moraines* (see Figure 12-31). Where the ice front remained stationary for long periods, and where the forward progression of ice was balanced by melting of the ice margin, large *end moraines* were formed, constituting linear belts of low hills, usually less than a hundred feet in height, although some several hundred feet high have been observed in Michigan and Wisconsin. Where there is a succession of end moraines, marking temporary halts in the ice-front retreat, the farthest is known as a *terminal moraine* and the others as *recessional moraines*. The former is not necessarily the highest or best developed.

The margins of the Pleistocene ice sheets were typically lobate; that is, they were in great tongues or lobes. Hence, the morainic ridges commonly are aligned in loops or arcs. Variations in the resistance of the underlying rock usually helped determine the channeling of the ice lobes. Moraines located between large, active lobes of ice are likely to be unusually high and are sometimes designated as *interlobate moraines*. Figure 12-32, which shows the position of morainic belts in the

Great Lakes area, clearly illustrates this lobate aspect. Note how the lobes roughly parallel the basins in which the Great Lakes are located.

As the ice sheets retreated gradually from their major morainic positions, a more even veneer of glacial till was deposited, forming what has been termed a *ground moraine.* If the resultant surface was essentially a flat plain, the term *till plain* is used. Much of the flat, fertile plains in Illinois, Iowa, and Indiana are typical till plains. Others are found in northern Germany and in the area east of the Baltic Sea. The till plains associated with the earlier Pleistocene ice advances in parts of southern Illinois, Iowa, and Missouri underwent considerable dissection by streams. The resulting rolling plains contrast markedly with the flat-to-undulating till plains of the last, or Wisconsin, glacial advance. Moraines can be found within the shield areas far to the north, but because of the lack of weathered surface material, which was carried far southward in earlier advances, the glacial deposits generally are low and stony (see Figure 12-33). Their major topographical role was to interrupt the prior drainage channels in the lowlands.

Drumlins are depositional landforms consisting of oval-shaped mounds of glacial till, aligned in the direction of ice motion and having steeper slopes on the side facing the direction from which the ice came. They are believed to be the result of overriding of former heavy clay till moraines by a readvance of the ice, especially in limestone areas, where thick deposits of heavy plastic till are likely to be found. The classic locality for them is in New York State, between Rochester and Syracuse, although they also are found

Figure 12-33 A stony moraine north of the Arctic Circle in Finland. Such stony moraines are common in the crystalline shield areas of Canada, the northern United States, and Scandinavia. Clearing and preparing such land for cultivation is extremely difficult. Note the lichens growing on the boulders, a characteristic feature of the high latitudes. [Van Riper]

in parts of Wisconsin, Michigan, southern Sweden, Ireland, and east of the Baltic Sea.

Outwash plains are perhaps the most common depositional landform type associated with glacial stream deposition. The meltwater from glaciers is heavily charged with rock waste, and much of this debris is dropped as the streams leave the steep front edge of the ice. Cobbles, sand, and gravel are the first to be deposited, and this material accumulates in broad sheets beyond the ice margin. These plains are especially well developed on the borders of prominent end moraines. The former vary greatly in size—from small, narrow valley fillings, where they are termed *valley trains,* to extensive plains several hundred square miles in area.

Blocks of ice, broken from the ice margin and buried in outwash material, later may result in depressions or *kettles* in the outwash plains as the ice melts away (see Figure 12-34). The term *kettle*, however, is often used to designate any depression without an outlet in a glacial depositional surface. A *pitted outwash plain* is one that has a large number of such depressions. Swamps or lakes may occupy the kettles. Outwash plains cover a larger percentage of the glacial depositional surfaces in Europe than in North America, occurring throughout much of the North German Plain and in the northwestern Soviet Union. However, large outwash plains also are found in the northern part of the United States, especially in Wisconsin, Michigan, and Minnesota.

Mounds and ridges of stratified outwash sands and gravels are often found in the vicinity of moraines, especially where there is evidence of ice stagnation with little or no ice movement. The mounds are known as *kames*, and the narrow, often sinuous gravel ridges are termed *eskers*. The latter probably were formed by water deposition behind the ice margins in stream channels confined within narrow ice walls. Eskers more than a hundred miles in length have been found in Maine and Scandinavia, where these forms are especially common features on the ground moraines. Kames have a variety of origins, the most usual being deposition by running water within crevasses, tubes, or wells in the ice mass. Later, as the ice melts, this material is let down as mounds upon the land surface. *Kame terraces* are found bordering the lower slopes of valleys in hilly terrain that underwent continental glaciation, as in the northern Appalachians, New England, parts of the northern British Isles and northern Norway. These are irregular benches or terraces that result from fluvioglacial deposition between the ice blocks in the centers of the valleys and the ice-free hill slopes.

Lakes are common features throughout glacial areas. Some are tiny, occupying small depressions in moraines or outwash plains. Others, like the Great Lakes, occupy huge basins excavated by ice scour. Still others are the result of clogging of former drainage channels by deposition. Glacial lake plains rimmed with low, sandy beach ridges border many of the Great Lakes in the United States (see Figure 12-32), representing higher lake levels during the waning stages of glaciation. Lakes also are often associated with mountain glaciation, some occupying former cirque depressions or lying behind recessional moraines.

Lake basins, terraces, and beach ridges formed during the Pleistocene were not always restricted to areas covered by the continental ice sheets. One of the major fea-

Figure 12-34 Kettles forming in the debris-clogged lower end of the Tasman Glacier, New Zealand. As the ice slowly melts, the gravelly debris is let down onto the underlying surface. Note the small pond that someday probably will become a typical kettle lake. [Van Riper]

tures of this unusual climatic period in earth history was a displacement equatorward of the normal climatic zones. This resulted in an appreciable increase in the precipitation of the subtropical desert areas. Many of the large interior basins in these regions were occupied by large lakes. Great Salt Lake is a last remnant of a much larger lake (Lake Bonneville) that once covered much of western Utah and northeastern Nevada. Similar Pleistocene lake basins with associated lacustrine features are found in many parts of what is now arid Africa, South America, and Australia. Other large temporary lakes were formed by ice sheets that dammed north-flowing rivers in Canada and northeastern Europe. Many of the large swamps in eastern Poland and the northern Soviet Union probably were former shallow, ice-dam lake basins that have become filled with sediment and vegetation.

Depositional features associated with mountain glaciation reproduce in miniature nearly all of the forms discussed under continental glaciation. Since ice confined within a valley flows more rapidly in the center than at the sides, there is a tendency for included rock waste to be diverted toward the valley sides, as shown in Figure 9-18. Weathering of the cliffs adjoining the glacial streams also causes rock debris to accumulate along the sides. After the melting of valley glaciers, ridges of morainic material are left behind along the valley flanks to produce *lateral moraines.* The union of ice streams results in the junction of lateral moraines to form *medial moraines* closer to the center of the valleys and downvalley from the valley junctions.

Glacial landforms may be highly intermingled in complex patterns. To illustrate, a typical small portion of an interlobate morainic area in east-central Wisconsin, with its associated glacial landforms, is shown in Figure 12-35. In this region, a farmer may have sand, gravel, clay, muck, or stony spots in many different slope conditions, associated with a large variety of glacial landforms, and within the confines of his own farm. Changes in soil composition also occur vertically. Bedrock may lie anywhere from 2 to 200 feet below the surface. Sand may underlie clay at variable depths, or vice versa, and a layer of sand or clay 3 feet below the surface may be entirely absent a short distance away. The great differences locally, as shown in Figure 12-35, can also be observed on continental scales. The fertile till plains of the American Corn Belt contrast markedly with the sterile, sandy, outwash plains of northern Michigan or the heath lands of northern Germany.

Landforms Associated with the Work of the Wind

Landforms associated with the work of the wind are mainly depositional. The erosive work of wind is not great mainly because of the limited size of the tools it can carry, the limited height above the surface these tools can be supported, and the inconstancy of the wind itself. Wind erosion usually is limited to a sandblasting effect that etches out weak parts of rocky surfaces. Soft sandstones, for example, tend to be especially susceptible to this etching effect, and unusual sculpture forms can be produced. The largest forms resulting from wind erosion are *deflation pits* or *blowouts,* produced by the removal of fine-textured surface materials, such as fine sands or silts, by wind deflation. These are shallow depressions of various sizes up to a mile or more across, 10 to 50 feet deep, and with gently sloping sides. Thousands of them can be found in the Great Plains of the United States. Many are associated with low dunes in the Sand Hills region of Nebraska.

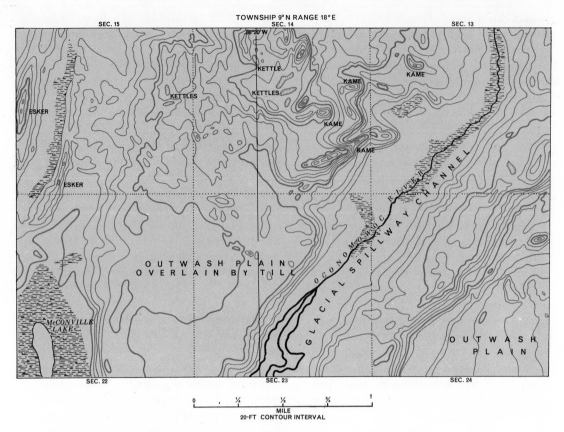

MILE
20-FT CONTOUR INTERVAL

Figure 12-35 Landform detail in a section of the interlobate moraine northwest of Milwaukee, Wisconsin. [Topography from Oconomowoc, Wis., 15-minute Quadrangle, U.S. Geological Survey].

The topographic forms resulting from wind deposition include those related to loess deposits and sand dunes. Loess rarely forms individual landform types because it is a regional depositional feature that mantles a broad surface with fine silt. However, its tendency to stand in near-vertical cliffs without slumping, owing to its columnar structure, produces unusually steep gullies and rectangular drainage patterns. Easily excavated, the steep, loessal valley sides of northwestern China are widely terraced for

cultivation (see Figure 12-36), and homes are excavated in the silty material. Similar excavations made in the loess or loesslike cliffs along the Mississippi River at Vicksburg were a factor in the stubborn defense of that city during the Civil War.

Many people believe deserts generally to be only vast expanses of billowing sand dunes, possibly because such settings invariably were chosen for many desert scenes in early motion pictures. The sandy deserts, or ergs, are by no means as widespread as often believed, although approximately one-fourth of all the desert areas are composed of ergs. A large supply of sand is required for their formation and is brought into large interior basins by networks of wadis that tap upland areas having a preponderance of sandstones or quartz-rich crystalline rocks. A fairly constant wind direction from the source areas is also helpful in establishing the erg dune complexes. Individual dunes invariably take one of two forms: a crescent shape or an elongated S or sigmoid curve (see Figure 12-37). The former is more characteristic of individual dunes and of areas where the wind direction is unusually constant. The latter is more common where the wind direction is more variable and where the dunes intersect each other by overriding.

Figure 12-36 Loess country in southern Shensi Province, northwestern China. Despite centuries of care in terracing hillsides, gullies continue to cut into these thick deposits of windblown dust. [U.S. Air Force (MATS)]

Figure 12-38 Erosional and depositional landforms along shorelines. [A after Stamp; B after Strahler]

Figure 12-37 A large sand dune in central Saudi Arabia. This is a typical sigmoid, or S-shaped, dune caused by reversals of wind direction. Note the extremely sharp crest of the dune and the alternation of the side containing the angle of rest. [Courtesy George Kuriyan]

Desert dunes vary in height, but normally small ones are between 10 and 50 feet high. Dunes 100 to 300 feet high are fairly common in the larger ergs, and individual ones 700 feet high and almost a mile long at the base have been observed in southern Iran. The largest erg deserts are those in the Sahara, southern Saudi Arabia, southeastern Iran, the Thar Desert of northwestern India, the Tarim Basin in central Asia, and parts of central Australia. Some of the individual ergs in the Sahara cover tens of thousands of square miles.

Dunes are not found exclusively in desert regions. They are found along windward, low, sandy coastlines nearly everywhere, although here they tend to be confined within a narrow strip less than 1 to 2 miles wide. Some of the large parabolic dunes along the eastern side of Lake Michigan reach heights

comparable to those in the large desert ergs. Some low dunes have formed on sandy outwash plains following glaciation and prior to the establishment of vegetation. The beach dunes of humid regions have profiles very similar to those of the deserts. The sand is rolled up a variable front slope of the dune, and, once past the sharp crest, rolls down the unstable back slope that invariably lies at an angle of 32°, the angle of rest of dry dune sand.

Erg deserts are rare in the arid portions of North America, but a small and highly unusual one is found in the United States in the White Sands National Monument near Alamogordo, New Mexico. It is unusual in that the dunes are composed largely of gypsum flakes, a salt that accumulated on the surface of an adjacent playa flat.

Landforms Associated with Waves and Currents

12-12 EROSIONAL LANDFORMS ASSOCIATED WITH WAVES AND CURRENTS Erosional landforms caused by wave and current action can best be observed along steep, rocky shorelines. A steep, rugged coast usually signifies deep water near shore, and since storm waves in deep water do not have to spend their energy through friction on the bottom, the waves are able to attack the shoreline directly (see Figure 12-38). Waves pounding against a cliffed shoreline often will undermine it, producing a *notched cliff.* They also etch rocky shores, concentrating on the weaker spots in the rocks, widening cracks and crevices, bringing into bold relief the

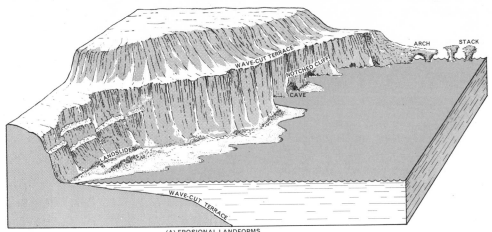

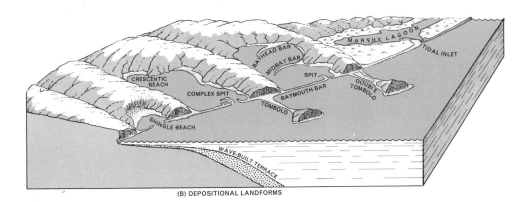

(B) DEPOSITIONAL LANDFORMS

hitherto hidden rock structure. *Wave-cut terraces* sometimes border sea cliffs, representing the erosional recession of the cliffs. These usually are mantled with a thin veneer of rock debris broken from the cliffs that is being rounded and reduced in size by the to-and-fro surges of wave motion.

Stacks, arches, caves, (see Figure 12-38), and other erosional forms can also be found along cliffed coasts. Caves are simply weak places in the shore cliffs that have been excavated by solution or hydraulic action. If a cave extends through a narrow point or promontory, an arch is formed. Collapse of

the roof of an arch, or the widening of a fissure, may result in a towerlike erosional remnant termed a stack. Erosion along rocky shorelines usually is slow enough not to be a serious problem with respect to land use, but where the material is loose and unconsolidated, storm waves may do an enormous amount of erosion within a single season or even a single storm. Several borders of the Great Lakes, for example, are experiencing unusually rapid recessions of their shorelines, such as along the west side of Lake Michigan north of Chicago and the south shore of Lake Erie. The material along these coasts is largely glacial or lacustrine deposits and erodes rapidly. Such lakefront property often is extremely valuable for residential purposes. Some of the most highly valued estate properties north of Evanston, Illinois, have lost 100 to 200 feet of their landscaped lawns during the past 50 years. Similar damage has occurred along the coasts of the English Channel, where unusually strong tidal currents help waves attack the soft rocks and sediments in this area (see Figure 12-39).

12-13 DEPOSITIONAL LANDFORMS PRODUCED BY WAVES AND CURRENTS
Several of the typical landforms produced along shorelines by depositional processes are illustrated in Figure 12-38. The material eroded from coasts by waves is distributed along the coast by the backwash of waves and by shore currents. Since the movement of water below the surface is weak, boulders,

Figure 12-39 System of groynes and seawall used to stabilize the coast of southern Britain against marine erosion. This view is near Dymchurch, Kent. [British Crown Copyright]

cobbles, and gravel usually remain near the surf zone where they originate. As they become reduced in size, they are more easily moved outward into deeper water. Sometimes the coarse material forms stony *shingle beaches* or *pebble beaches* along the shore, especially in small coves adjacent to cliffed headlands. Most of the beaches along the coasts of southern France and western Italy are of this type. The finer material, usually composed of sand, accumulates beyond the wave-cut terrace to form a *wave-built terrace,* thus extending the general gradual slope away from shore. Steep coastlines commonly are irregular, with pronounced headlands and bays. As erosion attacks the headlands, the eroded material is washed to one side, into the protected waters of the nearby bays. A protruding tongue of sand extending from the side of a promontory is termed a *spit.* Currents eddying around the end of a spit may bend it to form a *hook.* If the spit deposition moves completely across the bay or joins one from the opposite headland, it forms a *baymouth bar.* More frequently, such material is carried along the margins of the adjoining bays, rimming them to form *crescentic beaches.* A *tombolo* is a spit that has joined an offshore island with the mainland. An interesting example of a tombolo is the narrow neck of land that connects the ancient city of Tyre (Sur), Lebanon, with the mainland. During the time of the Phoenician Empire, Tyre was an island and utilized the shelter between the island and the mainland as a harbor. Soon after the Crusades it became tied to the mainland by sand deposition. The site of Carthage, in Tunisia, had a similar fate.

Low seacoasts with a gradually sloping bottom, where sands and gravels are abundant, often are fronted by a low ridge of sand that parallels the coast and extends above sea level. Such ridges are known as *barrier* or *offshore beaches.* They are built by large storm waves that pound against the gently sloping bottom, excavating it and piling the sandy debris into asymmetrical, low ridges. The wide marshy *lagoons* that lie back of barrier beaches are connected to the sea by narrow *tidal inlets,* or passageways separating the barrier islands. Offshore barrier islands and beaches are rare on inland-water shorelines, because storm waves rarely reach sufficient size to create them. Here they are represented instead by sand bars in shallow water. Barrier islands are prominent features along low coasts in many parts of the world. A succession of them borders much of the Atlantic and Gulf of Mexico coasts of the United States. The Frisian Islands along the south coast of the North Sea are well-known European examples. Attractions of ocean bathing, fishing, and cool ocean air in summer have helped to make barrier islands favorite summer resort areas.

Tidal mud flats are found wherever silts or clays accumulate along a seacoast faster than they can be removed by waves and currents. They occur along protected coasts where wave and current action is weak and where streams discharge large quantities of alluvium. Tides subject them to periodic inundations. They are not easy to reclaim for agricultural use because of their content of ocean salts. Diking to prevent flooding by seawater, draining by excavating ditches and canals, flushing with freshwater, and soil treatment using soil-conditioning chemicals must be done to transform tidal flats into productive agricultural lands. Such projects have been carried on for centuries in parts of the world where land is extremely valuable, as in northern China, the Netherlands, the Po delta, Japan, and northern Java. In low latitudes, such mud flats tend to be unusually difficult to reclaim because of the rank growth of mangroves and swamp palms.

Ice-rampart ridges are common features along shorelines that experience severe winters, especially on inland lakes. The expansion caused by ice crystallization on lakes where ice forms to thicknesses of 2 to 4 feet shoves the ice onto the shore, forcing rocks and soil material into low ridges a few feet back of the water's edge. On larger bodies of water, except in narrow bays, the ice buckles into ridges well out from shore; hence, the push shoreward does not have as much leverage.

Study Questions

1. Explain how it is possible to obtain a cinder cone during an eruption of highly basic, liquid basalt.
2. Explain why the volcanic-cone slopes of Java or the Philippines are choice sites for cultivation and why similar slopes in Japan are among the poorest in the country.
3. Explain with a diagram why grabens frequently are found near the summit of a large regional arch produced by compression.
4. Measure the ratio between maximum stream length and area of a small drainage basin on a U.S. Geographical Society topographic map. Compare this with two or three others. How much variation in the ratio do you find?

5. Explain the shape of the deltas shown in Figure 12-18.
6. How do you distinguish between a *bajada* and a *pediment*? Can pediments be found in humid regions? Explain.
7. Why are limestone caves often arranged in a series of levels, with narrow passageways separating broad rooms at different height levels?
8. Give five clear pieces of evidence for the occurrence of continental glaciation during the Pleistocene Period. Do *not* list five landforms.
9. How do you distinguish between a *kame*, an *esker*, and a *drumlin*? Is it possible to have a *kame moraine*? Explain.
10. Why are glaciers found at a much greater elevation in the northern Rocky Mountains of Montana than on the slopes of Mount Rainer, Washington, and at about the same latitude?
11. Explain the lack of continental glaciation in (a) most of Siberia; (b) central Alaska; (c) Australia.
12. Why do most lakes have sand bars paralleling the beaches but well offshore?

Part 3: References

Bertin, Leon, *Larousse Encyclopedia of the Earth*, G. P. Putnam's Sons, New York, 1961.

Birot, Pierre, *General Physical Geography*, John Wiley & Sons, New York, 1966.

Butzer, Karl W., *Environment and Archaeology; An Introduction to Pleistocene Geography*, Aldine Publishing Company, Chicago, 1964.

Chorley, R. J., "Geomorphology and General Systems Theory," *U.S. Geological Survey Professional Paper 500-B*, Washington, D.C., 1962.

Coulomb, J., and G. Jobert, *The Physical Constitution of the Earth*, English trans. by A. E. M. Nairn, Hafner Publishing Company, Inc., New York, 1960.

Dury, George H., *The Face of the Earth*, Penguin Books, Inc., Baltimore, Md., 1959.

Fairbridge, Rhodes W. (ed.), *The Encyclopedia of Geomorphology*, Reinhold Book Corporation, New York, 1968.

Flint, Richard F., *Glacial and Pleistocene Geology*, 2d ed., John Wiley & Sons, Inc., New York, 1961.

Keller, W. D., *The Principles of Chemical Weathering*, Lucas Brothers, Columbia, Mo., 1962.

King, C. A. M., *Introduction to Oceanography*, McGraw-Hill Book Company, Inc., New York, 1963.

Leet, L. D., and S. Judson, *Physical Geology*, 2d ed., Prentice-Hall Inc., Englewood Cliffs, N.J., 1958.

Leopold, Luna B., M. G. Wolman, and John P. Miller, *Fluvial Processes in Geomorphology,* W. H. Freeman and Company, San Francisco, 1964.

Lobeck, A. K., *Geomorphology, An Introduction to the Study of Landscapes,* McGraw-Hill Book Company, Inc., 1939.

Monkhouse, Francis J., *Principles of Physical Geography,* 5th ed., Philosophical Library, Inc., New York, 1964.

Pough, F. H., *A Field Guide to Rocks and Minerals,* Houghton Mifflin Company, Boston, 1953.

Rittmann, A., *Volcanoes and Their Activity,* trans. by E. A. Vincent, Interscience Publishers, Inc., New York, 1962.

Sharpe, C. F. S., *Landslides and Related Phenomena,* Columbia University Press, New York, 1934.

Strahler, Arthur N., *Physical Geography,* 3d ed., John Wiley & Sons, New York, 1968.

————, *The Earth Sciences,* Harper & Row, Publishers, Incorporated, New York, 1963.

Thornbury, William D., *Principles of Geomorphology,* John Wiley & Sons, Inc., New York, 1954.

Umbgrove, Johannes H. F., *Symphony of the Earth,* Martinus Nijhoff, The Hague, 1950.

Wright, H. E., and D. G. Frey (eds.), *The Quaternary of the United States,* Princeton University Press, Princeton, N.J., 1965.

Scientific American articles obtainable as offprints from W. H. Freeman and Company, San Francisco.

Anderson, Don L., "The Plastic Layer of the Earth's Mantle"	July, 1962
Bullard, Sir Edward, "The Origin of the Oceans"	Sept., 1969
Bullen, K. E., "The Interior of the Earth"	Sept., 1953
Ellison, W. D., "Erosion by Raindrop"	Nov., 1948
Emery, K. O., "The Continental Shelves"	Sept., 1969
Field, William O., "Glaciers"	Sept., 1955
Heirtzler, J. R., "Sea-Floor Spreading"	Dec., 1968
Heiskanen, Weikko A., "The Earth's Gravity"	Sept., 1955
Hurley, Patrick M., "The Confirmation of Continental Drift"	April, 1968
Janssen, Raymond E., "The History of a River"	June, 1952
Kay, Marshall, "The Origin of Continents"	Sept., 1955
Kuenen, Ph. H., "Sand"	April, 1960
Kurten, Bjorn, "Continental Drift and Evolution"	March, 1969
Leopold, Luna B., and W. B. Langbein, "River Meanders"	June, 1966
Menard, H. W., "The Deep-Ocean Floor"	Sept., 1969
Tuttle, O. Frank, "The Origin of Granite"	April, 1955
Urey, Harold C., "The Origin of the Earth"	Oct., 1952
Williams, Howel, "Volcanoes"	Nov., 1951
Williams, J. Tuzo, "Continental Drift"	April, 1963

PART 4
PLANT GEOGRAPHY

The great variety of plants that clothe the naked surface of the earth is one of the most interesting and most obvious features of physical geography. All around us can be seen the culmination of eons of plant selection: the forest giants; the lowly microscopic algae that form a scum on stagnant pools; cacti that thrive on meager desert moisture; lichens clinging to bare rock surfaces; carnivorous plants; plant parasites that consume the foods manufactured by other plants, animals, and even man; acid-tolerant plants and salt-tolerant plants; plants that live wholly underwater; plants that live wholly underground. These represent some of the different ways in which plants have reacted to widely differing environments.

Plants are not a simple, passive element in the earth environment. They serve an important function even in the operation of the great earth-atmosphere energy system. The position of plants in the carbon cycle has already been noted (see Figure 10-5). The processes of photosynthesis and transpiration by plants decrease the carbon dioxide content in the atmosphere and increase the content of oxygen and water. It has been estimated that if all of the carbon that is now tied up in vegetation and in the fossil fuels were to be burned or oxidized, there would be little if any free oxygen left in the atmosphere. We probably can thank the plant life on earth for most of our free oxygen. Some ecologists today are becoming alarmed at the increasing destruction of the vegetation cover by man, along with his increased consumption of coal, petroleum, and natural gas. They are not concerned so much with the resultant increase in carbon dioxide content as with a decrease in oxygen. There are several automatic controls other than vegetation to regulate the atmospheric content of CO_2, but man probably would have to manufacture oxygen synthetically if the natural supply were to decrease significantly.

Regional variations in vegetation also influence the maintenance of the global energy balance by plants. Transpiration of water, which is an important heat-exchange process, and the varying degrees of reflection of incoming solar radiation by vegetation are two examples of this role. One of the complexities of the global energy system is the variation in the net radiation balance both seasonally and geographically. The part played by vegetation in helping to reduce net radiation surpluses and deficiencies may be well illustrated by taking a stroll through a heavily wooded park on a hot summer day after leaving hot city pavements; or—at the other extreme—comparing the temperature differential between the center of a frozen lake or open field with that beneath a large spruce tree covered with snow. The biotic world, including plants and animals, constitutes a subsystem within the global energy system that is comparable functionally with that of the hydrologic cycle, the carbon cycle, or the geologic cycle, all mentioned earlier.

The two chapters included in this part of the text are devoted mainly to showing how differences in environmental conditions influence the associations of plants found in different parts of the world. The first chapter (Chapter 13) deals largely with the individual major factors that influence plant growth. The second treats the global associations that, while exhibiting an unbelievable variety owing to local conditions, have some generalized features that can be used for regional classification.

Man has become a highly significant independent variable in the geography of plants, and "natural" plant communities unaffected by man are becoming exceedingly rare. To emphasize this point, a separate division has been included in Chapter 13 on the role of man in plant geography.

CHAPTER 13
PRINCIPLES AND FACTORS IN PLANT GEOGRAPHY

13-1 PLANT COMMUNITIES An overall view of the earth and its plant covering reveals that, despite the almost infinite variety of plant forms, distinct groupings of plants appear at different levels of observation. Whether one views the arrangement of plants on an entire continent, within a state or province, on a casual stroll through a neighboring patch of forest, or on one's knees along a fence row bordering a field, distinct groupings of plants will be observed. These groupings of plants are not miscellaneous mixtures of plant types. Instead, they represent plant *communities*, whose individual members share a given environment and often have developed a mutual support in the successful occupation of this environment (see Figure 13-1).

Plant communities vary in size from small local groupings to broad agglomerations that span continents. They are as different from one another as a Chinese peasant village is from an English mill town. There is little in common between the riotous combination of tall grasses and herbaceous plants in a virgin prairie and the soggy, moss-covered, and shapeless forms of the cloud-zone forest on tropical mountain slopes; yet both are distinct communities of plants, developed under widely different environments.

Individual plant types or entire communities may compete with each other for a particular site and react differently to changes in environmental factors. Some groups resist change with stubborn persistence, whereas others are highly sensitive to slight external forces. Continued disruptions of communities often lead to their dissolution, until only traces of their former presence may be observed. Plant communities have a remarkable resilience, however, and after the disruptive factors have been removed, they often revert to their original form.

There is a decided hierarchy among the individual plant types within a community. In a forest community, for example, trees are the ruling members, and the ferns and low shrubs that dwell on

Figure 13-1 A forest community with hemlock and yellow poplar as the tree dominants. Rhododendrons and ferns form part of the characteristic ground flora. The photo was taken in the Joyce Kilmer Memorial Forest in the Great Smoky Mountains of North Carolina. [U.S. Dept. of Agriculture]

the forest floor represent the lesser members of the plant community. They require less light or less moisture and thus can live on the "crumbs from the table."

The broadest types of plant community are the plant *formations,* comprising the global forests, the grasslands, the desert shrubs, and the tundra. These cover large areas that are closely related to major climatic differences on a global scale. There are, however, great differences within each of these formations. The spruce-fir forests of Canada, the deciduous, oak-hickory forests of the east-central United States, and the broadleaf evergreen forests of the humid tropics are comparable only in that trees form the dominant vegetation forms. There are similar broad differences within the other formations. Both likenesses and dissimilarities of vegetation are present everywhere. Tree growth is a feature common to all forests; yet every local variation in slope, drainage, exposure, and microclimate results in localized differences. A timber "cruiser," in estimating the stand of merchantable timber in a particular forest, is acutely aware of local regional similarities and differences in the arrangement of tree species and ages; yet he usually has only a passing glance for the kaleidoscopic pattern of microflora that changes every foot or so as he strides along his route.

Vegetation communities invariably are arranged in tiers, or stories, reflecting the three-dimensional arrangement of plant environments. A forest, for example, consists of many plants other than trees. It may include the feathery fronds of ferns, thriving in the dampness and shade of the forest floor. Perhaps it also includes small flowering plants, such as anemones, trilliums, or mandrakes that find ideal conditions for flowering and fruiting in the short spring season of mid-latitude deciduous forests, before tree foliage reduces available sunshine. A forest may also contain the slender white threads of molds interfingering the matting of needles on a pine-forest floor, or clumps of mushrooms, which are able to feed on partially decayed forest litter and are not called upon to manufacture plant sugars by photosynthesis. In the depths of the tropical rain forests, the lower stories often include tree types that can thrive under different conditions of energy, moisture, or nutrient supplies. This arrangement of vegetation into tiers is not confined to forests. All environments inhabited by plants tend to have a tiered effect in the competition for survival.

Plant communities are never in a condition of complete stability, although the various elements within an environmental complex always tend to work toward mutual adjustments. The changes that take place in the environmental progression away from or toward relative equilibrium are accompanied by corresponding changes in plant communities. If, for example, a forest is destroyed by lumbering, the struggle for sunlight proceeds from the ground level. Under the open sun, the higher ground-surface temperatures and lower moisture conditions are advantageous to many plants that never had a chance in the competition of a mature forest environment. In time, if left undisturbed, the forest community will once again establish itself—perhaps at first with varieties of trees that are different from those of the relatively stable, or *climax* types. Eventually, however, these newcomers usually are replaced by the members of the ultimate community.

Not all environmental changes in a forest are as sudden as that associated with lumbering operations. There may be a change in hydrologic (drainage) conditions over several decades, a significant alteration of climate, or extensive gullying over still longer periods of time. Each of these changes, as well as many others of much shorter duration, involves alterations in the forest community, resulting in changes not only in the individual species of trees present but in the groupings of plants within the forest complex.

13-2 PLANT SUCCESSION There is an orderly sequence in the development of a vegetation cover. A bare rock surface anywhere on earth is not at balance with the energy flow system which seeks to establish a mixture of rock, water, gases, and life at the earth-atmosphere interface. At first, only *lichens*[1] may find a foothold on the base rock surfaces, but as soil develops, other plants begin to insert themselves competitively into the environment, eventually crowding out their predecessors. Change after change takes place, not only in the type of plant community but also in the environment itself. The end result is a "balanced" environment or subsystem through which energy flows—a deep, weathered soil mantle containing stable mineral compounds, water and associated solutes, and a soil atmosphere. This association tends to support the greatest possible *biomass,* or mass of living organisms, including both plants and animals. A forest has a greater biomass than a surface covered with shrubs. Hence, where the environment permits forests, trees eventually will replace shrubs in the equilibrium.

The sequence in the establishment of the ultimate maximum biomass is termed *plant succession.* The end result of succession is what has been termed a *climax vegetation.* It might be equated to the condition of equilibrium or a steady state in a system. It is doubtful if any vegetation association can be considered as climax because of the long-term climatic changes that alter energy inputs and outputs in biosystems throughout the world. The concepts of plant succession and relative equilibrium (climax) are useful, however in appraising man's use of the plant world for food and feed. To illustrate the principle of plant succession, the author observed the following sequences on a garden plot in southern New York State. The environment consisted of a well-drained soil on an underlying fluvioglacial gravel deposit.

[1]*Lichens* are interesting "double" plants that consist of fungi and algae living together in a mutually beneficial association. They are able to extract water directly from the vapor state, hence are not dependent on liquid sources.

Figure 13-2 Plant succession on an abandoned pasture in southern New York State. White pine (*Pinus strobus*) seedlings are among the first trees to follow the tall perennial herbs and grasses. The herbaceous plants shown in the foreground include goldenrod, Queen Anne's lace, and New England asters. [Van Riper]

The garden was plowed and harrowed but was not seeded. Within a week or so, the ground was carpeted with small annuals—the weeds that were so troublesome in cultivation the year before. Untouched by the hoe, they grew in profusion. At first the short types covered the ground, but in a few weeks taller varieties made their appearance. Here and there a few grasses took root and spread laterally. By the end of the first summer, the plot was a confused tangle of plants, ranging from lowly purslane to waist-high Compositae. A few scattered perennials appeared, such as dandelions. The perennial herbs and grasses had become more dominant by the end of the second year, and at the end of the third and fourth years they had taken over the garden. Now a definite, storied field community had been established, with goldenrod, Queen Anne's lace, and milkweed as the taller members, grass as the dominant intermediate story, and a ground flora of low vines, including field strawberry, dewberry, and cinquefoil.

Figure 13-3 Diagram showing typical plant succession on a section of the Atlantic coastal plain in the southeastern United States. The pine probably represents a fire-succession flora. [After Odum]

Trees and bushes appeared above the herbaceous plants during the sixth and seventh years (see Figure 13-2) and consisted mainly of scattered white pine (*Pinus strobus*), clumps of aspen, and willow. The pine rapidly outstripped the others, and from evidence elsewhere in the area, it will have established a pine forest within about 20 years. The pine, however, still is not a climax vegetation, and once shade is provided, maple, yellow birch, beech, and elm will become established as an understory, later destined to be the dominants in the climax community.

Widely dissimilar natural environments exhibit great differences in the sequence of plant succession. Not only are the climax communities at the end of the plant succession different, but so also are the stages in the progression toward that climax. A cultivated field, for example, if left to nature for regrowth, will have a much different plant succession in the humid tropics than in Aroostook County, Maine, or on the steppes of Alberta. Likewise, the succession that develops on a sandy soil often will be different from one on an adjacent heavy clay. The succession following a forest fire will be different from one following lumbering or cultivation, even within the same area (see Figure 13-3).

Stages in plant succession may be observed everywhere and are important in understanding the complexity of vegetation patterns. They may be seen along the sides of roads; along fence lines; at the edges of forests; bordering lakes, swamps, and marshes; in old abandoned fields; and on steep, rocky hillsides. If left alone, all these stages will progress toward a relatively long-term balanced state, which is determined principally by climate and surface configura-

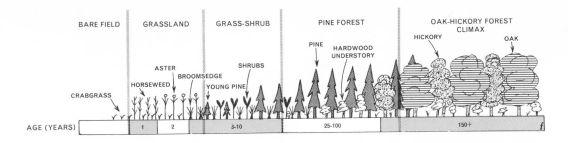

AGE (YEARS)

| BARE FIELD | GRASSLAND | GRASS-SHRUB | PINE FOREST | OAK-HICKORY FOREST CLIMAX |
| | 1 | 2 | 3-10 | 25-100 | 150+ |

tion. Some parts of the world show a greater dominance of stable vegetation communities than others, either because they have more stable physical environments or because man has not been as active in disturbing the vegetation (see Figure 13-4).

Man is especially effective in altering environments, not only the vegetation but other related elements of the landscape. Climax vegetation, therefore, is rare in populated areas. Botanists still are not in agreement as to what the prehuman regional vegetation was in North China, because there are no traces of undisturbed vegetation left in this crowded land, which has had such a long history of human occupancy. Perhaps the most extensive areas of climax vegetation are to be found in some of the great tropical forests or in the unglaciated sections of the northern coniferous forests of the Soviet Union. Both of these regions are now, and always have been, relatively sparsely inhabited.

The climatic changes of the Pleistocene

period produced drastic alterations in the vegetation patterns throughout the world (see Figure 13-5). Careful analyses of pollen grains obtained from swamps and bogs have verified the correlation between the wide swings of global temperatures and vegetation patterns. Evidence indicates that throughout most of the Mesozoic and Cenozoic eras, the global average annual temperature was around 75°F (in contrast to the present figure of 67°), with the polar areas at roughly 50°, the latter somewhat similar to the temperature of the present *Cfb* climates. Beginning about 10 million years ago, or in the late

Figure 13-4 An example of cultural alteration of vegetation forms. The shrubs shown are evergreen oaks (*Quercus calliprinos*). When undisturbed, this species of tree grows to a height of 30 feet, with leaves 6 inches long.

These oaks, located on a mountain slope in Lebanon, have been so continuously cropped by goats that dwarfing has taken place in branches, leaves, and fruit. [Van Riper]

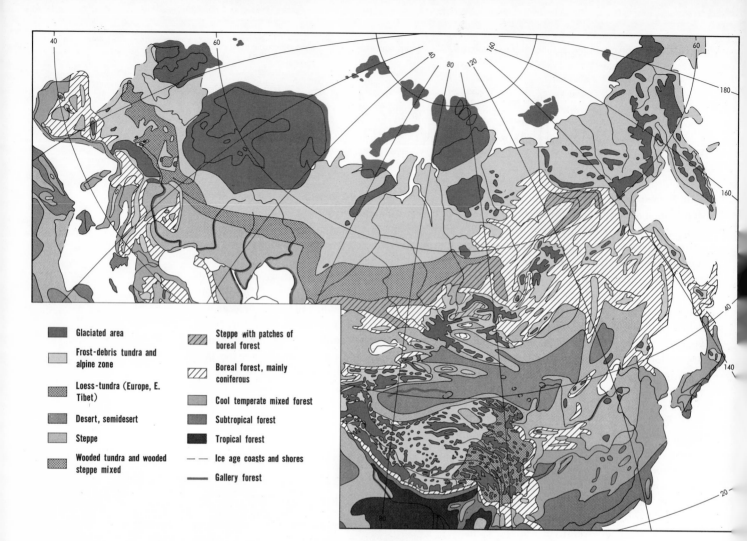

■	Glaciated area	▨	Steppe with patches of boreal forest
▨	Frost-debris tundra and alpine zone	▨	Boreal forest, mainly coniferous
▨	Loess-tundra (Europe, E. Tibet)	▨	Cool temperate mixed forest
▨	Desert, semidesert	▨	Subtropical forest
▨	Steppe	■	Tropical forest
▨	Wooded tundra and wooded steppe mixed	– – –	Ice age coasts and shores
		━━━	Gallery forest

480   *Plant Geography*

Figure 13-5 **Pleistocene vegetation boundaries.** Note that tundra conditions were found as far south as southern France and that in the area east of the Caspian Sea, steppe conditions prevailed in a region that now is a desert. [After Wissmann]

Tertiary, the global temperatures began to drop. At roughly 750,000 to 1,000,000 years ago, the global average had dropped to about 65°, and once the ice sheets began to form, this figure dropped quickly to a low of about 50°. Four major oscillations with an amplitude of about 18° in the global annual mean produced four major advances and retreats of the ice, but probably without complete removal during the interglacial stages. Minor oscillations with lesser amplitudes have occurred since the peak of the last glacial advance about 18,000 years ago.

The major effect of the glacial advances and retreats during the past million years has been to displace the normal climatic and vegetation patterns latitudinally several times over belts 15 to 20° wide. The low temperature periods greatly reduced the total quantity of vegetation on the earth surface as compared with the present, not only by extending periglacial (tundra) conditions almost into the subtropics but also by shifting the desert conditions equatorward into normally humid tropical savannas and forests. There was some increase in the volume of flora in parts of the present desert areas, but usually this was in the form of grasslands rather than forests.

The periodic shock of vegetation translocations during the Pleistocene and even the lesser fluctuations during the postglacial period produced major changes in the character of the total global flora as well as in regional differences. One effect was the increased mixture of species with a resultant increase in crossbreeding. Another was the increased number of *relict* forms, or those species that remained from previous but later displaced climax communities. Some plants, like some animals, are much more adaptable to climatic change than others. This appears to be especially true of some of the larger forms. Many plant ecologists believe, for example, that hemlock (*Tsuga*), which is a needle-leaf conifer commonly associated with the upland maple and beech forests of the northeastern United States, is a relict form left from a climax forest community of cooler periods in the postglacial period. Others believe that the marine west coast climates (*Cfb*) probably were influenced less by Pleistocene climatic changes than most parts of the world and contain relict forms not only from Pleistocene, but from the warmer pre-Pleistocene floras. *Sequoia*, the redwood genus, for example, is known to have been widely distributed across the entire North American continent during the late Tertiary period. These redwoods and other unusual plant forms within the west coast coniferous forest region, to be described later, are interesting examples of historical factors in the processes of vegetation change.

The tendency of plant communities everywhere to assume an equilibrium with the local environment is most useful to man when he seeks to exploit or control particular features of the vegetation. In the management of grazing lands, for example, different plants indicate various stages in the sequence toward, or away from, vegetation equilibrium. This information may be critical in maintaining the range at maximum carrying capacity for animals. Certain weeds in cultivated fields signify a need for drainage to improve crop yields, and others indicate that liming is necessary to check an increase in soil acidity. There are plants in irrigated deserts that indicate increasing or decreasing alkalinity in soils, a critical factor in this type of land utilization. In forest management programs, where selective logging practices are carried on, the appearance or disappearance of particular plant "indicators" may be critical to the maintenance of proper rates of cutting or to the regrowth of valuable species.

13-3 PLANT GEOGRAPHY AND THE PRINCIPLES OF NATURAL DISTRIBUTIONS Plants, as elements in the environmental complex, form no exception to the basic principles of natural distributions that were described in Chapter 1. The first principle, that of *infinite variety of forms and areal expression amidst graded likeness,* is beautifully illustrated by vegetation. Botanists still do not have a complete catalog of plant species on earth, and no perfect duplication has been found within any one species. The variety of plant types is mainly the result of *mutations,* those molecular rearrangements (now believed to be the result mainly of cosmic-ray bombardment) which occur spontaneously from time to time within plant reproductive cells and which are repeated in later generations of plants. Such alterations have taken place since plant life first appeared on earth. Mutations alone, however, do not explain the forms that are now present. The great variety of environments on earth and alterations in these environments create conditions for either the acceptance or rejection of the new forms. Mutations, plus natural selection, result in the kinds of plants that now mantle the earth surface. Because the environment of plants cannot be exactly duplicated in any two spots on the earth surface, neither can the association of plants.

It is likewise true that certain plant characteristics can be continued genetically for long periods of time, and natural selection tends to weed out poor competitors and to stabilize the highly adaptive species. Each plant form has its own *niche,* or contributory role in the community of living things. Similar environments tend toward similar groupings of plants, and the recognition and explanation of the processes that lead to recognizable spatial patterns of plant associations on earth lie at the central core of plant geography.

The second principle, that *regional similarities grade outward in all directions through transitional zones,* is revealed in plant arrangements everywhere. Examine, for example, the zonal sequence of plant communities bordering a swamp or marsh, and note how the gradual improvement of soil drainage is marked by successive changes in plant groupings. Even such sharp boundaries as a fence line separating a cleared field from a wood lot is not without its sequence of plant communities, reflecting the changes in microclimate and particularly the amount of sunlight available along such a boundary.

Some vegetation transitions can be observed much more easily than others, especially where the dominant types of one plant community are quite unlike those of another in size or form. Also, the place of areal change generally becomes less clear when we are dealing with major plant groupings. The boundaries of the great global vegetation formations, for example, represent unusually wide transitional zones. The long tongues of forest that extend along the watercourses far within the world grasslands and into the tundra, the islands of grass such as the *oak openings* that formerly were found in the forests adjacent to prairies in Indiana and Illinois, the sagebrush that hides the mergence of steppes and desert over hundreds of miles, not only in the western United States but also in the Soviet Union, are all examples of the broad width of the transitional zone between the formations. On the other hand, the transitions from reeds to sedges, from sedges to bushes, and from bushes to trees along a local marsh-forest boundary may take place within a few feet of one another. Furthermore, the zone which the eye classifies as a zone of sedges contains smaller plant forms that could be further classified to represent still smaller plant regions, and consequently would have sharper regional boundaries. The borders of vegetation regions, regardless of scale, exhibit all the characteristics of transitional patterns, including varying gradients of change, diffusion, interfingering, and enclaves.

The third principle indicates the *continuous alteration of areal associations with the passage of time.* In earlier chapters we noted that land-surface features are altered continuously through various geologic processes and at different rates, ranging from the suddenness of a landslide or volcanic explosion to the slow creep of soil down a slope mantled with forests. We also caught a glimpse of the day-to-day, year-to-year, and multicentury climatic changes that continually take place. Plant groupings are highly sensitive to changes in their environments, and both long- and short-term alterations can be observed. An early frost, an exceptionally dry year, a sudden windstorm—each of these claims its plant victims and thus aids other plants by freeing them from their formerly successful competitors for living space. To be sure, recovery is rapid, but the changes continue. The more lasting alterations are tied to long-range trends, where recovery is less certain. The erosion of an outlet to a swamp or marsh, which results in draining the area, will permanently alter the groupings of moisture-tolerant plants. No plant association in any area ever remains exactly the same with the passage of time.

The fourth and last principle of natural distributions states that *the elements in an environmental complex tend to develop mutual adjustments to each other and to aim toward a relative equilibrium.* This concept was discussed earlier in the preceding section. Plant succession and the idea of a climax vegetation are other ways of expressing this fourth principle as applied to plant distributions.

The concept of trends toward relative equilibria in vegetation is perhaps most justifiable in respect to man's relationship to vegetation. Man is extremely active in altering vegetation patterns because of his interest in utilizing plants for many purposes. From a utilitarian point of view, his own effects on vegetation are far more important than long-run climatic or geologic alterations. For his own welfare, he needs to know whether a particular method of changing vegetation will lead toward, or away from, relative stability.

To illustrate, suppose that a forest in the northeastern United States has been cut and that not a tree remains standing. If left alone, this environment will develop a forest cover, mainly because of climatic conditions and plant competition. Obviously, a forest here represents a more balanced and stable form of vegetation than the low ground cover that immediately follows deforestation. Proof of this is the rapid change in plants that compete with one another for occupancy of the bare site. Suppose, however, that man wishes to reestablish a forest cover without waiting for the natural progression toward the relatively balanced forest community. It is clearly useful for him to know that certain species of trees will thrive and that others will not. It is also important for him to know that, by altering an environmental factor such as drainage, he can produce conditions which will lead toward the establishment of a relatively stable type of forest that yields more valuable kinds of trees than the original forms that may have occupied the site.

13-4 WATER SUPPLY

13-4 WATER SUPPLY   Among the variety of factors that determine the gross features of vegetation differences on the earth, the most important is the availability of water.

Water plays a leading and decisive role in nearly every life process of plants, including germination, growth, and reproduction. It is more than the lifeblood of plants, more than a transporting medium for plant foods and waste products. It enters into the formation of most of the organic compounds of plants and acts as an important energy exchanger in maintaining plant temperatures within proper limits. Photosynthesis, or the utilization of solar energy in the manufacture of sugars and starches within green plants, could not take place without water, and even parasitic plants, which do not manufacture such compounds themselves but steal them from others, require water for cell construction and maintenance. When water leaves plant cells without being replaced, the cells wither and die.

The size of individual plants generally is related closely to the quantity of water that passes from the soil into the air via the plant. High actual transpiration rates usually favor large plants, such as trees. The growth of trees thus requires large quantities of available water in the soil, plus a ready supply of heat energy in the air which can be used to vaporize the water brought to the leaf surfaces. Competition for sunlight favors tall, large plants; a plentiful supply of sensible heat, water, and nutrients makes them possible. It might be noted here that some environments have extremely high water supplies and net radiation surpluses, but without tree growth. Canebrake marshes in the tropics and subtropics are examples. Evapotranspiration rates here may be just as great as in

Figure 13-6   World distribution of major plant formations.

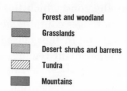

Forest and woodland
Grasslands
Desert shrubs and barrens
Tundra
Mountains

nearby forests, but the soft, deep mud or organic oozes, saturated with water, are not favorable for the physical support of trees. The biomass may be as great as in the forests, because the smaller size of plants is compensated by extremely dense plant populations.

The global plant formations (see Figure 13-6) represent major differences in moisture balances. Forests occur where plant moisture exchanges are large. Desert shrubs exhibit many characteristics, including small size, which enable them to adjust low intakes to high transpiration rates. The dwarfed vegetation forms of the tundra and alpine associations represent balances that occur between low transpiration rates, resulting from low air temperatures, and low intakes, resulting from long frozen periods. Grasslands appear to develop best under conditions where fairly large intakes of water occur for short periods during the year, particularly during hot periods, when transpiration rates are high and where soil moisture deficiencies frequently occur at other times of the year. Grasslands also are found in areas where the nutrient intake is insufficient to maintain a forest despite a favorable moisture balance. Many grasses require only a meager soil fertility for the completion of their life cycles.

Most of the major subdivisions of the global vegetation formations also reflect plant moisture balances. One of the major subdivisions of global forests, for example, is the *deciduous* types, which tend to shed

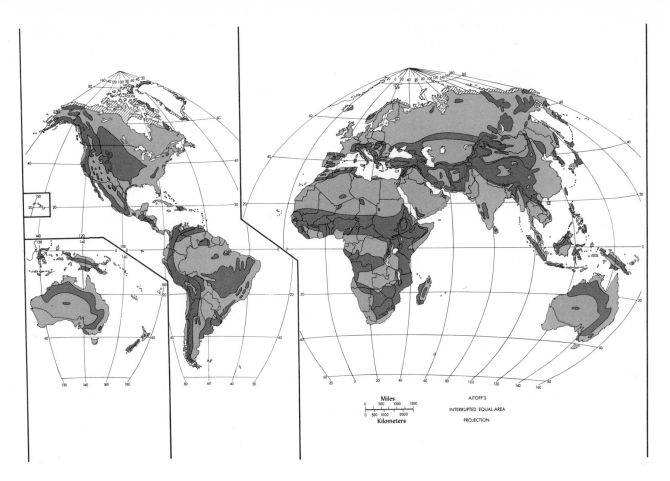

Miles
0   500   1000   1500

Kilometers
0   500 1000   2000

AITOFF'S

INTERRUPTED EQUAL-AREA

PROJECTION

their leaves during some season of the year. This is largely a result of pronounced seasonal climatic differences. In the tropics and subtropics, the deciduous nature of trees and shrubs is the direct result of alternate wet and dry seasons. Here the abundance of soil water for plant intake and the high transpiration potentials during the humid period favor luxurious tree growth. The dry period, in contrast, supplies insufficient soil water to maintain this lush growth. To compensate, the trees shed their leaves and maintain their life processes at a bare minimum during the dry season.

Winter freezing in the mid-latitudes also produces a deciduous reaction in many trees. In these areas temperature is a decisive factor, but its influence on the deciduous nature of plants operates through the availability of water. In such climates, trees lead a sort of double life. In summer, they pump water rapidly through their cells, behaving much like trees everywhere that have sufficient moisture and heat for rapid growth. In winter, on the other hand, they take on the characteristics of *xerophytes*, or drought-resistant plants. They reduce transpiration to a minimum by shedding leaves, retaining only a small amount of water within the plant cells, since freezing would cause death.

Needleleaf trees, such as pines and spruces, have many characteristics that enable them to withstand strong seasonal contrasts in moisture balances. The small surface area of the leaves tends to reduce transpiration; yet their large number enables the tree to perform photosynthesis effectively. The thick, pitchy sap of needleleaf trees resists evaporation and quickly heals wounds in the thick, spongy, and highly insulative bark. The shallow but well-developed root system assures that much rainwater percolating downward in the soil can be intercepted during the warm season, but since the roots lie at or near frost depth, they cannot absorb too much water from deep in the soil during winter seasons. Because of their moisture-retention capabilities, conifers are also adaptable to highly permeable, and thus droughty, sand and gravel soils of climatic regions where rainfall and temperature would appear to warrant luxurious broadleaf forests.

Moisture is such an important factor in determining plant characteristics that it is the primary basis for several vegetation classifications. The simplest such grouping includes the *xerophytes,* or moisture-deficient plants; the *hydrophytes,* water-loving plants that tolerate an excess of water; the *mesophytes,* which have moderate moisture preferences;

and finally the *tropophytes,* which undergo special changes in form to adjust to seasonal variations in moisture utilization. Representative examples of plants in each of these groups are given in Table 13-1.

13-5 TEMPERATURE Temperature alone, or rather, the availability of sensible heat, is not a significant factor in determining the major global vegetation patterns, although its indirect influence upon water availability through its effect on evaporation rates and the freezing of cell fluids is of primary importance. Cell destruction due solely to high temperatures rarely occurs below 122°F, and some plants thrive in temperatures not far below the boiling point, such as the algae that dwell in the hot springs of Yellowstone National Park. Similarly, low temperatures alone rarely kill plants without the aid of freezing water. This does not mean that temperature in itself plays only a minor role in plant characteristics; it signifies only that it is a less important factor than water supply in determining the boundaries of the major plant formations. One eminent plant ecologist has stated that, whereas water balances determine the major vegetation formations, temperature plays the major part in determining the floristic variations within the formations. The *direct* influences of temperature

Table 13-1 Classification of Plants According to Water Requirement

Type	Examples
Xerophytes (*xero* = dry; *phyto* = plant)	Cacti, desert junipers, sagebrush, certain bulb plants
Hydrophytes (*hydro* = water)	Water lilies, seaweed, sedges, bulrushes, mangroves, swamp cypress
Mesophytes (*meso* = middle)	Live oaks, mahogany, rubber trees, rhododendrons, and many common field and forest plants
Tropophytes (*tropos* = changing)	Maple, birch, ash, larch, teak, some acacias

affect rate of growth, fertilization, plant stature, maturing of tissues ("ripening off" in autumn), seed germination, and time of flowering, especially when the latter is combined with length of day.

Every plant has an optimum temperature for each of its various life processes, and in the competition for survival, the optimum temperatures for growth, germination, and reproduction are important assets. Plants also have maximum and minimum temperatures beyond which they cannot survive. Although some plants may survive below-freezing temperatures, they must curtail their normal life processes during such periods. Thus, protracted freezing temperatures weed out all plants that require extended germinating, growing, and reproduction periods. On the other hand, certain varieties of low plants thrive in the cold, inhospitable environments of the tundra, alpine meadows, or high bogs. Covered with snow for much of the year, they are protected against the extreme cold and find, in the cool but long days of summer, sufficient energy for rapid growth and reproduction processes (see Figure 13-7). The familiar crocus and snowdrop thrive in temperatures that are prohibitive to most garden flowers, and we welcome their bright blooms soon after the snows of midlatitude winters disappear. Although temperature has little direct influence on the boundary between forests and grasslands, it is a highly significant factor in the selection of the particular trees found in a forest and in the association of grasses found in a prairie or steppe.

The only plants to survive where temperature and moisture conditions show great seasonal contrasts are those that have highly specialized mechanisms for resisting such changes. This is why the number of species within forests increases as one proceeds toward the humid tropics. General growth conditions in the latter are adaptable to a

Figure 13-7 Close-up of a dense but low ground flora in a Norwegian upland bog. This dense matting of bog moss (*Sphagnum*), *Vaccinium* (foreground), and other low, acid-loving plants was observed in a high, treeless *fjeld* or glacial valley near Oppdal. [Van Riper]

wide variety of trees that differ only in minor respects. Relatively few species of trees, however, have been able to survive the vicissitudes of climatic fluctuations between the long summer days and long winter nights in the forests just south of the tundra.

Temperature differences may influence plants in many ways. The optimum temperature for germination is sometimes well below that of seed production or growth. Lettuce, for example, tends to grow tall and produce seed during high summer temperatures, but it germinates and produces leaves most effectively during cooler periods. Some plants require that their seeds be frozen prior to germination. Others require a special conditioning period in moist soils, with temperatures slightly above the freezing point. Many plants adjust their growth conditions to the fluctuations of temperature between night and day, developing a rhythm (termed *thermoperiodism*) which appears to be essential to their optimum growth.

Anyone who has done any gardening knows that plants differ widely in their sensitivity to low temperatures. Some plants actually suffer cell deterioration at temperatures somewhat above freezing, while others may tolerate temperatures far below zero for short periods.

The impact of sensible heat on plant geography is subtle and difficult to isolate. In general, however, it appears to have a more direct and determining influence the closer one looks at the vegetative pattern. On a topographic scale, it is ever present as a selector of the individual plant species that one sees about him. The attention paid to microtemperature patterns by large commercial horticultural producers is evidence of this fact. At the same time, neither the net radiation balance nor the quantity of sensible heat determines the major plant formations. These are outlined primarily by effective moisture balances, or what Thornthwaite termed the water budget (see Figure 5-33), and involve both the net radiation balance and the availability of water.

13-6 LIGHT Light is a form of energy, and most plants require it in their life processes. In burning a ton of coal, one releases the energy of sunlight once captured by plants, bound into complex organic molecules, and later concentrated by compression or consolidation into the free carbon of coal. It takes energy to break apart the bond between carbon and oxygen in a molecule of carbon dioxide. Yet plants must extract this carbon in order to build their tissues and cells and to synthesize their starches and sugars. This is photosynthesis, nature's alchemy, and since light availability is essential to this process, the competition for sunlight among plants plays an important part in the selection of plant types in any given community.

Whereas the intake and outgo of large quantities of water make it possible for plants to grow into tall trees, the competition for light provides the stimulus for vertical growth in forests. Availability of light, therefore, probably is a decisive factor in the arrangement of the storied, or vertical, zonation of plant groupings. Ferns, for example, thrive under conditions of low light availability, lower temperature, and high humidity that typify the deep shade of forests (see Figure 13-8). Vines have developed specialized mechanisms for climbing tree trunks in order to reach sunlight far above the ground and thus do not require massive size to compete with trees for light (see Figure 13-9). Certain fungi, such as mushrooms—since they have

Figure 13-8 Tree ferns as an understory in a eucalyptus forest in Victoria, Australia. These messmate and blue gum trees are more than 100 feet high. The moist, cool climate and dim light of the forest floor enable ferns to find an extremely favorable habitat for their growth. The tree ferns shown here are between 6 and 20 feet high. [Van Riper]

Figure 13-9   Scene in the tropical rain forest near Klamono, West New Guinea, showing vines and lianas clinging to the tree trunks.   By depending on large trees for support, vines are able to reach the sunlight far above the forest floor without building a massive structure. This is a good example of the adaptation of form to environmental conditions in the struggle for existence. [Standard Oil Company (N.J.)]

no *chlorophyll*[2] and do not require photosynthesis—thrive in dark places, free from the competition of green plants.

Latitudinal variations in the length of day and night result in periodic fluctuations in light availability. Many plants adjust their forms to match these seasonal and latitudinal differences, and these structural changes are termed *photoperiodism*. Other plants have the capability to move their leaves, branches or flowers either to seek more light or to avoid it. This is termed *phototropism* (Gr.: *photo* = light; *tropo* = turning). The turning of a sunflower's head to follow the sun's course across the sky and the manzanita's ability to keep its leaves edgewise to the sun are examples. Many deciduous (leaf-shedding) plants have their triggering mechanisms for budding, leaf shedding, flowering, etc., correlated with seasonal light changes, since they are a far more reliable signal than those supplied by weather or climate, which are subject to wide fluctuations. The flowering of chrysanthemums and dahlias, for example, is controlled by the length of daylight, and florists now are able to force these plants to bloom prematurely in greenhouses, at any time of year, by regulating the length of light-exposure periods.

The long daylight hours in higher mid-latitudes are responsible for abnormal size in some vegetables that can thrive in the moist, cool weather of the short summer season. Root or tuber crops, which are high in starch content, may have enormous yields in these areas because of the long hours of available light. It is partly for this reason that the major centers of potato cultivation in the United States lie near the northern boundary of the country. Visitors to the subarctic regions of Alaska, northern Canada, and northern Eurasia are always amazed at the enormous size of such hardy, common garden vegetables as cabbages, head lettuce, turnips, and radishes.

The influence of intense solar radiation on sugar manufacture by fruits in low-latitude deserts is well known. In these hot deserts, when water can be supplied by irrigation, sugars can be synthesized rapidly and efficiently. The date palm is a splendid example. Even in mid-latitudes, fruits vary in sugar content, depending largely on the number of sunny days. The ideal maple-syrup season is one that contains a high percentage of cloudless skies.

Like temperature, light is more important in the details of plant morphology and ecology than in the gross patterns. Yet it assumes great significance in some features of horticulture and is an ever-present factor in the total environment of plants. Other aspects of its influence are covered later, in the descriptions of global vegetation groupings.

13-7 HUMIDITY AND FOG The principal effect of atmospheric humidity is its influence on evaporation rates and plant transpiration rates, operating through the mechanism of fluctuations in vapor pressure (see Section 5-3). Directly, humidity plays a relatively insignificant role. Small quantities of water may pass into leaf or stem openings following condensation (dew) on plant surfaces, but except for a few desert plants which require only small amounts of moisture, the quantities absorbed are too small to influence plant distributions appreciably. Lichens and some mosses, however, which appear to be able to extract water directly from the air, vary in quantity in proportion to relative humidity. They are able, therefore, to thrive on bare rock surfaces free of any soil or liquid water.

[2]*Chlorophyll* is a complex organic compound that is essential to most plants because it acts as a catalytic agent in the formation of plant carbohydrates (starches and sugars). Being a strong pigment, it gives most plants their green color.

490 *Plant Geography*

Fog and clouds on mountain slopes have some noticeable influences on plant geography. The extent of the redwoods (*Sequoia sempervirens*) along the coastal area of northern California, for example, corresponds almost exactly to that of the coastal fogs that drift in from the sea. Water dripping from fog-enshrouded vegetation adds to groundwater supplies. It also tends to reduce incoming solar radiation and thus to lower the surface temperatures of plants. The fog belt along the western slopes of the Andes in Peru supplies water for a narrow zone of short, herbaceous vegetation in this largely rainless area. This vegetation, in turn, supports a distinct zone of pastoral activity. The cloud zones of mountain slopes in the humid tropics result in characteristic moss forests, a most distinctive forest type that appears in widely separated localities in low latitudes. For a description of this unique vegetation type, see Section 14-25.

The role of atmospheric moisture in relation to plant diseases is often a limiting factor in the growth of plants. Fungus diseases are especially common in humid areas or in regions where fogs are prevalent. Thus potatoes grown in warm rainy or foggy areas may suffer from late blight, whereas those grown in arid regions under irrigation may be comparatively free from disease infection.

13-8 SOIL Nearly all plants derive most of their moisture and plant foods from soil; hence, soil differences produce variations in plant distributions, particularly on local, or topographic, scales of observation. Soils themselves represent adjustments to a host of environmental conditions; thus the direct influence of soils on plant distributions can rarely be isolated from other factors, such as climate and drainage. Sharp changes in vegetation occur, however, where there are marked differences in underlying rock mate-rial and where insufficient time has elapsed for soils to develop an ecological balance with their total environments.

Among the most important soil characteristics influencing the vegetation cover are texture, structure, accessible water content, acidity, organic content, and the chemical composition of the soil particles. The influence of these and other soil properties on vegetation can be touched upon only briefly here. A fuller treatment of these properties is presented in Chapters 15 and 16.

Too little water or too much water for certain types of vegetation may be the result of soil texture. The influence of permeable sands and heavy, tight clays on water supply is clear; therefore this may become a plant geographical factor of considerable importance. Close correlations can be found between the distribution of pine forests and sandy or gravelly soils in the north-central part of the United States, the Atlantic-Gulf of Mexico coastal plain, and the Landes district of southwestern France (see Figure 13-10). In these areas, the pine is a fire-succession flora. The correlation with sandy soil is related to the fact that the sandy soils create drier soil conditions. During hot summer periods, the dry vegetation is more susceptible to ignition than that on moister, heavier soils. The texture of soils also may influence their temperature. A sandy soil, for example, warms much more quickly in spring than does a clay soil, and hence, by supplying optimum germination conditions, plays a part in the early selection of plants.

The sensitivity of many plants to soil acidity is recognized even by amateur gardeners. Some plants, such as the acid fruits (huckleberries, raspberries, etc.), many conifers, rhododendrons, and some varieties of oak, are acid lovers and require moderate-to-high acidity in soils. Others, such as hickory, maple, and many garden vegetables, do best

Figure 13-10 Northern scrub pine (Pinus Banksiana) with an understory of bracken and sweet fern (Myrica) near Grayling, Michigan. This scrub pine, locally termed "jack pine," is a true fire-succession tree. It covers thousands of square miles of sandy outwash plains from New Brunswick and Maine west to the Rocky Mountains. Much of its present acreage was once covered with magnificent stands of red pine (*Pinus resinosa*) and white pine (*Pinus strobus*). [Van Riper]

under conditions of low acidity. Table 13-2 lists the optimum acidity range of several well-known cultivated plants. In this table, increasing acidity is indicated by lower pH values. An explanation of pH, or hydrogen-ion concentration, is given in Section 15-17. A pH value of 7 denotes a neutral solution. Values above 7 indicate respective increases in alkalinity.

The physiological effects of acidity on plant growth are mostly correlated with the requirements of different plants for certain critical plant food elements and with the influence of soil acidity on the availability of these elements. The content of calcium and other bases in a soil, for instance, usually decreases as the soil becomes more acid.

Soil alkalinity, which refers to low hydrogen-ion concentration in soils, or low acidity, is a characteristic of dry regions. It occurs wherever soluble salts tend to accumulate in the soil, and, like acidity, it results in plant selectivity. Highly alkaline or saline conditions produce toxic effects on most plants, and only a relatively few highly specialized types are able to thrive in such environments. One property that appears to be typical of most salt-tolerant plants is *succulence,* or an unusually high water content within the plant cells (see Figure 14-30). The playa basins of arid regions, which accumulate salts because of poor drainage and high evaporation, show rapid changes in alkalinity about their borders. Concentric zones of plant communities, each with its own degree of tolerance for alkalinity, commonly ring these shallow topographic depressions. Much broader zones of plant forms, corresponding to broader alkalinity tolerances, mark the general transition from arid to humid regions, or from predominantly alkaline soils to acid soils. It is difficult, however, to correlate such broad zones directly to pH concentration, since other variables, such as moisture availability, may be as important.

Examples can be found throughout the world to illustrate the influence of underlying materials on the properties of soils derived

Table 13-2 The pH Preferences of Some Common Cultivated Plants	
Plant	*Optimum pH range*
Cranberry	4.2–5.0
Cotton	5.0–6.0
White clover	5.6–7.0
Sugar beets	6.5–8.0
Red pepper	7.0–8.5

SOURCE: C. H. Spurway, "Soil Reaction Preference of Plants," *Michigan Agricultural Experimental Station Special Bulletin* 306.

from them and, in turn, on the vegetation cover. This is an especially important factor in areas of young soils, which are only beginning their general progression toward relative environmental equilibrium. Examples of young soils include those recently developed on alluvium, marine sediments recently uplifted from the sea, recent volcanic deposits, and glacial detritus. Concentric belts of vegetation, corresponding to the belts of marls (limy muds), sands, and clays, are easily distinguished along the Gulf of Mexico coastal plain. Sharp boundaries between types of forest communities mark the change from sandy glacial outwash plains to the heavier soils of morainic belts and from narrow sandy beach ridges to heavy lake clays on former lake plains in many parts of the glaciated portions of North America and Eurasia. Although such borders are the direct result of differences in water availability, this in turn is related to the texture, structure, and composition of the underlying material.

In some areas, such as on the windswept rocky hills of the Scottish highlands, the rock is so near the surface that there is insufficient soil to furnish anchorage for tree roots. Forests are therefore absent, and small woody plants, such as broom and heather, furnish the permanent ground cover. In the same area, however, on the leeward side of the exposed hill slopes, trees can flourish on soils just as thin, because of the absence of wind stress.

As with the other factors affecting plant growth, the influence of soil can never be completely isolated from that of other variables. Sometimes it plays a dominant role in the selection of the plant cover; in other instances it exerts only a subtle influence and operates through other environmental forces. It is part of the environmental complex which seeks to develop an equilibrium among all its constituent parts.

13-9 DRAINAGE AND SLOPE Drainage and slope are treated together because they are so intimately related. The importance of water supply to plants has already been treated. Whereas water availability on a continental or global scale is mainly dependent on climatic conditions, local variations are largely the result of drainage and slope conditions. When there has been a heavy rain, less water is left in the soil for plant use on an extremely steep slope than on more gradual slopes. Erosion on steep slopes may remove much of the unconsolidated soil material above the underlying rock, speed runoff, and reduce available soil moisture for plant growth. Conversely, a flat land surface in a humid region may result in excessive water accumulation in soils and require consequent adjustments in the vegetation cover.

The impact on plants of excessive drainage, either horizontally or downward, is similar to that of decreased precipitation. Not all cacti are desert dwellers. The common prickly pear (*Opuntia*) has been observed by the author on thin soils covering steep, rocky hillsides in southern Illinois and Pennsylvania and on sand dunes of the New Jersey coast and the eastern shore of Lake Michigan. All these areas are in humid climates.

Inadequate drainage produces a number of harmful effects on plants that have not developed specialized forms to adjust to it. The most important factor is not the excess of water itself but the lack of aeration. Absorption and the discharge of CO_2 and oxygen are important aspects of plant maintenance, and these do not pass freely into and out of water. The intake of CO_2 and the release of oxygen in plant photosynthesis are well known. Less familiar is the reverse process. Plants require mechanical energy for many purposes—for example, the mechanical force of root penetration. They obtain this energy from the combustion of carbohydrates in the same way that animals do, except that the quantities are much smaller. Involved in this combustion are the intake of oxygen and the release of carbon dioxide. This process, termed *respiration,* is most noticeable at night, when owing to the lack of sunlight, photosynthesis is not taking place.

Hydrophytes, or plants that thrive under excessive water conditions, have many ways of overcoming inadequate aeration. Most such plants have spongy tissues, or abnor-

mally large intercellular openings, which are capable of storing large quantities of oxygen supplied by photosynthesis. This oxygen slowly decreases in the water during the night and is replenished during the day. In floating forms, these air passages further provide buoyancy, which keeps the top of the plant near the surface of the water (see Figure 13-11). This helps such plants to maintain an adequate source of light. Other plants, such as seaweeds, have extremely slow rates of photosynthesis, extracting dissolved carbon dioxide from the water and releasing oxygen directly. Different amounts of excess water result in varying degrees of adaptation in plants, ranging from completely submerged or floating forms to erect plant types that are occasionally or partially submerged.

The influence of slope on vegetation is not restricted to its effect on the water supply. Temperature and the availability of light may also be related to slope conditions. Mountain regions usually have distinctively different types of plant communities due largely to contrasts in exposure to sunlight. A sunny mountain slope with surface temperatures much higher than shaded slopes may influence plant selection through both light sensitivity and soil-moisture availability. The accumulation and duration of snow cover on mountain slopes may also be related to slope conditions, especially in connection with

Figure 13-11 Water lilies, typical hydrophytes. The fragrant white blooms of this water-loving plant, *Castalia odorata,* are among the most beautiful flowers of the north woods. Note the typical hydrarch succession along the edge of this lake in southern Quebec. [Van Riper]

494 *Plant Geography*

wind direction and sunshine availability. These, in turn, affect temperatures, moisture balances, and plant selection. Irregularity of slopes may produce noticeable differences in plant communities in dense forest areas. Sudden changes in slope cause breaks in the continuity of the forest canopy overhead, thus permitting greater light penetration and air circulation and producing a general alteration in humidity and soil-moisture conditions. Steep slopes tend to make the forest crown, or canopy, less regular than on plains. For this reason, forests on steep hill and mountain slopes are likely to have more undergrowth, or lower-story vegetation, than those on flat land, and the plant communities tend to be more complex and varied.

13-10 ELEVATION Some of the sharpest, most noticeable differences in major plant communities can be observed on mountain slopes because of differences in elevation. This is especially true of high mountains in low latitudes, where representatives of many of the great global plant formations can be found, arranged in successive zones up the mountainsides (see Figure 13-12). Such groupings are related primarily to the lapse rate, or the decrease in air temperature with increased elevation above sea level, but factors other than temperature change play a part in the vertical zonation. The timber line, or the upper limit of trees, does not always occur at the same elevation in similar latitudinal positions or even on different sides of the same mountain ridge. Moisture availability, wind direction and strength, and exposure of sunlight are other factors that influence the elevation of vegetation boundaries on mountain slopes. Although such elements result in many local exceptions, mountain vegetation boundaries generally follow the contour of the slopes.

Experiments in growing plants at different

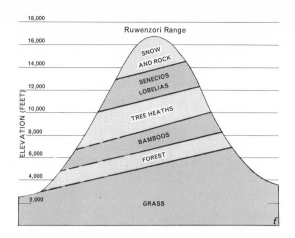

Figure 13-12 Zonation of vegetation on the slopes of the Ruwenzori Range in Uganda. The greater rainfall of the western (left) side is responsible for the lower elevation of the vegetation zones on that side. [After Woosnan]

altitudes have yielded much interesting evidence of changes in size and color associated with the alpine environment. The bright light combined with the cool air of high altitudes intensifies the color of flowers and promotes dwarfness. Conversely, alpine plants grown at low altitudes, such as some of the azaleas, fail to develop the brilliant colors seen in their natural high-altitude habitats, and most alpine plants, from the lowly dandelion to the spruces and firs, grow considerably larger and taller at lower altitudes.

One should not assume too close a correlation between the characteristics of mountain vegetation communities and the communities found in broad global zones having comparable temperature conditions. Although the flora high above the timber line may have many characteristics similar to those of tundra communities, the mountain environment produces many dissimilar features. Wind velocities and snowfall, both factors in plant selection, are likely to be much greater on mountain summits than at high latitudes, and the appreciable decrease in air pressure at high elevations leaves its mark on plant characteristics. In low latitudes, seasonal variations in temperature are no more pronounced at high elevations than they are at low elevations; therefore deciduous forests resulting from seasonal low temperatures do not appear at high elevations in the tropics. Conifers, on the other hand, are common members of high-elevation forests at all latitudes, because they can adjust moisture balances to depressed temperature conditions. Mountain slopes, particularly windward slopes, usually have much higher precipitation than do lowland areas, and thus plant relationships to temperature are not always comparable to those which exist elsewhere in the same latitudinal temperature zone. The cloud-zone environments of low-latitude mountains, for example, have few, if any, low-elevation equivalents. Since the totality of environmental conditions at high elevations cannot be reproduced at low elevations, the characteristics of mountain vegetation should be treated apart from those of the great lowland plant formations. Major characteristics of mountain vegetation patterns are discussed in the next chapter.

The Role of Man in Vegetation Balances

13-11 GENERAL It has been stated that there is no such thing as purely natural vegetation anywhere on earth, that man is found wherever plants will grow, and that his presence and abilities, no matter how primitive or sophisticated, have left their mark on the floral landscape. Lest this observation flatter human beings unduly, it should be remembered that the nonhuman members of the animal kingdom also exert their influence on the evolution, maintenance, and distribution of plant communities everywhere. Plant ecologists recognize that plant communities cannot be adequately studied apart from the animal life present in the same area, any more than they can be understood apart from the soil, air, or water conditions in the environmental complex. Microbes decay the forest litter; bees pollinate flowers; earthworms aerate the soil; grazing animals maintain balances between edible and nonedible plants; insects act as food for certain carnivorous plants; and there are many symbiotic, or mutually beneficial, relationships between plants and animals.

Man is a part of the environmental complex of plants, but he plays a much more dynamic role than other animals. He has forces at his disposal for altering vegetation that are far more effective than the feeding habits of animals. Furthermore, his abilities grow with knowledge, since they do not depend on instincts bred into the species of man by millions of years of selection in relatively stable environments. From the humble beginning of gathering wild fruits and roots and taming wild animals that thrived on broad, open grasslands, through the first deliberate planting, to the present-day applications of machine agriculture, controlled plant breeding, and the creation of synthetic growth conditions, man has come a long, long way.

In his attempt to make plants work for him, man has made many mistakes and has suffered unfortunate consequences. His most common mistake has been to demand too much of plants in highly balanced environments, without thought for the future. Sometimes this has been due to ignorance and at other times to pressures beyond his control. Nonliterate human groups, through trial and error, usually developed a mutually beneficial relationship with the particular vegetation communities in which they lived. This was not difficult to achieve so long as man's relationship to the land was direct; in fact, the environment generally forced human and plant life into equilibrium. In more complex societies, with powerful tools and inanimate energy sources to aid in satisfying human desires and with added knowledge of how to alter many environmental factors, the relationships between man and vegetation become more subtle and more difficult to recognize and to control. The worker on a banana plantation who purchases his food at the company store has a far different relationship to his environment than does a self-sufficient farmer who is living off a tiny patch of cleared land in the adjacent tropical forest.

The succeeding paragraphs of this section are concerned with some of the problems of maintaining natural balances in the face of human exploitation of plant life. Since such problems differ greatly with different environments and with different cultures, the subject cannot be fully treated here. The following discussion merely presents a few of the more prominent problems for illustrative purposes.

13-12 MAINTENANCE OF SOIL FERTILITY WITH CROP CULTIVATION In a well-balanced plant community, soil fertility is maintained by adding to the soil most of the inorganic plant foods used by the plants during their growth. This is accomplished largely through the decay of the plant material that falls upon the forest floor, but the weathering of inorganic soil particles also results in a small but steady addition of these same chemical materials. Whenever more nutrients are removed than are added by these two processes, soil fertility decreases. Organic material supplied by dead vegetation plays an additional role in fertility by acting as a sponge and retaining water and essential

plant food elements which, because of their solubility, would otherwise soon be leached out of the soil. When man removes some of the plant cover for his own food, he must either replace the loss in some way or face declining crop yields. Over the thousands of years of agricultural history, he has found several ways of replacing the nutrients that he takes, but in many situations he has found it impractical or impossible to do so and thus has eventually been forced to abandon his cultivated fields. In a few areas nature periodically restores lost fertility by adding fresh, unweathered material, as in volcanic regions or along the flood plains of rivers. Usually, however, man has had to use foresight in replenishing the fertility of the soil in order to maintain a stabilized system of plant exploitation.

13-13 MIGRATORY AGRICULTURE One of the simplest and oldest methods of maintaining agricultural production has been to permit natural regeneration of soil fertility

Figure 13-13 Typical migratory-agricultural clearing in the Philippines. Crops are planted haphazardly between the stumps and remaining trees following burning. [R. L. Pendleton; *Geographical Review*]

by simply abandoning the land following a decline in fertility. This is combined with a system of agriculture that imitates the cycling of nutrients in a forest, yet speeds the process and routes the vegetative growth into useful forms. This has been generally termed *migratory agriculture,* but it is also known under many other names, such as *swidden, ladang, milpa, caingin,* or *ray.* It is largely but not exclusively found in the humid tropics. The process involves forest clearing by girdling or felling trees and carefully burning as much of the dead plant material as possible (see Figure 13-13). A large variety of cultivated plants are used in a jumbled, tiered complex somewhat like the storied aspect of the forest. Planting is haphazard, usually consisting of inserting roots or shoots into holes made with a pointed stick. Yields are high at first because of the concentration of nutrients supplied by the fire ash, but these soon decline after a few years. Following the decline of fertility and the growing competition of unwanted plants, the clearing is abandoned and a new clearing is made (see Figure 13-14). Where population is sparse, where forest land is cheap or free, and where nutrient regeneration is comparatively rapid, this process can maintain a certain population density indefinitely and with an extremely low expenditure of human energy. Recommended population densities using this system are between 60 and 80 people per square mile. It is not an efficient system of agriculture because of the great loss of energy and materials in burning and weed growth,

Figure 13-14 Air photo showing an extensive area of migratory-agricultural clearings in the Annamese Mountains of North Vietnam. Different shades represent different stages of vegetation regrowth. Little of this tropical forest has remained without having been cleared at some time or other. [U.S. Air Force (MATS)]

but with care it can provide a stable system for cycling plant nutrients and solar energy. All too frequently, however, this system is abused, sometimes because of overburning and frequently because of overpopulation.

The density of population that can be adequately supported by this cultivation method depends on the initial soil productivity and the length and speed of the regenerative process. In the forest societies of the humid tropics, the balance between population density and soil capacity was usually kept near the optimum by the dual forces of a high birth rate and a high death rate, the latter involving intertribal warfare. This often was the result of territorial violations, which, in turn, were due to the pressures of a growing population and decreasing food supplies. Once law and order was established by colonial or national authorities and medical facilities were introduced, the older population controls were removed and population increased rapidly. This, in turn, led to greater attacks on remaining areas of primary forest, a shortening of the regenerative period, and a progressive decrease in crop yields. Erosion was accelerated, and many former productive areas degenerated into sterile grasslands whose tough sod was too difficult for the simple tools to handle. Humid tropical forest lands today contain many culturally-induced grasslands of little value and directly the result of migratory agricultural abuse (see Figure 13-15). The general tendency today is for governments to prohibit migratory agriculture and to attempt to force the forest cultivators to adopt a sedentary life and to practice the more controllable methods of crop rotation and fertilization. The principle of this type of agriculture, however, still is wholly valid. If population can be stabilized at an appropriate level without resorting to high death rates, and if deliberate, planned forest regeneration can be practiced without

Figure 13-15 Culturally induced grasslands in east-central New Guinea. This district has no pronounced dry season, and the climax vegetation is tall rain forest. Repeated burning has so depleted the soil that only coarse kunai grass (*Imperata*) occupies the interstream areas. Note how dense the vegetation is along the steep valley sides. [Royal Australian Air Force]

abuse and a more efficient set of crops established, migratory agriculture could provide a good means of maintaining agricultural stability, and one which may be far easier to control in the tropics than some of the complex fertilizer-oriented crop systems.

13-14   THE FIELD-FALLOW SYSTEM   A second method of maintaining agricultural stability, and one somewhat akin to migratory agriculture, is the *field-fallow* system. This is a form of sedentary agriculture; the farmer does not move to new areas from time to time, but he periodically retires a certain fraction of his acreage from cultivation. Usually the regenerative period is much shorter than in migratory agriculture, and thus regeneration is only partially complete. This system is much more adaptable to the midlatitudes than to the tropics because of the greater nutrient-storage capacity of the soils, and it was the basis of the ancient *dreifelder-schaft* (three-field system) that was characteristic of much of European agriculture until the seventeenth century. This involved raising grains and root crops on two-thirds of the cultivated land and allowing the other third to stand idle. The fields were rotated every 3 years. Later, with the introduction of animals into the European farm economy, hay crops, forage legumes, and rotation pasture took the place of the fallow period in the crop-rotation plan.

The field-fallow method is still widely used in parts of India and in some sections of Africa. With this system, agricultural stability can be maintained, but only on an extremely low-yield basis. It is interesting to note that a low-yield fallow system is employed in many of the great specialized sub-humid-to-semiarid wheat regions of the world. Here the fallow period not only provides some regeneration of plant foods but, more important, adds to the reserves of water in the soil. To compensate for low yields, mass-production methods, involving highly mechanized farm practices, are employed; thus per-unit costs are low. Frequently the fallow acres are cultivated but not cropped, in order to keep weeds away and to create a moisture-conserving, mulched surface (see Figure 7-15).

13-15 THE ORGANIC COMPOST SYSTEM
A third method of providing agricultural stability is the *organic compost system.* This is the principal method used in much of the Far East, although it is by no means restricted to this part of the world. Very simply stated, this system requires that man return to the fields what he removes—or, more accurately, as much as he can. This involves collecting and treating human and animal wastes, plus unused vegetation in the cultivation process, and utilizing this material as a partially oxi-

dized compost, which is added to the cultivated fields as fertilizer.

With this system, a high level of plant productivity can be maintained for long periods, as evidenced by the continuous cultivation of some parts of China for over 40 centuries. The main difficulty with this form of agriculture is the great expenditure of hand labor that is required, a demand that can be met only by dense populations having a comparatively low standard of living. Recently, however, some interesting experiments have demonstrated that organic compost can be collected and processed by special equipment on a mass-production basis. This new technique may prove to be particularly advantageous in utilizing the efficiency of vegetative growth in the humid tropics. Another disadvantage of the compost system is the spread of human diseases that may result when the compost is not adequately processed. With care, however, this can be prevented, and the danger apparently has been exaggerated.

13-16 INORGANIC FERTILIZATION The last major method of maintaining fertility is *inorganic fertilization.* In this system plant nutrients are added in the form of commercial fertilizers, mostly of inorganic origin. The most important constituents are soluble compounds of nitrogen, phosphorus, and potassium. Calcium, as a soil-corrective agent, and small quantities of other elements essential to plant growth are also added. This procedure usually is used in conjunction with selected crop-rotation plans and with *green manuring,*[3] a process designed to improve the physical state of the soil in order to promote ease of cultivation and the retention of water and nutrients. Legumes, which add to the

[3]*Green manuring* refers to the practice of plowing green plants directly into the soil. Such plants decay rapidly and add to the organic content of the soil.

nitrogen content of soils, also are an important part of the crop-rotation system.

Inorganic fertilization is the system that has received the greatest attention from agricultural scientists, and it is practiced today wherever Western culture has spread. Commercial fertilizers are well-adapted to an industrial economy, and agricultural systems have been able to become stabilized at high yield levels. Essentially the system creates artificial conditions for plant growth, at least with respect to soil environment (see Figure 13-16). The new, highly productive crop balances, however, are not without their disadvantages. Their security depends largely on the stability of the exchange economy on which they are based; and when maladjustments in the cycle of production and consumption take place, the agricultural stabilization program may likewise become precarious. With depressed prices, a commercial farmer often will begin his retrenchment by reducing his expenditures for fertilizers, hoping that his plant-food reserve in the soil will tide him over the unfavorable period. Obviously also, a nation dependent on commercial fertilizers for its food supply is more vulnerable in case of military attack.

The new environments created by the use of commercial fertilizers and insecticides present an interesting ecological problem. Although the dangers have been exaggerated by many proponents of organic compost agriculture, inorganic fertilization may upset the delicate balances between plants in their natural habitat and insects and microlife. Such relationships are not fully understood. The roles of soil molds and fungi in relation to plant resistance to disease, for example, are just beginning to be discovered through research into antibiotics. There appears to be no doubt that in some special cases the use of commercial fertilizers has led to an increase in plant enemies. Whether or not sci-

Figure 13-16   One of the new techniques for replenishing soil fertility.   Field equipment is being serviced with liquid anhydrous ammonia (a nitrogen compound) in Mississippi for use on a cotton farm. Increased nitrogen compounds in soils can also be harmful in that the groundwater, heavily charged with such compounds, changes the natural balance in the water bodies into which it passes. [U.S. Dept. of Agriculture]

ence keeps its present lead in its fight to understand and to control the new synthetic plant environments is a matter that remains to be seen.

Little has been included here concerning techniques of plant breeding and research into the behavior and pathology of a wide variety of useful plant species. Space permits only a mention of the fact that man is learning how to change not only the environments of plants but also the properties of many plants themselves, in order to make them more adaptable for use. He is discovering, for example, how to make the sugar cane sweeter; the wheat plant more tolerant of drought, rust, and frost; the potato less scabby; the rose more beautiful; the orange seedless; and the apple tree smaller.

Modern man promotes an ever-increasing specialization of the useful plants, and in this lie his greatest strength and his greatest danger. Unless he carefully learns the lessons of ecology as he proceeds, he is likely to find that he has opened a Pandora's box containing a set of unexpected horrors. The serious repercussions throughout the animal world of the use of certain types of insecticides, such as DDT, and the exploding growth of microflora in ponds and lakes following increased use of commercial fertilizers on adjacent fields are beginning to be realized. Harmful examples of ecological meddling by man are everywhere about us. The new synthetic environments set up to increase selected plant yields are new open systems in which the through-put of energy and materials has become greatly increased. Supplementing solar energy inputs are those indirectly supplied by the fossil fuels and hydroelectric power, for example, the electricity used to alter atmospheric nitrogen into soluble fertilizers (see Figure 13-16); the energy used by a chemical plant producing insecticides and pesticides; the gasoline or diesel fuel used in rapid and efficient cultivation practices; and the energy used in agricultural experiment stations everywhere to study the mysteries of plant growth. At the same time, certain aspects of the system are being badly neglected. We know little, for example, concerning the effects of the new agricultural systems on the nutritional balance; in fact, we know little about nutrition. Nature's systems for maintaining stable biomasses are subtle and compelling. We need to understand their complexities before we move energy inputs and outputs to higher and more dangerous levels.

13-17 NATURAL BALANCES AND FOREST MANAGEMENT PROBLEMS A large part of the world population lives in areas that once were forested. The reduction of forest land in the mid-latitudes, particularly in the eighteenth and nineteenth centuries, accompanied by a rapid increase in population, resulted in growing shortages of wood products. The shortages for a while appeared mainly in highly specialized products, such as tall spars for the masts of ships, oak beams for house timbers and the planking of warships, clapboards for siding, and special woods for wine barrels. Later the shortages became more general and began to invade the critical field of fuel. Men realized that even the welfare of nations depended on a regular supply of forest products, a supply which for a long time was taken for granted. Public forest reserves were established, and increasing attention was paid to learning the basic principles of forest conservation and management.

Today, forest management is a complex science which involves a thorough knowledge of plant ecology. Its objective should be *to adjust the plant community and forest practices in a forested area so as to produce a maximum yield adjusted to human needs and the productiveness of the site.* Many factors are involved in this objective, the most important of which are present in the following list of pertinent questions:

1. *Purpose.* Does the forest-land use have both long-range and short-range objectives? Is its purpose to supply forest products? If so, what kinds, and are the management practices designed to yield maximum efficiency in production? What role does the forest play in water storage, erosion control, and flood control? Is recreational use an important side function? Does it serve a multipurpose public function or profit-motivated private objectives?

2. *Stage of succession.* What appears to be the local climax or subclimax forest com-

munities in the area? How far along is the present forest in the successional stages? At what stage along the succession is the forest best adapted to its purposes? How can management practices shift the community toward the ideal successional stage?

3. *Alteration of the forest environment.* Can any of the environmental characteristics of the site be altered relatively simply to produce a new forest succession which would contain stages more consistent with forest purposes? Possible examples here are the use of fire to promote a pine-grass association versus an oak-hickory-rhododendron association in the southeastern United States, or the draining of a marsh to produce a more accessible and more valuable forest stand.

4. *The economic balance sheet.* If the forest is in private hands and the objective of the management program is to maximize profit, the following questions would be pertinent: What is the capital investment in land, forest, equipment, etc.? What is the annual increase in forest growth? What is the current return in the form of total economic input and output? What is the estimated return in the future? What are the current and future demands with respect to avowed purpose? How much investment should be made in ecological improvement? What is the most profitable time to cut particular trees?

Not all these factors are investigated by a single person. Some are studied by government agencies, some by corporation managers, and some by highly trained foresters operating in the field. Still others are determined by public sentiment and taste. A completely integrated forest management program can be developed only in a highly integrated society.

One of the major lessons learned from plant geography is that forest management programs that work well in one area may not be suitable for other areas. A few illustrations will emphasize this point. The periodic burning of dry litter and low brush in the longleaf pine areas of the southern United States has been found to be beneficial and even necessary for the reproduction of the desired species and for more rapid tree growth. It also makes possible an understory of grass on the forest floor which will support a cattle-grazing industry. The same practice in other parts of the United States would yield only worthless fire-succession shrubs. Reforestation of large areas by seedlings of the principal local lumber species may be totally unsuited to areas where such species usually are in a highly mixed, integrated plant community. On the other hand, it may be a recommended procedure in areas where the climax forest contains a high proportion of the same type of trees. Reforestation by planting seedlings normally should be used only in areas that will not reforest themselves naturally, areas that have been ecologically damaged by forest fires, floods, insect depredations, or the complete elimination of seed trees.

Selective logging[4] is necessary for a stabilized forest management program in most areas. It may, however, be uneconomical and undesirable in such forests as the Douglas fir forests of Washington and Oregon, since these trees commonly thrive best in pure stands of similar size (see Figure 13-17). Unlike most other climax forest trees (or, more correctly, subclimax), the Douglas fir does not require the shade of other trees for seed germination, but springs forth readily from bare ground. Having an extremely rapid growth, it eliminates most tree competitors other than its fellows of similar size. Standard practices of forest cutting in the Douglas fir areas, therefore, involve the removal of all

[4]*Selective logging* refers to the practice of cutting only scattered trees that have reached a desired size and quality.

Figure 13-17 Natural regrowth of Douglas fir in formerly cleanly cut area. Seed trees in the adjacent uncut stands provide the source for reproduction. Note the uniformity in height and density of stand typical of Douglas fir. [Weyerhaeuser Company]

the trees from a tract of land once they have reached proper size. A few scattered trees may be left as seed producers for natural reproduction, and a neighboring tract of land with trees near maturity is also recommended for reproduction (see Figure 13-18).

A final illustration involves the relationship between cultures or civilizations and forest management programs. It is customarily assumed that trees should be left to grow to maturity. Yet in many densely populated parts of the world, the need for narrow poles to be used in house construction, for thin firewood that does not require splitting, and even for leaves to be used for miscellaneous household purposes may be so urgent that the most beneficial forest management plan would be aimed specifically at a high productivity of young, immature forest stocks. Increases in the production of firewood may have far-reaching consequences. In some areas firewood may be able to replace animal dung as cooking fuel, and the dung, in turn, can be added to fields, thereby raising soil fertility and aiding in the stabilization of the agricultural economy.

13-18 FIRE AS A TOOL IN VEGETATION ALTERATION   Fire has always been a significant factor in the human alteration of vegetation patterns. Primitive people used it as a means of clearing agricultural land, as an aid in hunting, as a tactical weapon in both offensive and defensive warfare, as a means

Figure 13-18   Block cutting in an extensive stand of lodgepole pine.   This photo was taken in the Lewis and Clark National Forest in western Montana. [U.S. Dept. of Agriculture]

of stimulating the growth of new grass for animals, and perhaps merely as an enjoyable spectacle.

Fire has had its greatest impact on vegetation in three major vegetation regions: the grasslands, the deciduous forests, and the coniferous forests. Its abusive use by man, especially in primitive cultures, has led to some noteworthy shifts in the boundaries of some of the great global plant formations. Usually the shift has been at the expense of the more humid vegetation formations. For example, desert shrubs have increased at the expense of grasslands, and grasslands have moved into forested regions.

Grasslands have been the target for intentional burning for thousands of years. Except near the dry margins of the steppes, where grasses struggle with aridity for survival, grasslands are not especially harmed by burning at the end of the growth season. The tropical grasslands (savannas), which lie between the desert and the tropical forests, have a constant smoke haze hanging over them during the dry period (see Figure 7-5). The rank, coarse, dried savanna grass is symbolic of this hated time of year, when water holes dry up, the heat becomes unbearable, animals are scrawny and ill-tempered, and the grass itself tears the clothes and cuts the flesh. Better a charred landscape; at least it is easier to cross, the fire itself is wonderful to behold, and the people know that the new shoots that will appear a little later will be the greener for it. Near the forest-grassland border, the grass fires slowly nibble at the forest edges, aided in their conquest by the clearings for migratory agriculture or by the efforts of pastoral people who seek to extend their grazing territory. Continued burning in a tropical forest clearing may destroy the regenerative capacity of the forest, so that rank grasses become dominant. Useless for native agriculture, be-

cause of the tough sod cover, the grassy openings in the forest are abandoned but are periodically burned, as in the savannas, if even a short dry season is experienced. Thus the forest slowly becomes grassland, and the savannas migrate equatorward. Grasslands created by man, primarily through the use of fire, are found throughout the humid tropics, especially in areas where the population pressure is great.

Many authorities believe that most of the humid grasslands of the mid-latitudes, the tall-grass prairies, are the result of the use of fire by aboriginals. Near the forest-prairie border in Wisconsin and Minnesota, patches of forest cling to the lee sides of lakes, where the sweep of prairie fires was checked. Tongues of forest extend along the stream courses, which usually are entrenched below the general level of the prairies, where there was some protection from the sweep of prairie fires. Distinct ash layers have been found in the soil of the Argentine pampas, indicating that fire has been an active factor there for a long time.

Large areas covered with secondary scrub brush and grass tend to be self-perpetuating through the use of fire. Brush land is of little direct use to human beings, and it is usually resented by local inhabitants. In their desire to destroy it, they use fire without realizing that the brush itself may be a postfire succession. One of the worst areas for brush fires in the United States lies in southern California. The overlay for the grass- and brush-covered hills, so familiar to viewers of western movies, is largely a fire-induced vegetation. The brush reproduces mainly by the suckers of its extensive and well-developed root system. Fire destroys the tops, but not the roots, of these plants; thus the natural competition of other plants is reduced. The fires are unfortunate in many respects; but the brush vegetation helps protect vital watershed areas against too rapid runoff and floods, soil erosion, and silting of water reservoirs, and it provides for a greater retention of rainfall in subsurface reservoirs in this thirsty land.

Wide expanses of less valuable deciduous trees, including birch, aspen, and wild cherry, are found in the northern coniferous forests of Canada and Eurasia. These mark the sites of old fires and appear from the air like huge, elongated, light-colored patches on a dark-green quilt. The low branches, resinous leaves, and tangle of fallen trees that characterize the northern coniferous forests are highly inflammable in the late summer and fall months (see Figure 13-19). In this sparsely populated region, fires may continue to burn for days until quenched by a rainstorm. Lightning and spontaneous combustion, along with the matches of careless human beings, are responsible for igniting them. Burning with exceedingly high temperatures, these northern forest fires consume most of the organic material in the topsoils, leaving them unsuited for reproduction except by specialized fire-succession flora. These fires should not be thought of, however, as being entirely detrimental. The regrowth of grasses, herbaceous plants, berries, and shrubs that follows burning provides cover and food for wildlife, especially deer and other larger animals.

Many conifers are aided by fire in their competition with other plants. The lodgepole pine (*Pinus contorta*) of the Rocky Mountains and the jack pine, or northern scrub pine (*Pinus banksiana*), of the northern United States glacial outwash plains are the best known of several pines that have the unusual property of requiring the heat of a forest fire in order to open their tight cones; thus fire provides for their replacement and improves their competitive position within the postfire succession. The southern longleaf pine (*Pinus*

Figure 13-19 Charred remains following a forest fire near Hope, British Columbia. The timber loss following this huge fire was valued in millions of dollars. [Van Riper]

palustris) of the southeastern United States coastal plain has a highly insulative bark that is almost fireproof, and the peculiar arrangement of the terminal buds apparently enables the tree to be virtually defoliated by fire without undergoing serious harm. Fire also seems to assist pine reproduction by providing the bare ground surface necessary for seed germination and by destroying young competitive plants. For a long time pine was believed to be a dominant climax vegetation on coastal-plain sands, but evidence now indicates that oak will replace it if fire is kept away. The periodic burning of longleaf pine forests is standard practice in maintaining and improving this important source of naval stores (turpentine and resin) and rapidly growing timber.

Burning generally tends to exhaust the organic, biological, and chemical resources of the soil. It directly destroys most of the plant remains standing or lying on the surface and also a large part of the plant residues and old root fibers incorporated within the topsoil, leaving only ash and mineral soil material and reducing the soil capacity for retaining water and plant foods. Beneficial soil fungi, along with helpful bacteria, other microorganisms, and insect life, are also destroyed, or at least have their ecological equilibrium disturbed. Minerals and essential chemicals that are ordinarily made available slowly and are normally released only as needed by plants are suddenly rendered soluble by the high temperatures and are rapidly leached out of the soil by subsequent rains. The ultimate result is soil exhaustion and lifelessness, although an apparent effect of increased fertility is exhibited temporarily while the newly released soluble elements are made available to the postfire flora. Burning is generally a wasteful, destructive process, *except in certain special circumstances,* where factors other than soil nutrition seem to justify its use.

13-19 SUMMARY OF THE HUMAN FACTOR IN PLANT GEOGRAPHY Man is a comparatively new ecological factor in plant geography, because of his relatively recent appearance on this planet; yet the impact of his influence can be seen throughout the world. Despite the importance of plant life for food and feed, man's modifications of the vegetation cover have not always been to his own benefit. The marks of his abuse are encountered all too frequently. Only in a few regions has he achieved a stabilized balance in his utilization of vegetation growth. He has lowered the carrying capacity of many grasslands by burning and by overgrazing, and he has seriously depleted the world stock of valuable timberland. Eroded, bare, sterile slopes in many humid areas testify to abuses in hillside cultivation. Only in recent decades

has he set out to understand the natural balances of vegetation communities and to attempt to control and use them with regard to his future, as well as his present, welfare. Unfortunately, his greatest progress in this direction has been in areas where the pressures for increased plant productivity are not so great.

Notwithstanding the efficiency with which man uses the ax and the plow, the bulldozer and the dragline, his most effective tools for altering vegetation patterns have been fire and livestock. The reduction of the unprincipled use of fire has been the greatest achievement in forest conservation in the Western world. It is hoped that scientific management will bring our ranges to their optimum carrying capacity. The principal reason for the continued abuse of the vegetation cover in the world today is poverty—a poverty that is both a cause and a result of vegetation degeneration.

In the past, many peoples maintained a stabilized existence for a long time through a mutually beneficial relationship with the vegetation cover that furnished them food or feed. These older groups developed their balances, not through complete understanding, but through long periods of trial and error. The stabilized plant yields which they accepted were on a low level, far below environmental capabilities. Today we do not accept a position in the balance at such a low productivity level. Through knowledge, we are learning nature's secrets and are beginning to adjust the environmental forces to suit our needs. Until recently it might be said that the natural vegetation balances either accepted or rejected human intervention. Now we are learning how to make the plant world work for us. Plant scientists tell us that we have hardly begun to harness the potentialities of plants for feeding us and that the control of plant environments will eventually have as great an impact on man as did the first domestication of plants.

The Classification of Vegetation

13-20 PROBLEMS IN THE CLASSIFICATION OF VEGETATION PATTERNS As was stated earlier, the problems of describing the distribution of vegetation are similar to those encountered in depicting the geography of the other elements of the physical environment. Differences appear at every step, differences that grade into one another and that change with fluctuating conditions, ranging from the daily changes in weather to the slow action of geologic forces operating in cycles of millions of years. Yet, as with the other elements in the environmental complex, similarities can be seen as well as differences, and these make possible regional classifications of plant characteristics. As is pointed out many times in this volume, there are no universally applicable systems of geographic classification. Each system must be suited to the purpose for which it is to be used and to the observational scale that is appropriate to that purpose.

The most widely used systems of classifications of vegetation regions today are based on the plant-community concept, distinguishing between the communities which have reached a relatively long-term stability with the major environmental controls (the climax communities) and those which represent short-term balances (the successional communities). Subdivisions of the climax communities are based on a scale of generalization that ranges from the great global formations to local associations distinguished by the *dominant species*[5] present. Since this is the classification system to be used in describing global patterns of vegetation in the following chapter, no further elaboration of it will be given here.

In describing vegetation patterns on a local, or topographic, scale, classifications that recognize stages in plant succession are most useful. The reason for this is that local variations in slope and drainage are critical in the pattern of plant communities. Thus we may have a *hydrarch* succession, which refers to a sequence of plant communities that characterize the gradation from a bare surface covered with water (as in a lake or pond), through various degrees of poor drainage, to the well-drained upland climax (see Figure 13-20). Such a complete series is known as a *sere*. It will be noted that any area may undergo a successional shift in either direction; for example, in the hydrarch succession, a site that is poorly drained may become either better drained or more poorly drained. Keeping in mind that a sere represents the complete succession from a bare surface to a climax association, we may note that a simple classification of seres includes:

1. *Hydrarch* (a body of open water)
   a. Fresh-water sere: *hydrosere*
   b. Salt-water sere: *halosere*
2. *Xerarch* (a bare, waterless surface)
   a. Rock sere: *lithosere*
   b. Sand sere: *psammosere*

Other types of bare surface on which xerarch seres could develop include gravel, clay, and a burned-out soil following a forest fire. Seres differ widely with different types of climax vegetation, although the plants resulting from the excess or dearth of water may have comparable physiologic characteristics. For example, the swamp plants of the humid tropics often differ markedly in appearance and

[5]The term *dominant*, as here used, refers to the largest common members of a plant community. In a forest, the dominant species always include trees.

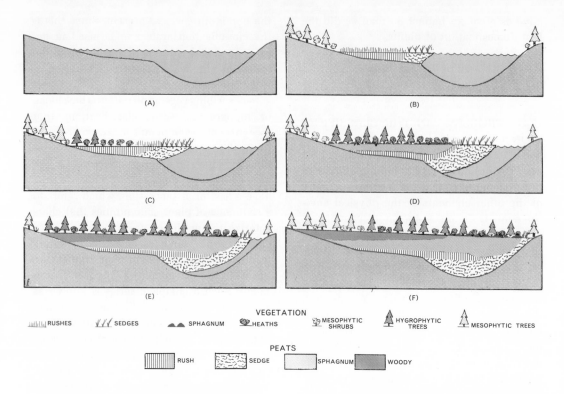

VEGETATION

⊥⊥⊥⊥⊥⊥ RUSHES   ∀∀∀ SEDGES   ▲▲ SPHAGNUM   🌿 HEATHS   🌿 MESOPHYTIC SHRUBS   🌲 HYGROPHYTIC TREES   🌲 MESOPHYTIC TREES

PEATS

RUSH   SEDGE   SPHAGNUM   WOODY

species from those of higher latitudes, but both may use many of the same mechanisms for overcoming water excesses.

A much simpler classification of plants based solely on water availability is sometimes used. This involves a primary division into four groups: the *xerophytes,* or drought-resistant plants; the *mesophytes,* plants that need a moderate amount of water; the *hydrophytes,* plants requiring an excessive water supply; and the *tropophytes,* plants that undergo periodic changes with respect to water availability. These categories can be divided into as many subgroups as are desired. Usually the subdivisions are based on the morphological characteristics of plants. For example, the xerophytes could be divided into cacti, perennial woody shrubs, needle-leaf trees, etc. The major disadvantage of such a classification is that it is difficult

to use in showing mixtures of plants, and it does not take into consideration the changes that individual sites may undergo.

General Perspective

If one realizes that nearly all the factors cataloged in this chapter operate more or less together to influence plant growth in one way or another, he should have some idea of the enormous scope of plant geography and the amazing complexity of nature. Not only is this subject material immense in breadth, and hence formidable to describe comprehensively, but it is never static. The plant world is in a constant state of change, for it is made up of a living, pulsating complex of unstable variants. The facts of plant life therefore are not of such a nature that they can all be categorically presented in a simple, brief

Figure 13-20 Diagram of a typical hydrarch succession. [After Clarke]

listing. Instead, the various factors seem to work together as a great dynamic system, filled with fluctuations and seeming contradictions, with opposite forces pushing and shoving one another. Plant geography might be pictured as a vast jungle in which science could lose itself; yet this is not the case.

Plant geography is a subject that has developed gradually, along with the various specialized plant sciences. At first a static, descriptive study, it assumed its dynamic character as the biological sciences uncovered some basic facts concerning the substance of living organisms, their habits and potentialities, and their relations with their environments. Plant taxonomy, physiology, pathology, genetics, and a myriad of other ramifications in botanical science are all involved, and it is the task of the plant geographer to bring together many of the contributions of these fields in an attempt to explain the complex pattern of plant life as it clothes the surface of the earth.

In an overall view, some of the basic points to be observed might be summarized as follows:

1. Plant growth is conditioned by numerous vital factors.

2. These factors are not static, but dynamic and variable.

3. Pressures are exerted upon the plant and upon each other by these factors.

4. One factor may become limiting and temporarily exert a controlling influence.

5. Whenever the pressure of one limiting factor is relieved, some other factor of the complex may assume control and limit the plant; this in turn may be superseded by others, one after another.

6. The factors themselves, however, are not without limitations, for each factor travels in its own special orbit.

7. The plants themselves, in relation to one another, serve as agents in creating part of their own environment.

To illustrate the dynamic nature of plant life, let us say that a plant is inhibited by lack of sufficient water. Such a water deficiency, then, would constitute a limiting factor. Later, with adequate water supplies, there might be inadequate light, or insufficient plant foods, or removal of organic material from the soil by burning—and so on through the long list of environmental factors that may impinge upon plant growth.

Despite all these variables, the circumstances affecting a plant's life are subject to certain laws and rules, enabling plants to grow, thrive, and react in ways that are known and predictable. The end result, therefore, is a fluctuating though orderly system of pressures and demands, controlled by checks and balances, which constantly operate to produce the harmonious but highly diverse results that we see in plant life everywhere.

If one makes a detailed study of the plant growth that can be found on almost any bit of undulating terrain in the humid midlatitudes, he will observe that each cubic inch of surface soil is occupied by plant life of one sort or another—usually several kinds, living together in a kind of society. In examining this variety at close hand, one can obtain a fair idea of the pressures which are present, the intense competition which exists, and the relative balance which is maintained. One can also recognize the various ecological agents which are focused upon any spot. Actual observation of such conditions in the field will be more effective than volumes of words in giving the student an understanding of these phenomena, which appear at first to be too complicated for the memory to grasp but which in actual operation perform as a successful and orderly process in the total scheme of things.

These complex interrelationships of plants with one another and with their environment are not unlike some situations that exist in human society; they might, in fact, be described by Plato's statement about democracy:

. . . a charming form of government, full of variety and disorder, and dispensing a sort of equality to equals and unequals alike.

Study Questions

1. What is a plant community? Give some examples of how different plants in a forest community assist each other.
2. Show how both plants and animals together form an open subsystem for the cycling of energy. Indicate some of the mechanisms for regulating the through-put of energy. What different types of energy can you identify in such a system?
3. What is *plant succession*? Find evidences of a succession in your local area and describe its stages. What happens to the biomass as the successional stages develop? Does biomass include animals?
4. What is a plant indicator? Find some plant indicators in your locality for each of the following: a. soil acidity (or alkalinity), b. poor drainage, c. inadequate soil moisture.
5. Give some specific illustrations indicating how vegetation patterns conform to the principles of natural distributions.
6. List as many properties as you can that show how plants adapt to seasonal deficiencies of water.
7. Why are desert fruits unusually sweet?
8. Distinguish between transpiration and respiration of plants and indicate their respective functions.
9. Why do tropical forests have so many more species than the forests of Finland?
10. What are the harmful effects of fire on tropical and subtropical soils and vegetation?
11. Explain why coniferous trees appear in nearly all climates except true deserts and polar regions.
12. What are the functions of the microflora in the soil? How do they derive their energy supplies?

CHAPTER 14

MAJOR VEGETATION REGIONS OF THE WORLD

The four primary types of vegetation regions of the world, termed *formations*, are the forests, and woodlands, grasslands, desert shrubs, and tundra (see Figure 13-6). There is sufficient variety within the first two formations to justify a further breakdown into subformations, each of which has sufficient area to be indicated on a world map.

The classification of forests, or the breakdown into subformations, is based on the forms that the vegetation assumes in order to adjust to particular moisture balances. The subformations are as follows:

1. Broadleaf evergreen forest and woodland
   a. Hygrophytic (abundance of water) tropical rain forest, or *selva*
   b. Mesophytic (moderate water supply) scrub woodland[1]
2. Broadleaf deciduous forest and woodland
   a. Hygrophytic broadleaf deciduous forest
   b. Mesophytic broadleaf deciduous scrub woodland
3. Needleleaf evergreen forest and woodland
   a. Hygrophytic conifers with tall trees
   b. Mesophytic low forest (the *taiga*)
   c. Xerophytic (low water supply) scrub woodland
4. Needleleaf deciduous forest
5. Mixed broadleaf deciduous—needleleaf evergreen forest
6. Semideciduous forest (broadleaf evergreen—broadleaf deciduous forest)

The grasslands are divided into the steppe, prairie, savanna, and marshland subformations.

Each of the formations and subformations is described in this chapter. A separate division of the chapter is reserved for mountain vegetation, which includes characteristics of several of the formations and subformations but which differs enough from the lowland types to warrant separate

---

[1]The term *scrub woodland* refers to a sparse tree stand, 20 to 40 feet high, in which a continuous canopy overhead is usually absent but in which trees predominate over grass in the landscape.

Figure 14-1 World distribution of vegetation subformations.

Forest and woodland

Hygrophytic broadleaf evergreen forest (selva)

Mesophytic broadleaf evergreen scrub woodland (maquis, chaparral, mallee)

Hygrophytic broadleaf deciduous forest

Mesophytic broadleaf deciduous, scrub woodland and thorn forest

Hygrophytic evergreen coniferous forest

Mesophytic evergreen coniferous forest (taiga)

Xerophytic coniferous and broadleaf scrub woodland (forest-steppe)

Mesophytic deciduous coniferous forest (larch taiga)

Mixed broadleaf deciduous-coniferous hygrophytic forest

Mixed broadleaf deciduous-broadleaf evergreen hygrophytic forest (tropical semideciduous)

Grasslands

Steppe and prairie

Savanna

Desert shrubs and barrens

Tundra

Mountains (vertical zone)

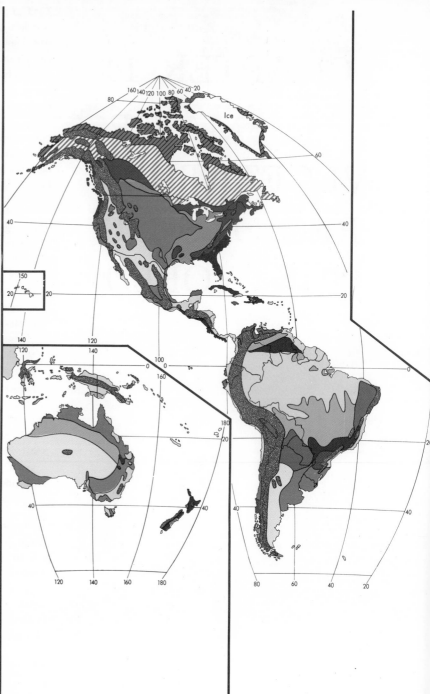

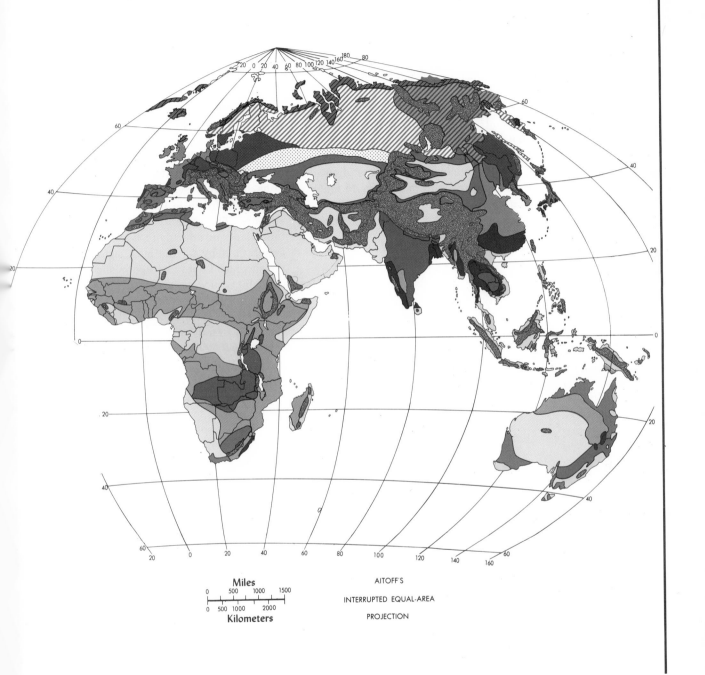

Miles

0   500   1000   1500

0   500 1000   2000

Kilometers

AITOFF'S

INTERRUPTED EQUAL-AREA

PROJECTION

treatment. Figure 14-1 shows the pattern of the global, vegetation subformations.

Descriptive generalizations of such immense areas as are covered by the vegetation regions described in this chapter must be subject to continual qualification. Exceptions to the generalizations can be found everywhere, if one looks closely enough. The descriptions here are those of climax or near-climax plant groupings. Little can be done with the characteristics of *plant succession* within the scale of our global presentation, but a few of the major implications of succession are noted from time to time to indicate that the processes of vegetation replacement and site alteration are present everywhere.

Our task is not only to portray and to interpret the vegetation patterns but also to indicate their significance to man; therefore, the discussion of each regional pattern is followed by a brief section on economic implications.

Broadleaf Evergreen Forest and Woodland

14-1 HYGROPHYTIC, TROPICAL RAIN FOREST (SELVA) Highly favorable conditions for photosynthesis and vegetative growth are found in the humid *Af* and *Am* climates. The quantity of the biomass, or total plant and animal life, is related to many factors, but by far the most significant are the surplus of net radiation balance and the availability of water and plant nutrients. The first two are reflected in the consistently high temperatures and rainfall in these climates. The third, or nutrient, factor can be built into a biotic subsystem through an appropriate and adequate circulation subsystem. We have already seen that high sensible heat and humidity favor rock decomposition through chemical weathering. This releases soluble bases into the soil solution and enables plants to use these bases in building plant tissue. However, such bases also are liable to be washed out of the soil via groundwater and

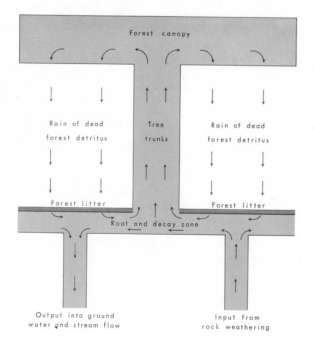

Figure 14-2 Schematic flow chart of plant nutrients in a forest. The flow of plant nutrients in a forest, like that of energy in a biotic system, is contained within a circulatory system. With continued circulation, and where the net outgo of nutrients via soil and stream drainage approximates the net input from rock weathering within the soil, the biomass can be maintained indefinitely.

Figure 14-3 The canopy of a typical selva, or tropical rain forest, as seen from the air. The varied composition of the forest is apparent. Location is in the upper Magdalena Valley in Colombia. [Standard Oil Company of N.J.]

stream drainage. A balanced, climax forest and its faunal assemblage is a kind of subsystem that maintains not only a balanced equilibrium of energy input and output but also a balanced input and output of materials, the nutrients that are diagrammed in Figure 14-2. The slow rain of leaves, twigs, branches, and other plant detritus onto the forest floor is rapidly consumed by microfloral and fauna assemblages in the tropics, such as bacteria, fungi, and termites. When decomposed, the nutrients are once more released into the soil solution and are taken up again by the broad, shallow root system of the forest. A circulation is thus formed that can be maintained indefinitely. Flora and fauna (the *biota*) thus are intimately related in the subsystem. While some nutrients always escape via groundwater, more are being released by rock decomposition far below the surface to be tapped by some deep-rooted trees and brought into the circulation system.

The high moisture and nutrient intakes and high transpiration rates make for a large biomass. The principal competition within the floral sector of the subsystem is for sunlight. Hence, tall trees are the rule, and each square inch of sunlight exposure is fought for. The typical *selva,* therefore, has an extremely tight overhead canopy (see Figure 14-3). The canopy sector not only contains the largest part of the active floral life, but it also is the main biotic realm for most of the larger faunal types, including birds,

monkeys, and other tree dwellers. The ground floor is the zone of the faunal forms that feed on dead organic material—mainly the decay-producing organisms. This part of the biotic subsystem is a zone of death rather than life, but it is as important to the continuation of the system as the teeming life in the treetops above.

There is a distinct storied aspect to the selva, or tropical rain forest, with two and sometimes three levels of tree growth. In the upper story, tree hights reach 150 to 200 feet, and branches begin far above the ground. A distinctive feature of many of the larger trees is their buttressed bases, often in the form of ribbonlike flanges that flare outward, so that from a distance the bases of the trees appear to have diameters of 15 to 20 feet. These buttressed bases are necessary to support the tall, slender trunks. The middle story consists of smaller trees, 30 to 60 feet high and only 4 to 8 inches in diameter. These smaller trees are not necessarily younger versions of their giant companions. Many of them are distinct species that have become adapted to this middle level, in which the conditions of light, heat, and moisture are intermediate between those of the forest floor and the canopy overhead. Leaves of trees within the middle story of the forest usually are considerably broader than those of the larger trees. Giant ferns (tree ferns) sometimes abound in the middle story in areas where moisture is especially abundant. The lowest story of the selva includes small seedlings, ferns, and often varieties of climbing bamboos (see Figure 13-8). Mushrooms and other forms of giant fungi are plentiful.

Vines are frequently found in the selva community, ranging from small tendrils of beanlike plants to giant lianas, 3 to 6 inches in diameter, which may hang down from the trees like crooked hawsers (see Figure 13-9). Equipped with tiny hooks, some of the vines constitute one of the more troublesome features for human beings traveling through the selva. For the most part, however, the relative sparseness of underbrush makes the selva fairly easy to penetrate. *Epiphytes*, or "plants upon plants," which include the orchids, abound in the upper story of the forest, where they find moisture and sustenance in tree crotches and on the more horizontal branches. The more showy members of this highly diverse group of plants are well known in florist shops, but most of them in the selva are small and inconspicuous. Large, brilliantly colored flowers generally are rare throughout the selva. The brilliant flowers seen by tourists in the tropics are more characteristic of human settlements than of natural forests. Another distinctive feature of most rain forests is the presence of many parasitic plants. The most widely distributed of these is the strangling fig (*Ficus*), which may slowly envelop the host tree and literally choke the life out of it.

The species of trees in the rain forest are extremely varied. Whereas ten or a dozen species of trees per acre are common in mid-latitude forests, a hundred or more will often be encountered in the selva. The properties of the trees are as diverse as the species they characterize. There are woods that are extremely hard and so dense that they will not float on water. Others, such as balsa, are light and spongy. There are dark woods and light woods, woods of different colors, straight-grained woods and gnarled woods. Some trees have a milky, thick sap; others contain fluids that are strong irritants and that may even be poisonous. Some trees bear nuts, and others produce fleshy fruits. Contrary to common belief, edible fruits are not abundant in a typical climax selva, despite the great variety of trees. Although there is a striking similarity of forest structure in the humid tropics, diversity is the outstanding characteristic of the trees themselves.

A walk in a climax selva is a unique experience for visitors. A heavy, rank, moldy odor permeates the dimly lit lower story of the forest. There is a surprising amount of bare soil, usually damp and reddish in color. Fallen trunks of dead trees are conspicuously absent. From far overhead comes the raucous clamor of birds and monkeys, interrupted by the shrill piping of tree frogs near at hand.

The largest expanse of undisturbed climax selva in the world is found in the upper Amazon Basin. Thinly populated even today, this area presents an almost unbroken expanse of rain forest 1,000 miles long and 750 miles wide. Elsewhere, however, man has been much more active in his penetration of the tropical rain forests, and, except for Brazil, primary forest occurs mainly in small, discontinuous areas and in the more inaccessible localities. This is particularly true of the tropical forests of southeastern Asia, where unaltered selva is rare.

Figure 14-4 Early stage in the development of a secondary forest, northeastern New Guinea. This is an abandoned clearing slowly reverting to forest. Note the scattered palms and the rank growth of vines covering some of the trees. The coarse, dense stand of kunai grass is an indication of soil depletion. The tangle in lower left would be extremely difficult to penetrate. [Royal Australian Air Force]

14-2 LOCAL VARIATIONS IN THE SELVA Typical selva vegetation, as described in the preceding section, is developed mainly on well-drained, rolling-to-undulating uplands that have not been occupied by man. Various types of modifications of the selva vegetation occur in the tropics, however, and are sufficiently widespread to warrant separate treatment. All these variations, however, are characterized by typical broadleaf evergreen trees.

Secondary rain forest. Much of the forest land in the humid tropics is in various stages of regrowth following destruction of the climax selva. There is fierce competition among many plants to occupy the site of a forest clearing. The length of time taken to establish the dominance of the giant trees has been estimated at between 100 to 250 years. As a result of migratory agriculture, there are large areas of secondary rain forest in various stages of regrowth in Africa and southeast Asia (see Figures 13-14 and 14-4).

At first the forest clearings are taken over by a tangled mass of low bushes, vines, ferns, briars, and young saplings. If the clearing was formerly cultivated and not too badly depleted in fertility or too heavily burned, the remains of garden fruits and vegetables often will be found in the tangle, including the spreading vine of the *yam*, the drooping leaves of the *plantain* (banana), and fruit trees such as the *mango, jack fruit*, and *guava*. In southeast Asia, bamboo thickets frequently crowd the assemblage. Gradually the trees become larger, undergrowth thins out, and eventually the selva reestablishes itself.

The vegetation in the early stages of a secondary forest is generally difficult to penetrate, and trails must be hacked out of the tangle. Such vegetation is usually avoided by human beings, and its wide distribution surrounding settled areas probably has helped to establish the erroneous conception that most tropical forests consist of dense jungle growth. A forest is replaced by coarse grasses instead of by secondary forest if annual burning is used consistently to clear away the entering vegetation (see Figure 13-15).

Influences of slope and elevation in altering typical selva vegetation. The true selva is typically developed in well-drained, level-to-rolling interstream areas. In hill and mountain country, steep slopes generally tend to increase the amount of undergrowth and to reduce somewhat the height of the mature forest. Vines and lianas become more plentiful, particularly some of the climbing varieties of bamboos. The sparsest vegetation is generally encountered along the crest lines of narrow ridges, probably because of the lower moisture content of the soil. Most of the native trails are found here. Narrow ravine bottoms have the densest tangles of vegetation.

Selva trees also tend to become shorter as elevation increases, and the general appearance of the forest changes noticeably as one ascends mountain slopes. Above 3,000 to 5,000 feet, tree heights are reduced to 50 to

Figure 14-5 Nipa palm thicket on the Menam delta, Thailand. The nipa palm, like the sago palm, is a common swamp palm that grows on poorly drained mud flats in the low latitudes. The nipa palm tolerates more salinity than sago, hence is especially common along tidal flats. Its leaves are widely used for thatching, and a fermented beverage is prepared from its sap. [Van Riper]

100 feet, the middle story tends to disappear, and buttressed trunks are much less usual. Vines are not so well developed, and tree ferns (tall, treelike ferns 20 to 40 feet high) become much more abundant.

Riverine and swamp forests. The rain forest that borders large rivers appears as a solid green wall of vegetation. Such areas contain an abundance of water and some of the most fertile soil in the tropics. The swath of the river breaks the dense forest canopy and thus permits sunlight to reach ground level. Vines grow in rank profusion, extending from the water's edge to the summits of the adjacent forest. No wonder that early explorers in the rainy tropics, penetrating along the waterways, which still are the easiest routes of access, believed that the tropical forest was virtually impenetrable!

The freshwater swamp forests that develop on the flood plains of rivers, away from the immediate stream banks, differ from the selva proper mainly in having a somewhat denser stand of taller trees and a much larger number of palms as a lower story (see Figure 14-5). Among the latter is the sago palm, particularly abundant in southeastern Asia and Indonesia. The pith contained in the trunk of this palm is high in starch and constitutes a basic element in the diet of some of the peoples in this part of the world.

Eucalyptus rain forest. An unusual variety of rain forest is found in southern Australia, including the extreme southwestern coast west of Albany, and also the rough hill and low mountain country of Victoria, New South Wales, and parts of Tasmania. Nearly all of the trees are *Eucalyptus,* but there are dozens of species. The mature trees are 100 to 150 feet tall, and some, including the giant *karri* (*Eucalyptus diversicolor*), reach up to 250 feet. Unlike the true selva, the eucalypts are *sclerophylls,* or narrow-leaf trees, but the trees are evergreen and the continuous canopy, the dark, damp ground floor, and the understory of lower trees make it similar in other ways to the selva. Tall tree ferns are a common feature of the middle-story flora (see Figure 13-8) and are accompanied by the low wattles, species of *Acacia*, whose bright yellow blooms seem to light up the forest during the flowering season. Several of the eucalypts are important lumber trees, and their tolerance of occasional frosts as well as protracted droughts have made them important shade trees in many of the world Cs climates, as in California and the Mediterranean Basin.

14-3 UTILIZATION OF THE SELVA Nonliterate, nonagricultural people have lived in the selva since early in human history. Some were gatherers, utilizing the great variety of trees to supply them with their needs. Some were hunters as well, such as the pygmies of Africa, who used blowguns with poison-tipped darts to bring down birds or monkeys. Occasionally they also found means of trapping or killing much larger animals, including elephants that found their way into the selva. Fish generally are available in the full-running streams, and the forest supplies unusual tree poisons that, when placed in a pool, have the property of stupefying fish.

A few of the more important tree fruits became early staple elements in the forest economy, and it is believed that the first deliberate planting, the first steps in agriculture, took place in the tropical forest and consisted of inserting roots or shoots of the wild forest plants into the ground, especially the fruits, such as the plantain, mango, papaya, and breadfruit. Some of the plants with tubers or starchy roots were domesticated later and incorporated into native agriculture. Among these were the *manioc* plant (mandioca, or cassava) and the yam. One may

wonder what accident led to the discovery that removing the poisonous prussic acid in the manioc root could yield a starchy staple food and the tapioca of commerce. Although there are only a few people in the world today who still obtain their food wild from the tropical forests, many still make a living by gathering certain products that appear in world markets. The best-known of these products include *brazil nuts*, *tagua nuts* (vegetable ivory), *rattan*, and *dammar*, a vegetable gum. Rubber was once among the important products gathered wild in the Brazilian selva, but it is now almost entirely produced on plantations or synthesized in the industrial centers of mid-latitudes.

The reserves of lumber in the tropical rain forest are vast, but some major disadvantages must be overcome before they can serve as an important source of the world timber supply. At present, the principal commercial woods in the tropics are the hard, durable cabinet varieties, which, because of their great value, justify the required high costs of extraction and transportation. Such woods include mahogany, ebony, rosewood, and many others less well known. The greatest disadvantage of the lumber industry in the tropics is probably the large number of tree species. Since an acre of forest may contain only two or three trees of the type desired, large-scale lumbering methods are hardly practicable. If the forest is clean-cut, there remains the difficulty of handling and marketing such a large variety of woods. Some success has been obtained in planting *djati* (teak) forests in southeastern Asia, but, in general, the slowness of replacement is a distinct handicap (see Figure 14-6).

14-4 MESOPHYTIC BROADLEAF EVERGREEN SCRUB WOODLAND The mesophytic type of broadleaf evergreen tree vegetation differs markedly from the tropical rain forest, or selva. The trees are relatively low,

Figure 14-6 Young teak plantation, Equatoria Province, southern Sudan.
This teak plantation southwest of Juba is approximately 20 years old. The diameter of the trees at waist height is about 10 inches; hence, about 30 to 40 years must elapse before the plantation will yield logs for timber. [Van Riper]

and a continuous canopy is absent. Such a woodland is found typically in climates that have a mild, moist winter season and a dry summer (Cs). The winter rainfall, plus low evaporation, replenishes soil-moisture reserves. The winters do not have sufficiently low temperatures to halt most native vegetation growth; therefore, the principal adjustment of the vegetation is to the heat and aridity of summer.

The distribution of this woodland is mainly poleward of the low-latitude deserts. It is typical of the Cs climates but extends somewhat into the BShs type (hot semiarid climate, with most of the rain in the winter months). The largest expanse is in the Mediterranean Basin, with extensions into Turkey and Iran along the lower slopes of mountains.

Smaller areas are found in southern California, middle Chile, southernmost Africa, and southern Australia.

Climax woodlands or even near-climax woodlands are exceedingly uncommon because of excessive alteration by humans, partly owing to cutting for fuel and partly because of overgrazing by sheep or goats. Much more characteristic of the vegetation within the Cs or Mediterranean climates is a low, evergreen, sclerophyllous (narrow-leaf) shrub growth. In the area bordering the Mediterranean Sea, where it is most widely distributed, this shrub growth is differentiated into *maquis,* where the dominant woody shrubs are from 3 to about 20 feet in height, and *garrigue,* where the dominants are mainly under 3 feet. Maquis shrubs are termed *chaparral* or *mesquite* in the southwestern United States and northern Mexico. The Mediterranean maquis has a strong dominance of evergreen oaks (see Figure 14-9), which also are frequently found in the chaparral and climax woodland in California, South Africa, and central Chile. Australia is the only continent where the evergreen oaks have not become dominants. Here various species of Eucalyptus take their place in a maquislike stand of woody shrubs termed *mallee* (see Figure 14-7). The intermixture of Acacia and fire-resistant plants, such as the blackboy (Xanthorrhoea), and palmlike cycads, which are direct descendants of some of the earliest plants on earth, give the mallee a highly exotic or alien appearance.

Probably the most distinctive feature of the maquis is the specialized plant mechanisms for resisting summer heat and drought. These include hard-surfaced, narrow, waxy leaves, a thick insulating bark, a well-developed and deep root system, and viscous plant fluids. Many of them are sclerophylls (Gr: *sclero* =hard; *phyllos* =leaf). The manzanita, a

Figure 14-7 Scrub woodland (mallee) southeast of Perth, Australia. The mallee scrub woodland tends to be located on alkaline soils with a rainfall of less than 18 inches per year. Like the evergreen oak of Mediterranean regions, the dominants reproduce quickly from rootstocks following fire. *Mallee* is a local word meaning *thicket.* The spiny-looking plants are "blackboys" (*Xanthorrhoea*), closely similar to cycads. [Van Riper]

common shrub of southern California, is also able to turn its leaves edgewise to the sun to avoid direct radiation. Nearly all of the maquis shrubs are able to tap deep-seated groundwater. This is a basic feature of the evergreen holm oaks in the Mediterranean Basin region (see Figure 14-8), and their roots are able to follow crevices in limestone downward to depths of 50 to 60 feet. Reproduction by shoot growth from rootstocks is another adaptive device of many of the maquis plants. The waxy leaves and stems of the maquis are susceptible to burning when dry, and shoot propagation provides the means for rapid recovery following burning or cutting. Such shoot growth is likely to take place in clusters; hence, the maquis typically is a copse or thicket. The understory vegetation in the maquis association has a wide variety of plants, in which many of the kitchen herbs have an important part. These include thyme, rosemary, myrtle, laurel, sage, oregano, basil, and the mints. Their aromatic oils serve two functions: to increase the density of the internal plant fluids to decrease transpiration rates, and to make the plants less palatable to grazing animals. The latter is extremely important in the Mediterranean area. Other low-level shrubs in this region include the colorful rock roses (*Cistus*), which have the unique property of greatly reducing the area of their thick, hairy, leaf surfaces by wilting during the dry summer season.

Figure 14-8 *Quercus calliprinos,* **one of the evergreen holm oaks that make up much of the maquis of the eastern Mediterranean Basin.** Note the hard, narrow, somewhat spiny leaves. *Quercus ilex,* or the holly oak, is a closely related species throughout the western Mediterranean Basin. [Van Riper]

The garrigue growth (see Figure 14-9) is developed largely by overgrazing. Here the principal shrubs to persist are those that are least palatable to animals. Thorn bushes (especially leguminous varieties), thistles, strongly scented herbs, and plants with small, hard leaves are examples. Garrigue occurs mainly in areas that are relatively unsuited for agricultural use. These include steep hill and mountain slopes, especially those underlain by limestones, highly porous and droughty marls and chalks (see Figure 10-7), and sandy areas along coastal beaches. Although the greatest areas of garrigue are found in the Mediterranean region, where man and his grazing animals have been present for thousands of years, it is by no means uncommon in the frequently burned-over hillsides of southern California and South Africa.

Figure 14-9 Typical garrigue, Lebanon.
Nearly all of the plants shown here are thorn bushes, mainly *Poterium spinosum* (the rounded tufts) and *Calycoteme villosa* (right foreground). Thistles and aromatic herbs also are common. After centuries of grazing by goats and sheep, these less palatable shrubs have become the dominants in the flora.

Flowering herbaceous (nonwoody) plants and grasses make up the lowest story of vegetation and are known collectively as the *batha*. Especially common members of this floral assemblage are the bulbous plants. These sprout and bloom during the late winter and early spring, and after storing water and nutrients in the bulbs beneath the ground, lie dormant through the dry summers. Especially common examples are the lilies, cyclamens, crocuses, hyacinths, wild onions, and arums. Grass always is present and is well adapted to the seasonal change in climate, but because it is among the most preferred plants to be eaten by grazing animals, its many species find survival a continual struggle. Thistles and other low-thorn bushes are extremely common. The small flowering annuals and low perennials also form part of the batha. Wild poppies, anemones (wind flowers), the *Compositae* (daisies), oxalis, tiny wild orchids, and a host of other flowers make the batha in early spring an unforgettable sight.

Coniferous forests, or cone-bearing, needle-leaf evergreens, are an important part of the upper mountain slopes in all of the Cs areas except Australia. Temperature decreases and precipitation rises with increased elevation, so that the floral associations in the mountains include larger and denser stands of trees than on the adjacent lowlands. Conifers have a remarkable tolerance of a wide variety of climates, and many of their structures are well adapted to the seasonal changes of the Cs climates. There are many species, including the western yellow pine, sugar pine, and redwoods of California; the stone, black, and umbrellalike Aleppo pines, cypresses, and cedars of the Mediterranean Basin; and the Auraucaria pine of the Chilean Andes. In South Africa, practically all of the original needle-leaf forest cover has been destroyed, but a few patches of trees similar to the cypress can be found in isolated places above 2,000 feet. Only fragments of a once extensive series of mountain conifers remain in the Mediterranean region (see Figure 14-10).

Some of the original woodland tree species have been domesticated. The olive, the pistachio, and the cork oak have become significant elements in local agricultural economies. The long pods of the *carob* (*Ceratonia Siliqua*)[2] are a concentrated animal feed that is an important part of the exports from Cyprus.

Centuries of overgrazing on the mountain slopes of northwest Africa, southern Europe, and in the Middle East have not only altered the vegetation associations but have also tended to alter the domestic fauna, mainly by substituting goats for sheep in many of the pastoral areas. Goats are far less selective in their diets than sheep and are able to sustain themselves even on the thorny garrigue if necessary. The goat has been blamed for much of the destruction of the vegetative cover, but much of this is unjust, because many of the stony limestone slopes probably could not support any type of land use other than goat browsing or the making of charcoal. Rather, the harm is the result of the intensity of browse, a condition that is caused more by the desperation of poverty and overpopulation than by the desire for food.

[2]The *carob* tree is unusual in often being spared the depredations of grazing flocks, so that it often appears as scattered trees in the midst of low garrigue. Shepherds in the Middle East tend to keep their flocks away from the carobs—first, because the harvest of the wild pods supplies valuable winter feed, and second, because the dense, heavy foliage provides scattered and welcome shade during the extremely hot sunny days of summer. A similar tree that also is protected and that even appears in the midst of cultivated fields in western Turkey is the *valonia oak* (*Quercus aegilops*), whose enormous, shaggy acorn cups 2-2½ inches in diameter are collected and exported as a source of tannin.

Broadleaf Deciduous Forest and Woodland

The broadleaf deciduous trees are those which have broad, flat leaves like those in the selva but which shed these leaves during a particular time of the year. Like the broadleaf evergreen subformation, the broadleaf deciduous vegetation has both true forest and scrub woodland subtypes. The former is found in the humid mid-latitudes, where the leaf shedding is the result of frosts during the winter season. The scrub woodland subtype is found mainly in the *Aw* climates, which are tropical but which have a pronounced wet and dry season.

Figure 14-10   Remnant of Lebanese cedar (*Cedrus libanica*) **forest near Barouk, Lebanon.** This particular grove of cedars is increasing slightly by natural reproduction. A more famous grove near Bcharre scarcely has an opportunity to do so because it is easily reached by automobile, and the ground beneath the trees is kept bare of any seedlings by thousands of picnic lovers every summer who admire the ancient enormous trees, many of which have diameters of 6 to 8 feet. The elevation of the grove shown here is about 5,900 feet.

A distinct middle story of trees, characteristic of the selva, is not present in the broadleaf deciduous forest, although younger members of the dominant tree species are found at various heights. A distinctive feature of this forest is the mixture of tree sizes, ranging from tiny saplings to fully mature trees 2 to 4 feet in diameter. The spacing between the trees increases as the trees become larger. The root systems of the deciduous trees are much less shallow than those of the selva, and some of them have taproots which descend far below the surface. For this reason, and also because of the lower tree heights, buttressed bases are not required for trunk support. Lower trunks, as in the selva, are largely free of branches. Decay is not so rapid in this forest as it is in the selva. Scattered windfalls are found throughout the forest, and mounds mark the sites of fallen, partially decomposed tree trunks.

Vines are not nearly so common as in the selva, although they tend to become more frequent toward the equatorward margins. Poison ivy (*Rhus*), Virginia creeper (*Parthenocissus*), and wild honeysuckle (*Lonicera*) are small vines that may cover the lower trunks of trees in some areas.

14-5 HYGROPHYTIC BROADLEAF DECIDUOUS FOREST Conditions for tree growth in the summer season of the humid, mild-winter mid-latitudes are similar to those in the humid tropics. High temperatures combine with high moisture content in the soils to promote rapid plant growth. Freezing temperatures, even for short periods during the winter season, however, are dangerous to trees that contain large quantities of water in their cells. Most trees in such climates, therefore, withdraw all but a small amount of moisture from their cells, shed their broad leaves, and curtail their life processes. The period of rest varies in length, depending on the length of the frost season. Thus have evolved the mid-latitude broadleaf deciduous forests, which reach their culmination in the humid climates with hot summers and mild, short winters (*Cfa*).

The trees of the humid deciduous forest are typically shorter than those in the selva, usually ranging from 50 to 100 feet in height. The canopy overhead, although continuous, is not so tightly closed as in the selva, and considerably more light filters through to lower levels. The variety of trees is much smaller, and commonly three or four species dominate. For this reason, this forest is frequently divided into communities, termed *associations*, distinguished by the combination of dominant species. The deciduous forest in the United States, for example, includes three major associations: the oak-hickory, the birch-beech-maple, and the oak-chestnut-yellow poplar (tulip tree). The distribution of these is shown in Figure 14-11.

The most spectacular feature of many of the broadleaf deciduous forests is perhaps the brilliant coloration in the early autumn, immediately prior to defoliation. The yellow pigment, *xanthophyll,* is always present in the leaf cell structure, but it becomes visible only when the green chlorophyll begins to deteriorate and disappear in the fall. The red pigment, *anthocyanin,* on the other hand, is produced in association with sugars in the tree sap. Like most plant sugars, it is better developed with greater intensity of sunlight; therefore, the scarlet foliage is more brilliant during dry, sunny autumns. It is also chemically influenced to some extent by sudden, chilling temperatures. Individual types of trees differ considerably in the dominance of yellows or reds. The sugar maple (*Acer saccharum*) usually has much anthocyanin in its sap, associated with the high sugar content; thus, reds and orange colors predominate in the autumn leaves. The birches, beeches,

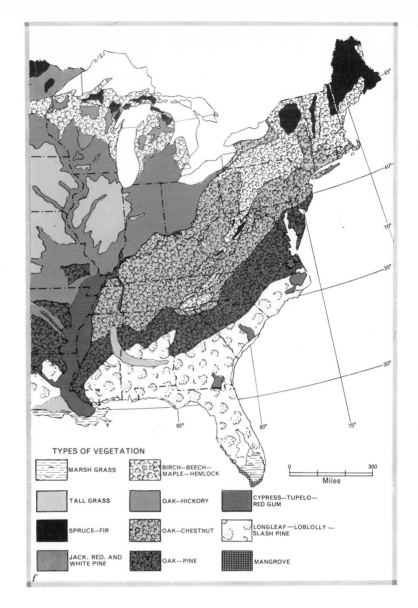

TYPES OF VEGETATION

MARSH GRASS

TALL GRASS

SPRUCE—FIR

JACK, RED, AND WHITE PINE

BIRCH—BEECH—MAPLE—HEMLOCK

OAK—HICKORY

OAK—CHESTNUT

OAK—PINE

CYPRESS—TUPELO—RED GUM

LONGLEAF—LOBLOLLY—SLASH PINE

MANGROVE

0   300
Miles

poplars, and elms, on the other hand, develop mainly yellow hues. Oaks and hickories usually show shades of brown. Certain other deciduous trees, such as the willows, do not develop bright coloration, but merely turn a grayish green before falling from the tree. The more brilliant coloration that usually is seen as one proceeds poleward is the result of the greater numerical dominance of sugar maples in the forest associations. The west coast (*Cfb*) deciduous forests usually are not nearly so brilliantly colored as those in the eastern parts of the continents.

The deciduous broadleaf forests of both Europe and Asia have been so modified by man's occupation of the land that few characteristics of the climax vegetation remain. Cultivation has removed much of the natural vegetation on the plains. In Europe, the forests that remain on the hill slopes are carefully tended, and all but the most valuable species of trees have been eliminated. Conifers have been planted in many areas to supply firewood and lumber, because of their relatively rapid growth, as compared to the deciduous broadleaf trees (see Figure 14-12). In densely populated China, the need for fuel has resulted in the removal of even young trees, except in the most isolated parts of the country.

14-6 HYDROSERE SUBCLIMAX[3] ASSOCIATIONS IN THE BROADLEAF DECIDUOUS FOREST The last stages of the hydrosere in the mid-latitude broadleaf deciduous forest are characterized by vegetation that shows many of the typical features of the

[3]The term *subclimax* refers to a plant community that has nearly reached the end of the successional stages in a sere.

climax type, but that differs mainly in the species of trees that are present. In the birch-beech-maple association of the United States, for example, swamp-forest associations usually include red maple, elm, and ash in areas where the groundwater has a neutral-

Figure 14-12 A cultivated coniferous tree plantation in the Black Forest (Schwartz-wald), southwestern Germany.   [Van Riper]

14-7 UTILIZATION OF THE BROADLEAF DECIDUOUS FOREST Broadleaf deciduous forests are utilized principally as a source of lumber, charcoal, and firewood. Most of the trees have a dense, hard wood; therefore, the lumber is best suited for furniture, flooring, the interior trim of buildings, and for such miscellaneous uses as wooden utensils, tool handles, and barrel staves. The special qualities of these hardwoods, combined with their nearness to markets and their relatively slow growth, make their lumber expensive. Today most of the furniture hardwoods are thinly sliced and used for veneering on softer, cheaper woods. Particularly prized are walnut, cherry, maple, and oak.

The Appalachian uplands contain the greatest reserves of deciduous hardwood timber in the United States, but farm woodlots supply a considerable quantity throughout the eastern part of the country. The high value of this timber has led to selective logging practices throughout most of the deciduous forest. Large areas are required to yield a steady supply of high-volume production. Thus, the hardwood lumber industry in the United States is mainly in the hands of jobbers, who contract for the removal of mature timber from a large number of local woodlot owners, mainly farmers. It has been estimated that between 75 and 125 years is required to replace a clean-cut deciduous forest; so there is little incentive for private ownership of cut-over climax forests. It is likely that there are only a few areas of full-grown hardwood forests left in the United States, and these are either in government forest reserves or in inaccessible localities. Most of the forest region is in various stages of regrowth, with the mature trees removed as they reach saw-timber size. Culling and thinning of the secondary forest, which aid in speeding the replacement of lumber, afford further income from firewood or charcoal.

to-slightly-acid reaction. Where it is highly acid, however, swamp conifers may be included, such as arbor vitae, larch, or spruce. A dominant swamp tree in the southern part of the United States is the bald cypress, unusual in being, like the larch, a deciduous conifer, or a cone-bearing tree that sheds its needles during the winter season (Figure 14-22). Associated with the cypress are the gum trees and swamp oaks. Sometimes the subclimax forms of the hydrosere are found on rolling uplands, as in the northern Appalachian uplands, because of poor internal drainage above hardpan soils. Red maple and ash are illustrative dominants in these areas.

14-8 THE MESOPHYTIC BROADLEAF DECIDUOUS SCRUB WOODLAND As one travels away from the rainy tropics, rainfall decreases, becomes more erratic, and tends to show a distinct seasonal deficiency mainly in the winter or low-sun period. Because of this, conditions for tree growth also tend to become more unfavorable. Where a pronounced dry season occurs, daytime surface temperatures become extremely high, and evaporation rates are similar to those in desert areas. Trees become stunted, as compared with those in the selva, and most of them shed their leaves during the dry season. Also, as conditions become less suited for tree growth, grass begins to cover the surface, aided in its competitive struggle with tree growth by man's use of fire. The mesophytic broadleaf deciduous scrub woodland represents a transition between the tropical grasslands (savannas) on one hand and the true tropical forests on the other.

The general appearance of the deciduous scrub woodland is somewhat similar to the evergreen scrub woodland of the Cs climates. The principal difference is that nearly all the trees are deciduous. The soils of the Aw regions retain less moisture than those of the Cs climates, and there is no cool winter season with fairly reliable rainfall and low evaporation rates to store water in the soil for plant use the following summer. The trees cannot obtain enough moisture to support continued tree growth during the dry season, even with various specialized methods of reducing transpiration.

The typical stand of trees in the deciduous scrub woodland varies from an open park or orchardlike cover to dense tangles of brush containing occasional grassy openings. The transitions from the deciduous scrub woodland to the savanna, in one direction, and to the true forest, in the other, are gradual. The boundary between the open woodland and the savanna is especially difficult to define, since the only mark of separation is the relative dominance of trees or grass in the landscape. The distinction is somewhat clearer with respect to the forest border, for the continuous canopy of a true forest is absent in a woodland.

The more open stands of the deciduous scrub woodland are composed of trees that are usually between 20 and 40 feet high. There are relatively few young trees and little undergrowth in this area. The influence of fire is evident in the absence of lower branches on the trees and the spreading, flat-topped crowns. Grass fires can sweep through such a forest without having much effect on mature trees. In such a woodland, one seems to be always in the midst of a clearing, with trees becoming closer together in the distance. Usually one can see through the woodland for a distance of about 100 to 200 yards. One characteristic genus of trees in the more open stands of the deciduous scrub forest is the *Acacia,* a flat-topped, sometimes deciduous, thorny tree. The remarkable adaptability of this hardy scrub tree, however, gives it a range throughout the subhumid to semiarid tropics, and subtropics almost to the desert margin. In Africa and southeastern Asia, clumps of bamboos are found.

The denser stands of this scrub woodland contain small trees and many low, woody, thorny shrubs. The smaller bushes, requiring less water, are more likely to be evergreen *sclerophylls,* which are like the chaparral or maquis of the evergreen scrub woodland of Cs climates. They vary in height but are usually under 20 feet. This denser, thornbush variety of the deciduous scrub woodland is found in the drier parts of the subformation. Sometimes xerophytic, leafless plants such as the Euphorbia are encountered in this area.

The world map of vegetation subformations (Figure 14-1) does not adequately indicate the distribution of this woodland type,
mainly because the details of the transitions
between the humid forests and the grasslands
never have been worked out carefully. Only
the larger recognized areas are shown. Unfortunately also, little correlation is possible
between climate and this woodland subformation, because in many places dry-
season burning has destroyed much of the
woodland, replacing it with savanna grasslands. For this reason, the selva passes directly into the savannas, particularly in much
of central Africa and in Brazil.

In South America, the two areas of this
woodland type shown in Fig. 14.1 include the
caatinga scrub woodland of northeastern
Brazil and the Chaco area of Paraguay and
northern Argentina. For the most part, the
caatinga consists of smaller thorn trees than
usually occur in this subformation. The erratic rainfall of this part of Brazil is shown
by the presence of cacti among the thorn
trees and bushes. The Chaco area is divided
between marshy grasslands in the low, flat
depressions and typical open scrub woodland just above flood level. Much of this area
contains the *quebracho* tree as a member of
the woodland association.

In Africa, the largest area of scrub woodland lies in a broad belt extending from central Tanganyika southwestward through
Northern Rhodesia into eastern Angola.
Patches of it can also be found on the broad
interstream areas bordering the Congo Basin
and in the northern parts of the Guinea coast
countries (Nigeria, Ghana, and the Ivory
Coast). Acacias dominate the African open
woodland. A considerable portion of peninsular India is in deciduous scrub woodland,
and the interior plains of Burma, northeastern Thailand, and west-central Indochina
have large areas of this subformation. In
Australia, the deciduous scrub woodland
corresponds to what is locally termed the
Brigalow scrub (see Figure 14-13), a variety
of *Acacia*, which extends along the northern
coast and southward within Queensland,
excluding a narrow coastal zone of tropical
forest.

The deciduous scrub woodland is not a
source of lumber because of the small size
of the trees. It does, however, yield a number
of valuable products, the principal ones being
waxes and gums. In South America, the
carnauba palm supplies a hard wax which
is used in the manufacture of phonograph
records. Gum lacquer and waxes are also
important products of this woodland in the
lowlands of Laos in Indochina. Sandalwood,
an aromatic wood that is especially prized
for fine cabinets, has long been sought in the
scrub woodlands of the southwest Pacific
Islands and in Australia, although little of it
remains. Another important product is quebracho bark, from the Chaco area of South
America. It has an unusually high tannin
content and is used in the tanning of leather.

Needleleaf Evergreen Forests

The needleleaf evergreen forests are composed largely of conifers, or cone-bearing
trees. These trees are the most widely distributed of the major types of trees and extend from the tropics to subarctic regions.
They also are found in widely differing conditions of moisture availability, from the
humid west coasts of upper mid-latitudes to
semiarid high plains, such as the Colorado
Plateau, and from thin mountain soils barely
a foot thick to water-saturated swamp soils.
The feathery casuarina tree borders the
beach strands of many tropical beaches, and
the slender spires of alpine firs abut against
the timber line of high mountains thousands

of feet above sea level. For this reason the conifers, as a type of vegetation, cannot be correlated with any particular climate. For one reason or another, however, large areas of the world are made up almost exclusively of these needleleaf evergreen trees.

The principal features of the conifers include, besides their unique evergreen leaf form and seed cones, a thick, insulative bark; a thick, pitchy sap; a broad, shallow root system; and a tolerance for wide ranges of soil acidity. These properties enable the conifers to become dominant under varying conditions. Their dominance in subarctic regions is mainly due to their resistance to low winter temperatures and highly acid soils. In other areas, tall pines, free of lower branches, are dominant because of their resistance to fire. Sandy and gravelly soils in humid regions sometimes contain conifers as a dominant type of vegetation in an early stage of plant succession because these trees can stand irregular periods of almost complete soil dryness. Their shallow root system enables them to obtain anchorage and nutrient solutions from thin, rocky soils on mountain slopes.

The subdivisions of the needleleaf evergreen forests are based on moisture availability. The first, or hygrophytic, group includes the forests where humid conditions exist for nearly the entire year. These forests are composed of tall trees ranging from 80 feet to 300 feet. The second, or mesophytic, group is found where water is available for only part of the year. This second group also is limited to the forests of high latitudes, where the plant forms are ideally suited to the long periods of low temperatures and

frozen soils, snow cover, and oblique solar radiation. The trees in the mesophytic group are not so large as the hygrophytic types, and they are usually less than 80 feet tall. The xerophytic, or drought-resistant, conifers do not cover a large total area, but patches of woodland containing them are found in the semiarid lands of most continents, especially at elevations over 3,000 feet.

All three groups of the needleleaf evergreens are found in mountain areas, and mountain conifers are discussed later under "Mountain Vegetation."

14-9 HYGROPHYTIC NEEDLELEAF EVERGREEN FORESTS The tall forests of conifers found in the warm humid regions of the world cover broad areas in the west coast *Cfb* climates of North America (see Figure 14-14) and Europe; in the Atlantic–Gulf of Mexico coastal plain of the eastern and southern United States; and in widely scattered patches in the low latitudes, including upland plains in southeastern Asia, Luzon, northern Sumatra, New Guinea, and southeastern Brazil. The sandy and gravelly

Figure 14-14 Mature stand of trees in a typical west coast coniferous forest. Note the ground flora of large-leafed plants and ferns, adapted to minimum light conditions and high humidity on the forest floor. [Weyerhaeuser Company]

coastal plains of Honduras and Nicaragua also have a pine flora, and similar pines are found in places in the West Indies.

The west coast conifers. Conifers are abundant in many parts of the west coast marine climates (*Cfb*) of the world, including western North America, southern Chile, southwestern France, coastal Norway, and parts of New Zealand.

In North America, the western conifers extend from northern California into Alaska along the coast. At higher elevations, they extend far southward in the Sierra Nevada. Inland, this forest extends to the lower eastern slopes of the Cascade Mountains in the United States (see Figure 14-15). In central British Columbia and northward, however, the same coniferous forest prevalent along the coast extends eastward to the Canadian Rockies.

The conifers of northwestern North America include several species that are highly important as sources of construction lumber. The most important is the Douglas fir (*Pseudotsuga*), whose tall, straight trunks frequently rise over 200 feet above the ground and have diameters up to 15 feet. Sitka spruce, western hemlock, red and white cedar, and the famous redwoods (*Sequoia sempervirens*) are other important varieties. Farther south, the conifers merge with the mountain conifers of California's *Cs* climate, and western yellow pine and sugar pine become dominant. In the central Sierra Nevada the world's largest trees, the giant redwoods (*Sequoia gigantia*), are found. Throughout the Pacific Northwest region, the conifers are huge. The forests usually are comparatively free of underbrush, although along the western slopes of the Olympic Range in Washington, a rather dense stand of low broadleaf evergreen shrubs and ferns is found, reflecting the constantly wet, dripping, fog-bound environment. Here the larger trees are covered with moss.

The distinct zonation of different coniferous associations in this region reflects a number of environmental contrasts (see Figure 14-15). The marine belt along the coast supports several species, including Port Orford cedar, red cedar, Douglas fir, the redwoods, and Sitka spruce. Most of the western slopes of the Cascade Mountains and the Canadian Coast Range have a climax of western hemlock and white cedar, with Douglas fir as a fire-succession subclimax (see Figure 14-16). The broad extent of Douglas fir indicates that extensive fires must have occurred frequently throughout this region for centuries. Because of its extraordinarily thick bark, the Douglas fir remains relatively undamaged by ground fires. The long mountain slopes provide updrafts of air and favorable conditions for forest fires, which are extremely difficult to check in these areas, even with modern fire-fighting facilities. Mountain areas also are susceptible to lightning, which was undoubtedly responsible for many of the fires of past centuries that, once started, burned over large areas. Ravine bottoms are less vulnerable to fires which tend to sweep up or along slopes, and these contain the densest stands of the hemlock-cedar association.

Elevation, exposure, and precipitation also play distinctive parts in producing the zonation of associations in the Pacific Northwest. An entirely different climax association of western white pine, yellow pine, and larch becomes dominant on the drier slopes east of the crest of the Cascades. Lodgepole pine replaces Douglas fir in the fire succession. Spruce and fir are found at high elevations throughout the mountains and, because of the decreased temperatures in such areas, the vegetation resembles that of the northern coniferous forest.

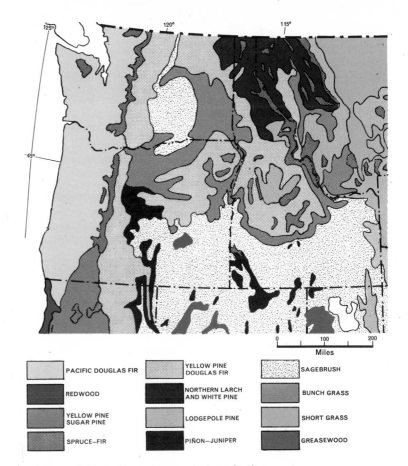

PACIFIC DOUGLAS FIR	YELLOW PINE DOUGLAS FIR	SAGEBRUSH
REDWOOD	NORTHERN LARCH AND WHITE PINE	BUNCH GRASS
YELLOW PINE SUGAR PINE	LODGEPOLE PINE	SHORT GRASS
SPRUCE—FIR	PIÑON—JUNIPER	GREASEWOOD

Figure 14-15 Forest associations in the northwestern United States. [U.S. Dept. of Agriculture; *American Atlas of Agriculture*]

It should be noted that the humid west coast climates, characterized by mild winters and cool summers, do not result exclusively in coniferous forests. Some deciduous forests can be found in the Pacific Northwest, frequently on alluvial soils. These forests also were predominant over most of northwestern Europe centuries ago. Maples (*Acer macrophyllum* and *circinatum*), red alder (*Alnus oregona*), and black cottonwood (*Populus trichocarpa*) are the principal species along the coasts of Oregon, Washington, and British

Columbia. Alder in most parts of the world is a small tree, but in the Pacific Northwest it becomes a true forest tree and reaches diameters of up to 2 feet. Deciduous forests also are encountered in southern Chile, where the conifers are restricted mainly to the slopes of the Andes.

Some broadleaf evergreens are encoun-

Figure 14-16 Understory of western hemlock developing beneath a mature stand of Douglas fir in the Willamette National Forest, Oregon. The Douglas fir is a fire-succession tree in the western associations. [U.S. Dept. of Agriculture]

tered in the west coast marine climates. The mild, humid winters undoubtedly are responsible for the fact that these usually tropical forms are found so far poleward. They are important parts of the local flora along the immediate coastal area in Oregon, Washington, and southern British Columbia. One of the most interesting members of this evergreen flora is the madrone (*Arbutus*), a broadleaf tree with a scaly trunk, somewhat similar to the sycamore but reddish in color. Usually from 20 to 30 feet high, it may reach 70 to 80 feet and have diameters of up to 2 feet. Giant rhododendrons are representative of this flora in the coastal areas of northwestern Europe, and a still more unusual and distinctive feature is the sprinkling of palms that can be found as far north as Scotland.

The wide variety of tree types in the humid west coast climates is caused by the availability of sensible heat and water throughout the year. Winter temperatures are not low enough to halt plant growth. Highly specialized mechanisms for resisting low temperatures or rainfall thus are not required.

The west coast conifers afford some of the most concentrated reserves of softwood timber, which is used mainly for construction lumber. Redwood and cedar have unusual resistance to weathering, hence are important for house siding and shingles. Regrowth is unusually rapid in the mild, humid climate of the west coast, and although a great many years are required to replace the full-grown forest giants, trees of saw-timber size generally develop in 25 to 50 years. The large size of the trees and the rough terrain make lumbering in this region a large-scale industry. Scientific forest management is the rule, largely because there is big business at stake. Lumber companies have to be large to operate, and the general long-view approach of corporation management aids in promoting stability and efficiency. The large area in public-owned forest reserves, in which leased cutting under supervision is permitted, is also helping to stabilize the forest economy of the Pacific Northwest region. Today, despite large-volume cutting, the region is replacing most of the forest wealth that is removed.

The southeastern United States conifers. One of the largest areas of hygrophytic needleleaf evergreen forests in the world is found on the Atlantic–Gulf of Mexico coastal plain, stretching from Cape Cod, Massachusetts, and northern New Jersey, southward into Florida, and thence westward into Alabama and Mississippi. After a gap of about 100 miles at the Mississippi River, it reappears in Texas. This forest area consists almost entirely of species of pine, including southern longleaf and shortleaf pine, pitch pine, loblolly pine, and Jersey scrub pine. The most widely distributed and most valuable species for lumber and naval stores is the southern longleaf, or yellow, pine (*Pinus palustris*) (see Figure 14-17).

The pines are closely related to the distribution of sandy or gravelly coastal plain soils. The high summer evaporation rates of this region, combined with the loose, permeable soils, are suited to the xeric (drought-resistant) characteristics of conifers. As was indicated in Chapter 13, burning may have been a major influence in effecting the dominance of the pines. A mixed oak-pine forest borders the coastal plain coniferous forest on somewhat heavier soils, and this indicates that the deciduous oaks may be part of the climax vegetation for this region. The same varieties of pine that are dominant on the coastal plain and in the mixed-forest zone can also be found at higher elevations in the Appalachian uplands. Regardless of their successional position, the United States coastal plain pines cover a sizable area, even on a global map. A comparable area of coni-

fers is found in the Landes district of southwestern France, south of Bordeaux. Here too the pines are closely related to the distribution of sandy soils.

The pine forests of the southern and southeastern United States constitute one of the most valuable timber areas in the country. The southern longleaf pine is an important source not only of turpentine and resin but also of lumber for both flooring and construction, and its regrowth is rapid. The region has also been important as a major supplier of paper pulp to be used in making kraft paper ever since the introduction of the Herty process for utilizing these resinous woods.

14-10 THE MESOPHYTIC NEEDLELEAF EVERGREEN FOREST (TAIGA) The mesophytic needleleaf evergreen forest, otherwise known as the *taiga,* or northern coniferous

Figure 14-17 Mature southern yellow, or longleaf pine (*Pinus palustris*) in Bronson State Park, Texas. Note the absence of an understory of shrubs. Such open forests, controlled by light burning of ground plants, make fair grazing land for beef cattle in the southern United States. [U.S. Dept. of Agriculture]

Figure 14-18 Birch-pine association in the taiga near Joensu, Finland. This association of white birch with Scotch pine (*Pinus sylvestris*) is extremely widespread in Scandinavia. [Van Riper]

forest, constitutes the largest subformation of all the global forests. It extends across the continents of the northern hemisphere from Alaska to Labrador and from Scandinavia to the Pacific coast of the Soviet Union. Although the general characteristics of the taiga are similar on both continents, the species of trees are different. In Canada, spruce (*Picea*) and fir (*Abies*) are dominant throughout the taiga. The pines (*Pinus*), including white pine, red pine, and jack pine, are found

principally in the southern part of the region and appear to be localized types that occur on unusually sandy soils or in fire-succession communities. Arbor vitae (*Thuja*) and larch (*Larix*) are associated swamp conifers. The latter is unique in being a deciduous needle-leaf tree. The taiga of Eurasia differs in having a much wider latitudinal distribution of pine, with Scotch pine predominating in Europe and the stone pines in the east. As in Canada, the pine tends to be found on sandier soils or on thin, stony knolls. Spruce is common throughout the Eurasian taiga, as is the fir, and their slender, tapering spires give the northern coniferous forest one of its most characteristic landscape features.

The taiga conifers are not large trees, usually being about 30 to 50 feet in height and having diameters of from 8 to 12 inches at maturity. Tree heights progressively decrease poleward, and those of saw-timber size are found mainly along the extreme southern portions. The forest is remarkably homogeneous with respect to the types of trees present, one or two species dominating over wide areas (see Figure 14-18). The density of the foliage intercepts most of the sunlight; therefore, the dark, humid forest floor is suitable only for small herbaceous plants and fungi that require little light. Among these forest-floor dwellers are wintergreen (*Gaultheria*), wood sorrel (*Oxalis*), trailing arbutus (*Epigaea repens*), ground pine (a club moss), and ferns. There are a number of terrestrial orchids, including the beautiful "lady's slipper" (*Cypripedium*). Most of these epiphytes are small and are confined to the ground surface. Lichens and mosses are abundant and can be found on nearly every tree. The extremely slow decay of dead wood results in a dense tangle of fallen trees, crisscrossed together. Cross-country travel on foot through the taiga is extremely difficult. This is true especially in spruce and arbor vitae swamps, where the trees are close together. Branches begin a short distance above the ground, although some of the larger pine forests have long lower trunks that are branchless.

As is characteristic of forests that have only one or two dominant species, the typical taiga is uniform in height and in density of stand. Seedlings have little chance to grow unless the canopy is broken by the death or destruction of any of the mature trees. Regrowth is unusually slow, being retarded by the long winter season, and it becomes progressively slower toward the north. In the south, about 30 to 40 years are required to develop trees with diameters of 8 to 10 inches. In the far north, along the border of the tundra, fully mature spruce and birches only a few feet high can be found.

Broad interruptions in the continuity of the taiga are formed by fire-succession forest communities, which include such broadleaf deciduous trees as white or paper birch (*Betula alba*), quaking aspen (*Populus tremuloides*), and fire or pin cherry (*Prunus pennsylvanica*). These postfire deciduous enclaves may persist for years because of the slow growth of trees in the northern climates and also because of the destruction of humus in the topsoils. In the Soviet Union, these patches of broadleaf deciduous trees among the conifers are referred to as the *white taiga* because of the light-colored leaves and trunks that contrast so markedly with the darker conifers. Spring and fall are threatening seasons in the taiga because of fire. When dry, the resinous leaves and branches of the conifers are virtual tinder sticks. Unlike the pine or Douglas fir forests in lower latitudes, the typical taiga conifers cannot withstand a ground fire, because their branches begin so close to the ground. In isolated areas, forest fires may burn for weeks until they are extinguished by heavy rains.

The taiga bogs. Hydroseres in the form of bogs are common throughout a large part of the taiga of both Canada and Eurasia. Much of the taiga was glaciated, excluding only the Soviet Union section east of the Yenesei River and parts of central Alaska. Poor drainage, caused by glaciation, resulted in many small swamps, lakes, and ponds. Perhaps the most frequent hydrosere is that associated with the gradual filling of small lakes and ponds with vegetation, producing a typical *muskeg bog* (see Figure 14-19). All stages of succession can be found in such bogs—the early stages, in which the mosses and feathery cranberry (*Vaccinium*) adjoin the margins of open water; the middle stages, characterized by leatherleaf (*Chamaedaphne*) and Labrador tea (*Ledum*), both of which are coarse, hard-

Figure 14-19 A taiga bog north of Rovaniemi, Finland. This small glacial lake is gradually being filled with vegetation in a typical hydrarch succession. Note that spruce is dominant north of the Arctic Circle in Lappland instead of the pine that prevails farther south. [Van Riper]

leaf, woody shrubs; and finally the later stages, in which the subclimax spruces, firs, and larches begin to establish the forest.

Sphagnum moss usually is present in all the successional stages of the taiga hydrosere. This is a tough, highly acid, stringy moss that is usually grayish in color but is sometimes reddish or light green (see Figure 13-7). Its high acidity makes it virtually sterile to most bacteria; hence, it decomposes slowly. Years ago it used to be gathered and sold as anti-septic dressing material for wounds. The growth of a sphagnum moss cover on some of the flat taiga plains following forest fires or clearing often has resulted in "water-logging," or the establishment of a retrogres-sive hydrosere. The reason for this is that the moss absorbs precipitation and not only keeps it from evaporating but also prevents it from running off into the subsoil. In time

Figure 14-20   A cross section of an Arctic bog in northern Norway.   The peat is cut, dried, and used as a low-grade fuel. Note the cotton grass (*Eriophorum*) in the foreground, which is widespread on poorly drained, acid soils throughout the arctic and subarctic regions of Canada and Eurasia. [Van Riper]

the water content may increase to such an extent that a typical bog is formed. The greater water content and moss growth in the center of the bog cause a slight doming or convex shape to the bog surface. Continued heightening at the center eventually leads to the broadening of the poorly drained area, an invasion of adjacent well-drained forests, tree destruction, and a reversal of the normal successional direction. In some parts of northern Europe, these bogs are sources of peat, which when dried provides a low-quality fuel (see Figure 14-20).

The taiga riverine meadows.   Flat flood plains of rivers which are flooded in the spring, due to ice jams, are not usually en-countered in the taiga of North America, because glaciation interrupted the normal stages in the erosional cycle of streams. In the eastern Soviet Union, however, where glaciation did not take place, such plains frequently are present. The seasonal flooding prevents the establishment of forest vegeta-tion, and usually coarse marsh grasses and sedges form strips of meadowland along the stream flats. Narrow equivalents of these can be found in the North American taiga along small streams which are flooded as the result of damming by beavers. Frequently the meadow grasses may grow as tall as a man. When drainage is restored, such meadows usually are replaced by thickets of speckled alder (*Alnus*), prior to the reestablishment of the forest. These alder thickets are notori-ously dense and difficult to penetrate, as many fishermen know who have followed trout streams in the "north country."

Utilization of the taiga.   Pines constitute the major source of lumber in the taiga, but they reach sizes suitable for saw timber only along the extreme southern margins. The white and red pine lumber of northern New England

and the northern Great Lakes states was the source of the massive lumber supplies that were needed in building homes during the settlement of the central part of the United States between 1840 and 1900. Some of these pine forests were magnificent stands of tall, straight-grained softwood trees, 3 to 5 feet in diameter. The indiscriminate cutting of this forest resource and, later, the repeated burning of the pine plains, until their capacity for regrowth was almost destroyed, constitute one of the most unfortunate episodes in the history of American forest utilization. The removal of the fire hazard over wide areas is beginning to produce signs that some of these magnificent forests might once again replace themselves, although probably not for another 75 to 100 years. The greatest reserves of taiga pine timber today lie in the southeastern Soviet Union, northern Korea, and northeastern Manchuria.

The greatest resource of the taiga today is pulpwood, or wood which is destined for paper manufacture. Nearly all the conifers can be utilized in this way, as well as many of the deciduous secondary species, such as aspen. An interesting potential resource of the taiga lies in the possibility of using the softwoods for the production of edible wood sugars by chemical processing. In Norway and Sweden these trees have already become an important source of animal feed, in the form of a molasses syrup produced as a by-product of the paper industry.

14-11 XEROPHYTIC NEEDLELEAF EVERGREEN SCRUB WOODLAND An open stand of low, scrubby needleleaf evergreens frequently is encountered at high elevations (3,000 to 7,000 feet) in semiarid regions. Because of the lower temperatures associated with the higher elevations, evaporation is not so great as in the lowlands in such climates, and drought-resistant varieties of scrub con-

Figure 14-21 Typical juniper-sagebrush association near Montrose, Colorado.
The juniper is a type of xerophytic conifer widely distributed in the higher elevations of dry lands throughout the northern hemisphere. [U.S. Dept. of Agriculture]

ifers develop a parklike type of cover (see Figure 14-21). The tree most frequently encountered in these areas is the juniper (*Juniperus*), and it is widely distributed in the dry uplands of both North America and Asia. Most varieties of junipers are low trees 15 to 30 feet high, but occasionally they are bushlike and form dense thickets. In parts of central Asia, juniper trees may reach heights of 50 to 60 feet. In the southwestern United States, juniper often is associated with piñon pine, another gnarly, low conifer. Low, xerophytic shrubs, like sagebrush and greasewood, also accompany the scrub conifers. In the eastern United States, a variety of juniper (*Juniperus virginiana*), inaccurately termed red cedar in some areas, is often found in pastures and hillsides where bedrock lies a few inches below the surface.

This coniferous scrub woodland is not found over broad areas. Instead, its distribution is patchy, corresponding to intermediate slopes of mountains in dry regions and to some parts of the high plains. In the United States, the largest areas are immediately east of the Cascades and Sierra Nevada in southeastern Oregon and California and in some of the higher sections of the Colorado Plateau.

Because of their low, gnarly, small trunks, these trees are of little value as lumber. The berry of the juniper supplies an oil which is used as a flavoring extract in the manufacture of gin. Piñon pines supply a small edible nut that is in some demand locally in the western United States. Probably the most important use of these scrub trees is as shade for range animals during warm summer days. The juniper-pine areas are favored as summer pasturage for sheep, goats, and cattle, which graze or browse on the low grass and shrub vegetation between the trees.

14-12   BROADLEAF CONIFEROUS FORESTS   A large percentage of the conifers of the southern hemisphere differ from those north of the equator in having broad, straplike leaves or scaly protuberances from the branches instead of needles. The most widely distributed genus is *Auraucaria,* which comprises most of the conifers in the Andes but is also found in parts of northeastern Australia and eastern Indonesia. The kauri pines of New Zealand and Australia (*Agathis spp*) are tall, straight trees 3 to 6 feet in diameter; originally they were widespread in New Zealand but have been almost removed by cutting. A similar genus (*Dammara*) is scattered throughout Indonesia and Malaya. *Dammar gum,* a resin that is widely used in varnishes, is derived from this tropical broadleaf conifer. Another interesting example within the Auraucaria genus is the

"monkey puzzle" tree (*Auraucaria imbricata*), one of the Peruvian Andes "pines," whose leaves consist of sharp bracts that point downward and grow outwards from the stem. Its name was derived from the reported claim that monkeys left them alone and never climbed them. Their odd appearance has caused them to be carried over the world into mild, frostless climates for use as ornamental trees. They are found in many front lawns along the west coast of the United States from Los Angeles to Vancouver.

The conifers were among the first seed-bearing plants to evolve from the lower plant forms; they date back to the Permian period. The different evolutionary direction taken by the southern hemisphere conifers in developing their leaf structures may well be related to the continental drift that is believed to have taken place in late Mesozoic times. The lesser continentality or lower seasonal range of temperature of the southern continents may have been a factor in favoring the broader leaf form, in that the needle leaf was an adaptation to heavy snowfall and severe winter temperatures. None of these forests has sufficient area to be distinguished on a world map.

Needleleaf Deciduous Forest

Not all needleleaf trees are evergreen. A few of them shed their needles during the winter season. The principal tree of this type is the larch (*Larix*), which has a wide distribution in the northern forest region of North America and Eurasia. In North America, however, the most common variety in the northern coniferous forest is the black larch, or tamarack (*Larix laricina*). It is a swamp tree and is usually associated with spruce in bog associations. Western larch (*Larix occi-*

Figure 14-22 Bald cypress (*Taxodium distichum*) near Lafayette, Louisiana. Note the "knees" that are extensions of the roots above water and the festoons of Spanish moss, an epiphyte that thrives on the branches and trunks of the trees. This tree, not a true cypress, is a common swamp tree on the Mississippi bottomlands and in poorly drained sections of the Atlantic and Gulf Coast coastal plains. It sheds its needles during the winter season and is thus a deciduous conifer. [Standard Oil Company of N.J.]

dentalis), however, is an upland tree of magnificent proportions, growing up to 200 feet or more in height, with a trunk diameter of 3 to 4 feet. It belongs to the giant coniferous forest associations of the Pacific Northwest region. A smaller alpine larch also is found in the western mountains above 8,000 to 9,000 feet. Another deciduous needleleaf tree of North America, and one that grows in an entirely different environment, is the bald cypress (*Taxodium distichum*), a stately tree of the swamps within the Mississippi Valley south of Illinois and along the Atlantic–Gulf of Mexico coastal plain south of Delaware (see Figure 14-22). This cypress is able to thrive in standing swamp water and is unique in having protuberances on the roots termed "knees," which rise several feet above the ground or water level.

The above deciduous conifers of North America are only subsidiary members of climax forests, or successional species in hydroseres, and hence are not indicated on the global map of vegetation. The only area of this subformation indicated on the map is a large expanse immediately south of the tundra in the eastern Soviet Union. Two varieties of larch (*Larix sibirica* and *L. dahurica*) are dominant species in mature forests within this region, and the evergreen conifers here, such as spruce and pine, are subsidiary in the climax association.

The dominance of the larches in the Soviet Union probably is related to the extremely low winter temperatures and to permanently frozen subsoils (permafrost). The larches, by shedding their needles, are able to withstand temperatures from 50° to 80° below zero. These trees also have an extremely shallow root system, which enables them to derive nourishment from the narrow soil zone which lies above permafrost. Spruce (*Abies*), which has a similar root system, is found along with the larches, but it is more restricted to fertile soils, such as those developed on alluvium in the region. It is noteworthy that the Siberian larch region broadens toward the east, as do the areas where permafrost is found.

The wood of the larches is remarkably heavy and coarse. It does not make good construction lumber, but it is a valuable source of railroad ties, mine props, etc., because of its resistance to decay in the presence of water. Also, except for the giant larches of northwestern North America and those which thrive on some of the more fertile Russian soils, the larches do not become large trees of saw-timber size. The bald cypress wood of the southern United States is especially prized for use in house siding and shingles because of its resistance to weathering.

The Mixed Broadleaf Deciduous-Coniferous Forest

The northern coniferous forest, or taiga, merges gradually with the broadleaf deciduous forest along its equatorward margins, producing a mixed broadleaf deciduous-coniferous transitional forest, which is indicated on the world map of vegetation. The mixture of trees may consist either of a combination of broadleaf deciduous and needleleaf evergreen trees in the same forest or of homogeneous patches of one intermingled with areas of the other. Strong contrasts in local environmental conditions, such as soil texture and drainage, generally are responsible for the latter type of mixture. Sandy soils in this transitional subformation,

for example, usually support a pine forest cover, whereas the clay loam soils of ridges support a deciduous broadleaf cover with a sprinkling of conifers. Spruce and fir are found mainly in swamps. In the northern United States, the dominant deciduous trees in this mixed forest zone include maple, yellow birch, and beech. The oaks, beech, and hornbeams (*Carpinus*), on the other hand, are the most common in Eurasia. An interesting tree in the northern United States is the hemlock (*Tsuga*), which is interspersed with the maple-beech-birch association, particularly on cooler, more moist, north-facing slopes. It appears to be a relict of a cooler climatic era.

The world map of vegetation formations and subformations (Figure 14-1) indicates that the major areas of the mixed deciduous-coniferous forest are found in the northern hemisphere.

The utilization of this mixed subformation of forests depends on the dominance of the types of trees present. Except for the swamp conifers, the needleleaf evergreens grow much larger than their counterparts in the taiga, and their soft woods are valuable for construction lumber. The deciduous broadleaf hardwoods have uses similar to those described in Section 14-7.

The Semideciduous Broadleaf Forest

The semideciduous broadleaf forest is a transitional subformation that borders the broadleaf evergreen forests of low latitudes in many places. Most of the trees in this subformation are broadleaf evergreens, but some are broadleaf deciduous varieties.

Rainfall is the principal climatic variable in the tropics, and wherever a dry period occurs, distinct adjustments are made in the vegetation. If the dry season is short and excessive amounts of rain fall during the rainy period, most plants can pass through the dry period without requiring a leaf-shedding, or deciduous, period. In such areas, however, there are usually some members of the rain-forest community that are more sensitive to drought than others; hence, the forest generally contains a sprinkling of broadleaf deciduous varieties. The principal effect of such deciduous trees is to produce interruptions in the forest canopy, so that undergrowth becomes much more dense because of the greater availability of light at some times of year. These semideciduous jungle forests are located on the windward coasts and mountain slopes of monsoon Asia, along the eastern edge of the Brazilian uplands, and in parts of the Guinea coast of Africa. The bamboos are especially abundant in the understory vegetation of these forests (see Figure 14-23). Teak is one of the representative deciduous broadleaf trees in the forests of southeastern Asia and is a valuable timber tree (see Figure 14-6).

In the poleward extensions of the humid tropics and along the eastern margins of the continents, broadleaf deciduous trees may be mixed with the broadleaf evergreen types, not because of a dry season but because of the occasional low temperatures during the winter season. Some of these subtropical mixed forests are found in southern China, Formosa, southern Japan, and southern Brazil. A sprinkling of broadleaf evergreen trees in a few areas containing dominantly deciduous forests along the Gulf of Mexico coast in the southern United States indicates that the transition from the rain forest to the subtropical deciduous forest is not a sharp one. Here the live oaks and southern magnolias are green throughout the year and represent the more resistant exotic outliers of the tropical broadleaf evergreens. Were it

Figure 14-23   A clump of bamboos in the bush savanna country west of Juba, southern Sudan.   Bamboos belong to the grasses but are normally most common in tropical semideciduous forests. The savanna in this area undoubtedly once was forest but now has reverted to savanna because of cultivation and seasonal burning. Some wild bamboos may reach diameters of a foot or more. [Van Riper]

not for the Gulf of Mexico and the Caribbean Sea, this general area would undoubtedly have a much larger extent of this transitional subformation, such as exists in southeastern Asia. The broadleaf semideciduous forest has sometimes been termed a *monsoon* forest because of its dominance in southeastern Asia.

The Grasslands

14-13 THE GRASSES AS PLANT COMPETITORS Simple, unspecialized forms of life generally tend to have the longest lifespan, the widest distribution, and the greatest resistance to the cataclysms of nature. Although the grasses are neither the simplest nor the least specialized form of plant life, they are far less complex than trees. Probably the greatest specialization that they have developed is the property of providing mobility for their seeds. Some of the seeds are tiny and light, enabling them to be carried aloft on wind currents.[4] Others have tuftlike appendages which act as tiny parachutes, aiding in their air travel. Still others are equipped with small hooks, which cling to animals and human beings, thus "hitchhiking" rides to distant sites. The grasses, therefore, have an unusually wide geographic range. Combined with the mobility of the seeds is the great durability of the plants, due primarily to the relatively large percentage of the plant structure that lies below ground level, where it can be protected somewhat against the variable conditions of the atmosphere and the exploitive activities of both animals and men.

Grasses have a large variety of forms and, with the possible exception of the *Compositae* (asters) and orchids, have a greater number of species than any other family of plants. This is to be expected because of the varied kinds of environments in which they are found. This variety can be observed even among our selected cultivated grasses, ranging from food grasses such as oats, wheat, corn, rice, and sugar cane to the bamboos. Among the properties common to all of them are bladelike, opposite leaves attached to a jointed stem by means of a sheathlike junc-

ture. The roots are finely divided but well developed.

The mobility of grass seeds and the general durability of the plants themselves explain why grasses are found in nearly all environments, from the polar wastes to the equator, from deserts to periodically flooded tidal marshes. Even in the depths of the selva, which offers the most ideal conditions for the dominance of trees, members of the grass family, the bamboos, are found. Only the lichens and mosses, simpler than the grasses, have a wider distribution among the many forms of macroplant life.

Not only are grasses found nearly everywhere, but they dominate the plants covering about 40 percent of the land area of the world. The regions included in this 40 percent are far from uniform with respect to climate, soil, or any other environmental factor. Grass tends to dominate areas that are unsatisfactory for tree growth but that provide favorable conditions for plant growth, at least in the topsoils, for part or all of the year. The conditions that most usually prevent tree growth include dry subsoils, saturated or "drowned" soils, and periodic burning. Grass can thrive under each of these conditions. Where subsoils are dry, grasses can grow and reproduce by utilizing the moisture that penetrates into the topsoil following occasional rains. In marshes, sedges and reeds, which represent the grasses, are among the first plants to send their roots into the saturated soils. Fire destroys most tree growth, but a rapid grass fire during the dry season only destroys the dead grass tops. In climates that will support both trees and grass, periodic burning usually favors the dominance of the grasses.

The grasslands of the world, or the regions in which grass is the prevailing plant form, include the natural grasslands and the culturally induced grasslands. The boundaries

---

[4] Grass seeds have been found in air samples taken from an airplane at an elevation of 4,000 feet above the ground.

between these two types cannot be defined accurately because of lack of historical evidence, but we at least know that the global natural grasslands have their centers in areas that straddle the boundary between arid and humid regions. Smaller areas where grass dominates are related to early stages in plant succession. Grasses are among the first plants to clothe naked soil surfaces, including (1) the muds that border water bodies; (2) the loose sand at the back of beaches; (3) the first crumbled rock fragments of weathered rock surfaces (where grasses follow lichens and mosses); (4) the fresh ash of volcanic eruptions; and (5) fresh alluvium or glacial detritus. Marshes represent the early, or grass, phase in a hydrosere (see Figure 13-11). Once the marsh grasses have accumulated enough organic detritus, plants of a higher and more complex order inhabit the graveyard of their predecessors. Similarly in the humid tropics, when plant foods are depleted from tropical soils, grass is the main plant form that comes to the sterile, abandoned fields. Asking few favors if left alone, it prepares the way for the reestablishment of the forest cover. It may persist for centuries if periodic burning prevents the restoration of the forest.

An additional environmental feature that favors the dominance of a grass cover is underlying material that consists of soft, pure, permeable calcium carbonate, usually chalks or marls. Trees apparently do not tolerate soils that are abnormally high in lime, but grass thrives on them. Such soils are relatively rare, but examples include the black prairie soils of north-central Alabama and southeastern Texas and the chalk "downs" of southern England. These are grassy areas, surrounded by forests on all sides, and their boundaries conform closely to those of the chalks and marls found below the surface.

Excluding the small successional grass-lands mentioned above, the global grasslands can be subdivided for study into three main subformations: the prairies, steppes, and savannas (see Figure 14-24). The prairies include the tall-grass communities that straddle the boundary between semiarid and humid regions in the mid-latitudes; the steppes are short-grass areas located mainly in semiarid regions; and the savannas include the tropical grasslands that are found in low latitudes. These subformations, along with marshlands, are discussed in turn in the remaining divisions of this section. A fifth grassland type, which includes alpine meadows, is treated later, under mountain vegetation.

14-14 THE PRAIRIES It should be admitted that discussing the prairies of the world today is an academic exercise, because so little of them remains. Man has appropriated these humid, tall-grass lands as some of his most productive agricultural land. Strips of prairie nevertheless can be observed throughout the "prairie" regions, along the edges of roads and railroads, along fences and hedgerows, and occasionally in areas that are too rough for cultivation. These narrow strips should not be considered identical with the original prairies. Many new species of plants have been introduced, such as bluegrass (*Poa*), which was carried to the United States from Europe. In general appearance, however, these strips are similar to the original cover.

In describing the plant life characteristic of prairies, the same caution must be used as in dealing with any other vegetation generalization. Differences can be found anywhere, and undoubtedly the varieties of plants that inhabit the prairies in Illinois are considerably different from those of the Argentine pampas. Despite such differences, some common characteristics may be noted.

First, the prairies are not, and never have

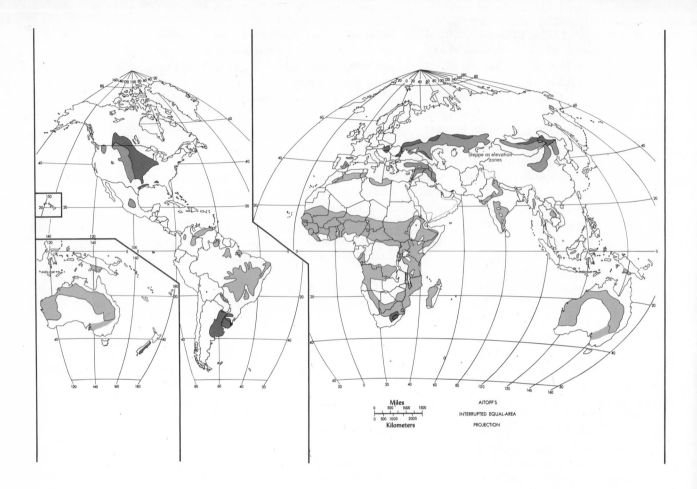

Miles
0 500 1000 1500

Kilometers
0 500 1000 2000

AITOFF'S
INTERRUPTED EQUAL-AREA
PROJECTION

been, great expanses of grasses alone. Although they were without trees, they contained an unusually large variety of non-woody, herbaceous plants, many with showy flowers. A good example of these in the American Midwest is the wild sunflower (*Helianthus*), appropriately the state flower of Kansas, whose yellow flowers brighten the highway borders during the late summer and early fall months. As in the deciduous broadleaf forest, the flowering herbs are arranged in seasonal cycles of dominance. Each spring, summer, and autumn presents its own aspect, with the prevailing vegetation differing in color and height at each season.

Although the flowering plants are most noticeable, the tall and varied grasses rule the communities. Heights at maturity are generally between 2 and 3 feet, although they occasionally rise to between 5 and 8 feet. A wide variety of grasses are included, but some of the genera have a worldwide distribution, including feather or needle grass (*Stipa*), bearded grass (*Andropogon*), June grass (*Koelaria*), wheat grass (*Agropyron*), and the fescues (*Festuca*). All of these have representatives in the prairies of the United States, Argentina, and the Soviet Union.

Figure 14-24 World distribution of major grassland subformations.

■ Savanna
■ Prairie
■ Steppe

The prairies of the world have been without trees during historic times, although stream courses usually were bordered by trees, commonly of deciduous broadleaf varieties. Trees are seen frequently in the prairies today, however, having been planted for shade or for farm woodlot use. These indicate that the environment is not entirely unsuited to forest growth.

If trees will grow in prairie environments, why is it that these regions were not forested within historic times? It is likely that there is no single answer to this question. Among the many hypotheses offered by students of the problem are the following: (1) periodic burning by primitive peoples, which enabled the grasslands to intrude into the forest from the adjacent steppes; (2) development of a subclimax grass community perpetuated on glacial detritus in the postglacial period either by burning or by animal browsing; (3) perpetuation of a subclimax grass community by the same processes in a hydrosere succession developed on poorly drained, flat plains; (4) a grass dominance on soil parent material unusually high in lime; (5) maintenance of a grass-herbaceous flora by rare cycles of protracted droughts; (6) a general climatic change toward more moist conditions, which, however, have not yet enabled forests to replace the grasslands, since trees have been held back by animal browsing or by burning. It is likely that all these factors help to explain the existence of some of the prairies of the world. Evidence exists for the influence of fire in most parts of the world, but undoubtedly burning is not the only explanation.

The generalized climatic location of the prairies in the world is along the semiarid-humid boundary in mid-latitudes, usually between lat 25° and 50° and in a transitional position between the steppe grasslands and the deciduous broadleaf forests.

In North America, the main area of prairies is shaped more or less like a triangle, with its apex, pointing east, in the vicinity of west-central Indiana. The western base of the triangle extends roughly along the 98th meridian, from central Texas into Canada, where it trends northwestward as far as central Alberta province. Small outlying areas include the Central Valley of California, the coastal prairies of Texas, and the Palouse country of eastern Washington and Oregon.

The largest prairie region of Eurasia comprises an east-west belt extending from the northern Ukraine, north of the Black Sea, eastward to approximately long 85° E. This is the famous chernozem, or black-soil, belt that is the major granary of the Soviet Union. The Hungarian Plain and parts of the Romanian Plain were prairies prior to cultivation. Much farther east, the central portion of the Manchurian Plain and part of the middle Amur Valley originally had a prairie vegetation. Prairies in South America include the pampas of Argentina, Uruguay, Paraguay, and southern Brazil. The *high veldt* region of South Africa, including all of the Orange Free State and much of southern Transvaal, is a prairie region. No true prairie occurs in Australia, although a large part of eastern New Zealand had a tall-grass–herbaceous cover, similar to the prairies of central California.

Today the prairie regions of the world are utilized mainly for agriculture, especially for grain cultivation. The prairies probably have the most inherently fertile soils in the world. Deep, often stoneless, black with humus, rich in plant nutrients, and highly capable of retaining water and plant foods, they are ideal soils for modern-day agriculture.

14-15 THE STEPPE GRASSLANDS The steppe is synonymous with the short-grass region that borders the desert or lies between the prairies and the areas of desert shrubs in the mid-latitudes. The term *steppe* has been used in various ways. The Russians often use it to designate any grassland of appreciable extent. It also has been used in geographic literature to include all areas of short-grass vegetation, including those within the tropics. The definition used here is convenient and less confusing.

The more or less continuous sod cover of the steppe consists mainly of short grasses 6 to 12 inches high but includes occasional representatives of the taller prairie grasses (see Figure 14-25). In the United States the short-grass dominants are grama grass (*Bouteloua*) and buffalo grass (*Buchloe*). As in the prairie, herbaceous annuals and perennials are mixed with the grasses, forming cyclical groupings during the year. The variety and conspicuousness of the flowering herbs, however, are not so great as in the prairie.

The ecological position of the steppes is not entirely understood. Plant ecologists are not in agreement about whether the short-grass dominance is a climax vegetation or whether it has resulted from overgrazing by animals that have shown a preference for the taller grasses and have left the shorter varieties behind to establish dominance. If grazing is responsible for the short-grass dominance, was this necessarily the work of domesticated flocks? Most of the steppe regions had large numbers of wild grazing animals, and many of the native rodents, such as the prairie dog, have a preference for the taller varieties of grass. It is likely that the steppes went through cycles of expansion and contraction of short-grass dominance long before man appeared, corresponding in some cases to changes in animal populations and in other cases to changes in climate. Regardless of their successional position, the steppes apparently have been grasslands for a long, long time.

The largest area of short-grass steppe is in the United States, located between the western boundary of the prairie and the foot of the Rocky Mountains. In Eurasia, there are three steppe areas: a narrow strip along the southern edge of the Russian prairies, the region along the eastern margin of the Gobi Desert in Mongolia and northwestern China, and a strip in western Manchuria. Smaller strips of steppe are found along the middle slopes of the great mountain ranges in central Asia.

There are no extensive steppes in the southern hemisphere. Although short-grass areas occur, they are invariably accompanied by desert shrubs. Illustrative of this is the *monte,* a region which borders the Argentine pampas on their dry, western margin—a location where one would ordinarily expect steppes to be found. The dominance of low, xerophytic shrubs gives the monte an appearance quite unlike that of the northern hemisphere steppes.

The utilization of the steppes is confined mainly to grain cultivation and grazing. This clearly is a marginal region for agriculture, and cultivated land has alternately expanded and contracted within the region in accordance with cycles of greater and lesser precipitation, respectively. There does not appear to be any reason for prohibiting agriculture in the steppe, provided that reasonable precautions are taken. Long-range weather forecasting, when improved, will aid immeasurably in planning farming schedules. Farm practices also should be geared to a combination of grazing and dry-farming tech-

Figure 14-25 Typical short-grass steppe vegetation on the Great Plains near Sioux Pass, Montana. [Standard Oil Company of N.J.]

niques. In the latter, the emphasis is continually on adding to ground-moisture supplies by checking runoff, reducing evaporation from the soil, and destroying weeds in order to minimize transpiration. Some areas probably will never be cultivated, because of surface configuration or uniquely unfavorable soil characteristics, and should remain as grazing lands, carefully managed so that the density of animal population is geared to fluctuating carrying capacities. Successful utilization of the steppes demands skills and equipment to combat rapidly soil water deficiencies and impending soil erosion.

14-16 THE SAVANNAS The savannas are tropical grasslands. They differ markedly from the prairies and steppes of mid-latitudes. First, savannas without trees or bushes are exceptionally rare. Second, savannas seldom have a continuous sod cover, but rather a variable spacing of bunch grasses with patches of bare ground between them. Third, nonwoody herbaceous plants are much less common in the savanna plant communities, the grasses themselves are much coarser, and only a few of the dominant genera of the mid-latitude grasslands find representation there.

Figure 14-26 Palmyra palms in the savanna east of Wau in southern Sudan. The coarse, harsh grass in this photo is between 4 and 8 feet high. The photo was taken in late December, toward the beginning of the dry season. [Van Riper]

Savannas differ considerably in general appearance. In some, the grass grows to heights of 8 to 12 feet, but in others it may be only about a foot high. In some areas, the grasses are surmounted by scattered palms, such as the Palmyra palm (*Borassus*) (see Figure 14-26), whereas in others the low, umbrellalike *Acacia* thorn trees are found. Patches of scrub thickets are another type of nongrassy vegetation on the savannas. Stream courses in savanna regions frequently are lined with *galeria*[5] forests or dense thickets of tall canes or reeds.

Nearly all savannas have pronounced wet and dry seasons, and present a markedly different appearance in each. In the dry season the grass is parched, tough, and brown, and the trees shed their leaves and stand

[5]The term *galeria* signifies a gallery, or corridor, so named because a dense tangle of trees meets over the stream, producing a tunnellike opening below.

naked under the scorching sun. The advent of the rainy season produces a dramatically sudden change in the landscape that is even more marked than the coming of spring in higher latitudes, because the infirmities of the dead season are not hidden under a snow cover. Trees and shrubs appear to burst into leaf, and the green shoots of grass quickly color the land surface. A general appearance of lushness grows as the rainy season continues, and this has often deceived travelers into making extravagant statements regarding the potentialities of agriculture or animal husbandry in these tropical grasslands.

Students of tropical vegetation have uncovered as many problems in attempting to interpret plant succession in the savannas as have been raised with respect to the prairies and steppes. Some researchers claim that nearly all the savannas were once similar to tropical open woodland or semideciduous rain forest and that they have been transformed into savannas by periodic burning. Others admit that some of them were produced by burning but insist that, over wide areas, the seasonal droughts could never support more than a sparse stand of trees and that a dominance of the coarse grasses is to be expected. All are agreed, however, that the area of savannas is growing at the expense of the tropical forests and woodlands and that savannas have existed for thousands of years.

The distribution of savannas today is confined largely to South and Central America, Africa, and Australia. Because of their gradual transition to desert-shrub lands on one side and to tropical woodland on the other, maps of savanna distribution may show considerable differences, depending on the judgment and measurements of the compilers. The major savanna areas, nevertheless, have been known and recognized for many years; these include the *llanos* of Ven-

Figure 14-27 Tree savanna along the Magdalena River, Colombia. Note the scattered palms that are found on savannas in many parts of the world. The savanna here is nearly a tropical scrub woodland. [Standard Oil Company of N.J.]

ezuela (see Figure 14-27), the *campo* of the Brazil-Guiana uplands, and the Sudan region of Africa. Africa contains the largest single area of savannas, and Shantz, a noted student of African vegetation, estimated that they comprise some 37 percent of the continental area. Along with the large areas shown on the world vegetation map, there are scattered small areas of savannas throughout the low latitudes, representing abandoned clearings in the rain forest or tropical woodland in which grasses are maintained by burning.

The savannas are utilized in somewhat the same way as the steppes, with grazing as the principal industry and with dry-farming agriculture making some inroads. There are many handicaps to both of these activities, and, in general, the savannas have low population densities. Grazing is hindered by the low nutritional value of the coarse grasses, which are particularly low in phosphorus. Only certain types of animals are capable of digesting these grasses during the dry season and of resisting the high temperatures of the tropics. The major products of the grazing industry are hides and skins. Animal diseases, particularly those transmitted by insects, are common and virulent.

Variations in the length and severity of the dry season are a major disadvantage to agriculture. Yields are also likely to be low because of the poor soil fertility. Here grass does not have the same relationship to soil fertility that it does in higher latitudes. In climates that have a frost season, grass tends to improve the quality of soils for agriculture by increasing the humus content of soils, by adding to the nutrient content (especially calcium), and by generally increasing the retention of water and plant foods. These benefits are related largely to the slowness of decay and to low leaching. In the tropics, on the other hand, fertility is maintained more by a tree cover than by grass, especially if the grass is burned. Trees are able to tap a much greater depth for plant foods than grass and thus are able to return more plant foods to the surface via the decay of leaves and other forest litter. Surface soils are impoverished because of excessive washing during the rainy season; thus, grass cannot return much nutrient content to the surface. The rapidity of decay and oxidation prevents the grass roots from adding as much agriculturally valuable organic material to the soil as they do in cooler climates. When the dead surface grass in the tropics is burned, it concentrates its former nutrients into ash which can be made immediately available to plants. The plant nutrients in the ash, however, are subject to leaching by rainfall. Thus the burning of savanna grasses produces a temporary stimulus to plant growth but, over a period of time, results in a progressive impoverishment of soils that were not fertile to begin with.

The grasslands of the tropics do not have the inherently high fertility of the steppes or prairies, where water availability is the critical item in productivity. This does not mean that the savannas cannot be made productive, but if agriculture in these areas is to raise itself above the level of providing a meager subsistence, it must develop specialties that can support the cost of soil improvement.

14-17 MARSH GRASSLANDS *Marshes* are poorly drained areas that have a cover predominantly of grasses or other plants closely related to the grasses, such as bulrushes, reeds, canes and sedges. The groundwater table in marshes stands either slightly above or below the surface. Although marsh grasslands are too small in individual extent to be shown on the text map of global vegetation regions, they are widely scattered throughout the world and comprise a large area collectively.

Marsh grasses thrive when there is some environmental factor that inhibits the growth of woody plants, shrubs, or trees. The most common factor is an alkaline rather than an acid reaction in groundwater. This, in turn, results from a variety of reasons, including (1) impregnation by seawater, as along tidal mud flats (note that in the tropics, mangrove and palm thickets are exceptions); (2) ground seepage through glacial till that contains a large amount of finely ground limestone; (3)

seepage from adjacent areas underlain by chalk, marl, or highly soluble limestones; (4) arid or semiarid areas, where soluble salts tend to be concentrated in the soil by evaporation.

Marshes occur also in areas with acidic groundwater when some other factor prevents the woody plants from developing. Many of the great rivers that flow into the Arctic Ocean in the Soviet Union and Canada are bordered by riverine meadows and marshlands (see "The Taiga Riverine Meadows," p. 538). Marshes are also common in poorly drained portions of the tundra regions, mainly because grasses and sedges are much more adaptable than shrubs or trees to the extremely short growing season, the shallow permafrost zone, and the cold-wind stress of winter.

The ecology of the great marshes on the African continent, such as those of the Sudd along the upper Nile above Khartoum, bordering Lake Chad, the Okavango Basin in northwestern Bechuanaland, and bordering some of the lakes in East Africa has not been carefully worked out. Fire, submergence through subsidence, and climatic changes since the Pleistocene all could have played a part. The areas mentioned are among the largest individual marshes in the world. Others are found in the Orinoco Basin of Venezuela, the Gran Chaco area of Paraguay, Bolivia, and Brazil, and along the lower portion of the Tigris-Euphrates Basin in Iraq.

Marshes have little more than academic or recreational interest to industrial-commercial cultures, but they have provided a unique ecological habitat for some highly specialized cultural groups, such as the Marsh Arabs of Iraq and the Nilotic tribesmen of the Sudd, who have developed an isolated and wholly adaptive life based on the limited resource base within these great marshes. For these groups, aquatic life is more important than the marsh vegetation itself, although some grazing animals are used. Marsh grasses are not highly nutritious, and should not be considered as potential areas for large-scale animal grazing.

Desert Shrubs

Nearly all desert surfaces contain some vegetation. There are, of course, a few exceptions, such as active sandy ergs, where the blowing dunes do not remain in place long enough for plants to develop, and windswept, bare rock surfaces, which have no soil to nourish plants; but such areas constitute only a small part of arid regions. In the struggle for life, plants have developed remarkable mechanisms for existing and reproducing under conditions of limited moisture supply. Desert plants consist of five major groups, each of which has a different method of overcoming the handicaps of desert environments; (1) leafless evergreen herbaceous plants, such as cacti; (2) the sclerophylls, or evergreen hardleaf plants; (3) deciduous bushes and shrubs; (4) the ephemerals, or short-lived plants; and (5) the halophytes, or salt-tolerant plants. All five are found in deserts, and their relative proportions depend largely on local conditions.

14-18 THE LEAFLESS EVERGREEN HERBS The evergreen herbs of the desert, which include the *cacti* and *Euphorbia,* are succulent, leafless, nonwoody plants that use their stems as breathing surfaces and are often equipped (as in the cacti) with spines to protect them from browsing animals. The lack of leaves results in low rates of photosynthesis; therefore, these plants grow extremely slowly. Their slow growth and low metabolism thus require only shallow and poorly developed root systems. Outer surface cells

Figure 14-28 Saguaro cactus near Phoenix, Arizona. The fragrant white, waxy bloom of this giant cactus is the state flower of Arizona. The saguaro are not widely distributed in the deserts of southwestern North America, but they are protected and can be observed in the Saguaro National Monument near Tucson, Arizona. [U.S. Bureau of Reclamation]

are hard, waxy, and waterproof, thus protecting the water-storage cells in the interior plant tissues. The spiny or cactus forms are mainly confined to the northern hemisphere, whereas the *Euphorbiae* are found mainly in Africa. Both groups of plants have a wide variety of species and form. Varieties of cactus range from the low supine types, such as the prickly pear (*Opuntia*), to the slender, branching *Ocotilla* and the giant *Saguaro* of the Mojave Desert, which may reach 20 to 30 feet in height (see Figure 14-28). Giant *Euphorbiae* may be as tall as the saguaro, and their candlebralike forms produce an unusual landscape feature over large areas of western Ethiopia and southeastern Sudan. Not all of the *Euphorbiae* are leafless, and small leafy species have extended out of the Saharan floral assemblage into the more humid regions of southwest Asia. For protection against browsing animals, the *Euphorbiae* universally are equipped with an extremely bitter white latex, or plant fluid. They thus do not need needles.

14-19 THE SCLEROPHYLLS The *sclerophylls* include many of the same plant forms that are dominant in the *Cs* climates. They are shrubs and low trees that have leaves and thus are able to grow faster and taller than most cacti, but they must also have elaborate devices for reducing transpiration. Among various of these devices are the following:

1. Dense cell fluids, frequently a milky latex.
2. Hard, evergreen, waxy leaf surfaces.
3. Insulative and waterproof cork cells in the bark of stems.
4. Complex *stomata* (breathing pores) with adjustable openings and various protective devices.
5. Phototropism, or the movement of leaves under the influence of light, so as to present edges to the sun.

Root systems are unusually well developed in the sclerophylls, and much of the water is stored there. The most characteristic location for the sclerophylls is along the pole-ward margins of the low-latitude deserts, adjacent to the Cs climates, as in the south-western United States, northern Mexico, and along the northern borders of the Sahara Desert.

An important exception to the generalized pattern is Australia, where about 95 percent of *all* native trees, without regard to climate, are sclerophyllous eucalypts (see Figure 13-8).

14-20 THE DECIDUOUS SHRUBS OF THE DESERT The deciduous shrubs are the most widely distributed of the conspicuous desert vegetation. Their leaf-shedding property makes them adaptable to droughts in both high and low temperatures. Growth can be rapid in this group of plants, since they have leaves during moist periods. Root systems are well developed. Many of the various protec-tive properties of the sclerophylls are also found in the deciduous shrubs.

The two most widely distributed repre-sentatives of the deciduous desert shrubs are sagebrush (*Artemisia*) and *Acacia*. The former is found mainly in the cooler arid and semiarid regions (see Figure 14-15), and the latter is a widely adaptable thorn tree and shrub of the low latitudes. In the United States, sagebrush and the creosote bush (*Larrea*) are the most common representa-tives of the desert deciduous vegetation. The sagebrush ranges from northern Arizona and New Mexico northward into central British Columbia and eastern Washington and Oregon, whereas the creosote bush is found in the warmer, drier region to the south (see Figure 14-29). The creosote bush differs from sagebrush in having an unusually widespread but shallow root system. Like most desert shrubs, these plants become shorter and more widely spaced with in-creasing aridity. The deciduous desert shrubs encounter considerable fluctuations in their

Figure 14-29 Creosote bush (*Larrea mexi-cana*), in southern Utah. This low, sclero-phyll (hard-leaf shrub) takes the place of sage-brush in the southwestern deserts of the United States. A desert free of vegetation is a rarity, not only in the southwestern United States but also in the world. Note the sparse grass cover between the desert shrubs. This country is graz-ing land, but the carrying capacity is extremely low. [Van Riper]

geographic range, interchanging desert envi-ronments with steppe and savanna grass-lands. A decrease in the grass cover on steppes because of overgrazing, for example, leads to increased soil dryness, even without climatic change, and thus fosters the intro-duction of desert shrubs. The shrub vegeta-tion, once fixed, is not easily replaced by grass and may persist for many years. This is why sagebrush covers such a large area of semiarid (BS) climates rather than true desert (BW).

14-21 THE EPHEMERALS The ephemerals, or short-lived plants, comprise many diverse types, including grasses, flowering annual herbs, and many tuberous or bulb plants. Most of them are small plants that can exist in dry areas, primarily because they are able to complete their life cycles in a short time. Their seeds, tubers, or corms remain dormant until a moist period stimulates them into germination and growth. After a rainy period, deserts are frequently mantled with a spectacular display of flowers, most of them having brilliant colors and a heavy fragrance. The colors and fragrance aid in insect pollination during the short life-span of these plants. A large number of our common garden flowers originated in desert environments; among them are the peonies, some of the irises, lilies, and poppies.

14-22 THE HALOPHYTES The halophytes, or salt-tolerant plants, are found in association with poor drainage and consequent saline or alkaline soils in dry regions, al-

Figure 14-30 Halophytes on the salt-encrusted edge of Tuz Gol, a large playa lake in central Turkey. Note the puffy, succulent "leaves" and branches of these low, salt-tolerant plants. Highly specialized structures enable these plants to thrive in a habitat that contains few competitors. [Van Riper]

though some of them are also encountered along seacoasts within humid regions. Most of them are succulent plants; that is, they have many water-storage cells in their stems and leaves, resulting in a thick, fleshy appearance (see Figure 14-30). The fluids in some of these have a high salt content and density, thus balancing the groundwater of their environments. When the outgo of water by transpiration exceeds intake, the salt solutions within the plants increase in density to the point where salt crystals are formed. These are forced outward from the plant surface, coating it and giving rise to the descriptive term *salt bush.*

14-23 THE UTILIZATION OF DESERT VEGETATION Desert vegetation is far from valueless as food for animals, although the carrying capacity of a desert is much lower than that of the grasslands. Feed for animals consists of (1) the bunch grasses, which are associated frequently with the low bush vegetation; (2) bushes and low trees, particularly the tender new tips; and (3) the smaller herbaceous plants. Cattle, goats, and sheep tend to prefer each of these three types of feed, respectively. In many parts of the lower mid-latitudes, desert range is assigned as winter pasturage, supplementing summer pasturage at higher elevations. Since the grasses often have cured into natural hay by this time, summer growth and reproduction are not hindered. Because of the low carrying capacity of desert range, successful ranching requires mobility of herds and large areas for pasturage. Fire is not the destructive agent that it is in the grasslands, because the plants are so widely spaced that there is a large amount of bare ground between them. Overgrazing, however, has seriously altered vegetation balances in many areas, decreasing the vegetative portion of the biomass and accelerating erosion by both wind and water.

The Tundra

The tundra is a vast, treeless expanse bordering the polar margins of the North American and Eurasian continents (see Figure 14-1). It extends farther south along the eastern margins of these continents, reaching almost to the Gulf of St. Lawrence along the eastern coast of Labrador and as far south as the peninsula of Kamchatka in eastern Asia. Plant growth on the tundra must adapt itself to the following environmental limitations:

1. A short, cool, growing season, usually less than 2 months long, with occasional frosts throughout this season.
2. Shallow soils lying above permafrost, or permanently frozen ground.
3. Poor internal soil drainage.
4. Strong, steady, drying winds.
5. Soils having a low nitrogen content.

Compensating somewhat for these adverse conditions are the long days of summer and the low rate of soil leaching.

The most outstanding characteristic of the tundra vegetation is dwarfing, or the tendency to produce much smaller varieties of plants than those common elsewhere. This is best exemplified by the trees and shrubs, such as arctic birches, willows, and heaths, which may be only 6 to 24 inches tall at maturity. Wind stress and the shallow soils rarely permit tall trees anyway, but the dwarfing appears to be more a result of extremely low rates of growth, which in turn are due to low transpiration rates and the slowness of protein synthesis for plant tissue development. Except for the first stages of leaf development in early summer, this slowness of growth holds true even during the long summer days. Burial beneath the snow is another factor in favoring dwarfing.

The most conspicuous plant forms of the tundra are low, cushionlike woody or herbaceous perennials that reproduce by budding, or from root shoots. Fruiting, which would be difficult in the short growing season, is thus unnecessary. By remaining short and keeping terminal buds at or near the ground, these plants can take advantage of the protection by snow in the cold winters and the radiation of heat from the ground during the summer months. Wind velocities near the surface also are much reduced.

Lichens are found throughout the arctic tundras. They thrive where there is plenty of light and a high relative humidity; they require little else. Free from the close competition of higher life forms, the lichens find a good home in the long daylight summers of the Arctic. They reproduce by disjunction; that is, pieces of the colony break off and are carried to a new site. Arctic lichens are distributed widely and take many different forms. *Cladonia,* one of the largest varieties in the tundra, is between 2 and 4 inches high and has a preference for somewhat better drained, sandy soils. Widely misnamed as *reindeer moss,* this pale-green plant is one of the principal foods for arctic browsing animals (see Figure 14-31).

True mosses, such as the sphagnums, are encountered far less frequently in the tundra than in the taiga. According to Berg, a Russian biologist, tundra bogs are by no means so common as was formerly believed, and those which do exist consist generally of sedges (*Carex*) rather than of the sphagnum moss typical of taiga bogs. The sedges are grassy plants with narrow leaves that are triangular in cross section. The water saturation of the tundra subsoils almost everywhere during the warm season, along with an acid environment, is revealed by the wide distribution of plants that are associated farther south with true bogs, such as the arctic

Figure 14-31 Cladonia, or reindeer "moss," a giant lichen that is common throughout the Arctic. This photograph was taken on the subarctic plains north of Alta, Norway. *Cladonia* forms an important item in the diet of the reindeer during the winter season. The object on the right is the end of a pen used for scale. [Van Riper]

cranberry, or cowberry (*Vaccinium*), and the ledums.

The tundra does not always have its customary drab, colorless appearance. Early in the spring, usually in late June and early July, before all the snow has disappeared, the arctic flowers appear. The display does not have the riot of color that a desert blooming presents, but it is impressive and even more welcome. Botanists who have visited the Arctic have often been astonished at the large variety of flowering herbs, particularly those which have tight rosettelike clusters of leaves close to the ground. Among these are the *saxifrages* and *arctic poppies*. Many such plants also have hairy, rather thick leaves that help retard transpiration and insulate against occasional frosts. Grasses are present in the tundra, as they are nearly everywhere

on earth. Their most common representatives are the sedges.

The ecological pattern of vegetation is almost entirely related to minute local variations of slope and drainage (see Figure 14-32). There is little difference between the general features of the Alaskan tundra and those of the tundras of Lapland or the Soviet Union. Extremely sharp contrasts in types of plants, however, are found within a few feet throughout the tundra, representing different moisture and temperature regimes associated with variations in snow retention, soil permeability, permafrost depth, or exposure to wind or sun. Since most of this region was covered with glacial ice a relatively short time ago, environmental balances have not been established except on a local microlevel. Plants normally occur only in patches, and usually there is more bare ground than vegetation cover on the true tundra. Some evidence of plant succession is present, but it is not nearly so clear as in regions to the south.

The southern boundary of the tundra merges with the northern edge of the taiga along a highly irregular line. Isolated clumps of low conifers, mainly larches or spruces, act as outliers of the taiga in the treeless tundra, and long, narrow strips of forest may penetrate the tundra for 100 to 200 miles along stream valleys. The presence or absence of trees along this boundary is closely related to the depth of the permafrost layer in the subsoil. Where the permafrost zone lies within 1 or 2 feet of the surface, the water-saturated topsoils will not support trees. Sufficient drainage to avoid soil saturation is therefore a requirement for tree growth in this harsh climate. Trees are encountered along stream valleys, not because of a greater water supply but actually because of less water. Streams can remove excess surface water, and permeable sands and gravels are more common in valleys. For the same reasons, permafrost consistently lies farther below the surface in valleys.

Except for its utilization in connection with reindeer herding and the gathering of tundra berries such as the cloud berry of northern Scandinavia, the tundra vegetation has no direct economic value. There is considerable potential for increasing the small number of reindeer now herded on the tundra, but a limited market, scarcity of trained herders, and high transportation costs are serious handicaps.

Mountain Vegetation

Local relief in mountainous areas is sufficient to produce a vertical zonation of vegetation and a consequent contrast with the adjacent lowlands. A global map utilizing a classification that distinguishes mountain vegetation signifies only that local differences occur with elevation; no adequate areal generalization can be made here. A few vegetation communities occur in mountain areas that appear to be sufficiently widespread and distinctive to warrant separate description. These include the tropical moss forests, the mountain conifers, and the alpine meadows.

14-24   THE TROPICAL MOSS FORESTS
Moss forests are found throughout the humid tropics, but always within the cloud zone of mountains, which is usually located on the windward slopes and which is nearly always enclosed in cloud mists. It cannot be related to any particular elevation, although it is rarely encountered below 3,500 feet.

The tropical moss forest is a weird type of plant community, quite unlike any other on earth. Compared with their giant neighbors farther down the mountainsides the trees are small, rarely reaching heights over

Figure 14-32   Typical tundra near Pond Inlet, Baffin Island.   Microrelief features are extremely important in the distribution and composition of the tundra vegetation. The low mound of earth in the foreground, somewhat better drained than elsewhere, is covered with a profusion of grasses and wild flowers. [National Film Board of Canada]

50 feet and usually remaining under 20 feet. The branches grow straight and stiff, like outflung arms, in order to resist the strong winds that frequently sweep these slopes during thunderstorms. Aerial roots sometimes spring out from the tree trunks and are somewhat reminiscent of those of the mangrove swamps of tropical tidal flats. Tree ferns are scattered throughout. Branches, tree trunks, and the ground itself are heavily mantled with mosses, liverworts, various orchid epiphytes, and ferns of many kinds. The mantle of moss and other plants varies in thickness but is sometimes more than a foot thick. In such cases, it seems as though

ground, sky, branches, roots, and moss all merge together. This forest is always wet, and water continually drips from the leaves and branches. Adding to the unearthly atmosphere of the landscape is its quietness. Few birds or monkeys live in this soggy, gloomy sepulcher, and even insects seem to avoid it.

14-25 MOUNTAIN CONIFERS Conifers not only constitute a large part of the midlatitude mountain forests but are frequently present at high elevations throughout the world. They are suited to the wide fluctuations in temperature caused by alternate sun and shade in the thin air, as well as to heavy snows and the soil droughts resulting from rapid runoff on steep slopes. Also, because of their wide-spreading, shallow roots, conifers can anchor themselves more adequately in thin soils than can other types of trees. Near the upper timberline appear the firs and spruces (see Figure 14-33), similar to the

dominants in the taiga, but in more open stands. Pines are usually found at lower elevations, below the spruce-fir association. Many of the pines, such as the lodgepole pine (*Pinus contorta*) in the Rocky Mountains and the southern yellow pine of the Great Smoky Mountains of the eastern United States, are the result of old forest fires. Lightning undoubtedly has been responsible for fires and for the establishment of many of the mountain conifers in these latitudes.

Conifers, especially the pines, are by no means uncommon at fairly high elevations in the tropics. The high plains of northern Sumatra, central New Guinea, and southeastern Brazil contain pine forests; pines are found in areas above 3,000 to 4,000 feet in Mindoro and northwestern Luzon in the Philippines and in northern Indochina. The highland pines of the tropics are most frequently located in regions where the adjacent lowlands have a distinct dry season. This is also true of the pines of the Atlas Mountains and the various other mountain areas of Cs climates, although the latter locations are more subtropical than tropical.

14-26 ALPINE MEADOWS Grasses and low, herbaceous woody plants, such as the heaths, often form meadows just above the upper timberline of mountains and form glades in the forests farther down the moun-

Figure 14-33 Engelman spruce (*Picea engelmanni*) near the upper timberline in the vicinity of Monarch Pass, east of Gunnison, Colorado. Spruce and fir are the dominant trees at high elevations throughout the mountains of the western United States. Note how much smaller they are than the tall pines at lower elevations. These are almost pure stands, because the harsh environment requires special adaptability to withstand environmental stress. [Standard Oil Company of N.J.]

tain slopes. This high-elevation environment has low temperatures, high wind velocities, and soils that alternate between water saturation and drought. Melting snow keeps the ground saturated for much of the earlier part of the summer season, but thereafter the winds dry the thin soils rather rapidly. The wet environment is illustrated by the types of grasses and herbs that are present. The sedges (*Carex*) are by far the most common grasses, and many of the flowering herbs are those which are typical of damp soils at lower elevations, such as the buttercups, bluebells, marigolds, violets, saxifrages, and poppies. In the upper forest zone, which is usually coniferous, swales and depressions usually are in meadow rather than forest, and this too reflects the influence of water saturation.

The alpine meadows long have been important as summer pastures for animals, especially where a pastoral economy prevails on the adjacent lowlands. This is less often true in tropical regions because of the difficult forest zones that must be crossed in order to reach the upper timber line. Also, the alpine meadows of the tropics lie at a much higher elevation than do those of higher latitudes.

Above the alpine meadows, and in more exposed sites, the alpine vegetation begins to take on many of the features of the arctic tundra. Cushionlike plants, streamlined by the wind, are frequent, and lichens abound on patches of thin soils and even on bare rock surfaces. Here and there a dwarf tree can be found, bent and blasted by the wind. Wherever snow accumulates, it helps to protect the low vegetation, and shrubs are denser and higher in the hollows, where they are sheltered from the wind. Tiny flowering plants are found growing up to the edges of permanent snow fields and glaciers. The existence of plants in this harsh environment illustrates the general principle that competition for survival among plants is so keen that special adaptations can be found suitable for almost any environment on earth— from the high temperatures of hot springs to the frozen edges of glaciers, from bare rocky surfaces to loose dune sand, from deserts to swamps and marshes, from mountaintops to the continental-shelf zone far beneath the surface of the sea.

Study Questions

1. If tropical soils generally have a low, inherent fertility, at least compared with most soils elsewhere, why is the biomass so great there?
2. What are some of the main handicaps to utilizing the Amazon forest as a world lumber-resource area?
3. Explain why fire has a much more serious effect on soil fertility than it does on the prairies.
4. The Cs and *Aw* climates both are characterized by much scrub woodland. Explain the essential difference between the two woodlands in terms of climate.
5. Why are there so many brush fires in southern California?
6. Using the concept of systems analysis, suggest a connection between savanna grasslands in East Africa and a dense animal population prior

to human depredations. Might animals replace trees in the savanna biomass?

7. Contrast the subformations of vegetation encountered as one passes poleward from the equator on the western side of the continents to those encountered progressing poleward on the eastern side. Explain the difference.

8. Look up in the Appendix some comparable climatic statistics for the northwestern U.S. coast and the west coast of Norway. What connection do you see between this and the size of trees in the two areas?

9. What indications are there in the broadleaf deciduous forest of the southern Appalachians as to the characteristic subformation that probably would have occupied the Gulf of Mexico area had the gulf not been there?

10. Examine the vegetation in Figure 12-26 and determine the vegetation subformation that is present there. What evidence do you have?

11. Contrast and explain the difference between the lumbering operations in West Virginia and Oregon.

12. What is the natural home for most of our common kitchen herbs? Explain.

13. Name two parts of the world *outside* of the tropics where the principal trees are broadleaf evergreens.

14. Where is there an area of deciduous conifers sufficiently large to be distinguished on the world map of subformations? Explain its presence there.

15. Explain the process by which overgrazing may turn a steppe into a desert shrub vegetation.

16. Give several examples of the relationship between geographic range of plants and plant specialization. Relate this to species counts in various climates.

Part 4: References

Atlas of American Agriculture, Section E, "Natural Vegetation," U.S. Government Printing Office, Washington, D.C., 1924.

Berg, L. S., *Natural Regions of the U.S.S.R.,* The Macmillan Company, New York, 1950.

Braun-Blanquet, Josias, *Plant Sociology, the Study of Plant Communities,* tr. by George D. Fuller and Henry Conrad, McGraw-Hill Book Company, New York, 1932.

Cain, Stanley A., *Foundations of Plant Geography,* Harper and Brothers, New York, 1944.

Dansereau, Pierre, *Biogeography: An Ecological Perspective,* The Ronald Press Company, New York, 1957.

Daubenmire, R. F., *Plants and Environment,* John Wiley & Sons, Inc., New York, 1947.

Eyre, S. R., *Vegetation and Soils; A World Picture,* Aldine Publishing Company, Chicago, 1963.

Farb, Peter, *Living Earth,* Harper and Brothers, New York, 1959.

Forestry and Timber Bureau, *Forest Trees of Australia,* Department of the Interior, The Government Printing Office, Canberra, Australia, 1957.

Gleason, Henry A., and Arthur Cronquist, *The Natural Geography of Plants,* Columbia University Press, New York, 1964.

Good, Ronald D., *The Geography of Flowering Plants,* Longmans, Green & Co., Ltd., London, 1947.

Grass, The Yearbook of Agriculture, 1948, U.S. Department of Agriculture, U.S. Government Printing Office, Washington, D.C., 1948.

Haden-Guest, Stephen, John K. Wright, and Eileen M. Teclaff (eds.), *A World Geography of Forest Resources,* The Ronald Press Company, New York, 1956.

Küchler, August W., "A Geographic System of Vegetation," *Geographical Review,* vol. 37, pp. 233–240, 1947.

McDougall, W. B., *Plant Ecology,* 4th ed., Lea & Febiger, Philadelphia, 1949.

Neil, Wilfred T., *The Geography of Life,* Columbia University Press, New York, 1969.

Oosting, Henry J., *The Study of Plant Communities; An Introduction to Plant Ecology,* 2d ed., W. H. Freeman and Company, San Francisco, 1956.

Polunin, Nicholas V., *Introduction to Plant Geography and Some Related Sciences,* McGraw-Hill Book Company, New York, 1960.

Richards, P. W., *The Tropical Rainforest: An Ecological Study,* Cambridge University Press, New York, 1952.

Trees, The Yearbook of Agriculture, 1949, U.S. Department of Agriculture, U.S. Government Printing Office, Washington, D.C., 1949.

Weaver, John E., and F. E. Clements, *Plant Ecology,* 2d ed., McGraw-Hill Book Company, Inc., New York, 1938.

PART 5
SOILS

The term *soil* has many meanings, all of them legitimate and many of them contradictory. To a construction engineer, soil may be merely the loose material that lies above solid rock, to be used either as construction material or as a construction site. To a housewife, the term may be synonymous with dirt. Even the soil scientists are not in total agreement as to what constitutes *soil* and *nonsoil*. Rather than become too involved with precise definitions in this introductory volume, we shall simply define soil as a *mixture of fragmental material that will support a vegetation cover and that usually is a three-dimensional natural body exhibiting some vertical difference with depth.*[1] Most soils evolve slowly as a result of the interaction between the atmosphere and its weathering functions, the solid crust of the earth or the fragmental rock debris lying on it, the water contained within the soil or moving through it together with its included solutes, and finally, the complex biological world that dwells within or upon the surface. The latter includes both microflora and microfauna as well as the larger life forms, such as trees, animals, and humans.

The most significant part of our definition is that soils support a vegetation cover. This signifies more than individual plants. A bare rock surface will support individual lichens or mosses, but soil is required to supply an association of plants with nutrients and water. Recent river alluvium will support a rich floral assemblage; hence, it should not be excluded from consideration as soil because it may not exhibit any vertical differentiation. Many of the cultivated fields in Europe and Asia have been under cultivation for so long that the dominant characteristics of the underlying material

[1] A more precise definition is suggested in the Soil Survey Manual, Handbook No. 18, U.S. Dept. of Agriculture, G.P.O., Washington, D.C., p. 7, August, 1951. It reads as follows: ". . a collection of natural bodies on the earth's surface, supporting plants, with a lower limit at the deeper of either the unconsolidated mineral or organic material lying within the zone of rooting of the native perennial plants; or where horizons impervious to roots have developed, the upper few feet of the earth's crust having properties differing from the underlying material as a result of interactions between climate, living organisms, parent material, and relief."

are the direct result of human cultivation practices, such as manuring, spading, and draining, rather than natural soil-forming processes. Such materials should not be excluded as soils because they are not natural bodies. Over the major part of the earth surface, however, the dominant features of soils have resulted from nonhuman processes.

The lower limit of soils in the progression downward toward unaltered or fragmental rock is taken as the lower limit of plant feeding or root penetration. The presence of crusts or hardpans in soils complicates this limitation somewhat, but in general, the association between soil and the covering of vegetation should be emphasized. Soil formation is not synonymous with geologic rock weathering. Rock weathering in the humid tropics may extend to depths of 100 feet or more, but the interaction between the weathered rock material and the surface vegetation generally is much less.

A further difficulty in defining soil arises in determining how small a unit of soil can be. A careful soil morphologist may observe differences horizontally and vertically every few inches. Obviously, some arbitrary lower limit of generalization must be established in classifying soils. By general agreement among soil morphologists, an arbitrary lower unit has been set at the minimum lateral extent in which the study of *horizon*[2] shapes and relations can be made. This has worked out in practice to have a minimum diameter of about a meter. This minimum unit is termed a *pedon.* Units below the pedon no longer are considered as soils. There is no upper limit to the size of pedons, and some have been observed that extend over several square miles. For mapping purposes, it is clear that individual pedons may be much too small to be shown; hence, larger units consisting of generalized agglomerations of pedons are shown in their areal extent. The pedon thus sets the level of generalization of form in studying and classifying soils. The classification of soils at different levels of generalization is discussed in Chapter 16.

No soil is an inert body, regardless of its origin or its age. Like its covering of vegetation, it reacts to changes in the total environment. Every rain that falls on it changes it in some minute way. The winter frosts, the high temperatures at ground level on a warm summer day, the activities of living creatures of all kinds, continually modify soil at varying rates, in different ways, and in different places. The breakdown of rock into soil was indicated earlier as part of the modification of the earth surface by exogenic energy flows. Bare rock is not a reflection of equilibrium at the juncture of the earth crust and the atmosphere. When such rock is exposed, energy is applied to alter it. Part of that energy is contained in the complex alteration of organic and inorganic material in soils. In time, perhaps measured in millions of years, a flat plain near sea level could be covered by a soil that had reached some state of equilibrium, a balanced mixture of air, water, minerals, and life. Until that time— and the earth crust rarely remains stationary that long—changes continue, governed by the processes of soil formation.

Like landforms, climate, and vegetation, soils are subject to the basic principles of natural distributions discussed in Chapter 1. That is, they (1) develop an infinite variety of forms areally, yet show degrees of similarity that permit regional comparisons; (2) show transitional changes between one soil and another; (3) alter with the passage of time; and (4) progress toward a relative equilibrium within the total environment.

[2]A *horizon* is a depth zone within soil that exhibits a certain degree of homogeneity in properties and contrasts with that in zones above and below.

CHAPTER 15

SOILS:

THEIR PROPERTIES AND PROCESSES OF FORMATION

The most important differences between soils include variations in one or more of the following properties: color, texture, structure, horizons, composition, and depth. Each of these is discussed briefly in the following sections.

15-1 SOIL COLOR Color is one of the more recognizable properties that distinguish soils. Strong contrasts in the color of surface soils are significant features in the comparative geography of continental regions. The bright red and yellow soils of the southeastern United States and the black prairie soils of the Ukraine are prominent regional characteristics even to the casual visitor. Like the other properties of soil, color varies not only in different areas but also at different depths within the same soil. Nearly all the colors of the soil are various mixtures of white, black, red, yellow, and brown. The diagram in Figure 15-1 shows an arrangement of the dominant soil colors and suggests possible results of color combinations.

The identification of color frequently presents problems of differences in personal judgment, especially in transitional zones. In order to remove such subjective differences, a standard color unit chart, termed the *Munsell Color Chart*, has been developed for use by soil surveyors. This consists of a set of about 175 different colors printed on bits of paper and mounted on cards. These represent all the gradations of color that have been observed in soils. They have been so arranged that the color index indicates not only the predominant hue (red, brown, yellow, etc.) but also the intensity, or *chroma*, of that color and its *value*, or the degree of lightness or darkness. The color cards are mounted in a looseleaf notebook and can be laid alongside a particular soil in the field for a rapid and accurate qualitative designation.

A wide variety of organic and inorganic compounds are found in soils, but only a few of them are responsible for the various color combinations. Among these, the most important are the iron

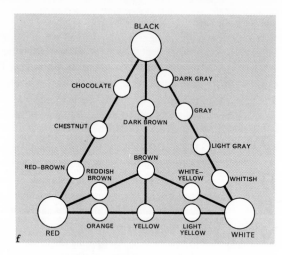

Figure 15-1   Soil color combinations.
[After Zakharov]

oxides and organic material. Table 15-1 lists the basic colors of soils and the constituents that are mainly responsible for them.

Greenish and bluish shades slightly tint heavy gray clays in the soils of humid, poorly drained regions. As shown in Figure 15-1, combinations of the constituents listed in Table 15-1 yield various color mixtures. For example, iron (red) and organic material (black to brown), both common soil materials, result in reddish-brown shades, such as chestnut. The effectiveness of the coloring agent varies. A small percentage of iron, for example, may produce a bright red color in soils that contain much higher percentages of aluminum or silica, because the latter are white to colorless. The coloring may also be more noticeable in fine-textured soils, because the dispersal is greater. This is some-

Table 15-1   Origin of Soil Colors

Color	Constituent
Black	Carbonate ions, usually Ca^{++} or Mg^{++}, plus highly decomposed organic material; other cations (Na^+, K^+), plus highly decomposed organic material; sulfur compounds;* manganese oxide*
Red	Ferric iron oxides: *hematite* (Fe_2O_3); *turgite* $2(Fe_2O_3) \cdot H_2O$; *goethite* (Fe_2O_3) $\cdot H_2O$
Yellow (ocherous or light yellowish-brown)	Hydrous ferric oxides: *limonite* $2(Fe_2O_3) \cdot 3(H_2O)$
Brown	Partially decomposed, acid, organic material; combinations of iron oxides, plus organic material
White to colorless	Aluminum oxides and silicates (*kaolinite, gibbsite, bauxite*); silica (SiO_2); alkaline earths ($CaCO_3$, $MgCO_3$); gypsum ($CaSO_4 \cdot 2H_2O$); highly soluble salts (chlorides, nitrates, borates of sodium and potassium); certain organic colloids*
Bluish	*Alloysite*, a hydrous aluminum oxide;* and *vivianite*, a hydrous ferrous phosphate.*
Green	Ferrous (incompletely oxidized) iron oxides*

*Relatively minor soil constituents that require unusual conditions for their accumulation.

what similar to the small amount of coloring agent needed to tint glass. An increase in moisture content also tends to sharpen the intensity of soil colors.

Many soils exhibit spots or streaks of color that stand out markedly against the background of the general soil mass. These often are shades of yellow or brown and represent the oxidation of iron along root channels or within large soil openings. When noticeably present, such soils are said to be *mottled.* The degree of mottling is a good index of the amount of alternate wetting and drying. Other mottling may be formed as the result of manganese concretions, which appear as irregular nodules with a dark red to black coloration.

While color is an important property by which soils may be distinguished, care must be taken not to draw too close a correlation between color and other properties, such as soil fertility or organic content. While some black soils are indicative of a high content of lime-saturated humus, some of the black soils of low latitudes have humus contents that are much less than those of adjacent red ones. It is generally true, however, that most white surface soils, with very few exceptions, tend to be relatively infertile.

15-2 SOIL TEXTURE Soil texture refers to the relative portion of different sizes of individual soil particles. It usually does not include the material over very coarse sand (+2 mm). Texture is a useful criterion in distinguishing soil differences vertically and horizontally. Table 15-2 presents a classification of textural classes as adapted by the U.S. National Cooperative Soil Survey.[1]

[1]The National Cooperative Soil Survey is a cooperating group of individuals who map, classify, and interpret soils in connection with their professional work. They represent many agencies, such as the Soil Conservation Service, the land grant colleges, the Forest Service, and the State Highway departments.

Table 15-2 Soil Textural Classes

Class	Particle size (mm)
Very coarse sand or fine gravel	1–2
Coarse sand	0.5–1
Medium sand	0.25–0.5
Fine sand	0.1–0.25
Very fine sand	0.05–0.1
Silt	0.002–0.05
Clay	Below 0.002

Since soils rarely consist of particles of only one size group, an additional classification is necessary to describe mixtures of particle sizes. The diagram in Figure 15-2 presents in graphic form the quantitative definitions of the various textural classes used by the National Cooperative Soil Survey. The textural-class triangle shown in Figure 15-2 also can be adapted to indicate differences in textural grouping with depth, as this figure illustrates. Points A, B, and C give the quantitative mixtures of particle sizes for the surface, lower soil, and underlying parent material, respectively, for a Miami silt loam from Indiana. The mechanical analysis is tabulated in Table 15-3.

Many factors influence the texture of soils, among them the mineral composition of the parent rock material, the nature and rate of the local soil-forming processes, and the relative age of the soil.

Unqualified statements regarding relationships between the texture of soils and soil fertility should not be made, although texture may be significant in the relative ease of tilling the soil. Sandy soils in humid climates usually are rapidly leached of soluble bases and hence are low in plant foods. They are also likely to have a low capacity for holding soil moisture; therefore they tend to be

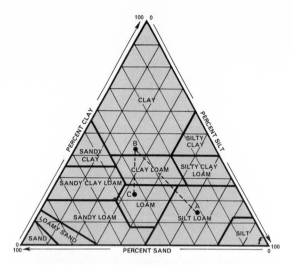

Figure 15-2 Soil texture diagram. The points *A*, *B*, and *C* represent samples taken from three levels in a soil profile in Indiana. The topsoil (*A*) is thus a silt loam, the lower horizon (*B*) a clay, and the parent material (*C*) a loam. The mechanical analysis of this soil profile is given in Table 15-3. [U.S. Dept. of Agriculture]

cultivation and fertility, depending on their chemical and structural arrangement. Differences in the behavior of the tiny clay particles are discussed later (see Section 15-5). Some clays are plastic, sticky, and almost unworkable when wet, and hard and cloddy when dry. Others, however, have a high degree of water permeability and crumble readily when dry.

From the standpoint of tillage, or the workability of the soil, silts and loams are desirable, since they rarely are too loose or too tight for plowing or harrowing. Their fertility varies widely, however, depending on the quantity and availability of plant foods and soil moisture.

15-3 **SOIL STRUCTURE** Soil structure refers to the physical arrangement and grouping of individual soil particles. It is developed to some extent in nearly all soils, although it may be absent in loose dune sands or in extremely plastic clays. The behavior of a soil is closely related to the size, shape, and arrangement of the structural *aggregates* (see Figure 15-3). An individual, natural soil aggregate is termed a *ped*. A *clod* is an aggregate caused by some outside disturbance, such as plowing. Natural, irreversible cements sometimes will unite soil particles into irregular lumps termed *concretions.* Some of the most common types of soil structures have been briefly classified in Table 15-4.

Soil structure has an important influence on the direction and ease of movement of water in the soil. Many of the usual methods of field cultivation are designed to produce a more desirable type of structure in order to improve conditions of soil drainage, aeration, and availability of plant foods. The addition of lime to a heavy, plastic clay soil, for example, flocculates the clay particles into larger structural units. *Flocculation* is

droughty. By adding much organic material, however, some sterile, humid, sandy soils have been changed into highly productive ones. Although sandy soils in arid regions have a high nutrient content, because of the small amount of precipitation and leaching, their low capacity for water retention reduces their agricultural value.

Clay soils vary considerably in ease of

Table 15-3 Mechanical Analysis of a Miami Silt Loam			
Horizon	Sand, %	Silt, %	Clay, %
A	21.5	63.4	15.0
B	31.1	25.0	43.4
C	42.4	34.0	23.5

the process of uniting the clay particles into aggregates and is somewhat comparable to the process of curdling milk. Rolling a freshly seeded field breaks up the aggregates into smaller sizes, thus increasing the number of soil openings and bringing capillary water closer to the surface to aid in seed germination. One of the main reasons for fall plowing is to expose the soil to alternate freezing and thawing, processes which tend to break up the larger clods into smaller structural units and to aerate the soil more thoroughly.

Structure in most soils is better developed in the lower part of the soil than it is at or near the surface, where, indeed, it may be missing entirely. The reason for this is partly that fine particles can be washed out of the upper part of the soil and also because the bases and other flocculating agents are more likely to be found below. Many sandy soils are so deficient in small-sized particles as to prevent the formation of structural aggregates. The sandy soils of desert regions, beaches, and dunes are illustrative.

15-4 HORIZON DIFFERENTIATION Soil-forming processes begin to act at once upon any exposed mass of unconsolidated rock material. Since these processes usually operate from the surface downward, differences in color, texture, structure, composition, and other properties eventually appear at varying depths. These vertical differences, which generally become more pronounced as the soil ages, may divide the soil into zones, known as *horizons.* A vertical section through all the soil horizons constitutes the soil *profile.*

Although most soils exhibit a profile—or morphological, chemical, and biological differences with depth—the separation of the vertical differences into distinct horizons does not always take place. This is illustrated

Figure 15-3 An open, crumb, soil structure in a prairie soil of McLean County, Illinois. This is representative of the prairie soils (Mollisols) of the American Corn Belt. Note the deep black color, indicative of a high content of calcium humates. [Standard Oil Company of N.J.]

by many humid tropical soils, which may show only a gradual transition, through perhaps dozens of feet, between the surface and the underlying material from which the soil is made.

Early investigators of soil profiles suggested a simple threefold division of soil horizons. These were designated as *A, B,* and *C horizons.* The *A* horizon included the topsoil, containing the vegetation litter and the *zone of eluviation* or leaching zone, from which material was shifted downward, by

Table 15-4 Classification of Soil Structural Units

	Type (shape and arrangement of peds)						
		Prismlike; horizontal dimension much less than vertical. Vertical faces		Blocklike; polyhedronlike, or spheroidal; 3 dimensions fairly even			
				Blocklike; blocks or polyhedrons having surfaces that are casts of the molds formed by ped faces		Spheroids or polyhedrons having plane or curved surfaces not related to faces of nearby peds	
	Platelike; vertical dimension much less than length or width	Without rounded caps	With rounded caps	Flat faces; most vertices angular	Mixed rounded and flat faces; rounded vertices	Nonporous peds	Porous peds
	Platy	Prismatic	Columnar	Blocky or angular	Subangular blocky	Granular	Crumb
Very fine or very thin	1 mm	−10 mm	−10 mm	−5 mm	−5 mm	−1 mm	−1 mm
Fine or thin	1–2 mm	10–20 mm	10–20 mm	5–10 mm	5–10 mm	1–2 mm	1–2 mm
Medium	2–5 mm	20–50 mm	20–50 mm	10–20 mm	10–20 mm	2–5 mm	2–5 mm
Coarse or thick	5–10 mm	50–100 mm	20–100 mm	20–50 mm	20–50 mm	5–10 mm	
Very coarse or very thick	+10 mm	+100 mm	+100 mm	+50 mm	+50 mm	+10 mm	

SOURCE: Soil Survey Staff, *Soil Survey Manual,* U.S. Dept. of Agriculture, Handbook No. 18, p. 228, U.S. Govt. Printing Office, Washington, D.C., August, 1951.

either the mechanical movement of fine particles or chemical solution by percolating rain water. The *B* horizon was designated the *zone of illuviation,* or zone of accumulation, located at a variable depth below the surface, and receiving material brought to it from above. The *C* horizon was simply designated the parent material from which the soil was derived. Such a simplified system, with a few refinements, worked reasonably well in Europe and in parts of North America, but with more extensive studies of soils through-out the world, this classification became increasingly difficult to apply. Frequently *A* horizons were absent entirely, or *B* horizons did not exist. Obvious changes in parent material (such as alternate layers of till, outwash, and loess) and combinations of transported and nontransported fragmental material caused uncertainty as to what to designate as *C.* Furthermore, the differences in types of accumulations in the illuvial zone appeared too great to be lumped together under a simple *B* designation. Classifications are still being changed, and there is a growing tendency to shift to diagnostic terms, but a highly modified version of the *ABC* classification is presently in use by soil surveyors of the National Cooperative Soil Survey. This version utilizes four types of code symbols: (1) a series of capital letters, *O, A, B, C,* and *R,* representing master horizons; (2) Arabic numerals 1, 2, and 3, following the capital letters, representing distinguishable subdivisions of the master horizons, in which the uppermost in the soil is number 1; (3) lowercase letter symbols representing special alterations of the master horizons, as compared with the parent material, and appended as suffixes to the Arabic numerals; and (4) a set of Roman numerals that precede the capital letters of the master horizons and are used when more than one kind of *lithology* (rock composition) is involved in the profile. A brief summary of these coded symbols as applied to horizons is presented in the following subsections.[2]

The master horizons and their subdivisions. The *O horizon* refers to the organic layer or layers that overlie the major mineral portion of the soil profile. Therefore, it is not used to designate any part of an organic soil, such as a thick peat or muck deposit. The *O1* subdivision refers to the fresh-litter portion of the *O,* in which the dead organic remains of either plants or animals still are clearly recognizable. The *O2* subdivision differs in that it includes some inorganic soil material. Also, the organic debris has been so decomposed that the original plant structures are no longer recognizable. The organic portion of the soil mass exceeds 30 percent in clay soils and 20 percent in coarse-textured soils. Since organic material weighs much less than

the mineral portion, 30 percent by mass indicates a pronounced predominance by volume. There should be no doubt that the organic portion dominates within the horizon.

The *A horizon* retains its older designation as an *eluvial* zone from which material (mainly clay, aluminum, and iron) is removed, with a resultant proportionate increase in resistant fragmental materials, such as quartz and mica. The horizon may contain some finely divided organic material, but this is never more than 30 percent of the soil mass. Boundaries above and below may or may not be sharp; hence, the subdivisions 1, 2, and 3 are used to designate transitions, in which 1 and 3 represent transitions with the horizons above and below, respectively. If both boundaries are sharp, no subdivision numeral is used.

The *B horizon,* like the *A,* is retained from the older classification, but it has taken on additional meanings besides that of a simple *illuvial* zone. The horizon shows a noticeable increase in the percentage of such materials as colloidal silicate clays (see Section 15-5), iron and aluminum oxides, and colloidal silica, compared with the parent material, yet not all of them need be transported into the horizon from above. Some of them may be created as the result of the weathering of feldspars and ferromagnesium minerals within the soil horizon. This is especially active in the deep subsoils of the humid tropics. The B horizon usually also is colored with redder and deeper shades than the parent material because of the concentration of iron oxides. Structural aggregates unrelated to those within the parent material also are generally present. When the deposition of material from above is concentrated into tight cemented layers and *fragipans,*[3] these

---

[2]The material in the following subsections has been condensed from "Soil Classification: A Comprehensive System—7th Approximation," Soil Survey Staff, Soil Conservation Service, U.S. Dept. of Agriculture, U.S. Govt. Printing Office, Washington, D.C., pp. 24–29, 1960.

[3]A *fragipan* (L.: *fragilis* = brittle) is a compact, fine-textured layer that is almost cementlike when it is dry but breaks or crumbles relatively easily when moist. Its material is closely packed, and it frequently features vertical blocks or prisms as structural units.

layers are labeled separately within the *B* horizon. As in the *A* horizon above, the subdivision numbers 1 and 3 represent zones in which subsidiary properties of the horizon above and below are present. Usually the gradation with the *C* horizon is more gradual than that with the *A*.

The *C horizon* is a mineral layer that does *not* include bedrock but yet has not been influenced by normal soil-forming processes. It may have undergone rock weathering, as in the deep tropical rock mantle, and it can be influenced by water table conditions, such as the reduction (decrease in oxygen) of iron oxides, the formation of fragipans, and the deposition of soluble salts, alkaline earth carbonates, colloidal silica, and iron and aluminum oxides. It definitely lacks the properties normally attributed to the *A* and *B* horizons. It is material that lies outside the normal reach of the biotic activities of soils. Plant roots and soil microorganisms are not found in this zone, and it might be considered a geological product rather than soil itself. It may, however, support plant life if exposed at the surface, and for that reason it is considered part of the soil. It should be noted that the *C* horizon is not necessarily the parent material from which the soil is derived, although in most cases it is.

The *R horizon* is the consolidated underlying bedrock.

The horizon suffix symbols denoting special conditions within the master horizons.

b (buried soil horizon): Climatic changes, a change in hydrologic conditions, and several other causes may result in relict horizons that belong to some previous soil-forming environment. Such horizons are more common in the lower latitudes.

ca (calcium): Accumulation of carbonates of the alkaline earths, usually lime ($CaCO_3$) or dolomite ($MgCO_3$). The content must be greater than the parent material. The *ca* zone may appear in the *A*, *B*, or *C* horizons.

cs (calcium sulphate): Accumulation of gypsum ($CaSO_4$). If present, this usually is located immediately below a *ca*.

cn (concretions, nodules): Accumulations of lumps or crusts composed of iron, aluminum, or manganese oxides.

f (frozen): A frozen layer below the surface. This usually is permafrost, although it need not be.

g (gley): Prevalent *gleying*. This is a reduction process caused by the removal of oxygen from ferric oxides, mainly (but not always) under stagnant water conditions. Ferrous compounds are produced that are shades of dark gray to slate blue or green. The interiors of soil structural units are highly mottled.

h (humus): Illuvial humus. This consists of distinct coatings of colloidal humus or organic material on the sand and silt particles of a *B* horizon, having washed down from above.

ir (illuvial iron): This is a coating or cementing (bridging) of sand or silt particles in a *B* horizon caused by the downward migration of colloidal iron oxides.

m (massive): Strong cementation into a distinct layer. *Plinthite, duripan,* or *caliche* (desert hardpan) are examples.

p (plowing): Human disturbance of the soil through cultivation, pasturing, manuring, etc. This symbol is the only letter suffix to precede the horizon subdivision numerals, and it is used only with the *A* horizon (*Ap1*).

sa (salt): Accumulation of soluble salts that are more soluble than calcium sulfate and where there is more salt than the parent material.

si (silica): Cementation into layers, crusts, or nodules by colloidal silica (silica soluble in alkaline solutions). This is restricted to *C* horizons and is usually found in dry regions

Figure 15-4 Typical profile of a Miami loam.

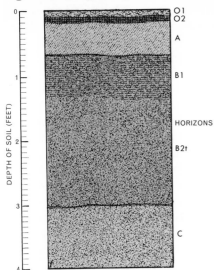

O1 A thin accumulation of forest litter.

O2 From 1 to 2 inches, dark, mellow loam or humus, containing a high percentage of organic matter, much decomposed and thoroughly incorporated in the soil. The reaction is medium acid.

A From 2 to 8 inches, light-gray, floury, loose loam with a high content of silt and slight development of a platy or laminated (layered) structure. The material crumbles easily into a structureless mass. The reaction is strongly acid.

B1 From 8 to 16 inches, light-yellow or grayish-yellow loose friable loam, platy in the upper part, becoming granular below. Very acid.

B2t From 16 to 36 inches, clay loam breaking into irregular angular particles about $\frac{1}{2}$ inch in diameter. The structure particles are brown, but when they are crushed, the resulting soil is yellowish brown. A thin coating of very fine textured brown material on the structure aggregates accounts for their color. When wet, the material is sticky; when dry, it is difficult to crush between the fingers. The reaction is acid.

C Imperfectly weathered, heavy, pale-grayish-yellow calcareous sandy loam or loamy glacial till containing a few stones. The material is variable in color, structure, and texture. It is hard when dry.

or in the alternate wet and dry climates of the tropics.

t (Ger.: tot = clay): Illuvial clay. It consists of accumulations of colloidal silicate clays as masses, pore fillings, and coatings on sand and silt particles within B horizons only, and carried there from above.

x Fragipan formation (see footnote 3): These brittle pans can be found in $A2$, B, and C horizons.

Lithologic discontinuities. Roman numerals preceding the capital letter of the master horizons are used when the depth of the soil profile encompasses more than one kind of parent material. The numbers are arranged from the top downward. Buried soils should not be so designated unless there is clear indication that the material above was transported and is different from the older soil.

Combinations and gaps in horizon sequences. It is possible to combine two horizons into a BC horizon, where the transition is so broad that it is difficult to determine which horizon is dominant. In this case, it would represent a zone between $B3$ and $C1$. It is not necessary that all the master horizons be represented in the soil profile. Soil erosion frequently will remove topsoils, in which case only the C may be exposed at the surface. A and C combinations are fairly common along side hills. The B horizon may be absent under conditions of continual poor drainage, as in a shallow swamp, where the gley development reaches to the surface of the inorganic material. It also is possible to have more than one B or A horizon, in which case the lower ones are indicated as buried soils. Figure 15-4 shows a typical sequence of horizons in a Miami loam, a rather common soil type in the American Midwest; it is developed under a forest cover, and the parent material consists of a highly calcareous (limey) glacial till. Note that the A horizon is sharply defined, since no Arabic numeral is used. The B horizon has an upper transition, but the contrast with the C horizon is clear. The $B2t$ indicates that some illuvial clay has been moved downward, coating the soil aggregates.

Soil horizons are treated further in Chapter 16, where specific diagnostic horizons are discussed in connection with soil classification (see "Soil Classification").

15-5 INORGANIC SOIL CONSTITUENTS

The inorganic portion of soils represents that portion which was derived from the parent rock or mineral material. This material is chemically similar to the common rocks of the earth crust, as indicated in Table 10-1. Soils from widely separated parts of the world have oxygen, silicon, aluminum, and iron as their leading chemical elements, with the common bases of magnesium, calcium, sodium and potassium as minor elements. Unusually high percentages of the minor elements appear only in dry land soils, where they can be concentrated by the evaporation of water. The inorganic soil particles can be divided into two groups: (1) the silts and sands, which are merely the small pieces of the original rock material that have not yet undergone chemical decomposition; and (2) the tiny clay fractions that have resulted from the chemical alteration by weathering processes.

The fragmental constituents. These particles form the skeleton of any soil and are the residue of rock and mineral disintegration. They tend to be composed mainly of the more resistant minerals found in the crustal rocks, but almost any type of mineral found on earth may be present. Especially common, however, are sand grains composed of quartz, mica, magnetite, garnet, etc., minerals that are highly resistant to weathering. Particles of less-resistant minerals, such as the feldspars, may be present in fairly young soils, or where the particles may have been protected by a coating of some kind. Weathering of the inorganic particles does not stop when soils are formed, and nearly all soils still retain many mineral fragments that are susceptible to further alteration by weathering. This helps to explain the natural regeneration of soil fertility. Such regeneration is more rapid in young soils.

The inorganic clay colloids. The processes of chemical weathering supply most of the clay fractions in soils. These are colloidal[4] and have unique properties that influence the form and behavior of soils.

Recent investigations indicate that the clay colloids in soils, instead of being irregular lumps of material, microscopic in size, have a characteristic crystal form, usually consisting of tiny bundles of flakes or sheets and resembling the arrangement of sheets in a crystal of mica (see Figure 15-5). This flake-like property tends to increase the interfacial characteristics of the mass, since atoms, ions, and molecules are able to work their way between the tiny sheets of the clay colloids. The colloidal chemistry of soils is a relatively new field of soil science, but it has already indicated that many of the processes responsible for the differences in soil productivity throughout the world are closely related to the makeup of inorganic and organic colloids. One of the most significant factors in soil

[4]The peculiar physical and chemical properties of colloids are essentially the result of combining the electrical forces of different molecules along *interfaces* (the boundaries of liquid-solid, liquid-liquid, liquid-gas, etc.). A colloidal system is simply a mass which contains enough interfaces to produce a dominance of interfacial properties. If a subdivision of any mass is carried far enough, it may result in a colloidal system. When particles of a substance enter a colloidal system, they often assume physical and chemical properties that are distinctly different from those of the same substance in larger-sized units. Fat globules in milk, for example, rise to the surface and become cream because of their lower specific gravity. When these globules are broken into much smaller droplets of colloidal size (homogenization), however, they become stabilized throughout the liquid. Milk itself is an irreversible system of solid colloidal particles within a liquid. The process of coagulation (flocculation) associated with souring or the addition of an acid destroys the former colloidal stability and unites the solids into masses, the curds. The process cannot, however, be reversed.

Figure 15-5 Soil inorganic colloids and their properties. The sandwichlike arrangement of an individual clay-colloid particle is shown in (1). The adsorption and absorption of cations and water molecules is diagramed in (2). The base-exchange function of silicate clays is shown in (3), where calcium and magnesium ions in the soil solution replace hydrogen and sodium ions in the clay complex. The flocculating action of calcium and magnesium is indicated in (4), where the positive charge of these ions acts as a bridge between adjacent clay particles whose surface charge usually is negative. The peptizing action of hydroxyl (OH^-) ions is shown in (5), where the negative charge helps to keep the particles separated. A typical clay floccule or soil aggregate is shown in (6).

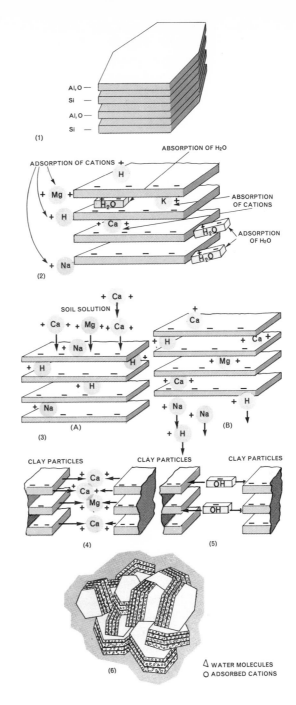

fertility, for example, is soil *cation-exchange capacity.*[5] This is a measurement of the capacity of the soil colloids to exchange cations held by the colloidal particles for others that are found in the soil solution. A high cation-exchange capacity indicates that the colloidal soil "bank" is well supplied with various types of cations and can exchange them for different types of ions from the soil solution. The plant foods that roots derive from the soil solution are metallic cations or bases that have been displaced from the colloidal interfaces by less useful ions, such as hydrogen.

The capacity of the inorganic clay colloids to absorb, adsorb, or exchange cations and water depends much on the chemical makeup of the lattice layers within the colloidal particles. Each layer in the lattice, as shown in Figure 15-5(1), consists of a horizontal sheet of oxygen atoms combined with those of

[5]A *cation* is a positively charged atom or one that has lost one or more electrons, thus having a net positive charge. Many of the cations in the soil are metallic ions that also act as soil nutrients, such as calcium, potassium, sodium, and iron. *Monovalent cations* are those that have lost a single electron (K^+), while *bivalent cations* have lost two (Ca^{++}, Mg^{++}, or Fe^{++}). The metallic cations often are referred to as *bases* to distinguish them from some of the nonmetallic ones, such as hydrogen (H^+).

silicon, aluminum, or iron. The silica lattice sheet invariably is in the form of SiO_2 molecules, while the iron and aluminum tend to unite with oxygen in the sesquioxide form (Fe_2O_3 or Al_2O_3). The most effective layer in the lattice structure appears to be that of combined silica. The greater the number of silica layers, the more stable the colloidal complex and the greater its absorptive or adsorptive capacities. Clays differ in their arrangement and in the proportion of the silica layers to those of the sesquioxides.

There seems to be a relationship between the composition of the clays and the general acidity of the prevailing soil environment. In highly acid soils, the silica layers tend to remain in the lattice, thus preventing further breakdown of the colloidal particles. In neutral or slightly alkaline environments, on the other hand, the silica tends to be removed from the lattice. Although the clays, like the feldspars, have a wide range of chemical composition, they have been classified into three major groups: (1) the *montmorillonites*, which have the greatest amount of combined silica layers in the lattice and whose capacity for absorption is extremely high; (2) the *illites*, sometimes termed *mica clays* because of their similarity to tiny flakes of the potash mica *muscovite*; their combined silica tends to be intermediate; and (3) the *kaolinites*, which have only a small amount of combined silica and hence low absorptive and adsorptive capacities. Kaolinites might be considered clays that are nearing the end of the process of clay decomposition, that of removing the silica layers from the lattice.

It is interesting to note that the montmorillonite clays appear to be found either in young soils, as in newly weathered basalts and glacial tills, or in the older soils of cool, humid regions. The kaolinites, on the other hand, are found mainly in the old soils of the humid tropics. The desilication of clays beyond that of the kaolinites apparently requires a long period of time in warm, humid climates. The low fertility of humid tropical soils is related in large measure to the dominance of kaolinite in the clay fraction of the soils.

15-6   ORGANIC SOIL CONSTITUENTS
The organic content of most soils is low, usually less than two or three percent of the total soil mass. Only in places where organic material can accumulate, as in marshes or swamps, does it constitute an appreciable amount, and even here, silts and other fine-textured inorganic material often form a surprisingly large part of the total mass.

Despite its proportionately small amount in well-drained soils, dead organic material plays an extremely important and active role in the soil-forming processes, as well as constituting a significant link in the carbon cycle. Here in the decay process of altering dead organic material is the connecting link between the biosphere, hydrosphere, atmosphere, and lithosphere. The conversion of organic compounds into their inorganic constituents is also a point of energy exchange; thus this becomes one of the control points in the global energy system. The mineralization of organic material involves a host of chemical and biological processes. The end results are the elementary building blocks which once were taken from the soil, water, and air to form living organic tissue. These include CO_2, water, oxygen, nitrogen, some of the simpler hydrocarbons, and various bases, the so-called plant nutrients. The last sometimes are collectively referred to as *ash*. There is, however, a wide variety of intermediate decay products. In classifying such intermediate products, two general groupings can be used: (1) the bits and pieces of the original plant tissue, ranging from freshly fallen leaves and twigs to durable resins and pollens that may remain within the soil relatively unchanged for centuries; and (2) the colloidal and usually gelatinous material that is termed *humus*. This material is by far the most active part of the entire soil mass.

Organic colloids, as a rule, are much more efficient than the silicate clay colloids in absorbing and adsorbing water molecules and bases (metallic cations). The strong negative charge on the surfaces of the organic colloids, and thus their great attraction for cations, tends to produce a surplus of positively charged hydrogen ions in the soil solution. Some of the colloids, moreover, become so highly dispersed under the acid environment that they move downward with the soil solution. For this reason, such solutions are sometimes termed *humic acids*. There are many types of these, partly because of the variety of material from which they can be derived and also because of the many stages and types of decay processes. Such organic acids, or more properly, solutions of dispersed organic colloids, differ greatly in properties and behavior. The brown color of the water in the lakes, streams, and swamps of the taiga is derived from the organic colloidal "solutions."

Organic colloids may play an important role in agriculture through their influence on the cation-exchange capacity and their ability to store water and nutrients. Some extremely poor, sandy soils can be made highly productive by plowing green plant material (*green manuring*) into the soil. This quickly decomposes into humus. Saturated with applied lime and fertilizer, the humus flocculates and provides the needed storage for water and nutrients. Not all organic colloids aid in fertility, however. Some of them are irreversible gels that decompose extremely slowly and coat soil particles, making them chemically inert.

The type and amount of soil organic material depend mainly on three factors: (1) the addition of new supplies, dependent in turn on the quantity and type of the biomass; (2) the rate of decay; and (3) the amount of bases in the soil that tend to flocculate the humus into gels, protecting the colloids within from decay and leaching. Humid tropical areas, although having a large biomass in the luxuriant growth of tropical rain forests, generally have a low humus content within the soil. This is because of rapid decay and the relatively lower content of bases in the soil. The luxuriance of the forest, as described earlier, is based on the rapid mineralization of the forest litter, the high surplus of net radiation, the abundance of water, and the addition of new nutrients to the cycle by the weathering of inorganic soil particles through great depth (see Figure 14-2). In contrast to the humid tropics, cold regions such as the tundras, which have a small biomass, owing to the low net radiation, have a relatively high organic content in their shallow soils. Much of this organic content has not reached the colloidal state, owing to the slowness of decay, which takes place only during a short period of the year. The soil humus, however, tends to be highly acid and highly dispersed, and the amount of exchangeable cations is small.

The soils that have the greatest amount of humus, other than the organic accumulations of some marshes, are those found in the transitional zone between wet and dry, and cold and warm climates. These areas over much of the world are grasslands. Despite a low grass cover above the surface, all of it dies each year to add to the litter, and the dense root system adds a further amount annually to the soil. Decay is not so great as in more humid regions, yet is sufficient to convert the nonwoody vegetation rather quickly into colloidal form after death. The plentiful supply of alkaline earth carbonates (bivalent cations), also resulting from the arid-humid transition, maintains the flocculated, crumblike structure of the organic colloids and the associated mineral portion of the dark topsoils (see Figure 15-3). The black, mid-latitude grassland soils are among the most naturally fertile soils in the world because of their high amounts of organic colloids, well saturated with bases.

Closely related to the organic content of the soil is the amount of nitrogen in both soluble and insoluble form. This element, so abundant in the atmosphere and so essential to the manufacture of proteins for tissues, does not easily unite with other elements to form soluble compounds. Most soluble nitrogen compounds in soils are of organic origin. They result mainly from the activity of special bacteria that require organic material on which to feed or living plants on which they may be parasitic, such as the legumes. Most soils thus have a soluble nitrogen content that is roughly proportional to their contained organic material. This explains why desert soils, which are rich in most other soluble nutrients, are generally deficient in soluble nitrogen ions. Some also believe that the extremely low height of plants in the tundra may be related to the low availability of soluble nitrogen in the soils. Nitrogen-fixing bacteria do not operate in the frozen soils during the long arctic winters.

Swamps and marshes are favored places for the accumulation of organic material. When the vegetation there dies, it is buried below the water level and thus is partially protected from decay, since most decay-producing microorganisms require free oxygen. There are some forms of microlife, however, that are able to extract oxygen from compounds instead of requiring the gaseous form. These can feed on submerged organic material, but do so slowly and show a decided preference for swamp water of low acidity. Under high acidity, such as is found in many bogs, the swamp organic material is brown, fibrous, and not greatly decayed, and the original plant material usually can be easily identified. Such material is known as *peat.* Where the groundwater is not acid, the resulting organic material tends to be a black, finely divided humus, termed *muck.* When drained and fertilized, muck forms one of our most productive soils for intensive cultivation.

15-7 THE SOIL SOLUTION   This is a dilute solution of acids, bases, and salts and may also carry colloidal particles. It enters the soil profile from either the top or the bottom and either passes through it or remains behind in the small soil openings. Despite its transient nature, the soil solution is a vital part of the soil body. Water itself is a poor solvent for most earth materials, but when combined with acids or bases, it becomes an important agent of rock decay and soil formation. Were it not for the mobility and chemical activities of the soil solution, soil profiles would not develop. The characteristics and the amount of the soil solution are strongly influenced by climate, and it is partly for this reason that many of the major soil-forming processes can be correlated with climatic differences. Plants depend on the soil solution as a source of mineral foods and of the water which is necessary for life.

The quantity of fluids contained within the soil may produce many different physical traits in the soil mass. Plasticity, frost heaving, swelling and shrinking, soil flowage down slopes, and many other properties are closely related to water content and its interaction with soil texture and structure. One of the most important factors influencing the quantity of the soil solution is *porosity,* or the total amount of space between the soil solids. This influences the capacity of soil to hold water. *Permeability,* or the relative ease with which the soil solution passes through the soil, is directly related to the size of the soil openings. In general, fine-textured soils have greater porosity but less permeability than those of coarse texture; thus, they not only hold more water but retain it longer. Colloidal particles, particularly the organic colloids, likewise greatly increase the water-storage capacity of soils.

Not all of the soil solution is available to plants; some of it is bound to the tiny surfaces of soil colloids by a powerful force, at times reaching the magnitude of 15,000 pounds per square inch. In an aggregate of colloidal particles, as in a clay aggregate or micelle, there is a gradual transition between the available water on the outside and the tightly held solution near the center. The colloids are the storage vaults that keep the soil solution from passing completely out of the soil by percolation or evapotranspiration and thus provide plants with a supply of water and nutrients to draw upon between rains.

15-8 THE SOIL ATMOSPHERE The soil atmosphere, another component of the soil, is found in all of the openings that are not filled with water. It is a necessary constituent, not only because plant roots and microorganisms depend on it for respiration and transpiration, but also because the included gases play important roles in the processes of inorganic and organic decomposition. Soil air differs from the air just above the ground in several ways. The most important may be summarized as follows:

1. Soil air is not continuous; hence, wide differences in composition are possible.

2. The carbon dioxide content of soil air usually is much greater than that above-ground, and the oxygen content tends to be somewhat less.

3. Soil air tends to have a higher relative humidity, frequently being close to saturation. The lack of continuity with the surface air and the constant exposure to water molecules held along soil interfaces are the forces responsible.

Proper soil aeration, or the exposure of the soil interstices to surface air, is an important part of cultivation. The harmful effects of water-saturated soils are related not so much to the presence of water as to the shortage of suitable soil air. It can be readily seen that as drainage conditions alter, so too do the conditions of the soil atmospheres.

The Variables in Soil Formation

Soils are a product of their environment, and the different characteristics they show from place to place and from time to time are the result of changes in any one or more of the variables that make up the total environment. Although many factors affect soil development, the basic determinants are parent material, climate, vegetation, slope, hydrography and drainage, microorganisms, man and time. Any one of these eight elements may exert a dominant influence on the soil properties by which areal differences and similarities are recognized.

15-9 PARENT MATERIAL The parent material of a soil consists of the fragmental rock debris that masks the solid rock below. It may consist of transported material, such as alluvium, colluvium, loess, or glacial drift, or it may be the partially weathered material derived from the underlying rock. It sometimes is referred to as the *regolith* or rock mantle. It is incorrect to consider the soil parent material as unweathered. Some weathering occurs wherever air and water come in contact with the lithosphere. The major difference between the parent material and the soil is that the latter involves the biosphere and all of the processes connected with life and death, ranging from the translocation of nutrients by plant roots to the effective cation-exchange mechanisms on the part of the decomposing humus.

The importance of the parent material in conditioning the properties of the soil above—that part of the profile known as the *solum*—varies greatly. In young soils, or those in which the soil developmental processes have not worked for long, many of the soil properties resemble closely those of the parent material, such as texture, color, and mineral composition. In others, such as on the Russian erosional interior plains, which have remained in their present condition for millions of years, adjacent soils from widely dissimilar parent materials may be almost indistinguishable.

Some kinds of parent materials have a more lasting influence on soils than others. Siliceous sandstones or granites with a large quartz content will produce sandy soils that could persist for millions of years because quartz is almost indestructable, while fragments of porous basaltic rock weather quickly and become far different when worked on by the soil-forming processes above. It is interesting to note that the soils derived from many sedimentary rocks often are much more closely related to their parent materials than those derived from igneous rocks. This is because the sedimentary material has already been through part of the weathering cycle, and materials resistant to further breakdown tend to predominate—quartz sand from sand-stones and the colloidal clays obtained from shales. In the case of limestones, the overlying soil may be related much more to the properties of the insoluble impurities within the limestone than to the limestone itself.

15-10 CLIMATE The three variables of climate that are most significant in soil-forming processes are precipitation, temperature, and evaporation. These factors, related closely to net radiation surpluses, may influence soils directly through the rate of chemical reactions, the type of rock weathering that is dominant, and the characteristics of the soil solution. They may also affect soils indirectly by influencing the quantity and type of vegetation and the microorganisms that are present in the soil environment. The results of climatic influences are more apparent than are the exact processes by which these results are produced.

We have accurate climatic records for relatively a few score years. We know that weather is constantly changing, yet no one has been able to isolate climate as a single variable in soil formation, partially, of course, because climate itself is a generalization of many variables. Most of our measurements of climate are inadequate for detailed studies of soil processes. The climate of the soil atmosphere is quite different from that above the ground surface, where meteorological observations customarily are taken. Despite these difficulties, the influence of climate upon soils is unmistakable, and properties that are largely the result of climate have generally been used to classify soil differences on chorographic or global scales of generalization. The global patterns of soils, which correspond closely to climatic patterns, are discussed in the following chapter.

15-11 VEGETATION Vegetation affects soils, first, by supplying most of their organic content. Especially important is the organic colloidal material that may be present in the soil profile. Second, the chemical composition of the dead plant remains that become incorporated into soils influences the acidity and composition of the soil solution. Finally, the dead organic residue supplies the food for the rich microlife that dwells in the soil and plays a critical role in maintaining soil fertility and determining the characteristics of profile development. Unlike parent material, vegetation is not an independent variable. It is largely dependent on climate for its most important characteristics; yet, as has been demonstrated, it has features of its own that help influence the details of soil geography.

Some of the principal influences of vegetation on soils can be summarized as follows (see Figure 15-6):

1. Forests tend to supply more organic material to the surface of the ground than does any other major type of vegetation.

2. Grasslands, although they do not supply as much organic material to the surface as forests do, supply large quantities to the soil immediately below the surface. This is much less true in the rainy tropics.

3. The content of bases in the leaves and stems of plants varies with different types of plants. Broadleaf trees generally have a higher base content than do conifers (needleleaf trees); hence, the derived humus is likely to be less acidic in reaction. Some oaks, however, produce highly acid humus.

4. Forests tend to reduce soil erosion on slopes and hence help to stabilize the soil-forming processes there.

5. The rate and amount of translocation of bases and other plant foods from the subsurface to the surface vary with different types of plants.

6. Forests, by decreasing the velocity of winds, tend to reduce the evaporation of soil moisture and to retain snow cover; these effects, in turn, influence the activities of the soil solution.

7. Forests, as compared with grasslands, tend to have a larger proportion of fungi and a lesser proportion of bacteria among the microlife of their soils—other factors, such as climate, remaining the same. The different

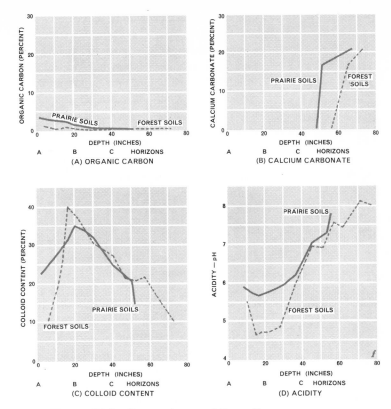

Figure 15-6 **Comparisons of the soil properties in typical prairie and forest soils in the American Midwest.** Note the higher organic and colloidal contents in the prairie topsoils, the higher pH (lower acidity), and the appearance of lime (calcium carbonate) at somewhat shallower depths in the subsoil, as compared with the forest soils. [After Jenny]

activities of such organisms are an important part of soil development.

Where man has appreciably altered the native vegetation of an area, significant changes in soil properties have been observed, often after a remarkably short period of time. Careful measurements of soils in New England, in areas where white pine has followed the abandonment of cultivated fields that originally were covered with virgin, deciduous hardwoods, have indicated noticeable changes in soil profiles within 10 years following the change in vegetation. Sharp local contrasts between prairie and forest soils, each having similar parent material or slope conditions, are observable in many places along the forest-prairie border in Minnesota, Wisconsin, and Illinois.

15-12 LOCAL RELIEF AND SLOPE CONDITIONS The rate and type of soil formation are influenced greatly by local conditions of surface irregularity and slope. Four major categories of local relief are recognized as influencing soil characteristics. Each one, however, is subject to qualifications, depending on special conditions.

Normal relief. This can be considered as comprising moderately sloping upland surfaces in which the runoff of surface water is neither excessive nor low. Under such conditions, soils are fairly well stabilized, erosion of surface material is compensated by the normal deepening process of soil formation, and a kind of dynamic equilibrium is established. This is the typical location for normal, mature soils.

Subnormal relief. This is found on upland surfaces that are flat to gently sloping. Drainage is slow, and the soils often exhibit characteristic features connected with alternate water saturation and drying. Water tables are likely to be shallow, and the accumulations of clay and sesquioxides tend to form hardpans of various kinds. Both the A and B horizons tend to be relatively fixed in position and do not move downward with age, as in better-drained soils.

Excessive relief. This includes terrain where steep slopes predominate, as in hilly and mountainous areas. Rapid runoff of water and erosion prevent the normal soil-forming processes from developing diagnostic horizons, and the soils are often closely related to the features of the parent material. Soil profiles exhibit stripped or truncated horizons in some cases, and the A and/or B horizons may be missing. The greater amount of runoff of surface water than that entering the soil means less water to form soil solutions and a much slower rate of soil development than on gentler slopes.

Flat or concave relief. This occurs on nearly level lowlands, where runoff is extremely slow or absent. Water accumulates, therefore, from both rainfall and upland drainage. Deposition of material prevails over removal both within and upon the soils. Buried horizons, saline conditions within arid regions, and various effects of accumulated groundwater are present, ranging from claypans to the accumulation of organic soils in poorly drained depressions.

The meaning of topographic factors cannot be expressed in terms of slope alone. The texture and structure of soil can either strengthen or mitigate the slope factor. Runoff and soil stripping are likely to be much less on a slope where the underlying material is highly permeable than where the material has a high clay content. The soils even on steep slopes under conditions of almost continuous rain are likely to exhibit characteristics normally found with poor drainage. The implications of this are important, because particular slope classes alone cannot be used as limits for slope cultivation. The influence of vegetation on erosion rates has been treated earlier in this text. On some soils, cultivation may be a stimulus to serious erosion on slopes of two to three percent, while on some of the permeable kaolinitic soils of the humid tropics, serious erosion following cultivation may not occur on slopes

Figure 15-7 Topographic influences on soil types near Dover, Delaware. The dark patches are organic half-bogs, developed in the slight depressions on an undulating land surface. Field ditches have been dug to drain these depressions. [U.S. Dept. of Agriculture]

of up to 40 percent. Nevertheless, slope is a factor that influences soil formation and stability. One need only fly over a cultivated area in most rolling country to observe the impact of slope and relief. In Figure 15-7, the dark, moist topsoils in the depressions contrast sharply with the lighter, less weathered material exposed on the swells or rises.

15-13 SOIL DRAINAGE Soil drainage refers to the rate at which water passes through the soil. Different aspects of it already have been mentioned in the discussions of slope conditions, horizon formation, texture, and structure. The most important factors influencing soil drainage are soil permeability, which in turn is influenced by the size and arrangement of the soil openings; runoff, which is affected mainly by slope and surface texture; precipitation and evapotranspiration; and various attitudes of the groundwater table. Any or all of these factors may be significant in influencing the characteristics and development of a particular soil.

Recognizing that drainage conditions can vary over a broad range of conditions, a few of the major influences are described under an extremely generalized classification within three groups: poor drainage, adequate to good drainage, and excessive drainage.

Poor drainage. Under poor drainage, the soil openings are largely filled with water during a large part of the year, the water table stands above, at, or slightly below the surface, and general anaerobic conditions prevail within

Figure 15-8 The development of hardpans in soils that have a high water table.

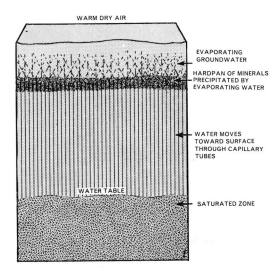

WARM DRY AIR

EVAPORATING GROUNDWATER

HARDPAN OF MINERALS PRECIPITATED BY EVAPORATING WATER

WATER MOVES TOWARD SURFACE THROUGH CAPILLARY TUBES

WATER TABLE

SATURATED ZONE

the soil environment. If the water table lies above the surface, organic material accumulates to various thicknesses. Reduction processes dominate to produce the resultant compounds that yield typical shades of gray, bluish or greenish gray, yellowish brown, or white within the mineral soils. Where the water table fluctuates in depth within the soil profile, typical mottling is common. When the water table remains fairly stationary at some point within the profile, its upper margin may be marked by a distinct depositional zone formed by the precipitation of material that is brought to it from above or below (see Figure 15-8).

Adequate to good drainage. This may be presumed to represent conditions where the water moves freely but slowly through the soil in any direction, where it stands on the surface for only short periods of time at or following rains, where a moderate amount leaves via evapotranspiration or into groundwater outlets, and where there is a balance between soil formation and soil loss under normal cultivation. The soil remains moist for much of the year. Fairly uniform coloring is found throughout most of the individual horizons, and if mottling occurs, it is found deep within the lower *B* or *C* horizons. This is the condition most sought after in cultivation, with the exception of those crops (such as rice) which require adequate lateral drainage but poor vertical drainage.

Excessive drainage. Water moves through the soil rapidly, and soil openings are large and contiguous. There is little retention of water anywhere within the soil, and supplies for plant growth may not be adequate for much of the year. Such soils usually show little horizon differentiation and differ little from the rock detritus and regolith from

which they are formed. Such conditions can be found in shallow soils on steep slopes or on deep, coarse-textured, permeable textures. They usually are unsuited for cultivation.

Like slope, poor or excessive drainage conditions rarely can be treated significantly in interpreting the pattern of soil distributions on a global scale. On a topographic scale, however, variations in soil properties may be more closely linked with drainage and slope conditions than with climate, vegetation, or any of the other variables. It is for this reason that most detailed soil surveys on a topographic scale use a classification system in which the properties determined by slope and drainage conditions play a leading role.

15-14 SOIL ORGANISMS Wherever organic material is added to soils as the result of death, it immediately becomes potential food for a great variety of living organisms. Were it not for the transformation of dead plant tissue by plants and animals, many of the active soil processes could not occur. There would be no humus, and once the base ions were removed from the soil by plant roots, they would be lost from the soil forever. The importance of soil organisms, ranging from the active roots of giant trees to submicroscopic bacteria, is reflected in the importance given to the biosphere as the determinant in fixing the lower boundary of soils and in distinguishing the solum from the parent material.

The casual observer has little conception of the amount of biological activity that goes on in the soil, and in general, the more productive the soil is in maintaining a plant cover on the surface, the greater is the biological activity that goes on below. To illustrate, Charles Darwin estimated that earthworms alone can pass through their bodies an amount of soil equal to 15 tons per year on a single acre. While the capacity of the soil to support living organisms varies greatly, the total biomass within the soil may become as high as 3 to 5 tons per acre. The number of individual living organisms may vary from 13,000 to a million earthworms per acre, while estimates of protozoa, fungi, and bacteria run up to millions of individuals per gram of soil.

The end products of soil-organism activities differ little from the end products of the metabolic activities of the larger plants and animals. First, energy is released, often in the form of heat. Experiments at the Rothamsted Experimental Station in Britain[6] indicated that an unfertilized, low-producing soil lost a million kilocalories of energy per acre per year through organic decomposition, while heavily manured soils yielded about 15 times as much. The latter figure is about equal to the normal heat loss from the interior of the earth! The heat exchange involved in organic decay in the soils thus constitutes a small but noticeable figure in the total global energy balance.

Second, as a result of organic decay, relatively simple substances, including carbon dioxide, oxygen, water, and various soluble proteins, are released into the soil, altering the soil solution and the soil atmosphere. These substances are quickly converted from the sugars, starches, and simple proteins in the organic residue. Eventually, also, all the bases that once were plant foods are returned to the soil.

Third are the more complex organic products, including the higher proteins, cellulose, lignins, and waxes, which require special digestive treatment. These materials tend to be worked on by the microorganisms, including fungi, bacteria, and protozoa, and

become the gelatinous masses that are termed *humus*. As indicated in Section 15-6, this colloidal organic material has a wide range of properties and composition.

The organisms are not independent variables in the soil. As indicated above, their numbers and activity increase with the general suitability for plant growth or soil fertility. Climate also is a contributing factor. Bacteria, for example, tend to find their optimum temperature range in humid climates between 70 and 100°F, while fungi are much more prevalent and find fewer competitors in cool, damp climates. Lifting the matting of needles on the floor of a northern coniferous forest invariably will reveal the delicate tracery of the *mycelium*, or threads of the molds that dwell there, feeding on the forest litter. The more complete mineralization of humus in the humid tropics by bacteria may well be a factor in preventing the formation of the highly acid organic colloids that are such a prominent part of the soil solution in the cool, humid climates.

The nature of humus, therefore, is the result of a combination of many organic factors. Despite its organic origin and composition, it can be remarkably durable, especially when it develops partial mobility and enters into colloidal bonds with some of the more absorptive inorganic clay colloids. Whether its greater durability in the cool, humid climates is due more to the less digestive waxes and lignins of the needleleaf, coniferous litter or to the difference in the type of microorganisms present in such climates is not entirely clear. It is likely that, as in all other environments, the energy-exchange mechanisms within the soil, which involve the biota, constitute subsystems in which there is a constant adjustment of materials and forms to provide as smooth a flow of energy as possible within a kind of dynamic equilibrium. The soils of the humid tropics must

[6] E. J. Russell and E. W. Russell, *Soil Conditions and Plant Growth*, Longmans, Green, and Company, New York, p. 194, 1950.

have a rapid return of bases via effective organic decay to ensure the maintenance of the forest, which in turn helps in the removal of net radiational surpluses. The flow of energy through the soil can take place at a much slower rate beneath the firs and spruces of the taiga.

15-15 MAN As human beings increase in number on earth and modify the land surface more and more in serving their needs, they are becoming increasingly significant as an independent variable in soil modification. Cultivation, deforestation, reforestation, irrigation and drainage projects, erosion control, erosion acceleration, and many other human activities produce significant alterations in soil properties and hence change the patterns of soil distribution. So much of the earth surface has been modified by man that completely natural soil conditions are exceptional. A thorough analysis of the human factor in the interpretation of soil differences is far beyond the scope of this volume. It is sufficient to note that, with his increasing knowledge of soil technology, man is now virtually able to alter the soil of any environment to produce whatever properties or qualities he desires, provided he is willing and able to pay the price. Unfortunately, his investment in soil conditioning and improvement depends on economic, political, and social feasibility.

Furthermore, despite his technical ability to alter soil properties, man is slowly learning that his uses of the soil are still subject to the inexorable progressions of the environment toward, or away from, relative equilibrium. He has learned, to his sorrow, that interference in the normal development of soils through his increased production of food or feed has sometimes led to unfavorable long-term effects. These effects often are subtle, and by the time the damage is done,

it may be too expensive to repair. Thousands of square miles of fertile, dry land soils have been made sterile through the encroachment of salinity resulting from mismanaged irrigation. The harmful effects of clearing and burning vegetation on the soils of the humid tropics have already been mentioned. Soil is a vital part of the nutrient cycle upon which the rapidly growing world population must depend. Soil management methods that work extremely well in one part of the world produce disastrous results in others. Man is continuing to increase the demands on soils by increasing the flow of energy and materials into them. At the same time, the nutrients consumed by him rarely are returned to the soil in the more sophisticated cultures. They enter sewers and flow to the sea. The cultivated soils in countries with high living standards are becoming human artifacts rather than natural bodies, dependent on a man-guided input of materials for the maintenance of the soil-fertility cycle. In the less privileged parts of the world, increased demands on the soil without compensating by inputs of nutrient material have led to serious soil deterioration.

The immense productivity of the reclaimed *polder* lands of the Netherlands and the sterility of the millions of acres of thin, stony, erosion-scoured hillsides in Korea, North China, and many parts of Mediterranean Europe stand as impressive testimonials to the potentialities and destructiveness, respectively, of human intervention in the natural processes of forming soil.

15-16 TIME All factors that influence soils, as well as soils themselves, are subject to alterations with time. Some soils on ancient erosional plains are subject to change only over long periods of time, whereas others exhibit noticeable alterations in soil properties within a few years. Soils may be considered young if they are just beginning to develop the properties that represent relative equilibrium with the total environmental conditions existing in a particular area. In young soil, changes are rapid. It does not take long for horizons to develop on freshly exposed parent material, and some of the properties of the ultimate soil balance may begin to be recognizable within 50 to 100 years. Studies made of soils that have accumulated on old, abandoned fortress walls in Europe indicate that, over periods of 250 to 300 years, these soils have developed characteristics remarkably like those of the soils overlying the local rock out of which the fortress walls were constructed. Investigations made in the Harvard forest, an experimental area in the Berkshire Hills of Massachusetts, to determine the influence of vegetation alterations on soil properties show significant readjustments of profile characteristics within 20 to 30 years following a change in vegetation.

It is not possible to express soil youth or maturity in terms of years. The rate of change varies with soil texture, with climate, and with age. Because of rapid leaching, loose-textured soils may reach a relative stability or balance in far less time than soils containing a high clay content. The rates of rock weathering, decay of organic material, and chemical leaching of the soil are much greater in the humid tropics than elsewhere.

The task of classifying and mapping soils on a topographic scale is extremely difficult in areas that have undergone the four successive major advances and retreats of continental ice sheets during the past million years. In these areas there is a wide range of parent materials, the slope and drainage patterns in morainic zones are complex and jumbled, and the differences in age of the successive glacial deposits result in appreciable variations in soil depth. In southwestern Wisconsin, for example, the weathered loess associated with the post-Illinoian glacial period (about 200,000 years ago) has reached depths of approximately 8 feet. On similar local parent material of post-Wisconsin age (10,000 to 18,000 years ago), unweathered loess is found about $2\frac{1}{2}$ to 3 feet below the surface.

The Climatic-Biotic Soil-forming Processes

The previous division of this chapter treated the major factors that operate with varying degrees of importance to influence soil properties. Now closer attention will be given to the major processes themselves, which are related to climate and to the soil biota closely associated with it. Three major processes will be discussed: (1) the process that predominates on well-drained soils in cool, humid regions, termed *podzolization;* (2) the process that prevails on well-drained soils in warm, humid regions, termed *latosolization* or *ferrallization;* and (3) the process that is related to well-drained soils in dry regions, or *calcification.* Note that all of these represent conditions that are most likely to be found on rolling-to-undulating uplands. We will also assume that there is no unusual outside factor operating, such as a recent deposition of material from the outside or a highly abnormal parent material.

The processes to be described in this section also are somewhat hypothetical. Some of them may take thousands of years to evolve, and the activities that occur in the colloidal portion of the soil can only be inferred; they cannot be observed directly. We have not been studying soils long enough to be certain of their processes of formation. We are on more certain ground when we describe the morphology of soils, and hundreds of soil investigators are pooling their observations in order to arrive at usable generic classifications that continue to yield clues as to the knowledge of processes. Close attention will be paid in Chapter 16 to the description of diagnostic horizons, which represent the end products of soil-forming processes. Soil scientists are beginning to realize that soil-forming processes cannot be neatly cataloged as a few simple types, and many former assumptions as to climatic influences have had to be changed. One of the most important considerations that interfered with previous assumptions was the effect of climatic change. This has not been much of a problem in the relatively young profiles of the northeastern United States and northwestern Europe, where most of the early soil investigations took place. In many other areas, however, erosional surfaces may have remained in place for tens of millions of years, and some of the morphological features, instead of being related to present conditions, may be the soil relics of ancient climates far different than those at present. The simplified generalizations of the three climatic-biotic processes described in this part of the chapter seem to be valid, but they still are likely to be subject to many qualifications.

15-17 PODZOLIZATION, THE COOL, HUMID PROCESS The word *podzolization* is derived from the Russian term *podzol*, meaning ash soil. The principal operating process of podzolization involves soil leaching, or internal washing by highly acid soil solutions. The high acidity, sometimes reaching a pH[7] of 4 to 3.5, is especially prevalent in the *Dfc-Dfb* climates, and the process appears to become gradually less effective toward warmer and drier climates. The high acidity of the soil solution in cool, humid climates is mainly related to the prevalence of dispersed organic colloids (organic acids) that have an extraordinary affinity for bases, leaving the soil solution dominated by hydrogen ions. The organic litter from the northern coniferous forests also has a low base content, thus favoring acid soil solutions. Acid soils, however, are not restricted to cool, humid climates, and there are many occurrences of podzolized soils in the humid tropics. Parent materials high in quartz sand, such as beach deposits, under a sparse, coniferous tree cover in the tropics may have such a low content of bases that the resultant soil solution has a high predominance of the only cation that can be found—that of hydrogen (H^+). Sometimes also in the humid tropics, extremely ancient soils or plinthites may be so deficient in bases, especially following human abuse by exploitive agriculture and burning, that the only cations present are hydrogen ions. Such acid soils may take on many of the characteristics of podzolized soils, and some of them have been referred to as *pseudopodzols*.

The morphological effect of podzolization, or leaching by highly acid solutions, may be summarized as follows:

1. Soluble bases (metallic cations) are displaced from the soil colloidal interfaces and removed via groundwater from the solum.

2. Aluminum and iron sesquioxides become mobile and are transported downward in the soil, thus making the *A2* horizon

[7] The pH of a solution refers to the concentration of hydrogen ions (H^+) within it and is therefore a measurement of acidity. A pH of 7 signifies a proportion of 10^{-7}, or the equivalent of 1/10,000,000 by volume. Such a solution is neutral in reaction, because the concentration of hydrogen ions is exactly balanced by the concentration of hydroxyl ions (OH^-), which is also 10^{-7}. A pH of 5, however, signifies a hydrogen ion concentration of 10^{-5}, or 1/100,000, whereas the hydroxyl-ion concentration would be 10^{-9}. Note that the sum of the negative powers must equal -14. With a pH of 5, there is an excess of positive hydrogen ions over the negative hydroxyl ions, and the concentration of the former is 100 times that of a neutral solution. In summary, as the pH drops below 7, the solution becomes acidic; as it rises above 7, it becomes alkaline in reaction.

somewhat lighter in color than it had previously been. Silica is not so affected and tends to remain in the *A2* horizon.

3. Inorganic clays tend to be peptized and susceptible to dispersion and downward movement.

4. *B* horizons tend to develop at variable depths, in which deposition of sesquioxides, migrating clays, and humus takes place. This illuvial zone is primarily the result of a reduction in soil acidity, related in turn to contact with more bases in the underlying material, decomposition of the humus, or importation of bases via groundwater.

In summary, podzolized soils are acid soils that generally have pronounced *A* and *B* horizons, in which the *A2* is lighter in color, more acid, has fewer bases, and a higher silica/sesquioxide content than the parent material. The *B* horizon tends to have almost the opposite effect, in contrast to the parent material.

It was formerly believed that the typical podzol soils, or those with pronounced ash-gray (*albic*) *A2* horizons and dark brown *B* horizons, represented podzolization in its most ideal setting. Now such soils are known to be related almost entirely to soils with an unusually high quartz content combined with a coniferous forest cover (see the discussion of Spodosols in Section 16-13). They are by no means restricted to cool, humid climates and could not develop on soils with a high clay content, no matter what the climate might be.

15-18 LATOSOLIZATION[8] (FERRALLIZATION), THE WARM-HUMID PROCESS In passing from cool-humid to warm-humid

regions, a different soil-forming process, *latosolization*, gradually takes precedence over podzolization. The transition is extremely gradual, and there is much evidence that both processes can operate simultaneously within the transitional belt, with one or the other gradually becoming dominant. Certain conditions of texture and parent material can favor one over the other.

Latosolization involves soil leaching by mildly acid (pH 5 to 7) to mildly alkaline solutions. The lesser acidity is believed to be the result of more rapid and more complete mineralization of humus in the soils. When humus is completely decomposed, the bases that were utilized originally to build plant or animal tissues are again released into the soil, thus helping to reduce the general acidity of the soil solution. Furthermore, chemical rock weathering under warm, humid conditions is likely to release bases from the mineral fragments still remaining in the soil more efficiently than in cool climates.

A vital chemical process takes place within the silicate clay colloids when the pH rises to values between 5 and 8. The silica layers in the colloidal lattices tend to become loosened or separated from the adjacent sesquioxide layers (see Figure 15-5). Once separated, they are taken into the groundwater and carried out of the soil. This process of *desilication*, or the removal of combined silica, lies at the heart of latosolization. The process essentially involves the decomposition of clays or the alteration of the high-silicate clays into low-silicate clays such as kaolinite. Or it may progress to the point where nothing is left but a residue of iron and aluminum sesquioxides under various degrees of hydration, with small inclusions

---

[8] The term *latosolization* is derived from *later* (L. = brick). For many years a general term used for red tropical soils was *laterite*. This term now refers specifically to *plinthite*, an iron hardpan. To avoid confusion, a newer term, *latosol*, was suggested to refer to soils in which the humid, tropical

process predominated. Another term, British in origin, is *ferrallization*, from *ferrallite*, derived from *fer* (Fr. = iron) and *al* from aluminum. Although the author prefers the more descriptive term *ferrallization*, he bows to the more commonly used *latosolization*.

of insoluble impurities and kaolin. The time factor is critical here, because the alteration tends to decrease with increasing age. Kaolinite clays are not easily decomposed further, and pure deposits of bauxite or hematite resulting from this process are exceedingly rare, even in the tropics. Soils that have begun the accumulation of desilicated iron and aluminum sesquioxides are known as

Oxisols. They are found throughout the rainy tropics or humid subtropics, where the land surface has been stable for a long time, perhaps millions of years.

Unlike podzolization, there is little migration of the sesquioxides in latosolization. Although the combined silica in the colloids tends to leave, what sesquioxides remain in the soil tend to be stabilized there, coating the soil particles and giving these soils their common red and yellow coloration. As might

Figure 15-9 Silica/sesquioxide (alumina) ratios in clay colloids, representative of podzolization and latosolization. A silica/Al_2O_3 (alumina) ratio of 2 appears to be a reasonable boundary between the two processes of soil development. In proceeding from cool-humid to warm-humid climates, the combined-silica percentage in the clays tends to decrease, and that of the sesquioxides (both Al_2O_3 and Fe_2O_3) tends to increase. [After Jenny]

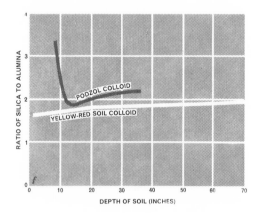

be expected, the inorganic clay colloids show various amounts of combined silica, depending on the age of the soil and the characteristics of the parent material.

In summary, latosolization consists largely of a breakdown of the silicate clay colloids, the removal of silica and bases from the soil, and a gradual increase in the amount of iron and aluminum sesquioxides throughout the soil. The fundamental cause appears to be the pH range, mostly between 5 and 7.

A graphic portrayal of the difference between podzolization and latosolization is shown in Figure 15-9. The bottom curve represents a soil dominated by the latter, although it is of a marginal type. A more advanced latosol would have its curve placed lower on the chart. The top curve represents a soil in which podzolization is the main soil-forming process. Note the sudden break in the top curve, representing the illuvial horizon, where the aluminum sesquioxide, contained in the translocated silicate clays, accumulates. No such accumulation zone appears in the bottom curve. The entire profile shows little change in the silica/alumina ratio, with only a slight decrease near the surface. The entire profile, however, shows a lower ratio than that in the upper profile, although the two are rather near each other in the *B* horizon of the podzolized soil.

The major soil characteristics that tend to form under latosolization are as follows:

1. A lack of strong horizon contrasts where drainage is adequate.

2. A low cation-exchange capacity and a low base saturation in the colloids, owing to excessive hydration, rapid removal of bases, and low absorption and adsorption capacity of the clays.

3. Greater amounts of sesquioxides *throughout the profile* than in the parent material.

4. A smaller amount of combined silica throughout the profile than in the parent material.[9]

5. A predominance of the kaolins among the clays present, or at least a progression toward this.

6. Accumulations of silica-free iron and aluminum oxides in crystals or amorphous masses where the soils are extremely old.

7. A tendency toward bright red and yellow colors in the soil, due to iron oxides. It is possible, however, to have white or pink soils in which the iron content is unusually low in the parent material.

The physical properties of *latosols* (the collective term for a wide variety of soils where latosolization has been operative) can vary widely. With respect to cultivation, however, several generalizations can be made:

1. They require heavier applications of fertilizer to maintain fertility.

2. They dry rather rapidly, especially near the surface, and low field crops are susceptible to wilting.

3. Soils of texture comparable to those in the higher latitudes do not erode by mass wastage or gullying as readily on equivalent slopes.

4. The humus content tends to be low, and green manure, although less durable, has a markedly beneficial influence on the retention of water and plant foods.

15-19 CALCIFICATION The dominant process that operates on the well-drained soils of dry regions, termed *calcification,* consists of the formation of calcic horizons

[9]An exception to this sometimes occurs in the *Aw* climates, which feature pronounced seasonal differences in rainfall. Intense heat and drought may cause evaporation of groundwater sufficient to cause a precipitation of colloidal silica. Such gels, when dried, form irreversible crusts, nodules, and even *duripans.*

(usually within the *B* or *C* horizon) and a *mollic epipedon* at the surface. The latter is a soft (L.: *mollis* = soft), dark, friable surface horizon containing a high percentage of organic colloids well saturated with calcium ions. The prevalence of lime in these soils led them formerly to be termed collectively *pedocals* (soils of calcium).

Evapotranspiration exceeds precipitation in the dry climates. This means that even in well-drained soils, there will be some net loss in the amount of water as it passes through the soil. As they evaporate, therefore, soil solutions have an increasing concentration of soluble salts and bases as they pass downward through the soil and even when they are below the water table. As they become concentrated, a point will be reached at which the solution can no longer hold them as dissolved ions, and they will precipitate out of solution. Not all the soluble salts and bases are precipitated at the same time, and there is a progressive listing in order of solubility. Calcium and magnesium bicarbonates lie at the bottom of the list, thus are the first to be precipitated, in which case they are altered into the carbonate form. These are followed by calcium sulfate and a long list of water-soluble salts.

The predominance of lime or calcium carbonate in the well-drained soils of dry climates is related not only to its position in terms of solubility but also to the abundance of calcium in the earth crustal rocks. Magnesium is another rather common element and tends to be a common companion to lime when in the carbonate form. The calcic horizons tend to be thickest and deepest along the boundary between dry and humid climates and to become gradually shallower, thinner, and more compact with increasing aridity. Desert calcic horizons may have the consistency of concrete and carry the French term *crôute calcaire* (lime crust).

Whereas only calcium and magnesium carbonates or sulfates dominate the lower horizons in the semiarid well-drained soils, the water-soluble salts, such as sodium bicarbonate, sodium chloride, and potassium bicarbonate occasionally appear in the B horizons of drier climates, where they are precipitated during long dry periods. With adequate drainage, however, they are soon flushed out of the soil following even a single heavy rain. While these water-soluble salts become more common in the soils with increasing aridity, they do not constitute important parts of the permanent horizons in well-drained soils, as do the alkaline earth carbonates. *Salic,* or saline, horizons are almost always the result of poor drainage, where the groundwater has no outlet except by evaporation.

Another important aspect of the calcification process is the translocation of calcium ions from the lower horizons to the surface epipedon. This is brought about largely by the feeding habits of grasses and earthworms. Grass, unlike trees, has great affinity for soils high in lime and utilizes much calcium when available throughout its plant tissue. This is why the surface grass cover of the steppes and prairies, when it dies each year, supplies not only a large amount of material destined to become humus but also the calcium ions to saturate and flocculate humus and inorganic clay colloids, thus helping to protect the former from rapid decay by the soil biota and keeping the epipedon friable.

The thickness and darkness of the mollic epipedon in the pedocals are related largely to evapotranspiration and precipitation. As one passes toward areas of decreasing precipitation, the mollic epipedon becomes lighter in color and thinner. The same transition takes place with increasing temperature and evapotranspiration, given the same amount of precipitation.

It should be kept in mind that the presence of a mollic epipedon does not in itself indicate a dominance of calcification. The black prairie soils of Illinois are not pedocals because they lack a calcic horizon. The climate is humid enough to prevent the precipitation of calcium carbonate from the soil solution or the water table.

Iron and aluminum silicates (the inorganic clay colloids) form as a result of rock weathering in dry regions, but the rate of their formation decreases rapidly with increasing aridity. They are much less important as coloring agents in dry soils than they are in humid soils, mainly because of the dominance of organic material in the epipedons which may actually extend into the B horizons near the humid borders. The clays are relatively unimportant constituents of true desert soils. If desert soils are red, the color is likely to be the result of iron in the original parent material, not of soil-forming processes. One exception might be noted: With high temperatures in the low latitudes, considerable organic material may oxidize without producing soil humus. Therefore, the A horizons of tropical grasslands are usually not so dark as those of cooler climates, and the iron present tends to produce shades of chestnut or reddish brown. The physical and chemical activities of the soil colloids show a similar relationship. In the cooler, more humid grasslands, organic colloids greatly exceed the clays in water retention and cation exchange. In the tropical grasslands, on the other hand, the clay colloids perform a larger share of such functions. Both organic and inorganic colloids are rare in true desert soils.

In summary, calcification, as a soil-forming process, develops under good drainage conditions in areas where evapotranspiration exceeds precipitation. It is characterized by the removal of some of the bases from the

A horizon, the formation in the subsoil of a depositional layer of alkaline earths—mainly calcium carbonate—and a stabilization of organic colloids in the epipedon. The process is best developed near the humid margins of the dry lands and weakens toward the desert.

In the latter, mechanical weathering processes, rather than the solvent action of percolating water, are responsible for determining the major characteristics of the fragmental material that is found mantling the earth surface.

Study Questions

1. Explain why white soils almost always are infertile, and why black soils are not always an indication of fertility.
2. Using the textural chart in Figure 15-2, determine the textural class of a soil with 60 percent sand, 25 percent silt, and 15 percent clay.
3. What condition would tend to peptize clays, and what would tend to flocculate them?
4. In Figure 15-4, what properties are indicated by the symbol t in the B2 horizon?
5. What is a colloid? Give some examples of colloidal properties. Distinguish between a reversible and an irreversible colloid.
6. Compare and contrast the properties and functions of organic and inorganic colloids in the soil.
7. What is the principal variant within the inorganic clay colloids? Is there a correlation with climate? Explain.
8. Explain the pH of soils. What is the difference between a pH of 8 and one of 5?
9. Why do most dry-land soils require nitrogen applications for cultivation?
10. How does the effect of a grass cover on soils in mid-latitudes differ from that in the tropics?
11. Explain the processes of podzolization, latosolization, and calcification.
12. Illustrate and explain why these soil-forming processes usually are linked but not necessarily directly correlated with climate.
13. Why is lime ($CaCO_3$) such a common compound in the B horizon of well-drained semiarid soils?
14. Why do red colors prevail in well-drained humid, tropical soils?

CHAPTER 16
SOIL CLASSIFICATION, GLOBAL SOIL ORDERS, AND SOIL FERTILITY

This chapter discusses the importance of scale in classifying the differences in soil forms, summarizes the global soil orders and the diagnostic horizons on which they are based, and treats briefly the variables that are most important in soil fertility. Maintaining soil fertility and preventing soil destruction have repercussions throughout the world in the face of constantly increasing demands on the soil to provide food and feed for ever larger generations of humans. It is perhaps appropriate that the text end at this point. If we examine the global energy system and the many subsystems that help guide the flow of energy and material across the face of the earth, sooner or later we realize that the drain on the world reserves of fossil fuels cannot continue to rise at a constantly increasing rate as population increases. We are now tapping "bank reserves" resulting from the maladjustments of the normal operation of the global energy system that took place during relatively rare periods in earth history. The normal operation of the system does not provide for a continual detouring of energy into the renewal of fossil-fuel deposits. Except for certain small exceptions, these are nonrenewable resources.

Eventually, the energy to maintain human life must be derived from the sun, and mainly through plant and animal food *via the soil*. The focal point, therefore, in tapping the energy of the sun for food lies within the soil. The proper care of our soil resource, then, is one of the most vital requirements for maintaining the human species on earth. The history of man's use of the soil generally has not been to his credit, although the enormous capabilities for making the soil more efficient as a base for plant growth have been demonstrated in many areas where the force of modern agronomic technology has been applied—again, admittedly, with the assistance of fossil fuels.

Soil Classification

Attention has been given throughout this text to the constant interplay of phenomena that produce an infinite variety of differences from place to place. Soils are no exception, and minute differences can be observed, inch by inch, that could never be completely integrated into a rational, systematic study. However, as the first principle of natural distributions stated (Chapter 1), there are regularities that appear out of the chaos of infinite differences—regularities that permit generic classifications (those of kind) and genetic classifications (those of origin) in which certain similarities are recognized and differences discarded as irrelevant to the purpose at hand. There are millions of farmers who have never had an academic course in soil morphology but who recognize that some areas of their workable fields behave differently from others with respect to plant growth. Some portions dry out sooner; others are "sour"; some have lost their topsoils; some have gained the topsoils washed from adjacent knolls or rises; still others may have unusually stony surfaces, are steep and subject to erosion, or have shallow hardpans near the surface that interfere with deep ploughing. Every good farmer who has learned from his own experience or from that of his predecessors uses the knowledge of these differences in planning the operation of his farm.

Recognition of soil differences at scales far smaller than those used and recognized by farmers or those suitable to areas far larger than those consisting of a fraction of an acre also are important. Planning for intelligent use of land over a state, province, or nation could not possibly utilize soil maps that indicate differences down to the minimum size of a pedon. Utility, therefore, determines the scale at which soil classifications are made. The history of the development of soil classification systems has been marked by the attempt to create flexible systems that recognize soil differences at various scales of observation but at the same time are linked together within a consistent generic classi-

fication. This has not been an easy task because of the many variables that influence soil characteristics (see Chapter 15), and as surveys in newly developing parts of the world are being made, new soil varieties are being discovered and changes being suggested for world soil classifications.

The history of soil classification systems developed by the soil-survey specialists of the U.S. Department of Agriculture illustrates the problems of the geography of soils so well that a short résumé is presented in the following subsections.

16-1 THE EARLY SOIL CLASSIFICATION BASED ON TEXTURE AND PARENT MATERIAL The U.S. government did not undertake organized programs of soil surveying until the middle of the nineteenth century. At that time, by far the most important classification criteria were the properties that appeared to be significant in the use of virgin soils for crop growth. This was long before the advent of modern agronomic science, through which soils have been altered synthetically to produce enormous yields of selected crops. The emphasis on new lands was logical, because this was a period when the vast plains in the central and western parts of the United States were being developed for agriculture. Nearly all of the soils west of the Appalachian uplands were relatively uninfluenced by human use, even when under cultivation.

Two major characteristics formed the main variables in the early U.S. soil surveys: (1) texture of the topsoil and (2) type of parent material. The texture of the topsoil was used primarily because the general ease of cultivation was closely related to it. The parent material was considered the principal variable in the fertility of virgin soils. Soil mapping in its earliest stages was done on a topographic base, and the soil maps published by counties were generally at a scale of an inch per mile. Today the preliminary field investigations are made on photomaps or photomosaics at scales of three to eight inches per mile. The details of roads, fences, buildings, and even individual trees enable the soil surveyor to locate his soil boundaries quickly and easily. The completed published soil maps today also tend to be on larger scales than previously. The basis for most soil mapping in the United States is the separation of the soil into *series* and *type*. The former is related mainly to the arrangement and characteristics of the horizons within the soil profile, and the latter represents variations of the soil series based on surface texture. Soil types are especially important considerations in farming. In the United States, the names of the soil series are derived from local proper names, such as a river, a city, or a county—e.g., Miami, Caribou, Cecil, Chenango, and Saugatuck. The soil types are named after the surface textural class, such as the Miami silt loam, Miami loam, or Miami clay loam. The name *Miami silt loam,* representing a particular soil type, whose textural and horizon descriptions are given in Figures 15-2 and 15-4, refers, first, to the Miami soil series that originally referred to a parent material consisting of a glacial till high in lime. The designation *silt loam* identifies the soil type. Figure 15-2 shows that a typical Miami silt loam has a clay texture in its *B* horizon and a loam texture in its *C* horizon. Originally, little attention was given to the lower horizons and to internal drainage characteristics. Modern surveys, on the other hand, while they may have retained the same soil series and type names, often have subdivided the original series into several new ones derived from more precise definitions based on the complete profile.

The individual soil maps of series and types are supplemented by a soil-capability classification, which rates land areas according to their capabilities and limitations for major land uses. A rating scale of eight land-capability classes is used, numbered from I through VIII. Land class I has few limitations and represents the greatest potential for almost any kind of use. Class VIII, at the other extreme, has only slight possibilities for productive uses and has many limitations. Such areas are recommended to be left alone and diverted into a wild state for watershed protection, wildlife sanctuaries, nature trails, or simply scenic attractions. The use of subclasses recognizes certain special kinds of limitations within each of the land classes. The land-classification system, begun in the 1930s, is a useful attempt to arrive at programs for sound land management.

As the number of soil series multiplied into the hundreds, and once the soil surveys began to be conducted in other parts of the world, several attempts were made to develop a classification system into which all soils could fit. Since the problem of classifying soil characteristics on a global or regional basis is geographic in nature, two systems that displaced the early texture-parent material one will be presented in the next sections of this chapter.

16-2 THE GREAT SOIL GROUPS: THE RUSSIAN GENETIC CLASSIFICATION
Late in the nineteenth century, a soil-classification system was established, based largely on soil characteristics primarily the result of climatic and biological soil processes. Essentially, therefore, it was a genetic rather than a generic classification and evolved from some pioneer work in soils by a Russian scientist, V. Dukuchaev. His ideas were expanded and applied in Russia by Glinka, one of Dukuchaev's students, by Ramann in Germany, and by Marbut in the United States.

The emphasis given by the Russians to climate in soil genesis and classification is understandable. The soils of the flat-to-undulating plains south of the limit of Pleistocene glaciation in the Soviet Union have remained relatively undisturbed for long periods of time (even measured geologically), and some of these soils have properties quite unrelated to those of the parent rocks beneath them. The gradual change in soil properties as one passes from the desert, across the steppes, to the taiga, or from arid to cool, humid climates, is clearly observable everywhere. The Russian scientists recognized, however, that the climatic emphasis in soil classification could not be all-inclusive, and they suggested separate categories to include soils whose major properties are the result of local extraclimatic conditions.

The late Dr. Curtis Marbut,[1] of the former U.S. Bureau of Chemistry and Soils, was one of the first outstanding American soil scientists to recognize the importance of studying the evolution of soils as natural bodies, as products of dynamic natural processes, and not as static substances whose significance is related solely to their productivity. He was largely responsible for initiating modern soil science in the United States, and he generously acknowledged the contributions and leadership of the Russians in this field. In the course of his research, writing, and teaching, he introduced the Russian soil-classification system to the United States, with appropriate modifications. The first application of this classification to U.S. soils is contained in the monograph *Soils of the United States,* published in 1913 as Bulletin 96 of the Bureau of Chemistry and Soils, under the direction of Marbut and several other soil scientists.

[1]The culmination of Marbut's work on soil classification appeared in 1935 in the section on soils of the *Atlas of American Agriculture,* one of the classic volumes prepared by the U.S. Department of Agriculture.

The classification system introduced by Marbut and indicated (with modifications) in Tables 16-1 and 16-2 comprised three major orders: the zonal, intrazonal, and azonal soils. *Zonal* soils included those whose characteristics are mainly the result of climatic or biological conditions, and these had two major subdivisions: the *pedocals,* the lime soils of dry regions, and the *pedalfers,* or the aluminum and iron soils of humid regions. *Intrazonal* soils were considered to be those soils whose properties are related more to local conditions of drainage or parent material than to climate or vegetation. *Azonal* soils included miscellaneous unconsolidated materials that have not yet developed distinct soil horizons.

The suborders of this classification system included a rough division of zonal soils according to color and vegetation, and differences between saline, swamp, marsh, and high-lime soils within the intrazonal order. The divisions of the suborders were included in what was termed the *Great Soil Group* and are shown on the right-hand side of Table 16-1. The soil-mapping units, represented by soil series, types, and phases, were not placed within the classification system and were largely applicable only at the topographic scale of observation.

Marbut's original classification was modified several times. One of the early modifications is the one shown in Table 16-1. About 10 years later, this was followed by a further modification, shown in Table 16-2. Note that the differences between the two classifications are not marked.

16-3 THE PRESENT SOIL-CLASSIFICATION SYSTEM OF THE UNITED STATES
After several decades of use, the soil investigators in the United States recognized that the Great Soil Group classification had many shortcomings and needed to be modified.

Order	Suborder	Great Soil Group
Table 16-1 Soil-classification System		

Table 16-1 Soil-classification System

Order	Suborder	Great Soil Group
ZONAL SOILS	Soils of the cold zone	Tundra soils
Pedocals — Light-colored soils of arid regions	Desert soils Red desert soils Sierozem soils Brown soils Reddish-brown soils
Pedocals — Dark-colored soils of the semiarid, subhumid, and humid grasslands	Chestnut soils Reddish chestnut soils Chernozem soils Prairie soils Reddish prairie soils
Pedalfers — Soils of the forest-grassland transition	Degraded chernozem soils Noncalcic brown soils
Pedalfers — Light-colored podzolized soils of the forest	Podzol soils Brown podzolic soils Gray-brown podzolic soils Yellow podzolic soils Red podzolic soils
Pedalfers — Lateritic forest soils	Yellowish-brown lateritic soils Reddish-brown lateritic soils Laterite soils
INTRAZONAL SOILS	Saline soils of poorly drained arid regions	Solonchak soils Solonetz soils Soloth soils
	Poorly drained soils of humid regions	Wiesenboden (meadow soils) Alpine-meadow soils Bog soils Half-bog soils Planosols (upland hardpan soils) Podzolized groundwater hardpan soils Lateritic groundwater hardpan soils
	Calomorphic (high-lime) soils	Brown forest soils Rendzina soils
AZONAL SOILS		Lithosols (rocky soils) Alluvium Dry sands

SOURCE: Mark Baldwin, Charles E. Kellogg, and James Thorp, "Soil Classification," *Soils and Men*, Yearbook of Agriculture, 1938, U.S. Dept. of Agriculture, pp. 993–995, Washington, D.C., 1938.

Table 16-2 Classification of Soils into Orders, Suborders, and Great Soil Groups

Order	Suborder	Great Soil Group
ZONAL SOILS	Soils of the cold zone	Tundra soils
	Light-colored podzolized soils of timbered regions	Podzol soils Brown podzolic soils Gray-brown podzolic soils Red-yellow podzolic soils Gray podzolic or gray wooded soils
	Soils of forested warm-temperate and tropical regions	A variety of latosols are recognized. They await detailed classification
	Soils of the forest-grassland transition	Degraded chernozem soils Noncalcic brown or Shantung brown soils
	Dark-colored soils of semiarid, subhumid, and humid grasslands	Prairie soils (semipodzolic) Reddish prairie soils Chernozem soils Chestnut soils Reddish chestnut soils
	Light-colored soils of arid regions	Brown soils Reddish-brown soils Sierozem soils Red desert soils
INTRAZONAL SOILS	Hydromorphic soils of marshes, swamps, flats, and seepage areas	Humic-glei soils (includes wiesenboden) Alpine meadow soils Bog soils Half-bog soils Low-humic glei soils Planosols Groundwater podzols Groundwater latosols
	Halomorphic (saline and alkali) soils of imperfectly drained arid regions, littoral deposits	Solonchak soils (saline) Solonetz soils (alkali) Soloth (solodi) soils
	Calcimorphic (high-lime) soils	Brown forest soils (braunerde) Rendzina soils
AZONAL SOILS	No suborders	Lithosols Regosols (includes dry sands) Alluvium

SOURCE: J. Thorp and G. D. Smith, "Higher Categories of Soil Classification: Order, Suborder, and Great Soil Groups," *Soil Science*, Vol. 67, 1949, pp. 117–126.

Some of the deficiencies were:

1. There was no place in the classification for soils with distinct characteristics that were related directly to human utilization. This was not so serious in the western hemisphere as in Europe and in parts of Asia.

2. There was no provision for multiple soil profiles, such as result from long-run climatic changes.

3. The typical zonal soils were considered to have reached maturity of profile development on undulating, well-drained upland surfaces, too often an exceptional rather than a customary condition.

4. Many tropical and arid-land soils had features that did not fit the logic of the classification—example, the soil that combines the properties of podzolization and calcification, processes that theoretically should not operate in the same areas.

In short, the Great Soil Group classification system became inadequate as new information was discovered. In order to prepare a complete revision, the National Cooperative Soil Survey undertook to develop an entirely new system in which soils throughout the United States—and, it was hoped, anywhere in the world—could be placed in their proper relation to each other. After a number of progress reports, the first major published report appeared in 1960 as a publication by the Soil Survey Staff of the Soil Conservation Service, entitled *Soil Classification; A Comprehensive System (7th Approximation)*. Further supplements appeared later. The project is nearing completion, and it is anticipated that the final revised system will be published sometime in 1971 or 1972. Since 1960, the system has been referred to as the *Seventh Approximation.*

The new system, hereafter to be referred to as the *U.S. Soil Classification System*, is based on the morphological characteristics of the soil profile, whatever they may be. Particular attention is given to horizon sequences and characteristics. Many of the features of the Great Soil Group classification have been retained, such as the recognition of stages in the development of profiles, and the principles of zonal, intrazonal and azonal characteristics have been interwoven into the system, but considerably expanded. None of the names have been retained, and an entirely new system of nomenclature has been developed for taxonomic purposes, based on classical language roots in the tradition of botany and zoology.

The U.S. Soil Classification System recognizes ten major global orders. These, in turn, are subdivided into *suborders,* and the latter into *great groups* and *subgroups.* Some progress is being made in subdividing further, roughly on the order of the former *soil family,* which has been a loose grouping of similar soil series. A separate division of this chapter will be devoted to the various diagnostic horizons, the most fundamental bases for classifying soils under the new system.

Diagnostic Soil Horizons[2]

The diagnostic horizons within the soil can be divided into two main groups: the *epipedons,* or surface horizons, and the *subsurface horizons.* The latter greatly outnumber the former. Section 15-4 has described the code system by which soil horizons can be identified. Now we shall treat more fully their properties and formation.

[2]The substantive material contained in this and the following part of the chapter was derived almost wholly from the following source: Soil Survey Staff, Soil Conservation Service, U.S. Dept. of Agriculture, *Soil Classification; A Comprehensive System (7th Approximation)*, U.S. Government Printing Office, Washington, D.C., 1960, and the supplements to it dated March, 1967, and September, 1968.

16-4 DIAGNOSTIC SURFACE HORIZONS

A diagnostic surface horizon is not any horizon at the surface, but rather, one that forms at the surface. It is termed an *epipedon* (Gr: *epi* = over; *pedon* = soil). It includes the upper portion of the soil, which may be darkened by organic material even if this darkened zone extends downward into the B horizon. Six major epipedons that have been recognized, as follows:

Mollic epipedon (L: *mollis* = soft). This is a dark brown to black soft epipedon, normally between 4 and 30 inches thick, that contains at least 1 percent of organic matter. The colloids present are well saturated (+ 50 percent) with bivalent cations (Ca^{++}, Mg^{++}), and the soil preserves its soft, crumbly structure even when dry. Figure 15-3 shows a good example of a typical mollic structure. The phosphate (P_2O_5) content soluble in citric acid cannot be more than 250 parts per million to distinguish it from some of the dark, mellow farm soils, where centuries of manuring has yielded a similar epipedon (see "Anthropic Epipedon"). According to the Munsell Color Chart, the colors must have values darker than 3.5 when moist and 5.5 when dry, and chromas less than 3.5 when moist.

Umbric epipedon (L: *umbra* = shade). This epipedon may, at a casual glance, look remarkably similar to the mollic because of its dark coloration, but its colloids are saturated with ions other than the typical bivalent bases of the mollic. In terms of color, thickness, organic content, and structure, the two epipedons are closely alike. The base saturation is less than 50 percent; hence, the soil has a tendency to become much harder than the mollic when dry. The carbon/nitrogen ratios also tend to be somewhat farther apart in the umbric epipedon. The prevalence of hydrogen ions in the colloids tends to make these topsoils somewhat more acid in reaction than the mollic. Umbric epipedons are not widespread in occurrence. They occur on some of the *ando* soils that overlie young, basic, volcanic material, some of the tundra and highly acid gley soils, and over some rubruzems or red latosols.

Histic epipedon (Gr: *histos* = tissue). A histic epipedon is predominantly dominated by a thin organic horizon. It is not thick enough to be classed as an organic soil (Histosol), but at least at the surface, it is dominated by raw, acid peat. Invariably it is poorly drained in its natural state, which accounts for the surface organic layer. It typically forms the surface horizon for humic gleys or half-bog soils. In order to distinguish it from the organic soils, the thickness of the histic epipedon should be less than 12 inches drained and 18 inches undrained. There should be at least 30 percent of organic matter if the mineral part of the layer is clay or 20 percent if there is no clay. These may drop to 28 and 14 percent, respectively, if the soil has been drained and cultivated.

Ochric epipedon (Gr: *ochros* = pale). This surface horizon is light in color, hence its name. It thus contrasts markedly with all the other epipedons, which are generally dark in color because of their organic content. Or they tend to be hard and caked when dry. They may possess a dark surface layer, but this must be so thin as not to qualify as a mollic or umbric epipedon. Probably they are most commonly found in the dry land regions, where there is insufficient vegetation to form a dark epipedon, or in azonal soils, where insufficient time has elapsed to develop the darker epipedons.

Anthropic epipedon (Gr: *anthropos* = man, human). This epipedon is almost indistinguishable in appearance from the mollic, but its properties are the result entirely of heavy manuring and cultivation, and the distinguishing feature is its high content of phosphate (P_2O_5), which averages more than 250 parts per million. This epipedon is almost unknown in the United States but can be found on some of the old, intensively cultivated soils of Europe and Asia.

Plaggen epipedon (Ger: *plaggen* = sod). Like the anthropic epipedon, the plaggen horizon is a man-made horizon more than 20 inches thick. It also is dark in color, ranging from dark brown to black, and frequently contains spade marks and bits of human artifacts—e.g., pieces of brick or pottery. The field borders are clearly noticeable in the horizontal extent of these soils. The name was derived from the ancient practice of using sod as bedding for animals, then spading this into the topsoil as manure. Note that this epipedon is not synonymous with any tilled surface soil. The soils in the United States have not been cultivated long enough to develop distinct evolutionary epipedons. Instead, they are merely mixtures of the normal epipedons.

16-5 DIAGNOSTIC SUBSURFACE HORIZONS The diagnostic subsurface horizons are those that develop beneath the epipedons at various depths, some of them immediately below the surface organic litter or *A1* horizon and others deep within the *B* horizon. Although formed below the surface, some of them may be exposed at the surface as the result of erosion. This does not make them epipedons. It should be noted that since they are *soil* diagnostic horizons, they are confined to the solum proper and do not extend into the *C* horizon or parent material.

Argillic horizon (L: *argilla* = clay). This is sometimes also termed an *illuvial clay horizon*. The most distinctive feature is, first, the high content of clay colloids, which is greater than elsewhere in the profile and tends to form a tight, relatively impervious claypan. Second, there is evidence of movement of this clay into the horizon, mainly from above.

Figure 16-1 Flow structure of a typical argillic horizon in a solodized solonetz soil, Oregon. When the soil solution is either highly alkaline or highly acid, the silicate clay complex becomes mobile and is carried downward in the soil to be deposited in an argillic (illuvial clay) horizon. The argillic horizon in this highly alkaline soil shows the typical flow structure caused by the deposition of the mobile clays. The small strip of paper used to indicate scale is 2 mm in length. [W. M. Johnson, Soil Conservation Service, U.S. Dept. of Agriculture].

This evidence is not easy to see with the naked eye, but a small hand lens with a magnification of 10 to 20 times will easily reveal the typical flow structure where the clay has moved along the soil openings, coating the adjacent soil particles, the channel sides, and generally clogging the pore spaces. Figure 16-1 shows a good example of the flow structure typical of the argillic horizon.

The argillic horizons appear to have little relationship to climate and occur in a wide range of environments. The principal requisites for development are: (1) a parent material that has a high clay content or that can result in a high clay content with weathering; (2) conditions that tend to promote the dispersion of the clay colloids (a high-lime content tends to check clay migration through flocculation); (3) a plentiful water supply for soil washing; and (4) a long period of time. The last is extremely important, because argillic horizons apparently take a long time to develop. Soils that have a predominance of hydrogen and sodium ions are especially favorable to clay dispersion. Hence, heavily irrigated soils of arid regions, where sodium ions tend to be common, and the highly acid conditions of cool, humid regions appear to favor argillic horizons. Many of the clay soils in glaciated areas, however, are not sufficiently old to have developed pronounced argillic horizons.

Natric horizon (L.: *natrium* = salt). A natric horizon is a special variety of argillic horizon in which the dispersion of the clays leading to the development of the illuvial clay horizon is caused by sodium ions. The colloids have a relatively high content of these sodium ions, amounting to at least 15 percent of the base saturation. The illuvial horizon develops a typical columnar or prismatic structure with rounded tops. Figure 16-2 shows a segment of such a horizon from a *solonetz* soil. Natric horizons develop where desalinization has taken place, as where irrigation has flushed soluble salts from the solum to make them suitable for agriculture. Reclaimed salt marshes along coastlines may

Figure 16-2 Typical solonetz soil (black alkali) structure. Note the typical columnar structure of a natric horizon, with columns that have rounded tops. Such soils represent early stages in the desalinization of solonchaks. [U.S. Dept. of Agriculture]

also develop such a horizon. The formation of a natric horizon presents some problems to the reclamation of saline soils, mainly because the horizon is relatively impervious and could result in perched water tables, inadequate interior drainage, and a reestablishment of saline conditions in the soil above.

Spodic horizon (Gr.: *spodos* = wood ash). This horizon is one in which humus and active amorphous aluminum sesquioxides have been precipitated. It is thus an illuvial horizon. Iron sesquioxides may or may not be involved. The colloidal sesquioxides involved in the spodic horizon illuviation should not be confused with the silicate clay colloids of argillic horizons, in which sesquioxides of aluminum and iron also are present, but in combined form. The sesquioxides brought to the spodic horizon probably have been derived from the complete decomposition of certain primary minerals within the horizons above, or brought from below by plant translocation. Although they do not have the absorptive capacities of the silicate clay colloids, they still can retain water and cations. The spodic horizons rarely develop where the parent material has a high silicate clay content, and they also are restricted entirely to humid regions, being most common in the cool, humid areas. The color of the horizon normally is reddish brown, whether or not iron is present. Although the sesquioxides in the horizon help cement the soil particles together, filling the pore spaces and making the entire zone relatively impervious to water, the typical flow structure of the argillic horizon is missing. The horizon may or may not be found just below a bleached *A* horizon, as in the typical podzols. The precipitation of humus and sesquioxides in the horizon is believed to be caused by

a rather sudden decrease in the acidity of the soil solution. The derivation of the name is somewhat misleading, because the horizon does not have any of the properties of wood ashes. The spodic horizon, however, occurs most commonly in the podzol soils that have the typical ash-gray (albic) eluvial horizon just above the spodic zone, hence are associated with it.

Agric horizon (L.: *agricolum* = farming). This is an illuvial horizon of clay and humus that results from long-continued cultivation. It normally occurs immediately below plow depth. Plowing maintains a constant mixture of humus and inorganic soil material, and rainfall slowly washes the finer constituents downward. Generally the illuviated material collects in more or less horizontal sheets or strings. It is noticeably darker in color than the adjacent material, and coatings of the illuviated material tend to collect on the surfaces of peds and the sides of worm holes and larger soil openings. The volume of the illuvial material should be at least 15 percent of the soil in the horizon.

Cambic horizon (It.: *cambiare* = to change). The cambic horizon consists of fine-textured material which has been altered appreciably from what it formerly was. The alteration may be physical, such as dislocation by freezing and thawing, extensive animal activity or root growth or it may have resulted from some chemical change, such as the reduction of ferric oxides or oxidation of ferrous oxides, the hydrolysis (addition of water) of primary minerals, or the solution and redistribution of carbonates. The characteristics of such horizons may therefore be highly variable, and they can be found almost anywhere. The important distinction between this and the other subsurface horizons is that change has taken place without any addition of mineral material from the outside. The cambic horizon is not an illuvial horizon. Probably it occurs most frequently in connection with the process of reduction and oxidation, due to a fluctuating groundwater table.

The ability of frost action in periglacial regions to alter soil structure internally is well illustrated by Figure 16-3. The churning action of alternate freezing and thawing prevents stable soil profiles from developing, and the entire soil profile beneath the surface organic layer is typically cambic in nature.

Oxic horizon.   An oxic horizon is one in which desilication of clays has taken place to the point where only insoluble materials—such as quartz, kaolin clays, and free hydrated sesquioxides of iron and/or aluminum—remain. The sesquioxides form at least 12 percent of the clay volume. Frequently, if iron is abundant, some crystallization takes place into hematite or turgite. While red coloration is common, it is by no means necessary. The oxic horizon is thick (at least 12 inches) and relatively porous, and there is little tendency for the contained clays to be flocculated into distinct structural units. A few clay skins or coatings may be present, but these are thin and indistinct. The cation-exchange capacity is low, and there are not many unweathered rock fragments left in the soil. Oxic horizons are extremely old—probably at least a half-million years. They truly represent surface materials that cannot progress much further in terms of weathering. Bases are almost nonexistent except in transit through the soil, and unless organic colloids are added, such soils are extremely infertile after the normal forest supply of bases has been removed. They are rarely found outside the tropics and subtropics. The oxic horizon probably represents the end product of latosolization.

Figure 16-3 "Roller coaster" railroad near Strelna, Alaska, caused by differential subsidence resulting from thawing permafrost. This irregular surface is caused by differential freezing and thawing of soil material lying above permafrost. It clearly illustrates the active mechanical disruption of soil profiles within periglacial regions. [U.S. Geological Survey]

Calcic horizon.   This is the calcium carbonate layer that was discussed earlier under the process of calcification (see Section 15-19). While the precipitation of calcium carbonate may take place in the parent material far below the solum, it may, and often does, appear in the B horizon and thus qualifies as a diagnostic soil horizon. It normally appears as white nodules, as irregular masses of lime, or as a compact layer. It may be distinguished generally from other types of white (albic) horizons by its free effervescence with applications of hydrochloric acid. To qualify as a true calcic horizon, it must be at least six inches thick and have at least 5 percent more lime than the parent material. The distinction between it and the parent material may not be easy, because many C horizons contain high percentages of calcium carbonate, such as calcareous glacial tills derived from limestone, or soft marls or chalks.

Albic horizon (L.: *albus* = white). Albic horizons are those that have been bleached as the result of the removal of coloring material so that the color is that of the primary soil particles and not that of any coatings. They commonly occur within the *A2* horizon of a podzol soil developed on highly siliceous sand. Here the highly acid solutions have removed all of the iron oxides as well as any humus coatings, resulting in a typical ashgray horizon. Highly acid groundwater also can bleach quartz sands that lie below the water table and where anaerobic bacteria reduce the small quantity of iron oxides, making them soluble in the ferrous form. The albic horizons generally are underlain by an illuvial horizon of some kind and are especially common where the parent material is sandy.

Salic horizon. This is a horizon 6 inches or more in thickness that contains precipitated water-soluble salts. The salt must be at least 2 percent for a horizon 12 inches thick and progressively higher in percentage with decreasing thickness. It tends to form in soils without a groundwater outlet and where evapotranspiration exceeds precipitation. The salic horizon may occur at any level and may be found on the surface on some of the salinas bordering playa lakes. Soils having such salic horizons have been termed *solonchak* in the older classifications (see Figure 16-4).

Gypsic horizon. A gypsic horizon is one in which there has been precipitation of hydrous calcium sulfate (gypsum). The horizon must be at least 6 inches thick. Since gypsum has a water solubility slightly greater than, but very close to, that of calcium bicarbonate, it is found just below a calcic horizon.

16-6 HARDPAN (PAN) HORIZONS Unusual conditions may result in the formation of unusually durable, relatively impervious horizons. Most of them are associated with water table conditions, hence are not strictly diagnostic of processes related directly to soil development. Nevertheless, they can appear within the solum, hence are included here. The term *pan* implies induration or hardening, which occurs as the result of compaction or cementation.

Figure 16-4 A saline soil near Delta, Utah. The white salts in the foreground, mainly sodium chloride and sodium sulfate, were concentrated in the topsoil as the result of inadequate drainage and the use of too much water in an irrigated area. This typical *solonchak* is a common result of mismanagement in an irrigation project. Removal of the salts from the surface as well as the salic horizons is difficult and costly. [U.S. Dept. of Agriculture]

Duripans (L.: *durus* = hard). The term *duripan* is restricted to extremely durable, massive, indurated horizons whose cement is mainly a form of colloidal silica. Associated cementing media may be present, such as calcium carbonate or iron. Duripans tend to be found mainly in dry regions or in regions that experience strong seasonal extremes of precipitation. In the United States, for example, they occur mainly in older soils along the west coast, especially in southern California. Some extremely thick duripans are found in Australia and in parts of South Africa.

Fragipans (L.: *fragilis* = brittle). These are loamy hardpans that are extremely hard and durable when dry but become fragile and brittle when moist, crumbling fairly easily between the fingers. The pans have a strong platy structure within prismlike blocks. They seem not to be related to any particular kind of soil material, and tend to lie at about the same depth wherever found—that is, roughly 18 to 24 inches below the surface. Their origin is obscure. Some authorities believe them to have formed as the result of compaction by the continental ice sheet. Usually they are found below the major diagnostic subsurface horizons. They impede normal interior soil drainage when found and often are surmounted by mottling.

Plinthite (Gr.: *plinthos* = brick). Plinthite is a highly weathered, mottled mixture of sesquioxides with inclusions of quartz, kaolinite, and other impurities. As long as it remains moist, it does not appear to be much more than a fairly tight oxic horizon. When it dries repeatedly, however, the hydrated iron oxide dehydrates to form an irreversible mass of iron stone that is almost indestructible. When exposed by erosion and washed by rain, the softer accessory mate-

Figure 16-5 Closeup of a typical plinthite or laterite. This is an exposed outcrop of plinthite where rain has removed the softer claylike material. Note the typical slaggy or clinkerlike appearance of the ferric oxide material. [R. L. Pendleton, *Geographical Review*]

rials are removed, leaving a vesicular (full of holes) mass not unlike that of a large furnace clinker or a piece of volcanic scoria (see Figure 16-5). Plinthite development is related to the concentration of sesquioxides near the top of the water table in soils where the latosolization process is taking place. Water saturation for at least part of the year seems to be required for its formation. It is largely restricted to the tropics and subtropics. Thick, massive ironstone formations are well known in what are now dry regions of Africa and Australia, and may represent a pronounced climatic change, perhaps dating from the Pleistocene or older. Plinthite formerly was termed *laterite*. Its properties of being fairly easy to cut into with a spade when moist and becoming stone hard when dry have given it considerable utility. In India and southeastern Asia, it is widely used for building material and road construction.

Ortstein (Ger.: *ort* = place; *stein* = stone). Ortsteins are extremely indurated spodic horizons that contain an unusual quantity of ferric oxide. Generally they occur at the top of the water table, within the capillary fringe, and may represent the evaporation of groundwater that contains a high percentage of ferrous compounds in solution as well as humic acids. Flat, fluvioglacial outwash plains seem to favor their development because of their usually high water table and extremely sandy textures. The sands above the water table dry out thoroughly during short, dry summer periods—enough to produce irreversible ferric oxides in the ortstein horizons. Their depth depends largely on the height of the water table. They are responsible for the soils that formerly were termed *groundwater podzols.*

The Global Soil Orders

This part of the chapter is concerned with the ten major global orders of the U.S. Soil Classification System, their distribution, characteristics, and general utility.

16-7 THE WORLD MAP OF SOIL ORDERS Figure 16-6 shows the generalized world distribution of the ten major soil orders. It was prepared in 1969 by the World Soil Geography Unit of the U.S. Soil Conservation Service from data available in published form from sources throughout the world. In assessing this map, the reader should recognize the many limitations of its scale. Within most areas of 100 square miles, which scarcely would be shown on a map of this scale, at least four or five of the soil orders normally would appear. The map indicates in an ex-

Figure 16-6 The global soil orders according to the U.S. Soil Classification System. [World Soil Geography Unit; U.S. Soil Conservation Service; U.S. Dept. of Agriculture]

tremely general way the predominant soil order in each of the soil regions. To illustrate, Florida is shown as a region entirely composed of Spodosols, yet included in this designation are extensive areas of Entisols on the sandy plains of central and coastal Florida and Inceptisols in the Everglades. The map scale simply does not permit much of a regional division within the state; hence, Spodosols were selected as covering the entire area. The map should be considered a highly generalized one that seeks to "paint with a broad brush." Nevertheless, the major distinguishing regional properties of the world are retained. Some of the qualified exceptions to the map generalizations will be indicated in the sections to follow.

16-8 THE ENTISOLS (*ent* = no root meaning; can be associated with re*cent*, and *sol* = soil) The *Entisols* include those soils that have little or no profile development. Because of this, they have no direct climatic relationship and can be found in all parts of the world. Examples of typical Entisols include dune sands, rock regolith, tidal mud flats, recent alluvium, and completed gleyed soils without a histic epipedon. The Entisols shown on the world map include some of the poorly drained tundra soils that are little more than masses of wet mud during the short summer periods when they are not frozen. They tend to be found mainly outside of deep bog depressions. Another area of Entisols includes the plains of central and northern China, which are especially susceptible to periodic flooding and deposition of river silts. The great erg deserts of the world appear as Entisols, and the broken rock rubble that covers so much of the

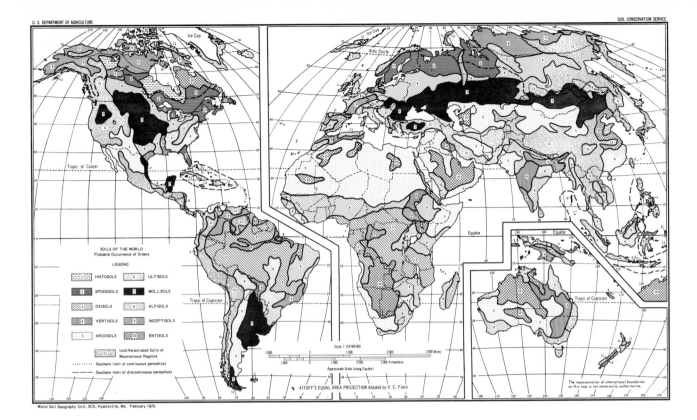

SOILS OF THE WORLD
Probable Occurrence of Orders

LEGEND

HISTOSOLS	6	ULTISOLS
2 SPODOSOLS	7	MOLLISOLS
OXISOLS	8	ALFISOLS
4 VERTISOLS	9	INCEPTISOLS
5 ARIDISOLS	10	ENTISOLS

Undifferentiated Soils of Mountainous Regions
Southern limit of continuous permafrost
Southern limit of discontinuous permafrost

Scale 1:154 000 000

Approximate Scale (along Equator)

AITOFF'S EQUAL AREA PROJECTION Adapted by V. C. Finch

The representation of international boundaries on this map is not necessarily authoritative.

World Soil Geography Unit, SCS, Hyattsville, Md. February 1970

Kalahari Desert in southwest Africa is included.

The Entisols are generally among the poorest soils in the world, having either too much or too little water, no structure, and not yet weathered sufficiently to supply plant nutrients. Yet, there are those who cultivate the desert sands where water is available, and the Entisols of central China support some of the densest rural populations on earth.

16-9 THE VERTISOLS (L.: *verto* = to turn)

Vertisols include those clayey soils that have extremely high bulk densities when dry and that shrink and develop wide cracks during fairly regular dry periods. Their name is derived from a unique property, that of self-mulching, or the capacity of the soil to mix its surface material mechanically into the lower horizons. Such soils have an unusually high absorptive capacity for water and change greatly in volume between wet and dry seasons, swelling when wet and shrinking when dry. The self-mulching results from the crumbling of the edges of the soil cracks, which shifts surface material downward into the soil profile. Rainwash also shifts surface material into the cracks until soil expansion closes them. The expansion of soil when wet sometimes produces internal-displacement, slippage planes or *slickensides* along hill slopes that in microdetail are similar to the smoothed surfaces along fault planes in rocks. Where the differential swelling is unusually great on flat-to-undulating surfaces, a hummocky surface somewhat similar to that on permafrost ground is found, termed

Soil Classification, Global Soil Orders, and Soil Fertility   **611**

gilgai relief. The Vertisols are known under various other names, such as *grumusols, tropical black clays, regur,* or *tir.*

The most clearly identifiable characteristic of most Vertisols, apart from their vertical cracking, is their dark color, which is unique in the tropics and subtropics with a wet-dry climate where they are found. This color ranges from black (when wet) through browns and reddish browns to dark gray. The dark color is not related to a high organic content, as in the black grassland mollic epipedons of the mid-latitudes. Instead, the surface horizon of Vertisols has a high percentage of clay, usually above 40 percent. The umbric epipedons result from a black, irreversible, organic colloid that coats the clay particles, keeping them peptized. The total organic content of some Vertisols may be lower than in some nearby red, tropical

The high absorptive capacity of the Vertisols, which accounts for their uniqueness, apparently is related to two factors: (1) an unusually high percentage of *montmorillonite* (see "The Inorganic Clay Colloids," pp. 576–578) among the colloidal fraction, possibly related to an early stage of weathering in basic rocks; and (2) the maintenance of a peptized state, in which the clay particles remain separate instead of flocculating. The Vertisols appear to have been formed from parent materials that are high in calcium, such as chalks, limey shales, or basalts, although the calcium ion does not form a significant part of the exchangeable cations. They also tend to be somewhat more common in climates that have a seasonal variation in precipitation, such as the *Aw, Cwa,* and *Csa* climates. They are not found outside the tropics and subtropics.

The major global areas of Vertisols, shown in Figure 16-1, include the famous black *regur* soils on the basalts of central India; the recent alluvium of the central Sudan, which is derived from the erosion of basaltic soils in highland Ethiopia; a broad area of grass-covered, calcareous soils in eastern Australia; and a small area in the United States, comprising some of the prairies of the south Texas coastal plains. Small patches of Vertisols are often encountered around the Mediterranean Basin, where they appear to be related to poor drainage in areas underlain by limey parent materials.

The Vertisols are among the better soils found in the humid tropics. Compared with most humid tropical soils, they absorb and hold fertilizers and irrigation water well because of their relatively highly absorptive montmorillonite clays. They are not easy to cultivate, however, being extremely plastic and heavy when wet. Their tendency to crack when dry makes them vulnerable to wind erosion.

16-10 THE INCEPTISOLS (L.: *inceptum* = inception, beginning) The *Inceptisols* include those soils that are usually moist, and whose surface and/or subsurface horizons exhibit some alteration of parent materials, but without any distinct illuvial zones, or zones of accumulation. There is no indication of translocation of material vertically within the soil. Since the main identifying characteristic of the Inceptisols is an extreme youthfulness of profile development, they can be found in nearly all climates except arid and semiarid regions. There is no limitation as to the kind of parent material from which these soils can be derived.

The world map (Figure 16-6) shows that some of the largest areas in the world containing Inceptisols are the arctic plains. Permafrost and the periglacial phenomena tend to prevent the development of distinct soil profiles. It is likely that Entisols, Inceptisols, and the organic Histosols all are found throughout the tundra regions, with the

Histosols occupying the poorly drained bogs (see Figure 16-7), the Entisols prevailing in the shallow half-bog soils, and the Inceptisols developing on the slight rises, where some oxidation can take place in the mineral soil.

Several types of young parent materials have Inceptisol soils, such as relatively recent loess, recent glacial material (especially those that contain appreciable quantities of lime, which tends to slow up the translocation of clay material), and some of the alluviums. The older alluvial plains along the lower Ganges, the periodically flooded alluvium of the Sudd marshes in the southern Sudan, the central Congo Basin, and the flooded llanos of Colombia are other examples. One of the most unique types of Inceptisols are the *ando* soils, which have a umbric epipedon, owing to rapid weathering of basic volcanic ash materials. Many of the dark soils on the flanks of the pyroclastic volcanoes in the southwest Pacific area are examples. They make fine soils for cultivation.

16-11 THE ARIDISOLS (L.: *aridus* = dry) The *Aridisols* are soils that have no water available for mesophytic plants and that have one or more clearly defined diagnostic horizons. The dominant soil-forming processes are calcification on well-drained soils (see Section 15-19) and the development of saline and alkaline soils in poorly drained depressions. Formerly, the Aridisols were included in the *reddish-brown, red desert, gray desert, sierozem, solonchak, solonetz,* and *soloth* soils (see Table 16-2 and 16-3).

Despite their dryness, Aridisols have distinct diagnostic horizons, especially below the epipedon. Among the most important are eluvial horizons of various types, such as the natric and argillic horizons. The argillic horizon shown in Figure 16-1 was derived from a desalinized soil, a typical solodi (soloth). These two eluvial horizons are common in

Figure 16-7 Problems of summer cross-country travel in regions of tundra soils. This is a tractor trail used for cross-country hauling during the winter season. It is located near Canning River, Alaska. The road becomes a quagmire in the summer owing to thawing permafrost. Note the *soil polygons,* low ridges formed by frost wedging. Such microlandforms, illustrating the churning action in soils overlying permafrost, are often encountered in unconsolidated soil material. If such soils could develop the beginnings of a profile, such mechanical action soon would destroy it. [U.S. Geological Survey]

the dry lands for two reasons: (1) A large amount of clay is brought to desert basins, because streams have no river outlets to carry away these fine colloidal materials; and (2) peptizing agents are dominant in the soil, especially some of the sodium and potassium ions, which maintain the clays in a highly dispersed state. Calcic, gypsic, and salic horizons also are common, having been precip-

deserts of the world. Their continuity there is broken only by the occurrence of Entisols, which make up the desert dunes and stretches of rock rubble.

The Aridisols can be productive agricultural soils when irrigated. Their principal deficiencies, apart from the availability of water, are a low cation-exchange capacity and a low nitrogen content. The former is related to a relatively low organic and inorganic colloidal content and indicates a low capacity for absorbing and holding both water and applied fertilizers. Exceptions to the latter occur in the clay deposits of desert basins, where salinity is a problem.

16-12   THE MOLLISOLS (L.: *mollis* = soft) The *Mollisols* derive their name from their typical mollic epipedons. They include those soils in which there has been alternate accumulation and decomposition of relatively large amounts of organic matter in the presence of calcium. The description and development of the mollic epipedon were given on page 603. The Mollisols are undoubtedly the finest natural soils for farming on earth and underlie the prairies and tall-grass steppes of the mid-latitudes. They were formerly included with the chernozem, brunizem, chestnut, and red prairie soils. Some of the humic gley soils and the rendzinas are also included.

The Mollisols differ widely with respect to

itated in the soil from the soil solution as the result of evaporation. Duripans and fragipans are sometimes encountered.

The distribution of the Aridisols, as shown in Figure 16-6, corresponds closely with the

their lower horizons, but their epipedons are closely similar. The relatively large number of suborders of Mollisols is an indication of how widely the subsurface horizons can vary. In the chernozems, for example, a pronounced calcic horizon is found (see Figure 16-8). Natric horizons, salic horizons, and others all indicate this variety.

The Mollisols, almost without exception, are found beneath a grassy vegetation cover. Even some of the poorly drained half-bog or humic gley soils that feature a muck surface, rather than one of peat, have a marsh cover of sedges or reeds rather than bushes or trees. Experiments indicate that when forests are established on a Mollisol, the mollic epipedon usually disappears in time. The preference of grass for high-lime material, the fine network of grass roots within the soil, and the movement of bases through the plant each year from the root zone to the surface are factors in the formation of this unique horizon. Only under exceptional conditions, where the soil contains an unusually rich fauna of earthworms and where the subsoil is high in lime, can enough lime be brought to the surface regularly to maintain a mollic epipedon under a forest cover. The mollic epipedon of the Mollisols gradually decomposes, but radioactive-carbon dating of the material in some mollic epipedons has indicated that there is a replacement cycle that varies from 100 to 600 years. The role of calcium in protecting the humus gels from decomposition is noteworthy.

The largest areas of Mollisols are generally found on both sides of the dry-humid climatic border in the mid-latitudes. Central North America, the Argentine pampas, the Russian "black belt" that stretches from the northern Romanian plains eastward across the Ukraine into central Asia, and the Mongolian and Manchurian grasslands of eastern Asia are good examples (see Figure 16-6).

Smaller areas, however, are found in alpine meadows, over chalks, and in shallow muck lands in many parts of the world. They have not been observed in high latitudes, and none of the tundra soils would qualify because of their acidity.

The extremely high inherent fertility of the Mollisols is related to their very great cation-exchange capacity, their calcium-saturated colloids, their high nitrogen content, their ability to retain water and fertilizers, and their general ease of cultivation. They include the major commercial granaries in the world and produce a major share of the world's supply of wheat, corn, and rye.

16-13   THE SPODOSOLS (Gr.: *spodos* = wood ash) The *Spodosols* are soils that feature the presence of a spodic horizon and typify the process of podzolization when acting on highly siliceous parent materials. A typical Spodosol profile is shown in Figure 16-9. The soils that formerly were termed podzols are typical Spodosols. As indicated earlier, however, not all Spodosols have the ash-gray, albic, eluviated A2 horizon of the typical podzol, and many of the spodic horizons would not be recognized as such if they did not have the albic horizon above them. True Spodosols require parent materials that have a high content of siliceous material, especially quartz sand. They are located mainly in the northern coniferous forest region, or taiga, on or near the great crystalline shields of Canada and northern Eurasia. They can be found in lower latitudes, although the tolerance of clay material becomes less in the warmer, more humid regions. In the humid tropics they generally can be found only on sandy parent material that is almost entirely composed of quartz particles, as along coastal beach deposits. Spodosols are poor agricultural soils in their natural state but can be conditioned to be-

Figure 16-9   A Spodosol profile on sandy glacial till in eastern Maine.   The bleached albic horizon is shown directly beneath thick O and A1 horizons. The top of the spodic horizon, shown at 6 inches on the scale marker, forms a sharp contrast because of its dark color, usually reddish brown. Note that typically the diagnostic horizons are not of even thickness. The horizons rarely are perfectly horizontal. Note also how the plant roots do most of their feeding within the B horizon. The spodic or B horizon here gradually merges with the C horizon below. [Chas. E. Kellogg, Soil Survey, U.S. Dept. of Agriculture]

come highly productive. Some of the sandy, outwash-plain Spodosols of northern Germany and Denmark have been made into fine farmland, but only by dint of many years of deep plowing, the use of fertilizers, and heavy manuring to decrease acidity and to increase the organic content. Except in the

tropics, the Spodosols usually are found beneath a coniferous or heath vegetation cover. A typical Spodosol profile analysis is given in Table 16-3. Note the highly acid organic O and A horizons, the increase in the amount of iron oxide in the B horizon, and the extremely low clay content. The symbols ir and h in the B horizons refer to the deposition of iron oxide and humus, respectively, in the spodic horizon. The II in the C2 horizon indicates a sudden change in parent material from a very silty material above to almost pure sand below.

16-14   THE ALFISOLS (alf = combined form from aluminum and iron)   The Alfisols include those soils with (1) ochric epipedons; (2) argillic horizons; (3) moderate to high base

Table 16-3   Profile Analysis of a Spodosol from Bardufoss, Norway (Data from U.S. Dept. of Agriculture)

Parent material: sandy glacial outwash

Depth (in.)	Horizon	Sand	Texture Silt	Clay	Org. mat. % (wt.)	Free Fe$_2$O$_3$	Base-exch. cap.	pH
6–0	O	—	—	—	100	—	116.3	4.3
0–3	A2	30.6	66.9	2.5	1.38	0.8	14.5	4.5
3–11	B2 ir, h	33.2	66.0	.8	1.26	2.7	16.0	5.5
11–14	B3 ir, h	29.3	68.9	1.8	.62	1.4	8.8	5.5
14–26	C1	52.9	46.1	1.0	.14	.7	3.8	5.3
26+	II C2	98.0	1.9	.1	.10	.3	1.1	5.1

saturation; and (4) a period during the year when evapotranspiration exceeds precipitation and one or more of the horizons become sufficiently dry to interfere with normal plant growth. The dry period need not be long, perhaps no more than a couple of weeks.

Alfisols and Spodosols may be found in the same area, the latter on coarse-textured sandy parent materials and the former on clays and clay loams. Such mixtures are often encountered in areas of glacial deposition. Some of the Alfisols exhibit features of calcification or salinization, yet lack the mollic epipedon that usually overlies such soils outside the deserts. Their principal diagnostic horizons are argillic or natric. Water does not move easily through the Alfisols, and there rarely are strong color contrasts between the horizons. They are best identified by the structural features of the B horizons. Probably the most widespread mid-latitude representatives are those that formerly were *gray-brown* podzolic soils and the *gray forest* or *degraded chernozem* soils. The latter are believed to be former Mollisols on which forests were established, thus destroying the mollic epipedon.

The Alfisols have a wide global distribution, as shown in Figure 16-6. Outside the upper mid-latitudes, they occur most widely in the tropical and subtropical areas that experience pronounced dry seasons, such as the Mediterranean (*Cs*) climates and the *Aw* or savanna climate. In the Mediterranean Sea region, much of the clay forming the argillic horizon in the Alfisols was derived from the solution of limestones. This soil formerly was termed *terra rossa*. In southwestern Australia, the Alfisols appear to feature a massive A horizon which is a typical duripan. Like terra rossa, it is soft and workable when wet but becomes hard as a rock when dry. It may represent a former buried horizon exposed by erosion. The argillic horizon is below it.

The broad expanse of Alfisols corresponding to the savanna country south of the Sahara, as well as those in India and South America, features massive claypans and poor internal drainage. Some of them also have indurated plinthite deposits overlain by loesslike silts, indicating climatic changes. Note that the African Alfisols lie between the desert and the humid tropics. Natric horizons can be formed in this manner, that is, by an increase in rainfall following salinization.

To summarize, the Alfisols essentially are soils with light-colored epipedons and feature argillic or natric horizons below.

A typical Alfisol profile analysis is shown in Table 16-4. The sudden increase in the illuvial clay in the B2t (argillic) horizon is diagnostic of this soil order, especially when it is combined with a slight podzolization effect (note the migration of iron sesquioxides into the illuvial horizon along with the clay). The high clay content in the B2 horizon precludes its being considered as a spodic horizon, despite an increase in iron oxide and humus. When podzolization takes place in high-lime parent materials, the result usually is the development of an Alfisol.

16-15 THE ULTISOLS (L.: *ultimus* = ultimate) The *Ultisols* include those soils that have illuvial horizons in which silicate clays have accumulated (argillic horizons) with a low cation-exchange capacity within and below this horizon. This is in contrast to the considerably higher base-exchange capacity of the Alfisols. They were formerly included among the *red-yellow* podzolic soils, *latosols,* some of the humic gleys, and groundwater laterites. The lower base-exchange capacity is related to the process of latosolization, which tends to remove some of the silica from the clays. Plinthite hardpans often are associated with poor drainage conditions in these areas. Red and yellow coloration is rather common but not always

Table 16-4 Profile Analysis of a Udalf (Gray-Brown Podzolic) Alfisol from Cayuga County, N.Y. (Data from U.S. Dept. of Agriculture)

			Parent material: calcareous glacial till						
Depth (in.)	Horizon	Sand	Texture (%) Silt	Clay	Org. mat. % (wt.)	Free Fe$_2$O$_3$	Base-exch. cap.	pH	CaCO$_3$ equiv (%)
0–4	A1	39.5	49.7	10.8	3.17	1.1	23.2	5.4	—
4–8	A21	41.6	48.5	9.9	1.17	1.1	12.5	4.8	—
8–16	A22	44.9	46.7	8.4	.38	1.0	6.7	5.2	—
16–18	A&B	43.2	45.4	11.4	.30	1.4	7.9	5.1	—
18–21	B&A	41.0	43.0	16.0	.35	1.8	10.4	5.4	—
21–27	B2t	37.9	38.7	23.4	.48	2.6	9.7	6.9	—
27–42	C1	43.2	49.0	7.8	.10	.8	calc.	8.0	31.4
70–80	C2	43.8	46.7	9.5	.12	.8	calc.	8.2	34.6

present. The Ultisols are found in two principal types of areas: (1) humid subtropics, as in the areas of *Cfa* climates, where the latosolization process is weakly developed; and (2) tropical areas, where the latosolization process has not reached maturity because of the youth of the parent material but where profile alteration is clearly indicated. In the former, the Ultisols have some of the characteristics associated with podzolization and latosolization. This is illustrated by Table 16-5. Note, for example, the argillic horizons (B1 and B2), the iron oxides in the illuvial zone, the mild acidity, and the color. The profile has many similarities to the Alfisol profile shown in Table 16-4, but two differences can be noted: (1) the yellowish-red color in the argillic horizon and (2) the low base-exchange capacity throughout the profile. The latter is a common feature of most of the Ultisols and, along with the low content of organic material, helps explain why such soils require heavy applications of fertilizers and green manure to maintain fertility, and why they tend to erode easily. Note also in Table 16-5 that the content of free iron

Table 16-5 Profile Analysis of an Ultisol from Covington County, Miss. (Data from U.S. Dept. of Agriculture)

			Parent material: coastal plain sandy loam						
Depth (in.)	Horizon	Sand	Texture (%) Silt	Clay	Org. mat. (%-wt.)	Free Fe$_2$O$_3$	Base-exch. cap.	pH	Color
0–3	A1	47.8	46.6	5.6	2.92	0.4	9.2	4.9	gray–brn.
3–5	A21	49.1	45.5	5.4	.82	.5	3.8	5.2	yel.–brn.
5–10	A22	49.4	44.7	5.9	.22	.5	2.0	5.2	brn.–yel.–brn.
10–13	B1t	40.7	47.4	11.9	.17	1.0	3.4	5.1	brn.–yel.–brn.
13–24	B2t	40.0	37.0	23.0	.18	2.3	8.3	5.0	yel.–red
24–32	B3	62.3	27.9	9.8	.08	1.0	3.0	5.0	yel.–red–brn.
32–41	II A2x	71.7	20.8	7.5	.04	.7	2.3	5.0	yel.–red–brn.
41–56	II B2x	61.7	19.8	18.5	.06	1.8	4.1	5.0	brn.–pale brn.

Table 16-6 Profile Analysis of an Oxisol from Puerto Rico (Data from U.S. Dept. of Agriculture)

Parent material: serpentine (a ferromagnesium rock)

Depth (in.)	Horizon	Sand	Texture (%) Silt	Clay	Org. mat. % (wt.)	Free Fe$_2$O$_3$	Base-exch. cap.	pH
0–11	A1	9.2	36.3	54.5	6.34	15.3	26.7	5.1
11–18	A3	7.4	34.9	57.7	2.04	14.3	12.1	5.0
18–28	B21	9.8	30.6	59.6	1.33	17.5	8.2	5.0
29–38	B22	23.3	21.0	55.7	.86	20.4	6.4	5.4
38–48	B23	17.0	23.3	59.7	.72	21.5	5.3	5.7
48–62	B3	19.2	27.2	53.6	.56	22.8	3.8	5.9
62–70+	C1	17.1	45.3	37.6	.19	28.4	1.4	6.1

oxide is low, even in the horizons that are colored by iron stain. The latosolization process has not progressed far.

16-16 THE OXISOLS (*oxi* from oxide) The *Oxisols* are relatively easy to define, because they are the only soils that contain an oxic horizon. For a description of this unique horizon, see page 607. As a general rule, the dark red shades associated with ferric oxides are typical of many of the Oxisols. Some, however, may be white or pink on granites or syenites, where the iron content is low or absent. Where the soil almost never dries out because of continuously high rainfall, the iron oxide may take the form of yellowish limonite, a hydrous ferric oxide. Argillic horizons are rare but not unknown in the Oxisols, indicating that the clays are not dispersed out of the topsoils, as they are under podzolization. Instead, they simply disappear through desilication. The Oxisols represent old soils in areas where the processes of chemical weathering are rapid because of high temperatures and rainfall. For this reason, the profiles are much deeper than in the other soil orders.

The Oxisols are restricted to ancient land surfaces in the humid tropical or subtropical regions. The global map of the soil orders (see Figure 16-6) shows Oxisols covering extensive areas in South America and Africa, but none elsewhere. This does not mean than Oxisols are restricted to these two continents. They are well represented on many of the well-drained upland surfaces in parts of India and on the erosional plains in southeastern Asia, such as in central Thailand and Cambodia, but their areal extent is limited, and rough topography and the relatively recent origin of many of the underlying rocks in this part of the world, especially in Indonesia, indicate much too young a surface for Oxisol development. Formerly the Oxisols were included among the *latosols, groundwater laterites,* or *laterites.* A profile analysis for one of the Oxisols is shown in Table 16-6. The color of the profile was dark red throughout.

16-17 THE HISTOSOLS (Gr.: *histos* = tissue) The *Histosols* are organic soils in which the thickness of the organic horizon exceeds that of the histic epipedons, or greater than 12 inches when undrained and 18 inches or more when drained. The Histosols are associated with swamp and marsh deposits, in which decay is retarded because of water saturation and anaerobic conditions. The organic soils always contain some mineral matter, and the limitations of it in order to qualify as an organic soil are the same as those in the histic epipedon (see p. 603). The type of organic material that accumulates in the Histosols varies widely, and some indication has been given of the difference between *peat* and *muck* (see Section 15-6), two types of Histosols.

Only one large area of Histosols is shown on the world soil order map; it is indicated for the area southwest of Hudson Bay. Such areas are, however, found in relatively small, poorly drained depressions throughout the world. They are especially common in the bog depressions of northern Canada and northern Eurasia. Dried peat from Histosols has been a low-grade but easily accessible fuel for centuries in northern Europe (see Figure 14-20).

Histosols can be important as agricultural soils. When drained and fertilized heavily, muck soils are ideal for truck garden vegetables, especially those that grow beneath the ground, such as onions, celery, carrots, beets, and potatoes. Flower bulbs also grow well there, because the friable, open texture of the organic soils permits easy bulb expansion beneath the surface. The mass of organic colloids, saturated with bases and water, also permits plants to feed as rapidly as possible. Some of the highly acid peat bogs have also been used for the cultivation of acid-loving fruits, such as cranberries and huckleberries (*Vaccinium*). Drying and baling peat for sale as lawn dressing and mulching has become a major industry in the northern United States and southern Canada.

One important feature should be noted with respect to the utilization of the Histosols. Unlike any of the mineral soils, they may be destroyed in place by fire, by removal for sale as soil-conditioning material, or by oxidation and normal decay once drained. Unless replenished by new increments of organic material, the Histosols are doomed

to eventual destruction when artificially drained and used for agriculture. Experience on some of the cultivated muck lands of the southern U.S. coastal plains indicates that these organic soils may lose as much as a foot of soil within 15 to 20 years.

16-18 SUMMARY OF THE U.S. SOIL CLASSIFICATION SYSTEM Table 16-7 presents the essential features of the new classification, including the orders and equivalents under the older classification systems.

Soil Fertility

Ever since man learned to domesticate plants to provide himself with a dependable food supply, he often has been more consciously concerned with soil differences than with the other features of his physical environment. It is likely that approximately two-thirds of the families in the world today obtain their livelihood by cultivating the soil, and they either bless or curse their fate, depending on how they appraise the particular piece of ground they cultivate. This appraisal of soil, wherever and whenever it is made, is based on human needs and customs; it has no necessary relationship to any one or any combination of the physical properties of the soil itself. Nature is indifferent to the desires of mankind, and all it says is, "Use me as you will, only remember that the rewards or punishments will be earned, not granted." Soil resources and soil resistances are created not by the properties of soils but by what man desires from it.

It can be noted by way of illustration that the soils having the greatest natural potential for agriculture, the Mollisols, were among the last to be cultivated. The tough grass sod that covered them was extremely difficult to re-

Table 16-7 Summary of the Global Soil Orders

Soil order	Diagnostic properties	Equivalent soils
Entisols	No diagnostic horizons or profile development	Azonal soils; tundra, lithosols, regolith, alluvium, sands and gravels
Vertisols	Dark topsoils that crack badly when dry; slanted columnar structure	Grumusols, tropical black clays, regur, tirs
Inceptisols	Young soils, with profile features just beginning; no clear-cut diagnostic horizons, usually dark epipedons	Subarctic brown forest, brown forest, ando, lithosols, regosols, some humic gleys
Aridisols	Desert soils that have one or more diagnostic horizons but no spodic or oxic horizons; light-colored epipedons	Desert, red desert, gray desert (sierozems), reddish brown, lithosols, regosols, solonetz, solod, solonchak
Mollisols	Dark brown to black, mellow epipedons; one or more diagnostic horizons, except spodic or oxic horizons; often a *ca* horizon	Chernozems, prairie (brunizem), chestnut, red prairie, humic gleys, planosols, black rendzinas, some brown forest, reddish chestnut soils, solonchak, solonetz.
Spodosols	Spodic horizon; usually also an albic horizon; usually coniferous forests over sandy soils	Podzols, brown podzolic, groundwater podzols
Alfisols	Pronounced argillic horizon without a spodic or oxic horizon; may have a natric horizon; no pronounced color changes vertically; high base exchange	Gray-brown podzolic, gray forest, noncalcic brown soils, planosols, half-bogs, solods, and terra rossas
Ultisols	Red or yellow argillic horizon; no oxic horizon; low base-exchange throughout	Red-yellow podzolic, reddish-brown lateritic, rubruzem, humic gleys, low humic gleys, groundwater laterites
Oxisols	Oxic horizon, with many free sesquioxides; little horizon differentiation; deep soils	Latosols, ferrallites, groundwater laterites, laterites
Histosols	A surface organic layer at least 12–18 inches thick	Peat, muck

move with hand tools, and, except for the Russian Ukraine, awaited the invention of the steel plow before being used for agriculture. The earliest American pioneers who moved into the Midwest chose forested alluvial bottomlands along the streams instead of the interstream prairies, partly because of the tough sod but partly also because they believed that if a soil could not grow trees, it was not worth much for farming.

The kinds of soil properties which man seeks are clearly related to the state of his technology. The dense populations of southeastern and southern Asia have been maintained and increased through the cultivation of paddy (wet-field) rice. Per-acre yield of such rice has only a slight relationship to soil base-exchange capacity, the normal index used for measuring soil fertility in agriculture. It is related much more to the hydrologic aspects of the soil, to the amount of paddy water, its rate and direction of movement, its degree of aeration, its acidity, solutes, and microorganic life. There is considerable evidence that paddy soils improve with age, that the laborious and unpleasant task of puddling the fields prior to planting creates an artificial illuvial horizon that resembles a fragipan separating the cultivated portion from the lower horizons. This is undesirable in most crop practices, but with rice, the important ties are with the surface lateral drift of water, not with the characteristics of the subsoil.

In our own cultural world, monetary profitability is the principal standard for appraising soils, and it may or may not be related to their capacity for producing food or feed. Soils may have value solely as places on which to erect buildings. In such cases, the abstract factor of position may be more significant than soil texture or structure. The engineer erecting an earth-filled dam may be far more interested in a soil's plasticity or porosity than in its acidity or organic content, factors that may be critical in agricultural appraisals. Obviously, an appraisal of soil differences based on standards of profitability alone involves complexities more appropriate to a textbook in economic geography than to the present study in global physical geography. The appraisal here is confined largely to conditions of fertility in agriculture.

16-19 THE ELEMENTS OF SOIL FERTILITY Soil fertility results from a large number of individual factors which, taken together, might be thought of as a chain. There must be a sufficient number of links in the chain to provide for each plant growth requirement, and the total strength of the chain is the strength of its weakest link. A soil may have an abundance of many essential elements of fertility, but if it lacks just one (for example, water supply), such elements have little value for plant growth. Among the most important factors in soil fertility are the following:

1. Dry land soils have a plentiful supply of most plant foods except nitrogen, but are generally subject to periodic deficiencies in soil moisture and, in poorly drained areas, to toxic salinity or alkalinity.

2. The best-balanced conditions of soil fertility, at least in a natural state, are found in transitional climatic areas, where the favorable factors of one climate tend to balance the unfavorable factors of another. Excluding the young soils of alluvial and volcanic material, which often are fertile because they have yet not been leached of their plant foods, the best soils for general agriculture are mid-latitude Mollisols that lie near the humid-dry boundary. In this intermediate position between cold and hot regions and between humid and arid conditions are

found the famous granaries of the world, including the Corn and Wheat Belts of the United States and Canada, the eastern European plains of Hungary and Romania, the Ukraine of the Soviet Union, the plains of northern China and Manchuria, the pampas of Argentina, and smaller areas in southeastern Africa and Australia.

3. There is no soil on earth whose fertility cannot be improved, and a wide range of corrective practices has been offered by the soil scientists. The problem of soil improvement is largely one of cost accounting. Experience in the United States seems to indicate that, with a few exceptions, investments in improvement pay off more on good soils than on poor ones. We have not yet reached the point of diminishing returns on the application of technology toward increasing soil productivity on inherently fertile soils. Until this point is reached, more and more food will continue to be produced on fewer acres of farm land, and more marginal land will be retired to other, less competitive uses.

One spectacular exception to the above generalization is worth noticing. The sandy coastal-plain soils of northern New Jersey have perhaps the lowest inherent fertility of any soils in the United States, aside from arid regions. Chemical analyses indicate that the Lakewood sand, a common Spodosol in this area, is composed almost entirely of quartz (SiO_2), one of the most inert minerals of the earth crust. The amount of quartz in the total soil mass in these regions may run well over 90 percent. Only traces of organic material and mineral colloids are present. Despite such obvious handicaps, some of these sands have been developed into some of the most productive and valuable market garden areas in the country, by gradually increasing the organic content and by supplying immense quantities of commercial fertilizers. This is man-made fertility, and the natural soil is useful only as a physical support for plant roots. The motivation for such expensive practices has been the proximity of metropolitan markets. Farmers in these areas are able to truck fresh garden produce directly to the metropolitan markets of Philadelphia and New York within a few hours. The market gardens are concentrated on sands rather than on adjacent heavier soils, which in their native state are much more fertile, because the former are easier to cultivate and the quantities of water and nutrients supplied to the plants can be better regulated. Such sections represent vegetable factories rather than farms. They illustrate the fact that soil fertility is not entirely a natural resource but can be created and modified by man, provided that he is given sufficient incentive. The human potential for increasing the productivity of soils is enormous, and a treatment of soil fertility solely as a function of nature is incomplete.

16-20 THE ECOLOGICAL PROBLEMS OF INCREASING SOIL PRODUCTIVITY The rapid increase in world population, resulting mainly from an increase in life expectancy, produces vast demands for increased production of food and feed. Under such pressures, many of the older agricultural systems, such as migratory agriculture, the fallow-field system, and the rice-compost system, are breaking down with tragic results in soil destruction and lowered per capita calorie intakes. The only areas that have produced spectacular increases in per-acre yields have been those in which heavy investments in the application of inanimate energy have been made. These investments in energy applications take many forms, including a complex chemical industry producing fertilizers, pesticides, insecticides, plant hormones, and soil conditioners; the development, use, and maintenance of a wide range of agricultural machinery using gasoline, diesel fuel, or electricity and designed for both large-scale and small-scale agriculture; the extension of irrigation into humid areas to ensure optimum intakes when needed; and an elaborate processing, transportation, and marketing system making possible regional specialization of crop production, or growing crops in the areas best suited to them. All of these require massive investments of capital and increasingly heavier drains on our nonrenewable reserves of fossil fuels. The lesson is clear: The production of food and feed from the soil, if it is to be increased to feed the world's population adequately, must require additional applications of inanimate energy. Unfortunately, the reserves of fossil fuels are irregularly distributed; hence, soil fertility becomes more a question of capital allocation than the distribution of soil characteristics.

There is a further important consideration in the problem of diverting more energy into selected plant growth. This relates to the ecological implications. Changing the energy distribution pattern produces reactions throughout the entire energy system. Harnessing the reserves of coal, petroleum, and natural gas to increase the world food production to needed levels could result in several important global ecological problems: (1) decreasing the atmospheric supply of free oxygen; (2) increasing the atmospheric content of pollutants on such a global level as to increase seriously the earth albedo; (3) increasing the supply of food for unwanted plant and animal life; and (4) decreasing the qualitative function of food varieties through selected standardized breeding for maximum yields. A few examples of harmful ecological effects resulting from increasing food pro-

duction are illustrated below as follows:

1. Increases in toxic salinity resulting from the extension of irrigation in dry lands.

2. Explosive effects on reproductive rates throughout the lower levels of the animal kingdom in heavily fertilized areas.

3. Serious curtailment of the reproduction rates of some useful birds and animals, resulting from unusually stable elements contained in certain pesticides, such as chlordane and DDT.

4. Reduction of the protein intake of many tropical agriculturalists, resulting from the easier and greater production of certain high-starch staples, such as corn and manioc.

5. Decrease in the level and quality of groundwater, resulting from more efficient withdrawal.

6. Reduction of plant resistivity to fungus and bacteriological infections, resulting from synthetic soil environments.

Little attention has been paid to the long-run implications of altering the entire *ecosystem,* or the entire association of environmental factors within which man earns his living. The unfortunate results of the use of chlordane and DDT as pesticides are only beginning to be recognized. The unity of physical and human geography comes into clear focus in any consideration of soil fertility. The way in which man solves the world food problem may well determine his fate on this planet. There must be a recognition that the physical environment of man places certain constraints on his demands upon it, and the greater these constraints are moved back through technology, the greater becomes the potential hazard of environmental deterioration. Soil fertility lies near the focus of this problem.

Study Questions

1. What is the difference between zonal and intrazonal soils?
2. Explain why the early American soil classification system emphasized intrazonal factors while the Russian soil scientists emphasized zonal properties.
3. What were some of the reasons for setting up the new U.S. Soil Classification System? What does the term "Seventh Approximation" mean?
4. Identify *soil series* and *soil type*.
5. Explain why the soil orders included in the tundra are Entisols, Inceptisols, Histosols, and no others.
6. Briefly summarize the major identifying characteristics of the following diagnostic horizons:
   a. spodic
   b. oxic
   c. albic
   d. calcic
   e. argillic
   f. mollic
   g. natric
7. What is an epipedon? Which of the soil orders is identified best by the characteristics of the epipedon?
8. Are Entisols zonal soils? Why is it possible to find Entisols virtually everywhere on earth?
9. What is the main diagnostic property of a Vertisol? What produces this characteristic? Locate some of the main areas where Vertisols occur.
10. What kinds of diagnostic horizons might be found in Aridisols, and why?
11. Why are the Mollisols that have mollic epipedons more than 24 inches thick *not* included within the Histosols?
12. Explain why mollic epipedons invariably have a grass vegetative cover. Where are they generally located with respect to global climates?
13. Explain the coexistence of Spodosols and Alfisols within the same square mile in Sweden. What are their major identifying characteristics?
14. Where would one look for a Spodosol in North Carolina?
15. The ferrallization process is represented in both the Ultisols and Oxisols. What is the major distinction between these two orders?
16. Why are there no Oxisols shown in southeast Asia or Indonesia on the global map of soil orders?
17. What is the principal difference between an Alfisol and an Entisol? Explain the difference.

Part 5 References

Bennett, Hugh H., *Elements of Soil Conservation*, 2d ed., McGraw-Hill Book Company, New York, 1955.

Buckman, Harry O., and Nyle C. Brady, *The Nature and Property of Soils: A College Text of Edaphology*, 7th rev. ed., The Macmillan Company, New York, 1969.

Bunting, B. T., *The Geography of Soil*, Aldine Publishing Company, Chicago, 1965.

Clarke, George R., *The Study of the Soil in the Field*, The Clarendon Press, Oxford, England, 1961.

Eyre, S. R., *Vegetation and Soils: A World Picture*, Aldine Publishing Company, Chicago, 1963.

Gibson, J. Sullivan, and James W. Batten, *Soils; Their Nature, Classes, Distribution, Uses, and Care*, University of Alabama Press, University, Ala., 1970.

Joffe, Jacob S., *Pedology*, Pedology Publishers, New Brunswick, N.J., 1949.

Marshall,Charles E., *Colloids in Agriculture*, Edward Arnold (Publishers) Ltd., London, 1935.

———, *The Physical Chemistry and Minerology of Soils,* John Wiley & Sons, Inc., New York, 1964.

Mohr, Edward C. J., and F. Z. Van Baren, *Tropical Soils; a Critical Study of Soil Genesis as Related to Climate, Rock, and Vegetation,* Interscience Publishers, Inc., New York, 1954.

Robinson, Gilbert W., *Soils: Their Origin, Constitution, and Classification; An Introduction to Pedology*, 3d ed., John Wiley & Sons, Inc., New York, 1951.

Russell, Sir Edward J., *Soil Conditions and Plant Growth,* 9th ed., John Wiley & Sons, Inc., New York, 1961.

Soil, Yearbook of Agriculture, 1957, U.S. Department of Agriculture, U.S. Government Printing Office, Washington, D.C., 1957.

Soils and Men, Yearbook of Agriculture, 1938, U.S. Department of Agriculture, U.S. Government Printing Office, Washington, D.C., 1938.

Soil Survey Staff, *Soil Survey Manual,* Soil Conservation Service, U.S. Dept. of Agriculture, U.S. Government Printing Office, Washington, D.C., 1951.

———, *Soil Classification; A Comprehensive System; (7th Approximation)*, Soil Conservation Service, U.S. Dept. of Agriculture, U.S. Government Printing Office, Washington, D.C., 1960. Also Supplements, March, 1947 and September, 1968.

APPENDIX A

DETAILS OF A SIMPLIFIED KOEPPEN-GEIGER SYSTEM OF CLIMATIC CLASSIFICATION

A Climates

Humid or subhumid, with no winters; no month with a mean temperature below 64.4°F (18°C).

For examples of the A climates, see the list of climatic stations in Appendix C.

SUBTYPES

f = at least 2.4 inches (60 millimeters) of rain in every month.

m = at least 1 month having less than 2.4 inches of rain; dry season compensated (see Table A-1 below for boundary between m and w or s).

w = uncompensated dry season during the low-sun period.

s = uncompensated dry season during the high-sun period.

g = highest temperature occurring before summer solstice.

w' = rainfall maximum in autumn.

w'' = double maximal rainfall periods.

B Climates
(Semiarid and Arid)

SUBTYPES

$BS = semiarid.$ The BS/humid climatic boundary corresponds to the following formulas:

1. Precipitation mainly in the summer (ten times as much rain in the rainiest month of summer as in the driest month of winter).	*English system* (°F and inches) $R = .44T - 3.0$
2. Precipitation mainly in the winter (three times as much rain in the rainiest month of winter as in	

Table A-1 Determination of *Am-Aw* or *As* Boundary	
Annual rainfall (in.)	Rainfall of driest month (in.)
40	2.34
41	2.30
42	2.26
43	2.22
44	2.18
45	2.14
46	2.10
47	2.07
48	2.02
49	1.98
50	1.94
51	1.90
52	1.86
53	1.82
54	1.78
55	1.75
56	1.70
57	1.66
58	1.63
59	1.58
60	1.55
61	1.51
62	1.47
63	1.42
64	1.38
65	1.34
66	1.30
67	1.26
68	1.22
69	1.18
70	1.13
71	1.10
72	1.06
73	1.02
74	0.98
75	0.94
76	0.90
77	0.86
78	0.81
79	0.78
80	0.74
81	0.70
82	0.66
83	0.61
84	0.58
85	0.54
86	0.50
87	0.46
88	0.42
89	0.37
90	0.34
91	0.29
92	0.26
93	0.22
94	0.18
95	0.14
96	0.09
97	0.06
98	0.02

the driest month of summer).

$$R = .44T - 14.0$$

3. Precipitation evenly distributed (neither #1 nor #2).

$$R = .44T - 8.5$$

NOTE: In the above formulas, R = mean annual precipitation in inches, and T = mean annual temperature in °F.

BW = *arid*. The BW/BS boundary is exactly one-half the figure for the BS/humid boundary, as determined from the formulas above or from Table A-2.

The following lowercase letters are used with the BW and BS climatic symbols to designate variations in the arid and semiarid climates:

h = mean average annual temperature above 64.4°F (18°C).[1]

k = mean average annual temperature below 64.4°F (18°C).

k' = warmest month below 64.4°F.

w = winter dry season; ten times as much rain in the rainiest month of summer as in the driest month of winter.

s = summer dry season; three times as much rain in the rainiest month of winter as in the driest month of summer.

For examples of the B climates, see the list of climatic stations in Appendix C.

C Climates

Humid or subhumid climates with mild winters; temperature of the coldest month above 26.6°F (-3°C) but below 64.4°F (18°C); temperature of the warmest month above 50°F (10°C).

[1] Some climatologists have recommended the use of 32°F for the coldest month as a better boundary between the hot and cold-dry climates (h and k), especially in North America.

628

Table A-2　Average Annual Precipitation along the BS/Humid Boundary (based on formulas 1, 2, and 3)

Average annual temp. (°F)	Summer dry season	Even distribution	Winter dry season
32	0.1	5.6	11.1
33	0.5	6.0	11.5
34	1.0	6.5	12.0
35	1.4	6.9	12.4
36	1.8	7.3	12.8
37	2.2	7.7	13.2
38	2.7	8.2	13.7
39	3.2	8.7	14.2
40	3.6	9.1	14.6
41	4.0	9.5	15.0
42	4.5	10.0	15.5
43	4.9	10.4	15.9
44	5.4	10.9	16.4
45	5.8	11.3	16.8
46	6.2	11.7	17.2
47	6.6	12.1	17.6
48	7.1	12.6	18.1
49	7.6	13.1	18.6
50	8.0	13.5	19.0
51	8.4	13.9	19.4
52	8.9	14.4	19.9
53	9.3	14.8	20.3
54	9.8	15.3	20.8
55	10.2	15.7	21.2
56	10.6	16.1	21.6
57	11.1	16.6	22.1
58	11.5	17.0	22.5
59	12.0	17.5	23.0
60	12.4	17.9	23.4
61	12.8	18.3	23.8
62	13.3	18.8	24.3
63	13.7	19.2	24.7
64	14.2	19.7	25.2
65	14.6	20.1	25.6
66	15.0	20.5	26.0
67	15.5	21.0	26.5
68	15.9	21.4	26.9
69	16.4	21.9	27.4
70	16.8	22.3	27.8
71	17.2	22.7	28.2
72	17.7	23.2	28.7
73	18.1	23.6	29.1
74	18.6	24.1	29.6
75	19.0	24.5	30.0
76	19.4	24.9	30.4
77	19.9	25.4	30.9
78	20.3	25.8	31.3
79	20.8	26.3	31.8
80	21.2	26.7	32.2
81	21.6	27.1	32.6
82	22.1	27.6	33.1
83	22.5	28.0	33.5
84	23.0	28.5	34.0
85	23.4	28.9	34.9

SUBTYPES

f = no pronounced dry season; difference between wettest and driest months less than required for s or w; precipitation in the driest month of summer more than 1.2 inches.

s = summer dry season; at least three times as much precipitation in the wettest month of winter as in the driest month of summer; precipitation in driest month of summer less than 1.2 inches.

w = winter dry season; at least ten times as much rain in the wettest month of summer as in the driest month of winter.

a = hot summers; temperature of the warmest month above 71.6°F (22°C).

b = cool summers; temperature of the warmest month below 71.6°F, but with at least four months above 50°F (10°C).

c = cool, short summers; one to three months above 50°F (10°C).

g = warmest month occurring before the summer solstice.

t' = warmest month occurring in the autumn season.

x = maximum rainfall in spring or early summer; dry period in the late summer.

For examples of the C climates, see the climatic statistics for selected world stations in Appendix C.

D Climates

Humid and subhumid climates with severe winters; coldest month below 26.6°F (−3°C); warmest month above 50°F (10°C).

SUBTYPES

f, s, and w defined as in the C climates; a, b, and c defined as in the C climates.

d = temperature of the coldest month less than −36.4°F (−38°C).

E Climates

No summer; warmest month below 50°F (10°C).

SUBTYPES

ET = periglacial climate; warmest month above 32°F (0°C) but below 50°F (10°C).

EF = icecap climates; warmest month below 32°F (0°C).

Details of a Simplified Koeppen-Geiger System of Climatic Classification

APPENDIX B

THE THORNTHWAITE CLIMATIC CLASSIFICATION

The Thornthwaite climatic classification[1] requires the following quantitative data:

1. Average monthly and annual precipitation.
2. Average monthly and annual temperatures.
3. Average monthly and annual potential evapotranspiration.
4. Average annual water surplus.
5. Average annual water deficit.

The last three require the determination of both actual and potential evapotranspiration, which in turn requires solving the empirical formulas that were determined by Thornthwaite from lysimeter readings. These equations have been solved for over 120 climatic stations, and the minimum classification data for these stations are given in Appendix C.

There are two primary and two secondary indices in the Thornthwaite classification. The two primary ones include the annual *moisture index* and the annual *thermal efficiency index*. The two secondary indices indicate the seasonal concentration of the primary indices.

Moisture index. The moisture index corresponds to the following equation: $M_i = 100(P/PE - 1)$, where P = annual precipitation and PE = annual potential evapotranspiration. The quantitative divisions of the moisture index by regions is given in Table B-1. To illustrate the classification of the moisture index regionally, determine the moisture index for Albany, N.Y. (see Appendix C). According to the data there, $P = 37.8$ inches and $PE = 27.1$ inches.

[1] The material in this appendix was derived from the 1955 modification of the Thornthwaite classification by Thornthwaite and Mather, described in Douglas B. Carter and John R. Mather, "Climatic Classification for Environmental Biology," *Publications in Climatology*, vol. 19, no. 4, C. W. Thornthwaite Associates Laboratory of Climatology, Elmer, N.J., 1966.

Table B-1 Moisture Regions, 1955 Thornthwaite Classification (after Thornthwaite and Mather)

Climatic symbol	Climatic type	Moisture index
A	Perhumid	100 and above
B4	Humid	80 to 100
B3	Humid	60 to 80
B2	Humid	40 to 60
B1	Humid	20 to 40
C2	Moist subhumid	0 to 20
C1	Dry subhumid	-33.3 to 0
D	Semiarid	-66.7 to -33.3
E	Arid	-100 to -66.7

Solving for M_i: $100(37.8/27.1 - 1) = 100(1.39 - 1) = 39$. According to Table B-1, a moisture index of 39 places the station in the $B1$ level of the humid climates.

Thermal efficiency index. This index uses potential evapotranspiration directly as an indicator of thermal efficiency. The appropriate divisions are shown in Table B-2.

Table B-2 World Thermal Efficiency Regions (After Thornthwaite)

Potential evapotranspiration (ann., in.)	Climatic symbol	Climatic type
	E'	Frost
5.61		
	D'	Tundra
11.22		
	$C'1$	Microthermal
16.83		
	$C'2$	
22.44		
	$B'1$	Mesothermal
28.05		
	$B'2$	
33.66		
	$B'3$	
39.27		
	$B'4$	
44.88		
	A'	Megathermal

Again using Albany as an example, its potential evapotranspiration (PE) for the year is 27.1 inches. When referred to Table B-2, this places the station in the B'_1 thermal efficiency region. Note that the thermal efficiency zone letters are marked with a prime mark (B') to distinguish them from the moisture-index code designations.

Seasonal concentration of moisture index. The third code index letter, shown in lower case, indicates the seasonal variation in precipitation effectiveness. The divisions and their numerical values are shown in Table B-3. Note that the concentration is based on the numerical values of the *aridity index* and the *humidity index.* The former, which is 100 times the annual deficit divided by the annual PE, is used only when P is greater than PE, or in the humid climates. The humidity index, or 100 times the annual PE, is used only in the dry climates (C_1, D, E), where PE is greater than P. Note that in each case, whenever there is a moderate or large deficit or surplus, the lowercase letters always refer to the drier of the two seasons.

Two examples: (1) Using the Albany example again, note that P is greater than PE. Thus the station is a humid climate, and the aridity index must be calculated to determine the seasonal concentration. The aridity index for Albany $= 100(1.26/27.1) = 4.6$. According to Table B-3, this would indicate little water deficit and would call for the use of the letter r.

(2) Use the data for Salisbury, Southern Rhodesia in Appendix C. This is a humid station. Hence the aridity index $= 100 \times (6.3/27.1) = 18.7$. Since this is a southern hemisphere station, the moderate deficit (between 10 and 20) comes during the winter half, hence the use of the letter w.

Seasonal concentration of PE. The fourth and last code index refers to the concentration of potential evaporation in the three summer months (June, July, and August north of the equator, and December, January, and February south of it). This is a rough index of seasonality and continentality. Low-latitude stations tend to have low summer concentrations, while high-latitude stations concentrate a large amount of their thermal efficiency into the short summer season with its long days and short nights. If the percentage drops below about 45 percent, no code letter is used. The zonal boundaries of the summer concentration of PE are shown in Table B-4.

To illustrate this last code index, we shall use the Albany data again. The total PE for the three summer months is 14.9 inches. The annual $PE = 25.4$ inches, Thus the summer concentration $= 14.9/25.4 = 58.6$ percent, which rates a $b'2$ according to Table B-4.

In summarizing the climatic classification for Albany, note that it carries a total code of $B1B'1rb'2$. Translated, this signifies a slightly humid, mesothermal climate that is near the microthermal boundary, with a fairly even distribution of precipitation effectiveness and a tendency toward pronounced seasonal variation in global thermal efficiency.

Table B-3 Seasonal Variation of Effective Moisture (1955 revision)

Moist climates (A, B, C_2)		Aridity index (D/PE) $\times$ 100
r	little or no water deficiency	0 to 10
s	moderate summer water deficiency	10 to 20
w	moderate winter water deficiency	10 to 20
s_2	large summer water deficiency	over 20
w_2	large winter water deficiency	over 20
Dry climates (C_1, D, E)		Humidity index (S/PE) $\times$ 100
d	little or no water surplus	0 to 16.7
s	moderate winter water surplus	16.7 to 33.3
w	moderate summer water surplus	16.7 to 33.3
s_2	large winter water surplus	over 33.3
w_2	large summer water surplus	over 33.3

Table B-4 Summer Concentration of Potential Evapotranspiration (after Thornthwaite)

Summer concentration of PE (%)	Summer concentration climatic symbol
	a'
—— 48.0 ——	
	$b'4$
—— 51.9 ——	
	$b'3$
—— 56.3 ——	
	$b'2$
—— 61.6 ——	
	$b'1$
—— 68.0 ——	
	$c'2$
—— 76.3 ——	
	$c'1$
—— 88.0 ——	
	d'

APPENDIX C
WORLD CLIMATIC RECORDS

The temperature and precipitation data included in this appendix were obtained from two sources:

1. U.S. Department of Commerce, Environmental Science Service Administration, *World Weather Records*, 1951–1960, 5 volumes, U.S. Government Printing Office, Washington, D.C., 1962–1968.

2. Great Britain, Meteorological Office, *Tables of Temperature, Relative Humidity and Precipitation for the World*, 6 volumes, H.M. Stationery Office, London, 1962–1969.

The averages represent the station records and, except for the stations within China (the Chinese People's Republic), include observations up to 1960. No station has less than a 20-year record.

The calculation of the Thornthwaite formulas yielding the figures for potential evapotranspiration and average annual water surplus and deficit was performed using an IBM 360/67 computer program developed by Dr. Nicolay Timofeeff and William MacArthur. Field-storage capacity for water at all locations was assumed at 8 inches (200 mm). The original Thornthwaite formula for the determination of the heat-index part of the equation was refined slightly to adjust evenly for variations in length of day and twilight energy inputs with increasing latitude up to 65°. It will be noted that the figures for actual evapotranspiration are not given, partly to save space and also because they are not needed in the determination of the Thornthwaite climatic classification, provided the annual water surplus and deficit are given. Letter symbols used in the station records that follow include:

PE = potential evapotranspiration, as determined by the Thornthwaite equations, in inches.

P = precipitation in inches.
T = temperature (°F).
S = average annual water surplus, inches.
D = average annual water deficit, inches.

K = Koeppen climatic classification symbol.
Th = Thornthwaite climatic classification symbol.

NORTH AMERICA

Albany, N.Y.: lat 42°45′ N; long 73°48′ W; elev. (ft) 288

		Jan.	Feb.	Mar.	Apr.	May	June	July	Aug.	Sept.	Oct.	Nov.	Dec.	Yr.	
S	14.8″	25.7	26.8	35.8	48.4	60.0	69.1	73.9	71.8	63.7	53.1	41.7	29.5	49.8	T°F
D	1.26″														
K	Dfa	2.5	2.2	2.9	2.9	3.6	3.7	4.3	3.3	4.0	2.8	2.9	2.7	37.8	P″
Th	B1B′1rb′2	0	0	0.2	1.5	3.4	5.0	6.0	5.2	3.4	1.8	0.6	0	27.1	PE″

Albuquerque, N.M.: lat 35°03′ N; long 106°37′ W; elev. (ft) 5,314

		Jan.	Feb.	Mar.	Apr.	May	June	July	Aug.	Sept.	Oct.	Nov.	Dec.	Yr.	
S	0	35.0	39.9	46.2	55.8	65.1	74.8	78.4	76.6	70.5	58.5	44.0	37.0	56.8	T°F
D	22.7″														
K	BWh	0.4	0.4	0.5	0.5	0.8	0.6	1.2	1.3	0.9	0.7	0.4	0.5	8.2	P″
Th	Eb′2db′2	0.1	0.3	0.8	2.0	3.6	5.5	6.3	5.6	4.0	2.1	0.5	0.1	30.9	PE″

Anchorage, Alaska: lat 61°10′; long 149°59′ W; elev. (ft) 105

		Jan.	Feb.	Mar.	Apr.	May	June	July	Aug.	Sept.	Oct.	Nov.	Dec.	Yr.	
S	0	12.4	18.0	23.4	35.8	45.9	54.5	57.0	55.6	47.8	35.1	22.1	14.4	35.2	T°F
D	7.29														
K	Dfc	0.8	0.7	0.5	0.4	0.5	1.0	1.9	2.6	2.5	1.9	1.0	0.9	14.7	P″
Th	C1C′2db′1	0	0	0	0.9	3.2	4.9	5.4	4.4	2.6	.6	0	0	22.0	PE″

Blue Hill obs., Mass.: lat 42°13′ N; long 71°07′ W; elev. (ft) 649

		Jan.	Feb.	Mar.	Apr.	May	June	July	Aug.	Sept.	Oct.	Nov.	Dec.	Yr.	
S	22.5	27.0	27.3	34.9	45.7	56.7	65.1	70.9	69.4	62.4	52.7	42.1	30.0	48.7	T°F
D	.79														
K	Cfb	4.5	3.7	4.5	4.0	3.5	3.7	3.3	4.1	3.9	3.7	4.6	4.0	47.5	P″
Th	B4B′1rb′2	0	0	0.2	1.4	3.1	4.5	5.5	4.9	3.3	1.9	0.7	0	25.4	PE″

Calgary, Alberta, Canada: lat 51°06′ N; long 114°01′ W; elev. (ft) 3,540

		Jan.	Feb.	Mar.	Apr.	May	June	July	Aug.	Sept.	Oct.	Nov.	Dec.	Yr.	
S	0	13.3	17.1	23.7	38.1	48.9	55.0	61.9	58.8	51.4	41.9	27.9	20.5	38.3	T°F
D	3.39														
K	Dfb	0.7	1.0	1.0	1.2	2.0	3.5	2.5	2.8	1.3	0.8	0.7	0.6	18.1	P″
Th	C1C′2db′1	0	0	0	1.1	3.0	4.0	5.1	4.2	2.7	1.3	0	0	21.5	PE″

Charlestown, S.C.: lat 32°54′ N; long 80°02′ W; elev. (ft) 59

		Jan.	Feb.	Mar.	Apr.	May	June	July	Aug	Sept.	Oct.	Nov.	Dec.	Yr.	
S	11.5″	50.4	51.4	56.7	64.2	72.0	78.3	80.1	79.7	75.6	66.2	55.9	50.0	65.1	T°F
D	.2″														
K	Cfa	2.6	3.3	3.9	2.9	3.6	5.0	7.7	6.6	5.8	2.8	2.1	2.8	49.1	P″
Th	B1B′3rb′4	0.7	0.7	1.4	2.6	4.4	5.9	6.5	6.1	4.6	2.7	1.2	0.7	37.5	PE″

Chesterfield Inlet, NWT, Canada: lat 63°20′ N; long 90°43′ W; elev. (ft) 13

		Jan.	Feb.	Mar.	Apr.	May	June	July	Aug.	Sept.	Oct.	Nov.	Dec.	Yr.	
S	0	−24.8	−25.6	−13.7	2.5	21.0	36.7	47.8	47.3	37.0	21.6	0.3	−15.3	11.3	T°F
D	2.84″														
K	ET	0.3	0.4	0.4	0.7	0.5	1.0	1.8	1.7	1.5	1.2	0.9	0.6	10.7	P″
Th	C1C′1dd′	0	0	0	0	0	2.6	5.0	4.2	1.7	0	0	0	13.4	PE″

Fairbanks, Alaska: lat 64°49′ N; long 147°32′ W; elev. (ft) 439

		Jan.	Feb.	Mar.	Apr.	May	June	July	Aug.	Sept.	Oct.	Nov.	Dec.	Yr.	
S	0	−11.0	−2.9	9.0	29.5	47.1	58.5	59.7	54.3	43.5	26.2	3.9	−7.8	25.9	T°F
D	11.41														
K	Dfc	0.9	0.5	0.4	0.2	0.7	1.4	1.8	2.2	1.1	0.9	0.6	0.6	11.3	P″
Th	DB′1dc′2	0	0	0	0	3.6	6.3	6.4	4.4	2.0	0	0	0	22.8	PE″

Kodiak Island, Alaska: lat 57°45′ N; long 152°30′ W; elev. (ft) 105

		Jan.	Feb.	Mar.	Apr.	May	June	July	Aug.	Sept.	Oct.	Nov.	Dec.	Yr.	
S	20.8″	30.0	31.5	31.3	36.5	42.8	49.6	54.1	55.0	49.8	41.0	35.8	30.2	40.6	T°F
D	.3″														
K	Cfc	4.8	3.7	3.9	3.5	5.2	3.6	3.5	3.8	5.0	5.9	6.5	4.8	54.2	P″
Th	AC′2rb′2	0	0	0	1.1	2.6	3.9	4.7	4.3	2.8	1.4	0.6	0	21.4	PE″

Sacramento, Calif.: lat 38°31′ N; long 121°30′ W; elev. (ft) 43

		Jan.	Feb.	Mar.	Apr.	May	June	July	Aug.	Sept.	Oct.	Nov.	Dec.	Yr.	
S	.24	46.2	50.2	54.3	59.9	66.0	72.5	77.4	76.1	73.6	64.9	54.3	47.5	62.0	T°F
D	20.0														
K	Csa	3.2	3.0	2.4	1.4	0.6	0.1	0	0	0.2	0.8	1.5	3.2	16.4	P″
Th	DB′3db′4	0.5	0.8	1.4	2.2	3.6	4.9	6.2	5.5	4.3	2.6	1.1	0.6	33.7	PE″

Sitka, Alaska: lat 57°03′ N; long 135°20′ W; elev. (ft) 66

		Jan.	Feb.	Mar.	Apr.	May	June	July	Aug.	Sept.	Oct.	Nov.	Dec.	Yr.	
S	63.7″	29.5	32.7	34.2	39.7	45.5	50.7	54.7	55.2	51.1	44.1	38.3	33.4	42.5	T°F
D	.1														
K	Cfb	6.0	7.9	7.5	5.9	5.1	2.8	4.8	7.3	10.1	13.7	11.7	12.1	94.9	P″
Th	AB′1rb′3	0	0.1	0.5	1.5	2.8	3.9	4.6	4.1	2.9	1.7	0.7	0.2	23.0	PE″

SOUTH AND CENTRAL AMERICA

Acapulco, Mexico: lat 16°50′ N; long 99°56′ W; elev. (ft) 10

		Jan.	Feb.	Mar.	Apr.	May	June	July	Aug.	Sept.	Oct.	Nov.	Dec.	Yr.	
S	13.5	79.5	79.9	80.6	82.0	83.7	83.5	83.5	84.2	82.6	82.9	82.0	80.0	82	T°F
D	30.4″														
K	Aww′i	.6	0	0	0	2.0	10.4	12.5	8.1	16.3	7.2	2.3	0.4	59.8	P″
Th	C1A′w	4.8	4.7	5.6	5.9	6.7	6.6	6.7	6.7	6.0	6.0	5.4	5.3	70.4	PE″

Antofagasta, Chile: lat 23°28′ S; long 70°26′ N; elev. (ft) 541

		Jan.	Feb.	Mar.	Apr.	May	June	July	Aug.	Sept.	Oct.	Nov.	Dec.	Yr.	
S	0	68.0	68.5	66.0	61.7	59.2	56.5	56.1	56.3	57.9	59.5	62.6	64.8	61.5	T°F
D	28.7″														
K	BWh	0	0	0	0	0	0	0	0	0	0	0	0	0	P″
Th	EB′2d	3.1	3.0	3.1	2.5	2.3	1.9	1.9	1.9	1.9	2.1	2.3	2.6	28.7	PE″

Lima, Peru: lat 12°04′ S; long 77°02′ W; elev. (ft) 449

		Jan.	Feb.	Mar.	Apr.	May	June	July	Aug.	Sept.	Oct.	Nov.	Dec.	Yr.	
S	0	69.8	72.0	71.8	68.4	64.2	61.3	59.7	59.2	59.4	61.0	63.3	66.6	64.8	T°F
D	31.5														
K	BWhs	0	0	0	0	0	.1	.1	.2	.2	.1	0	0	0.7	P″
Th	EB′2d	3.7	3.7	3.9	3.1	2.5	2.0	1.8	1.8	1.8	2.2	2.5	3.2	32.2	PE″

Managua, Nicaragua: lat 12°07′ N; long 86°11′ W; elev. (ft) 184

		Jan.	Feb.	Mar.	Apr.	May	June	July	Aug.	Sept.	Oct.	Nov.	Dec.	Yr.	
S	5.36″	78.3	80.0	81.1	82.8	82.9	80.6	80.1	80.8	80.6	79.2	79.5	79.3	80.4	T°F
D	22.4″														
K	Aww′gi	0.1	0.1	0.2	0.1	5.9	9.2	5.0	4.4	8.4	12.0	1.6	0.4	47.4	P″
Th	C1A′d	4.6	4.8	5.7	6.1	6.4	5.8	5.9	5.9	5.5	5.0	4.8	4.9	65.5	PE″

Maracaibo, Venezuela: lat 10°39′ N; long 71°36′ W; elev. (ft) 131

		Jan.	Feb.	Mar.	Apr.	May	June	July	Aug.	Sept.	Oct.	Nov.	Dec.	Yr.	
S	0	79.7	80.1	81.0	82.0	83.1	83.5	83.5	83.7	83.7	82.2	82.0	80.8	82.2	T°F
D	55.2″														
K	BWhw	0.1	0.1	0.1	1.0	2.9	1.7	1.1	1.6	1.6	3.9	0.9	0.2	15.2	P″
Th	EA′d	5.0	4.9	5.7	5.9	6.4	6.4	6.6	6.5	6.1	5.9	5.5	5.4	70.4	PE″

Monterrey, Mexico: lat 25°40′ N; long 100°15′ W; elev. (ft) 1,751

		Jan.	Feb.	Mar.	Apr.	May	June	July	Aug.	Sept.	Oct.	Nov.	Dec.	Yr.	
S	0	61.5	63.7	68.5	74.7	79.0	82.6	83.1	82.9	78.6	72.3	63.9	60.1	72.7	T°F
D	28.1″														
K	BSh	0.3	0.9	0.5	0.7	0.7	2.0	1.6	3.0	4.7	3.7	0.9	0.3	19.3	P″
Th	DA′d	1.4	1.6	2.7	4.2	5.8	6.7	7.0	6.6	5.0	3.4	1.7	1.3	47.4	PE″

Piarco Airport, Trinidad: lat 10°37′ N; long 61°21′ W; elev. (ft) 40

		Jan.	Feb.	Mar.	Apr.	May	June	July	Aug.	Sept.	Oct.	Nov.	Dec.	Yr.	
S	14.9	76.1	76.5	77.7	79.3	79.9	79.0	78.6	79.0	79.2	78.6	77.7	76.6	78.3	T°F
D	3.66″														
K	Awi	3.0	2.4	1.1	2.8	5.1	10.6	9.6	8.4	5.7	5.9	8.3	6.0	68.9	P″
Th	B1A′r	4.1	3.8	4.7	5.1	5.7	5.2	5.2	5.3	5.0	4.9	4.9	4.2	57.6	PE″

Punta Arenas, Chile: lat 53°10′ S; long 70°54′ W; elev. (ft) 92

		Jan.	Feb.	Mar.	Apr.	May	June	July	Aug.	Sept.	Oct.	Nov.	Dec.	Yr.	
S	0	51.3	50.9	48.0	43.5	39.2	36.1	36.1	37.4	40.1	44.2	48.4	50.4	43.9	T°F
D	6.26″														
K	Cfc	1.5	1.4	1.8	1.7	1.7	1.0	1.5	2.0	1.0	0.9	1.5	1.6	17.6	P″
Th	C1B′1da′	3.8	3.0	2.5	1.6	0.9	0.5	0.5	0.8	1.3	2.2	3.1	3.7	23.9	PE″

Quixeramobim, Brazil: lat 5°12′ S; long 43°10′ W; elev. (ft) 653

		Jan.	Feb.	Mar.	Apr.	May	June	July	Aug.	Sept.	Oct.	Nov.	Dec.	Yr.	
S	0	84.2	83.1	81.7	80.8	79.9	79.7	80.4	81.9	82.9	83.5	83.8	84.4	82.2	T°F
D	44.5″														
K	Awi	1.7	3.5	6.7	6.3	4.0	1.5	0.8	0.3	0.2	0.1	0.2	0.6	25.9	P″
Th	DA′d	6.5	5.7	5.9	5.5	5.4	5.5	5.5	5.9	5.9	6.3	6.2	6.6	70.4	PE″

Salvador (Bahia), Brazil: lat 12°57′ S; long 38°29′ W; elev. (ft) 30

		Jan.	Feb.	Mar.	Apr.	May	June	July	Aug.	Sept.	Oct.	Nov.	Dec.	Yr.	
S	11.1″	78.8	79.3	79.3	78.4	76.6	74.8	73.4	73.2	74.5	76.1	77.2	78.1	76.6	T°F
D	4.5														
K	Afgi	2.9	3.1	6.4	11.4	11.7	7.7	8.1	4.4	3.3	3.7	5.6	3.9	72.3	P″
Th	B1A′r	5.4	4.9	5.3	4.7	4.2	3.6	3.4	3.4	3.7	4.4	4.7	5.2	52.9	PE″

Santarem, Brazil: lat 02°25′ S; long 54°42′ W; elev. (ft) 66

		Jan.	Feb.	Mar.	Apr.	May	June	July	Aug.	Sept.	Oct.	Nov.	Dec.	Yr.	
S	28.0″	78.4	77.9	77.9	78.1	78.1	77.7	77.7	79.2	80.1	80.6	80.4	79.7	78.8	T°F
D	7.35″														
K	Amgi	7.3	10.8	14.1	14.3	11.5	6.9	4.4	2.0	1.5	1.8	3.4	4.8	82.8	P″
Th	B1A′w	5.0	4.3	4.8	4.6	4.8	4.5	4.7	5.1	5.4	5.7	5.5	5.4	59.7	PE″

Tacubaya, Mexico: lat 19°24′ N; long 99°11′ W; elev. (ft) 7,576

		Jan.	Feb.	Mar.	Apr.	May	June	July	Aug.	Sept.	Oct.	Nov.	Dec.	Yr.	
S	2.5	53.8	56.8	61.0	62.8	63.3	62.6	60.6	60.6	60.1	58.5	55.9	54.0	59.2	T°F
D	4.56														
K	Cwbi	0.3	0.2	0.4	0.9	2.2	4.6	6.3	5.7	5.1	1.9	0.7	0.2	28.5	P″
Th	C2B′1d	1.5	1.7	2.5	2.8	3.1	3.0	2.7	2.7	2.4	2.2	1.7	1.5	27.9	PE″

Uaupés, Brazil: lat 0°08′ S; long 67°05′ W; elev. (ft) 278

		Jan.	Feb.	Mar.	Apr.	May	June	July	Aug.	Sept.	Oct.	Nov.	Dec.	Yr.	
S	60.2″	61.7	61.2	58.3	52.9	49.5	46.2	45.7	46.4	48.4	52.9	57.2	61.2	53.4	T°F
D	0														
K	Afgi	4.0	2.0	3.8	7.0	17.3	14.6	17.8	12.5	8.9	3.9	3.2	2.8	98.0	P″
Th	AA′r	2.5	2.5	2.7	2.2	2.0	1.5	1.5	1.5	1.5	1.9	2.1	2.4	23.7	PE″

SOUTH AND CENTRAL AMERICA (continued)

Valdivia, Chile: lat 39°48′ S; long 73°14′ W; elev. (ft) 43													
		Jan.	Feb.	Mar.	Apr.	May	June	July	Aug.	Sept.	Oct.	Nov.	Dec.	Yr.
S	72.7″	61.7	61.2	58.3	52.9	49.5	46.2	45.7	46.4	48.4	52.9	57.2	61.2	53.4	T°F
D	.1													
K	Cfb	4.0	2.0	3.8	7.0	17.3	14.6	17.8	12.5	8.9	3.9	3.2	2.8	98.0	P″
Th	AB′1r	2.5	2.5	2.7	2.2	2.0	1.5	1.5	1.5	1.5	1.9	2.1	2.4	23.7	PE″

EUROPE

Athens, Greece: lat 37°58′ N; long 23°43′ E; elev. (ft) 351													
		Jan.	Feb.	Mar.	Apr.	May	June	July	Aug.	Sept.	Oct.	Nov.	Dec.	Yr.
S	0	48.7	49.8	52.3	59.5	68.0	76.3	81.7	81.3	74.3	66.2	58.5	51.8	64.0	T°F
D	20.9″													
K	Csa	2.4	1.4	1.5	0.9	0.9	0.5	0.2	0.3	0.6	2.0	2.2	2.8	15.7	P″
Th	DB′3db′3	0.6	0.7	1.1	2.0	3.8	5.7	7.2	6.6	4.4	2.7	1.5	0.8	36.6	PE″

Belgrade, Yugoslavia: lat 44°48′ N; long 20°28′ E; elev. (ft) 456													
		Jan.	Feb.	Mar.	Apr.	May	June	July	Aug.	Sept.	Oct.	Nov.	Dec.	Yr.
S	2.4″	31.6	34.9	43.2	54.0	62.8	68.9	72.7	71.6	64.9	54.5	44.2	36.5	53.2	T°F
D	6.2″													
K	Cfa	1.9	1.8	1.8	2.1	2.9	3.8	2.4	2.2	2.0	2.2	2.4	2.2	27.7	P″
Th	C1B′2db′3	0	0.1	0.8	2.1	3.8	5.0	5.8	5.2	3.5	1.9	0.7	0.2	28.7	PE″

Bergen, Norway: lat 60°24′ N; long 5°19′ E; elev. (ft) 144													
		Jan.	Feb.	Mar.	Apr.	May	June	July	Aug.	Sept.	Oct.	Nov.	Dec.	Yr.
S	47.9″	34.7	34.3	37.6	42.4	50.4	54.7	59.0	58.5	53.6	46.9	41.9	37.9	46.0	T°F
D	.1													
K	Cfb	7.1	5.5	4.3	5.5	3.3	5.0	5.6	6.6	9.0	9.3	8.1	8.0	78.3	P″
Th	AB′1rb′3	0.3	0.3	0.8	1.6	3.4	4.4	5.1	4.4	2.9	1.7	0.9	0.5	25.6	PE″

Bordeaux, France: lat 44°50′ N; long 0°42′ W; elev. (ft) 167													
		Jan.	Feb.	Mar.	Apr.	May	June	July	Aug.	Sept.	Oct.	Nov.	Dec.	Yr.
S	9″	41.4	42.6	48.7	53.1	58.5	64.4	67.3	67.1	62.8	54.9	47.1	42.3	54.1	T°F
D	2.3″													
K	Cfb	3.5	2.9	2.5	1.9	2.4	2.6	2.2	2.8	3.3	3.3	3.8	4.3	35.5	P″
Th	B1B′1rb′4	0.5	0.6	1.4	2.1	3.2	4.3	4.8	4.4	3.2	2.0	1.0	0.6	27.7	PE″

Budapest, Hungary: lat 47°31′ N; long 19°01′ E; elev. (ft) 426

		Jan.	Feb.	Mar.	Apr.	May	June	July	Aug.	Sept.	Oct.	Nov.	Dec.	Yr.	
S	1.1	30.0	33.8	42.4	53.2	62.2	68.4	72.0	70.5	63.3	52.3	42.4	34.7	52.2	T°F
D	7.7														
K	Cfa	1.6	1.7	1.5	1.8	2.8	3.0	2.1	2.0	1.3	2.2	2.7	1.9	24.6	P″
Th	C1B′2d	0	0.1	0.8	2.2	3.9	5.1	5.9	5.1	3.3	1.7	0.6	0.1	28.4	PE″

Edinburgh, United Kingdom: lat 55°55′ N; long 3°11′ W; elev. (ft) 439

		Jan.	Feb.	Mar.	Apr.	May	June	July	Aug.	Sept.	Oct.	Nov.	Dec.	Yr.	
S	2.5″	37.4	37.4	41.0	45.7	50.2	54.9	58.5	57.7	54.5	49.5	43.7	40.6	47.7	T°F
D	1.9″														
K	Cfb	1.9	1.4	1.3	1.3	1.9	1.8	3.5	3.6	1.9	2.0	2.4	2.9	25.9	P″
Th	C2B′1rb′4	0.5	0.5	1.1	1.9	3.0	4.0	4.6	4.0	2.9	1.9	1.1	0.7	25.5	PE″

Kharkov, Soviet Union: lat 49°56′ N; long 36°17′ E; elev. (ft) 499

		Jan.	Feb.	Mar.	Apr.	May	June	July	Aug.	Sept.	Oct.	Nov.	Dec.	Yr.	
S	0	19.2	20.3	29.1	46.0	59.0	66.0	70.0	67.8	57.4	45.1	33.3	24.8	44.8	T°F
D	7.1″														
K	Dfb	1.4	1.3	1.3	1.3	2.0	2.4	2.9	1.9	1.3	1.6	1.5	1.5	20.4	P″
Th	C1B′1db′1	0	0	0	1.6	3.9	5.1	5.9	5.0	2.9	1.2	0.1	0	25.4	PE″

Leningrad, Soviet Union: lat 59°58′ N; long 30°18′ E; elev. (ft) 13

		Jan.	Feb.	Mar.	Apr.	May	June	July	Aug.	Sept.	Oct.	Nov.	Dec.	Yr.	
S	1.9″	18.3	17.8	24.3	37.9	49.8	59.7	65.1	62.2	52.2	41.2	31.6	24.1	40.3	T°F
D	6.5″														
K	Dfb	1.4	1.3	1.0	1.3	1.6	2.1	2.7	3.0	2.3	2.0	1.8	1.4	21.9	P″
Th	C1B′1db′1	0	0	0	1.0	3.3	5.2	6.2	5.0	2.7	1.1	0	0	24.2	PE″

Lisbon, Portugal: lat 38°43′ N; long 09°09′ W; elev. (ft) 311

		Jan.	Feb.	Mar.	Apr.	May	June	July	Aug.	Sept.	Oct.	Nov.	Dec.	Yr.	
S	5.5″	51.4	52.9	56.5	60.1	63.0	68.2	72.0	72.5	70.2	64.8	57.9	52.7	61.9	T°F
D	13.6″														
K	Csa	4.4	3.0	4.3	2.1	1.7	0.6	0.1	0.2	1.3	2.4	3.7	4.1	27.9	P″
Th	C1B′2sa′	1.0	1.1	1.7	2.4	3.1	4.1	5.0	4.8	3.8	2.7	1.6	1.1	31.8	PE″

London, United Kingdom: lat 51°28′ N; long 0°19′ W; elev. (ft) 16

		Jan.	Feb.	Mar.	Apr.	May	June	July	Aug.	Sept.	Oct.	Nov.	Dec.	Yr.	
S	.2	39.6	39.9	43.9	48.7	54.3	60.4	63.7	63.0	58.6	51.4	45.0	41.4	50.9	T°F
D	4.7														
K	Cfb	2.1	1.6	1.5	1.5	1.8	1.8	2.2	2.3	2.0	2.2	2.5	1.9	23.4	P″
Th	C1B′1db′4	0.5	0.6	1.2	1.9	3.1	4.2	4.8	4.3	3.0	1.9	1.0	0.6	26.4	PE″

Madrid, Spain: lat 40°25′ N; long 03°41′ W; elev. (ft) 2,165

		Jan.	Feb.	Mar.	Apr.	May	June	July	Aug.	Sept.	Oct.	Nov.	Dec.	Yr.	
S	0	40.8	43.7	50.0	55.4	60.3	69.1	75.6	74.5	67.6	57.2	48.0	42.1	57.0	T°F
D	12.7														
K	Csa	1.5	1.3	1.8	1.7	1.7	1.1	0.4	0.6	1.2	2.1	1.9	1.9	17.2	P″
Th	DB′2db′3	0.4	0.6	1.3	2.1	3.1	4.6	6.0	5.4	3.7	2.0	0.9	0.5	29.9	PE″

Milan, Italy: lat 45°28′ N; long 9°17′ E; elev. (ft) 394

		Jan.	Feb.	Mar.	Apr.	May	June	July	Aug.	Sept.	Oct.	Nov.	Dec.	Yr.	
S	11.2″	34.3	37.2	46.2	54.0	62.6	69.3	73.4	71.6	65.1	54.3	43.9	37.4	54.1	T°F
D	2.8″														
K	Cfa	1.9	2.2	3.0	2.8	3.1	4.3	2.3	3.3	2.5	5.0	4.3	3.5	38.2	P″
Th	B1B′2rb′3	0.1	0.2	1.1	2.1	3.8	5.0	5.9	5.1	3.5	1.8	0.7	0.3	29.0	PE″

Munich, Germany: lat 48°05′ N; long 11°42′ E; elev. (ft) 1,736

		Jan.	Feb.	Mar.	Apr.	May	June	July	Aug.	Sept.	Oct.	Nov.	Dec.	Yr.	
S	13″	28.0	30.2	37.9	46.2	54.5	60.6	63.9	62.4	56.7	46.8	37.6	30.7	46.2	T°F
D	.1														
K	Cfb	2.3	2.2	2.0	2.4	4.2	4.9	5.5	4.1	3.4	2.6	2.2	2.0	37.8	P″
Th	B2B′1r	0	0	0.7	1.8	3.3	4.4	5.0	4.3	3.0	1.6	0.5	0	24.3	PE″

Nice, France: lat 43°39′ N; long 07°12′ E; elev. (ft) 33

		Jan.	Feb.	Mar.	Apr.	May	June	July	Aug.	Sept.	Oct.	Nov.	Dec.	Yr.	
S	9.2″	45.5	47.3	51.4	55.9	62.1	68.2	72.9	72.5	68.5	60.8	52.7	46.7	58.6	T°F
D	6.6″														
K	Csa	2.7	2.4	2.9	2.9	2.7	1.4	0.8	1.1	3.0	4.9	5.1	4.2	34.1	P″
Th	C2B′2s2a′	0.6	0.8	1.4	2.0	3.3	4.5	5.6	5.1	3.7	2.3	1.2	0.7	30.7	PE″

Paris, France: lat 48°58′ N; long 02°27′ E; elev. (ft) 174

		Jan.	Feb.	Mar.	Apr.	May	June	July	Aug.	Sept.	Oct.	Nov.	Dec.	Yr.	
S	.2	37.6	38.8	45.0	50.5	57.2	62.8	66.2	65.3	60.6	52.0	44.2	39.4	51.6	T°F
D	4.8″														
K	Cfb	2.1	1.7	1.3	1.5	2.0	2.0	2.2	2.4	2.0	1.9	2.0	1.9	23.0	P″
Th	C1B′1db′4	0.4	0.5	1.2	2.0	3.3	4.4	5.0	4.4	3.2	1.8	0.9	0.5	27.0	PE″

Prague, Czechoslovakia: lat 50°06′ N; long 14°17′ E; elev. (ft) 1,227

		Jan.	Feb.	Mar.	Apr.	May	June	July	Aug.	Sept.	Oct.	Nov.	Dec.	Yr.	
S	0	27.5	29.3	36.9	46.0	55.2	61.2	64.2	63.3	57.0	46.8	37.6	30.6	46.2	T°F
D	4.8″														
K	Cfb	0.9	0.9	0.9	1.3	2.4	2.6	3.2	2.6	1.4	1.6	1.0	1.0	19.8	P″
Th	C1B′1db′3	0	0	0.6	1.8	3.5	4.5	5.1	4.5	3.0	1.6	0.5	0	24.6	PE″

Rome, Italy: lat 41°48′ N; long 12°36′ E; elev. (ft) 429

		Jan.	Feb.	Mar.	Apr.	May	June	July	Aug.	Sept.	Oct.	Nov.	Dec.	Yr.	
S	5.1″	45.5	47.1	50.7	55.9	63.3	70.9	75.2	75.2	70.0	61.2	53.4	48.6	59.7	T°F
D	12.2″														
K	Csa	2.6	2.6	2.2	2.6	2.0	1.1	1.0	0.7	2.6	3.1	3.8	3.8	28.1	P″
Th	C1B′2db′4	0.6	0.7	1.2	1.9	3.4	4.9	5.9	5.5	3.9	2.3	1.2	0.8	31.7	PE″

Santander, Spain: lat 43°28′ N; long 03°49′ W; elev. (ft) 216

		Jan.	Feb.	Mar.	Apr.	May	June	July	Aug.	Sept.	Oct.	Nov.	Dec.	Yr.	
S	20.0″	48.6	48.4	53.1	53.6	57.9	62.2	65.5	66.4	64.8	59.7	53.8	51.1	57.0	T°F
D	.7″														
K	Cfb	4.8	3.6	3.3	2.9	2.8	2.4	2.5	3.5	4.7	6.2	4.6	6.1	47.4	P″
Th	B3B′2ra′	1.0	1.0	1.7	1.9	2.8	3.6	4.2	4.1	3.3	2.4	1.5	1.2	28.1	PE″

Stockholm, Sweden: lat 54°21′ N; long 18°04′ E; elev. (ft) 170

		Jan.	Feb.	Mar.	Apr.	May	June	July	Aug.	Sept.	Oct.	Nov.	Dec.	Yr.	
S	2.6″	26.8	26.4	30.7	39.9	50.2	58.8	64.0	61.9	54.0	44.8	37.0	32.2	43.9	T°F
D	5.7″														
K	Dfb	1.7	1.2	1.0	1.2	1.3	1.8	2.4	3.0	2.4	1.9	2.1	1.9	21.9	P″
Th	C1B′1db′1	0	0	1.0	1.2	3.1	4.6	5.5	4.6	2.9	1.5	0.5	0.1	23.5	PE″

Tromso, Norway: lat 69°42′ N; long 19°01′ E; elev. (ft) 78

		Jan.	Feb.	Mar.	Apr.	May	June	July	Aug.	Sept.	Oct.	Nov.	Dec.	Yr.	
S	30.0″	27.1	26.1	28.4	33.8	40.1	47.7	53.6	52.0	45.9	38.7	32.9	29.7	37.9	T°F
D	2.5″														
K	Dfc	4.7	3.7	4.4	2.9	2.6	2.2	2.2	3.3	4.5	5.2	3.8	4.5	44.0	P″
Th	B4C′2sc′2	0	0	0	0.8	2.7	4.7	5.7	4.5	2.7	1.3	0.2	0	22.5	PE″

Valentia, Eire: lat 51°56′ N; long 10°15′ W; elev. (ft) 46

		Jan.	Feb.	Mar.	Apr.	May	June	July	Aug.	Sept.	Oct.	Nov.	Dec.	Yr.	
S	28.5″	44.4	44.2	46.9	48.9	52.5	56.8	59.0	59.7	57.2	52.9	48.4	46.0	51.5	T°F
D	.1″														
K	Cfb	6.5	4.2	4.1	2.9	3.4	3.2	4.2	3.7	4.8	5.5	5.9	6.6	55.0	P″
Th	AB′1r	0.9	0.9	1.5	2.0	2.8	3.6	4.1	3.8	2.8	2.0	1.2	1.0	26.0	PE″

Warsaw, Poland: lat 52°09′ N; long 20°59′ E; elev. (ft) 350

		Jan.	Feb.	Mar.	Apr.	May	June	July	Aug.	Sept.	Oct.	Nov.	Dec.	Yr.	
S	0	27.7	26.1	33.1	45.1	55.2	63.1	65.7	64.0	55.6	46.8	37.4	32.7	46.1	T°F
D	6.2″														
K	Dfb	1.0	1.1	0.8	1.3	1.6	2.4	3.1	1.8	1.6	1.2	1.2	1.5	18.6	P″
Th	C1B′1dc′1	0	0.0	0.2	1.7	3.5	4.9	5.4	4.6	2.8	1.5	0.5	0.1	24.8	PE″

ASIA

Adana, Turkey: lat 36°39′ N; long 35°18′ E; elev. (ft) 66

		Jan.	Feb.	Mar.	Apr.	May	June	July	Aug.	Sept.	Oct.	Nov.	Dec.	Yr.	
S	7.0	48.4	50.4	55.2	62.4	70.2	77.0	81.7	82.4	77.4	69.4	59.9	51.6	65.5	T°F
D	22.4″														
K	Csa	4.4	3.7	2.6	1.8	1.8	0.7	0.2	0.2	0.7	1.6	2.4	4.0	24.1	P″
Th	DB′3sb′4	0.6	0.7	1.3	2.3	4.1	5.8	7.1	6.9	5.0	3.2	1.6	0.8	38.7	PE″

Alma-Ata, Soviet Union: lat 43°14′ N; long 76°56′ E; elev. (ft) 2,779

		Jan.	Feb.	Mar.	Apr.	May	June	July	Aug.	Sept.	Oct.	Nov.	Dec.	Yr.	
S	2.9″	17.8	22.8	34.9	51.4	60.8	68.7	73.9	72.1	63.3	50.0	31.8	22.3	47.7	T°F
D	9.8″														
K	Dfa	1.0	1.3	2.5	3.5	3.9	2.3	1.4	0.9	1.0	1.8	1.9	1.4	22.9	P″
Th	C1B′1db′2	0	0	0.2	2.0	3.6	5.0	6.0	5.3	3.4	1.6	0	0	26.8	PE″

Ankara, Turkey: lat 39°57′ N; long 32°53′ E; elev. (ft) 2,954

		Jan.	Feb.	Mar.	Apr.	May	June	July	Aug.	Sept.	Oct.	Nov.	Dec.	Yr.	
S	0	31.5	34.0	40.8	51.8	60.8	68.0	73.9	73.9	65.1	55.2	45.1	35.8	53.1	T° F
D	13.7″														
K	Csa	1.5	1.4	1.4	1.5	1.9	1.2	0.5	0.3	0.7	0.9	1.2	1.7	14.2	P″
Th	DB′1db′2	0	0.1	0.6	1.8	3.4	4.7	5.8	5.4	3.5	2.0	0.8	0.2	27.9	PE″

Baghdad, Iraq: lat 33°20′ N; long 44°24′ E; elev. (ft) 111

		Jan.	Feb.	Mar.	Apr.	May	June	July	Aug.	Sept.	Oct.	Nov.	Dec.	Yr.	
S	0	49.8	54.0	60.4	72.0	83.1	91.2	94.6	94.1	87.3	76.5	73.0	52.2	73.2	T° F
D	22.3″														
K	BWhs	1.0	1.1	1.1	0.7	0.3	0	0	0	0	0.1	0.8	1.0	6.1	P″
Th	EA′db′4	0.4	0.6	1.3	3.5	7.1	8.4	8.9	8.4	6.9	4.2	3.0	0.5	52.6	PE″

Balkash, Soviet Union: lat 33°20′ N; long 44°24′ E; elev. (ft) 111

		Jan.	Feb.	Mar.	Apr.	May	June	July	Aug.	Sept.	Oct.	Nov.	Dec.	Yr.	
S	0	49.8	54.0	60.4	72.0	83.1	91.2	94.6	94.1	87.3	76.5	73.0	52.2	73.2	T° F
D	46.5″														
K	BWhs	1.0	1.1	1.1	0.7	0.3	0	0	0	0	0.1	0.8	1.0	6.1	P″
Th	EA′db′4	0.4	0.6	1.3	3.5	7.1	8.4	8.9	8.4	6.9	4.2	3.0	0.5	52.6	PE″

Bangkok, Thailand: lat 13°44′ N; long 100°39′ E; elev. (ft) 39

		Jan.	Feb.	Mar.	Apr.	May	June	July	Aug.	Sept.	Oct.	Nov.	Dec.	Yr.	
S	.3	79.0	81.7	84.6	86.5	85.6	84.0	83.1	82.8	82.2	81.7	80.1	77.9	82.4	T° F
D	19.6″														
K	Awgi	0.3	1.1	1.3	3.5	6.5	6.7	7.0	7.5	12.1	10.1	2.2	0.3	58.6	P″
Th	C1A′d	4.8	5.2	6.4	6.7	7.0	6.6	6.6	6.4	5.9	5.8	5.2	4.4	70.5	PE″

Beirut, Lebanon: lat 33°39′ N; long 35°29′ E; elev. (ft) 79

		Jan.	Feb.	Mar.	Apr.	May	June	July	Aug.	Sept.	Oct.	Nov.	Dec.	Yr.	
S	1.3	57.0	57.4	59.5	64.6	69.8	75.4	79.2	80.8	78.3	73.4	65.8	59.9	68.4	T° F
D	24.6″														
K	Csa	4.4	3.1	3.0	1.0	0.4	0	0	0	0.3	0.8	3.1	4.1	20.2	P″
Th	DB′4da′	1.2	1.2	1.6	2.5	3.8	5.2	6.3	6.4	5.1	3.8	2.2	1.4	40.1	PE″

Calcutta, India: lat 22°32′ N; long 88°20′ E; elev. (ft) 20

		Jan.	Feb.	Mar.	Apr.	May	June	July	Aug.	Sept.	Oct.	Nov.	Dec.	Yr.	
S	10.3″	68.4	73.4	82.2	86.2	88.0	86.7	84.4	84.4	84.6	82.2	75.2	69.1	80.2	T°F
D	17.5″														
K	Awg	0.5	0.9	1.1	1.7	4.8	10.2	11.9	12.0	11.4	6.3	1.4	0.1	62.6	P″
Th	C1A′d	2.1	2.9	6.0	6.8	7.6	7.4	7.2	6.9	6.4	5.8	3.5	2.2	64.3	PE″

Damascus, Syria: lat 33°29′ N; long 36°14′ E; elev. (ft) 2,392

		Jan.	Feb.	Mar.	Apr.	May	June	July	Aug.	Sept.	Oct.	Nov.	Dec.	Yr.	
S	0	45.9	48.4	53.4	61.9	70.5	77.2	81.0	81.7	76.0	69.1	56.3	47.7	64.0	T°F
D	28.0″														
K	BShs	2.1	1.5	1.1	0.6	0.2	0	0	0	0	0.2	1.0	2.4	9.1	P″
Th	EB′2db′4	0.5	0.6	1.2	2.3	4.2	5.7	6.8	6.6	4.7	3.2	1.3	0.6	37.1	PE″

Darjeeling, India: lat 27°03′ N; long 88°16′ E; elev. (ft) 6,981

		Jan.	Feb.	Mar.	Apr.	May	June	July	Aug.	Sept.	Oct.	Nov.	Dec.	Yr.	
S	79.0″	43.5	45.9	52.2	57.7	60.3	62.6	63.5	63.5	63.0	59.4	52.3	46.6	55.9	T°F
D	.18″														
K	Cwb	0.9	1.1	2.1	4.3	7.4	20.5	27.7	22.5	16.1	4.6	0.5	0.2	107.9	P″
Th	Ab′1r	0.7	0.8	1.7	2.4	3.0	3.4	3.6	3.4	3.1	2.5	1.5	0.9	26.9	PE″

Dhahran, Saudi Arabia: lat 26°17′ N; long 50°09′ E; elev. (ft) 73

		Jan.	Feb.	Mar.	Apr.	May	June	July	Aug.	Sept.	Oct.	Nov.	Dec.	Yr.	
S	0	62.1	64.6	72.5	80.8	89.2	95.4	97.3	98.1	92.8	84.9	74.7	65.1	81.5	T°F
D	59.0″														
K	BWhs	1.1	0.6	0.4	0.1	0.1	0	0	0	0	0	0.1	0.9	3.3	P″
Th	EA′d	1.0	1.3	3.1	5.8	7.8	8.3	8.6	8.3	7.3	6.2	3.2	1.4	62.3	PE″

Fort Cochin, India: lat 9°58′ N; long 76°14′ E; elev. (ft) 10

		Jan.	Feb.	Mar.	Apr.	May	June	July	Aug.	Sept.	Oct.	Nov.	Dec.	Yr.	
S	64.5″	80.4	81.5	83.3	83.7	82.9	79.7	78.6	79.1	79.3	80.1	80.6	80.4	80.8	T°F
D	10.4″														
K	Awi	0.4	1.3	2.0	5.7	14.3	29.8	22.5	15.2	9.3	13.1	7.2	1.5	122.3	P″
Th	B4A′w	5.4	5.2	6.2	6.2	6.4	5.4	5.1	5.2	5.0	5.5	5.3	5.4	66.2	PE″

Hong Kong: lat 22°18′ N; long 114°10′ E; elev. (ft) 108

		Jan.	Feb.	Mar.	Apr.	May	June	July	Aug.	Sept.	Oct.	Nov.	Dec.	Yr.	
S	34.5″	59.7	60.4	64.8	71.2	78.1	81.5	83.1	82.2	81.1	76.5	70.2	63.3	72.7	T°F
D	1.3″														
K	Cwa	1.2	2.4	2.8	5.2	13.1	18.8	11.3	16.3	14.9	1.3	1.8	0.7	89.6	P″
Th	B4A′ra′	1.3	1.2	2.1	3.3	5.4	6.3	6.8	6.4	5.7	4.4	2.8	1.7	47.3	PE″

Hyderabad, India: lat 17°27′ N; long 78°28′ E; elev. (ft) 1,787

		Jan.	Feb.	Mar.	Apr.	May	June	July	Aug.	Sept.	Oct.	Nov.	Dec.	Yr.	
S	0	70.7	75.2	81.0	86.5	90.1	84.0	78.8	78.1	77.9	77.0	72.1	69.3	78.4	T°F
D	27.7														
K	BShwg	0.1	0.4	0.5	0.9	1.2	4.2	6.5	5.8	6.4	2.8	1.0	0.2	30.0	P″
Th	DA′d	2.7	3.5	5.7	6.8	7.7	6.7	5.5	5.1	4.7	4.4	3.0	2.4	57.7	PE″

Irkutsk, Soviet Union: lat 52°16′ N; long 104°21′ E; elev. (ft) 1,590

		Jan.	Feb.	Mar.	Apr.	May	June	July	Aug.	Sept.	Oct.	Nov.	Dec.	Yr.	
S	0	−5.4	0.0	15.3	34.9	47.8	59.7	64.2	59.2	46.8	34.0	12.6	−1.3	30.6	T°F
D	2.7														
K	·Dwc	0.5	0.3	0.3	0.6	1.1	3.3	4.0	3.9	1.9	0.8	0.7	0.6	18.0	P″
Th	C1C′2db′3	0	0	0	0.6	2.9	4.9	5.6	4.4	2.2	0.4	0	0	20.7	PE″

Istanbul, Turkey: lat 40°58′ N; long 25°05′ E; elev. (ft) 131

		Jan.	Feb.	Mar.	Apr.	May	June	July	Aug.	Sept.	Oct.	Nov.	Dec.	Yr.	
S	5.8″	41.9	42.1	44.6	53.3	61.2	69.1	73.9	74.1	67.5	60.1	52.7	46.0	57.0	T°F
D	12.6″														
K	Csa	3.5	3.1	2.4	1.5	1.3	1.1	1.1	0.9	1.9	2.4	3.4	3.8	26.4	P″
Th	C1B′2sb′3	0.5	0.5	0.8	1.7	3.2	4.7	5.7	5.4	3.6	2.4	1.3	0.7	29.8	PE″

Kagoshima, Japan: lat 31°34′ N; long 130°33′ E; elev. (ft) 16

		Jan.	Feb.	Mar.	Apr.	May	June	July	Aug.	Sept.	Oct.	Nov.	Dec.	Yr.	
S	57.7″	43.9	45.9	51.4	59.2	66.2	72.7	80.2	80.8	75.9	66.0	57.2	48.2	62.2	T°F
D	0														
K	Cfa	2.9	4.6	5.9	9.0	9.8	17.9	13.5	8.7	8.4	4.7	3.5	3.1	92.0	P″
Th	AB′3rb′4	0.4	0.5	1.1	2.1	3.5	4.7	6.5	6.4	4.8	2.8	1.5	0.7	34.4	PE″

Karachi, Pakistan: lat 24°55′ N; long 67°09′ E; elev. (ft) 73

		Jan.	Feb.	Mar.	Apr.	May	June	July	Aug.	Sept.	Oct.	Nov.	Dec.	Yr.	
S	0	63.9	68.4	76.1	82.8	86.9	88.5	86.2	83.8	83.5	81.7	75.0	67.3	78.6	T°F
D	51.7″														
K	BWhw	0.3	0.5	0.2	0.1	0	0.3	4.0	1.8	1.1	0.1	0.1	0.2	8.7	P″
Th	EA′d	1.5	2.0	4.2	6.2	7.5	7.7	7.6	6.8	6.2	5.7	3.5	2.1	60.4	PE″

Konya, Turkey: lat 37°52′ N; long 32°30′ E; elev. (ft) 3,366

		Jan.	Feb.	Mar.	Apr.	May	June	July	Aug.	Sept.	Oct.	Nov.	Dec.	Yr.	
S	0	28.4	34.9	41.0	51.8	60.6	67.6	73.6	73.2	64.8	54.5	43.7	34.9	52.7	T°F
D	15.0″														
K	Csa	1.6	1.3	1.2	1.2	1.5	1.0	0.2	0.2	0.4	1.1	1.2	1.5	12.4	P″
Th	DB′1db′2	0	0.2	0.6	1.9	3.4	4.6	5.7	5.3	3.5	2.0	0.7	0.2	27.4	PE″

Kweiyang, China: lat 26°35′ N; long 106°43′ E; elev. (ft) 3,468

		Jan.	Feb.	Mar.	Apr.	May	June	July	Aug.	Sept.	Oct.	Nov.	Dec.	Yr.	
S	16.4″	40.1	43.9	53.2	61.7	68.5	72.5	76.3	75.2	69.8	60.4	53.1	45.3	60.1	T°F
D	0														
K	Cwa	0.8	1.1	1.6	3.5	7.6	8.3	7.8	5.2	5.0	4.1	1.9	1.0	47.9	P″
Th	B2B′2rb′4	0.3	0.5	1.4	2.6	4.0	4.7	5.6	5.2	3.8	2.3	1.3	0.6	31.6	PE″

Lanchow, China: lat 36°03′ N; long 103°51′ E; elev. (ft) 4,941

		Jan.	Feb.	Mar.	Apr.	May	June	July	Aug.	Sept.	Oct.	Nov.	Dec.	Yr.	
S	0	20.7	29.5	41.9	53.6	63.1	69.6	73.0	70.6	61.4	50.5	34.9	23.0	49.3	T°F
D	14.3″														
K	BSkw	0.1	0.1	0.3	0.5	0.7	1.5	2.6	3.6	2.2	0.6	0.1	0	12.1	P″
Th	DB′1db′2	0	0	0.8	2.2	3.8	4.9	5.6	4.9	3.1	1.6	0.2	0	26.6	PE″

Lhasa, Tibet, China: lat 24°40′ N; long 91°07′ E; elev. (ft) 12,088

		Jan.	Feb.	Mar.	Apr.	May	June	July	Aug.	Sept.	Oct.	Nov.	Dec.	Yr.	
S	0	30.6	36.7	41.5	47.3	55.6	62.6	63.0	63.1	59.5	52.7	42.3	32.5	48.9	T°F
D	1.92″														
K	Cwb	0	0	0	0	0.7	2.8	6.2	6.0	2.7	0.2	0	0	18.6	P″
Th	C1B′1db′4	0	0.4	0.9	1.6	2.9	3.8	4.0	3.8	3.1	2.2	0.9	0	23.5	PE″

New Delhi, India: lat 28°35′ N; long 77°12′ E; elev. (ft) 710

		Jan.	Feb.	Mar.	Apr.	May	June	July	Aug.	Sept.	Oct.	Nov.	Dec.	Yr.	
S	0	57.7	63.1	73.2	84.4	92.3	94.1	88.2	85.8	84.7	78.6	68.4	60.3	77.6	T°F
D	29.65″														
K	BShw	1.0	0.9	0.7	0.3	0.3	2.6	8.3	6.8	5.9	1.2	0	0.2	28.2	P″
Th	DA′d	0.8	1.3	3.4	6.6	8.3	8.4	8.0	7.3	6.4	4.7	2.1	1.0	58.1	PE″

Omsk, Soviet Union: lat 54°56′ N; long 73°24′ E; elev. (ft) 308

		Jan.	Feb.	Mar.	Apr.	May	June	July	Aug.	Sept.	Oct.	Nov.	Dec.	Yr.	
S	0	−2.9	.1	11.5	36.1	52.3	63.0	65.8	61.2	50.7	36.0	15.3	2.3	32.7	T°F
D	10.0″														
K	Dwb	0.3	0.2	0.3	0.7	1.2	2.1	2.8	1.8	1.3	0.9	0.6	0.5	12.7	P″
Th	DB′1dc′2	0	0	0	0.7	3.4	5.3	5.8	4.5	2.5	0.5	0	0	22.7	PE″

Oymyakon, Soviet Union: lat 63°16′ N; long 143°09′ E; elev. (ft) 2,382

		Jan.	Feb.	Mar.	Apr.	May	June	July	Aug.	Sept.	Oct.	Nov.	Dec.	Yr.	
S	0	−53.0	−45.2	−29.6	4.3	34.5	52.9	58.6	51.6	34.9	2.8	−31.0	−47.2	2.7	T°F
D	10.23″														
K	Dwd	0.3	0.2	0.2	0.1	0.4	1.3	1.6	1.5	0.8	0.5	0.4	0.3	7.6	P″
Th	DC′2dc′2	0	0	0	0	1.2	5.3	6.2	4.3	0.9	0	0	0	17.9	PE″

Petropavlovsk, Soviet Union: lat 52°59′ N; long 158°39′ E; elev. (ft) 79

		Jan.	Feb.	Mar.	Apr.	May	June	July	Aug.	Sept.	Oct.	Nov.	Dec.	Yr.	
S	.47″	17.8	18.0	22.3	31.3	38.5	46.4	53.8	56.5	50.7	40.8	28.0	22.1	35.6	T°F
D	1.18″														
K	Dfc	2.4	1.2	1.8	1.7	1.6	1.8	2.9	2.9	5.2	3.3	2.7	4.7	32.2	P″
Th	B3C′2db′1	0	0	0	0	1.8	3.3	4.5	4.4	3.0	1.5	0	0	18.4	PE″

Rangoon, Burma: lat 16°46′ N; long 96°10′ E; elev. (ft) 76

		Jan.	Feb.	Mar.	Apr.	May	June	July	Aug.	Sept.	Oct.	Nov.	Dec.	Yr.	
S	49.5″	75.7	77.4	81.0	85.6	85.1	82.0	81.7	80.8	81.7	82.9	82.0	77.0	81.1	T°F
D	18.1″														
K	Amgi	0.3	0.2	0.2	0.7	10.2	20.6	19.4	22.6	15.7	8.2	1.3	0.1	99.5	P″
Th	B2A′w2	3.7	3.9	5.7	6.6	7.0	6.3	6.4	6.0	5.8	6.0	5.4	4.0	66.7	PE″

Saigon, Vietnam: lat 10°49′ N; long 106°40′ E; elev. (ft) 33

		Jan.	Feb.	Mar.	Apr.	May	June	July	Aug.	Sept.	Oct.	Nov.	Dec.	Yr.	
S	18.8″	78.4	79.3	82.0	83.8	82.8	81.3	80.8	80.8	80.1	79.7	79.0	78.3	80.6	T°F
D	13.3″														
K	Awgi	0.2	0.5	0.5	2.6	7.7	11.2	9.5	10.9	11.5	10.2	4.8	1.5	71.1	P″
Th	C2A′w2	4.6	4.5	5.9	6.2	6.4	5.9	6.0	5.9	5.4	5.2	4.7	4.6	65.4	PE″

Shanghai, China: lat 31°12′ N; long 121°26′ E; élev. (ft) 23

		Jan.	Feb.	Mar.	Apr.	May	June	July	Aug.	Sept.	Oct.	Nov.	Dec.	Yr.	
S	10.6″	38.1	40.1	47.3	57.3	66.4	73.9	81.3	81.5	73.5	64.0	53.6	43.2	60.1	T°F
D	.2″														
K	Cfa	2.0	2.6	3.3	3.3	3.8	7.8	6.1	5.2	5.4	2.9	2.4	1.6	46.4	P″
Th	B1B′2rb′3	0.1	0.2	0.8	1.9	3.6	5.0	6.7	6.4	4.3	2.6	1.2	0.4	33.2	PE″

Teheran, Iran: lat 35°41′ N; long 68°11′ E; elev. (ft) 3,907

		Jan.	Feb.	Mar.	Apr.	May	June	July	Aug.	Sept.	Oct.	Nov.	Dec.	Yr.	
S	0	38.3	41.4	50.4	59.7	70.2	79.0	85.1	83.1	76.3	64.9	51.1	40.8	61.7	T°F
D	28.9″														
K	BShs	1.5	0.9	1.4	1.2	0.5	0.1	0	0	0	0.2	1.1	1.1	8.0	P″
Th	EB′3db′2	0.1	0.2	0.9	2.1	4.2	6.3	7.8	6.9	4.8	2.6	0.8	0.2	36.9	PE″

Tobolsk, Soviet Union: lat 58°09′ N; long 68°11′ E; elev. (ft) 144

		Jan.	Feb.	Mar.	Apr.	May	June	July	Aug.	Sept.	Oct.	Nov.	Dec.	Yr.	
S	0	−1.3	0.9	12.6	34.3	49.3	60.8	64.6	60.3	48.9	34.9	15.1	1.8	31.9	T°F
D	28.9″														
K	BShs	0.7	0.6	0.8	1.0	1.6	2.4	2.9	2.5	2.1	1.5	1.4	1.0	18.5	P″
Th	C1B′3db′2	0	0	0	0.5	3.2	5.3	6.0	4.6	2.4	0.4	0	0	22.4	PE″

Tokyo, Japan: lat 35°41′ N; long 139°44′ E; elev. (ft) 20

		Jan.	Feb.	Mar.	Apr.	May	June	July	Aug.	Sept.	Oct.	Nov.	Dec.	Yr.	
S	27.56″	38.7	39.7	45.7	55.6	63.7	70.0	77.2	79.5	73.0	62.1	52.3	43.0	58.5	T°F
D	0														
K	Cfa	1.9	2.9	4.0	5.3	5.2	7.2	5.8	5.8	8.5	8.7	4.0	2.4	61.7	P″
Th	B4B′2rb′3	0.2	0.2	0.7	1.8	3.3	4.5	6.1	6.2	4.4	2.5	1.1	0.4	31.5	PE″

Ya-an, China: lat 30°0′ N; long 103°03′ E; elev. (ft) 2,132

		Jan.	Feb.	Mar.	Apr.	May	June	July	Aug.	Sept.	Oct.	Nov.	Dec.	Yr.	
S	38.1″	45.5	47.3	55.9	64.0	71.2	74.7	78.8	77.7	71.6	62.6	55.2	46.2	62.6	T°F
D	0														
K	Cwa	0.7	1.0	1.7	3.9	4.3	5.9	20.9	17.6	9.2	5.4	2.3	0.8	73.7	P″
Th	Ab′3rb′4	0.5	0.6	1.5	2.7	4.3	5.1	6.1	5.6	3.9	2.3	1.3	0.5	34.4	PE″

Vladivostok, Soviet Union: lat 43°07′ N; long 131°54′ E; elev. (ft) 452

		Jan.	Feb.	Mar.	Apr.	May	June	July	Aug.	Sept.	Oct.	Nov.	Dec.	Yr.	
S	10.0″	7.3	13.5	26.2	39.9	49.1	54.9	62.8	68.0	61.0	48.2	30.2	15.6	39.8	T°F
D	0														
K	Dfb	0.7	0.7	1.1	1.7	3.1	4.1	4.6	5.9	5.7	2.8	1.5	0.6	32.5	P″
Th	B2C′2rc′1	0	0	0	1.0	2.5	3.4	4.6	5.0	3.5	1.8	0	0	21.9	PE″

AFRICA

Accra, Ghana: lat 5°36′ N; long 0°10′ W; elev. (ft) 212

		Jan.	Feb.	Mar.	Apr.	May	June	July	Aug.	Sept.	Oct.	Nov.	Dec.	Yr.	
S	0	81.1	81.9	82.0	81.9	80.6	78.3	76.3	75.7	77.5	79.0	80.8	81.5	79.7	T°F
D	31.4″														
K	BShw	0.6	1.5	2.9	3.2	5.7	7.6	1.9	0.6	1.6	3.1	1.5	0.7	30.9	P″
Th	DA′d	5.6	5.3	5.9	5.8	5.8	4.8	4.4	4.1	4.5	5.0	5.4	5.6	62.4	PE″

Addis Ababa, Ethiopia: lat 9°02′ N; long 38°45′ E; elev. (ft) 7,899

		Jan.	Feb.	Mar.	Apr.	May	June	July	Aug.	Sept.	Oct.	Nov.	Dec.	Yr.	
S	20.7″	63.0	64.0	66.2	65.5	66.7	63.3	59.5	58.8	61.7	61.2	62.2	62.6	62.9	T°F
D	3.1″														
K	Cwbi	0.9	0.9	2.6	3.7	2.1	4.1	9.4	10.5	6.8	1.7	0.1	0.7	43.5	P″
Th	B2B′2w	2.5	2.4	3.0	2.9	3.2	2.6	2.2	2.0	2.3	2.3	2.3	2.4	30.1	PE″

Algiers, Algeria: lat 36°43′ N; long 3°15′ E; elev. (ft) 97

		Jan.	Feb.	Mar.	Apr.	May	June	July	Aug.	Sept.	Oct.	Nov.	Dec.	Yr.	
S	6.7″	50.5	51.4	55.4	59.4	64.4	71.2	75.9	77.2	73.6	66.0	58.8	53.1	63.1	T°F
D	16.1″														
K	Csa	4.6	3.0	2.2	2.6	1.4	0.5	0.1	0.2	1.1	3.3	3.6	4.6	27.2	P″
Th	C1B′3sa′	0.8	0.8	1.5	2.1	3.2	4.5	5.7	5.6	4.3	2.8	1.6	1.0	33.8	PE″

Cairo, U.A.R.: lat 30°08′ N; long 31°34′ E; elev. (ft) 311

		Jan.	Feb.	Mar.	Apr.	May	June	July	Aug.	Sept.	Oct.	Nov.	Dec.	Yr.	
S	0	56.8	59.4	63.3	70.0	76.5	80.4	82.4	82.4	78.8	74.7	66.7	59.5	70.9	T°F
D	44.0″														
K	BWhs	0.1	0.2	0.1	0	0	0	0	0	0	0	0.1	0.4	0.9	P″
Th	EA′da′	1.0	1.2	2.0	3.2	5.2	6.3	6.9	6.6	5.1	3.9	2.2	1.2	44.9	PE″

Capetown, Union of South Africa: lat 33°58′ S; long 18°36′ E; elev. (ft) 173

		Jan.	Feb.	Mar.	Apr.	May	June	July	Aug.	Sept.	Oct.	Nov.	Dec.	Yr.	
S	0	68.5	68.0	65.8	61.0	57.2	54.7	52.9	54.1	56.7	59.0	63.7	66.7	60.7	T°F
D	7″														
K	Csb	0.4	0.6	0.5	2.1	3.5	3.3	3.3	2.9	1.8	1.2	0.7	0.4	20.7	P″
Th	C1B′1d	3.0	2.8	3.1	2.5	2.2	1.9	1.7	1.7	1.9	2.0	2.4	2.7	27.8	PE″

Casablanca, Morocco: lat 33°34′ N; long 7°40′ W; elev. (ft) 190

		Jan.	Feb.	Mar.	Apr.	May	June	July	Aug.	Sept.	Oct.	Nov.	Dec.	Yr.	
S	0	54.3	55.4	58.5	61.0	64.4	68.9	72.5	73.2	71.2	66.9	61.3	56.1	63.7	T°F
D	16.14″														
K	Csa	2.6	2.1	2.2	1.5	0.8	0.1	0	0	0.2	1.5	2.2	3.4	16.6	P″
Th	DB′2d	1.1	1.2	1.8	2.3	3.1	3.9	4.8	4.7	3.8	2.9	1.9	1.3	32.8	PE″

Coquilhatville, Congo: lat 0°03′ N; long 18°16′ E; elev. (ft) 70

		Jan.	Feb.	Mar.	Apr.	May	June	July	Aug.	Sept.	Oct.	Nov.	Dec.	Yr.	
S	15.00″	76.3	76.8	77.0	76.8	77.0	75.6	74.7	74.8	75.0	75.6	75.6	76.5	76.0	T°F
D	0.1″														
K	Afi	3.2	4.0	6.1	5.5	5.2	4.7	3.9	4.3	8.1	8.4	7.7	4.8	65.9	P″
Th	B1A′r	4.4	4.1	4.6	4.4	4.6	4.1	4.0	4.0	3.9	4.2	4.1	4.4	50.7	PE″

Dakar, Senegal: lat 14°44′ N; long 17°30′ E; elev. (ft) 70

		Jan.	Feb.	Mar.	Apr.	May	June	July	Aug.	Sept.	Oct.	Nov.	Dec.	Yr.	
S	0	70.2	68.7	69.3	70.9	73.6	78.8	81.0	81.0	81.5	81.3	78.8	73.6	75.8	T°F
D	32.9″														
K	BShw	0	0.1	0	0	0	0.6	3.5	9.8	6.4	1.9	0.2	0.2	22.7	P″
Th	DA′d	2.7	2.3	2.7	3.1	3.9	5.3	6.2	6.0	5.7	5.7	4.7	3.4	51.7	PE″

Djibouti, French Somaliland: lat 11°36′ N; long 43°09′ E; elev. (ft) 23

		Jan.	Feb.	Mar.	Apr.	May	June	July	Aug.	Sept.	Oct.	Nov.	Dec.	Yr.	
S	0	77.2	77.7	80.2	83.3	87.1	91.9	93.6	91.9	90.0	85.3	81.3	78.6	84.9	T°F
D	68.2″														
K	BWhs	0.5	0.3	0.9	0.4	0.2	0	0.3	0.3	0.1	0.5	1.2	0.5	5.2	P″
Th	EA′d	4.1	3.9	5.5	6.1	7.1	7.6	7.9	7.3	7.0	6.4	5.4	4.5	73.3	PE″

El Golea, Algeria: lat 30°34′ N; long 2°53′ E; elev. (ft) 1,305

		Jan.	Feb.	Mar.	Apr.	May	June	July	Aug.	Sept.	Oct.	Nov.	Dec.	Yr.	
S	0	48.2	53.1	61.3	70.5	78.6	88.7	93.4	92.1	85.3	72.1	59.4	50.4	71.1	T°F
D	46.0″														
K	BWhs	0.3	0	0.2	0.1	0	0	0	0	0.1	0.2	0.2	0.6	1.7	P″
Th	EA′db′3	0.3	0.6	1.6	3.3	5.8	8.0	8.6	8.1	6.5	3.3	1.2	0.4	47.6	PE″

Fort Lamy, Chad: lat 12°8′ N; long 15°02′ E; elev. (ft) 984

		Jan.	Feb.	Mar.	Apr.	May	June	July	Aug.	Sept.	Oct.	Nov.	Dec.	Yr.	
S	0	74.3	78.6	86.2	90.9	90.1	86.9	81.5	79.2	80.8	83.5	80.8	75.4	82.2	T°F
D	42.7″														
K	BShw	0	0	0	0.2	1.4	2.6	6.1	10.1	4.1	0.9	0	0	25.4	P″
Th	DA′d	3.4	4.3	6.7	7.2	7.5	7.0	6.2	5.3	5.6	6.1	5.3	3.7	68.1	PE″

Johannesburg, Union of South Africa: lat 26°08′ S; long 28°14′ E; elev. (ft) 5,556

		Jan.	Feb.	Mar.	Apr.	May	June	July	Aug.	Sept.	Oct.	Nov.	Dec.	Yr.	
S	2.28″	66.0	66.2	64.4	60.4	54.7	49.8	50.5	54.9	58.6	63.5	64.6	64.6	59.9	T°F
D	1.42″														
K	Cwb	4.9	4.2	3.5	2.0	0.9	0.3	0.3	0.3	1.0	2.4	4.6	5.3	29.7	P″
Th	C2B′2r	3.6	3.2	3.1	2.2	1.5	1.0	1.1	1.6	2.1	3.0	3.2	3.4	28.9	PE″

Juba, Sudan: lat 4°52′ N; long 31°36′ E; elev. (ft) 1,502

		Jan.	Feb.	Mar.	Apr.	May	June	July	Aug.	Sept.	Oct.	Nov.	Dec.	Yr.	
S	0	81.3	83.1	83.3	81.5	79.2	77.5	75.6	75.4	76.8	78.3	79.3	79.9	79.3	T°F
D	21.9″														
K	Awgi	0.2	0.4	1.7	4.2	6.2	4.6	5.4	6.0	4.1	4.0	1.4	0.5	38.7	P″
Th	DA′d	5.7	5.5	6.2	5.7	5.2	4.6	4.2	4.1	4.3	4.8	4.9	5.4	60.6	PE″

AFRICA (continued)

Lagos, Nigeria: lat 6°35′ N; long 03°20′ E; elev. (ft) 124

		Jan.	Feb.	Mar.	Apr.	May	June	July	Aug.	Sept.	Oct.	Nov.	Dec.	Yr.	
S	10.66″	80.1	81.5	81.9	81.3	80.1	78.1	75.9	75.7	77.0	78.1	80.2	80.2	79.2	T°F
D	9.44″														
K	Amgi	1.6	2.2	3.9	4.5	8.5	13.2	5.9	2.3	8.4	8.7	3.0	1.6	63.8	P″
Th	C2A′w	5.4	5.2	5.9	5.7	5.7	4.8	4.3	4.2	4.3	4.7	5.3	5.4	61.0	PE″

Marrakech, Morocco: lat 31°37′ N; long 8°02′ W; elev. (ft) 1,529

		Jan.	Feb.	Mar.	Apr.	May	June	July	Aug.	Sept.	Oct.	Nov.	Dec.	Yr.	
S	0″	52.7	56.1	61.0	65.5	70.3	76.6	83.7	83.8	77.7	70.2	61.7	54.5	67.9	T°F
D	30.83″														
K	BShs	1.1	1.1	1.3	1.2	0.7	0.3	0.1	0.1	0.4	0.8	1.1	1.3	9.5	P″
Th	EB′4da′	0.8	1.0	1.8	2.6	3.9	5.4	7.3	7.0	4.9	3.2	1.6	0.9	40.3	PE″

Nairobi, Kenya: lat 1°18′ S; long 36°45′ E; elev. (ft) 5,898

		Jan.	Feb.	Mar.	Apr.	May	June	July	Aug.	Sept.	Oct.	Nov.	Dec.	Yr.	
S	8.61″	64.0	64.6	65.8	65.8	64.0	61.1	58.8	59.9	62.2	65.5	64.9	64.0	63.5	T°F
D	1.38″														
K	Cwbi	1.8	2.0	4.0	8.1	6.3	1.8	0.7	1.0	1.0	2.1	4.3	3.2	36.3	P″
Th	B1B′2r	2.7	2.5	3.0	2.9	2.7	2.2	1.9	2.1	2.3	2.9	2.7	2.7	30.6	PE″

Nova Lisboa, Angola: lat 12°48′ S; long 15°45′ E; elev. (ft) 5,594

		Jan.	Feb.	Mar.	Apr.	May	June	July	Aug.	Sept.	Oct.	Nov.	Dec.	Yr.	
S	20.54″	65.8	66.2	66.2	66.3	63.9	60.6	61.2	65.7	69.1	68.4	66.4	66.4	65.5	T°F
D	4.6″														
K	Cwbi	7.8	6.2	8.6	7.0	1.1	0	0	0	0.7	4.3	8.7	8.6	53.0	P″
Th	B3B′2w	3.0	2.7	2.9	2.7	2.4	1.8	2.0	2.7	3.2	3.3	2.9	3.1	32.8	PE″

Salisbury, Southern Rhodesia: lat 17°50′ S; long 13°12′ E; elev. (ft) 4,830

		Jan.	Feb.	Mar.	Apr.	May	June	July	Aug.	Sept.	Oct.	Nov.	Dec.	Yr.	
S	9.05″	69.1	69.3	67.8	66.6	62.1	57.0	57.0	61.3	66.4	71.2	70.5	69.6	65.7	T°F
D	6.3″														
K	Cwb	9.3	6.6	3.4	1.8	0.5	0.3	0	0.1	0.3	1.5	3.7	7.9	35.4	P″
Th	C2B′2w	3.6	3.2	3.2	2.7	2.1	1.4	1.4	2.0	2.8	3.9	3.7	3.7	33.6	PE″

Tripoli, Libya: lat 32°57′ N; long 13°12′ E; elev. (ft) 53

		Jan.	Feb.	Mar.	Apr.	May	June	July	Aug.	Sept.	Oct.	Nov.	Dec.	Yr.	
S	0	54.5	56.8	60.4	64.8	69.4	77.2	78.8	79.9	78.6	72.7	64.6	57.9	68.0	T°F
D	27.1″														
K	BShs	1.9	1.2	0.5	0.7	0.2	0	0	0	0.4	1.3	2.4	4.1	12.7	P″
Th	EB′4da′	0.9	1.1	1.7	2.5	3.7	5.6	6.1	6.1	5.2	3.6	2.0	1.2	39.7	PE″

Tunis, Tunisia: lat 36°50′ N; long 10°14′ E; elev. (ft) 13

		Jan.	Feb.	Mar.	Apr.	May	June	July	Aug.	Sept.	Oct.	Nov.	Dec.	Yr.	
S	0	51.8	53.1	56.1	60.3	66.4	74.1	78.6	79.9	76.3	68.7	60.6	54.3	65.1	T°F
D	18.2″														
K	Csa	2.8	1.8	1.7	1.6	0.9	0.4	0	0.4	1.5	2.2	2.2	2.8	18.3	P″
Th	DB′3da′	0.8	0.9	1.4	2.0	3.4	5.0	6.3	6.2	4.7	3.1	1.6	1.0	36.4	PE″

AUSTRALIA AND OCEANIA

Adelaide, Australia: lat 34°36′ S; long 138°35′ E; elev. (ft) 140

		Jan.	Feb.	Mar.	Apr.	May	June	July	Aug.	Sept.	Oct.	Nov.	Dec.	Yr.	
S	0	73.5	74.0	70.0	64.0	58.0	54.0	52.0	54.0	57.0	62.0	67.0	71.0	63.0	T°F
D	11.57″														
K	Csa	0.8	0.7	1.0	1.8	2.7	3.0	2.6	2.6	2.1	1.7	1.1	1.0	21.1	P″
Th	DB′2da′	5.1	4.4	3.8	2.4	1.6	1.1	0.9	1.2	1.6	2.5	3.5	4.6	32.7	PE″

Alice Springs, N. Terr., Australia: lat 23°38′ S; long 133°35′ E; elev. (ft) 1,901

		Jan.	Feb.	Mar.	Apr.	May	June	July	Aug.	Sept.	Oct.	Nov.	Dec.	Yr.	
S	0	83.5	82.0	76.5	67.5	59.5	54.0	53.0	58.0	65.0	73.0	78.5	82.0	69.4	T°F
D	32.8″														
K	BWh	1.7	1.3	1.1	0.4	0.6	0.5	0.3	0.3	0.3	0.7	1.2	1.5	9.9	P″
Th	EB′4da′	7.0	5.8	4.7	2.6	1.4	0.8	0.7	1.3	2.2	4.0	5.5	6.7	42.7	PE″

Auckland, New Zealand: lat 36°47′ S; long 174°39′ E; elev. (ft) 85

		Jan.	Feb.	Mar.	Apr.	May	June	July	Aug.	Sept.	Oct.	Nov.	Dec.	Yr.	
S	21.6″	66.5	66.5	65.0	61.5	56.5	53.0	51.0	52.0	54.5	57.5	60.0	63.5	59.0	T°F
D	.2″														
K	Cfb	3.1	3.7	3.2	3.8	5.0	5.4	5.7	4.6	4.0	4.0	3.5	3.1	49.1	P″
Th	B3B′2r	4.0	3.4	3.3	2.4	1.7	1.2	1.1	1.3	1.6	2.3	2.7	3.6	28.5	PE″

Canberra, N.S.W. Australia: lat 35°20′ S; long 149°15′ E; elev. (ft) 1,837

		Jan.	Feb.	Mar.	Apr.	May	June	July	Aug.	Sept.	Oct.	Nov.	Dec.	Yr.	
S	0	68.5	68.5	63.5	55.5	48.5	43.5	42.5	45.0	49.5	55.5	61.5	66.5	55.8	T°F
D	4.4″														
K	Cfb	1.9	1.7	2.2	1.6	1.8	2.1	1.8	2.2	1.6	2.2	1.9	2.0	23.0	P″
Th	C1B′1d	4.6	3.9	3.2	1.9	1.1	0.6	0.6	0.8	1.3	2.2	3.2	4.3	27.6	PE″

Christchurch, New Zealand: lat 43°32′ S; long 172°37′ E; elev. (ft) 32

		Jan.	Feb.	Mar.	Apr.	May	June	July	Aug.	Sept.	Oct.	Nov.	Dec.	Yr.	
S	.59″	61.5	61.0	58.0	53.5	48.0	44.5	42.5	44.0	48.5	53.0	56.5	60.0	52.6	T°F
D	3.89″														
K	Cfb	2.2	1.7	1.9	1.9	2.6	2.6	2.7	1.9	1.8	1.8	1.9	2.2	25.2	P″
Th	C1B′1da′	4.0	3.3	2.8	1.9	1.2	0.8	0.7	0.9	1.5	2.3	3.0	3.8	26.1	PE″

Mackay, Queensland, Australia: lat 21°09′ S; long 149°11′ E; elev. (ft) 35

		Jan.	Feb.	Mar.	Apr.	May	June	July	Aug.	Sept.	Oct.	Nov.	Dec.	Yr.	
S	27.16″	80.0	79.0	77.5	69.0	68.5	64.0	62.2	64.0	68.5	73.5	77.0	79.0	72.0	T°F
D	1.84″														
K	Cwa	13.8	11.8	12.2	6.0	3.8	2.7	1.7	1.0	1.7	1.7	3.1	7.0	66.5	P″
Th	B2B′4r	6.1	5.0	4.9	2.7	2.6	1.8	1.6	1.9	2.7	4.0	4.9	5.8	44.1	PE″

Macquarie Island, New Zealand: lat 54°30′ S; long 158°57′ E; elev. (ft) 20

		Jan.	Feb.	Mar.	Apr.	May	June	July	Aug.	Sept.	Oct.	Nov.	Dec.	Yr.	
S	19.5″	43.5	43.0	42.0	40.0	39.5	37.0	37.0	37.0	37.5	38.0	40.0	42.0	39.8	T°F
D	0														
K	ET	4.0	3.5	4.1	3.8	3.3	2.9	3.2	3.2	3.8	3.3	2.8	3.9	41.8	P″
Th	B4C′2ra′	3.3	2.6	2.3	1.6	1.3	0.8	0.9	1.1	1.4	1.8	2.4	3.1	22.3	PE″

Madang, N.E. New Guinea: lat 5°14′ S; long 145°45′ E; elev. (ft) 20

		Jan.	Feb.	Mar.	Apr.	May	June	July	Aug.	Sept.	Oct.	Nov.	Dec.	Yr.	
S	67.24″	81.0	80.5	80.5	81.0	81.5	81.0	81.0	81.0	81.0	81.5	81.5	81.5	81.1	T°F
D	0.1″														
K	Afi	12.1	11.9	14.9	16.9	15.1	10.8	7.6	4.8	5.3	10.0	13.3	14.5	137.2	P″
Th	AA′r	5.9	5.2	5.7	5.5	5.8	5.5	5.6	5.7	5.6	5.9	5.8	6.0	68.1	PE″

Malden, Line Island: lat 4°03′ S; long 155°01′ W; elev. (ft) 26

		Jan.	Feb.	Mar.	Apr.	May	June	July	Aug.	Sept.	Oct.	Nov.	Dec.	Yr.	
S	0	82.0	82.5	82.5	83.0	83.0	83.0	82.5	83.0	82.5	82.5	82.5	82.0	82.6	T°F
D	44.2″														
K	BSh	3.5	1.9	4.5	4.5	4.3	2.1	1.9	1.6	0.8	0.9	0.7	0.7	27.4	P″
Th	DA′d	6.1	5.5	6.1	5.9	6.1	5.9	6.0	6.1	5.9	6.1	6.0	6.1	71.6	PE″

Palau, Caroline Island: lat 7°14′ N; long 134°08′ E; elev. (ft) 104

		Jan.	Feb.	Mar.	Apr.	May	June	July	Aug.	Sept.	Oct.	Nov.	Dec.	Yr.	
S	88.3″	80.5	80.0	81.0	82.0	81.5	81.5	80.5	80.5	81.0	81.0	81.0	80.5	81.0	T°F
D	0														
K	Afi	15.3	9.4	6.8	7.6	15.5	12.4	19.9	14.0	15.7	14.8	11.8	12.7	155.7	P″
Th	AA′r	5.5	4.9	5.7	5.8	6.0	5.9	5.8	5.8	5.6	5.7	5.4	5.5	67.7	PE″

Papeete, Tahiti: lat 17°32′ S; long 149°34′ W; elev. (ft) 302

		Jan.	Feb.	Mar.	Apr.	May	June	July	Aug.	Sept.	Oct.	Nov.	Dec.	Yr.	
S	19.2″	80.5	80.5	80.5	80.5	78.5	77.5	77.0	77.0	77.5	78.5	79.5	80.0	79.0	T°F
D	4.5″														
K	Ami	9.9	9.6	16.9	5.6	4.0	3.0	2.1	1.7	2.1	3.5	5.9	9.8	74.1	P″
Th	B1A′r	6.1	5.4	5.7	5.3	4.6	4.1	4.1	4.2	4.4	5.1	5.4	6.0	60.6	PE″

Perth, S.W. Australia: lat 31°57′ S; long 115°51′ E; elev. (ft) 197

		Jan.	Feb.	Mar.	Apr.	May	June	July	Aug.	Sept.	Oct.	Nov.	Dec.	Yr.	
S	11.69″	74.0	74.0	71.0	66.0	61.0	57.0	55.5	56.0	58.5	61.5	66.5	71.0	64.4	T°F
D	12.00″														
K	Csa	0.3	0.4	0.8	1.7	5.1	7.1	6.7	5.7	3.4	2.2	0.8	0.5	34.7	P″
Th	C2B′2s	5.0	4.3	3.9	2.6	1.9	1.3	1.2	1.3	1.7	2.4	3.3	4.4	33.3	PE″

Port Darwin, N.W.T. Australia: lat 12°28′ S; long 130°51′ E; elev. (ft) 197

		Jan.	Feb.	Mar.	Apr.	May	June	July	Aug.	Sept.	Oct.	Nov.	Dec.	Yr.	
S	15.04″	83.5	83.5	84.0	84.0	82.0	78.5	77.0	79.5	82.5	85.0	86.0	85.0	82.5	T°F
D	29.8″														
K	Awgi	15.2	12.3	10.0	3.8	0.6	0.1	0.1	0.1	0.5	2.0	4.7	9.4	58.8	P″
Th	C1A′s	6.6	5.8	6.4	6.6	5.7	4.4	4.1	5.0	5.8	6.7	6.8	6.9	70.3	PE″

Sydney, N.S.W. Australia: lat 33°52′ S; long 151°12′ E; elev. (ft) 138

		Jan.	Feb.	Mar.	Apr.	May	June	July	Aug.	Sept.	Oct.	Nov.	Dec.	Yr.	
S	15.08″	72.5	72.5	69.5	64.5	58.5	54.5	53.0	55.5	59.0	63.5	67.0	70.0	63.4	T°F
D	0.63″														
K	Cfa	3.5	4.0	5.0	5.3	5.5	4.6	4.6	3.0	2.9	2.8	2.9	2.9	47.0	P″
Th	B2B′2r	4.8	4.1	3.7	2.5	1.6	1.1	1.0	1.3	1.8	2.8	3.5	4.3	32.5	PE″

Walgett, N.S.W. Australia: lat 30°02′ S; long 148°10′ E; elev. (ft) 436

		Jan.	Feb.	Mar.	Apr.	May	June	July	Aug.	Sept.	Oct.	Nov.	Dec.	Yr.	
S	0	82.5	81.0	76.0	67.5	59.0	53.0	51.5	54.5	66.5	69.0	76.0	81.0	68.0	T°F
D	23.5″														
K	BSh	2.1	1.9	1.6	1.2	1.5	1.6	1.3	1.1	1.0	1.2	1.5	1.7	17.7	P″
Th	EB′4da′	7.0	5.7	4.7	2.6	1.4	0.7	0.7	1.0	2.5	3.3	5.0	6.8	41.2	PE″

APPENDIX D
GLOSSARY
OF TERMS

Aberration   The failure of the zenithal extension of the earth axis to maintain a fixed position with respect to the background of stars.

Abrasion   The mechanical process of erosion by the friction of rock particles in transit.

Absolute humidity   A measurement of the weight or mass of water vapor contained within a given volume of air.

Absorption   The process by which substances penetrate the surfaces of solids.

Abyssal deeps   Deep ocean trenches that split the ocean floor, usually just beyond the foot of the continental rise.

Acacia   A thorny deciduous tree of tropical woodlands.

Acidic rocks   Igneous rocks containing a high percentage (above 50 to 55) of silica and a low percentage of metallic bases. See also *Sial*.

Acidity   The concentration of hydrogen ions in an aqueous solution. A concentration greater than 10^{-7} is considered acidic, and a concentration less than that is termed alkaline.

Actual evapotranspiration   The quantity of water that is evaporated from a land surface, including the amount transpired by plants. It differs from potential evapotranspiration in that it takes into account the increased difficulty of withdrawal with increased drain on soil-moisture reserves.

Adiabatic heating and cooling   Changes in temperature that do not involve an addition or withdrawal of heat. In the atmosphere, these are associated with the compression or expansion of air following an increase or decrease in atmospheric pressure. Also known as *dynamic heating and cooling*.

Adsorption   The process by which substances adhere to the surfaces of solids. The adsorbed substances may include ions, molecules, or colloidal particles.

Advection Heating or cooling resulting from horizontal airflow.

Aggradation The filling in of low areas by the deposition of sediment. It is the opposite of degradation.

Aggregates (soil) Soil structural units, consisting of many soil particles held together by some type of bond.

Agric horizon An illuvial horizon of clay and humus resulting from long-continued cultivation; usually immediately below plow depth.

A horizon The topmost horizon of the mineral portion of the soil from which material is removed by leaching or eluviation.

Air drainage The gravity flow of air into topographic depressions following radiational cooling.

Air mass An extensive body of air with more or less homogeneous characteristics within a horizontal plane.

Albedo The radiation that is lost to outer space by direct reflection or scattering.

Albic horizon A soil horizon that has been bleached as the result of the removal of clay and iron oxides.

Alfisols A mineral soil order whose soils have ochric epipedons, argillic horizons, and a moderate to high cation-exchange capacity, and that regularly undergo some seasonal drying.

Algae Single-celled plants, usually colonial.

Alkali Sodium or potassium carbonates that accumulate in poorly drained desert soils.

Alkaline earths Several of the bivalent oxides, such as those of calcium, magnesium, barium, and strontium.

Alkaline soil A soil that has a pH above 7, or one in which negative OH$^-$ ions predominate over positive H$^+$ ions.

Alluvial cone An unusually steep, cone-shaped deposit of alluvium deposited by streams as the result of a sudden decrease in slope.

Alluvial fan A fan-shaped alluvial deposit at the foot of a mountain stream or where the velocity suddenly drops.

Alluvial plain A plain that results from alluvial deposition by streams.

Alluvial terrace A river terrace composed of alluvium and marking a former higher level of stream deposition.

Alluvium Unconsolidated fragmental material recently deposited by streams. Does not include the material deposited in ponded water.

Alpine Pertaining to high altitudes near and above the timberline of mountains.

Amorphous Without a definite or distinctive form. When pertaining to rocks or minerals, the term implies the absence of a definite crystal structure.

Anaerobic Pertaining to life forms that are active despite a deficiency of free oxygen.

Analemma An asymmetrical figure, similar to the figure 8, which indicates the declination of the sun and the equation of time during the course of a year.

Anemometer An instrument used to measure wind velocity.

Angular momentum The tendency for a body moving in a curved path to maintain its angular velocity.

Angular velocity The velocity of movement along a curved path, measured in angles of arc per unit of time.

Anion A negatively charged ion, for example, OH$^-$.

Anomaly, gravimetric An abnormal gravity reading at the earth surface. A positive anomaly implies readings that are greater than normal; a negative anomaly indicates readings that are less than normal.

Anomaly, temperature A temperature that is higher or lower than expected for the position.

Anthracite A lustrous, black, hard coal containing a relatively high percentage (85) of fixed carbon. *Metaanthracite* consists almost entirely of fixed carbon.

Anthropic epipedon   A dark-colored epipedon similar to the *mollic* except that it has more than 250 p.p.m. of phosphoric acid, the result of long-continued manuring associated with cultivation.

Anticline   A rock upfold, or a rock fold in which the strata dip downward away from the center, or axis, of the fold.

Anticyclone (or *High*)   An area of relatively high atmospheric pressure, characterized by subsiding, diverging, rotating wind motion. The rotation is clockwise in the northern hemisphere and counterclockwise south of the equator.

Antipode   A point on the globe that is directly opposite any given point.

Aphanitic   A fine-grained igneous rock.

Aphelion   The position of the earth in its orbit around the sun when it lies farthest away. This occurs in July.

Apparent solar time   Local time that is based on the sun's apparent position in its course across the sky.

Aquifer   A permeable rock stratum from which flowing water may be obtained.

Arctic front   The line of discontinuity between very cold arctic air recently subsided within polar anticyclones and older, cool air in the upper mid-latitudes.

Arête   A jagged, narrow mountain-ridge crest resulting from glacial cirque excavation. Also known as a *comb ridge*.

Argillic horizon   An illuvial soil horizon in which fine-textured silicate clays have accumulated.

Aridisols   An order of mineral soils with no water available for mesophytic plants and including one or more clearly defined diagnostic horizons other than oxic or spodic.

Arkose   A sandstone that contains a high percentage of feldspar fragments rather than the more common quartz.

Association   (See *Plant associations*.)

Astrology   The pseudoscience that attempts to forecast events by noting various positions of the sun and planets against the stellar background.

Astronomical latitude   The angular distance of a particular star north or south of the celestial equator. It is synonymous with the *declination* of the star.

Atmospheric pressure   The mass weight of a column of air above a given point.

Atoll   (See *Coral reef*.)

Augite   One of the dark green, ferromagnesium, primary minerals belonging to the *pyroxene* group. It is especially high in calcium.

Aurora   A luminous phenomenon in the upper atmosphere caused by the ionization of gases. Termed *aurora borealis* in the northern hemisphere and *aurora australia* south of the equator.

Avalanche   A large mass of snow or ice that moves rapidly down a mountain slope. It is a form of mass wastage.

Axis   The line connecting the poles about which the earth rotates. Also, a line that marks the top of an anticline or bottom of a syncline.

Azimuth   The true direction of a meridian passing through a given point. Also, the horizontal direction of a line measured clockwise from the meridional plane. The azimuth of west, for example, is 270°.

Azimuthal   Pertaining to the property of map projections in which directions are true from a central point.

Azonal soils   Unconsolidated materials that have not yet developed distinct soil-profile characteristics.

Back marsh   The poorly drained areas on an alluvial floodplain, which are generally located well away from the river and its natural levees.

Back radiation   The radiation that passes back and forth between the earth and the atmosphere more than once.

Bacteria   The single-celled microorganisms

whose affinities for the plant or animal kingdom are debatable.

Badlands A highly dissected land surface, sculptured largely by rainwash and usually lacking an appreciable vegetation cover. It often is composed of fine-textured, unconsolidated sediments.

Bajada A sloping plain of detrital alluvial material lying along the flanks of a desert basin. Often also termed *piedmont alluvial plain.*

Barograph An instrument which continuously records barometric pressure.

Barometer An instrument used to record atmospheric pressure.

Barrier beach A ridge of sand paralleling the coast and thrown up by storm waves some distance off a gradually sloping coastline. Also known as an *offshore bar.*

Barrier reef A coral reef that lies some distance offshore and is separated from the mainland by a deep lagoon.

Basalt A dark-colored, fine-textured igneous rock. It usually is volcanic, or formed at or near the earth surface.

Bases Metallic cations, or the metallic atoms that have lost electrons, hence have a net positive charge. Among the more common ones are the plant nutrients Mg^{++}, Ca^{++}, Na^+, P^+, and Fe^{++}.

Base exchange (See *Cation exchange.*)

Base level The lower limit of effective erosion. This may be *ultimate,* or the extension of sea level into the continents; *temporary,* in which erosion is temporarily checked by some obstacle; or *local,* in which only a limited area is involved.

Basic rocks Igneous rocks that contain a relatively low percentage (below 50 to 55) of silica and a high percentage of metallic bases (iron, calcium, magnesium, etc.). See also *Sima.*

Basin (structural) A downwarped section of the earth crust produced by folding or faulting. An elongated basin would be termed a trough.

Batha The ground cover of herbaceous (nonwoody) plants that is found in the maquis and garrigue of Mediterranean regions.

Batholith A large, irregular intrusive mass of igneous rock that intersects the rock structure and has no apparent floor. It usually is greater than 40 square miles in area.

Bauxite The principal ore of aluminum. It is composed of hydrated oxide containing various impurities.

Baymouth bar A sandbar which extends across the mouth of a bay from one coastal promontory to another.

Bearing The horizontal angular measurement of a line from a cardinal point—for example, N 30° E.

Bed load The unconsolidated material that is moved along the bed of a stream channel without being carried in suspension.

Bench mark A cement marker and plate that records the exact geographic location and elevation of a point. A reference point for surveying.

Bergschrund The large crevasse left in the back of a cirque after a slippage of ice and snow away from the back wall.

B horizon That horizon of the mineral-dominant portion of the soil that has experienced illuviation, or the addition of clays and other colloidal material, or chemical precipitates. Sometimes referred to as the *zone of illuviation.*

Biomass The total mass of living organisms, large and small, that occupy a given area.

Biosphere The zone of life at or near the earth surface.

Biota The flora and fauna of a region.

Bituminous coal An intermediate coal, classified according to percentage of fixed carbon and volatile material. Sometimes loosely termed *soft coal.*

Black body A solid that is able to absorb 100 percent of the radiation it receives.

Block mountains A mountain mass bordered by faults. Also termed *fault-block mountains.*

Bog A thick deposit of acid peat, found in poorly drained depressions and covered with typical acid-loving, hydrophytic plants, such as the bog mosses (*Sphagnum*), sedges, and cotton grass. Especially common in periglacial regions.

Bolson A desert basin rimmed by mountains.

Bottoms (river) Alluvial terraces that are found along a river floodplain.

Braided stream A stream that divides into many interlocking channels when at grade.

Breccia A conglomerate whose constituent pebbles have sharp, angular edges.

Broadleaf coniferous forest A type of coniferous forest, mainly in the southern hemisphere, whose leaves are straplike rather than of the more usual needle form.

Brunizem (See *Prairie soils.*)

Butte A prominent, isolated, cliffed, erosional remnant in dry regions, usually bordered by talus and fall-face slopes; often turret-shaped.

Caatinga A scrub, thorny, open woodland in northeastern Brazil.

Cacti A group of desert plants having fleshy stems and branches with scales or spines instead of leaves.

Calcareous Limey, or containing a high content of calcium carbonate.

Calcic horizon A soil horizon with accumulations of calcium carbonate and possibly magnesium carbonate. It may become permanently indurated into a true cemented horizon, known as a *petrocalcic* horizon, or *croute calcaire.*

Calcification The soil-forming process which results in the formation of calcic horizons, or the deposition of alkaline earth carbonates within the soil profile.

Calcite The crystalline form of calcium carbonate ($CaCO_3$).

Caldera A large, basin-shaped depression associated with volcanoes, the diameter of which is many times greater than the included volcanic vent. Usually formed by foundering into the magma reservoir.

Caliche (See *Calcic horizon.*) A durable cemented horizon of precipitated sodium nitrate in the Atacama Desert of Chile is also thus termed.

Calorie The amount of heat energy necessary to raise the temperature of one gram of water one degree centigrade at sea level when the air temperature is 15 degrees centigrade. The *k calorie,* or kilocalorie, is equal to 1,000 calories and is used mainly in measuring food intakes.

Cambic horizon A fine-textured diagnostic soil horizon (usually subsurface) in which there has been noticeable alteration of the soil material by mechanical and/or chemical weathering.

Campo The Brazilian savanna.

Capillarity The action by which the surface of a liquid in an opening or a tube is raised or lowered because of surface tension.

Carbon 14 dating The process of dating organic material using the proportion of the carbon 14 isotope to other forms of carbon. It is effective up to 50,000–70,000 years.

Cartograms Maps that have proportionately distorted areas or scales to show quantitative values.

Cartography The art and science of map making.

Cation A positively charged ion, such as Na^+, that is, an atom that has lost one or more electrons.

Cation exchange The replacement of one cation by another along a colloidal surface or interface.

Cation-exchange capacity The capacity of soil colloids to exchange cations with the soil solution, or a measure of the storage capacity of the soil for plant nutrients.

Celestial equator The projection of the earth equatorial plane against the stellar background.

Celestial poles The points about which the stars in the sky appear to rotate during the period of earth rotation, or the intersection points of the extensions of the earth axis and the background of stars.

Center point The central point from which a map projection is geometrically based.

Centigrade (Celsius) The system of measuring temperature in which 0° is designated the freezing point of water at sea level and 100° is designated the boiling point.

Centrifugal An apparent force directed outward by a body moving in a curved path.

Centripetal A force directed inward by a body moving in a curved path.

Chalk A soft, amorphous rock composed of calcium carbonate.

Chaparral (See *Maquis*.)

Chernozem A dark-colored-to-black zonal soil having a lime-accumulation layer in the *B* horizon. Typically developed under a tall-grass vegetation in the mid-latitudes. A type of *Mollisol*.

Chert An amorphous rock composed mainly of hardened colloidal silica. Usually occurs as an impurity in limestones.

Chinook (or Foehn) A dry, warm wind that descends a mountain slope from high elevations and has been warmed by compression.

Chlorophyll A complex organic molecule that is a necessary catalyst in photosynthesis. Its bright green color is distinctive.

C horizon That portion of the soil profile in which the rock material has undergone some chemical and physical alteration, but which lies below the normal biotic activities. Normally the parent soil material.

Chorographic (medium-scale) maps Regional or subcontinental in scale. More specifically, the term usually refers to a scale of reduction of 1:500,000 to 1:5,000,000.

Chroma A measurement of the intensity of color.

Chronometer An accurate timepiece, generally used to determine longitude.

Cinder cone A cone-shaped mound of volcanic, pyroclastic material accumulated at the vent or opening of an explosive volcano. Usually contains a small crater at its summit.

Circle of illumination (See *Terminator*.)

Cirque An amphitheaterlike depression in mountain terrain that is produced by the sapping effect of mountain glaciers.

Cirrus High-elevation, wispy cloud forms that do not cast a shadow.

Clay Inorganic particles below .002 mm in diameter. Chemically, most clays are hydrous iron and aluminum silicates in colloidal form containing various absorbed and adsorbed cations.

Cleavage The property of some minerals and rocks to split along definite lines (usually parallel to the crystal faces).

Climate The overall, or aggregate, weather conditions of an area.

Climatology The scientific study of climates.

Climax vegetation A plant community that represents a relatively long-term equilibrium with respect to its constituent plant forms and their environment.

Clouds Air spaces that contain a sufficient number of condensed, suspended droplets of water to be opaque to sunlight.

Coal A simple mixture of carbon in a solid, noncrystalline form, containing various organic impurities.

Coastal plain Any plain which has its margin on the shore of a large body of water, especially the sea. Generally represents a strip of recently emerged sea bottom.

Cobbles Detrital depositional material, usually rounded, above 1½ to 2 inches in diameter.

Col A pass or saddle along a mountain ridge. Usually formed by the intersection of cirques.

Cold front A steep frontal surface between cold and warm air masses in which the cold air is displacing warm air in its path.

Colloid A material so finely divided that interfacial physical and chemical properties predominate.

Colloidal soil dispersion The dispersion of soil colloidal particles from aggregates into separate particles. This is sometimes referred to as *peptization*.

Colloidal soil flocculation The coagulation of soil colloidal particles into structural units or *micelles*.

Colluvial foot slope The accumulation of colluvial material at the base of a slope dominated by mass wastage.

Colluvial slope The sloping surface of material that is moving downslope by mass wastage.

Colluvium The unassorted detrital mass of soil and rock debris that has been transported by mass wastage.

Color value A measurement of the lightness or darkness of a color.

Comb ridge (See *Arête*.)

Comet A mass of dispersed solids and gases in the solar system that is in a highly elliptical orbit around the sun.

Compaction The hardening of unconsolidated material by the weight of overlying material.

Compensated dry period A dry period which results in no major adaptation by the dominant vegetation because of an abundance of rain during the remainder of the year.

Compost Partially decayed organic matter that is used as fertilizer.

Compromise A property of map projections in which all distortion is kept as low as possible, at the expense of having no characteristic of the surface truly represented.

Conchoidal fracture The curved breakage surface of certain materials similar to that of glass or quartz.

Condensation A change in phase from a gas to a liquid.

Condensed projections Map projections that have had portions removed and the remaining parts compressed to save space.

Conditional instability A condition in which air contains sufficient water vapor and is located where the lapse rate is steep enough to sustain lifting once the condensational level is reached. A condition in which the lapse rate lies between the values of the wet and dry adiabatic rate.

Conduction The transmission of heat from a warmer to a cooler body by direct contact.

Conformality A property of map projections in which shapes are true for limited areas.

Conglomerate A sedimentary rock composed of cemented gravel or water-worn pebbles. Also known as *pudding stone*.

Conifer A plant (usually a needleleaf tree) belonging to the order *Coniferae*. Such plants carry their naked seeds in cone-shaped structures.

Consequent stream A stream whose trajectory has been directed by the original slope of the land surface.

Constellation A clustering of major stars in the sky.

Continental margins Those parts of the earth crust that form a transition between the ocean floor and the edge of the land areas.

Continental rise That part of the continental margin that lies below the steep continental slope.

Continental shelf A gradually sloping submarine plain or terrace that borders the continents. Its steep outer edge generally lies at a depth of about 350 feet but may extend to about 600 feet.

Continental slope The steep edge of the continental blocks just beyond the continental shelf.

Contour line A line connecting points of equal elevation.

Convection A localized upward or downward air current or other fluid material involving a transfer of heat away from a heating source.

Convergence A condition characterized by a net horizontal inflow of air or water toward a common center or line.

Coral reef A large mass of calcium carbonate formed mainly by corals that grow upward from the sea floor. *Fringing reefs* border the coasts of islands or continents, while *barrier reefs* lie some distance offshore, roughly paralleling it. *Platform reefs* are large, broad reefs located on the continental shelf. *Atolls* are more or less circular reefs enclosing a lagoon.

Cordillera A mountain chain comprising several separate ridges or ranges that may or may not be parallel. The entire chain, however, has a general directional trend.

Coriolis effect The apparent deflective force of the earth's rotation. See *Ferrel's law.*

Corrasion The mechanical erosion of bedrock resulting from the abrasion of detritus and removal by transporting agents, such as streams, glaciers, and ocean currents.

Corrosion The chemical erosion of bedrock and its removal. Solution by weak acids is the principal agent.

Crater A steep-sided pit or depression at the summit or on the flanks of a volcano.

Cretaceous period The third and last period of the Mesozoic era, which featured a long period of crustal stability, rapid evolution of the flowering plants, the rise of the mammals and the decline of the reptiles.

Crevasse A fissure or cleft in glacial ice caused by stresses resulting from differential movement.

Cross channel A lateral drainage way cut by glacial drainage across interstream areas.

Cross grading The process of developing a dendritic pattern by rills and gullies as they begin to dissect a slope.

Crust That portion of the outer part of the solid earth that lies above the *Mohorovicic Discontinuity.*

Crystalline rock A rock whose constituent minerals have developed crystal forms as the result of cooling from a molten state or from excessive heating and compression.

Cuesta An asymmetrical ridge with a steep slope or cliff on one side and a gentle slope on the other. The latter generally conforms to the slope, or dip, of the underlying rock strata.

Cumulus Cloud types that typically have domed summits and flat bases.

Curve of equilibrium The longitudinal profile of a stream at *grade,* or the balance between deposition and erosion. Normally it is a concave curve that flattens downstream.

Cycle of erosion A sequence of stages in the erosion of a land surface (following an initial change in state) to the ultimate stage of stability or equilibrium. Usually the initial change implies uplift, but it might also involve a change in climate.

Cyclone (or Low) An area of relatively low atmospheric pressure, characterized by converging, ascending, rotating winds. The rotation is counterclockwise in the northern hemisphere and the reverse south of the equator.

Cyclonic precipitation   Precipitation in which the initial updraft of air is caused by cyclonic convergence.

Deciduous   Pertaining to plants that shed their leaves during a particular period or season, usually because of low temperatures or drought.

Declination   (See *Astronomical latitude.*)

Decomposition   The chemical breakdown of compounds. The term frequently is used specifically to signify chemical rock weathering.

Deflation   The removal of surface detrital material by the wind.

Deflation pit   A shallow surface depression caused by wind deflation.

Degradation   (See *Denudation.*)

Degraded chernozem   A chernozem soil in which the lime has been removed from the *B* horizon. Usually develops where a forest vegetation has encroached upon a prairie.

Degree-day   The number of degrees' difference between the mean daily temperature and an arbitrary base point of 65°F. The degrees below 65° are termed *heating degree-days,* and those above 65° are termed *cooling degree-days.*

Degree slope   The angular measurement of a slope from the horizontal plane, expressed in degrees. *Percentage slope* is the percentage of slope between 0° and 45°. A 45° slope is a 100 percent slope.

Delta   An accumulation of sediment in ponded water at the mouth of a stream.

Dendritic   Pertaining to an arborescent, or treelike, pattern. The term usually refers to a drainage pattern, with tributaries joining the main valley at acute angles.

Density   The ratio of mass, or weight, to volume.

Denudation   The removal of detrital material through erosion and mass wastage on the earth surface.

Desert pavement   A thin mantle of wind-polished pebbles that covers a desert land surface.

Desert varnish   A dark-red-to-black glaze which forms on exposed rock surfaces in low-latitude deserts. This generally consists of iron and manganese oxides.

Dessication   The process of drying out, or of losing water.

Detritus   Fragmental material resulting from weathering and erosion.

Dew   Water that condenses on a surface that is cooler than the adjacent air.

Dew point   The temperature at which condensation takes place in a given air mass.

Diabase   A basic igneous rock intermediate in texture between aphanitic (basalt) and granitic (gabbro).

Diagnostic horizon   A soil horizon within the profile whose characteristics clearly indicate the soil-forming processes that operate in the area.

Differential erosion   Irregular erosion resulting from differences in resistance of surface materials.

Diffuse radiation   The shortwave solar radiation that is scattered by being reflected off atmospheric particles.

Dike   A tabular body of igneous rock that crosscuts the grain, or structure, of the surrounding rock.

Diluvium   Fragmental material deposited by ocean waves or currents.

Dip   The angle made by a rock stratum or any other plane surface, measured from the horizontal.

Discontinuity   A relatively sharp difference in state or properties between two contiguous volumes, such as adjacent air masses, ocean currents, or earth interior zones.

Disintegration   (See *Weathering.*)

Dissection   The erosional process of sculpturing the land surface.

Distributary   A discharge channel of a river on its delta.

Diurnal   Of or pertaining to a 24-hour period.

Divergence   A condition of horizontal, divergent outflow of a fluid from a particular area.

Doldrum   A zone of calms or light variable winds located between the hemispheric trade-wind zones.

Doline   (See *Sinkhole*.)

Dolomite   A calcium-magnesium carbonate.

Dome   A roughly symmetrical rock upfold, in which the included rock strata dip outward in all directions from the center.

Dome (*salt*)   A dome structure in rocks caused by the upward thrust of salt columns under great rock pressure.

Dominant plants   The largest common members of a plant community.

Drainage basin   The area drained by a particular stream and its tributaries.

Drift (glacial)   A collective term for all detrital material derived from a glacier. Includes both till and fluvioglacial material.

Driftless area   A section in southwestern Wisconsin and adjacent parts of Illinois, Iowa, and Minnesota that was surrounded, but not covered by, the Pleistocene continental ice sheets.

Dripstone   Collective term for deposits of calcium carbonate in caves, resulting from the evaporation of dripping water carrying lime in solution.

Drumlin   An oval-shaped mound of compact glacial till that has its long axis oriented in the direction of ice motion and forms beneath the ice.

Dry adiabatic rate   The rate at which an ascending body of unsaturated air cools because of adiabatic expansion: approximately 5.6°F per 1,000 feet.

Dry farming   A system of farming designed primarily to conserve and retain moisture in the soil for crop use.

Dune   A deposit of windblown sand, usually having a gentle windward slope and a steep lee slope.

Duripan   A subsurface horizon that is cemented by indurated silica gels.

Dust devils   Small whirling vortices produced by mechanical instability of air over a highly heated surface. Usually found in dry regions.

Dynamic equilibrium   A condition within a dynamic system in which the form and arrangement of materials are at equilibrium with respect to the flow of energy through the system.

Earthquake   An earth tremor or elastic wave caused by rock rupturing or displacement along a fault.

Easterly wave   A wavelike undulation that moves from east to west within the isobars of a pressure gradient. Usually found on the equatorial flanks of subtropical anticyclones.

Eccentricity   The variation of an ellipse from a circle.

Ecliptic   The plane of an elliptical orbit.

Ecology   The science of interrelationships between organisms and their environments.

Ecosystem   The total environmental system within which an organism operates. In the case of man, it includes his economic, social and cultural environments, as well as the natural environment.

Edaphic   Pertaining to soil influences.

Eddy diffusion   The vertical transfer of fluid matter and energy from the surface by a spiral vortex or eddy.

Electromagnetic waves   Waves of radiant energy that are susceptible to electromagnetic influences.

Eluvial horizon   A soil horizon from which

material has been washed by eluviation. The *A* horizon of soil profiles.

Eluviation The physical movement of soil material within the solum, usually downward.

Endogenic energy Energy that originates within the earth interior (e.g., terrestrial heat flow).

Entisol A mineral soil order whose soils have no other diagnostic horizon than an ochric and anthropic epipedon and an albic or agric horizon.

Epipedon The horizon that forms at the surface of a soil, including the portion darkened by humus, which may include the upper eluvial horizon. It is *not* necessarily synonymous with the *A* horizon.

Epiphyte A plant that uses other plants for support (e.g., some mosses and orchids).

Equation of time The variation of apparent solar time ahead of or behind mean solar time during the course of a year.

Equator A great circle of the earth that is equidistant from the two poles, or the ends of the earth axis.

Equilibrium The concept relating to systems in which process, material, and pattern tend to form a balanced, self-correcting state with respect to the flow of energy.

Equinox The date on which the vertical rays of the sun are on the equator and the days and nights are of equal length at all latitudes. It occurs on or near March 22 (*vernal equinox*) and September 22 (*autumnal equinox*).

Equivalence A property of map projections in which the ratio between the area on the map and the corresponding area on earth are equal. Also referred to as *equal-area* maps.

Erg A desert surface composed largely of sand dunes.

Erosion Any or all of the processes that loosen and remove earth or rock material.

Esker A long, narrow ridge composed of stratified sand and gravel, formed by a stream in association with glacial ice.

Euphorbia A genus of succulent herbaceous, xerophytic plants, mainly leafless and having a milky latex sap. Dominant in dry African environments.

Eustatic fluctuations Changes in sea level.

Evapotranspiration (See *Actual evapotranspiration.*)

Exfoliation The peeling off or spalling of the outer, curved layers of rock.

Exogenic energy Energy derived from beyond the earth, such as that from the sun and from the motion of other heavenly bodies.

Explosion cone (See *Cinder cone.*)

Extratropical cyclone A cyclonic low-pressure center originating outside the tropics.

Extrusive rock An igneous rock that has cooled and solidified rapidly at or near the earth surface.

Fahrenheit An English system of measuring temperature in which 32° is designated the freezing point of water at sea level and 212° is designated the boiling point.

Fall face The steep, cliffed side of a valley or mountain from which rock fragments fall immediately upon being loosened by weathering.

Fault A fracture in the crust of the earth along which there has been displacement of the two sides relative to one another. A *normal fault* is one in which displacement has shifted the overhanging side downward with reference to the footwall. A *thrust* or *reverse fault* is one in which the displacement has moved the overhanging wall upward with reference to the footwall. A *strike-slip fault* is one in which the displacement is parallel to the strike of the fault.

Fault escarpment A steep slope or cliff that conforms closely to a line of faulting and has resulted from differential movement along the fault plane.

Fault scarp A steep slope or escarpment that has resulted from differential erosion along a fault rather than from displacement, as in a fault escarpment.

Feldspar A group or family of aluminum silicates carrying varying amounts of bases, such as potassium, sodium, calcium, and magnesium. A common constituent of *granite*.

Felsite A light-colored, fine-textured volcanic or extrusive rock containing much quartz and feldspar. Frequently has different-sized but small crystals.

Ferrallite (See *Oxisol*.)

Ferrallitic Pertaining to soils that have characteristic features of ferrallization. Sometimes defined as soils that have a silica-sesquioxide ratio of less than 2.

Ferrallization The soil-forming process of warm, humid regions which is characterized by the removal of combined silica and bases by mildly acidic-to-neutral soil solutions. It is also known as *latosolization, laterization,* or *kaolinization.*

Ferrel's law The law that all moving bodies of air or water tend to be deflected toward the right in the direction of motion in the northern hemisphere and to the left in the southern hemisphere. Also known as the *Coriolis effect.*

Ferric Iron compounds in which the iron combines in a monovalent bond, hence is stable (Fe_2O_3), (H_2FeO) with respect to oxidation.

Ferrous Iron compounds in which the iron combines in divalent bonds, hence may be oxidized. Examples are FeO, Fe_2O_2, FeS.

Ferruginous Containing a relatively large amount of iron oxide.

Field-fallow A form of agriculture in which fertility is replenished by the natural decomposition of minerals within the soil. It usually involves a period of rest, without cultivation.

Fire succession A plant succession that follows a fire.

Fissure-flow An outpouring of highly fluid lava from a fissure or crack in the earth crust.

Fjord A glaciated U-shaped mountain valley that opens onto the sea.

Flint A dark mass of hardened silica gel, usually found in calcareous shales or chalks. A dark variety of chert.

Flocculation (See *Colloidal soil flocculation.*)

Floodplain A flat plain adjacent to a river and composed of sediments derived from stream deposition during times of flood.

Fluvioglacial Of or pertaining to streams originating in glacial meltwater.

Fog A cloud at ground level.

Folded mountains Mountains having a folded rock structure.

Foraminifera Small, marine, one-celled animals that have a lime body casing.

Forest An association of plants dominated by large, woody perennials generally more than 20 feet high and having a continuous canopy overhead.

Fossil fuels The mineral fuels resulting from organic carbon, such as petroleum, coal, and natural gas.

Fractional land-use mapping A system of land-use mapping in which each digit of a fractional code refers to some land quality.

Fragipan A compact soil horizon whose soil aggregates are hard and durable when dry, but brittle and crumbly when moist.

Front (*air mass*)   The surface boundary between two adjacent air masses of unlike characteristics.

Frost   Ice crystals formed by the freezing of vapor on terrestrial objects.

Frost action   The weathering processes associated with alternate freezing and thawing.

Fungi   A grouping of spore-bearing plants without chlorophyll, such as molds and mushrooms.

Furlong   An ancient English basic field length of 660 feet.

Gabbro   A coarse-textured or granitic, basic igneous rock composed mainly of pyroxenes, hornblende and basic feldspars.

Galaxy   An island universe, or supergrouping, of stars and star clusters.

Galeria (*gallery forest*)   A forest growth along streams in grasslands, in which the treetops interlace over the streams.

Garrique   A low, shrubby vegetation consisting mainly of thorny sclerophylls less than 3 feet in height. It is found mainly in overgrazed areas within the Mediterranean climates.

Geanticline   A large regional anticlinal structure that may have minor folds along its flanks.

Gel   An amorphous colloidal mass that has not yet hardened.

Generic   Pertaining to things of the same kind or class.

Genetic   Pertaining to the genesis, or natural origin, of things.

Geographic grid   The system of parallels and meridians used to locate points on the earth surface.

Geography   The science of earth space; the descriptive and analytical study of the arrangement of spatial patterns on the surface of the earth.

Geoid   The exact shape of the earth.

Geologic cycle   The rock cycle whereby rocks are weathered, eroded, deposited, founder, are metamorphosed, and altered back into rock again.

Geology   The science of earth history.

Geomorphology   The study that deals with the form of the earth, the general surface configuration, and the evolution of landforms.

Geostrophic winds   Winds that parallel the isobars.

Geosyncline   A large sedimentary trough formed as the result of progressive downward warping and deposition.

Glacial lakes   Lakes formed as the result of glaciation, including cirque lakes, morainic-dam lakes, ice-dam lakes, glacial-scour lakes, kettle lakes, and pluvial lakes (the last are found in desert areas).

Glacial pavement   A polished rock surface produced by glacial abrasion.

Glacial spillway   A valley or drainage way that was eroded by meltwater flowing from the margins of a continental ice sheet.

Glacial striae   Grooves or scratches on a stone surface caused by glacial abrasion.

Glaciation   The various processes of erosion and deposition associated with the work of glaciers. Also, a period in which glaciers were active.

Glacier   A flowing mass of ice.

Glaze   An icy coating formed by the sudden freezing of rain after it strikes terrestrial objects.

Gley horizon   A diagnostic soil horizon featuring properties associated with alternate oxidation and reduction of iron compounds due to periodic flooding. Red, yellow, and brown spots or streaks between soil masses having grayish to bluish shades produce the typical mottling of gley horizons.

Gleying   A reduction process in soils in

which oxygen is removed from metallic oxides in the soil, especially from the iron and aluminum sesquioxides, by anaerobic microorganisms producing ferrous aluminum silicates.

Global energy balance   The general balance between total input and output of energy within the earth-atmosphere system.

Global energy system   The earth-atmosphere system for regulating and distributing the input and output of radiation.

Global scale   A scale of areal generalization involving all or a major part of the earth surface. Maps at this level have scales smaller than 1:5,000,000.

Gneiss   A banded, crystalline, coarse-grained metamorphic rock.

Gnomonic projection   A geometric projection derived from the projection of light from the center of a basic globe.

Gondwanaland   The hypothetical southern continent that preceded the proposed rupture and drifting apart of South America, Africa, peninsular India, Australia, and Antarctica during the late Mesozoic.

Gore   A triangular or lune-shaped piece forming part of the surface of a globe.

Graben   An elongated fault block that has been depressed with reference to the blocks on either side.

Gradation   The general tendency to reduce surface slopes by denudation and deposition.

Grade   The gradient of a stream that represents a steady state or equilibrium between downward erosion and deposition.

Granite   A light-colored, coarsely crystalline igneous rock, composed largely of quartz and feldspar.

Granitic   Pertaining to coarse texture in an igneous rock.

Gravity wind   A density airflow down a topographic slope under the influence of gravity.

Graywacke   A fine-textured sedimentary rock midway between shale and sandstone. Sometimes termed *siltstone*.

Great circle   All or part of an outer circumference of the earth. The shortest surface distance between any two points on earth.

Great Soil Group   A genetic classification of soils on a global basis, related mainly to climatic and biological soil-forming processes.

Greenhouse effect   The selective screening or absorbing of longer wavelengths of terrestrial radiation by the atmosphere, and the transmission of shortwave solar radiation.

Green manuring   Plowing green plants into the soil to increase the content of organic colloids (humus).

Grid   Any orderly, systematized series of lines.

Groundwater   The portion of the subsurface water that is below the water table.

Groundwater latosol   A plinthite horizon of precipitated iron lying at the top of the groundwater table in subtropical to tropical climates.

Groundwater podzol   A soil in cool, humid regions that features an iron hardpan immediately below a spodic horizon and that results from mineral precipitation at the top of a groundwater table.

Groundwater table   The upper limit of the zone of water saturation in the soil.

Grumusol   (See *Vertisols*.)

Guyot   (See *Sea mount*.)

Gypsic horizon   A soil horizon more than 6 inches thick that has been enriched by calcium sulphate. It usually lies just below a calcic horizon.

Gypsum   A crystalline mineral of hydrous calcium sulfate. Sometimes precipitated from groundwater in the soils of dry regions.

Gyre   A balanced, circular movement in a

fluid medium. Examples are the circular motions of ocean currents in the major oceans of the world.

Habitat   The effective areal range of action of an organism.

Hachures   Short lines on a map tending in the direction of slope and used to portray relative relief.

Hail   Round or irregular lumps of ice formed by the concentric layering of ice or snow derived from collisions with supercooled drops of water during thunderstorms.

Half-bog soils   A shallow layer of organic soil overlying a gray mineral soil. Usually associated with a swamp-forest type of vegetation. Sometimes also termed *humic gley* soils.

Halophytes   Plants tolerant of high salinity in soils.

Hamada   A dry, rocky, desert plain.

Hanging valley   A tributary valley that enters the main valley at a level well above the floor of the latter.

Hardpan   A collective term for soil horizons that are tight and relatively impermeable, owing to cementation or compaction.

Haystack topography   A land surface consisting of knobby, cone-shaped limestone hills with rounded summits. Such landforms are the result of solution and are found largely in the tropics or subtropics.

Heat   The kinetic energy of molecular motion.

Heath   An association of low, woody plants 1 to 2 feet high, including many belonging to the genus *Erica*, bog mosses, and other hydrophytic plants that are found on flat, poorly drained areas in cool, humid regions.

Heat low (Thermal low)   A low-atmospheric-pressure area caused by abnormal surface heating.

Heat, sensible   The heat energy that is expressed in terms of temperature.

Heat sink   An area in which radiation output exceeds input.

Hematite   The principal ore of iron, usually brick-red in color. Has the chemical formula Fe_2O_3, hence is the stable crystalline sesquioxide of iron.

Herbaceous   Pertaining to plants whose stems do not develop woody tissue.

High plains   Plains that lie at elevations above 2,000 feet. The term is also used specifically to refer to a relatively undissected section of the American Great Plains.

High-sun season   The summer season. The term is used primarily in referring to the tropics, where temperature seasons are not pronounced.

Hills   Areas where local relief is between 200 and 2,000 feet and where slopes of over 5 percent predominate.

Hill station   A high-elevation climatic refuge in the tropics.

Histic epipedon   A largely organic soil horizon lying at or near the surface, less than 12 inches thick when undrained and 18 inches when drained. Normally it is water-saturated for part of the year unless artificially drained.

Histosols   An organic soil order in which deep accumulations of organic material resulting from poor drainage have taken place. See also *Peat* and *Muck*, which are examples of Histosols.

Homolosine projection   An equivalent world map projection which uses the sinusoidal from 0 to 40° latitude and the homolographic between 40° and 90°.

Hook   (See *Spit*.)

Horizon   The intersection of a horizontal plane with the sky. For the use of the term in connection with soils, see *Soil horizon*.

Horn   A steep-sided pyramidal mountain peak resulting from the intersection of three or more cirques.

Hornblende A dark mineral found in basic rocks. A complex iron and calcium aluminum silicate belonging to the amphibole group of minerals.

Horse latitudes The belt of calms, light variable winds, and subsident air that is found near the center of the subtropical anticyclones.

Horst The upfaulted blocks along the side of a graben.

Humid acids Dispersed organic colloids that in solution act like acids, since they attract cations.

Humic gleys (See *Half-bog soils*.)

Humidity Water vapor in the air. *Absolute humidity* is the weight of water vapor per unit of air volume. *Relative humidity* is the ratio between the amount actually held in the air and the amount it could contain at saturation. *Specific humidity* is the mass of water vapor in relation to a given mass of air.

Humus A gelatinous, colloidal mass resulting from the decomposition of organic tissue.

Hurricane A severe tropical storm with winds exceeding 75 miles per hour.

Hydration The absorption of water molecules into a crystal lattice or within a colloidal complex.

Hydrocarbon A chain compound of carbon and hydrogen atoms.

Hydrogen ion The nucleus of a hydrogen atom, or an atom of hydrogen that has lost its outer electron, hence has a net positive charge.

Hydrography A description of distributional patterns of drainage.

Hydrology The scientific study of the hydrologic cycle.

Hydrologic cycle The course taken by water in moving from the oceans to the land via evaporation and precipitation and returning via stream flow. The subsystem of the global energy system that regulates the flow of energy through the heat-exchange property of water.

Hydrophyte A plant that thrives on an abundance of water.

Hydrosere The succession of plants that tend to fill in a body of water.

Hydrosphere The zone of liquid water at or near the earth surface.

Hydroxyl ion A union of oxygen with a hydrogen atom that contains an extra electron; hence, the molecule has a net negative charge (OH^-).

Hygrometer An instrument used to measure relative humidity.

Hygrophyte A plant that thrives in a humid environment.

Hygroscopic particles Particles of water-soluble salts in the atmosphere that attract water condensation below vapor pressure saturation.

Hypsographic curve The curve that relates area to elevation above and below sea level.

Ice cap An ice mass that moves outward fairly evenly from a single center.

Ice fog A fog composed of ice particles resulting from the direct sublimation of water from the vapor to the solid state in the air. It requires temperatures below $-40°F$.

Ice rampart ridge A ridge of soil or stones formed by lateral compression caused by ice expansion on frozen inland lakes.

Ice sheet An ice cap that mantles all of the land surface in its path, except the highest mountains, and whose general direction of flow is largely independent of underlying terrain features.

Igneous rock Rock that has solidified from a molten state.

Illite A group of clays (hydrous aluminum silicates) that contain an appreciable amount of potassium. They are sometimes referred to as hydrous micas and have a base-exchange capacity intermediate between that of the montmorillonites and kaolinites.

Illuvial horizon   A soil horizon that features soil material washed into it, usually from above. Also termed *B horizon*.

Illuviation   The process by which soil material is washed into a soil horizon, usually from above.

Immature soil   A young soil lacking well-developed diagnostic horizons.

Inceptisols   A mineral-soil order which includes soils that are usually moist, where there has been some alteration of parent material, but where there has been no distinct accumulation of soil-forming products, as in a spodic, argillic, natric, or oxic horizon.

Induration   The process of hardening rocks through the cementing of detrital material by heating or recrystallization.

Infrared radiation   Radiation having wavelengths slightly longer than those within the visible light range.

Inorganic fertilization   The process of increasing soil fertility by adding inorganic fertilizers.

Inselberg   An erosional hill or remnant standing above the general level of an erosional plain, usually in the tropics. It is sometimes also referred to as a *bornhardt*.

Insolation   Shortwave solar radiation received at the earth surface.

Instability, general   An atmospheric state in which the air, if started either upward or downward, tends to accelerate away from its source position.

Instability, absolute   Air instability that results from a lapse rate greater than the adiabatic rate.

Instability, conditional   An instability resulting from the addition of latent heat.

Instability, mechanical   An instability which requires no initial impetus. Usually a lapse rate above 19°F per 1,000 feet.

Interface   A boundary between solid and gas, solid and liquid, or solid and solid, liquid and gas, etc., in a colloidal system.

Interglacial stages   Intervals separating glacial advances during a glacial period.

International date line   The irregular line at or near the 180° meridian, at which there is an arbitrary change in time of one calendar day.

Interrupted projections   Projections that interrupt the continuity of the meridians and parallels.

Intertropical convergence zone (ITC)   The general zone of convergence between the northern and southern hemisphere trade winds.

Intertropical front   A front that may develop within the zone of intertropical convergence.

Intrazonal soils   Soils whose dominant characteristics are derived largely from local topographic conditions of slope, drainage, and parent material.

Intrusive rocks   Igneous rocks that have cooled slowly beneath the earth surface and consequently have a coarsely crystalline texture.

Ion   An atom that has either an excess or a deficiency of electrons, thus bearing a net electrical charge.

Ionic balance   The tendency for certain fluids and gases to maintain a given proportion of different types of ions—as, for example, the proportion of sodium ions in sea water.

Ionization   The process of creating ions from atoms through the addition or removal of electrons.

Ionosphere   A thick zone in the outer atmosphere in which the atmospheric gases become ionized by incoming solar energy.

Variations in the intensity of ionization separate this zone into distinct layers, labeled the *D, E, F,* and *F1* layers.

Irreversible colloid A colloidal system which is not reversible once it has undergone a change in physical state. Curdled milk is an irreversible colloidal system.

Isarithm A line connecting points of equal numerical quantity.

Isobar A line connecting points of equal atmospheric pressure.

Isobath A line connecting points of equal depth below sea level.

Isohyet A line connecting points of equal precipitation.

Isoline A line connecting points of equal value.

Isopleth A line connecting points of equal ratio.

Isostasy The tendency to maintain an even gravity balance at the earth surface despite denudation and deposition.

Isostatic movement An adjustment of the earth crust upward or downward following the removal or addition of weight on a section of the surface.

Isotherm A line connecting points of equal temperature.

Isothermograph A diagram that indicates average hourly temperatures by months throughout the year.

Jet streams Unusually swift geostrophic air streams in the upper troposphere, usually found near breaks in the tropopause. Most of them are westerly, but a weak easterly jet is sometimes found in low latitudes at high elevations.

Joints A geometric pattern of more or less vertical cracks or fractures in rocks caused by regional rock stresses.

Kame A mound, variously shaped, composed of stratified glacial drift. A *kame terrace* is a terrace of stratified fluvioglacial material formed between an ice mass and the side walls of a valley.

Kaolinite A variety of clay consisting mainly of aluminum silicates. It is usually the product of tropical weathering and represents other clays which have undergone desilication. It has a low cation-exchange capacity and a low retention of water and nutrients.

Karst topography An irregular land surface produced mainly by solution in a limestone upland.

Kelvin scale A scale of temperature having intervals equivalent to those of the centigrade (Celsius) scale, but beginning at absolute zero.

Kettle A depression in stratified glacial drift, resulting principally from the melting of included ice blocks.

Khamsin A sudden continental tropical air mass that blows into the eastern Mediterranean Basin from the deserts to the south, mainly during the spring months. Known also as a *ghibli, samiel,* or *leveche.* The *sirocco* of the western Mediterranean is similar but stronger and warmer, owing to the descent from the high plains of North Africa.

Kinetic energy The energy of motion.

Knot (See *Nautical mile.*)

Laccolith An intrusive rock mass injected between rock layers and doming the overlying rocks.

Lacustrine Of or pertaining to lakes.

Lagoon A former portion of the sea which has been sheltered from direct wave attack by some depositional feature, such as a reef or barrier beach.

Land-and-sea breeze A nightly landward wind and daily seaward wind frequently encountered along ocean coastlines.

Land creep The slow movement of the soil mantle under the influence of mass wastage.

Landform A recognizable feature of the earth surface, with characteristic form and composition.

Landscape The sum total of the features that give a topographic area or region its distinctiveness.

Landslide The sudden, rapid movement of a large mass of soil and rock debris down a steep slope.

Land slump The sudden downward movement of a section of the land surface.

Lapse rate The rate of decrease in free-air temperature encountered when passing from a lower to a higher elevation. It averages about 3.56°F per 1,000 feet.

Larch A deciduous, needleleaf, coniferous tree belonging to the genus *Larix*.

Latent heat The amount of heat energy necessary to produce a change in phase—for example, from a solid to a liquid or from a liquid to a gas. The former is termed the *latent heat of fusion*; the latter, the *latent heat of vaporization*.

Laterite (See *Plinthite*.)

Latitude The angular distance measured north and south of the equator.

Latosol (See *Oxisols*.)

Lattice The geometric arrangement of atoms in a crystalline solid.

Lava Molten rock at the surface of the earth.

Leaching The removal of soluble material from a soil through the action of aqueous solutions.

Legume A pea or beanlike plant whose roots contain nodular colonies of nitrogen-fixing bacteria. A member of the order *Leguminosiae*.

Lichen A mutually supporting plant combination of a fungus and an alga.

Lightning A flash of light produced in the air by the flow of electrons across a gap between two oppositely charged positions.

Light-year The distance traveled by light in a year at the rate of 186,000 miles per second.

Limestone An amorphous sedimentary rock composed largely of calcium carbonate.

Limonite A hydrous iron oxide, usually yellowish brown in color; has the formula $Fe_2O_3 \cdot 2H_2O$. Sometimes termed *bog iron*.

Linear scale The scale of a map that expresses the comparative length of distance on a map with distance on the earth.

Line squall A sudden and violent windstorm of short duration that often precedes large thunderstorms.

Lithologic Pertaining to rock characteristics.

Lithosols Soils which are largely unweathered rock fragments. Also know as *regosols*.

Lithosphere The solid portion of the earth, as contrasted with the atmosphere and hydrosphere.

Littoral Pertaining to the shore zone.

Llanos The savanna grasslands located in the Orinoco drainage basin of Venezuela.

Local relief The general difference in elevation within an area between the summits of hills or mountains and the adjacent valleys or lowlands.

Loess A nonstratified deposit of yellowish-brown silt, having a uniform texture and frequently exhibiting a vertical, columnar structure. Most loess is believed to have been deposited by wind at some time or other.

Longitude The angular distance measured east and west of a prime meridian.

Longitudinal profile The profile that shows the comparative elevation of points along a stream course.

Long waves Parallel sinuosities, or waves, occurring in the upper-air isobars and having exceptionally long axes.

Loran (Long-range radar) A directional device utilizing long-range radar beams.

Low (See *Cyclone*.)

Low-sun season The winter season in the low latitudes, where seasonal temperature differences are not pronounced.

Loxodrome A line connecting points that have similar bearing. Such a line is a chord of a great circle. Short sections of loxo-dromes are termed *rhumb lines*.

Magma A mass of molten rock beneath the earth surface.

Magnetic declination The deviation of the magnetic compass from true north.

Mallee A low scrub woodland in Australia dominated by Eucalyptus and Acacia.

Mangrove A tropical tree or shrub, belonging to the genus *Rhizophora*, that is adapted for growth in the saline waters of muddy coastlines.

Mantle The main solid portion of the earth, lying below the crust and above the fluid core.

Map projection The orderly or systematic arrangement of the earth grid on a plane surface.

Map scale The ratio between the total size of a map and the area on the earth it represents.

Maquis A scrub woodland with low trees 3 to 20 feet high. The trees are usually broadleaf, sclerophyllous evergreens. Known also as *chaparral* or mesquite.

Marble Metamorphic limestone in which the calcium carbonate has crystallized, owing to heat and pressure.

Marl A white-to-gray calcareous clay.

Marsh An essentially treeless area of grass, sedges, reeds, or low shrubs, in which water stands slightly above or below the surface.

Mass wastage The mass movement of material down a slope by the direct action of gravity.

Mature soil A well-drained soil which has clearly marked characteristics produced by soil-forming processes and which is in relative equilibrium with its environment.

Meadow A low, herbaceous plant association consisting of grasses and other woody flowering plants, found in humid regions. Meadows are especially common above the timber line of mountain regions, where they are known as *alpine meadows*.

Meander A loop in a river as it winds back and forth across its floodplain.

Mean solar day The average period of earth rotation.

Mean solar time The average at any one place of the fluctuations in time during the year that the sun reaches the meridian or the highest point in its apparent course across the sky.

Mechanical instability (See *Instability, mechanical*.)

Mechanical weathering (See *Weathering*.)

Meridian An imaginary line connecting points of equal longitude; half of any great circle that passes through the poles. Also, an imaginary line in the sky bisecting it from north to south and crossing the zenith.

Meridional Pertaining to a north-south or a south-north orientation.

Mesa A broad, flat-topped, erosional remnant flanked on at least one side by a steep cliff and associated talus.

Mesopause The temperature discontinuity separating the mesosphere above from the stratosphere below.

Mesophyte A plant with moderate water preferences.

Mesosphere The zone in the atmosphere above the stratosphere which is marked by a fairly steep lapse rate.

Mesozoic One of the great eras of geologic time; including the Triassic, Jurassic and Cretaceous periods.

Mesquite (See *Maquis*.)

Metabolic Pertaining to the synthesis or the destruction of protoplasm in living cells.

Metamorphic rock A rock that has been altered in form when in a solid state, mainly as the result of pressure, heating, or a change in chemical environment.

Meteorology The science of weather.

Mica   A group or family of silicate minerals having a tendency to split into thin sheets.

Micelle   A soil colloidal aggregate that features a net negative charge and various absorbed positive ions (cations).

Mid-oceanic ridge   The broad, high mountain range that extends through the middle of the ocean basins except that of the North Pacific.

Migratory agriculture   A type of agriculture which involves progressive land abandonment following a decline in fertility.

Millibar   A unit of force equal to one-thousandth of a *bar*, or 1,000 *dynes*. A millibar also is equivalent to 0.014 pound per square inch.

Mineral   A naturally occurring aggregate of inorganic substances that has a definite chemical composition and more or less characteristic physical properties.

Mirage   An optical image at or near the horizon resulting from the refraction of light. In a *superior mirage*, refraction due to rapidly decreasing air density with altitude causes an object below the horizon to be seen above the horizon and to appear inverted. In an *inferior mirage,* the density pattern is reversed; the object again appears to be inverted, but it is seen below its true position.

Mistral   A cold northern wind that descends into the western Mediterranean basin from higher elevations to the north. A similar wind in the Adriatic Sea is known as a *bora.*

Mogote   A limestone knob.

Mohorovicic Discontinuity Line (Moho)   A rock discontinuity separating the crust from the mantle. It generally lies about 12 miles below the continental surface and 5 miles below the general ocean floor.

Mollic epipedon   A dark, mellow, surface mineral layer in a soil which has a high percentage of base saturation (+50) and between 1 and 17.5 percent of organic matter.

Mollisols   An order of mineral soils that feature a mollic epipedon and result from the partial decomposition and accumulation of relatively large amounts of organic material in the presence of abundant calcium.

Monadnock   A residual hill or mountain rising above a plain that has been beveled by erosion.

Monsoon   A seasonal wind reversal; usually a land-and-sea wind in which the change in direction is sudden.

Monte   A region of low xerophytic shrubs bordering the Argentine pampas.

Montmorillonite   A group of clays with an unusually loose crystal lattice permitting easy absorption of water and cations. They represent early stages in the decomposition of clays and have an extremely high cation-exchange capacity.

Moraine, end   An irregular ridge of morainic material deposited at the outer end of a glacier.

Moraine, ground   A fairly even layer of morainic material deposited during gradual recession of an ice sheet.

Moraine, interlobate   A moraine deposited between the protuberances, or lobes, of a glacier.

Moraine, lateral   A morainic ridge deposited along the sides of a mountain glacier.

Moraine, medial   A morainic ridge deposited in the midportion of a mountain valley. Results from the joining of lateral moraines.

Moraine, recessional   An end moraine marking a temporary halt in the recession of an ice mass.

Moraine, terminal   An end moraine marking the farthest extent of glacial movement.

Morphology   Pertaining to the form and structure of things.

Moss forest   A low, humid forest characteristic of the cloud zone on high mountain slopes in the tropics. It features an unusually thick growth of mosses and other epiphytes.

Mottling   The variegated coloration of soils that undergo periodic oxidation and reduction of iron compounds.

Mountain-and-valley breeze   Nightly downvalley and daily upvalley winds frequently encountered in mountain valleys.

Mountainous   An area with a local relief greater than 2,000 feet and narrow summits.

Muck   The granular or finely divided black mass of partially decayed organic material found in bogs. Usually slightly acidic or weakly alkaline in reaction.

Mud cracks   The roughly polygonal cracks that develop in fine-textured muds upon drying.

Mud flow   A large stream of fine-textured soil material which, when sufficiently water saturated, becomes fluid and moves downslope. A form of mass wastage.

Muskeg   A type of acid bog found in northern regions. Represents an intermediate stage in a hydrarch succession.

Mutation   A sudden alteration in chromosome structure (the "blueprint" of organic development), which is hereditary.

Nappe   A series of flat, parallel, overturned folds.

Natric horizon   A variety of argillic (illuviated clay) horizon that has a distinctive columnar or prismatic structure and that shows somewhere in the horizon some appreciable amount (+15 percent) of base saturation with sodium.

Natural levee   A ridge of alluvium that adjoins the river channel on a floodplain.

Nautical mile   The mean length of one minute of earth arc, or 6,076.1 feet. When used in measuring speed, it is termed a knot.

Neap tide   (See Tide.)

Niche   The contributary role played by a living organism within the community in which it lives.

Nomograph   A graphic chart showing statistical correlations between two or more sets of data.

North celestial pole   The zenith point in the heavens above the North Pole.

Norther   A severe winter windstorm or blizzard.

Notched cliff   An undercut cliff resulting from wave erosion along a rocky coastline.

Nuclear fission and fusion   Nuclear fission is the process of splitting the nuclei of atoms. Nuclear fusion is a thermonuclear reaction in which energy is liberated by the fusion of neutrons or nuclei of hydrogen isotopes.

Nutrients   Food substances necessary for the sustenance of living organisms. Plant nutrients are soluble plant foods derived mainly from the soil solution.

Oblate spheroid   Earth shaped, or a near-sphere that is slightly flattened.

Obsequent stream   A stream that flows in a direction opposite to the slope of the original land surface.

Obsidian   Pertaining to a dark volcanic glass.

Occluded front   The front formed when a cold front overtakes a warm front or another cold front. The process is known as an occlusion. If the intruding mass is colder than the original mass, a cold-front occlusion develops. If the newer mass is warmer, a warm-front occlusion is formed.

Occlusion   An overtaking of one front by another.

Ocean basin floor   The general level of the ocean basins, roughly between 14,000 and 18,000 feet below sea level.

Ocean trenches   (See Abyssal deeps.)

Ochric epipedon   A surface soil horizon that features light-colored soil material, usually hard and massive when dry.

Offshore beach   (See *Barrier beach.*)

O horizon   The organic zone that overlies the dominantly mineral or inorganic portion of the soil profile.

Olivine   A light-green magnesium and iron silicate commonly found in basic rocks.

Open systems   An energy system into which energy flows at one or more points and from which an equivalent amount of energy leaves after performing work. The earth-atmosphere system is an open system.

Orbit   The path of a body in circular or elliptical revolution about a focal point.

Orogenic   Of or pertaining to major warpings of the earth crust.

Orogeny   A mountain-building period.

Orographic precipitation   Precipitation in which the initial lifting of air is caused by a topographic obstacle.

Orthoclase   A pink potash feldspar, one of the most common mineral constituents of granite.

Orthographic projection   A geometric map projection in which the position of the projecting light source is at infinity; hence, the projection beams are parallel.

Ortstein   A durable, illuviated layer of ferric oxide and humus that collects near the top of a water-saturated, sandy soil in cool, humid regions.

Outwash   (See *Fluvioglacial outwash.*)

Outwash plain   A plain resulting from the deposition of fluvioglacial sands and gravels. A *pitted outwash plain* is one that contains many irregular depressions.

Overturned fold   An almost horizontal fold.

Oxbow lake   A crescentic lake or slough found on a floodplain; marks the former channel of a meandering river.

Oxic horizon   A diagnostic soil horizon at least 12 inches thick, containing mixtures of hydrated iron and aluminum oxides, often in crystalline form, low-silica clays, and highly insoluble residues, such as quartz sand.

Oxidation   The addition of one or more oxygen atoms to a molecule.

Oxisols   Soils resulting from the latosolization process, which contain an oxic horizon.

Ozone   A triatomic molecule of oxygen.

Ozonosphere   A zone in the upper atmosphere with an unusually high concentration of ozone. It is generally between 15 and 35 miles above the earth surface and is sometimes referred to as the *G layer.*

Paleozoic   An era of geologic time, from late Precambrian time through the Permian period.

Palynology   The science of dating sediments from plant pollen analysis.

Pampa   A prairie grassland of Argentina.

Parallel   An imaginary line connecting points of equal latitude.

Parent material   The unconsolidated but weathered portion of the soil profile below the level of normal biological activity. See also *C horizon.*

Peat   A fibrous, brown, acidic mass of partially decayed plant remains found in bogs.

Pebble beach   (See *Shingle beach.*)

Pedalfer   A humid soil which contains a higher percentage of iron and aluminum oxides in the subsoil than is found in the parent material. Such a soil has no accumulated lime.

Pediment   A gently sloping erosion rock plain bordering mountains in dry regions. Usually veneered with sand and gravel.

Pediplain   (See *Pediment.*)

Pediplanation   The process of pediment formation.

Pedocal A soil that contains a clearly identifiable calcic (lime) horizon.

Pedology The science of soil formation.

Pedon The minimum areal or mapping unit within which soil comparisons can be made.

Peneplain A beveled, flat-to-undulating surface of considerable extent, resulting from erosion.

Penman equation A theoretical equation for determining potential evapotranspiration.

Peptization (See *Colloidal soil dispersion.*)

Percentage base saturation The percentage of the total exchange capacity of soil colloids that is occupied by exchangeable bases. Since the remainder is made up largely of hydrogen ions, there is an inverse relationship between percentage base saturation and soil acidity.

Peridotite A course-textured, basic igneous rock consisting largely of olivine and pyroxene.

Periglacial A climatic condition which features frequent alternation between freezing and thawing. Usually also is characterized by the presence of *permafrost.*

Perihelion The shortest distance between the sun and the earth during the latter's annual orbit. It occurs in July.

Permafrost The zone of permanently frozen ground below the surface.

Permeability The property of rock, sediments, or soils that permits liquids to pass through them.

Petrocalcic horizon (See *Calcic horizon.*)

pH An index of soil acidity or alkalinity; specifically, the logarithm of the reciprocal of the hydrogen-ion concentration.

Phase A uniform physical state of matter within a variable system (e.g., the three phases of water).

Phase equilibrium The tendency for a solid, gas, or liquid to remain in the same phase within a given temperature range and at a given pressure.

Photo periodism The periodic changes in plant form that are the result of changes in availability of light.

Photosynthesis The manufacture of starches and sugars by plants, utilizing air, water, sunlight, and chlorophyll.

Phototropism The ability of some plants to turn their leaves or flowers toward the sun.

Physiographic diagram A stylized portrayal of surface-relief forms through line sketches.

Piedmont alluvial plain (See *Bajada.*)

Plaggen epipedon A man-made epipedon more than 20 inches thick, produced by long-continued manuring. It differs from the anthropic epipedon in having a lower content of phosphoric acid, less mellow qualities, and more artifacts.

Plain An extensive area characterized by a local relief of less then 200 feet and slopes of generally less than 5 percent.

Planimetric map A topographic map without the representation of local relief.

Planosol A clay hardpan soil resulting from the poor drainage associated with shallow soils in areas without appreciable relief.

Plant associations Localized plant communities characterized by one or more dominant species of plant forms.

Plant community Any ecological (interrelated as to site) grouping of plant forms.

Plant formation The major global plant communities, generally considered to be the forests, grasslands, desert shrubs, and tundra.

Plant indicators Plants that can be used as evidence for a specific associated environmental characteristic.

Plant succession The sequence of plant communities that evolves on a bare surface and proceeds toward the climax stage.

Plateau A broad plain that drops to lower elevations on at least three sides.

Playa A shallow enclosed desert basin which collects water following a rain. In northern Africa, known also as a *shott*.

Pleistocene The last geological epoch, characterized by continental glaciation. Approximately 1 million years in duration.

Plinthite A soil horizon high in iron and sesquioxides, low in humus, containing low-silica clays as red mottles. With successive drying, the iron sesquioxide turns to irreversible ironstone. Sometimes termed *laterite*.

Plucking The quarrying action of glaciers in removing blocks of fractured rock by ice wedging.

Podzol A group of acid soils featuring an ash-gray *A2* horizon. Included among the Spodosols in the Seventh Approximation.

Podzolic Soils which evolve under leaching by acid soil solutions (podzolization).

Podzolization A soil-forming process resulting from leaching by highly acid soil solutions. Found mainly in cool, humid regions.

Polar front A line of discontinuity at the surface between air from polar sources and air from tropical sources.

Polar highs High-atmospheric-pressure centers in high latitudes, believed to be thermally induced, at least in part.

Polje A large, flat-bottomed basin formed as a result of the coalescence of sinkholes in a limestone region. Also known as *uvala*.

Porosity The degree to which the total volume of a rock, sediment, or soil is made up of empty space.

Porphyry An igneous rock with large crystals set in a finely textured groundmass.

Potential energy The energy that exists by virtue of its position with respect to gravitation.

Potential evapotranspiration The maximum amount of water that could be evaporated or transpired from a given area, provided plants could obtain as much water as they needed.

Prairie A humid-climate grassy vegetation containing a large percentage of herbaceous plants.

Prairie soils Soils of prairie regions that feature a thick mollic epipedon and without a calcic horizon. Also termed *brunizems* and *udolls*.

Precambrian A collective term for all of earth history prior to the Paleozoic era or Cambrian period.

Precession of the equinoxes The slow annual variation in the position of the sun within the zodiac at a particular time of year. The completion of the zodiacal circuit requires about 28,500 years.

Pressure gradient The slope of pressure change. In the atmosphere, it is at right angles to the isobars.

Prevailing westerlies The general prevailing winds on the polar side of the subtropical anticyclones.

Primary forest A forest consisting mainly of climax or subclimax species.

Primary mineral or rock Materials that have remained unchanged from the time they were formed out of molten rock.

Prime meridian An arbitrary meridian selected as the base line from which to measure longitude. The meridian conventionally used for this purpose is the one that passes through Greenwich, England.

Process The expenditure of energy to produce change.

Profile The shape of a surface along a line and in vertical section. A *longitudinal profile* of a stream valley is its profile along its length. A *transverse profile* of a valley is its profile along a line at right angles to the valley. As applied to soils, see *Soil profile*.

Pumice An extremely lightweight, acidic, porous, extrusive igneous rock.

Pyroclastic Fragmental material ejected from an explosive volcano, including ash and cinders.

Pyroxene A dark-colored ferromagnesium silicate mineral having associated calcium and a common constituent of basic igneous rocks.

Quartz The crystalline mineral of silicon dioxide (SiO_2). Also, a general term for a variety of noncrystalline minerals having the same chemical composition as quartz.

Quartzite A metamorphic rock composed largely of quartz. Usually metamorphosed sandstone.

Radar Directional shortwave radio impulses.

Radial drainage A radial drainage pattern such as develops on the flanks of a structural dome.

Radiation (*radiant energy*) A general term for radiant energy.

Radiation (lowland) fog Fog produced by air-drainage chilling.

Radioactivity The gradual breakdown of atomic nuclei through fission in certain elements.

Radiogenic energy The energy released as the result of atomic nuclear fission.

Radiosonde The automatic radio transmission of weather data from airborne instrument packages; usually carried by balloon.

Rain Drops of liquid water falling from the atmosphere.

Rainfall effectiveness The degree to which precipitation is made available for plant growth.

Rain forest A mesophytic forest association of tall trees, mainly of broadleaf, evergreen species.

Rain shadow A dry zone on the lee side of a topographic obstacle, usually a mountain range.

Reduction The removal of one or more oxygen atoms from a molecule.

Refraction The bending of light as it passes from one medium into another having a different density.

Region An area that exhibits a certain degree of likeness or homogeneity in contrast to the surrounding territory.

Regolith A collective term for the unconsolidated inorganic material mantling the earth surface.

Relative humidity (See *Humidity*.)

Relict plants Plants which have survived an environmental change and which were members of earlier climax communities.

Relative relief The general collective unevenness of the land surface.

Relief inversion A land surface condition resulting from erosion where a previously low structural position has become a high topographical position (e.g., a synclinal mountain).

Rendzina Soils whose properties are due mainly to calcareous parent material. They may be black, red or white.

Representative fraction The fraction that presents the linear ratio between a map and the area it represents.

Residual soils Soils that have developed in place from the underlying rock, without having been transported.

Respiration The inhalation of oxygen for metabolic purposes and the exhalation of waste products.

Reversible colloid Colloids that may alternate between a flocculated (coagulated) and a peptized (dispersed) state. *Irreversible colloids* are those whose change in state is only in one direction.

Revolution and rotation The motions of the earth spheroid. *Revolution* is the movement of the earth around the sun; *rotation* is the turning of the earth about its axis.

R horizon The consolidated bedrock that lies beneath the soil.

Rhumb line or Loxodrome (See *Loxodrome*.)

Ridge-and-valley topography A succession of parallel ridges and valleys resulting from the differential erosion of highly folded rock layers of varying resistances.

Rift An elongated structural valley formed by normal faulting.

Rilling The erosion of soil through the development of shallow trenches.

Rime A coating of ice crystals on solid objects resulting from contact with supercooled fog droplets.

Rock decomposition The chemical weathering of rocks.

Rock disintegration The mechanical weathering of rocks.

Rock glacier A stream of rocks in periglacial areas which slowly moves downslope through alternate freezing and thawing of included snow, ice and meltwater.

Rotation (See *Revolution*.)

Runoff The discharge of water from surface streams.

Salic horizon A soil horizon more than 6 inches thick that has been enriched by water-soluble salts (other than gypsum).

Salina A salt-encrusted flat, resulting from the evaporation of a playa lake.

Saline soils Soils that contain an excess of water-soluble salts but are not excessively alkaline. The pH range is roughly from 7.3 to 8.5.

Salinization The process of salt accumulation resulting from poor drainage in dry regions.

Salt The product, other than water, of the reaction between a base and an acid. Also, often refers specifically to sodium chloride.

Saltation The movement of fragmental material in a series of irregular leaps above the surface, such as along the bed of a stream or in a strong wind.

Sand Inorganic particles having diameters between .25 and 2 millimeters.

Sandstone A sedimentary rock formed by the cementation of sand.

Saturation vapor pressure That part of the atmospheric pressure composed of the mass of water vapor molecules at saturation, or when the number of water molecules passing into the air equals the number leaving it.

Savanna A tropical grassland, usually with scattered trees and shrubs.

Schist A foliated (having fine, leaflike layers) metamorphic rock, containing micalike minerals.

Sclerophyll A woody, xerophytic plant with shiny, hard-surfaced, and often waxy, narrow leaves.

Scree (See *Talus*.)

Scrub woodland A plant association dominated by low trees 20 to 40 feet high and without a continuous canopy overhead.

Sea arch A rock arch or natural bridge along a cliffed coast formed as the result of wave erosion.

Sea cave A cave excavated in rock as the result of wave erosion.

Sea mount A steep, volcanic pinnacle or peak that rises from the ocean floor. Sea mounts truncated by wave erosion are termed *guyots*.

Sebkha An Arabic term for a salt-encrusted flat. See *Salina*.

Secondary forest An immature stage of forest regrowth.

Sedges Marsh grasses characterized by leaf blades that are triangular in cross section.

Sedimentary platform A broad area of continental plains underlain by relatively thin strata of sedimentary rock.

Sedimentary rock A rock resulting from the compaction, cementation, or induration of depositional sediments.

Seismology The science of earthquakes.

Selective logging A lumbering procedure whereby only trees of a selected size are removed from a forest.

Selva The tropical rain forest, composed of broadleaf evergreen trees.

Semideciduous forest A forest in which some but not all of the trees shed their leaves seasonally.

Sensible heat Heat energy in the air that is sensed and measurable as temperature.

Sensible temperature The temperature that is felt by the human body. It is influenced by wind and humidity as well as by sensible heat.

Sere A complete series of successional stages leading to the establishment of a climax vegetation.

Sesquioxides Metallic oxides in which the ratio of the metal to oxygen is $1:1\frac{1}{2}$. Iron and aluminum oxides are the most common soil sesquioxides.

Seventh approximation A worldwide classification of soils based largely on morphology and currently in use by the U.S. Department of Agriculture.

Sextant An instrument used to measure the angle of elevation of heavenly bodies, such as the sun or the stars.

Shale A layered sedimentary rock in which the constituent particles are largely of clay size. Sometimes termed *mudstone*.

Shield volcano A broad shieldlike or domelike volcano resulting from successive outpourings of highly viscous, basaltic lavas; usually several tens or hundreds of square miles in extent.

Shingle beach A beach composed of flattened, rounded stones.

Shoran (Short-range radar) A directional device utilizing short-range radar beams.

Shott The local term for *playa lake* in North Africa.

Sial Acidic, continental, igneous rocks; light in color and high in silica.

Sidereal time (Star time) Time that is based on the apparent motions of the stars in their courses across the heavens.

Sierozems Gray desert soils.

Silica Silicon dioxide (SiO_2). It may exist in a pure crystal lattice form (*quartz*), as an amorphous colloidal mass (as in *silica gels* or *chert*), or as *combined silica,* in which the silica occurs as separate layers in a crystal lattice. Compounds containing combined silica are termed *silicates*. Rocks high in silica of any kind are termed *siliceous*.

Sill An intrusive rock mass of more or less uniform thickness that is injected parallel to the enclosing rock strata. It is relatively thin, in comparison to its lateral extent.

Silt Inorganic particles having diameters between .05 and .002 millimeter.

Siltstone (See *Graywacke*.)

Sima Dark, basic igneous rocks.

Sinkhole A funnel-shaped depression in the land surface caused by solution of limestone by underground water. Also known as a *doline*.

Sirocco A dry, hot wind entering the Mediterranean Basin from the Sahara. Known also as *khamsin, leveche,* and *samiel*.

Slate A fine-grained metamorphic rock that tends to split in one direction.

Sleet Frozen or partly frozen raindrops that lack the concentric layering of hail.

Slipoff slope The gradual side of an asymmetrical valley or stream channel.

Slumping The sudden mass movement downslope of soil and regolith when saturated with water.

Slump slide A sudden slumping of fine-textured material on a steep slope, often with a curling or rotational movement. Also termed *rotational slip*.

Smog A dense mixture of water droplets and solid particulates in suspension in the air at the surface.

Snow Crystalline water that has sublimated in free air around a nucleus directly from a vapor phase.

Snow line The limit, or height, beyond which snow is present throughout the year.

Soil A mixture of fragmental materials that will support a vegetation cover and that usually is a three-dimensional natural body exhibiting some vertical differentiation with depth.

Soil creep The slow, almost imperceptible movement of soil downslope by mass wasting.

Soil horizon A depth zone in soils that exhibits a certain degree of recognizable homogeneity in properties and contrasts with that in zones above and below.

Soil order A primary global soil grouping, the broadest classification in the Seventh Approximation.

Soil phase A subdivision of soil-mapping types based on a unique surface property other than texture (e.g., steepness, stoniness).

Soil polygons Roughly polygonal macro-structures in soils produced by alternate freezing and thawing above permafrost.

Soil profile The genetically related vertical differences in a soil body, taken as a unit.

Soil series A soil-mapping unit which distinguishes soil-profile characteristics.

Soil solution The water present in the soil, together with its dissolved salts.

Soil structure The arrangement and grouping of soil particles.

Soil texture The relative proportions of different sizes of soil particles.

Soil type Subdivisions of soil series that are based on surface texture.

Solar constant The number of calories received per square centimeter per minute on a standard flat surface held perpendicular to the sun.

Solar declination The latitudinal position at which the sun is vertical or at the zenith.

Solar radiation Radiant energy from the sun.

Solar system The sun and its associated orbital bodies or those that have fixed orbits around it.

Solifluction The movement of soil material downslope under the influence of frost action.

Solodi Soils in which the desalinization of saline soils has been complete. They exhibit many features common to podzols.

Solonchak Soils that contain observable accumulations of water-soluble salt crystals.

Solonetz Black alkali soils resulting from the recent desalinization of solonchak soils. The black color is the result of small amounts of humus combined with sodium and potassium carbonates.

Solstice The longest or shortest day of the year. The two solstices occur on or about June 21 and December 21, respectively.

Solum The total mass of the soil horizons, from the surface to the parent material, or down to the C horizon.

Solute A substance held in solution.

Spatter cone A small cone of pyroclastic volcanic ejecta.

Specific gravity A measurement of material density or ratio between mass and volume based on water = 1.

Specific heat The number of calories needed to raise one cubic centimeter of a substance one degree centigrade at sea-level atmospheric pressure.

Specific humidity The weight of water vapor in a given weight of air.

Sphagnum A highly acidic fibrous mass, (*Sphagnum*, sp.) commonly found in northern bogs.

Spit A tongue of sand or gravel extending out from a promontory along a coastline. A curved spit is termed a *hook*.

Spodic horizon A soil horizon of humid climates in which there has been a precipitation of organic material, plus aluminum and/or iron sesquioxides in a noncrystalline (amorphous) form.

Spodosols An order of mineral soils that have a spodic horizon resulting from the accumulation of free sesquioxides removed from an eluviated horizon.

Spring tides (See *Tides.*)

Stability As applied to air, a tendency to resist upward motion. *Absolute stability* occurs when the lapse rate is less than the dry adiabatic rate.

Stack A column of rock found as an erosional remnant along rocky coastlines.

Stage as used in connection with processes of various kinds, the degree to which progression is made toward a steady state. *Youth* and *maturity* are adjectives that refer to stage.

Stalactite Stone icicles of calcium carbonate that hang from the roofs of limestone caverns.

Stalagmite A column of dripstone extending upward from the floor of a limestone cavern.

Standard meridian Any meridian that is selected as a standard on which to base a map grid.

Standard parallel Any parallel of latitude that is selected as a standard on which to base a map grid.

Standard time Mean solar time at the control meridian for time zones.

Standard time zones Longitudinal zones approximately 15° wide which uniformly base local time on the mean solar time at the central or control meridian.

Statute mile An English linear measurement consisting of 8 *furlongs,* or 5,280 feet.

Steppe A mid-latitude, short-grass vegetation cover that mantles the ground with a fairly continuous sod.

Stereographic projection An azimuthal geometric map projection in which the point of a projecting light source for the grid is placed at the antipode of the point of tangency.

Stomata Breathing pores on leaves and stems of plants.

Stone rivers Linear belts of boulders following shallow drainage ways in arctic regions.

Stone stripes Alternate strips of coarse rock fragments and finer material found along hillsides and aligned at right angles to the contour. Periglacial phenomena.

Strata Distinct rock layers, or beds.

Stratopause The boundary between the stratosphere and mesosphere, or the upper boundary of the stratosphere.

Stratosphere An isothermal zone of the atmosphere lying above the tropopause. It is about 10 to 15 miles in thickness and is characterized by the absence of a distinct lapse rate.

Stratus Pertaining to sheet- or layerlike cloud forms.

Stream load The total amount of material carried in suspension and solution by a stream.

Stream order The ranking of a stream according to the grouping of tributaries. Originally, according to Horton, first-order streams have no tributaries, second-order streams have first-order tributaries, etc.

Striations Fine parallel scratches or grooves found on the surfaces of rocks.

Strike The bearing of the intersection of a fault, fracture or dipping rock strata with the horizontal.

Strike-slip fault A fault whose displacement is roughly parallel to the strike.

Strip cropping Cultivation of alternate strips of land, leaving intervening spaces fallow to conserve moisture or nutrients.

Structure The arrangement or attitude of the rocks underlying an area.

Subaerial In the open air, as contrasted with *submarine.*

Subarctic A general term for the tundra and *Dfc* climatic regions in North America and Eurasia.

Subclimax community A plant community that represents a stage in a sere or plant succession near the climax.

Subhumid A climate that is transitional between semiarid and humid climates.

Sublimation The passage of a substance directly from a gaseous to a solid phase or vice versa (e.g., water vapor turning into snow or ice crystals, or snow evaporating without melting).

Subpolar lows Centers of low atmospheric pressure located over the ocean areas of high latitudes. These occur only during the winter season in the northern hemisphere.

Subsoil Normally, that portion of the soil below plow depth.

Subsystem A set of interrelated objects, together with their attributes, that are designed to regulate the flow of energy and materials into and out of a system.

Subtropical Pertaining to the subtropics, a zone straddling the outer margin of the tropics generally between lat 25 and 35°.

Subtropical highs or anticyclones Permanent or semipermanent anticyclonic centers of air subsidence and divergence, located roughly between lat 25 and 35°.

Succulence The tendency of some plants to have an unusually high water content in their cells.

Sudd A large area of marshland along the Nile River upstream from Khartoum.

Sun compass A compass used to determine direction, utilizing solar observation.

Sun time Time based on the position of the sun in its apparent course across the heavens.

Supercooling The cooling of drops of liquid water below the freezing point without a change in phase.

Supersaturation The crowding of water-vapor molecules into a given mass of air beyond its normal capacity.

Surface configuration Qualitative properties of the general lay of the land, such as slope, relief, and elevation.

Surf depth The depth below which there is no direct work done by wave action on the bottom.

Swamp An area of tree or bush vegetation, where water stands slightly above or below the surface.

Syenite An igneous intrusive rock composed largely of feldspar.

Symbiosis A state of mutual support between two unlike organisms that are juxtaposed.

Syncline A rock downfold, or a rock fold in which the strata dip downward toward the center, or axis, of the fold.

System A set of interrelated objects, together with their attributes, that are linked together by a flow of energy.

Tableland A broad, flat upland area.

Taiga The northern coniferous forest.

Talus The unconsolidated mass of rock rubble accumulated at the foot of a fall face. Known also as *scree*.

Tarn A small mountain lake located in a cirque.

Tectonic Pertaining to crustal movements of the earth.

Temperature A measurement of the degree of sensible heat present in a substance.

Temperature anomaly A temperature of air or water that appears to be unusually high or low, considering its position.

Temperature gradient The slope or rate of temperature change.

Temperature inversion A sudden change from a decrease in temperature with elevation to an increase in temperature with elevation. Warmer air overlying colder air.

Terminator The line separating daylight

from darkness on the earth or moon. Also referred to as the *circle of illumination*.

Terracette A small terracelike ridge paralleling the slope and caused by solifluction in periglacial regions.

Tertiary The geologic period that immediately preceeded the Pleistocene, or Ice Age.

Theodolite A surveying instrument used to measure vertical and horizontal angles.

Thermodynamic Pertaining to the mechanical movements associated with heating and cooling.

Thermoperiodism Rhythmic patterns of plant growth and development that conform to variations in temperature.

Thermosphere The zone of the atmosphere above the mesosphere in which there is a steep lapse rate.

Thornthwaite equation The empirically derived equation devised by C. W. Thornthwaite which is used to determine potential evapotranspiration.

Thunderstorm Any thermodynamic storm accompanied by thunder and lightning.

Tidal bore An unusually steep, cresting wave that marks the onset of high tide, caused by the sudden constriction of the oceanic tidal bulge.

Tidal currents The localized flow of water caused by topographic constriction of the tidal flow.

Tidal inlet A passageway for incoming and outgoing tidal currents through a coastal obstruction, such as a coral reef or barrier beach.

Tidal marsh or mud flat A marshland that borders a low coast and is subject to tidal inundation.

Tide The alternate rising and falling of sea level caused by the gravitational attraction of the moon and the sun. *Daily tides* are the daily pattern of high and low tides caused by earth rotation. *Neap tides* have a low tidal range that occurs fortnightly when the moon is in its first- or third-quarter phases. *Spring tides* have a high tidal range that occurs fortnightly when the moon is in its new or full phases.

Tierra caliente The frostless zone along lower mountain slopes in the tropics.

Tierra fria The mountain zone between the upper timber line and the zone of perpetual frost.

Tierra templada The intermediate elevation zone in the humid tropics between the frostless zone below and the upper timber line.

Till Unassorted glacial debris deposited directly from glacial ice.

Till plain A flat-to-undulating plain composed of glacial till.

Timber line The boundary between forested and nonforested land owing to natural conditions.

Tombolo An island that has been connected to the mainland by marine deposition.

Topographic map A map that portrays the surface features of an area at a scale greater than 1:500,000.

Topographic scale A scale of reduction greater than 1:500,000.

Topography The surface content and characterization of an area.

Topsoil A general term for the surface portion of the soil, either the *A* horizon or the zone above plow depth.

Tornado A violent, localized cyclonic vortex, containing a visible, pendulous, funnel-shaped cloud.

Toxic Pertaining to a condition that hinders the maintenance of good health in an organism.

Trace elements Chemical elements utilized in minute quantities in the development of organic cell structures.

Trade winds Winds with an easterly component that are located on the equatorward sides of subtropical anticyclones.

Transpiration Exhalation of water vapor by plants.

Transverse profile The profile of a valley across or at right angles to the general stream flow.

Trellis drainage A drainage pattern characterized by parallel main streams intersected at or near right angles by their tributaries.

Trench (See *Abyssal deeps*.)

Tropical easterlies The general flow of air from easterly quadrants along the equatorial sides of the subtropical anticyclones.

Tropical year The number of days (earth rotations) it takes to complete one revolution of the earth around the sun, as observed on earth. Its value is 365.2422 mean solar days.

Tropic of Cancer The parallel of lat $23\frac{1}{2}°$ N. Indicates the northern limit of the sun's vertical rays.

Tropic of Capricorn The parallel of lat $23\frac{1}{2}°$ S. Indicates the southern limit of the sun's vertical rays.

Tropics A general term for the low latitudes or the latitudinal zone between the Tropic of Cancer and Tropic of Capricorn.

Tropopause The level in the atmosphere at which the decline in temperature associated with increasing elevation (lapse rate) suddenly stops.

Tropophytes Plants that undergo seasonal changes in form while adjusting to seasonal differences in moisture availability.

Troposphere The lower portion of the atmosphere, from the earth surface to the tropopause. That portion of the atmosphere in which temperature decreases fairly regularly with elevation.

Tundra A treeless expanse of low vegetation in arctic and subarctic areas.

Turbidity The thickness or opaqueness of a gas or fluid due to suspended material.

Turbulence High irregularity of fluid flow, with eddying owing to differential friction.

Turbulent flow The type of flow of gases or liquids in which the stream lines are thoroughly confused because of irregular mixing.

Ultisols A mineral soil order whose soils feature argillic horizons which have a generally low cation-exchange capacity at or below the zone of illuviation.

Ultraviolet Radiation that has wavelengths slightly shorter than those of visible light.

Umbric epipedon A soil surface horizon that is dark in color, and whose dominant exchangeable ion is hydrogen rather than calcium, as in the dark mollic epipedon. The acidity therefore is much higher.

Universal time Mean solar time at the Greenwich meridian. Also termed *Zulu time*.

Uvala A broad depression in limestone country, produced as the result of coalesced sinkholes.

Valley train A narrow outwash plain occupying the floor of a valley.

Vapor pressure That part of the total atmospheric pressure that is due to the included water vapor.

Vernal equinox The equinox that occurs after the winter season. It falls on or about March 22 north of the equator.

Vertisols A mineral soil order normally dark and fine-textured, whose soils feature extensive cracking during seasonal droughts and pronounced swelling when wet.

Volcanic field A large cluster of volcanic vents.

Volcanic neck A pinnacle or erosional remnant formed by the resistant crystallized lava in the conduit of a volcanic ash core.

Volcanic cones Cone-shaped volcanic vents built by lava ejection.

Volcano A vent at the surface of the earth for the extrusion of molten lava.

Wadi A dry stream channel in arid regions.

Warm front A gently sloping frontal surface between two air masses, in which there is an active movement of warm air over cold air.

Water balance Precipitation minus evapotranspiration and groundwater discharge equals runoff.

Water budget The quantitative disposition of precipitation in an area.

Waterspout A tornadolike vortex occurring over water.

Water table The upper surface of the zone of water saturation in the soil, except where that surface is formed by an impermeable body.

Wave-built terrace A sloping surface below water level formed as the result of deposition by waves or currents.

Wave-cut terrace A sloping plain below water level resulting from erosion by waves and currents.

Weather The general condition of the atmosphere at a given time and place.

Weathering The breakdown of rocks caused by the action of atmospheric agents.

Westerlies (See *Prevailing westerlies*.)

Wet adiabatic rate The rate of cooling in ascending air which has already reached the condensation point, in which the lapse rate is being increased as the result of the release of latent heat of condensation.

White taiga A successional forest association, usually following fire within the general region of the taiga (northern coniferous forest), in which the tree dominants are broadleaf deciduous varieties, such as *Betula* (birch) or *Populus* (poplar).

Wind Horizontal or nearly horizonal air movement.

Wind chill The depression of sensible temperature owing to wind velocity.

Woodland An open stand of trees without a continuous canopy of leaves overhead.

Xerophyte A drought-resistant plant. *Xerophytic* = the quality of drought resistance by plant life.

Zenith A point directly overhead, or a point at right angles to the horizontal and opposite the direction of earth's gravity.

Zodiac The belt of star constellations about 12° wide through which the sun appears to move during the course of a year.

Zonal soils The mature soils of well-drained uplands, whose properties are related to climatic and biological processes.

INDEX

Climates of the World
Koeppen-Geiger classification

- Af-Am
- Aw-s
- BS
- BSkw
- BW
- Cfa
- Cwa
- Csa
- Csb
- Cfb
- Cwb
- Dfa
- Dfb
- Dwa
- Dfc
- Dwb
- Dwc
- Dfd
- E
- Mountains

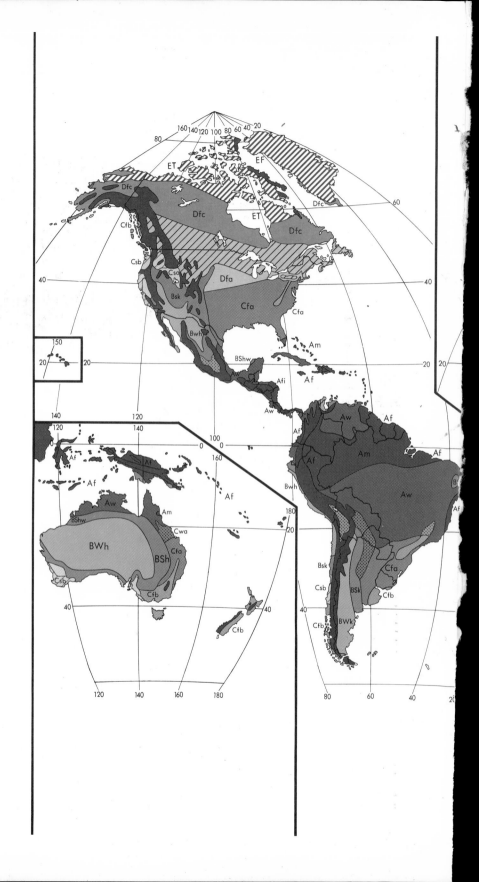